UNIVERSE

UNIVERSE

Third Edition

William J. Kaufmann, III

W. H. Freeman and Company
New York

To Mom and Leo, with love

Cover images: The Eta Carina Nebula, NGC 3372, is a huge cloud of interstellar gas and dust illuminated by radiation from a young cluster of massive, hot stars at its center. The nebula is located in the Carina spiral arm of our Galaxy, about 8000 light years from Earth. The star cluster, called the Carina OB1 association, is only about 1 million years old.

Eta Carina, which is visible to the naked eye from southern latitudes, is erratically variable. In 1677, the British astronomer Edmund Halley noted that its brightness was 4th magnitude. By 1843, however, it had become the second brightest star in the sky, surpassed only by Sirius. During the following decades it dimmed somewhat, but it is now brightening again. Many astronomers believe that these fluctuations are caused by variations in huge dust clouds surrounding Eta Carina.

The front cover photograph, taken with the 4-m telescope at the Cerro Tololo Inter-American Observatory in Chile, is a comparatively short exposure that displays the brightest regions of the core of the Eta Carina Nebula. The wide-angle photograph on the back cover is a much longer exposure and encompasses an area of the sky large enough to show the full extent of the nebula. Eta Carina has an angular diameter of 3°, which corresponds to a linear size of 400 light years. Delicate features seen on the front cover photograph are severely overexposed at the center of the picture on the back cover. A photograph of intermediate exposure, which clearly exhibits a feature called the "Keyhole," is shown in Figure 20-3 on page 398.

Illustration credits are listed on page 635.

Copyright © 1985, 1988, 1991 by W. H. Freeman and Company

Library of Congress Cataloging-in-Publication Data

Kaufmann, William J.
    Universe / William J. Kaufmann, III.—3d ed. p. cm.
    Includes bibliographical references and index.
    ISBN 0-7167-2094-9
    1. Astronomy. 2. Cosmology. I. Title.
QB43.2.K38 1990
520—dc20   90-42021

Printed in the United States of America

1 2 3 4 5 6 7 8 9 0   KP   9 9 8 7 6 5 4 3 2 1

Contents Overview

Contents

Preface

I wrote this book for the student who wants to explore the mysteries of the universe and probe some of the most profound questions people have ever asked. Since earliest times humanity has been fascinated by topics that astronomers examine today: the creation of the universe, the formation of the Earth and other planets, the motions of the stars, the structure of space and time. Armed with the powers of observation, the laws of physics, and the resourcefulness of the human mind, astronomers survey alien worlds, follow the life cycles of stars, and probe the dim and distant reaches of the cosmos.

I wrote this book for the instructor who wants to use the inherent intrigue of astronomy to inspire students to explore science and the world around them. One of the wonderful things about astronomy is that it can open our minds to the previously unimaginable and expand our awareness of the familiar. In studying astronomy, we investigate phenomena and explore realms far removed from our daily experience. Many of the objects that astronomers study are far too vast, distant, or intangible to ever sample directly; indeed, many of the phenomena that are observed today occurred very long ago. When viewed in the context of the evolution of the universe, even the dimensions of space and time take on new meaning.

To show students how scientists reason must be an objective of any first science course. Consequently, while discussing the fascinating nature of our physical universe, I describe also how astronomers have come to know what they know. By studying the methods that astronomers have used in exploring the universe, we can learn about the nature of scientific inquiry.

Organization

Because this book is designed to be used both for one- and for two-term courses, the level of coverage is generally more comprehensive and contains a little more mathematics than that found in the sister text *Discovering the Universe*. While comprehensive, this third edition of *Universe* has a flexible, modular structure that permits instructors to teach the topics in the order desired.

The book's twenty-nine chapters are divided into two nearly equal parts, the first half dealing with introductory material and planetary astronomy, and the second half treating stars, galaxies, and cosmology. Instructors may emphasize either half of the book according to preference, covering more or less of the detail as time and the preparation of the students permit. For instance, a course on stellar astronomy could begin with the introductory chapters on gravitation and light (Chapters 4 and 5) and proceed directly to Chapter 18, which introduces the Sun and stars, without any loss of continuity.

The traditional Earth-outward organization of this text emphasizes how our understanding of the universe developed, inviting the reader to share in the excitement of astronomical discovery. The first celestial objects to be examined are those that were observed by the ancient astronomers. Moving outward from the planets to the stars and galaxies, older Earth-based observations and the questions that they provoke are augmented or even supplanted by newer observations, including those from outside the visible range and those made from space. These new observations in turn raise new questions, which draw the reader on to the outer limits of our universe and our understanding.

The first six chapters introduce the foundations of astronomy, including descriptions of such naked-eye observations as eclipses and planetary motions and such basic tools as Kepler's laws, the fundamental properties of light, and the optics of telescopes. A discussion of the formation of the solar system in Chapter 7 prepares the reader for the next ten chapters, which cover the planets in outward order from the Sun. One of these chapters deals with the Galilean satellites, which are terrestrial worlds in their own right.

Chapter 18 on the Sun concludes coverage of the planets while introducing stellar astronomy. The properties of stars in general are described in Chapter 19. In Chapters 20 through 24, stellar evolution is described chronologically from birth to death. Molecular clouds, star clusters, nebulae, neutron stars, black holes, and various other phenomena are presented in the sequence in which they naturally occur in the life of a star, thus unifying the wide variety of objects that astronomers find scattered about the heavens.

A survey of the Milky Way introduces galactic astronomy in Chapter 25; this is followed by two chapters on galaxies and quasars. The final two chapters, on cosmology, emphasize exciting recent developments in our understanding of the physics of the early universe. All of the more "standard" cosmological issues are discussed in Chapter 28. It would thus be quite appropriate to end an astronomy course with that chapter. Chapter 29 was created for those who wish to go on to consider the interface between cosmology and particle physics with such speculative and exciting ideas as cosmic strings and eleven-dimensional spacetime. This challenging, optional chapter explores topics in which research is today proceeding at a fevered pitch. The student will learn that physicists are at the brink of tackling questions like Why is there only one dimension of time, yet three dimensions of space?

Major changes in this edition

My first step in planning the third edition of *Universe* was to compile and evaluate reactions to the second edition. Response to a questionnaire sent by my publisher to hundreds of instructors who had used *Universe* in both the United States and Canada gave me a most enlightening and inspiring impetus for this revision. Many of the improvements made in the third edition reflect the good judgment and class experience of numerous text users, reviewers, and colleagues.

Two major organizational changes were made in this edition. The first involves the treatment of the planets. After an overview of the solar system in Chapter 7, individual planets and satellites are discussed, beginning with the Earth (Chapter 8) and the Moon (Chapter 9). This permits the student to become familiar with the basic planetary properties by studying our own planet before moving on to alien worlds. The second significant reorganization involves the placement of the chapter on solar astronomy immediately following the planetary chapters. Many reviewers suggested that this arrangement offers a smoother transition from the solar system to the stars.

Studying the heavens inspires people to ask about the possibility of life elsewhere in the universe, and so an afterword on extraterrestrial life has been added. This speculative afterword utilizes what the student has learned about planets and stellar evolution to estimate the probability of extraterrestrial civilizations in the Galaxy and includes a brief discussion of SETI programs.

Many smaller, yet equally important, revisions enhance the rest of the text. For instance, the chapter on the Moon now includes the history of manned lunar exploration. The text was, of course, updated throughout. The planets chapters have been streamlined and strengthened. Coverage of Neptune and Triton reflects our latest findings. A discussion of close binary systems has been further enlarged. In light of SN 1987A, the treatment of Type I and Type II supernovae has also been expanded. And, finally, the discussion of galactic evolution has been greatly amplified. Finally, the glossary has been indexed to facilitate its use as reference material. With these changes, I have ventured to create a comprehensive yet entertaining text that gives a fair and accurate picture of the full scope of astronomy.

Pedagogical emphasis

Helping students to easily gain an understanding of—and an appreciation for—astronomy continues to be a major objective of this text. Each chapter begins with a brief, one-paragraph abstract that gives the reader a clear idea of the chapter's contents. The chapter headings, in the form of declarative sentences, highlight main concepts. Coupled with a list of key words, a formal summary outlines the essential facts addressed in each chapter. Each chapter concludes with a series of questions grouped into three categories: review, advanced, and discussion. The advanced questions are preceded by a short box of tips and tools that give students just enough of a hint to get them started on solving these problems. In response to instructors' requests, the number of questions and exercises has been significantly enlarged in this edition. Answers to questions that require computation (marked by an asterisk) appear at the end of the book, while the Instructor's Manual contains full step-by-step solutions that can be posted for students at the instructor's discretion. Care has been taken to include some questions whose answers require reasoning rather than just memorization. Hence the range of questions is considerable. Both basic and challenging questions are so noted, with most questions falling somewhere in between. Roger Culver of Colorado State University deserves special recognition for his many contributions to the questions. Also, to enhance student' interest in astronomy, observational activities are included at the end of the chapters and attendant star charts are located at the back of the book. An annotated list of further readings rounds out each chapter's presentation.

Boxed inserts set aside the text's more technical and/or mathematical material. Many review key formulas and the calculations that lead to astounding discoveries. Others contain technical reference information, such as the orbital and physical data for each planet. Still others present arguments, taking an idea raised in the text a step further. Boxes may be omitted without loss of continuity. This segregation permits greater flexibility in an instructor's use of the text and allows students to easily locate and review such topics.

Illustrations

Color illustrations greatly enhance the pedagogical effectiveness of this book. Because color photographs are indispensable to a truly modern view of astronomy, they are incorporated throughout. Color is now routinely used by astronomers in a wide variety of circumstances. One glance at a color photograph of a planet's cloudtops or of the glowing gases of a nebula reveals significant details about the object that cannot be gleaned from a black-and-white view. X-ray, infrared, and radio views of the sky are now very comprehensibly displayed in extraordinary computer-generated false-color images. The incorporation of color pictures in the main body of the text not only makes reading and learning more pleasurable and easier, but reflects how astronomy is practiced today.

Essays

It is again a great pleasure to include the essays of six renowned astronomers, who offer their personal views on topics of especially current interest. These exceptional essays round out the book and give it a depth and quality that could not have otherwise been possible:

Why Astronomy?	Sandra M. Faber
Astrology and Astronomy	Owen Gingerich
New ESO Telescopes	Richard M. West
The Evolution of Close Binary Stars	Icko Iben, Jr.
The Great Attractor	Alan Dressler
The Edge of Spacetime	Stephen W. Hawking

Supplementary materials

It is a pleasure to announce the availability of the following outstanding supplements to the second edition of *Universe*: A *Computerized Test Bank* has been prepared by T. Alan Clark, The University of Calgary, Alberta, Canada. Included on this IBM-PC-based testing program are multiple-choice questions, indexed by chapter and topic category, and a flexible edit feature that allows instructors to include their own questions. A printed test bank is also available.

Both overhead transparencies and slides are available to accompany *Universe*. They contain a selection of color drawings from the text. The test banks, overhead transparencies, and slides are free to adopters of 100 or more copies of the text.

An *Instructor's Manual* has been prepared by Thomas H. Robertson of Ball State University and Andrew Fraknoi of the Astronomical Society of the Pacific. It contains chapter synopses, hints for teaching and discussion, a current list of resources for teaching, and the fully worked-out solutions to the computational problems in the book.

For more information and to request copies of these supplements please contact:

Marketing Department
W. H. Freeman and Company
41 Madison Avenue
New York, NY 10010

Acknowledgments

I would like to begin by thanking the many instructors who responded to the questionnaire about the second edition of *Universe* and especially the dozens of people who sent in unsolicited detailed comments and suggestions for the second edition. Foremost among those deserving of thanks are the following seven people who scrutinized the second edition, making numerous specific suggestions that profoundly affected the writing of the third edition:

David Van Blerkom	University of Massachusetts
Austin F. Gulliver	Brandon University
Icko Iben, Jr.	Pennsylvania State University
Laurence A. Marschall	Gettysburg College
Kenneth S. Rumstay	Valdosta State College
Virginia Trimble	University of California, Irvine
Bruce A. Twarog	The University of Kansas

I am also deeply grateful to the following people who carefully reviewed the manuscript for the third edition:

Alice L. Argon	Harvard-Smithsonian Center for Astrophysics
Bradley W. Carroll	Weber State University
Paul F. Goldsmith	University of Massachusetts
Douglas P. Hube	University of Alberta

I also wish to thank the many people whose advice on the first and second editions has had an ongoing influence:

Robert Allen	The University of Wisconsin, La Crosse
John M. Burns	Mt. San Antonio College
Bruce W. Carney	The University of North Carolina
Roger B. Culver	Colorado State University
James N. Douglas	The University of Texas at Austin
David S. Evans	University of Texas
George W. Ficken, Jr.	Cleveland State University
Andrew Fraknoi	Astronomical Society of the Pacific
Owen Gingerich	Harvard University
J. Richard Gott, III	Princeton University
Bruce Hanna	Old Dominion University
Paul Hodge	University of Washington
John K. Lawrence	California State University of Northridge
Dimitri Mihalas	University of Illinois

Laurence A. Marschall	Gettysburg College
L. D. Opplinger	Western Michigan University
John R. Percy	University of Toronto
Terry Retting	University of Notre Dame
Richard Saenz	California Polytechnic State University
Thomas F. Scanlon	Grossmont College
Richard L. Sears	University of Michigan
David B. Slavsky	Loyola University of Chicago
Joseph S. Tenn	Sonoma State University
Gordon B. Thomson	Rutgers University
Donat G. Wentzel	University of Maryland
Nicholas Wheeler	Reed College
Raymond E. White	University of Arizona

I am especially grateful to the following individuals who were most helpful with my requests for photographs:

David F. Malin	Anglo-Australian Observatory
Rudolph E. Schild	Center for Astrophysics
Richard Schmidt	U.S. Naval Observatory

Many other people have participated in the preparation of this book, and I thank them for their efforts. Foremost among them is my developmental editor, Elizabeth Zayatz. I also thank my editor, Jerry Lyons, and W. H. Freeman's president, Linda Chaput, for their support and encouragement of this project. It was once again a pleasure to work with Georgia Lee Hadler, my project editor; Mike Suh, who managed both the design and art programs; Bill Page, the illustration coordinator; and Ellen Cash, the production coordinator. They all deserve special thanks for their unfailing concern for quality. I also acknowledge the fine efforts of my copy editor, Ruth Davis, who has significantly improved the readability of the manuscript and of Margot Getman, who oversaw the development of the supplemental materials; of the people at Vantage Art Studio, who drafted many of the illustrations; and of Tomo Narashima and George Kelvin, whose marvelous airbrush artistry makes their drawings jump off the pages of the book.

Although we have made a valiant effort to make this an error-free edition, some mistakes may have crept in. I would appreciate hearing from anyone who finds an error or who wishes to comment on the text. You may write to me in care of the publisher. I will respond personally to all correspondence.

William J. Kaufmann, III

Department of Physics
San Diego State University

The Horsehead and Orion Nebulae *New stars are forming in the clouds of interstellar gas and dust shown in this photograph, which covers an area of the sky approximately 4° × 6°. The gases glow because of the radiation emitted by newborn, massive stars. Clouds of interstellar dust block light; they appear as dark regions silhouetted against glowing background nebulosity. The Horsehead Nebula is in the lower left; the Orion Nebula is toward the upper right. Both nebulae are about 1500 light years from Earth. (Royal Observatory, Edinburgh)*

CHAPTER

Astronomy and the Universe

Astronomy, the study of the universe, addresses profound questions that people have pondered since the dawn of civilization. In this chapter, we see that by exploring the planets, astronomers learn about the formation and evolution of the solar system. By examining stars and nebulae, they fathom the life cycles of stars. By observing galaxies, they uncover important clues about the creation of the universe. Over the course of centuries of such delv-

ings into the mysteries of the universe, we humans have discovered that the physical world obeys certain rules called the laws of physics. Such discoveries owe as much to systems of measurement as they do to the creativity and insight of the human mind. In addition to the deep satisfaction each new discovery brings, our understanding of the universe gives us a sweeping perspective from which we can appreciate our existence on Earth.

Figure 1-1 The starry sky *The star-filled sky is a beautiful and inspiring sight. This photograph, taken from northern Mexico, shows Halley's Comet and a portion of the Milky Way (upper right). To get a good view of the heavens, you must be far from any city lights. (Courtesy of D. L. Mammana)*

The splendor of the star-filled night sky is one of the central experiences of life. Gazing out into the heavens, we see thousands of stars scattered from horizon to horizon. The delicate mist of the Milky Way traces a faerie path across the starry firmament and the entire spectacle swings slowly overhead from east to west as the night progresses. No light show, no artist's brush, no poet's words can truly capture the beauty of this breathtaking panorama (see Figure 1-1).

For thousands of years people have looked up at the heavens and found themselves inspired to contemplate the nature of the universe. Like our ancestors, we find our thoughts turning to profound questions as we gaze at the stars. How was the universe created? Where did the Earth, Moon, and Sun come from? What are the planets and stars made of? And how do we fit in? What is our place and our role in the cosmic scope of space and time?

To wonder about the nature of the universe is one of the most characteristic of human endeavors. Our curiosity, our desire to explore and discover, are unique qualities that distinguish us from lower creatures. The study of the stars transcends all boundaries of culture, geography, and politics. In a literal sense, astronomy is a universal subject—its subject is the entire universe.

1-1 Astronomers use the laws of physics and construct testable theories and models to understand the universe

Astronomy has a rich heritage that dates back to antiquity. In many ancient societies, myths and legends dominated astronomy. The heavens were thought to be populated with demons and heroes, gods and goddesses. Astronomical phenomena were explained as the result of supernatural forces and divine intervention.

The course of civilization was greatly affected by the realization that the universe is comprehensible. This first glimpse of the power and potential of the human mind is one of the great gifts to come to us from ancient Greece. Greek astronomers learned that by observing the heavens and carefully thinking about what they saw, they could discover something about how the universe operates. For example, they measured the size of the Earth and understood and predicted eclipses.

The approach of using observation and logic to explore physical reality evolved into the **scientific method,** which has become a dominant force in science and society over the past four hundred years. In essence, the scientific method requires that our ideas about the world around us be in agreement with what we actually observe. More specifically, a scientist trying to understand some phenomenon begins by proposing a **hypothesis,** which is an idea or collection of ideas that seems to explain the phenomenon. The hypothesis should not be in conflict with existing observations and experiments, because a discrepancy with known facts implies that the hypothesis is wrong. The hypothesis should also be capable of being tested with new observations and experiments. This crucial aspect of the scientific method requires that the scientist carefully examines the hypothesis to see what it predicts or in what way the implications of the hypothesis might be tested. Only after the hypothesis has withstood such tests, by having accurately forecast the results of new experiments and observations, does the scientist feel confident that the hypothesis is on firm ground.

It is important to realize that science does not discover absolute truth. Scientists instead describe reality in terms of **models** and **theories,** which are two commonly used names for hypotheses that have withstood observational tests. For example, the model of the atom, which scientists picture as electrons orbiting a central nucleus, is a descriptive representation of nature. Similarly, Einstein's general theory of relativity describes gravity in terms of its effects on space and time. The scientific method requires that the scientist be an open-minded person, willing to discard even the most cherished ideas if they fail to agree with observation and experiment. As new discoveries are made, old models and theories must be modified or discarded in favor of more comprehensive explanations of the world around us.

Some people think that astronomy deals only with faraway places of no possible significance to life here on Earth, but nothing could be further from the truth. For example, the seventeenth-century scientist Isaac Newton succeeded in describing how the planets orbit the Sun. The motions of the planets, unhampered by air resistance or friction, reveal some of the most fundamental laws of nature in their simplest form. From Newton's work we obtained our first complete, coherent description of the behavior of the physical universe. The resulting body of knowledge, which is called **Newtonian mechanics,** deals in concrete terms with the concepts of force, mass, acceleration, momentum, and energy. This understanding had immediate practical application in the construction of machines, buildings, and bridges. It is no coincidence that the Industrial Revolution followed hard on the heels of these theoretical and mathematical advances inspired by astronomy.

Astronomers use Newtonian mechanics, along with other physical principles (usually called the **laws of physics**), to interpret their observations and to understand the processes that occur in the universe. Light and its relationship to matter are of particular importance to astronomers. By trying to understand how objects emit radiation and how this light interacts with matter, we acquire the skills needed to analyze and interpret the wealth of information coming to us from the stars and galaxies. Astronomers use these skills to obtain fundamental information about stars, galaxies, and the evolution of the universe.

The laws of physics, particularly those involving optics and light, can also be used to develop new tools and techniques with which to examine and explore the universe. Until recently, everything we knew about the distant universe was based on visible light. Astronomers would peer through telescopes to observe and analyze visible starlight. By the end of the nineteenth century, however, scientists had begun discovering such nonvisible forms of light as X rays and gamma rays, radio waves and microwaves, and ultraviolet and infrared radiation.

Astronomers have recently constructed telescopes that can detect these nonvisible forms of light (see Figure 1-2). Whether located on Earth or in orbit above the obscuring effects of the atmosphere, these astronomical instruments give us views of the universe vastly different from anything our eyes can see. This new information is crucial to our understanding of familiar objects like the Sun and also gives us important clues about such exotic objects as neutron stars, pulsars, quasars, and black holes.

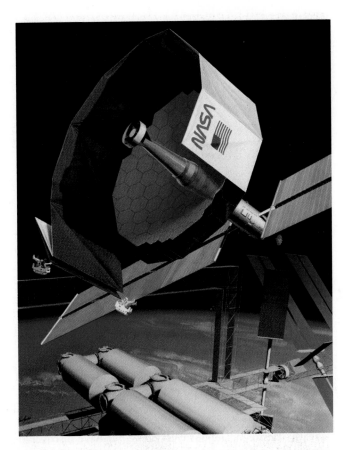

Figure 1-2 *A telescope in space* *This artist's painting shows astronauts constructing a telescope, called the Large Deployable Reflector, at an Earth-orbiting space station. Astronomers who are planning this ambitious project hope that the telescope will be launched around the year 2000. (NASA)*

1-2 *By exploring the planets, astronomers uncover clues about the formation of the solar system*

This book presents the substance of modern astronomy in three segments, corresponding to three major steps out into the universe: the planets, the stars, and the galaxies. The star we call the Sun and all the celestial bodies (including the Earth) that orbit it make up the **solar system**. We will explore the solar system, beginning with the Earth and the Moon, and then the other planets, moving outward toward the frigid depths of space where comets spend most of their time.

Most of the up-to-date information we have about the planets comes from numerous Soviet and American space flights, primarily those during the 1970s and 1980s (see Figure 1-3). Spacecraft have visited all the planets except Pluto, usually the outermost world in the solar system. We have flown over Mercury's cratered surface, we have peered

Figure 1-3 *Saturn and its satellites This composite photograph shows the planet Saturn along with several of its moons. Space probes to the planets have provided us with a wealth of information about other worlds. This new knowledge gives us important insights about the formation and evolution of the solar system, as well as an appreciation of the variety of which nature is capable. (NASA)*

beneath Venus's poisonous cloudcover, we have discovered enormous canyons and extinct volcanoes on Mars. We have visited the moons of Jupiter, we have seen the rings of Saturn and Uranus, we have looked down on the active atmosphere of Neptune. Along with the moon rocks brought back by the Apollo astronauts, this wealth of new and exciting information has produced a major revolution in how astronomers think about the solar system. In particular, we have come to realize that collision between objects in space is a fundamental process that has shaped many of the planets and their satellites. Craters on the Moon and on many other worlds stand in mute testimony to innumerable impacts by interplanetary rocks. But more importantly, the Moon itself may be the result of a catastrophic collision between the Earth and a planet-sized object shortly after the formation of the solar system. As we shall see, such a collision could have torn sufficient material from the primordial Earth to create the Moon.

Throughout our journey across the solar system, we shall find that our discoveries are relevant to the quality of human life here on Earth. Until recently, our knowledge of such subjects as geology, weather, and climate was based on data from only the planet Earth. Since the advent of space exploration, however, we have had a range of other worlds to compare and contrast with our own. As a result, we have not only made important progress in understanding the creation and evolution of the Earth and the solar system, but have also gained many valuable insights into the origins and extent of our natural resources.

1-3 By studying stars and nebulae, astronomers discover how stars are born, grow old, and eventually die

In studying the stars, including the Sun, we see again the surprising impact of astronomy on the course of civilization. In the 1920s and 1930s, physicists figured out how the Sun shines. At its center, thermonuclear reactions convert hydrogen into helium. This violent process releases a vast amount of energy, which eventually makes its way to the Sun's surface and escapes as light (see Figure 1-4). By 1950 physicists had learned how to reproduce these thermonuclear reactions here on Earth (see Figure 1-5). Hydrogen bombs operate on the same basic principles as do the thermonuclear reactions that produce energy at the Sun's center. In the decades following the invention of the hydrogen bomb, thermonuclear weapons were deployed around the world and offer the specter of dramatically influencing life on our planet. In contrast, research into peaceful applications of controlled thermonuclear reactions may provide a clean source of electrical energy early in the twenty-first century.

As we look deeper into space, we find star clusters and clouds of glowing gas, called **nebulae** (singular, **nebula**), scattered across the sky. These beautiful objects can tell us much about the lives of stars. Stars are born in huge clouds of interstellar gas and dust such as the Orion Nebula, shown in Figure 1-6. After many millions or billions of years, stars eventually die. Some end their lives with a spectacular detonation called a **supernova** that blows the star apart. The Crab Nebula seen in Figure 1-7 is a striking example of a supernova remnant.

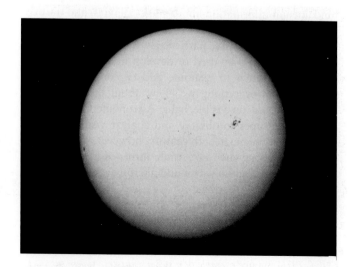

Figure 1-4 *Our star—the Sun The Sun is a typical star. Its diameter is about 1.39 million kilometers (roughly a million miles), and its surface temperature is about 5500°C (10,000°F). The Sun draws its energy from thermonuclear reactions occurring at its center, where the temperature is about 15 million degrees Celsius. (NOAO)*

Figure 1-5 A thermonuclear explosion *Understanding the Sun's source of energy has given humanity the ability to build thermonuclear weapons. The hydrogen bomb and the thermonuclear reactions at the Sun's center both operate under the same basic physical principle: the conversion of matter into energy. This thermonuclear detonation on October 31, 1952, had an energy output, or "yield," equivalent to 10.4 million tons of TNT. (Defense Nuclear Agency)*

During their death throes, stars return gas to interstellar space. This gas contains some of the original hydrogen and the heavy elements created by thermonuclear reactions in the stars' interiors. Interstellar space thus becomes enriched with newly manufactured atoms and molecules. The situation is analogous to a forest in which decaying leaves and logs enrich the soil for future generations of trees. In the same sense, the Sun and its planets were formed from enriched interstellar material. We therefore arrive at the surprising realization that virtually everything we touch, including the atoms in our bodies, was created deep inside ancient, now-dead stars.

Dying stars can produce some of the strangest objects in the sky. Some dead stars become **pulsars** that emit pulses of radio waves or **bursters** that emit powerful bursts of X rays. Massive dead stars become **black holes**, surrounded by such incredibly powerful gravity that nothing—not even light—can escape. Many of these bizarre stellar corpses have been discovered in recent years with Earth-orbiting telescopes that detect nonvisible light, like the X rays emitted by gases falling toward a black hole.

Figure 1-6 The Orion Nebula *This beautiful nebula (also called M42 or NGC 1976) is a fine example of a stellar "nursery" where stars are born. Intense radiation from the newly formed stars causes the surrounding gases to glow. Many of the stars embedded in this nebula are less than a million years old. The Orion Nebula is 1300 light years from Earth, and the distance across the nebula is about 5 light years. (U.S. Naval Observatory)*

Figure 1-7 The Crab Nebula *This nebula (also called M1 or NGC 1952) is a fine example of a supernova remnant. A dying star exploded, and this beautiful funeral shroud was caused by the gases blasted violently into space. In fact, these gases are still moving outward, at about 1000 km/s (roughly 2 million miles per hour). The Crab Nebula is 6300 light years from Earth, and the distance across the nebula is about 6 light years. (Lick Observatory)*

Figure 1-8 The galaxy M83 *This spectacular galaxy (also called NGC 5236) contains about 200 billion stars. The galaxy's spiral arms are outlined by many nebulae, which are sites of active star formation. This galaxy has a diameter of about 35,000 light years and is at a distance of 12 million light years from Earth. (Courtesy of R. J. Dufour)*

1-4 By observing galaxies, astronomers learn about the creation and fate of the universe

Stars are not spread uniformly across the universe but are grouped together in huge assemblages called galaxies. Galaxies are the largest individual objects in the universe. A typical galaxy, like our own Milky Way, contains several hundred billion stars.

The Milky Way Galaxy has beautiful, arching spiral arms like those of M83 in Figure 1-8, which are active sites of star formation. The center of the Galaxy is emitting vast quantities of energy. Some astronomers suspect that this energy output is caused by gas falling into an enormous black hole at the galactic center.

Other galaxies come in a wide range of shapes and sizes. Some galaxies are quite small, containing only a few hundred million stars. Others are veritable monstrosities that devour neighboring galaxies in a process called "galactic cannibalism."

Some of the most intriguing galaxies are those that appear to be in the throes of violent convulsions. The centers of these strange galaxies, which may also harbor extremely massive black holes, are often powerful sources of X rays and radio waves. In many such cases, the galaxy is rapidly shedding matter.

Even more dramatic sources of energy are to be found still deeper in space. At distances of billions of light years from Earth, we find the mysterious quasars. Although quasars look like stars (see Figure 1-9), they are probably the most distant and most luminous objects in the sky. A typical quasar shines with the brilliance of a hundred galaxies. Data suggest that quasars draw their awesome energy from enormous black holes.

Finally, we shall see how the motions of clusters of galaxies reveal that we live in an expanding universe. Extrapolating into the past, we learn that the universe must have been born from an incredibly dense—perhaps infinitely dense—state nearly 20 billion years ago.

Most astronomers believe that the universe began with a cosmic explosion, known as the Big Bang, that occurred throughout all space at the beginning of time. Shortly after the Big Bang, events happened that dictated the present nature of the universe. Astronomers are making significant

Figure 1-9 The quasar 3C48 *Quasars, believed to be the most distant objects in the universe, are the most luminous objects that astronomers have ever seen. At first glance, a quasar is easily mistaken for a faint star. This quasar is thought to be at a distance of 4.8 billion light years from Earth. (Palomar Observatory)*

progress in understanding these cosmic events. Indeed, they may be about to discover the origin of some of the most basic properties of the universe. In addition, the motions of the most distant clusters of galaxies may tell us the ultimate fate of the universe: whether it will continue expanding forever or someday stop and collapse back in on itself.

The foregoing discoveries and theories of today's astronomy owe a great debt to astute observers and thinkers spanning several centuries and to the first stirrings of the human imagination that led to the development of systems of measurement. Curiosity, insight, and creativity together with observation, measurement, and disciplined inquiry fostered the development of all science. Given the human need to know and understand, it seems inevitable that we would begin to unravel even the deepest mysteries of the universe. The basic tools we need for this bold endeavor in astronomy were developed long ago and are quite straightforward.

1-5 Astronomers use angles to denote the apparent sizes and positions of objects in the sky

Astronomers have inherited many useful concepts from antiquity. For example, ancient mathematicians invented angles and a system of angular measure that is still used to denote the positions and apparent sizes of objects in the sky.

An **angle** is the opening between two lines that meet at a point. Various kinds of angles are shown in Figure 1-10. **Angular measure** provides a more exact description of the shape or size of an angle. The basic unit of angular measure is the **degree**, designated by the symbol °. A full circle is divided into 360°. A right angle measures 90°. As shown in Figure 1-11, the angle between the two "pointer stars" in the Big Dipper is about 5°.

Astronomers use angular measure in a wide range of situations, including sizing a celestial object. For example, imagine looking up at the full moon. The angle covered

Figure 1-11 *The Big Dipper* *The Big Dipper is an easily recognized grouping of seven bright stars. The angular distance between the two "pointer" stars at the front of the Big Dipper is about 5°. For comparison, the angular diameter of the Moon is about ½°.*

by the Moon's diameter is nearly ½°. We therefore say that the **angular diameter**, or **angular size**, of the Moon is ½°. Alternatively, astronomers say that the Moon **subtends** an angle of ½°. Ten full moons could fit side by side between the two pointer stars in the Big Dipper.

To talk about smaller angles, we subdivide the degree into 60 **minutes of arc** (abbreviated 60 <u>arc min</u> or 60'). A minute of arc is further subdivided into 60 **seconds of arc** (abbreviated 60 <u>arc sec</u> or 60"). Thus,

$$1° = 60 \text{ arc min} = 60'$$
$$1' = 60 \text{ arc sec} = 60''$$

The *Astronomical Almanac* for 1990, for example, states that on January 5, Venus had an angular diameter viewed from Earth of 28.27 seconds of arc. That is a very convenient and precise statement of how big the planet appeared in Earth's sky on that date.

The angular size of an object can be converted into a linear size (measured in kilometers or miles, for example) if we know the distance to the object. A method for making this conversion is described in Box 1-1.

1-6 Powers-of-ten notation is a useful shorthand system of writing numbers

Astronomy is a subject of extremes. As we examine various environments, we find an astonishing range of conditions, from the incredibly hot, dense centers of stars to the frigid,

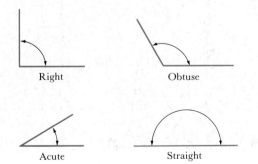

Figure 1-10 *Angles* *An angle is the opening between two lines that meet at a point. A right angle measures 90°. An acute angle has less than 90°, an obtuse angle more than 90°. A straight angle equals two right angles.*

Box 1-1   The small-angle formula

Astronomers usually are concerned with objects that subtend tiny angles in the sky. If you know the distance to an object, you can convert its angular size into a linear size (in kilometers or miles, for example). This conversion is accomplished with the **small-angle formula.** Suppose that an object subtends an angle α (measured in seconds of arc) and is at a distance d from the observer, as in the diagram below. The small-angle formula tells us that the linear size (D) of the object is given by the expression

$$D = \frac{\alpha d}{206,265}$$

Example: On July 3, 1981, Jupiter was at a distance of 824.7 million kilometers from Earth. Jupiter's angular diameter on that date was 35.72 seconds of arc, so we can calculate the planet's diameter as follows:

$$D = \frac{35.72 \times 824,700,000}{206,265} = 142,800 \text{ kilometers}$$

Incidentally, the number 206,265 is equal to $360 \times 60 \times 60/2\pi$, the total number of arc seconds in 360° divided by the circumference of a circle whose radius is 1.

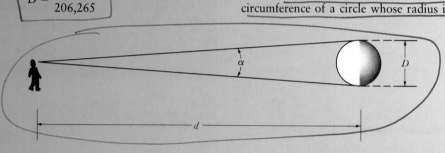

near-perfect vacuum of interstellar space. To describe such divergent conditions accurately, we need a wide range of both large and small numbers. Astronomers avoid such confusing terms as "a million billion billion," by using a standard shorthand system: All the cumbersome zeros that accompany a large number are consolidated into one term consisting of 10 followed by an **exponent,** which is written as a superscript and called the **power of ten.** The exponent indicates how many zeros you would need to write out the long form of the number. Thus,

$$10^0 = 1$$
$$10^1 = 10$$
$$10^2 = 100$$
$$10^3 = 1000$$
$$10^4 = 10,000$$

and so forth. The exponent tells you how many tens must be multiplied together to give the desired number. For example, ten thousand can be written as 10^4 ("ten to the fourth") because $10^4 = 10 \times 10 \times 10 \times 10 = 10,000$.

With this notation, numbers are written as a figure between one and ten multiplied by the appropriate power of ten. The distance between the Earth and the Sun, for example, can be written as 1.5×10^8 km, which, once you get used to it, is more convenient than writing "150,000,000 kilometers" or "one hundred and fifty million kilometers."

This shorthand system can be applied to numbers that are less than one by using a minus sign in front of the exponent. A negative exponent tells you to divide by the appropriate number of tens, for example, $10^{-1} = \frac{1}{10}$. The location of the

decimal point is as follows:

$$10^0 = 1$$
$$10^{-1} = 0.1$$
$$10^{-2} = 0.01$$
$$10^{-3} = 0.001$$
$$10^{-4} = 0.0001$$

and so forth. Perhaps the easiest way of understanding negative exponents is to realize that the negative exponent tells you how many tenths must be multiplied together to give the desired number. For example, one ten-thousandth can be written as 10^{-4} ("ten to the minus four") because $10^{-4} = \frac{1}{10} \times \frac{1}{10} \times \frac{1}{10} \times \frac{1}{10} = 0.0001$. With this notation, the diameter of a hydrogen atom is 1.1×10^{-8} cm. That is more convenient than saying "0.000000011 centimeters" or "eleven billionths of a centimeter."

Using the powers-of-ten notation, one can write familiar numerical terms as follows:

$$\text{one thousand} = 10^3 = 1000$$
$$\text{one million} = 10^6 = 1,000,000$$
$$\text{one billion} = 10^9 = 1,000,000,000$$
$$\text{one trillion} = 10^{12} = 1,000,000,000,000$$

and also

$$\text{one thousandth} = 10^{-3} = 0.001$$
$$\text{one millionth} = 10^{-6} = 0.000001$$
$$\text{one billionth} = 10^{-9} = 0.000000001$$
$$\text{one trillionth} = 10^{-12} = 0.000000000001$$

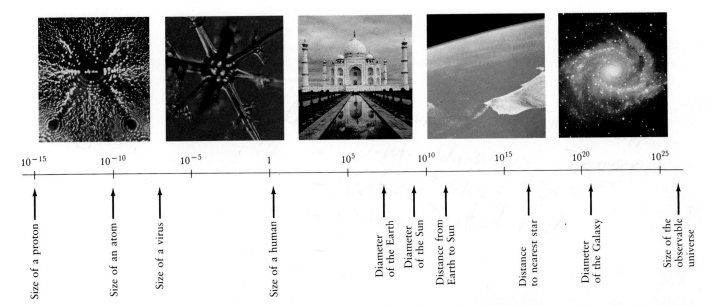

Size of a proton   Size of an atom   Size of a virus   Size of a human   Diameter of the Earth   Diameter of the Sun   Distance from Earth to Sun   Distance to nearest star   Diameter of the Galaxy   Size of the observable universe

10^{-15} 10^{-10} 10^{-5} 1 10^{5} 10^{10} 10^{15} 10^{20} 10^{25}

Figure 1-12 *Examples of powers-of-ten notation* *The scale gives the sizes of objects in meters, ranging from subatomic particles at the left to the entire observable universe on the right. The photograph at the left shows tungsten atoms, 10^{-10} meter in diameter. Second from left is a crystalline skeleton, 10^{-4} meter (0.1 millimeter) in size, of a diatom—a single-celled organism. At the center is the Taj Mahal, within reach of our unaided senses. On the right, looking across the Indian Ocean toward the South Pole, we see the curvature of the Earth, 10^{7} meters in diameter. At the far right is a galaxy, 10^{21} meters (100,000 light years) in diameter. (Courtesy of Scientific American Books; NASA; AAT)*

Powers-of-ten notation bypasses all the awkward zeros so that a wide range of circumstances such as those shown in Figure 1-12 can be described in a convenient fashion. Furthermore, by using powers-of-ten notation it is easy to multiply and divide numbers, as described in Box 1-2.

1-7 Astronomical distances are often measured in AUs, parsecs, or light years

As we turn toward the stars, in the second half of this book, we shall find that some of our traditional units of measure become cumbersome. It is fine to use kilometers or miles to give the diameters of craters on the Moon or the heights of volcanoes on Mars. But it is as awkward to use kilometers to express distances to stars or galaxies as it would be to talk about the distance from New York to San Francisco in millimeters or inches. Astronomers have therefore devised new units of measure.

When discussing distances across the solar system, astronomers use a unit of length called the **astronomical unit** (abbreviated AU), which is the average distance between the Earth and the Sun:

$$1 \text{ AU} = 1.496 \times 10^{8} \text{ km} = 93 \text{ million miles}$$

Thus the distance between the Sun and Jupiter can be conveniently stated as 5.2 AU.

When talking about distances to the stars, astronomers choose between two different units of length. One is the **light year** (abbreviated ly), which is the distance that light travels in one year:

$$1 \text{ ly} = 9.46 \times 10^{12} \text{ km}$$

or, alternatively,

$$1 \text{ ly} = 63,240 \text{ AU}$$

This distance is roughly equal to 6 trillion miles. Proxima Centauri, the nearest star other than the Sun, is 4.3 light years from Earth, for example.

The second commonly used unit of length is the **parsec** (abbreviated pc). Imagine taking a journey far into space, beyond the orbits of the outer planets. As you look back toward the Sun, the Earth's orbit subtends a smaller angle in the sky the farther you are from the Sun. The distance at which 1 AU subtends an angle of one second of arc, as shown in Figure 1-13, is defined as one parsec. The parsec turns out to be longer than the light year. Specifically,

$$1 \text{ pc} = 3.09 \times 10^{13} \text{ km} = 3.26 \text{ ly}$$

Thus, the distance to the nearest star can be stated as 1.3 pc as well as 4.3 ly.

Many astronomers prefer to use parsecs when describing interstellar distances because the definition of the parsec is closely related to a method of measuring the distance to the stars (you might want to glance ahead to Figures 19-1 and

Box 1-2   Arithmetic with exponents

Using powers-of-ten notation, it is easy to multiply numbers; simply add up the exponents.

Example:

$$10^2 \times 10^3 = 10^{2+3} = 10^5$$

To divide numbers, remember the rule

$$10^{-n} = \frac{1}{10^n}$$

In other words, dividing by 10^6, for instance, is the same as multiplying by 10^{-6}. You again add up the exponents.

Example:

$$\frac{10^4}{10^6} = 10^4 \times 10^{-6} = 10^{4-6} = 10^{-2}$$

Usually, a computation involves numbers multiplied by powers of ten. In such cases, to perform multiplication or division, you can treat the numbers separately from the powers of ten.

Example: The numerical example worked out in Box 1-1 can be done straightforwardly as follows:

$$D = \frac{35.72 \times 824{,}700{,}000}{206{,}265} = \frac{3.572 \times 10 \times 8.247 \times 10^8}{2.06265 \times 10^5}$$

$$= \frac{3.572 \times 8.247 \times 10^{1+8-5}}{2.06265} = 1.428 \times 10^5$$

Negative exponents are often used with units of measure. For instance, speed in kilometers per second can be written as "km/s" or as "km s^{-1}."

Example: The speed of light in a vacuum (usually designated by the letter "c") is written as

$$c = 3 \times 10^8 \text{ m/s}$$
$$= 3 \times 10^8 \text{ m s}^{-1}$$

Similarly, the average density of a typical rock (that is, its mass divided by its volume) is 3 grams per cubic centimeter, which can be written as

$$3 \text{ g/cm}^3 = 3 \text{ g cm}^{-3}$$

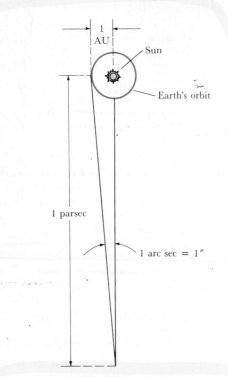

Figure 1-13   A parsec *The parsec, a unit of length commonly used by astronomers, is equal to 3.26 light years. The parsec is defined as the distance at which 1 AU perpendicular to the observer's line of sight subtends an angle of 1 second of arc.*

19-2). Physicists sometimes prefer to use light years because that unit of distance involves one of nature's most important numbers: the speed of light. Whether you choose to use parsecs or light years is a matter of personal taste.

For even greater distances, astronomers commonly use **kiloparsecs** and **megaparsecs** (abbreviated kpc and Mpc), in which the prefixes simply mean "thousand" and "million," respectively:

$$1 \text{ kpc} = 10^3 \text{ pc}$$
$$1 \text{ Mpc} = 10^6 \text{ pc}$$

For example, the distance from Earth to the center of our Milky Way Galaxy is about 8 kpc, and the rich cluster of galaxies in the direction of the constellation of Virgo is 20 Mpc away.

Some astronomers prefer to talk about thousands or millions of light years rather than kiloparsecs and megaparsecs. Once again, the choice is a matter of personal taste.

To the possible annoyance of some people, astronomers use whatever yardsticks seem best suited for the issue at hand and do not restrict themselves to one system of measurement. For instance, an astronomer might say that the supergiant star Antares has a diameter of 860 million kilometers and is located at a distance of 150 parsecs from Earth. A further discussion of units is contained in Box 1-3.

Box 1-3   Units of length, time, and mass

To understand and appreciate the universe, we shall need to describe a wide range of circumstances. Astronomers generally use units that are best suited to the topic at hand. For instance, interstellar distances are conveniently expressed in either light years or parsecs, whereas the diameters of the planets are more comfortably presented in kilometers.

Most scientists prefer to use a version of the metric system called the International System of Units, abbreviated SI. According to SI, length is measured in meters (m), time is measured in seconds (s), and mass is measured in kilograms (kg). Throughout this book, we shall try to conform to this widely accepted convention, although there will be notable exceptions because of the wide range of circumstances we encounter in studying the universe. It is therefore useful to see how the basic units of meters, seconds, and kilograms are related to other measures.

When discussing objects on a more or less human scale, sizes and distances are sometimes expressed in millimeters (mm), centimeters (cm), and kilometers (km). These units of length are related to the meter as follows:

$$1 \text{ millimeter} = 0.001 \text{ meter}$$
$$1 \text{ centimeter} = 0.01 \text{ meter}$$
$$1 \text{ kilometer} = 1000 \text{ meters}$$

To make conversions from the English system of inches (in.), feet (ft), and miles (mi), it is necessary to know that

$$1 \text{ inch} = 2.54 \text{ cm}$$
$$1 \text{ foot} = 0.3048 \text{ m}$$
$$1 \text{ mile} = 1.609 \text{ km}$$

Conversions are accomplished by using these equalities to cancel out the unwanted units while introducing the desired units.

Example: Suppose a NASA publication tells you that "the *Saturn V* rocket used to send astronauts to the Moon stands about 363 feet tall." You can convert this to meters as follows:

$$363 \text{ ft} \times \frac{0.3048 \text{ m}}{1 \text{ ft}} = 111 \text{ m}$$

For the convenience of readers used to the English system, we shall often give measurements in both systems; thus we might note that "the diameter of Mars is 6794 km (4222 mi)."

When discussing very small distances, such as the size of an atom, astronomers often use the micrometer (μm) or the angstrom (Å), which are related to the meter and centimeter as follows:

.00000l

$$1 \ \mu\text{m} = 10^{-6} \text{ m}$$
$$1 \ \text{Å} = 10^{-10} \text{ m}$$

Thus 1 μm = 10^4 Å.

The basic unit of time is the second (s). It is related to other units of time as follows:

$$1 \text{ minute} = 60 \text{ s}$$
$$1 \text{ hour} = 3600 \text{ s}$$
$$1 \text{ day} = 86{,}400 \text{ s}$$
$$1 \text{ year} = 3.16 \times 10^7 \text{ s}$$

According to SI, speed is properly measured in meters per second (m/s). Quite commonly, however, speed is also expressed in km/s and mi/hr. These units are related to each other in the following ways:

$$1 \text{ km/s} = 10^3 \text{ m/s}$$
$$1 \text{ km/s} = 2237 \text{ mi/hr}$$
$$1 \text{ mi/hr} = 0.447 \text{ m/s}$$
$$1 \text{ mi/hr} = 1.47 \text{ ft/s}$$

In addition to using kilograms, astronomers sometimes express mass in grams (g) and in solar masses ($M_\odot$), where the subscript $\odot$ is the symbol denoting the Sun. As we shall see, it is especially convenient to use solar masses when discussing the masses of stars and galaxies. These units are related to each other as follows:

$$1 \text{ kg} = 1000 \text{ g}$$
$$1 \ M_\odot = 1.99 \times 10^{30} \text{ kg}$$

On the Earth, a mass of 1 kg weighs 2.20 pounds. Thus we can relate the English pound (lb) to the gram and kilogram:

$$1 \text{ kg} = 2.20 \text{ lb}$$
$$1 \text{ lb} = 453.6 \text{ g}$$

These two equalities require a word of caution, because they are valid only on the Earth's surface. Elsewhere in the universe, where the strength of gravity is different from that on the Earth, a kilogram would weigh something other than 2.20 lb. For instance, in space far from any sources of gravity, a kilogram of matter would have no weight at all. It would be weightless. We will further examine the difference between mass and weight when we discuss gravity in Chapter 4.

Finally, it is often informative to describe a substance by its density, which is mass divided by volume. Because the density of water is 1 g/cm³, which is an easy number to remember, some people prefer to measure density in grams per cubic centimeter. Everyone is familiar with water and so, if we note that the density of a typical rock is 3 g/cm³, it is immediately obvious that rock is 3 times denser than water. According to SI, however, density should be measured in kilograms per cubic meter. It is therefore convenient to know that

$$1 \text{ g/cm}^3 = 1000 \text{ kg/m}^3$$

Box 1-4 Astronomy as a profession

Many thousands of people enjoy astronomy as a hobby. Small telescopes can be purchased in department stores, and most large cities have active astronomy clubs. Attending a "star party" on a cool, clear night to view planets and nebulae with friends can be a very enjoyable experience. But what does it mean to pursue astronomy as a career?

Astronomy is a physical science, and to be an astronomer you need a strong background in both physics and mathematics. It is advisable to major in physics at the undergraduate level. Formal training in astronomy occurs in graduate schools, where it typically takes six years to obtain the Ph.D. degree. A doctorate is usually the minimum requirement for employment as a professional astronomer. In addition, many newly graduated astronomers spend one to three years doing postdoctoral research to hone their talents. Finally, at an age of approximately thirty, the budding astronomer is ready to enter the job market.

The American Astronomical Society (AAS) is the primary professional organization for astronomers. Its membership totals approximately 5000 persons, about 4000 of whom are full-time professional astronomers in the United States. A recent survey of AAS members demonstrated that 50 percent are employed by colleges and universities. As a professor, an astronomer teaches classes and conducts research, often at major observatories in Arizona, Hawaii, or Chile.

About one-sixth of all astronomers are employed by government agencies, especially the National Aeronautics and Space Administration (NASA). As shown in the accompanying graph, industry and aerospace companies each also employ roughly one-sixth of all astronomers.

A star is a hot ball of gas with nuclear reactions occurring at its center, as is an exploding thermonuclear weapon. Thus, astronomers who have specialized in the structure and evolution of stars are ideally suited to work on the development of nuclear weapons. About one-seventh of all astronomers are employed at such facilities as Los Alamos National Laboratory and Lawrence Livermore National Laboratory.

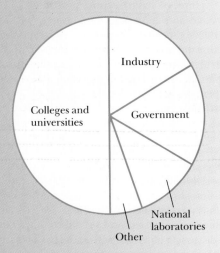

If you are thinking about a career in astronomy, you should know that the field is small and job opportunities are limited. As of 1987, the average income of all professional astronomers was approximately $43,000 per year. The starting salary of an assistant professor is considerably less than this. Women account for only 8 percent of the membership of the AAS, and surveys demonstrate that they are consistently paid less than their male colleagues. Perhaps this will change as the number of women in the sciences grows and they take an increasingly active role in their profession.

At present, there is a surplus of astronomers seeking academic positions at colleges and universities. Recent studies suggest, however, that this employment picture may soon change for the better. During the late 1960s more Ph.D.s in astronomy were awarded than in any comparable period before or since. The people who received these degrees will be reaching retirement age in the late 1990s, and the present rate of incoming new Ph.D.s will not suffice to replace them. Thus, around the year 2000 astronomers are likely to be in short supply.

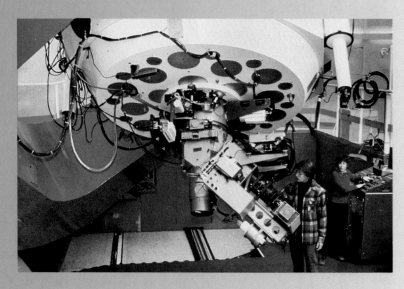

1-8 Astronomy is an adventure of the human mind

An underlying theme of this book is the idea that the universe is rational. It is not a hodgepodge of unrelated things behaving in unpredictable ways. Rather, we find strong evidence for the existence of fundamental laws of physics that govern the nature and behavior of everything in the universe. This powerful unifying concept enables us to explore realms far removed from our earthly experience. Thus a scientist can do experiments in a laboratory to determine the properties of light or the behavior of atoms, then use this knowledge to discover the life cycles of stars and the structure of the universe.

Such discoveries have had a direct and profound influence on humanity. The past four centuries of civilization clearly show that major scientific advances sooner or later make their way into our lives. The world around us is filled with examples of the impact of science in technology, commerce, medicine, entertainment, and transportation. In the near future we can look forward to enjoying the benefits of space technology. Weightlessness and the near-perfect vacuum of space will enable us to manufacture a wide range of exceptional substances, from exotic alloys to ultrapure medicines (see Figure 1-14).

The dreams of Jules Verne and H. G. Wells pale in comparison to the reality of today. Ours is an age of exploration and discovery more profound than any since Columbus and Magellan set sail. We have walked on the Moon, dug into the Martian soil. We have discovered active volcanoes and barren ice fields on the satellites of Jupiter. We have visited the shimmering rings of Saturn. Never before has so much

Figure 1-14 *An Earth-orbiting space station* *A new generation of high-technology materials could easily be manufactured in space. Exotic alloys, foam metals, ultrapure semiconducting crystals, and rare vaccines are among the obvious practical applications of zero-gravity industry. This artist's conception shows the first phase of the space station currently being planned by NASA. (Courtesy of Lockheed)*

been revealed in so short a time. As you proceed through this book, you will come to realize that one of the great lessons of modern astronomy is the awesome power of the human mind to reach out, to explore, to observe, and to comprehend, thereby transcending the limitations of our bodies and the brevity of human life.

Key words
Terms preceded by an asterisk are discussed in the optional boxes.

angle	*density	model	*SI
angular diameter	exponent	nebula (*plural* nebulae)	*small-angle formula
angular measure	galaxy	Newtonian mechanics	solar system
angular size	hypothesis	parsec	subtend
astronomical unit	kiloparsec	powers of ten	supernova
Big Bang	laws of physics	pulsar	theory
black hole	light year	quasar	
burster	megaparsec	scientific method	
degree	minute of arc	second of arc	

Key ideas

- The universe is comprehensible.

- Observations of the heavens have led to discovery of some of the fundamental laws of nature.

- The scientific method is a procedure involving the formulation of hypotheses. These are tested by observation or experimentation, in order to build consistent models or theories that accurately describe phenomena in the universe.

- Exploration of the planets provides information about the origin and evolution of the solar system, as well as the history and resources of the Earth.

- Study of the stars and nebulae provides information about the origin and history of the Sun and the solar system.

- Observations of galaxies provides information about the origin and history of the universe.

- Astronomers use angles to denote the positions and sizes of objects in the sky.

   The size of an angle is measured in degrees, minutes of arc, and seconds of arc.

- The powers-of-ten notation system is a convenient shorthand method of writing numbers.

- A variety of distance units, including the parsec and the light year, are used by astronomers.

Review questions

The answers to computational problems, which are preceded by asterisk, are at the end of the book.

1 What are degrees, minutes of arc, and seconds of arc used for? What is the relationship between these units of measure?

2 With the aid of a diagram, explain what it means to say that the Moon subtends an angle of $\frac{1}{2}°$.

*3 (Basic) How many seconds of arc equal 1°?

4 What is the advantage to the astronomer of using the light year as a unit of distance?

5 How is an AU defined? Give an example of where it would be convenient to use this unit of measure.

6 What is a parsec and how is it related to a kiloparsec and a megaparsec?

*7 (Basic) Write the following numbers using powers-of-ten notation: (**a**) ten million, (**b**) four hundred thousand,

(**c**) six one-hundredths, (**d**) seventeen billion, (**e**) your age, (**f**) the number of pages in this book.

Advanced questions

Tips and tools . . .
The size of an object is related to the angle it subtends by the small-angle formula given in Box 1-2. The method for converting from one unit of measure to another is illustrated in Box 1-3. The volume of a sphere is $\frac{4}{3}\pi r^3$, where r is the sphere's radius and π (pi) is a constant equal to 3.142, to an acceptable degree of accuracy.

*8 (Basic) According to information listed in the *Astronomical Almanac* for 1990, on August 7 the planet Uranus was at a distance of 18.610 AU from Earth. In an appendix at the end of this book, you find that the diameter of Uranus is 51,120 km. What was the angular size of Uranus as seen from Earth on August 7, 1990?

*9 (Basic) The bright star Sirius is 2.65 pc from Earth. (**a**) How long does it take for light emanating from Sirius to reach the Earth? (**b**) What is the distance to Sirius in kilometers?

*10 (Basic) The Sun's mass is 1.99×10^{30} kg, three-quarters of which is hydrogen. The mass of a hydrogen atom is 1.67×10^{-27} kg. How many hydrogen atoms does the Sun contain?

*11 (Basic) The average distance to the Moon is 384,000 km, and the Moon subtends an angle of $\frac{1}{2}°$. Calculate the diameter of the Moon.

*12 At what distance would a person have to hold a quarter (which has a diameter equal to about 2.5 cm) in order for the quarter to subtend an angle of (**a**) 1 degree? (**b**) 1 arc minute? (**c**) 1 arc second?

*13 The speed of light is 3×10^{10} cm/s. How long does it take light to go from the Sun to the Earth?

*14 The mass of the Earth is 5.98×10^{24} kg, and its diameter is 12,756 km. Divide the Earth's mass by its volume to find the average density of the Earth in kg/m^3. How does the Earth's density compare with the density of water, which is 1 g/cm^3?

*15 A hydrogen atom has a radius of about 5×10^{-9} cm. The radius of the observable universe is about 20 billion light years. How many times larger than a hydrogen atom is the universe?

*16 A certain fast-food chain advertises that they have sold over 20 billion hamburgers. If a typical hamburger has a diameter of 15 cm, how far out into space would all these hamburgers reach if placed in a straight line? Would they reach any celestial object?

*17 (*Challenging*) The diameter of the Sun is 1.4×10^{11} cm, and the distance to the nearest star, Proxima Centauri, is 4.3 ly. If the Sun were reduced to the size of a basketball (about 30 cm in diameter), at what distance would Proxima Centauri be from the Sun on this reduced scale?

*18 How many Suns would it take, laid side by side, to reach to the nearest star?

*19 When *Voyager 2* sent back pictures of Neptune during its historic flyby of that planet in 1989, the spacecraft's radio signals traveled for four hours to reach the Earth. How far away was the spacecraft?

*20 Suppose your telescope can give you a clear view of objects and features that subtend angles of at least 2 arc seconds. What is the diameter of the smallest crater you can see on the Moon?

*21 What is the age of the universe in seconds?

Discussion questions

22 How do astronomical observations and experiments differ from those of other sciences?

23 Scientists assume that "reality is rational." Discuss what this means and the thinking behind it.

Observing projects

24 Look up at the sky on a clear, cloud-free night. Is the Moon in the sky? If so, does it interfere with your ability to see the fainter stars? Why do you suppose astronomers prefer to schedule their observations on nights when the Moon is not in the sky?

25 Can you see the Milky Way from where you live on a dark, clear, moonless night? If so, briefly describe its appearance. If not, what seems to be interfering with your ability to see the Milky Way?

26 Look up at the sky on a clear, cloud-free night and note the positions of a few prominent, bright stars relative to such reference markers as rooftops, telephone poles, and treetops. Also note the location from where you made your observations. A few hours later, return to that location and again note the positions of the same bright stars that you observed earlier. How have their positions changed? From these

changes, can you deduce the general direction in which the stars appear to be moving?

For further reading

Asimov, I. *The Measure of the Universe.* Harper & Row, 1983 • An extensive journey through the universe in half powers of ten.

Chaisson, E. *Cosmic Dawn.* Little, Brown, 1981 • This excellent book presents the scientist's worldview by discussing the origins of matter and life.

Dickinson, T. *The Universe and Beyond.* Camden House, 1986 • A beautifully illustrated, nontechnical coffee-table book takes the reader on a tour of modern astronomy. Contrary to what the title suggests, the book stays firmly within our universe.

Gore, R. "The Once and Future Universe." *National Geographic,* June 1983 • A well-written, beautifully illustrated introduction to modern astronomy.

Graham-Smith, F., and Lovell, B. *Pathways to the Universe.* Cambridge University Press, 1988 • An introduction to astronomy by two distinguished British astronomers, consisting of 21 separate topics, from "How We Became Astronomers" to "Cosmology."

Hartmann, W., and Miller, R. *Cycles of Fire.* Workman, 1987 • An oversized, colorful album of paintings by two noted astronomical artists, accompanied by an excellent nontechnical tour of stars, nebulae, and galaxies.

Jastrow, R. *Red Giants and White Dwarfs,* 2nd ed. Norton, 1979 • An updated revision of a classic introduction to astronomy that offers one of the best expositions of cosmic evolution.

Morrison, P., Morrison, P., and the Office of Charles and Ray Eames. *Powers of Ten.* Scientific American Books, 1982 • A tour of the universe where each step corresponds to a power of ten.

Sagan, C. *Cosmos.* Random House, 1980 • A lavishly illustrated "personal voyage" through the universe written in conjunction with a popular television series of the same name.

Seielstad, G. *Cosmic Ecology.* University of California Press, 1983 • This eloquent survey traces evolutionary processes from the Big Bang to the development of intelligence on Earth.

Weiskopf, V. *Knowledge and Wonder: The Natural World as Man Knows It,* 2nd ed. MIT Press, 1979 • An articulate introduction to the world of science by a distinguished physicist.

Sandra M. Faber, Professor of Astronomy at the University of California, Santa Cruz, and Astronomer at the Lick Observatory, was intrigued with the origins of the universe when she entered Swarthmore College. She completed her doctorate in astronomy at Harvard University, but did her dissertation at the Carnegie Institution's Department of Terrestrial Magnetism, where she was influenced by the distinguished woman astronomer Vera Rubin.

Dr. Faber chaired the now-legendary group of astronomers called the "Seven Samurai," who surveyed the nearest 400 elliptical galaxies and discovered a new mass concentration, the "Great Attractor."

Sandra Faber is the recipient of many honors and awards for her research, including the Bok Prize from Harvard University in 1978. She was elected to the National Academy of Sciences in 1985 and won the coveted Dannie Heineman Prize from the American Astronomical Society, given in recognition of a sustained body of especially influential astronomical research, in 1986. Recently, she also has been appointed to the Board of Trustees of the Carnegie Institution.

Sandra Faber WHY ASTRONOMY?

As you study astronomy, it is appropriate for you to ask, why am I studying this subject? What good is it for human beings in general and for me in particular? Astronomy, it must be admitted, does not offer the same practical benefits as other sciences. So how can astronomy be important to your life?

On the most basic level, I like to think of astronomy as providing the introductory chapters for the ultimate textbook on human history. Ordinary texts start with recorded history, going back some 3000 years. For events before that, we consult archeologists and anthropologists, who tell us about the early history of our species. For knowledge of the time before that, we consult paleontologists, biologists, and geologists about the origin and evolution of life and the evolution of our planet, altogether going back some 5 billion years. Astronomy tells us about the vast stretch of time before the origin of the Earth, the 10 billion years or so that saw the formation of the Sun and solar system, the Milky Way Galaxy, and the origin of the universe in the Big Bang. Knowledge of astronomy is essential for a well-educated person's view of history.

Astronomy challenges our belief system and impels us to put our "philosophical house" in order. Take the origin of the world, for example. According to Genesis in the Bible, the world and all in it were created in six days by the hand of God. However, the ancient Egyptians believed that the Earth arose spontaneously from the infinite waters of the eternal universe, Nun. And Alaskan legends taught that the world was created by the conscious imaginings of a creator deity named Father Raven.

The modern astronomical story of the creation of the Earth differs from all these in that it claims to be buttressed by physics and by observational fact. The Sun, it is asserted, formed via gravitational collapse from a dense cloud of interstellar dust and gas about 5 billion years ago. The planets coalesced at the same time as condensates within the swirling solar nebula. This process altogether took not weeks or days but several hundred thousand years. Confidence in this theory stems from our looking out into the Galaxy and seeing young stars actually form in this way.

At issue here, really, is the deep question of how we are to gain information about the nature of the physical world—whether by revelation and intuition or by logic and observation. Where science stops and faith begins is a thorny issue for all human beings, but particularly so for astronomers—and for astronomy students.

Astronomy cultivates our notions about cosmic time and cosmic evolution. It is all too easy, given the short span of human life, to overlook the fact that the universe is a dynamic and changing place. When I first entered astronomy, I found myself handicapped by a conservative mindset that assumed, subtly, that celestial objects were unchanging. This might seem strange for someone whose avowed interest was in learning how galaxies formed. Such are the vagaries of the human mind! Fortunately I lost this bias as I grew older, mainly, I think, by observing evolution all around me. Many things that had seemed immutable to me as a child—social structures, customs, even the physical environment—turned out not to be so. With that realization, the scales fell from my eyes and I was able to accept cosmic evolution as a core concept in my thinking about the universe.

Following closely on the concept of evolution is the concept of fragility—if something can change, it might even actually disappear someday. The most obvious example is the limited lifetime of our Sun: in another 5 billion years of so, the Sun will enter its death throes, during which it will swell up and brighten to 1000 times its present luminosity, and, in the process incinerate the Earth.

Five billion years is long enough in the future that neither you nor I feel any personal responsibility for preparing the human race to meet this challenge. However, there are many other cosmic catastrophes that will befall us in the meantime. The Earth will be hit by sizable pieces of space junk; eroded impact craters show this happens all the time—every few million years or so. Debris thrown up into the atmosphere from these events could be so extensive as to totally alter the climate for an extended period of time, possibly leading to mass extinctions. There will be enormous volcanic eruptions that will make Mount St. Helens's look like a small firecracker. Another Ice Age, which will totally disrupt modern agriculture, is virtually certain within the next 20,000 years, unless we cook the Earth ourselves first by burning too much fossil fuel.

Closer to the human sphere, such common notions as the inevitability of human progress, the desirability of endless economic growth, the ability of the Earth to support its growing human population, are all based on limited experience and may not—probably *will* not—prove viable in the long run. Consequently, we must rethink who we are as a species and what our proper activity on Earth is. Contemplating cosmic time puts us into the appropriate frame of mind for grappling with these issues. These are problems that are very long range and involve the whole human race, yet are totally under human control and are vital to our long-term survival and well-being.

Astronomy is essential to developing a perspective on human existence and its relation to the cosmos. What question could be grander and at the same time more relevant to us? We can hardly contemplate human existence in general without taking at least a brief look at our own lives.

Earth is a mote in the vast cosmic sea, and our lives, in the evolutionary scheme of the universe, would seem at face value to be insignificant. Should that thought terrify us, de-priving our lives of any meaning? Should we seek to provide meaning by introducing an intelligent creator? Or should we be fully comforted merely to know that this great universe has given birth to us in an entirely natural, perhaps even inevitable, chain of events?

Not long ago I visited the southern hemisphere for the first time to observe the sky from Chile. At that latitude, the center of the Milky Way passes overhead, where it makes a grand show, not the miserable, fuzzy patch that is visible, low on the horizon, from North America. Stepping out on the observatory catwalk one morning before dawn, I saw for the first time the galactic center in all its glory soaring overhead. At that instant, my perceptions underwent a profound alteration. One moment I was standing in an earthbound "room" with the Galaxy painted on the "ceiling." The next instant, the walls of this earthly room fell away, and I was floating freely in outer space, viewing the center of the Galaxy from an outpost on its periphery. The sense of depth was breathtaking: sparkling star clusters and dense dust clouds were etched against the blazing inner bulge of the Galaxy many thousands of light years distant. I saw the Galaxy as an immense, three-dimensional object in space for the first time. It became "real." At that moment, my sense of place in the universe underwent a fundamental and irreversible change. I remained a citizen of the Earth, but I became also a citizen of the Galaxy.

Many astronomers, myself included, believe that the ultimate, proper concept of "home" for the human race is our universe. It seems more and more likely that there are a very large number of other universes. Virtually none of them would appear to resemble ours; they may have different numbers of space and time dimensions—perhaps even no space or time—and very different physical laws. The vast majority probably are incapable of harboring intelligent life as we know it.

The parallel with Earth is striking. Among the solar system planets, only Earth can support human life. Among the great number of planets which likely exist in our Galaxy, only a small fraction are apt to be such that we could call them home. The fraction of hospitable universes is likely to be smaller still. Our universe would seem, then, to be the ultimate instance of "home": a sanctuary in the middle of a vast sea of inhospitable universes.

I began this essay by talking about history and ended with questions that border on the ethical and religious. Astronomy is like that: it offers a modern-day version of Genesis—and of the Apocalypse, too. Astronomy stimulates and challenges us on our deepest, most human levels. I hope that, during this course, you will be able to take time out to contemplate the broader implications of what you are studying. This is one of the rare opportunities in life to think about who you are and where you and the human race are going. Don't miss it.

2

Knowing the Heavens

Before the dawn of recorded history, our ancestors divided the starry sky into constellations. Modern astronomers still use many of the constellations described by ancient Babylonians and Greeks. These starry patterns named for mythical figures help us find our way around the sky, which is conveniently described as the celestial sphere. The annual path of the Sun across the constellations of the zodiac, together with the Earth's equator and poles, serve as guideposts. We find that the seasons are related to the tilt of the Earth's axis of rotation and see that the axis itself is slowly changing its orientation. The motions of the Earth are crucial in timekeeping and the calendar, which are traditional, practical applications of astronomy.

Circumpolar star trails   This long exposure, aimed at the south celestial pole, shows the rotation of the sky. The foreground building is part of the Anglo-Australian Observatory, which houses one of the largest telescopes in the southern hemisphere. The telescope's primary mirror is 3.9 m (12.8 ft) in diameter. During the exposure, someone carrying a flashlight walked along the dome's outside catwalk. Another flashlight made the wavy trail at ground level. (Anglo-Australian Observatory)

2-1 Eighty-eight constellations cover the entire sky

The roots of astronomy reach back to the dawn of civilization. This ancient heritage is most apparent in the **constellations** (from the Latin words meaning "group of stars"). Perhaps it was shepherds tending their flocks at night, or families sitting by the fire, or priests studying the heavens who first imagined pictures among groupings of stars.

You may already be familiar with some of these patterns in the sky, such as the Big Dipper, which is actually part of a large constellation called Ursa Major (the great bear). Many of these constellations, such as Orion, in Figure 2-1, have names from the myths and legends of antiquity. Although some star groupings vaguely resemble the figures they are supposed to represent, most do not.

On modern star charts, the entire sky is divided into 88 constellations. Some constellations cover large areas (Ursa Major being one of the biggest), others very small areas (Crux being the smallest). The constellations are used to specify certain regions of the sky. For instance, we might speak of "the nebula M42 in Orion" much as we would refer to "the Ural Mountains in the Soviet Union."

Like constellations, many stars have ancient names. Star names and catalogues are discussed in Box 2-1.

People who spend time outdoors at night are very familiar with the apparent motions of the constellations. If you lack this knowledge, take the time to observe the basic facts of astronomy yourself. Go outdoors soon after dark, find a spot away from bright lights, and note the patterns of stars in the sky. A few hours later, check again. You will find that the entire pattern of stars (including the Moon, if it is visible) has shifted its position. New constellations will have risen above the eastern horizon and some will have disappeared below the western horizon. If you can check again before dawn, you will find low in the western sky the stars that had just been rising when the night began.

The constellations that you can see in the sky change slowly from one night to the next. This shift occurs as the Earth orbits the Sun (see Figure 2-2). The Earth takes a full year to go once around the Sun, which means that the darkened, nighttime side of the Earth is gradually turned toward different parts of the heavens. If you follow a particular star on successive evenings, you will find that it rises approximately four minutes earlier each night.

Constellations can help you orient yourself and find your way around the sky. For instance, if you live in the northern hemisphere, you can use the Big Dipper in Ursa Major to find the north direction by drawing a straight line through the two stars at the front of the Big Dipper's bowl, as shown in Figure 2-3. The first moderately bright star you come to is

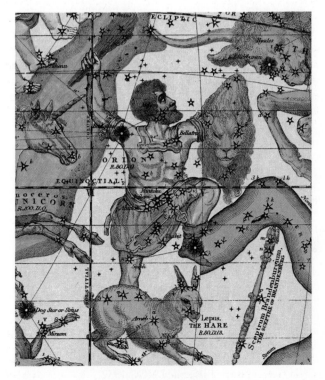

Figure 2-1 Orion *Orion is a prominent winter constellation. From the United States, Orion is easily seen high above the southern horizon from December through March. The photograph was a four-minute* exposure on Kodak film. Because of the time exposure, the colors of the stars are quite noticeable. The fanciful drawing of Orion is from an 1835 star atlas. (Courtesy of R. Mitchell and Janus Publications)

Box 2-1 Star names and catalogues

The customs and traditions for naming stars have changed over the centuries. Many of the brightest stars in the sky have Arabic names, assigned in medieval times when astronomy was widely studied among Islamic nations. The diagram shows the Big Dipper with the Arabic names for its seven brightest stars. The North Star received its formal name, Polaris, later, when Latin was the language used by European astronomers.

As you might suspect, memorizing exotic star names is an unwelcome burden for many astronomers. A simpler system, invented by Johann Bayer in 1603, uses the letters of the Greek alphabet and the constellation names.

The 24 lowercase letters of the Greek alphabet are as follows:

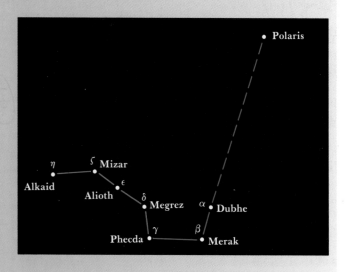

α alpha	ν nu
β beta	ξ xi
γ gamma	o omicron
δ delta	π pi
ϵ epsilon	ρ rho
ζ zeta	σ sigma
η eta	τ tau
θ theta	υ upsilon
ι iota	ϕ phi
κ kappa	χ chi
λ lambda	ψ psi
μ mu	ω omega

In constructing a star name, a Greek letter is used with the Latin possessive form of the name of the constellation in which the star is located. The 12 constellations of the zodiac with their Latin possessives are as follows:

Constellation	Possessive
Aries	Arietis
Taurus	Tauri
Gemini	Geminorum
Cancer	Cancri
Leo	Leonis
Virgo	Virginis
Libra	Librae
Scorpius	Scorpii
Sagittarius	Sagittarii
Capricornus	Capricorni
Aquarius	Aquarii
Pisces	Piscium

In most cases, the brightest star in the constellation is α, the second brightest is β, the third is γ, and so on. For example, the brightest star in Libra (the scales) is called α Librae. This name is more informative than its Arabic name, Zubenelgenubi.

The 24 letters in the Greek alphabet allow Johann Bayer's system to name only the brightest two dozen stars in a given constellation. However, astronomers are often interested in very faint stars, many of which are too dim to be seen with the naked eye. In referring to these fainter stars, astronomers make use of designations from standard star catalogues.

One of the first major star catalogues, called the **Bonner Durchmusterung** (Bonn Comprehensive Survey), was produced in Germany in the mid-1800s by F. W. Argelander at the Bonn Observatory. This catalogue lists the positions of 320,000 stars. Each star in the catalogue is designated by a "BD number." For example, the star BD+5 1668 is a dim star in the constellation of Monoceros (the unicorn).

Another commonly used catalogue is the **Henry Draper Catalogue**, which was compiled in the United States around 1920. (It is named after a physician whose widow financed the project.) Stars from this catalogue are listed by their "HD numbers." For example, HD 87901 is α Leonis (also called Regulus), the brightest star in Leo (the lion).

Tens of thousands of faint nonstellar objects such as galaxies, nebulae, and star clusters are scattered across the sky. As with the faint stars, astronomers refer to these objects by their catalogue designations. Roughly a hundred of the brightest nonstellar objects are listed in a well-known catalogue by the eighteenth-century French astronomer Charles Messier. Astronomers often refer to these objects by their "M numbers." For example, the Orion Nebula (see Figure 1-6) is the forty-second object on Messier's list and is therefore called M42.

During the nineteenth century, William Herschel and his son John Herschel extended this list to include nearly 5000 faint nonstellar objects. Their list, in turn, was enlarged by J. L. E. Dreyer and published in 1888 as the **New General Catalogue**. Further additions were made during the next 20 years so that this catalogue and its supplements ultimately listed nearly 15,000 objects. Astronomers frequently refer to galaxies and nebulae by their "NGC numbers." For example, the Crab Nebula (see Figure 1-7) is called NGC 1952.

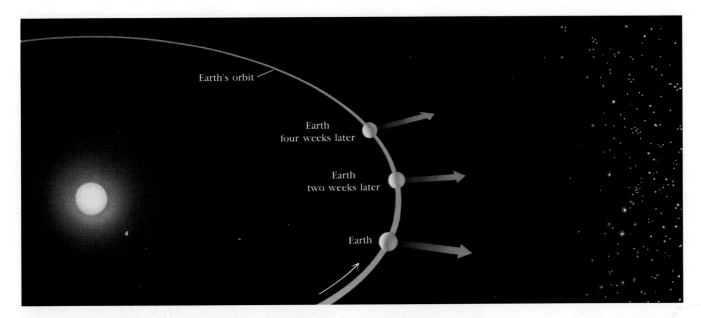

Figure 2-2 Our changing view of the night sky As we orbit the Sun, the nighttime sky of the Earth gradually turns toward different parts of *the sky. Thus, the constellations that we can see change slowly from one night to the next.*

Figure 2-3 The Big Dipper as a guide
This star chart shows how the Big Dipper can be used to point out the north star as well as the brightest stars in three other constellations. The angular distance from Polaris to Spica is about 101°, so a large fraction of the sky is covered.

Figure 2-4 The "Winter Triangle" This star chart shows the eastern sky as it appears during the evening in December. Three of the brightest stars in the sky make up the "winter triangle," which is about 26° on a side. In addition to the constellations involved in the triangle, Gemini (the twins), Auriga (the charioteer), and Taurus (the bull) are shown.

Polaris, also called the North Star because it is located almost directly over the Earth's north pole. If you draw a line from Polaris straight down to the horizon, you have found the north direction.

By drawing a line through the two stars at the rear of the bowl of the Big Dipper you can find Leo (the lion). As Figure 2-3 shows, that line points toward Regulus, the brightest star in the "sickle" tracing the lion's mane. By following the handle of the Big Dipper, you can locate the bright reddish star Arcturus in Boötes (the shepherd) and the prominent bluish star Spica in Virgo (the virgin). The saying "Follow the arc to Arcturus and speed to Spica" may help you remember these stars, which are conspicuous in the evening sky during the spring and summer months.

During the winter months in the northern hemisphere you can see some of the brightest stars in the sky. Many of them are in the vicinity of the "winter triangle," which connects bright stars in the constellations of Orion (the hunter), Canis Major (the large hunting dog), and Canis Minor (the small hunting dog), as shown in Figure 2-4. The winter triangle is nearly overhead during the middle of winter at midnight.

A similar feature, called the "summer triangle," graces the summer sky. This triangle connects the brightest stars in

Lyra (the harp), Cygnus (the swan), and Aquila (the eagle), as shown in Figure 2-5. A conspicuous portion of the Milky Way forms a beautiful background for these constellations, which are nearly overhead during the middle of summer at midnight.

A set of star charts for the evening hours of selected months of the year is included at the end of this book. You may find stargazing a surprisingly enjoyable experience, and these star charts will help you identify many well-known constellations.

2-2 It is often convenient to imagine that the stars are located on the celestial sphere

As you look at the sky on a clear, dark night, you might think that you can see millions of stars. Actually, the unaided human eye can detect only about 6000 stars over the entire sky. Because only half of the sky is above the horizon at any one time, you can see roughly 3000 stars. Of course, the Earth rotates from west to east once every 24 hours,

Figure 2-5 The "Summer Triangle"
This star chart shows the northeastern sky as it appears in the evening in June. The angular distance from Deneb to Altair is about 38°. In addition to the three constellations involved in the triangle, the faint constellations of Sagitta (the arrow) and Delphinus (the dolphin) are shown.

which is why we have day and night, so the stars rise in the east and set in the west, as do the Sun and Moon. This daily, or **diurnal**, motion of the stars is very apparent in time-exposure photographs such as Figure 2-6.

Many ancient societies believed the Earth to be at the center of the universe. They also imagined the stars to be attached to the surface of a huge sphere centered on the Earth. This imaginary sphere, called the **celestial sphere**, is still a useful concept.

Of course, the stars are actually scattered at various distances from Earth. Many of the brightest stars visible in the sky are 10 to 1000 light years away. These distances are so immense that all the stars appear to be equally remote, fixed to a spherical backdrop. We can use this spherical backdrop as a reference to specify the directions to objects in the sky.

The Earth is at the center of the celestial sphere, as shown in Figure 2-7. We can project the Earth's key geographic features out into space to establish directions and bearings on the celestial sphere. For example, if the Earth's equator is projected onto the celestial sphere, we obtain the **celestial equator**. The celestial equator divides the sky into northern and southern hemispheres, just as the Earth's equator divides the Earth into two hemispheres.

Figure 2-6 Star trails around the north celestial pole The Earth rotates once every 24 hours, and hence the stars appear to move across the sky. This time-exposure photograph is centered on the north celestial pole, which is directly over the Earth's north pole and is a point about which the heavens appear to revolve. The north celestial pole is slightly less than 1° from the moderately bright star Polaris. (U.S. Naval Observatory)

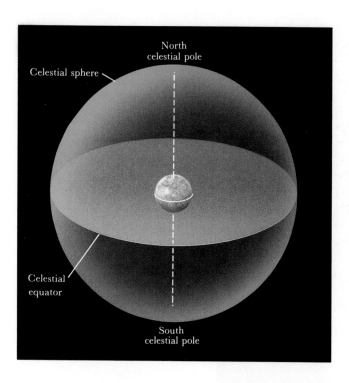

Figure 2-7 The celestial sphere *The celestial sphere is the apparent sphere of the sky. The celestial equator and poles are obtained by projecting the Earth's equator and axis of rotation out into space. The celestial poles are therefore located directly over the Earth's poles.*

We can also imagine projecting the Earth's north and south poles out into space along the Earth's axis of rotation. This gives us the **north celestial pole** and the **south celestial pole**. The two celestial poles are the points of intersection between the Earth's axis of rotation, extended out into space, and the celestial sphere (see Figure 2-7). The celestial equator and poles are extremely useful in setting up a coordinate system with which you can specify the positions of objects anywhere in the sky. The most commonly used coordinate system, which involves two angles (right ascension and declination) that are quite like longitude and latitude on Earth, is described in Box 2-2.

2-3 The seasons are caused by the tilt of the Earth's axis of rotation

In addition to rotating on its axis every 24 hours, the Earth revolves around the Sun every $365\frac{1}{4}$ days. The seasonal changes we experience on Earth during a year result from the way the Earth's axis of rotation is tilted with respect to the plane of the Earth's orbit around the Sun.

The Earth's axis of rotation is not perpendicular to the Earth's orbit but is instead tilted by $23\frac{1}{2}°$ away from the per-

Box 2-2 Celestial coordinates

To denote the positions of objects in the sky, astronomers use a system called "right ascension" and "declination," which is very similar to longitude and latitude. Declination corresponds to latitude. The **declination** of an object is its angular distance north or south of the celestial equator, measured along a circle passing through both celestial poles, as shown in the diagram.

Right ascension corresponds to longitude. Astronomers measure right ascension from a specific point, called the vernal equinox, on the celestial equator. This point is one of two locations where the Sun crosses the celestial equator during its apparent annual motion. In the Earth's northern hemisphere, the season of spring is said to begin when the Sun reaches the vernal equinox in late March. The **right ascension** of an object in the sky is the angular distance from the vernal equinox eastward along the celestial equator to the circle used in measuring its declination (see diagram). Following traditional practice, astronomers measure this angular distance in time units (hours, minutes, seconds) corresponding to the time required for the celestial sphere to rotate through this angle.

In catalogues of stars, galaxies, and nebulae, the positions of objects are given by their right ascension and declination. This valuable information tells the astronomer precisely where to point a telescope.

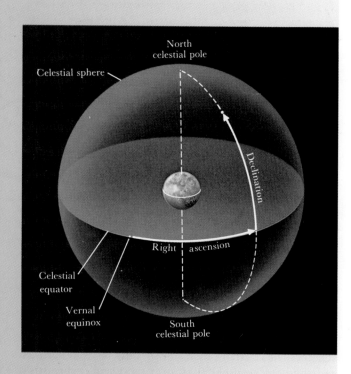

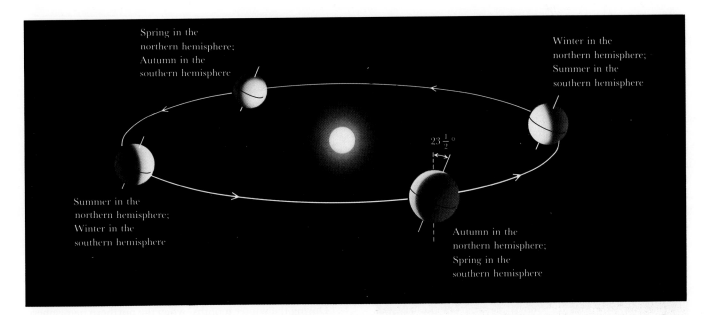

*Figure 2-8 The seasons The Earth's axis of rotation is inclined 23½°
away from the perpendicular to the plane of the Earth's orbit. The Earth
maintains this orientation (with its north pole aimed at the north celestial pole, near the star called Polaris) throughout the year as the Earth*
*orbits the Sun. Consequently, the amount of solar illumination and the
number of daylight hours at any location on Earth vary in a regular
fashion throughout the year.*

pendicular, as shown in Figure 2-8. As it orbits the Sun, the
Earth maintains this tilted orientation with the Earth's north
pole pointing toward Polaris. Thus, during part of the year
we find the northern hemisphere tilted toward the Sun and
the southern hemisphere tilted away. This produces summer
in the north and winter in the south. Half a year later the
situation is reversed, with winter in the northern hemisphere
(now tilted away from the Sun) and summer in the southern
hemisphere. As Figure 2-8 illustrates, the seasons of spring
and fall occur between winter and summer, when the two
hemispheres are receiving roughly equal amounts of illumination from the Sun.

We can represent this seasonal phenomenon on the celestial sphere by examining the Sun's apparent motion against
the background constellations. As the Earth moves along its
orbit, the Sun appears to shift its position gradually from
day to day, as viewed against the background stars. The Sun
therefore traces out a path in the sky during the course of a
year. This apparent path of the Sun on the celestial sphere is
called the **ecliptic**. Because of the tilt of the Earth's axis of
rotation, the ecliptic and the celestial equator are inclined to
each other by 23½° (see Figure 2-9). There are 365¼ days
in a year and 360° in a circle, so the Sun appears to move
along the ecliptic at a rate of approximately 1° per day.

As Figure 2-9 demonstrates, the ecliptic and the celestial
equator intersect at only two points, which are exactly opposite each other on the celestial sphere. Both points are called
equinoxes (from the Latin words meaning "equal night")

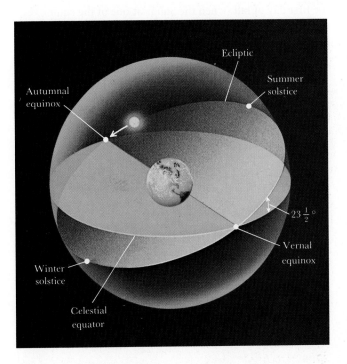

*Figure 2-9 Equinoxes and solstices The ecliptic is the apparent an-
nual path of the Sun as projected onto the celestial sphere. Inclined to
the celestial equator by 23½° because of the tilt of the Earth's axis of
rotation, the ecliptic intersects the celestial equator at two points called
equinoxes. The northernmost point on the ecliptic is called the summer
solstice, and the corresponding southernmost point, the winter solstice.
The Sun is shown in its position for approximately September 1.*

because, when the Sun appears at either of these points, day and night are each 12 hours long at all locations on Earth.

The **vernal equinox** marks the beginning of spring in the northern hemisphere, when the Sun moves northward across the celestial equator on about March 21. The **autumnal equinox** marks the moment when fall begins in the northern hemisphere (about September 22), when the Sun moves southward across the celestial equator.

Incidentally, we should always remember that the northern and southern hemispheres experience opposite seasons at any given point in time. For example, March 21 marks the beginning of autumn for people in Australia. Obviously, these astronomical terms come from a time when virtually all astronomers lived north of the equator.

Between the vernal and autumnal equinoxes, there are two other significant locations along the ecliptic. The point on the ecliptic farthest north of the celestial equator is called the **summer solstice** (from the Latin for "solar stand-still," denoting the fact that the Sun is as far from the celestial equator as it can get). It is the location of the Sun at the moment summer begins in the northern hemisphere (about June 21). At the beginning of winter, the Sun is farthest south of the celestial equator, at a point called the **winter solstice** (about December 21).

Seasonal changes in the Sun's daily path across the sky are diagrammed in Figure 2-10. On the first day of spring or the first day of fall (when the Sun is at one of the equinoxes), the Sun rises directly in the east and sets directly in the west. Day and night are of equal duration.

During the summer months, when the northern hemisphere is tilted toward the Sun, sunrise occurs in the northeast and sunset occurs in the northwest. The Sun spends more than twelve hours above the horizon, and it passes high in the sky at noon. It is hot during the summer, not only because of extended daylight hours, but also because sunlight at midday strikes the ground more nearly perpendicular than at any other time of the year. This nearly perpendicular impact of sunlight heats the ground more efficiently than when the Sun is lower in the sky, as it is during non-summer months.

Incidentally, the point in the sky directly overhead is called the **zenith**, as shown in Figure 2-10. At the summer solstice, the Sun is as far north of the celestial equator as it gets, giving the greatest number of daylight hours. There are even certain locations on Earth (north of the Arctic Circle, as discussed in Box 2-3) where the Sun does not set at all during summer nights.

During the winter months, when the northern hemisphere is tilted away from the Sun, sunrise occurs in the southeast. Daylight lasts for less than 12 hours as the Sun skims low over the southern horizon and sets in the southwest. Night is longest when the Sun is at the winter solstice. In fact, north of the Arctic Circle a winter night lasts for 24 hours, as described in Box 2-3.

2-4 Precession is a slow, conical motion of the Earth's axis of rotation

Ancient astronomers realized that the Moon orbits the Earth. It takes roughly four weeks for the Moon to go once around the Earth. The word "month" comes from the same Old English root as the word "moon."

As seen from the Earth, the Moon is never far from the ecliptic. In other words, the Moon's path through the constellations is close to the Sun's path. The Moon's path remains within a band called the **zodiac** that extends about 8° on either side of the ecliptic. Twelve famous constellations (listed in Box 2-1) lie along the zodiac. The Moon is generally found in one of these twelve constellations. As the Moon moves along its orbit, it appears north of the celestial equator for about two weeks, then south of the celestial equator for about the next two weeks.

Both the Sun and the Moon exert a gravitational pull on the Earth. We will discuss gravity in greater detail in Chapter 4. For now, it is sufficient to realize that gravity is the universal attraction of matter for other matter.

The gravitational pull of the Sun and the Moon affects the Earth's rotation because the Earth is not a perfect sphere. Our planet is slightly fatter across the equator than it is from pole to pole. The equatorial diameter of the Earth is

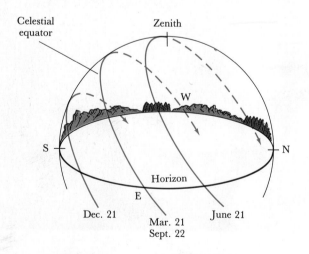

Figure 2-10 The Sun's daily path *On the first day of spring and the first day of fall, the Sun rises precisely in the east and sets exactly in the west. During summer, the Sun rises in the northeast and sets in the northwest. The maximum northerly excursion of the Sun occurs at the summer solstice. In the winter, the Sun rises in the southeast and sets in the southwest, with its maximum southerly excursion occurring at the winter solstice.*

Box 2-3 Tropics and circles

The inclination of the ecliptic to the celestial equator produces some noteworthy occurrences at particular locations on Earth. For example, on any date during the year the Sun appears directly overhead at "high noon" along a band of locations encircling the Earth. The northernmost band where this happens is called the **Tropic of Cancer**, which circles the Earth at a latitude of $23\frac{1}{2}°$ north of the equator. On the day of the summer solstice, the Sun is at the zenith at high noon, as seen from anywhere along the Tropic of Cancer.

The corresponding southern location is called the **Tropic of Capricorn**, $23\frac{1}{2}°$ south of the equator. On the day of the winter solstice, when the Sun has reached its southernmost declination, the Sun passes directly overhead at high noon, as viewed from any location along this tropic.

There are also regions on Earth near the two poles where the Sun spends many consecutive days either above or below the horizon. For example, as seen by someone standing at the north pole, the Sun rises on the day of the vernal equinox and stays above the horizon for six months, setting on the day of the autumnal equinox. During the next six months, the north pole is subjected to six continuous months of night, because the Sun is too far south to be seen.

The region around the north pole where you can see the Sun for 24 continuous hours on at least one day of the year is bounded by the **Arctic Circle**. The Arctic Circle lies $23\frac{1}{2}°$ south of the north pole (that is, at a latitude of $90° - 23\frac{1}{2}° = 66\frac{1}{2}°$ N).

The corresponding region around the south pole is bounded by the **Antarctic Circle**, as shown in the diagram below. While the Earth's north polar regions are being subjected to continuous night near the time of the winter solstice, antarctic explorers enjoy the "midnight sun."

The Earth's inclination and your own latitude conspire to keep some constellations permanently above your horizon and others forever hidden from your view. For example, as viewed from the United States, the north celestial pole is always above the horizon. In fact, the elevation of the north celestial pole above the horizon is exactly equal to your latitude on Earth. As the Earth turns, constellations near the north celestial pole revolve about the pole, never rising or setting. Similarly, as shown above, there are constellations around the south celestial pole that are too far south to be seen from northern latitudes.

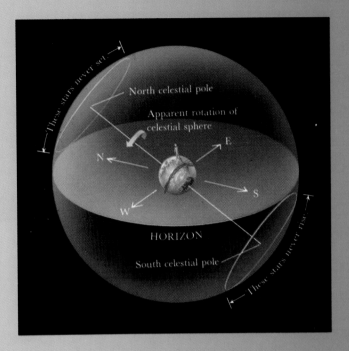

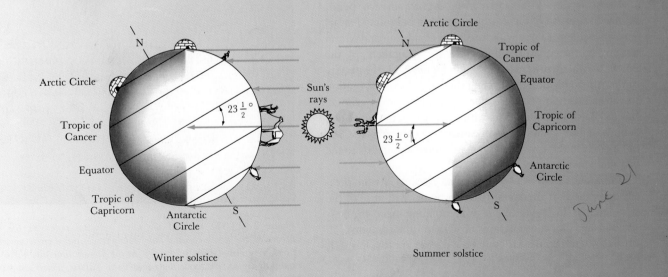

Winter solstice

Summer solstice

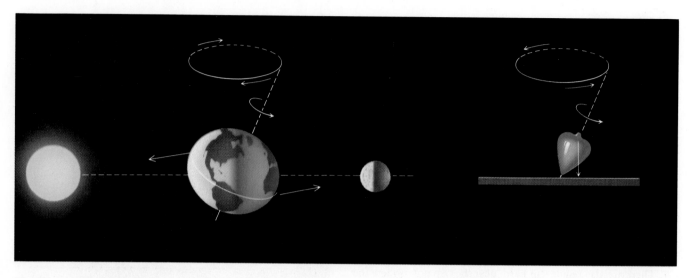

Figure 2-11 *Precession* *The gravitational pull of the Moon and the Sun on the Earth's equatorial bulge causes the Earth to precess. As the Earth precesses, its axis of rotation slowly traces out a circle in the* *sky. The situation is analogous to a spinning top. As the top spins, the Earth's gravitational pull causes the top's axis of rotation to move in a circle.*

43 kilometers (27 miles) larger than the diameter measured from pole to pole. The Earth is therefore said to have an "equatorial bulge." The gravitational pull of the Moon and the Sun on this equatorial bulge gradually changes the orientation of the Earth's axis of rotation, producing a phenomenon called precession.

The situation is analogous to a spinning toy top, as illustrated in Figure 2-11. If the top were not spinning, gravity would pull the top over on its side. When the top is spinning, the combined actions of gravity and rotation cause the top's axis of rotation to trace a circle, a motion called **precession**.

As the Sun and Moon move along the zodiac, each spends half its time north of the Earth's equatorial bulge and half its time south of it. The gravitational pull of the Sun and Moon tugging on the equatorial bulge tries to "straighten up" the Earth. In other words, as sketched in Figure 2-11, the gravity of the Sun and Moon tries to pull the Earth's axis of rotation toward a position perpendicular to the plane of the ecliptic. But the Earth is spinning. As with the toy top, the combined actions of gravity and rotation cause the Earth's axis to trace out a circle in the sky while it remains tilted about $23\frac{1}{2}°$ to the perpendicular.

The Earth's rate of precession is fairly slow. At the present time, the Earth's axis of rotation points within 1° of the star Polaris. In 3000 BC, it was pointing near the star Thuban in the constellation of Draco (the dragon). In AD 14,000, the "pole star" will be Vega in Lyra (the harp). It takes 26,000 years for the north celestial pole to complete one full precessional circle around the sky (see Figure 2-12).

Of course, the south celestial pole executes a similar circle in the southern sky.

As the Earth's axis of rotation precesses, the Earth's equatorial plane also moves. Because the Earth's equatorial plane defines the location of the celestial equator in the sky, the celestial equator precesses as well. The intersections of the

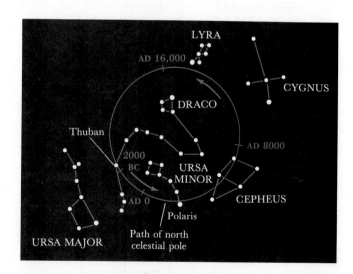

Figure 2-12 *The path of the north celestial pole* *As the Earth precesses, the north celestial pole slowly traces out a circle among the northern constellations in the sky. At the present time, the north celestial pole is near the moderately bright star Polaris, which serves as the "pole star."*

celestial equator and the ecliptic define the equinoxes, as we saw in Figure 2-9, so these key locations in the sky also shift slowly from year to year. In fact, the entire phenomenon is often called the **precession of the equinoxes**. Today the vernal equinox is located in the constellation Pisces (the fishes). Two thousand years ago, it was in Aries (the ram). Around the year AD 2600, the vernal equinox will move into Aquarius (the water bearer). Incidentally, astrological terms like the "Age of Aquarius" involve boundaries in the sky that are not recognized by astronomers and are generally not even related to the positions of the constellations.

As discussed in Box 2-2, the astronomer's system of locating heavenly bodies by their right ascension and declination is tied to the positions of the celestial equator and the vernal equinox. Thus, because of precession, the coordinates of stars in the sky are constantly changing. These changes are very small and gradual, but they add up over the years. To cope with this difficulty, astronomers always make note of the date (called the **epoch**) for which a particular set of coordinates is precisely correct. Star catalogues and star charts are updated periodically. Most new catalogues and star charts are now being prepared for the epoch 2000. The coordinates in these reference books, which are precise for January 1, 2000, will require very little correction over the next few decades.

2-5 Keeping track of time is traditionally a responsibility of astronomers

Astronomers have traditionally been responsible for telling time. The need for accurate timekeeping is universal. The Pharaoh wanted to know when the Nile would flood. The proper timing of seasonal religious events has always been extremely important in most cultures. Even today, we have many reasons to keep the calendar consistent with the cycle of the seasons. For example, many of our holiday traditions involve seasonal observances.

In measuring short time intervals, we want our system of timekeeping to reflect the position of the Sun in the sky. The Sun's position determines whether we are awake or asleep and whether it is time for breakfast or for dinner. Thousands of years ago, the sundial was invented to keep track of **apparent solar time**. An apparent solar day is the time it takes the Sun to go from one high noon to the next.

For more formal measurements, astronomers use the **meridian**, which is a north–south circle on the celestial sphere that passes through the **zenith** (the point directly overhead) and both celestial poles, as shown in Figure 2-13. Local noon occurs when the Sun crosses the meridian above the horizon. Local midnight occurs when it makes the opposite crossing below the horizon, although this crossing can-

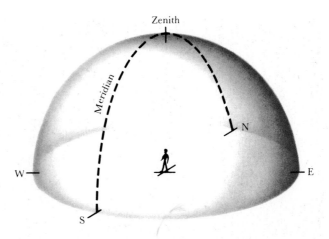

Figure 2-13 *The meridian* *The meridian is a circle that passes through the observer's zenith and the two celestial poles. The passing of celestial objects across the meridian can be used to measure time.*

not be observed directly. The crossing of the meridian by any object in the sky is called a **meridian transit**. If the crossing occurs above the horizon, it is an upper meridian transit. An **apparent solar day** is formally defined as the interval between two successive upper meridian transits of the Sun, as observed from any fixed spot on the Earth.

Early observers soon realized that the Sun is not a good timekeeper. The length of the apparent solar day (as measured by a device such as an hourglass) varies from one season to another. Similarly, the speed of the Sun's slow eastward progress against the background stars varies over the course of a year.

There are two main reasons for the Sun's nonuniform motion. First, the Earth's orbit is not a perfect circle. The orbit is actually an ellipse, as shown in exaggerated form in Figure 2-14a. As we shall learn in Chapter 4, the Earth moves more rapidly along its orbit when it is near the Sun than it does when it is farther away. The speed of the Sun along the ecliptic thus varies throughout the year. The Sun appears to move more than 1° per day along the ecliptic in January, when the Earth is nearest the Sun, and less than 1° per day in July, when the Earth is farthest from the Sun.

Even if the Sun did move along the ecliptic at a uniform rate, it would still not be a good timekeeper. Because the ecliptic is inclined by 23½° to the celestial equator, a significant part of the Sun's motion near the equinoxes is in a north–south direction (see Figure 2-14b). The net daily eastward progress in the sky is therefore somewhat foreshortened. But near the solstices, the Sun's motion is almost parallel to the celestial equator, so there is no comparable foreshortening at the beginning of summer or winter.

To cope with these difficulties, astronomers invented an imaginary object called the **mean sun** that moves along the

365.25

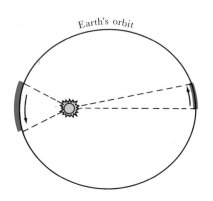

Earth's orbit

a A month's motion of
the Earth along its orbit

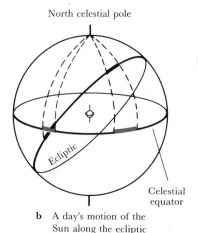

North celestial pole

Ecliptic

Celestial
equator

b A day's motion of the
Sun along the ecliptic

Figure 2-14 Why the Sun is a poor timekeeper
There are two main reasons why the Sun is a poor timekeeper. **(a)** *The Earth's speed along its orbit varies throughout the year. Hence, the apparent speed of the Sun along the ecliptic is not constant.* **(b)** *The projection of the Sun's daily progress along the ecliptic onto the celestial equator varies throughout the year. Hence, the Sun's net daily eastward progress against the stars is not constant.*

celestial equator at a precisely uniform rate. In this context, "mean" is a synonym for "average." The mean sun is sometimes slightly ahead of the real Sun in the sky, sometimes behind. Since the mean sun moves at a constant rate, it serves as a fine timekeeper.

A **mean solar day** is the interval between successive upper meridian transits of the mean sun. It is exactly 24 hours long, the average length of an apparent solar day. Our clocks and wristwatches measure mean solar time.

Apparent solar time and mean solar time can differ by as much as a quarter of an hour during certain seasons. The difference between apparent and mean solar time, called the **equation of time**, is graphed in Figure 2-15. With this graph, you can correct your sundial reading to get the mean solar time at your location on Earth.

Time zones were invented for convenience in commerce, transportation, and communication. In a **time zone**, everyone agrees to set all their clocks and watches to the mean solar time for a meridian of longitude that runs approximately through the center of the zone. Time zones around

the world are centered on meridians of longitude at 15° intervals. That is why going from one time zone to the next requires you to change the time on your wristwatch by exactly one hour. The time zones for North America are shown in Figure 2-16.

Although it is natural to want our clocks and method of timekeeping to be related to the Sun, astronomers are usually interested in observing the stars. Astronomers therefore find it convenient to use **sidereal time**, which is based on the apparent motion of the stars rather than that of the Sun. Sidereal time is usually used when aiming a telescope, and so most observatories are equipped with a sidereal clock. Box 2-4 discusses what sidereal time is and how it is used.

2-6 Astronomical observations led to the development of the modern calendar

Just as the day is a natural unit of time based on the Earth's rotation, the year is a natural unit of time based on the Earth's revolution about the Sun. Unfortunately, nature has not arranged things in a convenient fashion. The year does not divide exactly into 365 whole days. Ancient astronomers realized that the length of a year is approximately $365\frac{1}{4}$ days, and Julius Caesar established the system of "leap years" to account for this extra one-quarter of a day. By adding an extra day to the calendar every four years, Caesar hoped to ensure that seasonal astronomical events, such as the beginning of spring, would occur on the same date year after year.

Caesar's system would have been perfect if the year were exactly $365\frac{1}{4}$ days long and if there were no precession. Unfortunately, this is not the case. To achieve better accuracy, astronomers now recognize several different types of years. One of these years is defined to be the time required for the Sun to return to the same position with respect to the stars in

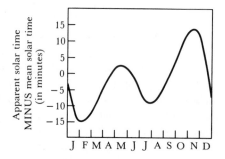

Figure 2-15 The equation of time The equation of time equals apparent solar time minus mean solar time. It is the amount by which a simple sundial is in error because the Sun's eastward motion against the background stars is not constant throughout the year.

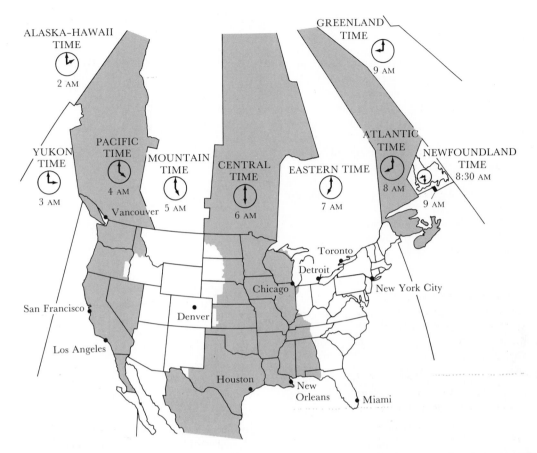

Figure 2-16 Time zones in North America   For convenience, the Earth is divided into 24 time zones centered on 15° intervals of longitude around the globe. Consequently, there are four time zones across the *continental United States, giving a three-hour time difference between New York and California.*

the sky. This year is called the <u>sidereal year</u> and is equal to 365.2564 mean solar days, or 365^d 6^h 9^m 10^s.

Although the sidereal year is the true orbital period of the Earth with respect to the stars, it is not the kind of year on which we base our calendar. Like Caesar, most people prefer seasonal events to happen on the same date year after year. For example, we want the first day of spring to occur on March 21. But spring begins when the Sun is at the vernal equinox, and the vernal equinox moves slowly against the background stars because of precession. Therefore, to set up a calendar we want to use a year equal to the time needed for the Sun to return to the vernal equinox. This period, which is called the <u>tropical year,</u> is equal to 365.2422 mean solar days, or 365^d 5^h 48^m 46^s. Because of precession, the tropical year is 20 minutes and 24 seconds shorter than the sidereal year.

This discrepancy was known to the ancient Greeks. In the second century BC, Hipparchus calculated the length of the tropical year as being only about six minutes different from the currently accepted value. (Hipparchus also became the

first person to detect the precession of the equinoxes when he compared his own observations with those of Babylonian astronomers three centuries earlier.) Caesar's assumption that the tropical year equals 365¼ years was off by 11 minutes and 14 seconds. This tiny error adds up to about three days every four centuries. Although Caesar's astronomical advisers were aware of the discrepancy, they felt that it was too small to matter. However, by the sixteenth century the first day of spring was occurring on March 11.

To straighten things out, Pope Gregory XIII instituted a calendar reform in 1582. He began by dropping ten days (October 5, 1582, was proclaimed to be October 15, 1582), which brought the first day of spring back to March 21. Next he modified Caesar's system of leap years.

Caesar had added February 29 to every calendar year that is evenly divisible by four. Thus, for example, 1980, 1984, and 1988 were all leap years with 366 days. But we have seen that this system produces an error of about three days every four centuries. To solve the problem, Pope Gregory decreed that only the century years evenly divisible by 400

Box 2-4 Sidereal time

If you want to observe a particular star or galaxy, it obviously must be above the horizon. It is best to view an astronomical object when it is high in the sky because the distorting effects of the Earth's atmosphere increase as the viewing direction approaches the horizon. Ideally, the object should be on or near the upper meridian.

To tell which objects are crossing the upper meridian, most observatories are equipped with a **sidereal clock** that tells sidereal time. Sidereal time is simply the right ascension of any object on the meridian. As new objects cross the meridian, sidereal time progresses.

For example, suppose you want to observe the bright star Regulus in the constellation of Leo (the lion). From reference books, such as the *Astronomical Almanac,* you find that the coordinates of this star are

$$R.A. = 10^h \ 07^m \ 28.0^s$$
$$Decl. = +12° \ 03' \ 03''$$

where R.A. and Decl. are commonly used abbreviations for right ascension and declination, respectively. Thus, since its right ascension is $10^h \ 07^m \ 28.0^s$, Regulus will be on the upper meridian when the sidereal time is 10:07:28. A sidereal clock is useful because it tells you the right ascension of those objects near the meridian and thus best suited for observation.

From this example, we can see why astronomers measure right ascension in units of time (hours, minutes, seconds). Because the Earth rotates once a day, it takes nearly 24 hours for all the stars to pass across the meridian. By measuring right ascension in units of time, astronomers have a convenient relationship between time and star positions. In contrast, declination is measured in units of angle (degrees, minutes of arc, seconds of arc) with a plus sign (or no sign at all) for north of the celestial equator and a minus sign for south.

It is sometimes convenient to be able to convert between angular and sidereal time. For example, you might want to know how many degrees correspond to one hour of right ascension. Because 24 hours of right ascension take you all the way around the celestial equator, we have the relationship that $24^h = 360°$. Therefore, the following equivalents hold:

$$1^h = 15°$$
$$1^m = 15'$$
$$1^s = 15''$$
$$1° = 4^m$$
$$1' = 4^s$$
$$1'' = 0.067^s$$

Sidereal time is useful only to people such as astronomers and navigators who deal with the stars. It is different from the time on your wristwatch. In fact, a sidereal clock and an ordinary clock even tick at different rates because they are based on

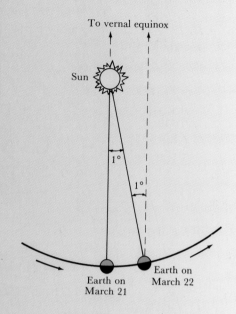

different astronomical objects. Ordinary clocks are related to the position of the Sun. Sidereal clocks are tied to the position of the vernal equinox, because that is the location from which right ascension is measured.

Regardless of where the Sun is, the sidereal day starts when the vernal equinox crosses the upper meridian. A **sidereal day** is the time between two successive upper meridian passages of the vernal equinox. In contrast, an apparent solar day is the time between two successive upper meridian crossings of the Sun. These two kinds of days are not equal, for reasons illustrated in the diagram. Because the Earth moves along its orbit around the Sun, the Earth must rotate through nearly 361° to get from one local noon to the next. This extra 1° of rotation corresponds to four minutes of time, which is the amount by which a solar day exceeds a sidereal day. To be precise,

$$1 \text{ sidereal day} = 23^h \ 56^m \ 4.091^s$$

where the hours, minutes, and seconds are in mean solar time. Of course, one day according to your wristwatch is one mean solar day and is exactly 24 hours of solar time long.

The sidereal clock measures sidereal hours, minutes, and seconds—dividing the sidereal day into exactly 24 sidereal hours. This explains why the sidereal clock "ticks" at a slightly different rate from your wristwatch. Measurements of right ascension are always expressed in sidereal hours, minutes, and seconds. All other time measurements given in this book are expressed in mean solar time unless otherwise stated.

should be leap years. For example, the years 1700, 1800, and 1900 (which would have been leap years, according to Caesar) are not leap years in the improved Gregorian system. But the years 1600 and 2000—which can be divided evenly by 400—remain leap years.

The Gregorian system is the one we use today. It assumes that the year is 365.2425 mean solar days long, which is very close to the true length of the tropical year. In fact, the error is only one day in every 3300 years. That won't cause any problems for a long time.

Key words

* Antarctic Circle
apparent solar day
apparent solar time
* Arctic Circle
* Astronomical Almanac
autumnal equinox
* Bonner Durchmusterung
celestial equator
celestial sphere
constellation
* declination

diurnal
ecliptic
epoch
equation of time
equinox
* Henry Draper Catalogue
mean solar day
mean sun
meridian
meridian transit
* New General Catalogue

north celestial pole
precession
precession of the
 equinoxes
* right ascension
* sidereal clock
* sidereal day
sidereal time
sidereal year
south celestial pole
summer solstice

time zone
tropical year
* Tropic of Cancer
* Tropic of Capricorn
vernal equinox
winter solstice
zenith
zodiac

Key ideas

- It is convenient to imagine the stars fixed to the celestial sphere with the Earth at its center.

   The surface of the celestial sphere is divided into 88 regions called constellations.

- The celestial sphere appears to rotate around the Earth once in each 24-hour period; in fact, of course, it is the Earth that is rotating.

   The poles and equator of the celestial sphere are determined by extending the axis of rotation and the equatorial plane of the Earth out to the celestial sphere.

   The positions of objects on the celestial sphere are described by specifying their right ascension (in time units) and declination (as an angle).

- The Earth's axis of rotation is tilted at an angle of $23\frac{1}{2}°$ from the perpendicular to the plane of the Earth's orbit.

   The seasons are caused by the tilt of the Earth's axis, which tips one hemisphere or the other toward the Sun at certain times of the year.

- Equinoxes and solstices are significant points in the sky, determined by the relationship between the Sun's path on the celestial sphere (the ecliptic) and the celestial equator.

- The orientation of the Earth's axis of rotation moves slowly in a phenomenon called precession.

   Precession is caused by the gravitational pull of the Sun and Moon on the Earth's equatorial bulge.

   Precession of the Earth's axis in turn causes precession of the equinoxes along the ecliptic.

   Because the system of right ascension and declination is tied to the position of the vernal equinox, the date (or epoch) of observation must be specified when giving the position of an object in the sky.

• Timekeeping is an important concern of astronomers.

> Apparent solar time is based upon the apparent motion of the Sun across the celestial sphere, which varies with the seasons.

> Mean solar time is based upon the motion of an imaginary mean sun along the celestial equator, which produces a uniform mean solar day of 24 hours.

> Sidereal time is based upon the apparent motion of the celestial sphere; the sidereal year is the actual orbital period of the Earth.

> The tropical year measures the period between vernal equinoxes; leap-year corrections in the calendar are needed because the tropical year is not exactly 365 days.

Review questions

1 Why is the ancient idea of the celestial sphere still a useful concept today?

2 Are there any stars in the sky that are not part of a constellation? Explain your answer.

3 Why can't a person in Australia use the Big Dipper to find the north direction?

*4 *(Basic)* On December 1 at 10:00 PM you look toward the eastern horizon and see the bright star Procyon rising. At approximately what time will Procyon rise two weeks later, on December 15?

5 At what location on Earth is the north celestial pole on the horizon?

6 Where do you have to be on the Earth in order to see the Sun at the zenith? If you stay at that location for a full year, on how many days will the Sun pass through the zenith?

7 Where do you have to be on Earth in order to see the south celestial pole directly overhead? What is the maximum possible elevation of the Sun above the horizon at that location? On what date can this maximum elevation be observed?

8 At what point on the horizon does the vernal equinox rise?

9 What causes the seasons? How does the daily path of the Sun across the sky change with the seasons?

10 What causes precession of the equinoxes? How long does it take for the vernal equinox to move 1° along the ecliptic?

*11 What is the sidereal time when the vernal equinox rises?

*12 On what date is the sidereal time nearly equal to the solar time?

13 How would you use the equation of time to correct your sundial readings?

14 Why is it convenient to have the Earth divided into time zones?

15 When is the next leap year? Will the year 2000 be a leap year?

Advanced questions

> **Tips and tools . . .**
> Take the time to become familiar with various types of star charts. For instance, you should compare the monthly star charts published in such magazines as *Sky & Telescope* and *Astronomy* with more detailed maps of the heavens, like those in Wil Tirion's *Sky Atlas 2000.0* (Sky Publishing Corporation and Cambridge University Press, 1981). You may also find it useful to examine a "planisphere," a device consisting of two rotatable disks, the bottom one showing all the stars in the sky (for a particular latitude), while the top one has a transparent, oval-shaped window through which only some of the stars can be seen. By rotating the disks about their pivot point you can immediately see which constellations are above the horizon on any date and at any time during the year.

*16 *(Challenging)* How would the sidereal and synodic days change **(a)** if the Earth's rate of rotation increased, **(b)** if the Earth's rate of rotation decreased, and **(c)** if the Earth's rotation were retrograde (i.e., if the Earth rotated about its axis opposite to the direction in which it revolves about the Sun)?

17 Consult a star map of the southern hemisphere and determine which, if any, bright southern stars could someday become south celestial "pole" stars.

18 Are there stars in the sky that never set? Are there stars that never rise? Does your answer depend on the observer's location on Earth?

19 With the aid of a diagram, demonstrate that your latitude on Earth is equal to the latitude of the north celestial pole above your northern horizon.

20 Go to a library that has recent issues of the *Astronomical Almanac*. In the section of these reference books entitled "Bright Stars," look up the coordinates of a particular star of your choice for two successive years. By how much did the right ascension and declination of that star change?

21 Using a star map, determine which bright stars, if any, could someday mark the location of the vernal equinox. Give the approximate years when this should happen.

*****22** What is the right ascension of a star that is on the meridian at midnight at the time of the autumnal equinox?

*****23** (*Challenging*) Suppose that you live at a latitude of 40°N. What is the elevation of the Sun above the southern horizon at noon at the time of the winter solstice?

Discussion questions

24 Examine a list of the 88 constellations. Are there any constellations whose names obviously date from modern times? Where are these constellations located in the sky? Why do you suppose they do not have archaic names?

25 Describe the seasons if the Earth's axis of rotation were tilted (**a**) 0° and (**b**) 90° to its orbital plane.

Observing projects

26 On a clear, cloud-free night, use the star charts at the end of this book to see how many constellations of the zodiac you can identify. Which ones were easy to find? Which were difficult?

27 Examine the star charts that are published monthly in such popular astronomy magazines as *Sky & Telescope* and *Astronomy*. How do they differ from the star charts at the back of this book? On a clear, cloud-free night, use one of these star charts to locate the celestial equator and the ecliptic. Note the inclination of the Milky Way to the ecliptic and celestial equator. What do your observations tell you about the orientation of the Earth and its orbit relative to the rest of the Galaxy?

28 Suppose you wake up before dawn and want to see which constellations are in the sky. Explain how the star charts at the end of this book can be quite useful, even though chart times are given only for the evening hours. Which chart most closely depicts the sky at 4:00 AM tomorrow morning? Set your alarm clock for 4:00 AM to see if you are correct.

For further reading

Allen, R. *Star Names: Their Lore and Meaning*. 1899. Dover reprint, 1963 • A classic book detailing the origins of constellations and star names.

Berry, R. *Discover the Stars*. Harmony, 1987 • An excellent manual featuring clear, beautiful star charts showing the whole sky, along with enlarged, detailed views.

Beyer, S. *Star Guide*. Little, Brown, 1986 • A useful guide for the beginner who wants to learn the sky.

Brown, H. *Man and the Stars*. Oxford University Press, 1978 • This book includes nice sections on the history of calendars, timekeeping, and navigation, as well as a discussion of the interaction between astronomy and philosophy.

Chartrand, M. *Skyguide*. Western Publications, 1982 • This pocket-sized book gives a concise introduction to observational astronomy and has superb illustrations and excellent star charts.

Gallant, R. *The Constellations: How They Came to Be*. Four Winds Press, 1979 • The stories of 50 constellations are accompanied by star charts and useful information.

Gingerich, O. "The Origin of the Zodiac." *Sky & Telescope*, March 1984 • A brief article on the constellations by a distinguished historian of astronomy.

Kunitzsch, P. "How We Got Our Arabic Star Names." *Sky & Telescope*, January 1983 • This fascinating article gives many historical insights about star names and Islamic astronomy during the Middle Ages.

Mayer, B. *Starwatch*. Putnam, 1984 • An unusual book with clever diagrams and a new way to learn the constellations using a coat hanger, some cellophane, and white ink.

Menzel, D., and Pasachoff, J. *A Field Guide to the Stars and Planets*, 2nd ed. Houghton-Mifflin, 1983 • This handy pocket-sized book introduces observational astronomy with sky maps, star charts, tables, and references.

Motz, L., and Nathanson, C. *The Constellations: An Enthusiast's Guide to the Night Sky*. Doubleday, 1988 • A useful guide to the constellations that gives mythology, observing information, and astronomical data.

Ridpath, I., and Tirion, W. *Universe Guide to Stars and Planets*. Universe, 1985 • All 88 constellations are covered in exquisitely drawn star charts along with sky maps for different seasons.

Whitney, C. *Whitney's Star Finder*. Knopf, 1981 • A useful guide for the novice who wants to identify stars and constellations.

The Caracol at Chichén Itzá *This ancient Mayan observatory in the Yucatán was built around* AD *1000. Its architecture is based on alignments with important celestial events. Mayan astronomers developed a very accurate calendar and measured the motions of celestial bodies with great precision. Special significance was associated with the planet Venus, which inspired sacrificial rites and other ceremonies. (Courtesy of E. C. Krupp)*

C H A P T E R

3

Eclipses and Ancient Astronomy

Ancient cultures made important astronomical observations and developed ideas that set the stage for modern astronomy. This chapter retraces the route to these discoveries, beginning with the phases of the Moon and their relationship to the Moon's motion about the Earth. Ancient astronomers made ingenious attempts to measure the size of the Earth and the distances from Earth to the Sun and the Moon. Eclipses of the Sun and the Moon, which are dramatic phenomena, played important roles in many early theories and measurements. Indeed, these first astronomers understood eclipses well enough to predict their occurrences with some degree of reliability. An enduring part of our legacy is the magnitude scale, handed down to us from a Greek astronomer in the second century BC and still used today to denote the apparent brightness of objects in the sky.

Figure 3-1 Stonehenge This astronomical monument was constructed nearly 4000 years ago on Salisbury Plain in southern England. The monument originally consisted of thirty blocks of gray sandstone each 4 meters (13 feet) high, set in a circle 30 meters (98 feet) in diameter. These stones were topped with a continuous circle of smaller stones. Inside this circle there still exists a geometric arrangement of other stones, most notably a horseshoe-shaped set of larger stones that open toward the northeast. (Courtesy of the British government)

The beauty of the star-filled night sky or the drama of an eclipse may well suffice to make astronomy fascinating. However, there are also practical reasons for having an interest in the universe. Even the early Greeks knew the connection between the seasons and the relative orientations of the Sun and Earth. And many early seafaring cultures were aware that the tides are influenced by the position of the Moon.

Ancient civilizations placed great emphasis on careful astronomical observation. Hundreds of impressive monuments, such as Stonehenge (Figure 3-1), that dot the British Isles provide evidence of this preoccupation with astronomy. Alignments of various stones point to the locations where the Sun and Moon rise and set at key times during the year, such as solstices. Similar monuments have also been found in the United States. For example, the Medicine Wheel in Wyoming, constructed atop a windswept plateau by the plains Indians, is a circular ring of stones. Certain stones and markers are aligned with the rising points of specific bright stars and of the Sun at the summer solstice.

Aztec, Mayan, and Incan architects in Central and South America also designed buildings to be astronomically oriented. For example, at the ruined city of Tiahuanaco, in Bolivia, the Temple of the Sun was built with its walls aligned north–south and east–west with an accuracy of better than one degree. A similar precision was exercised 5000 years ago by the builders of the great Egyptian pyramids, whose sides are likewise oriented north–south and east–west.

In addition to building temples and tombs, the ancients constructed what were apparently astronomical observato-

ries. One of the best examples is the Caracol in the Mayan city of Chichén Itzá on the Yucatán Peninsula. Built nearly a thousand years ago, the Caracol's cylindrical tower contains windows aligned with the northernmost and southernmost rising and setting points of both the Sun and the planet Venus. A similar four-story adobe building, probably constructed during the fourteenth century, is located at the Casa Grande site in Arizona. These structures all bear witness to careful and patient astronomical observations by the peoples of many cultures and civilizations.

3-1 Lunar phases are caused by the Moon's orbital motion

Ancient astronomers knew that the Moon shines by reflected sunlight and that the Moon orbits the Earth. These facts were deduced by observing the Moon's changing **phases** as varying amounts of its illuminated hemisphere became exposed to observers on Earth.

Figure 3-2 is an extraordinary photograph of both the Earth and the Moon from space. The Moon orbits the Earth every 27.3 days, during which time the Moon completes a full cycle of its phases, as shown in Figure 3-3.

The phase called **new moon** occurs when the dark hemisphere of the Moon faces the Earth. The Moon is not visible during this phase because it is in the same part of the sky as the Sun.

Figure 3-2 The Earth and the Moon The Moon orbits the Earth every 27.3 days, at an average distance of 384,400 kilometers (238,900 miles). This picture was taken in 1977 from the Voyager 1 spacecraft shortly after it was launched toward Jupiter and Saturn. (NASA)

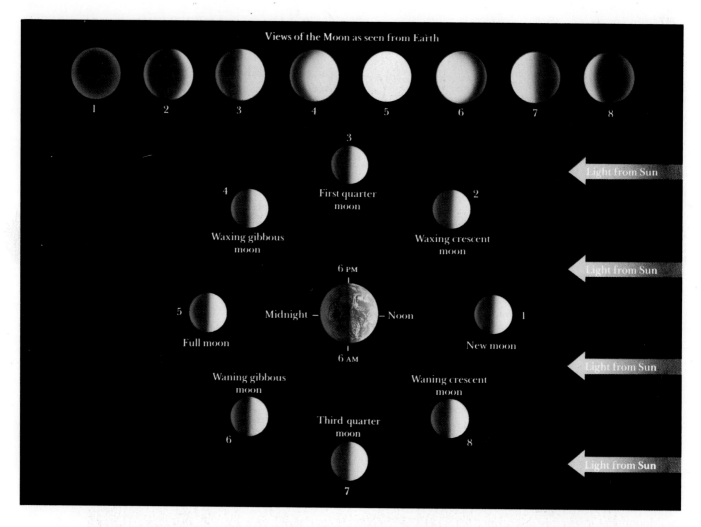

Figure 3-3 The phases of the Moon *Light from the Sun illuminates one half of the Moon, while the other half is dark. As the Moon orbits the Earth, we see varying amounts of the Moon's illuminated hemisphere. It takes 29½ days for the Moon to go through all its phases.*

During the next seven days, Earth-based observers see a phase called **waxing crescent moon**, in which progressively more and more of the Moon's illuminated hemisphere becomes exposed to our view. At **first quarter moon**, the angle between the Sun, Moon, and Earth is 90°, so we see half of the illuminated hemisphere and half of the dark hemisphere.

During the next week, still more of the illuminated hemisphere can be seen from Earth, giving us the phase called **waxing gibbous moon.** When the Moon stands opposite the Sun in the sky, we see the fully illuminated hemisphere in the phase called **full moon.**

Over the following two weeks, we see less and less of the illuminated hemisphere as the Moon continues along its orbit. This movement produces the phases called **waning gibbous moon, third-quarter moon,** and **waning crescent moon.**

Because the position of the Sun in the sky determines the local time, it is possible to correlate the Moon's phase and location with the time of day. For example, during the first quarter moon the Moon is approximately 90° east of the Sun in the sky, hence moonrise occurs approximately at noon. At full moon the Moon is opposite the Sun in the sky and thus moonrise occurs at sunset. Figure 3-4 is a series of photographs showing the various phases of the Moon.

The four-week interval required for the Moon to complete one cycle of its phases inspired our ancestors to invent the concept of a month. For various historical reasons, the calendar we use today has months of differing lengths, none of which has much to do with the heavens. In contrast, astronomers find it useful to define two types of months, depending on whether the Moon's motion is measured relative to the stars or to the Sun. Neither corresponds exactly to the months of our usual calendar.

The **sidereal month** is the time it takes the Moon to complete one full orbit of the Earth, measured with respect to the stars. This true orbital period is equal to 27.3 days.

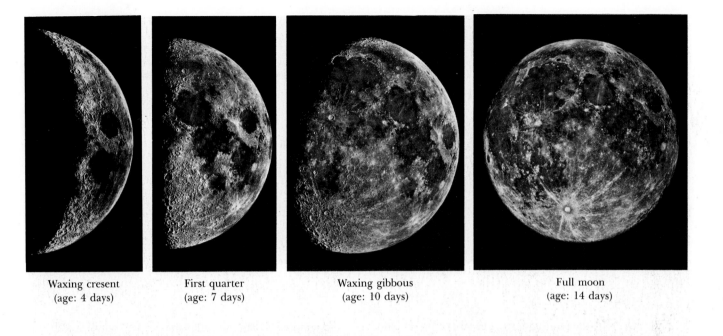

Waxing cresent
(age: 4 days)

First quarter
(age: 7 days)

Waxing gibbous
(age: 10 days)

Full moon
(age: 14 days)

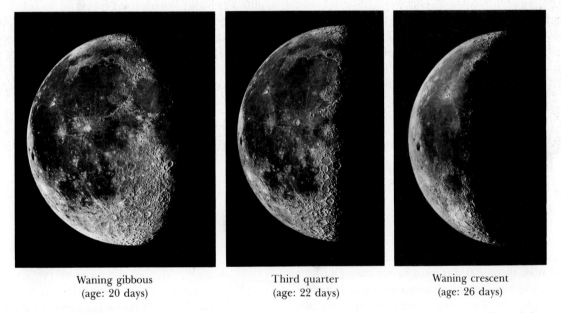

Waning gibbous
(age: 20 days)

Third quarter
(age: 22 days)

Waning crescent
(age: 26 days)

Figure 3-4 *The Moon's appearance* *The Moon always keeps the same side facing the Earth. Earth-based observers therefore always see the same craters and lunar mountains, regardless of the Moon's phase. (Lick Observatory)*

The **synodic month** is the time it takes the Moon to complete one cycle of phases (i.e., from new moon to new moon, or full moon to full moon), and thus is measured with respect to the Sun (rather than the stars).

Of course, while the Moon is going through its phases, the Earth keeps on orbiting the Sun. So, to complete a cycle of phases (e.g., from one new moon to the next), the Moon must travel *more* than 360° along its orbit as shown in Figure 3-5. Because of this extra distance, the synodic month is equal to about 29.5 days—about two days longer than the sidereal month.

The Moon stays in orbit about the Earth because of the gravitational attraction between these two bodies. (We shall discuss orbits and gravity in greater detail later, in Box 3-1 and Chapter 4.) In addition to the Earth's gravity, however, the Sun is also pulling on the Moon. The Sun's gravity sometimes causes the Moon to speed up or slow down slightly in its orbit. Thus, the Moon's orbital path is constantly being

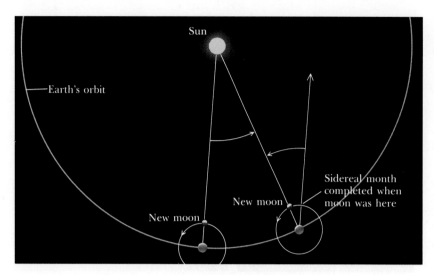

Figure 3-5 *The sidereal and synodic months* *The sidereal month is the time it takes the Moon to complete one full revolution around the Earth with respect to the background stars. However, because the Earth is constantly moving along its orbit about the Sun, the Moon must travel through slightly more than 360° to get from one new moon to the next. Thus the synodic month is slightly longer than the sidereal month.*

slightly altered by the Sun's gravity. The size of these minor variations depends on the relative configurations of the Sun, Moon, and Earth.

The final result is that both the sidereal and the synodic months are variable in length. The sidereal month (average length = 27^d 7^h 43^m 11^s) can vary by as much as seven hours, while the synodic month (average length = 29^d 12^h 44^m 3^s) can vary by as much as half a day.

3-2 Ancient astronomers measured the size of the Earth and attempted to determine distances to the Sun and Moon

More than two thousand years ago, Greek astronomers were fully aware of the Earth's spherical shape. Eclipses of the Moon convinced them that this must be the case. During a lunar eclipse, the Moon passes through the Earth's shadow (in other words, the Earth is between the Sun and the Moon). These Greek astronomers noted that the edge of the Earth's shadow is always circular. Because a sphere is the only shape that casts a circular shadow from any angle, they concluded that the Earth is spherical.

Around 200 BC, the Greek astronomer Eratosthenes devised a way to measure the circumference of the Earth. He had been intrigued by reliable reports from the town of Syene in Egypt (near the modern Aswan) that the Sun there shone directly down vertical wells on the first day of summer. Eratosthenes knew that the Sun never appeared at the zenith in his home in the Egyptian city of Alexandria, which is on the Mediterranean Sea almost due north of Syene. When he measured the position of the Sun at local noon on

the summer solstice in Alexandria, he found that it was about 7° south of the zenith, as sketched in Figure 3-6. Actually, Eratosthenes measured the angle as being one-fiftieth of a complete circle, so he concluded that the distance from Alexandria to Syene must be one-fiftieth of the Earth's circumference.

In Eratosthenes' day, the distance from Alexandria to Syene was said to be 5000 stades. Therefore, Eratosthenes found the Earth's circumference to be

$$50 \times 5000 \text{ stades} = 250,000 \text{ stades}$$

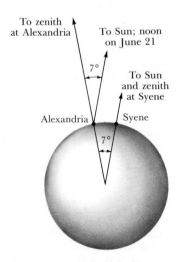

Figure 3-6 *Eratosthenes' method of determining Earth's diameter* *Eratosthenes noticed that the Sun is about 7° south of the zenith at Alexandria when it is directly overhead at Syene. This angle is about one-fiftieth of a circle, so the distance between Alexandria and Syene must be about one-fiftieth of the Earth's circumference.*

Unfortunately, no one today is sure of the exact length of the Greek unit called the stade. One guess is that the stade was about $\frac{1}{6}$ kilometer, which would mean that Eratosthenes obtained a circumference for the Earth of about 42,000 kilometers, which is within 5 percent of the modern value of 40,000 kilometers.

Eratosthenes was one of several brilliant astronomers to emerge from the so-called Alexandrian school, which by his time already had a distinguished tradition. For example, one of the first Alexandrian astronomers, Aristarchus of Samos, had devised a method of determining the relative distances to the Sun and Moon, perhaps as long ago as 280 BC.

Aristarchus knew the Sun, Moon, and Earth form a right triangle at the moment of first- or third-quarter moon, as diagramed in Figure 3-7. For reasons lost in antiquity, he also argued that the angle between the Moon and the Sun at first and third quarters is 3° less than a right angle (that is, 87°). Using the rules of Euclidean geometry, Aristarchus concluded that the Sun is about 20 times farther from us than the Moon is. We now know that the average distance to the Sun is about 390 times larger than the average distance to the Moon. It is nevertheless impressive that people were trying to measure distances across the solar system more than 2000 years ago.

Using lunar eclipses, Aristarchus also made an equally bold attempt to determine the relative sizes of the Earth, Moon, and Sun. From his observations of how long it takes for the Moon to move through the Earth's shadow, Aristarchus estimated the diameter of the Earth to be about three times larger than the diameter of the Moon. To determine the diameter of the Sun, Aristarchus simply pointed out that the Sun and the Moon have the same angular size in the sky. Therefore, their diameters must be in the same proportion as their distances. In other words, because Aristarchus believed

the Sun to be 20 times farther from the Earth than the Moon, he concluded that the Sun must be 20 times larger than the Moon.

We can now appreciate the significance of Eratosthenes' measurement of the Earth's circumference. The Greeks knew that the Earth's diameter is equal to its circumference divided by the constant called π (pi). Knowing the Earth's diameter, astronomers of the Alexandrian school could calculate the diameters of the Sun and Moon as well as their distances from Earth.

For ease of comparison, Table 3-1 summarizes some ancient and modern measurements of the sizes of the Earth, Moon, and Sun and the distances between them. Although some of these ancient measurements are far from the modern values, our ancestors' achievements still stand as an impressive exercise in observation and reasoning.

Table 3-1 A comparison of ancient and modern measurements

	Ancient measure (km)	Modern measure (km)
Earth's diameter	13,000	12,756
Moon's diameter	4,300	3,476
Sun's diameter	9×10^4	1.39×10^6
Earth–Moon distance	4×10^5	3.84×10^5
Earth–Sun distance	10^7	1.50×10^8

3-3 Eclipses occur only when the Sun and Moon are both on the line of nodes

A **lunar eclipse** occurs when the Moon, at full phase, passes through the Earth's shadow. This can happen only when the Sun, Earth, and Moon are in a straight line at full moon.

A **solar eclipse** occurs when the Earth passes through the Moon's shadow. As seen from Earth, the Moon moves in front of the Sun. Once again, this can happen only when the Sun, Moon, and Earth are aligned. However, the Moon must be between the Earth and the Sun for a solar eclipse. Therefore, a solar eclipse can occur only at new moon.

Both new moon and full moon occur at intervals of $29\frac{1}{2}$ days, but solar and lunar eclipses happen much less frequently. This situation occurs because the Moon's orbit is tilted slightly out of the plane of the Earth's orbit, as shown in Figure 3-8. The angle between the plane of the Earth's orbit and the plane of the Moon's orbit is about 5°. Because of this tilt, new moon and full moon usually occur when the Moon is either above or below the plane of the Earth's orbit.

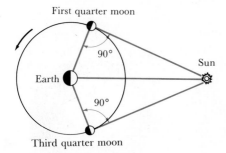

Figure 3-7 Aristarchus's method of determining distances to the Sun and Moon Aristarchus reasoned that the Moon should take longer to go from first quarter to last quarter than from last to first. By measuring this time difference, he determined the angles in the triangles formed by the Sun, Earth, and Moon at the first and last quarter phases. He was then able to calculate the relative lengths of the sides of these triangles and thereby obtain the distances to the Sun and Moon. The size of the Moon's orbit is greatly exaggerated in this diagram.

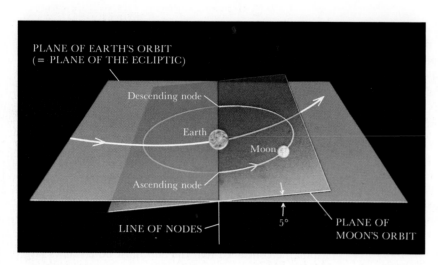

When the Moon is out of the plane of the Earth's orbit, a
perfect alignment between the Sun, Moon, and Earth is not
possible, and thus an eclipse cannot occur.

The plane of the Earth's orbit and the plane of the
Moon's orbit intersect along a line called the **line of nodes**.
The line of nodes passes through the Earth and is pointed in
a particular direction in space. Eclipses can occur only when
both the Sun and the Moon are on or very near the line of

nodes, because only then do the Sun, Earth, and Moon lie in
a straight enough line, as shown in Figure 3-9.

Knowing the orientation of the line of nodes is clearly
important to anyone who wants to predict upcoming
eclipses. However, determining the orientation of the line of
nodes is complicated by the fact that it gradually changes its
direction in space. The constant gravitational pull of the Sun
on the Moon causes the Moon's orbit gradually to shift its

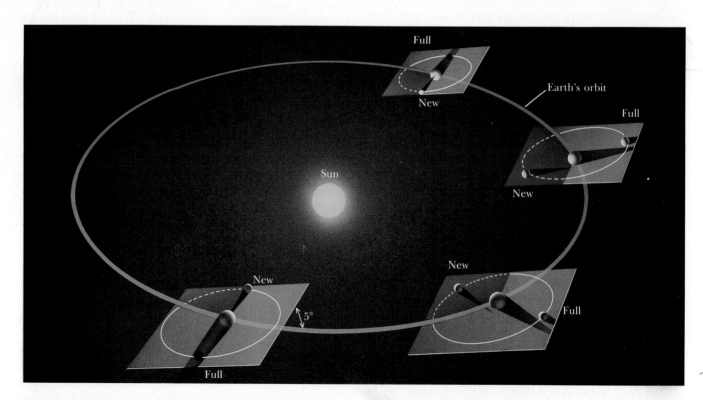

*Figure 3-9 **Conditions for eclipses*** *A solar eclipse occurs only if the
Moon is very near the line of nodes at new moon. A lunar eclipse occurs
only if the Moon is very near the line of nodes at full moon. These
conditions are met in two of the circumstances (upper right and lower*
*left) shown here. No eclipses would be possible in the other two circum-
stances (to the right of the Sun and at the lower right) because the Moon
is not near the line of nodes.*

Box 3-1 Some details of the Moon's orbit

The Moon's orbit about the Earth is an ellipse, as sketched in the accompanying diagram. The Moon is said to be at perigee when it is nearest the Earth and at apogee when it is farthest from Earth. The line connecting the points of perigee and apogee passes through the Earth and is called the **line of apsides** (see diagram *a*).

The center-to-center distance from the Earth to the Moon can vary by more than 50,000 km, from a minimum of 356,410 km (221,510 mi) at perigee to a maximum of 406,697 km (252,763 mi) at apogee. Consequently, the apparent angular size of the Moon as seen from Earth varies over the course of a month. At perigee, the Moon has an angular diameter of 33′ 31″, whereas at apogee the Moon's angular diameter is only 29′ 22″. The average apparent size of the Moon is 31′ 5″, which corresponds to the average Earth–Moon distance of 384,400 km (238,900 mi).

The average speed of the Moon along its orbit is 1.02 km/s (2280 mi/hr). As seen from Earth, the Moon appears to move eastward through the constellations from one day to the next. The Moon's daily eastward progress averages 13.2° (which is

360° divided by the 27.3 days in the sidereal month). In one hour, the Moon moves slightly more than $\frac{1}{2}$°, which is just a bit more than its own diameter. This rate of motion means that the time of moonrise is set back an average of about 50 minutes from one day to the next.

The mutual gravitational attraction between the Earth and the Moon keeps the Moon in orbit about the Earth. However, the Sun is also constantly tugging on the Moon. The Sun's gravitational pull on the Moon as it moves along its orbit causes both the line of nodes and the line of apsides to shift slowly over the years. The line of nodes gradually moves westward, as sketched in diagram *b*. This movement is called the **regression of the line of nodes**. It takes 18.61 years for the line of nodes to complete one full rotation.

While the line of nodes is moving westward among the constellations, the Sun's gravity causes the line of apsides to move eastward. It takes 8.85 years for the line of apsides to complete one full rotation. This effect, sketched in diagram *c*, is simply called the **rotation of the Moon's orbit**.

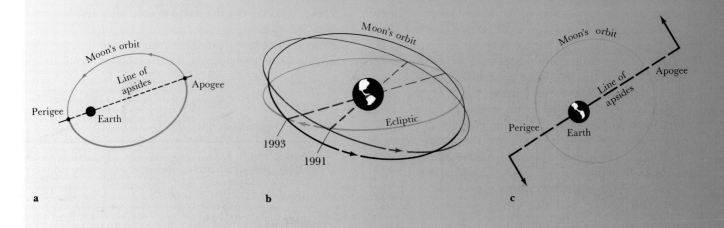

a   b   c

orientation in space, as described in Box 3-1. The resulting slow westward movement of the line of nodes is one of several details that astronomers calculate to fix the dates and times of upcoming eclipses.

There are at least two—but not more than five—solar eclipses each year. The last year in which five solar eclipses occurred was 1935. Lunar eclipses occur just about as frequently as solar eclipses. However, the maximum number of eclipses (solar and lunar combined) possible in a single year is seven.

3-4 Solar and lunar eclipses can be either partial or total, depending on the alignment of the Sun, Earth, and Moon

A lunar eclipse occurs when the Moon moves through the Earth's shadow. However, the Earth's shadow has two distinct parts, as diagrammed in Figure 3-10. The **umbra** is the

Figure 3-10 **The geometry of a lunar eclipse**
People on the nighttime side of the Earth see a lunar eclipse when the Moon moves through the Earth's shadow. The umbra is the darkest part of the shadow. In the penumbra, only part of the Sun is covered by the Earth.

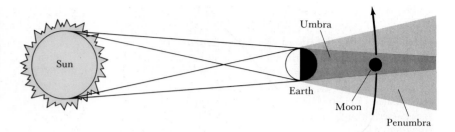

darkest part of the Earth's shadow, within which no portion of the Sun's surface can be seen. The **penumbra** is not quite so dark, since only part of the Sun's surface is covered by the Earth.

There are three kinds of lunar eclipses, depending on exactly how the Moon travels through the Earth's shadow. When the Moon passes only through the Earth's penumbra, we see a **penumbral eclipse**. During a penumbral eclipse, none of the lunar surface is completely shaded by the Earth, because in the penumbra only part of the Sun's light is blocked by the Earth. At mid-eclipse, the Moon merely looks a little dimmer than usual. There is no "bite" taken out of the Moon by the Earth's umbra. Since the Moon still looks full, penumbral eclipses are easy to miss.

Most people notice a lunar eclipse only if the Moon passes into the Earth's umbra. As the umbral phase of the eclipse begins, a "bite" seems to be taken out of the Moon when part of the lunar surface becomes immersed in the darkest portion of the Earth's shadow. If the Moon's orbit is oriented so that only part of the lunar surface passes through the umbra, we have a **partial lunar eclipse**. When the Moon travels completely into the umbra, as seen in Figure 3-11, a **total lunar eclipse** occurs. The maximum duration of totality occurs when the Moon travels directly through the center of the umbra. The Moon's speed through the Earth's shadow is roughly 1 kilometer per second (2280 miles per hour), which means that totality can last for as much as 1 hour 42 minutes.

As an example of the frequency of lunar eclipses, Table 3-2 lists all the eclipses, both total and partial, from 1991 through 1999. Penumbral eclipses are not included in this listing.

The Moon does not completely disappear from the sky during totality. A small amount of sunlight passing through the Earth's atmosphere is deflected into the Earth's umbra. Most of this deflected light is red, and thus the darkened Moon glows faintly in reddish hues during totality (see Figure 3-12).

It is a fortunate coincidence of nature that the apparent diameter of the Moon, as seen from Earth, is almost exactly the same as the apparent diameter of the Sun. Both the Sun and the Moon have angular diameters of about $\frac{1}{2}°$. For this reason, the Moon "fits" over the Sun during a **total solar eclipse**, blocking out the dazzling solar disk and not much else. This blocking is important to astronomers because it allows the hot gases (called the **solar corona**) that surround the Sun to be photographed and studied in detail during

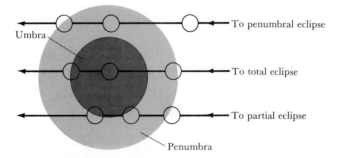

Figure 3-11 **Various lunar eclipses** *This diagram shows the Earth's umbra and penumbra at the distance of the Moon's orbit. Different kinds of lunar eclipses can be seen, depending on the Moon's path through the Earth's shadow.*

Table 3-2 Lunar eclipses, 1991–1999

Date	Percentage eclipsed (100% = total)	Duration of totality (hr min)
1991 December 21	9	——
1992 June 15	69	——
1992 December 10	100	1 14
1993 June 4	100	1 38
1993 November 29	100	0 50
1994 May 25	28	——
1995 April 15	12	——
1996 April 4	100	1 24
1996 September 27	100	1 12
1997 March 24	93	——
1997 September 16	100	1 06
1999 July 28	42	——

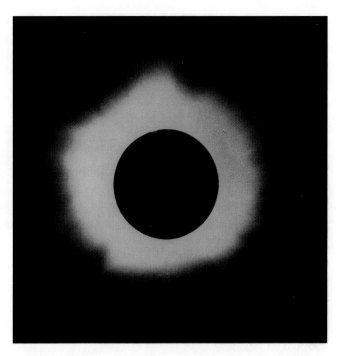

Figure 3-12 A total eclipse of the Moon *This photograph was taken by an amateur astronomer during the lunar eclipse of September 6, 1979. The distinctly reddish color of the Moon is caused by sunlight that is deflected into the Earth's shadow by the Earth's atmosphere. (Courtesy of M. Harms)*

Figure 3-13 A total eclipse of the Sun *During a total solar eclipse, the Moon completely covers the Sun's disk, allowing the solar corona to be seen. This halo of hot gases extends for thousands upon thousands of kilometers into space. Only the brightest, inner portions of the solar corona are seen in this photograph of a total solar eclipse on March 7, 1970. (NASA)*

the few precious moments when the eclipse is total (see Figure 3-13).

As with the Earth's shadow, the darkest part of the Moon's shadow is called the **umbra.** You must be inside the Moon's umbra in order to see a total solar eclipse, because that is the only region within which the Moon completely blocks the Sun. Because the Sun and the Moon have nearly the same angular diameter, only the tip of the Moon's umbra reaches the Earth's surface, as shown in Figure 3-14. As the Earth rotates, the tip of the umbra traces an **eclipse path** across the Earth's surface. Only the people within the eclipse path are treated to the spectacle of a total solar eclipse. Figure 3-15 shows the dark spot on the Earth's surface produced by the Moon's umbra.

Immediately surrounding the Moon's umbra is the region of partial shadow called the penumbra. From this area, the

Sun's surface appears only partially covered by the Moon. During a solar eclipse, the Moon's penumbra covers a large portion of the Earth's surface, and anyone standing inside the penumbra sees a **partial solar eclipse.**

The details of solar eclipses are calculated well in advance and published in reference books such as the *Astronomical Almanac.* Figure 3-16 shows a typical eclipse map, which displays the areas of the Earth covered by the Moon's umbra and penumbra. During a total eclipse, the Earth's rotation, coupled with the orbital motion of the Moon, causes the umbra to race along the eclipse path at speeds in excess of 1700 kilometers per hour (1060 miles per hour). Because of the high speed of the umbra, most people along the eclipse path will observe totality for only a few moments. In fact, totality never lasts for more than $7\frac{1}{2}$ minutes at any one location on the eclipse path. In a typical total solar eclipse, the

Figure 3-14 The geometry of a total solar eclipse *During a total solar eclipse, the tip of the Moon's umbra traces an eclipse path across the Earth's surface. People within the eclipse path see a total eclipse, whereas anyone within the penumbra would see only a partial eclipse of the Sun.*

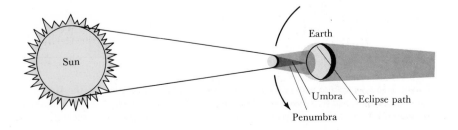

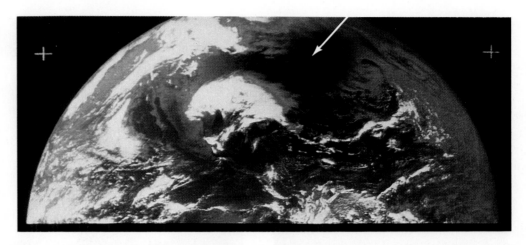

Figure 3-15 The Moon's shadow on the Earth *This photograph was taken from an Earth-orbiting satellite during a total solar eclipse on March 7, 1970. The Moon's umbra appears as a dark spot on the eastern coast of the United States. (NASA)*

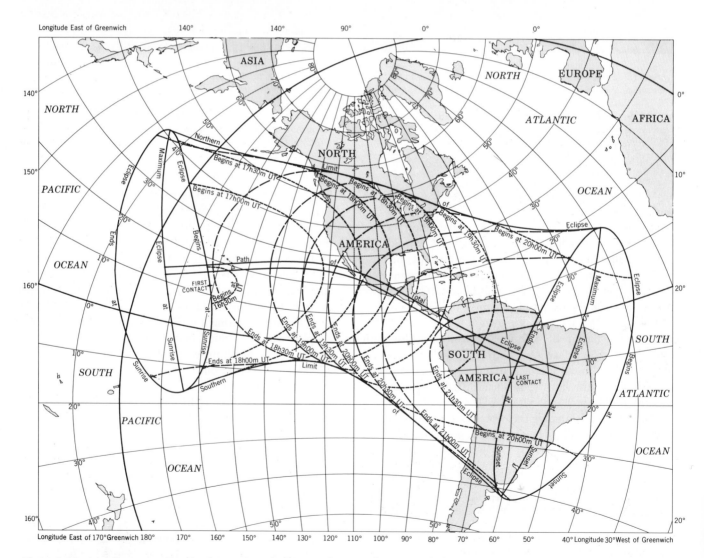

Figure 3-16 An eclipse map *Details of upcoming eclipses are published in astronomical reference books. Maps such as this one show the regions of the Earth that will experience either total or partial eclipse.*

The Moon's shadow travels generally eastward across the Earth's surface. This map shows details of a total solar eclipse scheduled to occur on July 11, 1991. (Reproduced from the Astronomical Almanac*)*

angle of the Sun and the Earth–Moon distance produce a duration of totality much less than the maximum of $7\frac{1}{2}$ minutes.

The width of the eclipse path depends primarily on the Earth–Moon distance during an eclipse. The eclipse path is widest if the Moon happens to be at **perigee**, the point in its orbit nearest the Earth (see Box 3-1). This nearness can produce an eclipse path up to 270 kilometers (170 miles) wide, though usually it is much narrower. In fact, sometimes the Moon's umbra does not even reach down to the Earth's surface. This happens because the Moon's orbit is slightly elliptical. If the alignment for a solar eclipse occurs when the Moon is at **apogee**, farthest from the Earth, then the Moon's umbra falls short of the Earth and no one sees a truly total eclipse. From the Earth's surface, the Moon appears too small to cover the Sun completely, and a thin ring of light is seen around the edge of the Moon at mid-eclipse. An eclipse of this type is called an **annular eclipse** (see Figure 3-17). The length of the Moon's umbra is nearly 5000 kilometers (3100 miles) shorter than the average distance between the Moon and the Earth's surface. Thus, the Moon's shadow often fails

Figure 3-17 An annular eclipse of the Sun This composite of six exposures taken at sunrise in Costa Rica shows the progress of an annular eclipse of the Sun that occurred on December 24, 1974. Note that at mid-eclipse the limb of the Sun is visible around the Moon. (Courtesy of D. di Cicco)

to reach the Earth, and annular eclipses are slightly more common than total eclipses.

To illustrate the frequency of solar eclipses, Table 3-3 lists all the total, annular, and partial eclipses from 1991 to 1999.

3-5 Ancient astronomers achieved a limited ability to predict eclipses

A total solar eclipse can be a dramatic event. The sky begins to darken, the air temperature falls, and the winds increase as the Moon's umbra races toward you. All nature responds: Birds go to roost, flowers close their petals, and crickets begin to chirp as if evening had arrived. As totality approaches, the landscape around you is bathed in an eerie gray or, less frequently, in shimmering bands of light and dark as the last few rays of sunlight peek out from behind the edge of the Moon. And finally the corona blazes forth in a star-studded midday sky. It is an awesome sight.

In ancient times, the ability to predict eclipses must have been very desirable. Archeological evidence suggests that astronomers in many civilizations struggled to predict eclipses, with varying degrees of success. The number and placement of certain holes in the ground around Stonehenge indicate some ability to predict eclipses more than 4000 years ago. One of three priceless manuscripts to survive the devastating Spanish Conquest shows that the Mayan astronomers of Mexico and Guatemala had a fairly reliable method of eclipse prediction. There are also numerous apocryphal stories such as the one about the great Greek astronomer Thales of Miletus, who is said to have predicted the famous eclipse of 585 BC, which occurred during the middle of a war. The sight was so awesome and unnerving that the soldiers put down their arms and declared peace.

In retrospect, it seems that what the ancient astronomers actually produced were eclipse "warnings" of various degrees of reliability, rather than true predictions. Working with historical records, these astronomers generally sought to discover cycles and regularities from which future eclipses might be anticipated.

To see how you might produce eclipse warnings yourself, suppose that you observe a solar eclipse in your home town and want to figure out when you and your neighbors might see another one. How would you begin?

First, remember that a solar eclipse can occur only if the line of nodes points toward the Sun at new moon (recall Figure 3-9). Second, you must know that it takes 29.53 days to go from one new moon to the next. This interval is called a **lunar month**, or synodic month. Because solar eclipses occur only at the time of the new moon, you must wait several whole lunar months for the proper alignment to occur again.

However, there is a complication. The line of nodes gradually shifts its position with respect to the background stars, as described in Box 3-1. It takes 346.6 days to go from

Table 3-3 Solar eclipses, 1991–1999

Date	Area	Type	Notes
1991 January 15/16	Australia, New Zealand, Pacific	Annular	
1991 July 11	Pacific, Mexico, Brazil	Total	Max. length 6 m 54 s
1992 January 4/5	Central Pacific	Annular	
1992 June 30	South Atlantic	Total	Max. length 5 m 20 s
1992 December 24	Arctic	Partial	84% eclipsed
1993 May 21	Arctic	Partial	74% eclipsed
1993 November 13	Antarctic	Partial	93% eclipsed
1994 May 10	Pacific, Mexico, United States, Canada	Annular	
1994 November 3	Peru, Brazil, South Atlantic	Total	Max. length 4 m 23 s
1995 April 29	South Pacific, Peru, South Atlantic	Annular	
1995 October 24	Iran, India, East Indies, Pacific	Total	Max. length 2 m 5 s
1996 April 17	Antarctic	Partial	88% eclipsed
1996 October 12	Arctic	Partial	76% eclipsed
1997 March 9	Russia, Arctic	Total	Max. length 2 m 50 s
1997 September 2	Antarctic	Partial	90% eclipsed
1998 February 26	Pacific, Atlantic	Total	Max. length 3 m 56 s
1998 August 22	Indian Ocean, East Indies, Pacific	Annular	
1999 February 16	Indian Ocean, Australia, Pacific	Annular	
1999 August 11	Atlantic, Europe, Turkey, India	Total	Max. length 2 m 23 s

one alignment of the line of nodes pointing toward the Sun to the next identical alignment. This period is called the **eclipse year.**

To predict when you will see another solar eclipse, you need to know how many whole lunar months equal some whole number of eclipse years. This information will tell you how long you will have to wait for the next virtually identical alignment of the Sun, the Moon, and the line of nodes. By trial and error, you find that the answer is 223 lunar months = 19 eclipse years, because

$$223 \times 29.53 = 19 \times 346.6 = 6585 \text{ days}$$

This calculation is accurate to within a few hours, and the interval is called the **saros.** A more accurate treatment gives the length of the saros as 6585.3 days. Eclipses separated by the saros interval are said to form an "eclipse series."

You might think that you and your neighbors would simply have to wait one full saros interval to go from one solar eclipse to the next. However, because of the extra one-third day, the Earth will have rotated by an extra 120° when the next solar eclipse of a particular series occurs. The eclipse path will thus be one-third of the way around the world from you. You must therefore wait three full saros intervals (54 years 34 days) before the eclipse path comes back around to your part of the Earth. Figure 3-18 shows a series of solar eclipse paths, each separated from the next by one saros interval.

There is evidence that some ancient astronomers knew about intervals such as the saros. The discovery of such intervals is more likely to have come from lunar eclipses than solar eclipses. If you are far from the eclipse path, yet still within the Moon's penumbra, there is a good chance that you could fail to notice a solar eclipse. Even if half the Sun is

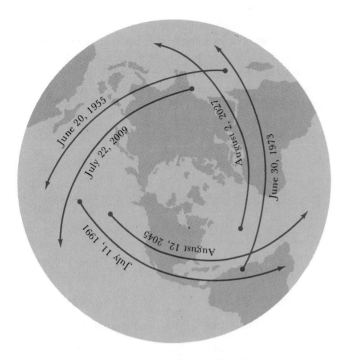

Figure 3-18 An eclipse series *Eclipses separated by the saros interval (18 years 11.3 days) are said to form an "eclipse series" because their geometric details, such as the orientation of the line of nodes, are nearly identical. Even the duration of totality is nearly the same for all members of a particular eclipse series.*

covered by the Moon, the remaining solar surface provides enough sunlight for the outdoor illumination not to be greatly diminished. In contrast, everyone on the nighttime side of the Earth can see an eclipse of the Moon, unless clouds block the view.

3-6 Ancient astronomers established traditions and invented systems that are still used

Ancient astronomers, particularly those of Greece, gave humanity a new and powerful way of thinking about the world. They gave the first clear demonstration that the tools of logic, reason, and mathematics can be used to discover and understand the workings of the universe. This general approach underlies all modern science. Aristarchus even went so far as to suggest that the motions of the planets in the sky could be explained simply if all the planets as well as the Earth orbit the Sun. As we shall see in the next chapter, this man was nearly 2000 years ahead of his time.

In addition to setting the stage for modern science, the early Greek astronomers also gave us ideas and established certain traditions that are still useful. For example, around

the year 160 BC, Hipparchus built an observatory on the island of Rhodes. Over a period of several years, he compiled the first comprehensive star catalogue, which listed the coordinates and brightnesses of some 850 stars. During the course of this pioneering work, Hipparchus compared his star positions with earlier records dating back to Aristarchus's time, a century earlier. Hipparchus soon noted systematic differences that led him to conclude that the north celestial pole had shifted slightly over the last hundred years. He had discovered precession.

While compiling his star catalogue, Hipparchus established a system to denote the brightnesses of stars. His system is the basis of the **magnitude scale** used by astronomers today. Quite simply, Hipparchus said that the brightest stars in the sky are "first magnitude." The dimmest stars visible to the unaided eye he called "sixth magnitude." To stars of intermediate brightness he assigned intervening numbers on the scale of 1 to 6.

With only a few refinements, this same system is used today. As shown in Figure 3-19, the magnitude scale has been extended to include even very faint stars. For example, with a good pair of binoculars you can see stars as faint as tenth magnitude. Through some of the largest telescopes it is

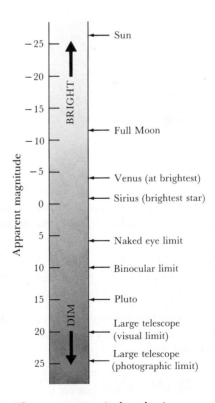

Figure 3-19 The apparent magnitude scale *Astronomers denote the brightness of an object in the sky by its "apparent magnitude." Most stars visible to the naked eye have apparent magnitudes between 1 and 6. Photography through a large telescope can reveal stars as faint as magnitude 24.*

possible to see stars as dim as magnitude 20. Photography with long exposure times reveals even dimmer stars.

Modern astronomers use negative numbers to extend Hipparchus's scale to include very bright objects. For example, Sirius, the brightest star in the sky, has a magnitude of $-1\frac{1}{2}$. At its brightest, the planet Venus shines with a magnitude of -4. The Sun is the brightest object in the sky. Its magnitude is $-26\frac{1}{2}$.

Later in this book, we shall discuss many details about the brightness of stars and galaxies. We shall see how careful observation and logical thinking can extend our knowledge and understanding far beyond our limited daily experiences. Although they have been refined and elaborated, many concepts like the magnitude scale have come down to us from the people who first discovered how the human mind can unlock the secrets of the universe.

Key words

annular eclipse	lunar month	phases (of the Moon)	third-quarter moon
apogee	magnitude scale	*regression of the line of nodes	total lunar eclipse
eclipse path	new moon		total solar eclipse
eclipse year	partial lunar eclipse	*rotation of the Moon's orbit	waning crescent moon
first quarter moon	partial solar eclipse		waning gibbous moon
full moon	penumbra (*plural* penumbrae)	saros	waxing crescent moon
*line of apsides		sidereal month	waxing gibbous moon
line of nodes	penumbral eclipse	solar corona	umbra (*plural* umbrae)
lunar eclipse	perigee	solar eclipse	
		synodic month	

Key ideas

- Thousands of years ago, many cultures were gathering useful observations about the apparent positions of the Sun, Moon, and planets against the background of the stars.

- The phases of the Moon occur because of the shifting relative positions of the Earth, the Moon, and the Sun.

  With respect to the stars, the Moon completes one orbit around the Earth in a sidereal month averaging 27.3 days.

  The Moon completes one cycle of phases (one orbit around the Earth with respect to the Sun) in a synodic month averaging 29.5 days, which is also called the lunar month.

  The lengths of the sidereal and synodic months vary slightly because of the Sun's gravitational pull.

- Ancient astronomers made great progress in determining the sizes and relative distances of the Earth, the Moon, and the Sun.

  Around 280 BC, Aristarchus attempted to measure the distances from the Earth to the Sun and the Moon, using the Moon's phases. He also estimated the relative sizes of the Earth, Sun, and Moon.

  Decades later, around 200 BC, Eratosthenes measured the Earth's size by comparing the position of the Sun in the sky at the same moment in two different locations.

  From these determinations, Greek astronomers were able to estimate the actual sizes of the Earth, Moon, and Sun and their actual separations.

- A lunar eclipse occurs when the Moon moves through the Earth's shadow. This happens when the Sun and Moon are both on the line of nodes at full moon. A solar eclipse occurs when the Earth passes through the Moon's shadow. This happens when the Sun and Moon are both on the line of nodes at new moon. The line of nodes is the line where the planes of the Earth's orbit and the Moon's orbit intersect.

The gravitational pull of the Sun causes the line of nodes gradually to shift its orientation with respect to the stars.

A fairly detailed knowledge of the Moon's orbit around the Earth is needed for an understanding of the intervals between eclipses.

· Depending on the exact relative positions of the Sun, Moon, and Earth, lunar eclipses may be penumbral, partial, or total. Solar eclipses may be partial, annular, or total.

The shadow of an object has two parts: the umbra, within which the light source is completely blocked; and the penumbra, where the light source is only partially blocked.

During a total solar eclipse, the Moon's umbra traces out an eclipse path over the Earth's surface, as the Earth rotates.

· Using an interval called the saros, it is possible to group total solar eclipses into series and to predict the time of the next eclipse in each series.

· Astronomers still use a system for denoting the apparent brightness of objects in the sky that was invented by Hipparchus around 160 BC.

Very bright objects have negative values of magnitude; very dim objects have large positive values of magnitude.

Review questions

1 Sketch a diagram of the relative positions of the Earth, Moon, Sun, and *Voyager 1* when the photograph in Figure 3-2 was taken.

*2 What is the phase of the Moon if it (**a**) rises at 3 AM? (**b**) is crossing the upper meridian at midnight? (**c**) sets at 9 PM? At what time does (**d**) the full moon set? (**e**) the first-quarter moon rise? (**f**) the third-quarter moon cross the upper meridian?

*3 What is the phase of the Moon if, on the first day of spring, the Moon is located at (**a**) the vernal equinox, (**b**) the summer solstice, (**c**) the autumnal equinox, (**d**) the winter solstice?

4 Describe the method that Eratosthenes used to measure the diameter of the Earth.

5 Describe the method that Aristarchus used to determine the distances to the Sun and Moon.

6 What is the difference between the sidereal and synodic months?

*7 How many more sidereal months than synodic months are there in a year? Why?

8 Why isn't there a lunar eclipse at every full moon and a solar eclipse at every new moon?

9 What is the line of nodes and why is it important to the subject of eclipses?

10 Which type of eclipse—lunar or solar—do you think most people have seen? Why?

11 What is the difference between the umbra and the penumbra of a shadow?

12 Is it possible for a total eclipse of the Sun to be followed three months later by a lunar eclipse? Why?

13 Can one ever observe an annular eclipse of the Moon? Why?

14 Why do you suppose that total solar eclipse paths fall more frequently on oceans rather than on land?

15 In his novel *King Solomon's Mines*, author H. Rider Haggard described a total solar eclipse that was reportedly seen in both South Africa and in the British Isles. Is such an eclipse possible? Why or why not?

Advanced questions

Tips and tools . . .
It is helpful to know that the saros interval of 6585.3 days equals 18 years 11⅓ days if the interval includes four leap years, but is 18 years 10⅓ days if it includes five leap years. The average angular speed of the Moon along its orbit can be estimated by dividing 360° by the length of a sidereal month.

16 During a lunar eclipse, does the Moon enter the Earth's shadow from the east or from the west? Explain why.

17 If the Moon revolved about the Earth in the same orbit but in the opposite direction, would the synodic month be longer or shorter than the sidereal month? Explain why.

*18 On July 11, 1991, residents of Hawaii were treated to a total solar eclipse. (a) When and over what part of the world will the next eclipse of that series occur? (b) When might the Hawaiians next expect to see an eclipse of that series?

*19 During an occultation, or "covering up," of Jupiter by the Moon, an astronomer notices that it takes the Moon's edge 90 seconds to cover Jupiter's disk completely. If the Moon's motion is assumed to be uniform and the occultation was "central" (i.e., center over center), find the angular diameter of Jupiter.

Discussion questions

20 Suppose that you were an ancient astronomer given the task of building a Stonehenge-type monument to be aligned with the apparent positions and phases of the Moon. Which directions and alignments might you consider important? What sort of observations would be needed to determine these directions? How long would it take to collect the necessary data?

21 Describe the cycle of lunar phases that would be observed if the Moon moved about the Earth in an orbit perpendicular to the plane of the Earth's orbit. Is it possible for both solar and lunar eclipses to occur under these circumstances?

22 How would a lunar eclipse look if the Earth had no atmosphere?

23 Examine a listing of total solar eclipses over the next several decades. What are the chances that you might be able to travel to one of the eclipse paths? Do you think you might go through your entire life without ever seeing a total eclipse of the Sun?

Observing projects

24 Observe the Moon on each clear night over the course of a month. On each night, note the Moon's location among the constellations and record that location on a star chart that also shows the ecliptic. After a few weeks, your observations will begin to trace the Moon's orbit. Identify the orientation of the line of nodes by marking the points where the Moon's orbit and the ecliptic intersect. On what dates is the Sun near the nodes marked on your star chart? Compare these dates with the dates of next solar and lunar eclipses.

25 It is quite possible that a lunar eclipse will occur while you are taking this course. Look up the date of the next lunar eclipse in a current issue of a reference such as the *Astronomical Almanac* or *Astronomical Phenomena*. You would also be well advised to consult such magazines as *Sky*

& Telescope and *Astronomy*, which generally run articles about upcoming eclipses the month before they happen. Make arrangements to observe the next lunar eclipse. If the eclipse is partial or total, note the times at which the Moon enters and exits the Earth's umbra. If the eclipse is penumbra, can you see any changes in the Moon's brightness as the eclipse progresses?

For further reading

Allen, D., and Allen, C. *Eclipse*. Allen & Unwin, 1987 • A well written history of our understanding of eclipses, including the mythology they engendered and the science behind them.

Cornell, J. *The First Stargazers*. Scribner, 1981 • This introduction to "archeoastronomy," the archeological study of ancient astronomy, includes excellent sections about the astronomy of Stonehenge, native Americans, China, and Egypt.

Daniel, G. "Megalithic Monuments." *Scientific American*, July 1980 • This article surveys stone monuments that were built by the thousands throughout prehistoric Europe, probably between 3000 and 1000 BC. Like Stonehenge, many of these structures seem to have been constructed with astronomical alignments in mind.

Gingerich, O. "Aristarchus of Samos: A Report on a Symposium." *Sky & Telescope*, November 1980 • This brief article gives interesting insights into Aristarchus and the island of Samos on which he was born.

Hadingham, E. *Early Man and the Cosmos*. Walker, 1984 • An introduction to prehistoric and early astronomy with nice sections on stone monuments in Britain and America. An excellent primer on archeoastronomy.

Hawkin, G. *Stonehenge Decoded*. Delta, 1965 • This classic book was one of the first to decipher and popularize Stonehenge as an astronomical monument.

Hetherington, N. *Ancient Astronomy and Civilization*. Pachart, 1987 • A slim volume that introduces early astronomy, much of it through translations of original writings.

Krupp, E. *Echoes of the Ancient Skies*. Harper & Row, 1983 • A superb introduction to archeoastronomy. Krupp skillfully brings together the mythology, rituals, and monuments of many cultures, demonstrating both similarities and differences.

Kundu, M. "Observing the Sun during Eclipses." *Mercury*, July/August 1981 • This article is a personal account of the trials and tribulations of observing the 1980 total eclipse of the Sun in India.

Menzel, D., and Pasachoff, J. "Solar Eclipse: Nature's Superspectacular." *National Geographic,* August 1970 • A beautifully illustrated article about observing a total solar eclipse.

Sagan, C. "The Shores of the Cosmic Ocean." In Sagan, C., *Cosmos.* Random House, 1980 • This chapter is an excellent recounting of Eratosthenes' experiment to measure the size of the Earth.

Stephenson, F. "Historical Eclipses." *Scientific American,* October 1982 • This fascinating article shows how reliable records of solar and lunar eclipses dating back to 750 BC can be used to see if the Sun is shrinking or if the Earth's rate of rotation is changing.

Williamson, R. *Living the Sky: The Cosmos of the American Indian.* Houghton-Mifflin, 1984 • This book gives a fascinating tour of the astronomical thought and monuments of native American Indian tribes.

Owen Gingerich is an astrophysicist at the Smithsonian Astrophysical Observatory in Cambridge and a Professor of Astronomy and the History of Science at Harvard University. His research interests have included modeling the solar atmosphere, interpretation of stellar spectra, and the re-computation of ancient Babylonian mathematical tables. He is a leading authority on the works of Johannes Kepler and Nicholas Copernicus. He is also actively involved with work on recent astronomy, having coedited A Source Book in Astronomy and Astrophysics, 1900–1975, *and serving currently as editor of the twentieth-century section of the International Astrophysical Union's* General History of Astronomy. *He is an active participant in such scientific organizations as the American Astronomical Society and the International Astronomical Union and has several hundred general interest, technical, and historical publications to his credit. The most accessible of these are in* Scientific American *and* Sky & Telescope.

Owen Gingerich ASTROLOGY AND ASTRONOMY

In 1226 the feared Mongol conqueror Genghis Khan, about to subdue the entire known world, suddenly called off his campaign. The planet Jupiter was about to overtake Saturn, and quite possibly Genghis Khan's astrologers warned of dangerous consequences heralded by this once-in-twenty-year planetary conjunction.

In 1634 the Bohemian general Wallenstein became more and more nervous. He carried close to his chest a horoscope prepared by the astronomer Johannes Kepler, and the predictions stopped in that year with the prophecy that "according to astrological lore" the month of March would be marked with "horrible disorder." Wallenstein became increasingly troubled, brought matters to a crisis, and then, in February, was assassinated by other officers.

What are we to make of these extraordinary events? Did the distant planets have some subtle influence on these military men? Or were these self-fulfilling prophecies, events destined to happen because the belief in their inevitability was so strong that they were subconsciously laid in place?

A few hundred years BC, when the earliest scientists began to appreciate the relationships between the Sun's noontime height and the seasons and between the Moon and the tides, it seemed logical to suppose that the planets might also have other terrestrial influences. This astrological thinking was far more scientific than the animistic notions that had preceded it, in which capricious gods in the natural world were continually carrying out unpredictable acts.

Gradually there grew up a systematic lore about the supposed planetary influences. Mars and Saturn were seen as generally evil, Venus and Jupiter as benign. Geometrical relationships between the Sun, Moon, and planets, such as conjunctions, oppositions (two objects 180° apart), and quadratures (90° apart), enhanced or countered these astrological effects. The 12 zodiacal signs acquired special properties in combination with the planetary movements. Furthermore, the sky was divided from horizon to horizon into a series of "houses," each of which controlled some aspect of human life such as marriage, friends, riches, or death. The configuration of the heavens and the placement of planets at the moment of birth (or conception) was believed to establish a person's personality and destiny. The part of the sky just below the eastern horizon, the ascendant, was considered the most powerful indicator in a person's horoscope.

The fourteenth-century poet Chaucer often used astrology to good effect in his *Canterbury Tales,* with lines such as

> Mine ascendant was Taurus and Mars therein.
> Alas, alas, that ever love was sin!
> I followed aye mine inclination
> By virtue of mine constellation.

Listeners knowledgeable in astrology understand the Wife of Bath's contradictory character through this speech because, while the sign of Taurus was particularly favorable to Venus, the planet of love, this happy circumstance was spoiled by the unfortunate presence of warlike Mars.

The continuously changing positions of the planets within the houses, as the daily rotation of the heavens carried the stars and planets across the sky, were thought to presage one's fortune from moment to moment. Altogether, astrology with its zodiacal signs and houses and planetary configurations was a marvelous system of folk psychology, but unfortunately there was little empirical evidence behind it.

At the same time that some astrologers tried to chart the course of human affairs through the stars, others tried to predict the weather. Those who worked on human affairs had better luck, probably because weather does not cooperate to conform to prophecies.

Like many astronomers of the early Renaissance, Kepler drew up many horoscopes for his friends and employers, although he was very skeptical about the prevailing opinions on celestial influences. For the benefit of other astrologers, Kepler gave the position of the planets at his own birth, but he added: "My stars were not Mercury rising in the seventh angle in quadrature with Mars, but Copernicus and Tycho Brahe, without whose observation books everything I have brought into light would have remained in darkness; my rulers were not Saturn predominating over Mercury, but the emperors Rudolph and Matthias; not a planetary house, Capricorn with Saturn, but Upper Austria, the house of the emperor."

In some respects, Kepler was the astrologer who destroyed astrology. By insisting on an astronomy based on physical causes, he furthered the development of a logically connected scientific framework. Astrology could find neither causes nor correlations between the planetary configurations and their supposed influences.

Perhaps the most extensive statistics on the effectiveness of astrological predictions have been compiled by the French psychologist Maurice Gauquelin. He failed to find any correlations between the birth horoscopes of 25,000 celebrities and the traditional qualities associated with the various zodiacal constellations. Gauquelin also conducted an interesting experiment in which he offered free computer-generated horoscopes with the provision that the recipients evaluate how well the astrological characterization fit them. Nearly 95 percent of the recipients said they recognized themselves in the "psychological portrait," and 80 percent of their family and friends shared the opinion. What they did not know was that exactly the same computer printout had been sent to everyone!

While astrology remains an amusing game comparable to, but rather more complicated than, Chinese fortune cookies, the fact that many people continue to take it seriously can only be attributed to wishful thinking and self-fulfilling prophecies.

Of all the evaluations pro and con, I particularly enjoy Michael Flanders' perceptive tongue-in-cheek view of astrology:

> Jupiter's passed through Orion
> And coming to conjunction with Mars.
> Saturn is wheeling through infinite space
> To its pre-ordained place in the stars
> And I gaze at the planets in wonder
> At the trouble and time they spend
> All to warn me to *be careful*
> *In dealings involving a friend.*

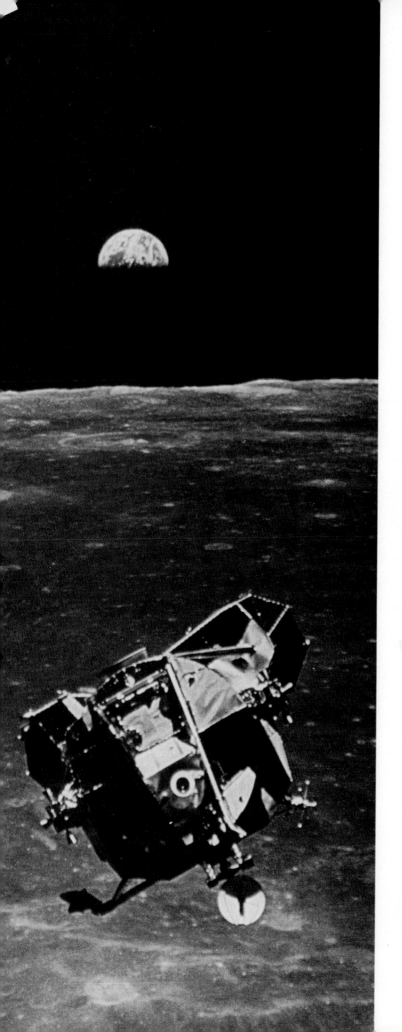

Gravitation and the Motions of Planets

Ancient astronomers believed that the heavens rotated around a stationary Earth. During the Renaissance, several brilliant thinkers spearheaded a revolution that dethroned the Earth from its central location and laid the foundation of modern science. First came the Copernican system—a workable model of planets going around the Sun along a system of circles. Later, observations by Tycho Brahe and Galileo clearly demonstrated inaccuracies in the earth-centered view. Kepler's profound discovery that planetary orbits are in fact ellipses set the stage for the inspired work of Isaac Newton, who formulated basic laws of physics. Newton's study of planetary motion led to a precise, mathematical description of the force of gravity, which holds the planets in their orbits about the Sun. Then in the early twentieth century, Albert Einstein proposed the radically new idea that gravity affects the curvature of space and the flow of time. The relationships between gravity, space, and time are central to all our modern ideas about the structure of the universe, including its creation and ultimate fate.

Apollo 11 *leaving the Moon   The lunar module Eagle returns from the Moon after completing the first successful manned lunar landing. This photograph was taken from the command module Columbia, in which the astronauts returned to Earth. All of the orbital maneuvers to dock the Eagle and the Columbia and then set course for Earth were based on Newtonian mechanics and Newton's law of gravity. These same physical principles are used by astronomers to understand a wide range of phenomena, from the motions of double stars to the rotation of an entire galaxy. (NASA)*

It is by no means obvious that the Earth moves around the Sun. Indeed, our daily experience strongly suggests that the opposite is true. The daily rising and setting of the Sun, Moon, and stars could lead us to believe that the entire cosmos revolves about an Earth that is at the center of the universe. This is just what most people did believe for thousands of years.

4-1 Ancient astronomers invented geocentric cosmologies to explain planetary motions

The ancient Greeks conceived of certain principles that even today guide modern scientists. For instance, around 550 BC Pythagoras and his followers put forth the idea that natural phenomena could be described with mathematics. About 200 years later, Aristotle asserted that the universe is governed by physical laws. These two concepts found their highest expression in the works of great astronomers and physicists who struggled to explain planetary motion.

These early Greek astronomers were among the first to leave a written record of their attempts to explain the motion of the planets. Most Greeks assumed that the Sun, the Moon, the stars, and the five planets revolve about the Earth, and thus their view of the universe is said to be "geocentric." A theory of the universe is called a **cosmology**, and thus Greek thinkers such as Pythagoras or Aristotle gave credence to a **geocentric cosmology**.

The Greeks, and other cultures of that time, knew of five planets: Mercury, Venus, Mars, Jupiter, and Saturn. These planets are quite obvious in the sky because from night to night they slowly shift their positions with respect to the background of the stationary, or "fixed," stars in the constellations. In fact, the word "planet" comes from a Greek term meaning wanderer. Furthermore, some of these planets are very bright in the night sky. Venus, for example, at its maximum brilliancy is 16 times brighter than the brightest star.

Observations of the positions of the planets against the stars from night to night make it clear that the planets do not move at uniform rates through the sky. Explaining these nonuniform motions of the five planets was one of the main challenges facing the astronomers of antiquity. It was not easy to develop a comprehensive geocentric theory of the universe.

As seen from Earth, the planets wander primarily across the 12 constellations of the zodiac. As mentioned in Chapter 2, these constellations encircle the sky in a continuous band centered on the ecliptic. If you follow a planet as it travels across the zodiac from night to night, you find that the planet usually moves slowly eastward against the background stars. This phenomenon is called **direct motion**. Occasionally, however, the planet will seem to stop and then back up for several weeks or months. This occasional west-ward movement is called **retrograde motion**. These motions are much slower than the apparent daily rotation of the sky caused by the Earth's rotation. Both direct and retrograde motions are best detected by mapping the position of a planet against the background stars from night to night over a long period. An example is the path of Mars from late 1992 through early 1993, shown in Figure 4-1.

The Greeks developed many theories to account for retrograde motion and the exact loops that the planets trace out against the background stars. One of the most successful and enduring ideas was expounded and embellished by the last of the great Greek astronomers, Ptolemy, who lived in Alexandria during the second century AD. The basic concept is sketched in Figure 4-2. Each planet is assumed to move in a small circle called an **epicycle**, whose center in turn moves in a larger circle called a **deferent**, which is centered approximately on the Earth. As viewed from Earth, the epicycle moves eastward along the deferent, and both circles rotate in the same direction (counterclockwise in Figure 4-2).

Most of the time, the motion of the planet on its epicycle adds to the eastward motion of the epicycle on the deferent. Thus, the planet is seen to be in direct (eastward) motion against the background stars throughout most of the year. However, when the planet is on the part of its epicycle nearest the Earth, the motion of the planet along the epicycle subtracts from the motion of the epicycle along the deferent. The planet therefore appears to slow down and then halt its usual eastward movement among the constellations, even to the extent of going backward for a few weeks or months. This concept of epicycles and deferents succeeds in providing a general explanation of the retrograde loops that the planets execute.

Using the wealth of astronomical data in the library at Alexandria, including records of planetary positions for hundreds of years, Ptolemy deduced the sizes of the epicycles and deferents and the rates of rotation needed to produce the recorded paths of the planets. After years of tedious work, Ptolemy assembled his calculations into 13 volumes, collectively called the *Almagest*. His work was then used to predict the positions and paths of the Sun, Moon, and planets with unprecedented accuracy. In fact, the *Almagest* was so successful that it became the astronomer's bible, and for over 1000 years, Ptolemy's cosmology endured as a useful description of the workings of the heavens.

Eventually, however, things began going awry. Tiny errors and inaccuracies unnoticeable in Ptolemy's day became compounded and multiplied over the years, especially those concerning precession. Thirteenth-century astronomers made some cosmetic adjustments to the Ptolemaic system. However, as more complicated and arbitrary details were added to keep it consistent with the observed motions of the planets, the system became less and less aesthetically satisfying.

One of the guiding beliefs of many scientists is that simple, straightforward explanations of phenomena are more

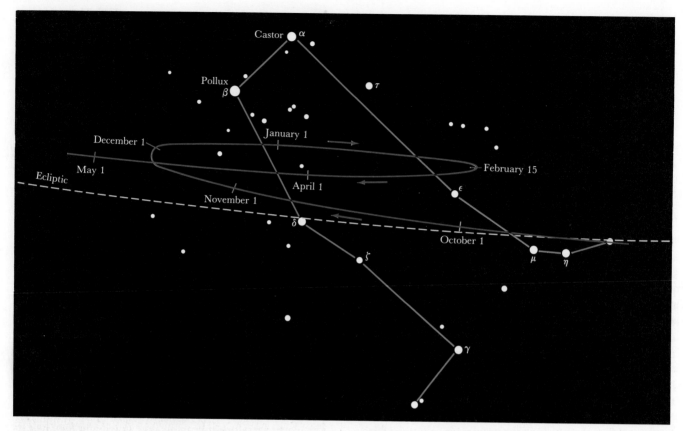

Figure 4-1 *The path of Mars in 1992–1993* *From the fall of 1992 through the spring of 1993, Mars will move across the constellation of* *Gemini. From November 30 through February 14, Mars's motion will be retrograde.*

likely to be correct than complicated, convoluted ones. This paraphrases an idea called **Occam's razor**, named after the fourteenth-century English philosopher who first expressed it. (The term "razor" refers to shaving an argument or explanation to its simplest terms.) Occam's razor has no proof or verification, but instead appeals to the scientist's sense of beauty and elegance.

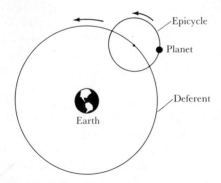

Figure 4-2 *A geocentric explanation of planetary motion* *Each planet moves along an epicycle, which in turn moves along a deferent centered approximately on the Earth. As seen from Earth, the speed of the planet on the epicycle alternately adds to or subtracts from the speed of the epicycle on the deferent, thus producing periods of either direct or retrograde motion.*

Half a century after the death of William of Occam, the cumbersome complexity of the Ptolemaic system led a Polish astronomer to doubt its validity. The Sun-centered system that emerged from his work offered a much simpler explanation of planetary motions and helped lay the foundations of modern physical science.

4-2 Nicolaus Copernicus devised the first comprehensive heliocentric cosmology

Imagine driving on a freeway at high speed. As you pass a slowly moving car, it appears to move backward, even though it is traveling in the same direction as your car. This sort of observation (probably with chariots instead of cars) inspired the ancient Greek astronomer Aristarchus to suggest a straightforward explanation of retrograde motion in which all the planets, including the Earth, revolve about the Sun. The retrograde motion of Mars, for example, occurs when the Earth overtakes and passes Mars, as shown in Figure 4-3. The occasional backward movement of a planet is the result of our changing viewpoint—an idea that is beauti-

Figure 4-3 *A heliocentric explanation of planetary motion* *The Earth travels around the Sun more rapidly than does Mars. Consequently, as the Earth overtakes and passes this slower-moving planet, Mars appears (from points 4 through 6) to move backward for a few months.*

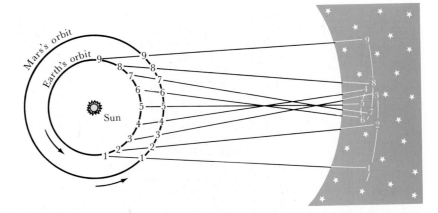

fully simple compared to an Earth-centered system with all its "circles upon circles."

Aristarchus had demonstrated that the Sun is bigger than the Earth (recall Table 3-1), which then made it sensible to consider the possibility that the smaller Earth might be orbiting the larger Sun. In Aristarchus's day, however, the idea of a moving Earth seemed quite incompatible with the apparent stillness and immobility of the Earth. Almost 2000 years elapsed before someone had both the insight and the determination to work out the details of a **heliocentric** (Sun-centered) **cosmology**. That person was a Polish lawyer, physician, economist, canon, and artist named Nicolaus Copernicus (see Figure 4-4). Especially gifted in mathematics, Copernicus turned his attention to astronomy in the early 1500s.

Copernicus realized that using a heliocentric perspective would enable him to determine which planets are closer to the Sun than is the Earth and which are farther away. Because Mercury and Venus are always observed fairly near the Sun in the sky, Copernicus concluded that their orbits must be smaller than the Earth's. The other visible planets—Mars, Jupiter, and Saturn—can be seen in the middle of the night, when the Sun is far below the horizon, which Copernicus realized could occur only if the Earth comes between the Sun and a planet. He therefore concluded that the orbits of Mars, Jupiter, and Saturn must be larger than the Earth's orbit.

Mercury and Venus are sometimes called **inferior planets**, because their orbits are smaller than the Earth's. Mars, Jupiter, and Saturn are known as **superior planets**, because their orbits are bigger than the Earth's. The other, dimmer superior planets—Uranus, Neptune, and Pluto—were discovered only after the telescope came into use.

It is often useful to specify various points on a planet's orbit, as shown in Figure 4-5. These points help us identify certain geometric arrangements, or **configurations**, involving the Earth, another planet, and the Sun. For example, when Mercury or Venus is between the Earth and the Sun, we say that the planet is at **inferior conjunction**. When Mercury or Venus is on the opposite side of the Sun, we say that the planet is at **superior conjunction**.

The angle between the Sun and a planet, as viewed from the Earth, is called the planet's **elongation**. At **maximum eastern elongation** an inferior planet is as far east of the Sun as it can be. At such times, the planet appears above the western horizon after sunset and is often called an "evening

Figure 4-4 *Nicolaus Copernicus (1473–1543)* *Copernicus was the first person to work out the details of a heliocentric system in which the planets, including the Earth, orbit the Sun. (E. Lessing/Magnum)*

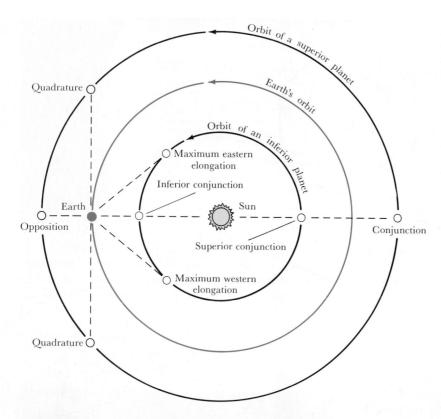

Figure 4-5 Planetary configurations *The key points along a planet's orbit are shown in this diagram. These points identify the specific geometric arrangements possible between the Earth, a planet, and the Sun.*

star." Similarly, at **maximum western elongation** Mercury or Venus is as far west of the Sun as it can possibly be and rises before the Sun, gracing the predawn sky as a "morning star."

When a superior planet is behind the Sun, the planet is at **conjunction** (with an elongation of 0°). When it is exactly opposite the Sun in the sky, the planet is at **opposition** (with an elongation of 180°). And when the planet's elongation is 90°, it is at **quadrature**.

It is not difficult to determine when a planet happens to be located at one of the key positions in Figure 4-5. For example, when Mars is at opposition, it crosses the upper meridian at midnight.

Although it is easy to follow a planet as it moves from one configuration to another, these observations do not immediately provide relevant data about the planet's actual orbit around the Sun. The Earth, from which we must make the observations, is also moving. Realizing this, Copernicus was careful to distinguish between two characteristic time intervals, or **periods**, of each planet. The **synodic period** is the time that elapses between two successive identical configurations, as seen from the Earth—from one opposition to the next, for example, or from one conjunction to the next. The **sidereal period** is the true orbital period of a planet, the time it takes the planet to complete one full orbit of the Sun, as measured relative to the stars.

The synodic period of a planet can be determined by observing the sky, but the sidereal period must be calculated.

Copernicus figured out how to do this (see Box 4-1), and his results are shown in Table 4-1.

With the planetary orbits now known, Copernicus devised a straightforward geometric method of determining the distances of the planets from the Sun. The mathematical details, which involve some trigonometry, are described in Box 4-2. His answers turned out to be remarkably close to the modern values (see Table 4-2). From Tables 4-1 and 4-2 it is apparent that the farther a planet is from the Sun, the longer the planet takes to travel around its orbit.

Copernicus compiled his ideas and calculations into a book entitled *De Revolutionibus Orbium Celestium* (On the Revolutions of the Celestial Spheres), which was published in 1543, the year of his death. Although Copernicus as-

Table 4-1 The synodic and sidereal periods of the planets

Planet	Synodic period	Sidereal period
Mercury	116 days	88 days
Venus	584 days	225 days
Earth	—	1.0 year
Mars	780 days	1.9 years
Jupiter	399 days	11.9 years
Saturn	378 days	29.5 years

Box 4-1 Synodic and sidereal periods

Consider an inferior planet orbiting the Sun as shown in the diagram. Let P be the sidereal period of the planet, S the synodic period of the planet, and E the sidereal period of the Earth (which Copernicus knew to be equal to nearly $365\frac{1}{4}$ days).

The rate at which the Earth moves around its orbit is $360°/E$ (nearly 1° per day). Similarly, the rate at which the inferior planet moves along its orbit is $360°/P$.

During the inferior planet's synodic period, the Earth covers an angular distance of $S(360°/E)$ around its orbit. In that same time, the inferior planet has covered an angular distance of $S(360°/P)$. Note, however, that the inferior planet has gained one full lap on the Earth. One lap corresponds to 360°. Thus, $S(360°/E) + 360° = S(360°/P)$. Dividing each term of this equation by $360°\,S$ gives

$$\frac{1}{P} = \frac{1}{E} + \frac{1}{S}$$

A similar analysis for a superior planet yields

$$\frac{1}{P} = \frac{1}{E} - \frac{1}{S}$$

Using these formulas, we can calculate a planet's sidereal period from its synodic period.

Example: Consider the superior planet Jupiter, which has an observed synodic period of 398.88 days, or 1.092 years. (When astronomers express a time interval in years, they mean Earth years of $365\frac{1}{4}$ days.) Of course, $E = 1$ year exactly. Thus

$$\frac{1}{P} = 1 - \frac{1}{1.092} = 0.084 = \frac{1}{11.9}$$

It takes 11.9 years for Jupiter to complete one full orbit of the Sun.

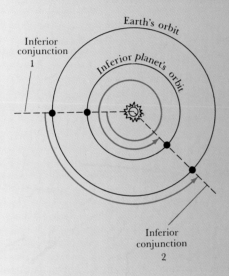

sumed that the Earth travels around the Sun along a circular path, he found that perfectly circular orbits could not accurately describe the paths of the other planets. He thus had to add an epicycle to each planet to account for its slight variation in speed along its orbit. Therefore, according to Copernicus, each planet moves along a small epicycle, which in turn traces out a circular path around the Sun.

Table 4-2 Average distances of the planets from the Sun (AU)

Planet	Copernicus	Modern
Mercury	0.38	0.39
Venus	0.72	0.72
Earth	1.00	1.00
Mars	1.52	1.52
Jupiter	5.22	5.20
Saturn	9.07	9.54

4-3 Tycho Brahe made astronomical observations that disproved ancient ideas about the heavens

In November of 1572, a bright star suddenly appeared in the constellation of Cassiopeia. At first, it was even brighter than Venus, but then it began to grow dim. After 18 months, it faded from view.

Modern astronomers recognize this phenomenon as a supernova explosion, produced by the violent death of a certain type of star (see Chapter 22). In the sixteenth century, however, the prevailing opinion was quite different. Classical teachings that dated back to Aristotle and Plato argued that the heavens are permanent and unalterable. Consequently, the "new star" of 1572 could not really be a star at all, because the heavens do not change; it must instead be some sort of bright object quite near Earth, perhaps not much further away than the clouds overhead.

A 25-year-old Danish astronomer named Tycho Brahe realized that straightforward observations might reveal the

Box 4-2 Copernicus's method of determining the sizes of orbits

To determine the size of an inferior planet's orbit, Copernicus measured the angle (α) between the Sun and the planet at its greatest elongation. As shown in diagram *a*, the triangle formed by the Earth, the inferior planet, and the Sun then contains a right angle. The hypotenuse of the triangle has a length of one astronomical unit (1 AU). Therefore, the radius (also measured in AU) of the inferior planet's orbit is equal to sin α.

Determining the size of a superior planet's orbit is slightly more complicated. First, note the date on which the planet is at opposition. (In diagram *b*, opposition occurs when the planets are in the positions designated by 1.) After a few months, note the date on which the planet appears at quadrature. (The plan-

ets have now moved to the positions marked 2.) Using the number of days that have elapsed, determine the angle β, which is the distance the Earth has traveled, moving at roughly 1° per day. Knowing the sidereal period of the planet (see Table 4-1), you can determine the angle γ in the same way.

The triangle formed by the Sun, the Earth, and the planet at quadrature contains a right angle. In addition, the short side of the triangle is exactly 1 AU long, and the angle ($\beta - \gamma$) is now known. The hypotenuse of the triangle, which is the radius of the superior planet's orbit, therefore has a length of 1/cos ($\beta - \gamma$), also measured in AU.

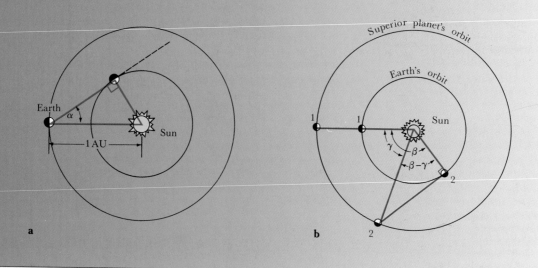

distance to the "new star." It is everyone's common experience that, when you walk from one place to another, nearby objects appear to shift their positions against the background of more distant objects. This phenomenon, whereby the apparent position of an object changes because of the motion of the observer, is called **parallax**. If the "new star" was nearby, then its position should shift against the background stars over the course of a night, because the Earth's rotation changes our viewpoint (see Figure 4-6). Actually, Tycho believed that the heavens rotate about the Earth, but the net effect is the same. Tycho's careful observations failed to disclose any parallax, and so the "new star" had to be quite far away, farther from Earth than anyone had imagined. Tycho Brahe summarized his findings in a small book *De Stella Nova* (On the New Star), published in 1573.

Tycho's discovery, which flew in the face of nearly 2000 years of wisdom, attracted the attention of the Danish king, who was so impressed that he financed the construction of a magnificent observatory for Tycho on the island of Hveen,

just off the Danish coast. The bequest included a lavish house with servants and coastal land that provided an income. Tycho designed and supervised the construction of astronomical instruments, like the wall quadrant shown in Figure 4-7, which permitted him to measure the positions of stars and planets with unprecedented accuracy. Tycho named his palatial observatory Uraniborg, which means *sky castle.*

Using his instruments at Uraniborg, Tycho Brahe attempted to test Copernicus's ideas about the Earth going around the Sun. Tycho argued that if the Earth was in motion, then nearby stars should appear to shift their positions with respect to background stars as we orbit the Sun. Tycho failed to detect any such parallax, and so he concluded that the Earth was at rest and the Copernican system was wrong. Actually, the stars are so far away that naked-eye observations cannot possibly detect the tiny shifting of stellar positions that have since been confirmed with telescopic observations. Nevertheless, Tycho Brahe's astronomical rec-

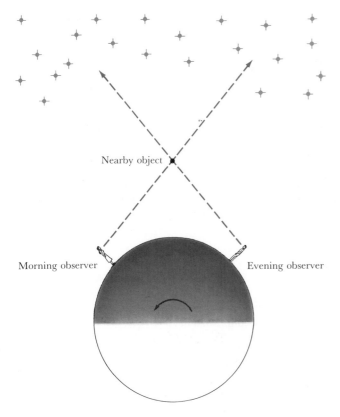

Figure 4-6 *The parallax of a nearby object* *Tycho Brahe argued that, if an object is near the Earth, its position against the background stars should appear to change over the course of a night. Tycho failed to measure such changes for a supernova in 1572 and a comet in 1577, and so he concluded that these objects are far from the Earth.*

Figure 4-7 *Tycho Brahe at his observatory* *During the sixteenth century Tycho Brahe measured the positions of stars and planets with unprecedented accuracy. His measurements were crucial to the development of astronomy in the seventeenth century. (Photo Researchers, Inc.)*

ords were destined to play an important role in the development of a heliocentric cosmology. Upon his death in 1601, many of his charts and record books fell into the hands of his gifted assistant Johannes Kepler (see Figure 4-8).

4-4 Johannes Kepler proposed elliptical paths for the planets about the Sun

Until Kepler's time, astronomers had assumed that heavenly objects moved in circles. A good part of the reasoning behind this assumption was that circles were considered the most perfect and harmonious of all geometric shapes. Since God was envisioned as being perfect and as residing in heaven along with the stars and planets, these astronomers concluded that a perfect God would use only perfect shapes to control the motions of the planets. Kepler questioned such arguments, and his first major contribution to astronomy was the suggestion that noncircular curves might fit the planetary orbits.

Kepler first turned from circles to ovals. For years he tried in vain to find ovals that would fit the observed positions of

Figure 4-8 *Johannes Kepler (1571–1630)* *Using Tycho Brahe's records of planetary positions, Kepler discovered that the planets orbit the Sun along ellipses. Kepler's three laws describe how the planets move about the Sun. (E. Lessing/Magnum)*

the planets against the background of the distant stars. Then he began working with a slightly different curve called an **ellipse**.

An ellipse can be constructed with a loop of string, two thumbtacks, and a pencil, as shown in Figure 4-9. An ellipse has two foci: each thumbtack is at a **focus**. The longest diameter of an ellipse passes through both foci and is called the **major axis**. Half of that distance is called the **semimajor axis**, whose length is usually designated by the letter a.

To Kepler's delight, the ellipse turned out to be the curve he had been searching for. He published this breakthrough along with other material in 1609 in a book known today as *New Astronomy*. This important discovery, now called **Kepler's first law**, is stated as follows:

The orbit of a planet about the Sun is an ellipse with the Sun at one focus.

Kepler also realized that planets do not move at uniform speeds along their orbits. A planet moves most rapidly when it is nearest the Sun, at a point on its orbit called **perihelion**. Conversely, a planet moves most slowly when it is farthest from the Sun, at a point called **aphelion**.

After much trial and error, Kepler found a way to describe how fast a planet moves along its orbit. This discovery, now called the **law of equal areas** or **Kepler's second law**, is illustrated in Figure 4-10. Suppose that it takes 30 days for a planet to go from point A to point B. During that time, a line joining the Sun and the planet sweeps out a nearly triangular area. Kepler discovered that the line joining the Sun and the planet also sweeps out an equal area during

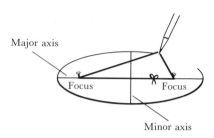

Figure 4-9 The construction of an ellipse *An ellipse can be drawn with a pencil, a loop of string, and two thumbtacks. If the string is kept taut, the pencil traces out an ellipse. The thumbtacks are located at the two foci. The ellipse's major axis is the longest diameter across the ellipse; the minor axis is the perpendicular bisector of the major axis.*

any other 30-day interval. In other words, if the planet also takes a month to go from point C to point D, then the two shaded segments in Figure 4-10 are equal in area. Kepler's second law, also published in *New Astronomy*, can be stated thus:

A line joining a planet and the Sun sweeps out equal areas in equal intervals of time.

Kepler was fascinated by the many harmonious relationships in the motions of the planets. His writing is filled with intriguing speculations, including musical scores intended to represent the celestial music that the planets make as they travel along their orbits. One of Kepler's later discoveries stands out because of its impact on future developments.

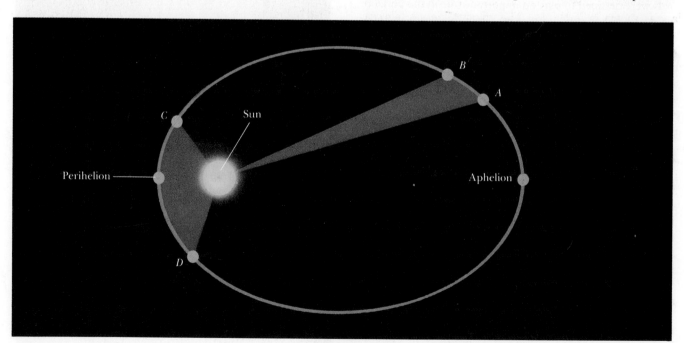

Figure 4-10 Kepler's first and second laws *According to Kepler's first two laws, every planet travels around the Sun along an elliptical orbit with the Sun at one focus in such a way that the line joining the planet and the Sun sweeps out equal areas in equal intervals of time.*

Table 4-3 A demonstration of Kepler's third law

Planet	Sidereal period P (years)	Semimajor axis a (AU)	P^2	a^3
Mercury	0.24	0.39	0.06	0.06
Venus	0.61	0.72	0.37	0.37
Earth	1.00	1.00	1.00	1.00
Mars	1.88	1.52	3.53	3.51
Jupiter	11.86	5.20	140.7	140.6
Saturn	29.46	9.54	867.9	868.3

Now called the **harmonic law** or **Kepler's third law**, it states a relationship between the sidereal period of a planet and the length of its semimajor axis:

The squares of the sidereal periods of the planets are proportional to the cubes of their semimajor axes.

If a planet's sidereal period (P) is measured in years and the length of its semimajor axis (a) is measured in astronomical units, then Kepler's third law is simply stated as

$$P^2 = a^3$$

The length of the semimajor axis may be thought of as the average distance between a planet and the Sun. Using data from Tables 4-1 and 4-2, we can demonstrate Kepler's third law as shown in Table 4-3. This relationship can also be displayed on a graph, as in Figure 4-11.

This third law was published in 1619. It is testimony to Kepler's genius that his three laws are rigorously obeyed in any situation where two objects orbit each other under the influence of their mutual gravitational attraction. Throughout this book, we shall see that Kepler's laws have a wide range of practical applications. Kepler's laws are obeyed not only by planets circling the Sun but also by artificial satellites orbiting the Earth, by two stars revolving about each other in a double-star system, by stars in their orbits within galaxies, and even by galaxies in their orbits about each other.

4-5 Galileo's discoveries with a telescope strongly supported a heliocentric cosmology

While Kepler was making rapid progress in central Europe, an Italian physicist was making equally dramatic observational discoveries in southern Europe. Galileo Galilei (see Figure 4-12) did not invent the telescope, but he was the first

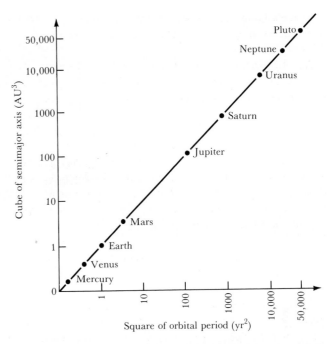

Figure 4-11 Kepler's third law *On this graph, the squares of the periods of the planets (P^2) are plotted against the cubes of their semimajor axes (a^3). The fact that the points fall along such a straight line is verification of Kepler's discovery that $P^2 = a^3$.*

Figure 4-12 Galileo Galilei (1564–1642) *Galileo used a telescope to observe the heavens. He discovered craters on the Moon, sunspots on the Sun, the phases of Venus, and four moons orbiting Jupiter. (Art Resource)*

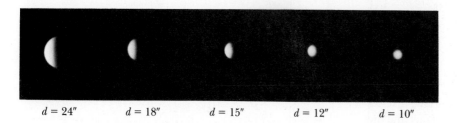

| d = 58″ | d = 56″ | d = 51″ | d = 42″ | d = 31″ |

Figure 4-13 The phases of Venus This series of photographs shows how the appearance of Venus changes as it moves along its orbit. The number below each view is the angular diameter of the planet in seconds of arc. (New Mexico State University Observatory)

| d = 24″ | d = 18″ | d = 15″ | d = 12″ | d = 10″ |

to point one of the new devices toward the sky and publish his observations. What he saw, no one had ever dreamed of: mountains on the Moon and sunspots on the Sun. He also discovered that Venus exhibits phases (see Figure 4-13).

After only a few months of observation, Galileo noticed that the apparent size of Venus as seen through his telescope was related to the planet's phase. Venus appears smallest at gibbous phase and largest at crescent phase. There is also a correlation between the phases of Venus and the planet's angular distance from the Sun. These relationships clearly support the conclusion that Venus goes around the Sun (see Figure 4-14).

In 1610, Galileo discovered four moons orbiting Jupiter (see Figure 4-15). He realized that they were orbiting Jupiter because they moved back and forth from one side of the planet to the other. Confirming observations made by Jesuits in 1620 are shown in Figure 4-16. Astronomers soon realized that these four moons obey Kepler's third law: The cube of a moon's distance from Jupiter is proportional to the square of its orbital period about the planet.

These telescopic observations constituted the first fresh influx of fundamentally new astronomical data in almost 2000 years. In contradiction to prevailing opinion, these discoveries strongly suggested adopting a heliocentric view

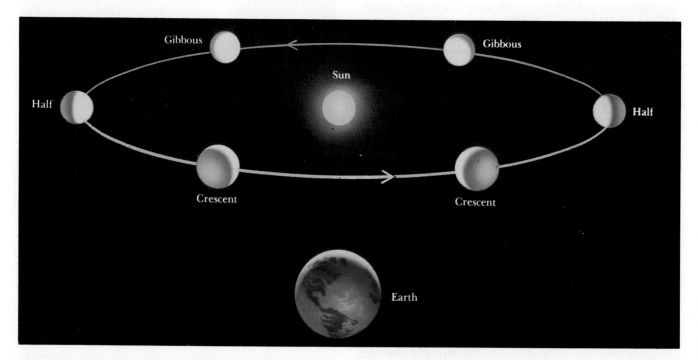

Figure 4-14 The changing appearance of Venus The phases of Venus are correlated with the planet's angular size and its angular distance from the Sun, as sketched in this diagram. These observations clearly support the idea that Venus orbits the Sun.

Figure 4-15 Jupiter and its largest moons *This photograph, taken by an amateur astronomer with a small telescope, shows the four Galilean satellites alongside an overexposed image of Jupiter. Each satellite is bright enough to be seen with the unaided eye were it not overwhelmed by the glare of Jupiter. (Courtesy of C. Holmes)*

of the universe. The Roman Catholic Church attacked Galileo's ideas because they were not reconcilable with certain passages in the Bible or with the writings of Aristotle and Plato. Nevertheless, there was no turning back. Although Galileo was condemned to spend his latter years under house arrest "for vehement suspicion of heresy," his revolutionary ideas were soon to inspire a sickly English boy who was born on Christmas Day of 1642, less than one year after Galileo died. The boy's name was Isaac Newton.

4-6 Isaac Newton formulated a description of gravity that accounts for Kepler's laws and explains the motions of the planets

Until the mid-seventeenth century, virtually all mathematical astronomy was entirely empirical, characterized by trial and error. From Ptolemy to Kepler, essentially the same approach was used. Astronomers would work directly from data and observations, adjusting ideas and calculations until they finally came out with the right answers.

Isaac Newton (see Figure 4-17) introduced a new approach. He made three assumptions, now called **Newton's laws of motion**, about the nature of reality. These are quite

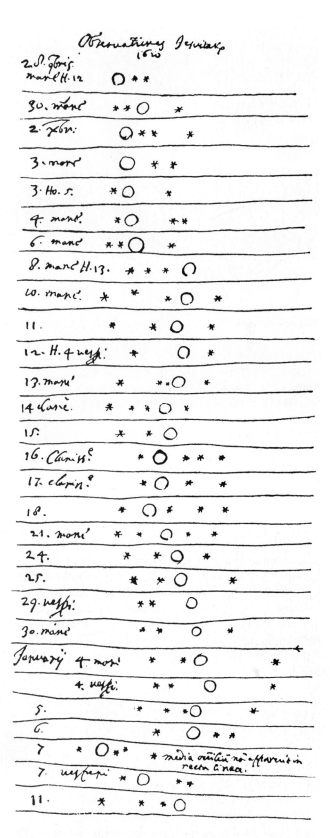

Figure 4-16 Early observations of Jupiter's moons *In 1610, Galileo discovered four "stars" that move back and forth across Jupiter from one night to the next. He concluded that these are four moons that orbit Jupiter much as our Moon orbits the Earth. This drawing shows observations made in 1620. (Yerkes Observatory)*

Figure 4-17 Isaac Newton (1642–1727) Using mathematical techniques that he invented, Isaac Newton formulated the universal law of gravitation and demonstrated that the planets orbit the Sun according to simple mechanical rules. (National Portrait Gallery, London)

general statements that apply to all forces and all bodies. Newton then showed that Kepler's three laws follow logically from the laws of motion and from a formula for the force of gravity that he derived from observations and his laws. He used this formula to describe the observed orbits of the planets, the Moon, and comets. His first assumption, known as **Newton's first law** or the **law of inertia**, reads as follows:

A body remains at rest, or moves in a straight line at a constant speed, unless acted upon by an outside force.

This idea tells us that a force must be acting on the planets. If there were no "outside" force acting on planets, they would leave their curved orbits and move away from the Sun along straight-line paths at constant speeds. Because this does not happen, Newton concluded that the continuous action of a force confines the planets to their elliptical orbits.

Isaac Newton did not invent the idea of gravity. An educated seventeenth-century person had a vague appreciation of the fact that some force pulls things down to the ground.

It was Newton, however, who gave us a precise description of the action of gravity. Using his first law, Newton proved mathematically that the force acting on each of the planets is aimed directly at the Sun. This discovery led him to suspect that the force pulling a falling apple straight down to the ground is essentially the same as the force on the planets that is always aimed straight at the Sun.

Newton's second assumption does two things; it defines the concept of **force** and it describes how a force changes the motion of an object. To appreciate these concepts, we must first understand the quantities that describe motion: quantities such as speed, velocity, and acceleration.

Imagine an object in space. Push on the object, and it will begin to move. At any moment, you can specify the object's motion by giving both its speed and its direction. Speed and direction of motion together constitute the object's **velocity**. If you continue to push on the object, its speed will increase— in other words, it will accelerate.

Acceleration is the rate at which velocity changes. Since velocity involves both speed and direction, acceleration can result from changes in either. Also note that, contrary to popular use of the term, acceleration is not restricted to increases in speed alone. A slowing down, a speeding up, or a change in direction all involve accelerations.

An apple falling from a tree is a good example of acceleration that involves an increase in speed only. Initially, at the moment the stem breaks, the apple's speed is zero. After one second, its downward speed is 32 feet per second. After two seconds, the apple's speed is 64 feet per second. After three seconds, the speed is 96 feet per second. Because the apple's speed increases by 32 feet per second for each second of free fall, the rate of acceleration is 32 feet per second per second, or 32 ft/s². In other words, the Earth's gravity produces a constant acceleration of 32 ft/s² (9.8 m/s²) downward, toward the center of the Earth.

A planet revolving about the Sun along a perfectly circular orbit is an example of acceleration that involves change of direction only. As the planet moves along its orbit, its speed remains constant. Nevertheless, the planet is continually being accelerated because its direction of motion is continually changing.

Newton's second law says that the acceleration of an object is proportional to the force acting on the object. In other words, the harder you push on an object, the greater is the resulting acceleration. This law can be succinctly stated as an equation. If a force F acts on the object, it will experience an acceleration a such that

$$F = ma$$

where m is the mass of the object.

The **mass** of an object is a measure of the total amount of material in the object and is usually expressed in grams or kilograms. For example, the mass of the Sun is 2×10^{30} kg. The mass of a hydrogen atom is 1.7×10^{-24} g. And the mass of the author of this book is 79 kg. The Sun, a hydrogen atom, and the author have these masses regardless of where they happen to be in the universe.

It is important not to confuse the concepts of mass and weight. **Weight** is the force with which an object presses down on the ground (due to gravity's pull), and like force, it is usually expressed in pounds, newtons, or dynes. These units of measure are related to one another as follows:

$$1 \text{ newton} = 10^5 \text{dyne} = 0.225 \text{ pound}$$

We can use Newton's second law to relate mass and weight. We have seen that the acceleration caused by the Earth's gravity is 32 ft/s^2, or 9.8 m/s^2. From the second law ($F = ma$), the force with which the author presses down on the ground is

$$79 \text{ kg} \times 9.8 \text{ m/s}^2 = 774 \text{ newtons} = 174 \text{ pounds}$$

Note that this answer is correct only when the author is standing on the Earth. He would weigh less on the Moon and more on Jupiter. Floating deep in space, he would have no weight at all; he would be "weightless." Nevertheless, under all these circumstances, he would always have exactly the same mass. Thus we see that mass is an inherent property of matter unaffected by details of the environment. Whenever we describe the properties of planets, stars, or galaxies, we speak of their masses, never of their weights.

Newton's final assumption, called **Newton's third law**, is the famous statement about action and reaction:

Whenever one body exerts a force on a second body, the second body exerts an equal and opposite force on the first body.

For example, if you weigh 165 pounds, you are pressing down on the floor with a force of 165 pounds. Newton's third law tells us that the floor is also pushing up against your feet with an equal force of 165 pounds. (If it were not, you would fall through the floor.) In the same way, Newton realized that, because the Sun is exerting a force on each planet to keep it in orbit, each planet must also be exerting an equal and opposite force on the Sun. However, the planets are much less massive than the Sun (for example, the Earth has only $\frac{1}{330,000}$ of the Sun's mass). So although the Sun's force on a planet is the same as the planet's force on the Sun, the planet's much smaller mass gives it a much larger acceleration, according to Newton's second law. This is why the planets circle the Sun, instead of vice versa. Newton's laws reveal the reason for our heliocentric solar system.

Using his own three laws and Kepler's three laws, Newton succeeded in formulating a general statement describing the nature of the force called **gravity** that keeps the planets in their orbits. Newton's **universal law of gravitation** is:

Two bodies attract each other with a force that is directly proportional to the product of their masses and inversely proportional to the square of the distance between them.

In other words, if two objects have masses m_1 and m_2 and are separated by a distance r, then the gravitational force F between these two masses is given by the equation

$$F = G \frac{m_1 m_2}{r^2}$$

If the masses are measured in kilograms and the distance between them in meters, then the force is measured in newtons. If the masses are in grams and the distance in centimeters, the force is calculated in dynes. In this formula, G is a number called the **universal constant of gravitation**. From laboratory experiments, G has been determined to have the following value:

$$G = 6.67 \times 10^{-8} \text{ dyne cm}^2/\text{g}^2$$
$$= 6.67 \times 10^{-11} \text{ newton m}^2/\text{kg}^2$$

We can use Newton's law of gravity to calculate the force with which any two bodies attract each other. For example, to compute the gravitational force between the Earth and the Sun, we substitute values for the Earth's mass ($m_1 = 5.98 \times 10^{24}$ kg), the Sun's mass ($m_2 = 1.99 \times 10^{30}$ kg), the distance between them ($r = 1 \text{ AU} = 1.5 \times 10^{11}$ m), and the value of G into Newton's equation to get

$$F_{\text{Sun–Earth}} = 6.67 \times 10^{-8} \left(\frac{5.98 \times 10^{24} \times 1.99 \times 10^{30}}{(1.5 \times 10^{11})^2} \right)$$
$$= 3.53 \times 10^{25} \text{ newtons}$$

Using his law of gravity, Newton found that he could mathematically prove the validity of Kepler's three laws. For example, whereas Kepler had to discover by trial and error that $P^2 = a^3$, Newton demonstrated mathematically that this equation follows logically from his law of gravity. (Some details of how he did so are presented in Box 4-3.) Newton also discovered new features of orbits around the Sun. For instance, his equations soon led him to conclude that the orbit of an object around the Sun could be any one of a family of curves called conic sections.

A **conic section** is any curve that you get by cutting a cone with a plane, as shown in Figure 4-18. You can get circles and ellipses by slicing all the way through the cone. You can also get two "open" curves called **parabolas** and **hyperbolas**. Comets hurtling toward the Sun from the depths of space sometimes follow parabolic orbits.

Newton's ideas and methods turned out to be incredibly successful in a wide range of situations. The orbits of the planets and their satellites could now be calculated with unprecedented precision. Using Newton's laws, mathemati-

Box 4-3 Ellipses and orbits

Using his universal law of gravitation and a lot of mathematics, Newton calculated many details of the planets' orbits around the Sun.

As shown in the diagram on the left, the largest diameter across an ellipse is called the **major axis** and has a length of $2a$. The shortest diameter through the center of the ellipse is called the **minor axis** and has a length of $2b$. The minor axis is the perpendicular bisector of the major axis.

The shape of an ellipse is determined by its **eccentricity**, e. The distance from the center of an ellipse to one focus is ae. Notice that, if $e = 0$, we have a circle. A few examples of ellipses of different eccentricities are shown in the middle diagram.

Kepler's second law tells us that the velocity of a planet varies as it orbits the Sun. Newtonian mechanics can be used to calculate the speed of the planet at various points along its orbit. If P is the planet's sidereal period, then the orbital speed v at perihelion is given by

$$v = \frac{2\pi a}{P}\left(\frac{1 + e}{1 - e}\right)^{1/2}$$

and at aphelion by

$$v = \frac{2\pi a}{P}\left(\frac{1 - e}{1 + e}\right)^{1/2}$$

Example: We have the following data for the Earth:

$$a = 1 \text{ AU} = 1.496 \times 10^8 \text{ km}$$
$$P = 1 \text{ yr} = 3.156 \times 10^7 \text{ s}$$
$$e = 0.0167$$

where the eccentricity of the Earth's orbit is determined from the annual variation of the Earth–Sun distance. Inserting these values into the above equations, we find that the Earth's orbital speed varies from 30.3 km/s at perihelion to 29.3 km/s at aphelion.

Although it is commonly said that the Moon goes about the Earth or that the planets revolve around the Sun, such statements are not entirely accurate. When one object seemingly orbits another, both objects are actually revolving about a common, stationary point called the **center of mass** as shown in the diagram on the right. To appreciate this concept, imagine placing a mass m_1 at one end of a seesaw or teeter-totter and a second mass m_2 at the other end. The center of mass is the point where you would put the fulcrum to balance the seesaw.

In the case of the Moon orbiting the Earth or a planet orbiting the Sun, one object is much more massive than the other, and so the center of mass is quite near the more massive object. For instance, the center of mass of the Earth–Moon system actually lies within the Earth, whereas that for the Sun–Jupiter system lies just outside the Sun's surface.

Consider the masses m_1 and m_2 orbiting their common center of mass as shown in the diagram on the right. The location of their center of mass is given by

$$m_1 r_1 = m_2 r_2$$

where $r_1 + r_2 = a$, the distance between the two masses. The gravitational force between these two masses is given by Newton's law:

$$F = G\frac{m_1 m_2}{a^2}$$

where G is the universal constant of gravitation. From these equations, it is possible to derive the following relationship between the total separation of the two masses a and the orbital period P with which they revolve about their common center of mass:

$$P^2 = \left[\frac{4\pi^2}{G(m_1 + m_2)}\right] a^3$$

This equation is actually the most general and complete statement of Kepler's third law. In the case of the solar system, the mass of the Sun is thousands of times greater than the masses of any of the planets. Consequently, for all practical purposes, $m_1 + m_2$ is equal to the mass of the Sun, $M_\odot$, which is 2×10^{30} kg. Thus, Kepler's third law becomes

$$P^2 = \left(\frac{4\pi^2}{G M_\odot}\right) a^3$$

If P is measured in years and a is measured in astronomical units, then the quantity in the parentheses equals 1 and we have

$$P^2 = a^3$$

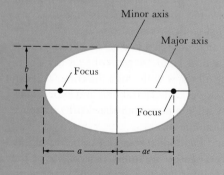

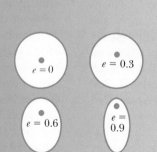

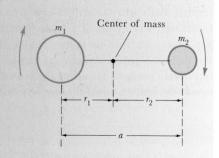

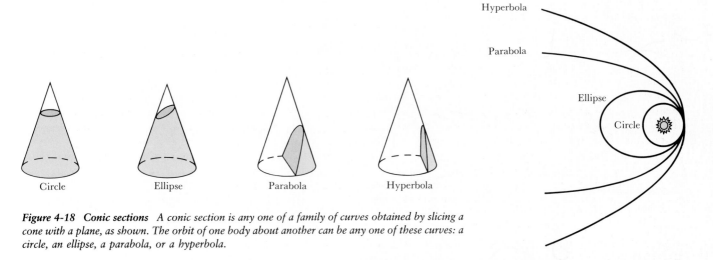

Figure 4-18 *Conic sections* *A conic section is any one of a family of curves obtained by slicing a cone with a plane, as shown. The orbit of one body about another can be any one of these curves: a circle, an ellipse, a parabola, or a hyperbola.*

cians proved that the Earth's axis of rotation must precess, because of the gravitational pull of the Moon and the Sun on the Earth's equatorial bulge (recall Figure 2-11). Similarly, all the details of the Moon's orbit (recall Box 3-1) could be demonstrated mathematically with a body of knowledge built on Newton's work that is today called **Newtonian mechanics.** A central feature of Newtonian mechanics is the idea that certain quantities—energy, momentum, and angular momentum—are conserved. In other words, in a system

of interacting objects, whatever the system may do and however it may evolve, these three quantities measured for the entire system remain unchanged. A brief discussion of these basic laws of nature is found in Box 4-4.

Newtonian mechanics not only explained in detail a variety of known phenomena, it was capable of predicting new phenomena. For example, one of Newton's friends, Edmund Halley, was intrigued by three similar historical records of a comet that had been sighted at intervals of 76 years. Assuming these records to be of the same comet, Halley used Newton's methods to work out the details of the comet's orbit and predicted its return in 1758. It was first sighted on Christmas night of 1757, and to this day the comet bears Halley's name (see Figure 4-19).

Perhaps the most dramatic success of Newton's ideas involved the discovery of the eighth planet from the Sun. The seventh planet, Uranus, had been discovered accidentally by William Herschel in 1781 during a telescopic survey of the sky. Fifty years later, however, it was clear that Uranus was not following its predicted orbit. Two mathematicians, John Couch Adams in England and U. J. Leverrier in France, independently calculated that the deviations of Uranus from its orbit could be explained by the gravitational pull of a yet unknown, more distant planet. They each predicted that the planet would be found at a certain location in the constellation of Aquarius. A brief telescopic search on September 23, 1846, revealed the planet Neptune within 1° of the calculated position. (Figure 4-20 shows photographs of both Uranus and Neptune.) Before it was sighted with a telescope, Neptune was really discovered with pencil and paper.

Because of the success of Newton's ideas in both explaining and predicting many important phenomena, Newtonian mechanics has become the cornerstone of modern physical science. Even today, as we send astronauts to the Moon and spacecraft to the outer planets, Newton's equations are used to calculate orbits and trajectories. These applications usually employ high-speed computers, called supercomputers,

Figure 4-19 *Halley's comet* *Halley's comet orbits the Sun with an average period of about 76 years. During the twentieth century, the comet passed near the Sun twice—in 1910 and again in 1986. This photograph shows how the comet looked in 1986. (Courtesy of J. Marling)*

Box 4-4   The conservation laws

One of the most powerful concepts in Newtonian mechanics is the idea that certain physical quantities remain unchanged as the universe evolves. The statements that express these constancies of nature are called **conservation laws**, and three of the most important conserved quantities are energy, momentum, and angular momentum.

According to the **conservation of energy**, the total energy of a physical system is constant. There may be different kinds of energy—chemical energy, thermal energy, radiation—in that system, and these different manifestations of energy may interconvert one into another, but their total remains unchanged. In order to use this law, however, you must know how what energy is and how to express it mathematically.

Energy is the ability to do work, and work is evaluated as the exertion of a force over a distance. From these definitions is it possible to derive mathematical expressions for certain kinds of energy. For instance, the energy associated with motion is called **kinetic energy**. The kinetic energy of a particle of mass m moving with a velocity v is $\frac{1}{2} mv^2$. Another example is the energy associated with position in a gravitational field, called **potential energy**. In the Earth's gravitational field, the potential energy of an object of mass m at a height h above some reference level is mgh, where g is the acceleration of gravity, 9.8 m/s^2.

Imagine dropping a ball of mass m from a height h above the floor, as shown in the top diagram. At the moment you release the ball, all its energy is entirely potential energy, because it has not yet begun to move. As the ball accelerates toward the floor, however, its potential energy is converted into kinetic energy. At the moment it strikes the floor, the ball's energy is entirely kinetic, because is height above the floor has shrunk to zero. Equating the energy of the ball when you released it (mgh) to the energy of the ball it strikes the floor ($\frac{1}{2} mv^2$) we get

$$mgh = \frac{1}{2} mv^2$$

Rearranging terms, we discover that the ball's velocity upon striking the floor is independent of the ball's mass and is given by

$$v = \sqrt{2gh}$$

The **conservation of momentum** says that the total momentum of the particles in a system remains unchanged. The momentum of an individual particle of mass m moving with a velocity v is mv. Keep in mind, however, that velocity has both size *and* direction, which means that momentum does also.

Angular momentum is a measure of the momentum carried by an object because of its rotation about an axis. To appreciate the definition of angular momentum, consider a mass m revolving about an axis, as shown in the bottom diagram. If r is the distance of the mass from the axis of rotation and v is the linear speed of the mass, then the angular momentum L is

$$L = mvr$$

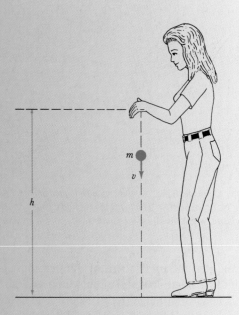

The **conservation of angular momentum** requires that both the size and the direction of the angular momentum never change. For a mass revolving about an axis, this means that the value of L must remain constant and the axis of rotation must remain fixed in space.

In the coming chapters we shall often see how these conservation laws play a crucial role in determining the behavior and outcome of many features of the universe.

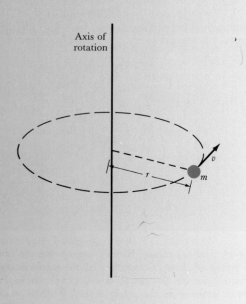

Figure 4-20 *Uranus and Neptune   The discovery of Neptune was a major triumph for Newtonian mechanics. The existence of Neptune (shown here with one moon) was deduced from deviations in the pre-* *dicted orbit of Uranus (shown with three moons). Both planets have more satellites than appear in these photographs. (Lick Observatory)*

that greatly facilitate up the repetitive computations that are often required (see Box 4-5).

There was one instance, however, in which Newtonian mechanics was not quite in agreement with astronomical observations. During the mid-1800s, Leverrier pointed out that Mercury was not following its predicted orbit. As the planet moves along its elliptical orbit, the orbit itself rotates (or precesses) as shown in Figure 4-21. Most of Mercury's precession is caused by the gravitational pull of the other planets, and so is explained with Newtonian mechanics. There is, however, an unexplained excess rotation of Mercury's major axis amounting to only 43 arc sec per century. Although the effect is very small, all attempts to ac-

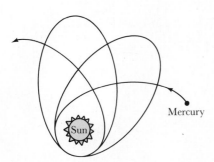

Figure 4-21 *The advance of Mercury's perihelion   As Mercury moves along its orbit, the orbit slowly rotates. Consequently, Mercury traces out a rosette figure about the Sun. The effect is greatly exaggerated in this diagram. The excess rotation of Mercury's orbit (beyond that predicted by Newtonian mechanics) amounts to only 43 arc sec per century.*

count for this phenomenon met with failure. Nevertheless, it is a testament to Newton's genius that his three laws were precisely the three basic ideas needed for a full understanding of the motions of the planets. Isaac Newton thus brought a new dimension of elegance and sophistication to our understanding of the workings of the universe. It remained for the insight, vision, and genius of Albert Einstein to develop a whole new perspective on the universe.

4-7 Albert Einstein's theory states that gravity affects the shape of space and the flow of time

In the cosmologies of most ancient civilizations, the Earth and its inhabitants occupied a special place at the center of the universe. Then Copernicus's cosmology made the Earth just one of a number of planets orbiting the Sun. We have seen that this new approach proved quite fruitful. Within less than a century, Kepler's accurate description of planetary orbits led directly to Newton's universal law of gravitation. Astronomers then began to explore the possibility that even the Sun may occupy no special location. Today we know that our Sun is merely one of the many stars scattered throughout the Milky Way Galaxy, which is only one of many galaxies in the universe.

Albert Einstein (see Figure 4-22) introduced a powerful extension of this perspective. Einstein believed that the

Box 4-5   Supercomputing—A new mode of science

The invention of high-speed electronic computers during the twentieth century is a revolutionary development that is dramatically changing the way scientists do science. Because of the speed with which they can perform repetitive mathematical operations, supercomputers have given us a new way of learning about the universe.

Previously, all scientific investigation fell into two broad categories: experiment and theory. The experimental/observational mode seeks to learn about the world through observation, measurement, and experimentation. Galileo Galilei, who showed us the heavens with his telescope and investigated gravitational acceleration by timing balls rolling down an inclined plane, can be considered the founder of this technique of science. The theoretical/analytical mode was conceived by Isaac Newton, who showed us that mathematics is the fundamental language of science. With this language the regularities we observe in nature can be written as universal laws.

Both of these traditional modes of science have flourished since the time of Galileo and Newton. But both have distinct limitations. Many of the phenomena that scientists would like to observe are too small, too far away, or too ephemeral to yield readily to scientific scrutiny. Similarly, the complexity of the real world poses insurmountable problems for the theoretician, who despairs at the monumental task of solving the myriad of equations that describe realistic situations, such as colliding galaxies or a thunderstorm.

In the 1940s, the Hungarian-American mathematician John von Neumann conceived of a third way of doing science: the computational mode. Von Neumann pointed out that, for the most part, we know all the laws of physics but we can solve them only in the very simplest, idealized cases. Realistic situations typically involve so many intertwined equations that a sci-

John von Neumann
(1903–1957)

entist with pencil and paper will not live long enough to work out their solution. Von Neumann therefore envisioned the day when high-speed computers would be able to tackle the arduous task of solving the equations that describe interesting and complex phenomena. That day has arrived.

A supercomputer is the fastest, most powerful computer that can be built. Examples include the Cray X-MP, shown in the

fundamental laws of the universe do not depend on a person's location or motion. In other words, the laws of physics should be the same, whether we happen to be sitting on Earth or moving through space at a high speed.

Soon after 1900, Einstein began developing a new approach to the phenomena of electricity and magnetism. The basic properties of electricity and magnetism had earlier been summarized in four equations formulated in 1865 by the great Scottish physicist James Clerk Maxwell. Maxwell's equations are the basis of **electromagnetic theory**, which today has a wide range of practical applications, from television sets to microwave ovens. Maxwell's electromagnetic

Figure 4-22 Albert Einstein (1879–1955) Albert Einstein demonstrated that gravity affects the curvature of space and the flow of time. These effects are noticeable only in extremely intense gravitational fields, like that found around a black hole. (L. Jacobi, Diamond Library, University of New Hampshire)

accompanying photograph, which can perform billions of mathematical operations per second. Every supercomputer enjoys its exalted status for only a few years, because continuing technological advances soon give rise to the next generation of supercomputers.

A supercomputer is a valuable research tool because the laws of physics can be written in a way that the supercomputer can solve them. In their basic form, the laws of physics are valid at every point in space and at every moment of time. Scientists seldom, however, need to apply the laws of physics on this fine a scale. Instead, they use a set of points in space that are separated by distances that are small compared to the size of the object under study. They also use intervals of time that are short compared to the duration of the process they want to explore. By programming a supercomputer with the laws of physics expressed only at selected points and time intervals, scientists can reduce a complicated problem to a level that the machine can handle.

For example, suppose that you want to calculate the orbit of Halley's Comet. If the Sun were the only object in the solar system, the comet's path would be an ellipse with the Sun at one focus. But all the planets are also pulling on the comet, and so its path will deviate slightly from a perfect ellipse. To calculate the true orbit, you start with the comet and planets in a particular configuration corresponding to their known locations in space on a certain day. Knowing the distances between the comet, the Sun, and the planets, you calculate the total gravitational force acting on the comet and, using Newton's second law, compute the acceleration that the comet experiences. This tells you where the comet is headed in the immediate future. Once you have deduced the location of the comet for a particular day, you must repeat the entire calculation to predict the comet's location on

the following day, using the data for the comet and the planets in their new positions. In this repetitive step-by-step fashion, you track the path of the comet as it moves across the solar system. Modern supercomputers greatly facilitate such tedious, repetitive computations.

Furthermore, using supercomputers, scientists can study phenomena that they otherwise would have no hope of observing. For example, later in this book we shall contemplate the Moon's creation by looking at a supercomputer simulation of a Mars-sized object striking the Earth. We shall also see what happens as gas is captured and swallowed by a black hole. We shall watch as two galaxies consisting of thousands of stars and gas clouds collide and interact with each other. These are but a few examples of the ways in which supercomputing is giving us a new, deeper understanding of the universe.

theory predicts various effects, depending upon the motion of electric charges and magnets. For example, a moving electric charge creates a magnetic "field," whereas a stationary electric charge does not. Such predictions imply some absolute or fixed spatial framework within which an object is either stationary or moving.

Einstein's goal was to eliminate this assumption of absolute space from electromagnetic theory. He wanted to rewrite Maxwell's equations so that they would depend only upon the relative motions of the observer and the electric charges or magnets. In 1905 he succeeded and published his results in a famous paper entitled "On the Electrodynamics of Moving Bodies."

Einstein's approach to electromagnetic theory introduced some revolutionary concepts to physics. To achieve his goal, Einstein had to abandon both the idea of absolute time and the idea of absolute space. This new perspective led to surprising conclusions: The length of an object depends upon its speed relative to the observer who measures it and, simi-

larly, the rate at which a clock ticks depends upon its speed relative to the observer who times it. The body of knowledge that describes all these effects is called the **special theory of relativity.**

The relativity of distances and time intervals is easily noticeable only at speeds approaching the speed of light. For most phenomena, Newtonian mechanics provides accurate predictions and explanations. However, for extremely high speeds, such as those of subatomic particles in particle accelerators, physicists found that special relativity is needed to explain and understand the results of their experiments. Using extremely accurate atomic clocks, physicists have even confirmed that a clock in a moving jet plane ticks slightly more slowly than an identical clock on the Earth. Einstein's theory has been supported by every experiment designed to test it.

Details of special relativity need not concern us here (they will be discussed in Box 24-1), but this newly discovered relativity of space and time was very important in Einstein's

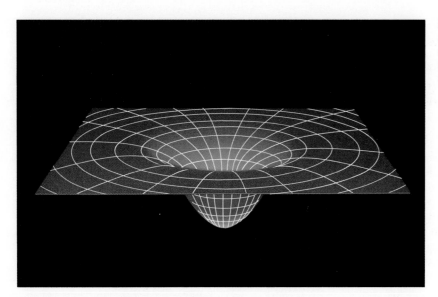

Figure 4-23 *The gravitational curvature of space* According to Einstein's general theory of relativity, space becomes curved near a massive object such as the Sun. This diagram shows a two-dimensional analogy of the shape of space around a massive object.

desire to understand gravity. Einstein found Newton's description of gravity disappointing because it assumes that space and time are absolute. However, gravity causes a falling object to accelerate and, according to the special theory of relativity, the resulting motion also affects rulers and clocks that are moving *with* the falling object. Thus Einstein argued that gravity must affect the shape of space and the flow of time.

In 1916, Einstein formulated a new theory of gravity that came to be called the **general theory of relativity**. The basic idea of general relativity is that the mass of an object alters the properties of space and time around the object. According to the general theory of relativity, gravity causes space to become curved and time to slow down. Einstein eliminated the idea of a "force of gravity" from his theory.

A useful analogy is to imagine that the space near a massive object such as the Sun becomes curved like the surface in Figure 4-23. Imagine a ball rolling along this surface. Far from the "well" that represents the Sun, the ball would move in a straight line since the surface is fairly flat. If it passes near the well, however, it would curve in toward the well. If it is moving at an appropriate speed with respect to the well, it might move in an orbit around the sides of the well. In this analogy, it is the curvature of the surface that makes the ball follow a curved path in the vicinity of the well.

One of the first things Einstein did with his new theory was to recalculate the orbits of the planets. Far from the Sun, space is almost flat and objects travel along nearly straight-line paths. Near the Sun, planets and comets travel along curved paths because space itself is curved. The general theory of relativity predicts that only strong gravitational fields will have easily detectable effects upon space and time in their vicinity.

With only one minor exception, general relativity gave almost exactly the same answers as Newtonian theory. Mercury is the only planet to pass close enough to the Sun

for the curvature of space to give rise to motions that differ noticeably from those predicted by Newtonian mechanics. Indeed, throughout most of the solar system the curvature of space is so slight that Einstein's equations are essentially equivalent to Newtonian calculations. However, unlike Newtonian mechanics, Einstein's theory could account for the excess precession of Mercury's orbit, thereby explaining a phenomenon that had frustrated astronomers for half a century.

To help validate his theory, Einstein made predictions that could be tested. With his calculations he showed that light rays passing near the surface of the Sun should appear to be deflected from their straight-line paths, because the space through which they are moving is curved. In other words, gravity would bend light rays, an effect not predicted by Newtonian mechanics, because light has no mass.

Figure 4-24 shows a beam of light from a star passing by the Sun and continuing on down to the Earth. Because the

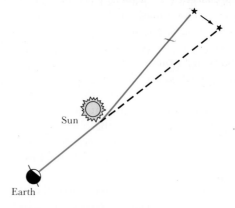

Figure 4-24 *The gravitational deflection of light* Light rays passing near the Sun are deflected from their straightline paths. By relativistic standards, the Sun's gravity is very weak. Therefore, the maximum angle of deflection is only 1.75 arc sec for a light ray grazing the Sun's surface.

light ray is bent, the star appears to be shifted away from its usual location. At most, though, the position of a star is shifted by 1.75 arc sec for a light ray grazing the Sun's surface.

This prediction was first tested in 1919 during a total solar eclipse. During the precious moments of totality, when the Moon blocked out the blinding solar disk, astronomers succeeded in photographing the stars around the Sun. Careful measurements afterward revealed that the stars were shifted from their usual positions by an amount consistent with Einstein's theory. General relativity had passed another important test.

Einstein made a third prediction. He stated that because gravity causes time to slow down, clocks on the first floor of a building would tick slightly more slowly than clocks in the attic, which are farther from the Earth, as shown in Figure 4-25. This prediction was tested by analyzing light from compact stars (whose surface gravity is strong) and by using extremely accurate clocks first developed in 1960. Again Einstein was proven correct.

The general theory of relativity now stands as our most precise and complete description of gravity. Einstein demonstrated that Newtonian mechanics is accurate only when applied to low speeds and weak gravity. If extremely high speeds or powerful gravitational fields (such as those of neutron stars and black holes) are involved, only a relativistic calculation will give correct answers.

Einstein's general theory of relativity evoked a sensation when first proposed. Newtonian gravitation had been overthrown by a radically different approach that worked better. After the initial excitement, however, interest in general relativity rapidly waned. No one could imagine places where the curvature of space might have a major effect. It was generally believed that Newtonian mechanics would suffice in virtually all circumstances and that relativistic theory offered only a little more accuracy, as in calculating Mercury's orbit. The complex mathematics of relativity seemed burdensome, and Newton's simpler approach was reasonably precise in most conceivable circumstances.

Although Newtonian mechanics is still widely used, in recent years there has been a reawakening of interest in general relativity, primarily because of dramatic advances in our understanding of the evolution of stars. As we shall see in

Figure 4-25 *The gravitational slowing of time* *A clock on the ground floor of a building is closer to the Earth than a clock at a higher elevation. According to general relativity, the clock on the ground floor ticks more slowly than the clock on the roof.*

later chapters, we now have a reasonably complete picture of how stars are born, what happens to them as they mature, and how they die. In particular, it appears that the most massive dying stars are doomed to collapse completely in upon themselves, thereby producing some of the most bizarre objects in the universe, black holes. The gravity around one of these massive stellar corpses is so strong that it in effect punches a hole in the fabric of space. Because of this intense gravitational effect, black holes can be described only in terms of general relativity.

As we set our sights on understanding the universe as a whole, we again must turn to general relativity. All of the matter in all the stars and galaxies is responsible for the overall curvature or shape of space. From the viewpoint of general relativity, it is therefore reasonable to ask about the actual shape of the universe. We shall see that the answer suggests the ultimate fate of the cosmos.

Key words

acceleration	*center of mass	conjunction	*conservation of energy
*angular momentum	configuration (of a planet)	*conservation of angular momentum	*conservation of momentum
aphelion	conic section		

cosmology

deferent

direct motion

*eccentricity

electromagnetic theory

ellipse

elongation

*energy

epicycle

focus (of an ellipse; *plural* foci)

force

general theory of relativity

geocentric cosmology

gravity

harmonic law

heliocentric cosmology

hyperbola

inferior conjunction

inferior planet

Kepler's laws

*kinetic energy

law of equal areas

law of inertia

*major axis (of an ellipse)

mass

maximum eastern elongation

maximum western elongation

*minor axis (of an ellipse)

Newtonian mechanics

Newton's laws of motion

Occam's razor

opposition

parabola

parallax

perihelion

period (of a planet)

*potential energy

quadrature

retrograde motion

semimajor axis (of an ellipse)

sidereal period

special theory of relativity

superior conjunction

superior planet

synodic period

universal constant of gravitation

universal law of gravitation

velocity

weight

Key ideas

- Ancient astronomers believed the Earth to be at the center of the universe. They invented a complex system of epicycles and deferents to explain the direct and retrograde motions of the planets.

- A heliocentric (Sun-centered) theory simplifies the general explanation of planetary motions.

    In a heliocentric system, the Earth is one of the planets orbiting the Sun. The other planets are divided into the inferior planets (with orbits smaller than Earth's) and the superior planets (with orbits larger than Earth's).

    The sidereal period of a planet, its true orbital period, is measured with respect to the stars. Its synodic period is measured with respect to the Earth and the Sun (for example, from one opposition to the next).

- Ellipses describe the paths of the planets around the Sun much more accurately than do circles. Kepler's three laws give important details about elliptical orbits.

- The invention of the telescope led to new discoveries that supported a heliocentric view of the universe.

- Newton based his explanation of the universe on three assumptions or laws of motion. These laws and his universal law of gravitation can be used to deduce Kepler's laws and extremely accurate descriptions of planetary motions.

    The mass of an object is a measure of the amount of matter in the object; its weight is a measure of the force with which the gravity of some other object pulls on it.

    In general, the path of one object about another, such as that of a comet about the Sun, is one of the curves called conic sections: a circle, an ellipse, a parabola, or a hyperbola.

- Although Newtonian mechanics adequately describes and predicts numerous phenomena, Einstein's relativistic theories are always more accurate and must be used where extremely high speeds or intense gravitational fields are involved.

    The special theory of relativity was designed to eliminate the notion of absolute space and time. The theory leads to the conclusion that measurements of distance and time intervals are affected by the motion of the observer.

    The general theory of relativity explains that gravity causes space to be curved and time to slow down. The curvature of space and slowing of time in turn give rise to accelerated motions.

Review questions

1 What is an epicycle, and how is it important in Ptolemy's explanation of the retrograde motions of the planets?

2 How did Copernicus explain the retrograde motion of the planets?

3 Which planets can never be seen at opposition? Which planets can never be seen at inferior conjunction?

4 At what configuration (superior conjunction, greatest eastern elongation, etc.) would it be best to observe Mercury or Venus with an Earth-based telescope? At what configuration would it be best to observe Mars, Jupiter, or Saturn? Explain your answers.

5 What is the difference between the synodic and the sidereal periods of a planet?

6 What are Kepler's three laws, and why are they important?

∗7 *(Basic)* A line joining the Sun and an asteroid is found to sweep out 5.2 AU^2 of space in 1990. How much area is swept out in 1991? Over a period of five years?

∗8 A comet with a period of 1000 years moves in a highly elongated orbit about the Sun. What is the comet's average distance from the Sun? What is the farthest it can get from the Sun?

∗9 The orbit of a spacecraft about the Sun has a perihelion distance of 0.5 AU and an aphelion distance of 3.5 AU. What is the spacecraft's orbital period?

10 What are Newton's three laws? Give an everyday example of each law.

∗11 *(Basic)* What is your weight in pounds and in newtons? What is your mass?

∗12 The mass of the Moon is 7.35×10^{22} kg, while that of the Earth is 5.98×10^{24} kg. The average distance from the center of the Moon to the center of the Earth is 384,400 km. What is the size of the gravitational force that the Earth and Moon exert upon each other? How does this compare with the force between the Sun and the Earth calculated in the text?

∗13 Suppose that the Earth were moved to a distance of 10 AU from the Sun. How much stronger or weaker would the Sun's gravitational pull be on the Earth?

14 What is the difference between weight and mass?

15 Why was the discovery of Neptune an important confirmation of Newton's universal law of gravitation?

16 Why is Einstein's general theory of relativity a better description of gravity than is Newton's universal law of gravitation?

Advanced questions

> **Tips and tools . . .**
> Details about sidereal and synodic periods are covered in Box 4-1. Various useful formulas involving orbital parameters are found in Box 4-3. Data about the planets and their satellites are found in the appendices at the back of this book.

∗17 *(Basic)* The synodic period of Mercury (an inferior planet) is 115.88 days. Calculate its sidereal period.

∗18 Is it possible for a planet's sidereal period to equal its synodic period? Would such a planet be closer to the Sun than the Earth is or farther away? Is there a planet that nearly fits this description?

∗19 *(Challenging)* A satellite is said to be in a "geosynchronous" orbit if it appears always to remain over the exact same spot on Earth. **(a)** At what distance from the center of the Earth must such a satellite be placed into orbit? **(b)** Explain why the orbit must be in the plane of the Earth's equator.

∗20 *(Basic)* Calculate the orbital speed of Pluto at **(a)** perihelion and **(b)** aphelion. Relevant orbital data are found in Appendix 1 at the back of this book.

∗21 Suppose you have discovered an alien solar system in which a planet circles a star once every two years at an average distance of 4 AU. How does the mass of this star compare with that of our Sun?

∗22 With what speed would a rock strike the ground if it were dropped from an altitude of **(a)** 1 m, **(b)** 100 m, **(c)** 1 km? Neglect air friction. Why do your answers *not* depend on the mass of the rock?

∗23 Suppose that you traveled to a planet whose mass is the same as Earth's but its diameter is twice as large as Earth's. Would you weigh more or less on that planet than you do on Earth? By what factor?

∗24 If you landed on Mars, would you weigh more or less than you do on the Earth? By what factor? See Appendix 2 at the back of this book for relevant data about Mars.

25 Look up the dates of the greatest eastern and western elongations of Mercury for any year you choose. Does it take longer to go from eastern to western elongation, or vice versa? Why do you suppose this is the case?

26 Look up orbital information for the four largest moons of Jupiter. Demonstrate that these data obey Kepler's third law.

Discussion questions

27 Which planet would you expect to exhibit the greatest variation in apparent brightness as seen from Earth? Explain your answer.

28 Use two thumbtacks, a loop of string, and a pencil to draw several ellipses. Describe how the shape of an ellipse varies as you change the distance between the thumbtacks.

Observing projects

29 It is quite probable that, within a few weeks of your reading this chapter, one of the planets will be near opposition or greatest eastern elongation, making it readily visible in the evening sky. Consult a reference book such as the current issue of the *Astronomical Almanac* or the pamphlet entitled *Astronomical Phenomena* (both published by the U.S. government) to select a planet that is at or near such a configuration. At that configuration, would you expect the planet to be moving rapidly or slowly from night to night against the background stars? Verify your expectations by observing the planet once a week for a month, recording your observations on a star chart.

30 If Jupiter happens to be visible in the evening sky, observe the planet with a small telescope on five consecutive clear nights. Record the positions of the four Galilean satellites by making nightly drawings, just as the Jesuit priests did in 1620 (see Figure 4-16). From your drawings, can you tell which moon orbits closest to Jupiter and which is farthest from the planet? Is there a night when you could see only three of the moons? What do you suppose happened to the fourth moon on that night?

31 If Venus happens to be visible in the evening sky, observe the planet with a small telescope once a week for a month. On each night, make a drawing of the crescent that you see. From your drawings can you determine if the planet is nearer or farther from the Earth than the Sun is? Do your drawings show any changes in the shape of the crescent from one week to the next? If so, can you deduce if Venus is coming toward us or moving away from us?

For further reading

Banville, J. *Kepler: A Novel*. Godine, 1981 • A fictionalized biography of Johannes Kepler.

Christianson, G. *This Wild Abyss*. Free Press, 1978 • A very readable book about the evolution of astronomy.

Cohen, I. B. "Newton's Discovery of Gravity." *Scientific American,* March 1981 • An intriguing account of how Newton compared the real world with a mathematical representation of it. Many historical insights about Newton and his contemporaries are given.

Dicks, D. *Early Greek Astronomy to Aristotle*. Cornell University Press, 1985 • A scholarly survey of some early Greek astronomers and their thoughts about the universe.

Drake, S. "Newton's Apple and Galileo's Dialogue." *Scientific American,* August 1980 • This article shows how Newton's thinking was influenced by the work of Galileo.

———. *Telescopes, Tides, and Tactics*. University of Chicago Press, 1983 • An engrossing book that includes a new translation of Galileo's *Starry Messenger*, illuminating the effects of his work on the worldview of his time.

Durham, F., and Purrington, R. *Frame of the Universe*. Columbia University Press, 1983 • A history of cosmology that begins with ancient ideas, then moves on to Copernicus, Kepler, Tycho Brahe, Galileo, and Newton. The book ends with Einstein and two chapters on modern cosmology.

Fauvel, J., et al., eds. *Let Newton Be: A New Perspective on His Life and Works*. Oxford University Press, 1988 • Thirteen historical essays on Newton and his contributions to science, seen from today's perspective.

Field, J. *Kepler's Geometrical Cosmology*. University of Chicago Press, 1988 • A scholarly study of Kepler's writings, emphasizing his worldview and his search for the geometrical plan of the universe.

Gardner, M. *The Relativity Explosion*. Vintage, 1976 • An excellent book about Einstein's ideas and relativity theory.

Gingerich, O. "Copernicus and Tycho." *Scientific American,* December 1973 • An article about two giants of astronomy and how they grappled with basic issues of the structure of the then-known universe.

———. "The Galileo Affair." *Scientific American,* August 1982 • This article gives many insights into the life and times of Galileo, with special emphasis on his relationship with the Church.

Kaufmann, W. *Relativity and Cosmology*. Harper & Row, 1977. • This slim book, which emphasizes relativity, describes the evolution of our understanding of gravitation.

Kuhn, T. *The Copernican Revolution*. Harvard University Press, 1957 • A classic book, with special emphasis on the paradigm shift associated with the development of a heliocentric worldview.

Lerner, L., and Gosselin, E. "Galileo and the Specter of Bruno." *Scientific American,* November, 1986 • The turbulent times of Galileo and the politics of the Roman Inquisition are the focus of this well written article.

Wilson, C. "How Did Kepler Discover His First Two Laws?" *Scientific American,* March 1972 • An excellent treatment of the steps that Kepler took toward his great discoveries.

Molecular hydrogen

Neon

Lithium

Iron

Barium

Calcium

The Sun

Incandescent lamp

Fluorescent lamp

Spectra and spectral lines When an electric spark passes through a gas, atoms of the gas emit radiation at certain wavelengths, and so a spectrum of that gas contains bright spectral lines. Here the spectra of various elements are displayed. Each element has a unique and distinctive pattern of spectral lines. Spectra of the Sun and two common household lamps are also shown. The spectrum of an incandescent light bulb is quite like the continuous spectrum of a blackbody. (Courtesy of Bausch and Lomb)

C H A P T E R 5

The Nature of Light and Matter

Astronomy is built on an understanding of light, and every scientific discovery about light has led to important discoveries about the universe. Light is a form of energy possessing both wavelike and particlelike properties. Because the light emitted by an object is dependent upon the object's temperature, we are able to determine the surface temperature of stars. With an understanding of the structure of atoms comes an understanding of why each chemical element emits and absorbs light at specific wavelengths. Astronomers use this knowledge to determine the composition of stars and galaxies. Even the motion of a light source affects its emitted wavelengths, permitting us to deduce its speed of approach or retreat. These are but a few of the reasons why understanding light is a prerequisite to understanding the universe.

In the early 1800s, the French philosopher Auguste Comte argued that humanity would never know the nature and composition of the stars. Because they are so far beyond our earthly reach, he thought that these remote celestial worlds would forever remain unfathomable and mysterious. But only a few years after Comte's bold pronouncement, scientists began discovering the basic properties of matter and radiation. It soon became clear that by analyzing starlight we can learn many things about the stars. We now know that the stars and galaxies are made of substances found here on Earth and that all atoms, regardless of their location in the universe, obey the same physical laws. Thus, laboratory experiments conducted here on Earth provide the insight and understanding needed to probe the distant reaches of the universe.

5-1 Light travels through empty space at a speed of 300,000 km/s

Does light travel instantaneously from one place to another, or does it move with a measurable speed? Whatever may be the nature of light, it does seem to travel swiftly from a source to our eyes. We see a distant event before we hear the accompanying sound.

Our modern understanding of light began with the pioneering work of Galileo and Newton. They were the first to ask basic questions about the properties and behavior of light. In the early 1600s, Galileo performed the first experiment to try and measure the speed of light. He and an assistant stood at night on two hilltops a known distance apart, holding shuttered lanterns. First Galileo opened the shutter of his lantern; as soon as his assistant saw the flash of light, he opened his own. Galileo used his pulsebeat to try and measure the time between opening his lantern and seeing the light from his assistant's lantern. From the distance and time, he computed the speed at which the light had traveled to the distant hilltop and back.

Galileo found that the measured time failed to increase noticeably, no matter how distant the assistant was stationed. Galileo therefore concluded that light travels too rapidly for our slow human reactions to measure its speed in this fashion.

The first accurate measurement of the speed of light was made in 1675 by Olaus Roemer, a Danish astronomer. Roemer had been studying the orbits of the moons of Jupiter by timing carefully the moments when the satellites passed into and out of the planet's shadow. To his surprise, the orbital motions of the satellites seemed to depend upon the relative positions of Jupiter and the Earth. As the Earth moved closer to Jupiter (during the months just before an opposition), the satellites seemed to speed up slightly in their orbits about Jupiter. Then as the Earth moved away from Jupiter, they seemed to slow down slightly.

Roemer realized that this puzzling effect could be explained if the light was taking a measurable time to travel from Jupiter to the Earth. When the Earth is closest to Jupiter, the image of a satellite disappearing into the Jovian shadow arrives at our telescopes a little sooner than it does when Jupiter and the Earth are farther apart, as shown in Figure 5-1. Roemer's measurements indicated that it takes about $16\frac{1}{2}$ minutes for light to travel across the diameter of the Earth's orbit (a distance of 2 AU). The size of the Earth's orbit was not accurately known in Roemer's day. Today, using the modern value of 150 million kilometers for the astronomical unit, Roemer's method yields roughly 300,000 km/s (186,000 mi/s) for the speed of light.

The first successful laboratory experiments to measure the speed of light were conducted in the mid-1800s in France. In 1850, Hippolyte Fizeau and Jean Foucault built an apparatus consisting of a light source, a rotating mirror, and a stationary mirror, as sketched in Figure 5-2. Light from the light source is reflected from the rotating mirror toward the stationary mirror 20 m away. While the light is making the round trip, the rotating mirror moves slightly. Thus the returning light ray is deflected away from the initial beam by a small angle. By measuring this angle and knowing the dimensions of their apparatus, Fizeau and Foucault could deduce the speed of light. Once again, the answer was very nearly 300,000 km/s.

The speed of light in a vacuum is usually designated by the letter "c." The modern value of the speed of light is

$$c = 299,792.458 \text{ km/s}$$

In most calculations, you can use $c = 3 \times 10^5$ km/s = 3×10^8 m/s. Note, however, that this value is the speed of light in a vacuum. Light traveling through air, water, glass, or any

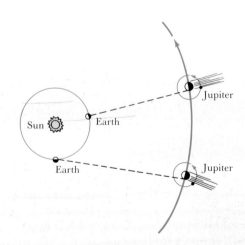

Figure 5-1 *Roemer's method of measuring the speed of light* When the Earth is near Jupiter, eclipses of Jupiter's moons occur slightly earlier than anticipated. When the Earth–Jupiter distance is large, the eclipses occur slightly later than anticipated. Roemer correctly attributed this effect to the travel time of light from Jupiter to the Earth.

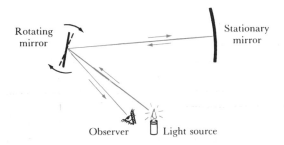

Figure 5-2 The Fizeau–Foucault method of measuring the speed of light Light from a light source is first reflected off a rotating mirror to a stationary mirror. The returning ray is deflected away from the path of the initial beam because the rotating mirror has moved slightly while the light was making the round trip. From the deflection angle and the dimensions of the apparatus, the speed of light can be determined.

other transparent substance always moves more slowly than it does in a vacuum.

The speed of light in empty space is one of the most important numbers in modern physical science. This value appears in many equations that describe atoms, gravity, electricity, and magnetism. Furthermore, according to Einstein's special theory of relativity, nothing can travel faster than the speed of light.

5-2 *Light is electromagnetic radiation and is characterized by its wavelength*

Light is energy. This fact is apparent to anyone who has felt the warmth of the sunshine on a summer's day. But what exactly is light? What is it made of? How does it move through space? These are profound questions that scientists have struggled with for the past four centuries.

The first major breakthrough in understanding light came from a simple experiment performed by Isaac Newton in the late 1600s. Newton was familiar with what he called "the celebrated Phenomenon of Colours": A beam of sunlight passing through a glass prism is spread out into the colors of the rainbow, as shown in Figure 5-3. This rainbow, called a **spectrum** (plural, **spectra**), suggested to Newton that white light must actually be a mixture of all the colors. Up to that time, it was erroneously assumed that the colors were somehow added to the light by the prism. Newton elegantly disproved this idea by passing the spectrum through a second prism inverted with respect to the first. Only white light emerged from this second prism. So it was clear that the second prism reassembled the colors of the rainbow to give back the original beam of sunlight.

From many experiments, Newton concluded that light is composed of undetectably tiny particles. In fact, as we shall see later in this chapter, the modern theory of quantum mechanics developed in the early 1900s also takes this viewpoint. Quantum theory postulates that light is composed of tiny packets of energy called **photons**. In the mid-1600s, however, a rival explanation was proposed by the Dutch astronomer Christian Huygens. He suggested that light travels in the form of waves rather than particles. Later experiments demonstrated that light does indeed act like a wave, at least in certain circumstances. Scientists now realize that light possesses both particlelike and wavelike characteristics.

The wavelike nature of light was convincingly demonstrated around 1801 by Thomas Young, an English physicist, physician, and general scholar. Young passed a beam of light through two thin, parallel slits in an opaque screen (see Figure 5-4a). On a white surface some distance beyond the slits, the light formed a pattern of alternating bright and dark bands. Young reasoned that if a beam of light was a stream of particles (as Isaac Newton had argued), the two beams of light from the slits should simply form bright images of the slits on the white surface. The pattern of bright

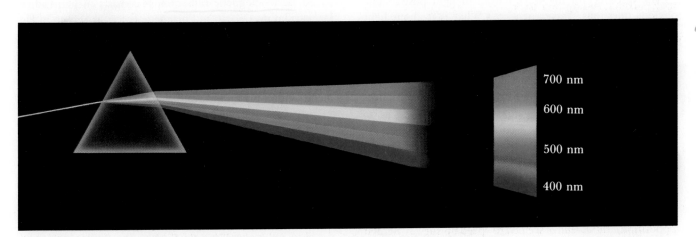

Figure 5-3 A prism and a spectrum When a beam of white light passes through a glass prism, the light is broken into a rainbow-colored band called a spectrum. The numbers on the right side of the spectrum indicate wavelengths.

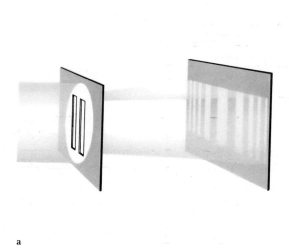

a

b

Figure 5-4 *Young's double-slit experiment* (a) *In the modern laboratory, Thomas Young's classic double-slit experiment is most easily repeated by shining light from a laser onto two closely spaced parallel slits. Alternating dark and light bands appear on a screen beyond the slits.* (b) *The intensity of light on the screen in the double-slit experiment*

is analogous to the height of water waves that strike a shore after passing through a barrier with two openings. In certain locations, ripples from both openings reinforce each other to produce extra-high waves. At intermediate locations along the shoreline, ripples and troughs cancel each other and produce still water.

and dark bands he observed is just what would be expected, however, if light had wavelike properties.

Imagine ocean waves pounding against a reef or breakwater that has two openings (see Figure 5-4b). A pattern of ripples is formed inside the barrier as the waves coming through the two openings interfere with each other. At certain points along the shore, crests arrive simultaneously from the two openings, producing high waves. At intermediate points along the shore, a crest from one opening meets a trough from the other opening, thus canceling each other and producing bands of still water. A similar explanation accounts for the pattern of bright and dark bands that Young observed on the white surface behind the slits in his experiment.

The discovery of the wave nature of light posed some obvious questions. Waves of what? What is it about light that goes up and down like water waves on the ocean? The answer came in the 1860s from a seemingly unlikely source: a comprehensive theory that described electricity and magnetism. Numerous experiments during the first half of the nineteenth century had demonstrated an intimate connec-

tion between electric and magnetic forces. As mentioned in Chapter 4, the Scottish mathematician and physicist James Clerk Maxwell succeeded in describing all the basic properties of electricity and magnetism in four equations. This mathematical achievement demonstrated that electric and magnetic forces are really two aspects of the same phenomenon, which we now call **electromagnetism**. By combining his four equations, Maxwell showed that electric and magnetic fields should travel through space in the form of waves, and calculated their speed at about 3×10^8 m/s. Maxwell pointed out that this is equal to the best available value for the speed of light. His suggestion that these waves do exist and are observed as light was soon confirmed by experiments.

Because of its electric and magnetic properties, light is called **electromagnetic radiation**. It consists of perpendicular, oscillating electric and magnetic fields, as shown in Figure 5-5. The distance between two successive wave crests is called the **wavelength** of the light and is usually designated by the Greek letter λ (lambda).

More than a century elapsed between Newton's experiments with a prism and confirmation of the wave nature of

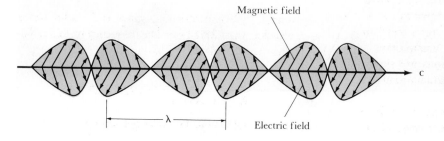

Magnetic field

Electric field

c

Figure 5-5 *Electromagnetic radiation* All forms of light consist of oscillating electric and magnetic fields that move through space at a speed of 3×10^5 km/s ($= 3 \times 10^{10}$ cm/s). The distance between two successive crests, called the wavelength of the light, is usually designated by the Greek letter λ.

light. One reason for this delay is that the wavelength of **visible light** is extremely short—less than a thousandth of a millimeter—and thus is not easily detectable. To express such tiny distances conveniently, scientists use a unit of length called the nanometer (abbreviated nm), where $1 \text{ nm} = 10^{-9}$m. Experiments demonstrated that visible light has wavelengths covering the range from about 400 nm for violet light to about 700 nm for red light. Intermediate colors of the rainbow, like yellow (550 nm) have intermediate wavelengths (see Figure 5-3).

Another unit of measure commonly used by astronomers to denote wavelengths is the angstrom (abbreviated Å and named after the Swedish physicist A. J. Ångström), where $1 \text{ Å} = 10^{-10}$ m. Thus 10 Å=1 nm, and visible light has wavelengths between 4000 and 7000 Å. As more and more people adopt SI units, however, they tend to prefer nanometers over angstroms.

Maxwell's equations place no restrictions, however, on the wavelength of electromagnetic radiation. In other words, electromagnetic waves could and should exist with wavelengths both longer and shorter than the 400- to 700-nm range of visible light. Consequently, researchers began to look for invisible forms of light: forms of electromagnetic radiation to which the cells of the human retina do not respond.

Around 1800, the British astronomer William Herschel had discovered infrared radiation, which is just beyond the red end of the visible spectrum of light. In 1888, the German physicist Heinrich Hertz succeeded in producing electromagnetic radiation a few centimeters long in wavelength, now known as **radio waves.** In 1895, Wilhelm Röntgen invented a machine that produces electromagnetic radiation of a wavelength shorter than 10 nm, now known as **X rays.** Modern versions of Röntgen's machine are today found in medical and dental offices.

Over the years, radiation has been discovered in many other wavelength ranges. Visible light occupies only a tiny fraction of the full extent of possible wavelengths collectively called **the electromagnetic spectrum.** As shown in Figure 5-6, the electromagnetic spectrum stretches from the longest-wavelength radio waves to the shortest-wavelength gamma rays. For example, at wavelengths slightly longer than those of visible light, we find **infrared radiation,** which covers the range from about 700 nm to 1 mm. From roughly 1 mm to 10 cm is the range of **microwaves,** beyond which is the domain of radio waves.

At wavelengths shorter than those of visible light, **ultraviolet radiation** extends from about 400 nm down to 10 nm. Next are X rays, which consist of wavelengths between about 10 and 0.01 nm, and beyond them are **gamma rays.** It should be noted that these arbitrary divisions are simply rough boundaries that allow us to identify certain broad sections of the electromagnetic spectrum.

Scientists working with the various kinds of electromagnetic radiation have adopted slightly different ways of de-

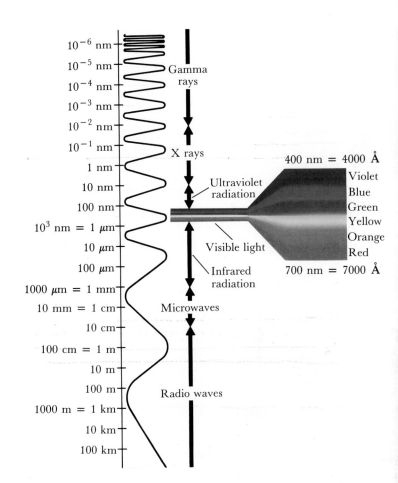

Figure 5-6 The electromagnetic spectrum *The full array of all types of electromagnetic radiation is called the electromagnetic spectrum. It extends from the longest-wavelength radio waves to the shortest-wavelength gamma rays. Visible light occupies only a tiny portion of the full electromagnetic spectrum.*

noting wavelengths. For example, astronomers interested in infrared radiation often express wavelength in micrometers (abbreviated μm), where $1 \text{ μm} = 10^3 \text{ nm} = 10^{-6}$ m. Astronomers who work with radio telescopes often have training in electrical engineering and prefer to speak of frequency rather than wavelength. The **frequency** of light is simply the number of wave crests passing a given point in one second. Because light travels at the speed $c = 3 \times 10^8$ m/s, the frequency (ν) of light is related to its wavelength (λ) by

$$\nu = \frac{c}{\lambda}$$

For example, hydrogen atoms in space emit radio waves with a wavelength of 21.12 cm. The frequency of this radiation is therefore calculated as follows:

$$\nu = \frac{c}{\lambda} = \frac{3 \times 10^8 \text{ m/s}}{0.2112 \text{ m}} = 1.42 \times 10^9 \text{ s}^{-1}$$

$$= 1,420,000,000 \text{ cycles per second}$$

Box 5-1 Temperatures and temperature scales

Three temperature scales are commonly used, for various purposes. It is useful to be able to convert temperature readings from one scale to another.

Temperatures are expressed throughout most of the world in degrees Celsius (°C). The Celsius temperature scale is based on the behavior of water, which freezes at 0°C and boils at 100°C at sea level on Earth. This scale is named in honor of the Swedish astronomer Anders Celsius, who proposed it in 1742.

For many purposes, however, scientists prefer to express temperatures in a unit called the kelvin (K), named after the British physicist Lord Kelvin (William Thomson), who made many important contributions to our knowledge about heat and temperature. On the Kelvin temperature scale, water freezes at 273 K and boils at 373 K. Because water must be heated through a change of 100 K or 100°C to go from its freezing point to its boiling point, you can see that the "size" of a kelvin is the same as the "size" of a degree Celsius. When considering temperature *changes*, measurements in kelvins and in degrees Celsius are the same. A temperature expressed in kelvins is always equal to the temperature in degrees Celsius plus 273. The scientific preference for the Kelvin scale arises from a physical interpretation of the meaning of temperature.

All substances are made of atoms. These atoms are very tiny (typical atomic diameters are about 10^{-10} m) and are constantly in motion. The temperature of a substance is directly related to the average speed of its atoms. If something is hot, its atoms are moving at high speeds. If a substance is cold, its atoms are moving much more slowly.

The coldest possible temperature is the temperature at which atoms move as slowly as possible (they can never quite stop completely). This minimum possible temperature is called absolute zero and is the starting point for the Kelvin scale. Absolute zero is 0 K, or −273°C. Since it is impossible for anything to be colder than 0 K, there are no negative temperatures on the Kelvin scale.

In the United States, many people still use the now-archaic Fahrenheit scale, which expresses temperatures in degrees Fahrenheit (°F). When the German physicist Gabriel Fahrenheit introduced this scale in the early 1700s, he intended 0°F to represent the coldest temperature then achievable (with a mixture of ice and saltwater) and 100°F to represent the temperature of a healthy human body. On the Fahrenheit scale, water freezes at 32°F and boils at 212°F. Because there are 180 degrees Fahrenheit between the freezing and boiling points of water, a degree Fahrenheit is only $\frac{100}{180} = \frac{5}{9}$ as large as either a degree Celsius or a kelvin.

The following equation is useful to convert from degrees Celsius to degrees Fahrenheit:

One cycle per second is also called a hertz (abbreviated Hz) in honor of Heinrich Hertz, the physicist who discovered radio waves. Often it is convenient to use the prefix *mega-* (meaning million and abbreviated M) or *kilo-* (meaning thousand and abbreviated k). Thus the frequency of hydrogen emission is 1420 MHz, or 1420 megahertz.

5-3 An object emits electromagnetic radiation with intensity and wavelengths related to the temperature of the object

The amount of energy radiated by an object depends upon its temperature. The hotter the object is, the more energy it emits in the form of electromagnetic radiation. The dominant wavelength of the emitted radiation also depends upon the temperature of the object. A cool object emits most of its energy at long wavelengths, whereas a hot object emits most of its energy at short wavelengths.

These basic phenomena are familiar to anyone who has ever watched a welder or blacksmith heat a bar of iron (see Figure 5-7). As the bar becomes hot, it begins to glow deep red. As the temperature rises further, the bar begins to give off a brighter, reddish-orange light. At still higher temperatures, it shines with a brilliant, yellowish-white light. If the bar could be prevented from melting and vaporizing, at extremely high temperatures it would emit a dazzling, blue-white light.

To appreciate the relationship between energy and temperature that describes these phenomena, we must understand the units that scientists use to measure energy and temperature. Energy is usually measured in joules (J), named after the nineteenth-century English physicist James Joule. A joule is the amount of energy contained in the motion of a two-kilogram mass moving at a speed of 1 m/s. (Recall from Box 4-4 that kinetic energy is $\frac{1}{2}mv^2$.) The joule is a convenient unit of energy because it is closely related to the familiar watt (W): 1 W = 1 J/s. For instance, the energy consumption of a 100-watt light bulb is 100 joules per second.

In discussing temperature, scientists usually prefer the Kelvin temperature scale, on which temperature is measured in kelvins (K) upward from absolute zero, the coldest possible temperature. On the familiar Celsius and Fahrenheit temperature scales, absolute zero (0 K) is −273°C and −460°F. Water freezes at 273 K and boils at 373 K, and "room temperature" is roughly 295 K. The relationship be-

$$T_F = \frac{9}{5}T_C + 32$$

where T_F is the temperature in °F and T_C is the temperature in °C. To convert in the opposite direction, a simple rearrangement of terms gives the relationship

$$T_C = \frac{5}{9}(T_F - 32)$$

Example: Consider a typical room temperature of 72°F. Using the second equation, we can convert this measurement to the Celsius scale as follows:

$$T_C = \frac{5}{9}(72 - 32) = 22°C$$

To arrive at the Kelvin scale, we simply add 273 degrees to the value in °C. Thus,

$$72°F = 22°C = 295 \text{ K}$$

The diagram displays the relationships between these three temperature scales.

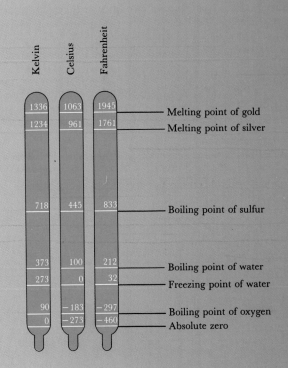

Kelvin	Celsius	Fahrenheit	
1336	1063	1945	Melting point of gold
1234	961	1761	Melting point of silver
718	445	833	Boiling point of sulfur
373	100	212	Boiling point of water
273	0	32	Freezing point of water
90	−183	−297	Boiling point of oxygen
0	−273	−460	Absolute zero

tween the Kelvin, Celsius, and Fahrenheit temperature scales is discussed in Box 5-1.

In 1879, the Austrian physicist Josef Stefan summarized the results of experiments he had done by stating that an object emits energy at a rate proportional to the fourth power of the object's temperature (measured in kelvins). In other words, if you double the temperature of an object (for example, from 500 to 1000 K), then the energy emitted from the object's surface each second increases by a factor of $2^4 = 16$. If you triple the temperature (for example, from 500 to 1500 K), the rate of energy emission increases by a factor of $3^4 = 81$.

The energy emitted from each m² of an object's surface each second is called the **energy flux** (F). In this context, flux means "rate of flow," and thus F is a measure of how much energy is flowing out of the object. Using this term, we can write Stefan's law as

$$F = \sigma T^4$$

where σ (sigma) is a constant, T is the object's temperature measured in kelvins, and F is the energy flux measured in joules per square meter per second (usually written as $J \, m^{-2} \, s^{-1}$). Since 1 watt equals 1 joule per second, F can also be expressed in watts per square meter.

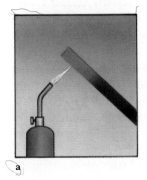
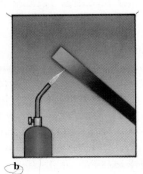

a   **b**   **c**

Figure 5-7 Heating a bar of iron This sequence of drawings (a→b→c) shows the changing appearance of a heated bar of iron as its temperature rises. As the temperature increases, the amount of energy radiated by the bar increases. The color of the bar also changes because, as the temperature goes up, the dominant wavelength of light emitted by the bar decreases. These effects are described by the Stefan–Boltzmann law and Wien's law.

Box 5-2 The Sun's luminosity and surface temperature

Using detectors above the Earth's atmosphere, astronomers have measured the average flux of solar energy arriving at the Earth; this value is called the **solar constant** and is equal to 1370 W m^{-2}. Imagine a huge sphere with a radius of 1 AU with the Sun at its center, as shown in the diagram. Each square meter of that sphere receives 1370 watts of power from the Sun, so we can compute the total energy output of the Sun by multiplying the solar constant by the sphere's area. The result, called the **luminosity of the Sun** (L$_\odot$), is

$$L_\odot = 3.90 \times 10^{26} \text{ W}$$

(Astronomers use the symbol $\odot$ to denote the Sun.)

We know the size of the Sun, so we can now compute the energy flux emitted from its surface. The radius of the Sun is $R_\odot = 6.96 \times 10^8$ m, and its surface area is $4\pi R_\odot^2$. Therefore, its energy flux is

$$F_\odot = \frac{L_\odot}{4\pi R_\odot^2} = 6.41 \times 10^7 \text{ W m}^{-2}$$

Now we can use the Stefan–Boltzmann law to find the Sun's surface temperature:

$$T_\odot^4 = \frac{F_\odot}{\sigma} = 1.13 \times 10^{15} \text{ K}^4$$

Taking the fourth root (the square root of the square root) of this value, we find the surface temperature of the Sun to be

$$T_\odot = 5800 \text{ K}$$

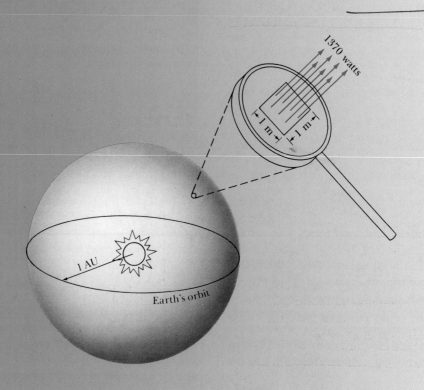

Five years after Stefan announced this law, another Austrian physicist, Ludwig Boltzmann, showed how it could be derived mathematically from basic assumptions about atoms and molecules. The law today is thus commonly known as the **Stefan–Boltzmann law.** The Stefan–Boltzmann constant σ is known from laboratory experiments to have the value

$$\sigma = 5.67 \times 10^{-8} \text{ W m}^{-2} \text{ K}^{-4}$$

Boltzmann showed that this law is best obeyed by an object called a **blackbody**. A blackbody does not reflect any light; it instead absorbs all radiation falling on it. Since it reflects no electromagnetic radiation, the radiation that it does emit is entirely the result of its temperature. Ordinary objects, like tables and textbooks, are rather poor blackbodies; they reflect light, which is actually the only reason they are visible. We shall see shortly that the wavelengths emitted by blackbodies at room temperature are far too long for our eyes to perceive.

It is important to realize that a blackbody does not necessarily look black. For instance, a star behaves like a blackbody because it efficiently absorbs almost all of the radiation falling upon it from the outside. A star does not look black, however, because it consists of hot, glowing gases. Since a star behaves like a blackbody, astronomers can use the Stefan–Boltzmann law to relate its energy output to its surface temperature. Box 5-2 shows how the Stefan–Boltzmann law can be used to calculate the surface temperature of the Sun.

An object emits radiation over a wide range of wavelengths, but there is always a particular wavelength (λ_{max}) at which the emission of energy is strongest. This dominant wavelength gives a glowing hot object its characteristic color.

In 1893, the German physicist Wilhelm Wien discovered a simple relationship between the temperature of a blackbody and the dominant wavelength (λ_{max}) of the energy it emits:

$$\lambda_{max} = \frac{2.9 \times 10^{-3}}{T}$$

where λ_{max} is measured in meters, and T is measured in kelvins. This relation is today called **Wien's law.**

Wien's law states that, for any blackbody, λ_{max} (in meters) times T (in kelvins) equals the constant value 0.0029. If the temperature of the blackbody increases, the dominant wavelength of its radiation must decrease so that the product remains constant. In other words, the hotter an object, the shorter the dominant wavelength of the electromagnetic radiation it emits.

From the Stefan–Boltzmann law, we see that any object with a temperature above absolute zero (0 K) emits some electromagnetic radiation. From Wien's law, we find that a very cold object with a temperature of only a few kelvins emits primarily microwaves. An object at "room temperature" (about 295 K) emits most of its radiation in the infrared portion of the spectrum. And an object with a temperature of a few thousand kelvins emits most of its radiation as visible light. An object with a temperature of a few million kelvins emits most of its radiation at X-ray wavelengths.

Consider energy coming from the Sun. The Sun emits energy over a wide range of wavelengths, but the maximum intensity of sunlight is at a wavelength of roughly 500 nm = 5×10^{-7} m. From Wien's law, we find the Sun's surface temperature ($T_{\odot}$) to be

$$T_{\odot} = \frac{2.9 \times 10^{-3}}{5 \times 10^{-7}} = 5800 \text{ K}$$

This result agrees with the value we computed using the Stefan–Boltzmann law (see Box 5-2). Wien's law is very use-

ful for computing the surface temperatures of stars, because it does not require knowledge of the star's size or luminosity. All we need to know is the dominant wavelength of the star's electromagnetic radiation.

5-4 A full explanation of blackbody radiation requires assuming that light has particlelike properties

The Stefan–Boltzmann law and Wien's law describe only two basic properties of **blackbody radiation**, the electromagnetic radiation emitted by an ideal blackbody. A more complete picture is given by **blackbody curves** like those in Figure 5-8. These curves show the relationship between wavelength and the intensity of the light emitted by a blackbody at a given temperature. The total area under a blackbody curve is proportional to the energy flux F. The wavelength corresponding to the peak of the curve is the dominant wavelength λ_{max}. Note that the blackbody curves clearly illustrate both of the laws we have discussed. A cool

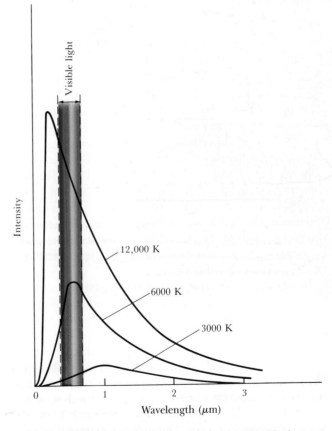

Figure 5-8 Blackbody curves *Three representative blackbody curves are shown here. Each curve shows the intensity of light at every wavelength that is emitted by a blackbody at a particular temperature. Note the range of wavelengths of visible light.*

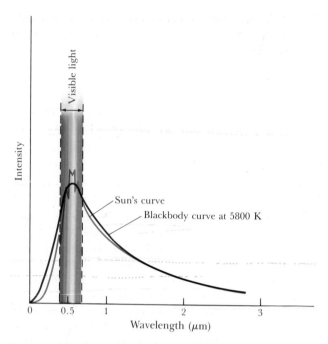

Figure 5-9 *The Sun as a blackbody* *This graph compares the intensity of sunlight over a wide range of wavelengths with the intensity of radiation coming from a blackbody at a temperature of 5800 K. Measurements of the Sun's intensity were made above the Earth's atmosphere. The Sun mimics a blackbody remarkably well.*

The apparatus to demonstrate the photoelectric effect consists of a glass container, called a vacuum tube, from which all the air has been pumped (see Figure 5-10). At either end of the vacuum tube is a metal plate called an electrode. Both electrodes are connected to a battery. In the dark or when the apparatus is illuminated by visible light, no current flows through the circuit. If a beam of ultraviolet light is focused on the negatively charged electrode, however, a current does flow through the circuit. An electric current is carried by tiny, negatively charged particles called **electrons**. (We will see below that electrons are one of the basic particles of which atoms are composed.) Somehow, the ultraviolet light liberates electrons from the negative electrode, allowing them to move across the tube to the positively charged electrode. It is surprising that a bright red light has no effect on the apparatus, whereas a dim ultraviolet light causes a current to flow.

Einstein explained the photoelectric effect by assuming that light exists in particlelike packets, or photons, of energy. The amount of energy in one **photon** of light is inversely proportional to the wavelength of the light. In other words, the longer the wavelength, the less energy there is in the photon. A single photon of red light, for instance, with

object has a low curve that peaks at a long wavelength, whereas a hot object has a high curve that peaks at a short wavelength.

Figure 5-9 shows how the intensity of sunlight varies with wavelength. This curve was obtained by measuring the intensity of sunlight at various wavelengths above the Earth's atmosphere. As mentioned earlier, the peak of the curve is at a wavelength of about 0.5 μm = 500 nm. The blackbody curve for a temperature of 5800 K is also plotted in Figure 5-9. Note how closely the observed intensity curve for the Sun matches the blackbody curve. This close correlation between the observed intensity curves for most stars and the blackbody curves is the reason astronomers are interested in the physics of blackbody radiation.

By the end of the nineteenth century, physicists realized that they had reached an impasse. All attempts to explain the characteristic shape of blackbody curves in terms of then known science had failed.

In 1900, however, the German physicist Max Planck discovered that he could derive a mathematical formula for blackbody curves if he assumed that electromagnetic energy is emitted in discrete packets. This important assumption was verified in 1905 by Einstein, who used the idea of the particlelike nature of light to explain a phenomenon, called the **photoelectric effect,** in which a metal gives off electrons when exposed to short-wavelength radiation. Einstein won a Nobel Prize for this explanation.

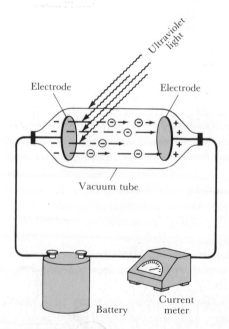

Figure 5-10 *The photoelectric effect* *The photoelectric effect can be demonstrated with a vacuum tube containing two electrodes attached to a battery. When no light falls on the negative electrode (at left), no electrons are liberated from its metal and no current flows. When long-wavelength red light is shined on the electrode, still no electrons are released and no current flows. But when short-wavelength ultraviolet light falls on the electrode, many electrons are liberated from the metal. These negatively charged electrons are attracted to the positively charged electrode at the opposite end of the vacuum tube and thus create an electric current.*

its relatively long wavelength, does not have enough energy to knock an electron out of the metal electrode. So, no matter how many photons of red light strike the metal, no collision of a photon with an atom can liberate an electron. A single photon of ultraviolet radiation, however, with its much shorter wavelength, does have enough energy to knock loose an electron. Thus even a very weak beam of ultraviolet light can liberate electrons and cause a current to flow.

The relationship between the energy E of each photon and the wavelength λ of the electromagnetic radiation can be expressed in a simple equation:

$$E = \frac{hc}{\lambda}$$

where c is the speed of light and h is a constant value now called Planck's constant. Because $c/\lambda = \nu$, we can express this relationship in another form, often called **Planck's law:**

$$E = h\nu$$

where ν is the frequency of the light. Note that both equations express a relationship between a particlelike property of light (the energy E of a photon) and a wavelike property (the wavelength λ or the frequency ν).

Laboratory experiments show that Planck's constant has the value

$$h = 6.625 \times 10^{-34} \text{ J s}$$

This is a tiny value; by the standards of our everyday experience, a single photon carries a very small amount of energy (see Box 5-3).

5-5 Each chemical element produces its own unique set of spectral lines

In 1814, the German master optician Joseph Fraunhofer repeated the classic experiment of shining a beam of sunlight through a prism (recall Figure 5-3), but this time he subjected the resulting rainbow-colored spectrum to intense magnification. To his surprise, Fraunhofer discovered that the solar spectrum contains hundreds of fine dark lines, now called **spectral lines.** Fraunhofer counted over 600 such lines; today we know of about a million. A photograph of the Sun's spectrum is shown in Figure 5-11.

Half a century later, chemists discovered that they could produce spectral lines in the laboratory. Chemists had long known that certain substances emit distinctive colors when sprinkled into a flame. Such experiments were often involved in the identification of **chemical elements,** the fundamental substances that are impossible to break down into more-basic chemicals. About 1857, the German chemist Robert Bunsen invented a special gas burner that produces a clean, colorless flame. This Bunsen burner was very useful for analyzing substances, because its flame had no color of its own to be confused with the color produced when a substance was sprinkled into it.

Figure 5-11 The Sun's spectrum *Numerous spectral lines are seen in this photograph of the Sun's spectrum. The spectrum is so long that it had to be cut into convenient segments to fit it on this page. (NOAO)*

Box 5-3 Photon energy and the electron volt

A beam of light can be regarded as a stream of tiny packets of energy called photons. The energy (E) carried by a photon is directly related to its frequency (ν) by Planck's law, as discussed in the text:

$$E = h\nu$$

or equivalently

$$E = \frac{hc}{\lambda}$$

where h is Planck's constant, λ is the wavelength of the light, and c is the speed of light in a vacuum. Laboratory experiments reveal that

$$h = 6.625 \times 10^{-34} \text{ J s}$$

The energy carried by a single photon is very small and is conveniently expressed in a unit of energy called the **electron volt** (eV), where

$$1 \text{ eV} = 1.602 \times 10^{-19} \text{ J}$$

Thus, Planck's constant may be written as

$$h = 4.135 \times 10^{-15} \text{ eV s}$$

Example: Consider a photon of wavelength 620 nm, visible to the human eye as orange light. The energy of that photon is calculated as

$$E = \frac{hc}{\lambda} = \frac{(4.135 \times 10^{-15}\text{eV s})(3 \times 10^8 \text{ m s}^{-1})}{6.2 \times 10^{-7} \text{ m}} = 2 \text{ eV}$$

Visible light consists of photons in the energy range of approximately 2 to 3 eV. X-ray photons carry thousands of times more energy than visible photons, so their energies are conveniently expressed in thousands of electron volts (keV), where the prefix k indicates one thousand. For instance, the energy carried by an X-ray photon with a wavelength of 1 nm is 1.24 keV. Similarly, the energies of gamma rays are often expressed in millions of electron volts (MeV), where the prefix M indicates one million.

The accompanying table gives the photon energy ranges of various portions of the electromagnetic spectrum, along with the corresponding blackbody temperature, computed using Wien's law.

	Wavelength (m)	Photon energy (eV)	Blackbody temperature (K)
Radio	$>10^{-1}$	$<10^{-5}$	<0.03
Microwave	10^{-1} to 10^{-4}	10^{-5} to 10^{-2}	0.03 to 30
Infrared	10^{-4} to 7×10^{-7}	0.01 to 2	30 to 4100
Visible	7×10^{-7} to 4×10^{-7}	2 to 3	4100 to 7300
Ultraviolet	4×10^{-7} to 10^{-9}	3 to 10^3	7300 to 3×10^6
X ray	10^{-9} to 10^{-11}	10^3 to 10^5	3×10^6 to 3×10^8
Gamma ray	$<10^{-11}$	$>10^5$	$>3 \times 10^8$

Note: The symbol $>$ means "greater than." The symbol $<$ means "less than."

A young colleague, the Prussian-born physicist Gustav Kirchhoff, suggested that light from the colored flames might best be studied by passing it through a prism (see Figure 5-12). Bunsen and Kirchhoff collaborated in the design and construction of a spectroscope, a device consisting of a prism and several lenses, by which the spectra of flames might be magnified and examined. They promptly discovered that the spectrum from a flame consists of a pattern of thin, bright spectral lines against a dark background. Later, they found that each chemical element produces its own unique pattern of spectral lines. Thus was born in 1859 the technique of **spectral analysis**, the identification of chemical substances by their unique patterns of spectral lines.

New chemical elements were discovered through spectral analysis. After Bunsen and Kirchhoff had recorded the prominent spectral lines of all the then-known elements,

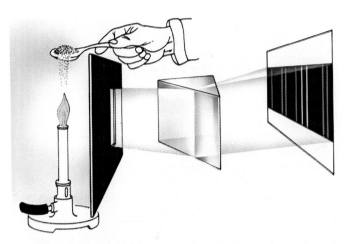

Figure 5-12 The Kirchhoff–Bunsen experiment *In the mid-1850s, Gustav Kirchhoff and Robert Bunsen discovered that when a chemical substance is heated and vaporized the resulting spectrum exhibits a series of bright spectral lines. They found also that each chemical element produces its own characteristic pattern of spectral lines. (In an actual laboratory experiment, lenses would be needed to focus the image of the slit onto the screen.)*

they soon began to discover other spectral lines in the spectra of mineral samples. In 1860 they found a new line in the blue portion of the spectrum of a sample of mineral water. After chemically isolating the previously unknown element responsible for making the line, they named it cesium (from the Latin *caesium*, gray-blue). The next year, a new line in the red portion of the spectrum of a mineral sample led them to discover the element rubidium (Latin *rubidium*, red).

During the solar eclipse of 1868, astronomers found a new spectral line in light coming from the upper surface of the Sun while the main body of the Sun was hidden by the Moon. This line was attributed to a new element that was named helium (from the Greek *helios*, sun). Helium was not discovered on the Earth until 1895, however, when it was located in gases obtained from a uranium mineral.

By the early 1860s, Kirchhoff's experiments had progressed sufficiently for him to formulate three important statements about spectra that today are called **Kirchhoff's laws of spectral analysis.**

Law 1 A hot object or a hot, dense gas produces a **continuous spectrum**—a complete rainbow of colors without any spectral lines.

Law 2 A hot, rarefied gas produces an **emission line spectrum**—a series of bright spectral lines against a dark background.

Law 3 A cool gas in front of a source of a continuous spectrum produces an **absorption line spectrum**—a series of dark spectral lines among the colors of the rainbow.

Figure 5-13 illustrates Kirchhoff's laws. The bright lines in the emission spectrum of a particular gas occur at exactly the same wavelengths as the dark lines in the absorption spectrum of the gas. Also note the importance of the relative temperatures of the gas cloud and its background. Absorption lines are seen if the background is hotter than the gas. Emission lines are seen if the background is cooler.

At the time of these discoveries, scientists knew that all matter is composed of atoms. An **atom** is the smallest particle (typical diameter = 0.1 nm) of a chemical element that still has the properties of that element. Thus, Kirchhoff's laws mean that the atoms in a gas somehow extract light of very specific wavelengths from the white light that passes through the gas. Thus, dark absorption lines are created in the continuous spectrum of the white-light source. The atoms then radiate light of precisely these same wavelengths in all directions, so that an observer at an oblique angle (without the white-light source in the background) detects bright emission lines.

Why does an atom absorb only light of particular wavelengths? Why does it then emit only light of these same wavelengths? Maxwell's theory of electromagnetism could not answer these questions. The answers did not come until early in the twentieth century, with the development of quantum mechanics and nuclear physics.

5-6 An atom consists of a small, dense nucleus surrounded by electrons

The first important clue about the internal structure of atoms came from an experiment conducted in 1910 by Ernest Rutherford, a gifted chemist and physicist from New Zealand. Rutherford and his colleagues at the University of Manchester in England had been investigating the recently discovered phenomenon of radioactivity. Certain radioactive elements, such as uranium and radium, were known to emit particles. One type of particle, called an alpha particle, has about the same mass as a helium atom and is emitted from a radioactive substance with considerable speed.

In one series of experiments, Rutherford and his colleagues were using alpha particles as projectiles to probe the structure of solid matter. They directed a beam of these particles against a thin sheet of metal (see Figure 5-14). Almost all the alpha particles passed through the metal sheet with little or no deflection from their straight-line paths. To the surprise of the experimenters, however, an occasional alpha particle bounced back from the metal sheet as though it had struck something quite dense. Rutherford later remarked, "It was almost as incredible as if you fired a fifteen-inch shell at a piece of tissue paper and it came back and hit you."

Rutherford concluded from this experiment that most of the mass of an atom is concentrated in a compact, massive lump of matter which occupies only a small part of the

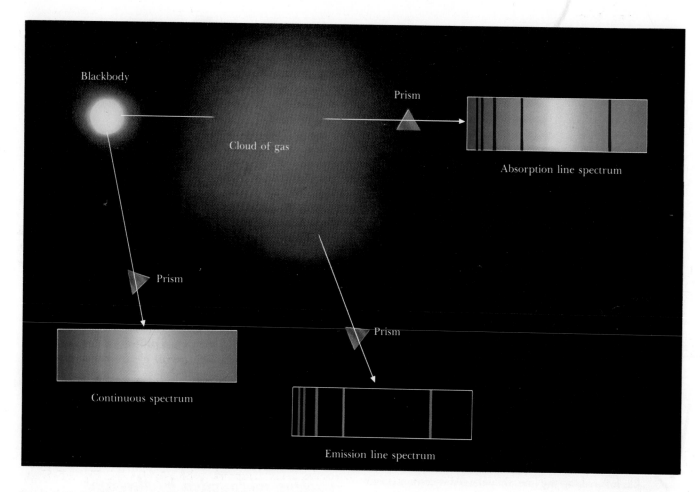

Blackbody

Cloud of gas

Prism

Absorption line spectrum

Prism

Continuous spectrum

Prism

Emission line spectrum

Figure 5-13 Kirchhoff's laws of spectral analysis *This schematic diagram summarizes Kirchhoff's laws. A hot, glowing object emits a continuous spectrum. If this source of light is viewed through a cool gas,* *dark absorption lines appear in the resulting spectrum. When a heated gas is viewed against a cold, dark background, bright emission lines are seen in the spectrum.*

atom's volume. Most of the alpha particles pass freely through the nearly empty space that makes up most of the atom, but a few particles happen to strike the dense mass at the center of the atom and bounce back.

Rutherford proposed a new model for the structure of an atom. According to this model, a massive, positively charged <u>nucleus</u> at the center of the atom is orbited by tiny, nega-

tively charged electrons (see Figure 5-15). Rutherford concluded that at least 99.98 percent of the mass of an atom must be concentrated in a nucleus whose diameter is only about one ten-thousandth the diameter of the atom.

We know today that the nucleus of an atom contains two types of particles: <u>**protons**</u> and <u>**neutrons**</u>. A proton has almost the same mass as a neutron, and each has about 2000

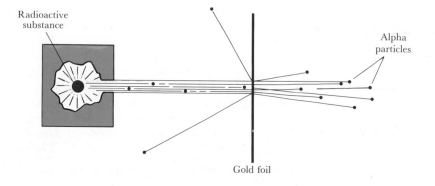

Radioactive substance

Alpha particles

Gold foil

Figure 5-14 Rutherford's experiment *Alpha particles from a radioactive source are directed against a thin metal foil. Most alpha particles pass through the foil with very little deflection. Occasionally, however, an alpha particle will recoil, indicating that it has collided with the massive nucleus of an atom. This experiment provided the first evidence that the nuclei of atoms are relatively massive and compact.*

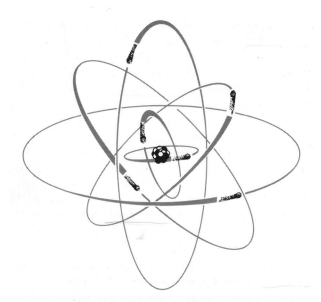

Figure 5-15 Rutherford's model of the atom *Electrons orbit the atom's nucleus, which contains most of the atom's mass. The nucleus contains two types of particles: protons and neutrons.*

times as much mass as an electron. A proton has a positive electric charge; a neutron has no electric charge. The electric force between the positively charged protons and the negatively charged electrons holds an atom together.

5-7 Spectral lines are produced when an electron jumps from one energy level to another within an atom

The task of reconciling Rutherford's atomic model with Kirchhoff's laws of spectral analysis was undertaken by the young Danish physicist Niels Bohr, who joined Rutherford's group at Manchester in 1911.

Bohr began by trying to understand the structure of hydrogen, the simplest and lightest of the elements. A hydrogen atom consists of a single electron and a single proton. Hydrogen has a simple spectrum consisting of a pattern of lines that begins at 656.3 nm and ends at 364.6 nm. The first spectral line is called H_α, the second spectral line is called H_β, the third is H_γ, and so forth, ending with H_∞. The closer

you get to the short-wavelength end of the spectrum at 364.6 nm, the more spectral lines you see.

The regularity in this spectral pattern was described mathematically in 1885 by Johann Jacob Balmer, an elderly Swiss schoolteacher. Balmer used trial and error to discover a formula from which the wavelengths (λ) of hydrogen's spectral lines can be calculated. Balmer's formula is usually written

$$\frac{1}{\lambda} = R\left(\frac{1}{4} - \frac{1}{n^2}\right)$$

where n is an integer (whole number) greater than 2 and R is a number called the Rydberg constant (R = 1096.77 m^{-1}) in honor of the Swedish spectroscopist J. R. Rydberg. To get the wavelength of H_α, you put $n = 3$ into Balmer's formula. To get H_β, use $n = 4$. To get H_γ, use $n = 5$. To get the short-wavelength end of the hydrogen spectrum at 364.6 nm, use $n = \infty$ (the symbol ∞ stands for infinity).

The spectral lines of hydrogen at visible wavelengths are today called the **Balmer lines**, and the entire pattern from H_α through H_∞ is called the **Balmer series**. Nearly two dozen Balmer lines, from H_{13} through H_{40}, are seen in the spectrum of HD 193182 in Figure 5-16.

Bohr realized that, if he were to succeed in understanding the structure of the hydrogen atom, he should be able to derive Balmer's formula using the laws of physics. He assumed first that the electron in a hydrogen atom orbits the nucleus in certain specific orbits. As shown in Figure 5-17, it is customary to label these orbits, called Bohr orbits, $n = 1$, $n = 2$, $n = 3$, and so on. Although confined to one of these allowed orbits while circling the nucleus, an electron can jump from one Bohr orbit to another.

For an electron to jump from one Bohr orbit to another, the hydrogen atom must gain or lose a specific amount of energy. For the electron to go from an inner orbit to an outer orbit, the atom must absorb energy. Conversely, if the electron goes from an outer orbit to an inner orbit, the atom must release energy.

The energy gained or released by the atom when the electron jumps from from one orbit to another is the difference in energy between these two orbits. According to Planck and Einstein, the packet of energy gained or released is a photon whose energy is given by $E = hc/\lambda$, where h is Planck's constant, c is the speed of light, and λ is the photon's wavelength. Bohr combined these ideas to prove mathematically

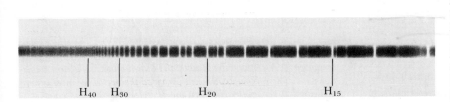

Figure 5-16 Balmer lines in the spectrum of HD 193182 *This portion of the spectrum of the star HD 193182 shows nearly two dozen Balmer lines. The series converges at 364.56 nm, just to the left of H_{40}. This star's spectrum also contains the first twelve Balmer lines (H_α through H_{12}), but they are located beyond the right edge of this photograph. (Mount Wilson and Las Campanas Observatories)*

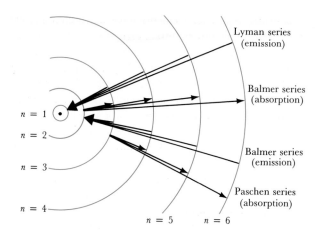

Figure 5-17 *The Bohr model of the hydrogen atom* *An electron circles the nucleus in orbits n = 1, 2, 3, and so forth. A photon is absorbed by the atom as the electron jumps from an inner orbit to an outer orbit. A photon is emitted by the atom as the electron falls from an outer to an inner orbit. These absorptions and emissions of photons, which occur only at specific wavelengths, produce characteristic patterns of lines in the hydrogen spectrum.*

that the wavelength (λ) of the photon emitted or absorbed as the electron jumps from one orbit to another is:

$$\frac{1}{\lambda} = \left(\frac{2\pi^2 Me^4}{ch^3}\right)\left(\frac{1}{m^2} - \frac{1}{n^2}\right)$$

where m is the number of the lower orbit, n is the number of the higher orbit, M is the mass of the electron, and e is the charge on the electron. As usual, h is Planck's constant and c is the speed of light.

Bohr's discovery elucidated the meaning of Balmer's formula: All the Balmer lines are produced by electron transitions between the second Bohr orbit ($m = 2$) and higher orbits ($n = 3, 4, 5$, etc.).

Bohr's formula not only gives the wavelengths of the Balmer series (when $m = 2$), but also correctly predicts the wavelengths of other series of spectral lines that occur at nonvisible wavelengths. For example, $m = 1$ gives the **Lyman series**, which is entirely in the ultraviolet range. Because $m = 1$, all the spectral lines in this series involve electron transitions between the lowest Bohr orbit and all higher orbits ($n = 2, 3, 4$, etc.). This pattern of spectral lines begins with L_α ("Lyman alpha") at 122 nm and converges on L_∞ at 91 nm. At infrared wavelengths is the **Paschen series**, for which $m = 3$. It begins with P_α ("Paschen alpha") at 1875 nm and converges on P_∞ at 821 nm. Additional series exist at still longer wavelengths.

Bohr's ideas give us an explanation of Kirchhoff's laws. Each spectral line corresponds to one particular transition

between the orbits of the atoms of a particular element. An absorption line occurs when an electron jumps from an inner orbit to an outer orbit, extracting the required photon from an outside source of energy, such as the continuous spectrum of a hot, glowing object. An emission line is produced when the electron falls back down to a lower orbit and gives up a photon. However, the wavelengths of the absorption and emission lines of a particular element are the same, since they involve the same electron orbits.

Physicists today use many features of the Bohr model of the atom, although they no longer picture electrons as moving in specific orbits about the nucleus. Instead, electrons are now said to occupy certain **energy levels** in the atom. An extremely useful way of displaying the structure of an atom is with an **energy-level diagram**, such as that shown in Figure 5-18 for hydrogen. The lowest energy level, called the ground state, corresponds to the $n = 1$ Bohr orbit. An electron can jump from the ground state up to the $n = 2$ level only if the atom absorbs a Lyman-alpha photon with a wavelength of 122 nm. The energy of a photon is determined by the relationship $E = h\nu = hc/\lambda$ and is usually expressed in electron volts (eV). As explained in Box 5-3, an electron volt is a tiny amount of energy ($1\ eV = 1.6 \times 10^{-19}$ J). The Lyman-alpha photon has an energy of 10.19 eV, so the energy level $n = 2$ is shown in the diagram as having an energy 10.19 eV above the energy of the ground state (conventionally assigned a value of 0 eV). Similarly, the $n = 3$ level is 12.07 eV above the ground state, and so forth up to the $n = \infty$ level at 13.6 eV. If the atom absorbs a photon of any energy greater than 13.6 eV, an electron from the ground state will be knocked completely out of the atom. This process, in which an electron is removed from the atom, is called **ionization**.

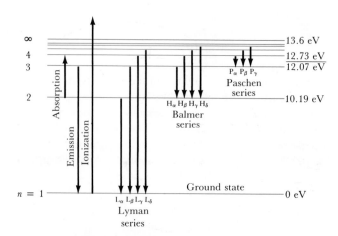

Figure 5-18 *The energy-level diagram of hydrogen* *The structure of the hydrogen atom can be conveniently displayed in a diagram showing the energy levels above the ground state. A number of electron transitions are shown, including those that produce some of the most prominent lines in the hydrogen spectrum.*

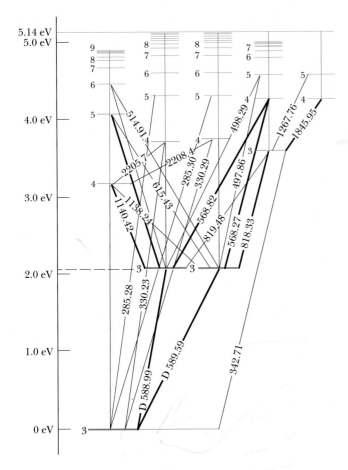

The atoms of heavier elements have more complex energy-level diagrams. For example, Figure 5-19 shows such a diagram for sodium. Astronomers find it useful to refer to energy-level diagrams to understand specific details of the lines they observe in the spectra of stars and nebulae.

With the work of people like Planck, Einstein, Rutherford, and Bohr, the interchange between astronomy and physics came full circle. Modern physics was born when Newton set out to understand the motions of the planets. Two and a half centuries later, physicists in their laboratories had probed the properties of light and the structures of atoms. The fruits of their labors had immediate and direct applications for astronomy.

5-8 The wavelength of a spectral line is affected by the relative motion between the source and the observer

Christian Doppler, a mathematics professor in Prague, pointed out in 1842 that wavelength is affected by motion. As shown in Figure 5-20, light waves from an approaching light source are compressed. The circles represent the peaks of waves emitted from various positions as the source moves along. As each successive wave is emitted from a position slightly closer to you, you see a shorter wavelength than you would if the source were stationary. All the lines in the spectrum of an approaching source are shifted toward the short-wavelength (blue) end of the spectrum. This phenomenon is called a **blueshift**.

Conversely, light waves from a receding source are stretched out, so that you see a longer wavelength than you would if the source were stationary. All the lines in the spec-

Figure 5-19 The energy-level diagram of sodium Complex atoms have complicated energy-level diagrams. The energy-level diagram of sodium is displayed here, along with the wavelengths of photons absorbed or emitted in some of the major electron transitions. At visible wavelengths, the sodium spectrum is dominated by two strong lines, called the sodium D lines, at 588.99 and 589.59 nm. These two lines are strong because they correspond to two transitions that are the primary avenue through which electrons cascade from high orbits down to the ground state.

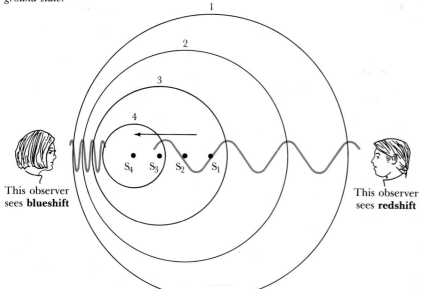

This observer sees **blueshift**

This observer sees **redshift**

Figure 5-20 The Doppler effect This diagram shows why wavelength is affected by motion between the light source and an observer. A source of light is moving toward the left. The four circles (numbered 1 through 4) indicate the location of the light waves that were emitted by the moving source when it was at points S_1 through S_4, respectively. Note that the waves are crowded together in front of the moving source but spread out behind it. Consequently, wavelengths appear shortened (blueshifted) if the source is moving toward the observer and lengthened (redshifted) if the source is moving away from the observer. Motion perpendicular to an observer's line of sight does not affect wavelength.

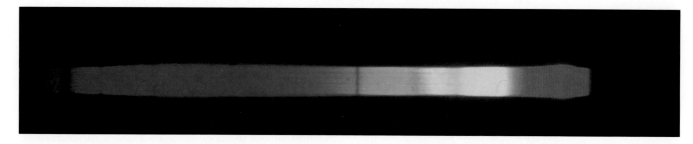

Figure 5-21 **The spectrum of Vega** *Several Balmer lines are seen in this photograph of the spectrum of Vega, the brightest star in the constellation of Lyra (the harp). The fact that all the spectral lines are shifted slightly toward the blue side of the spectrum indicates that Vega is approaching us. (NOAO)*

trum of a receding source are shifted toward the longer-wavelength (red) end of the spectrum, producing a **redshift**. In general, the effect of relative motion on wavelength is called the **Doppler effect**.

Suppose that λ_0 is the wavelength of a particular spectral line from a source that is not moving. It is the wavelength that you might look up in a reference book or determine in a laboratory experiment for this spectral line. If the source is moving, this particular spectral line is shifted to a different wavelength (λ). The size of the wavelength shift is usually written as $\Delta\lambda$, where $\Delta\lambda = \lambda - \lambda_0$. Thus $\Delta\lambda$ is the difference between the wavelength listed in reference books and the wavelength that you actually observe in the spectrum of a star or galaxy.

Christian Doppler proved that the wavelength shift $\Delta\lambda$ is governed by the following simple equation:

$$\frac{\Delta\lambda}{\lambda_0} = \frac{v}{c}$$

where v is the speed of the source measured along the line of sight between the source and the observer. As usual, c is the speed of light (3×10^5 km/s). (This equation is valid only for cases where v is small in comparison to c.)

For example, the spectral lines of hydrogen appear in the spectrum of the bright star Vega (also called α Lyrae) as shown in Figure 5-21. The prominent hydrogen line H_α has a normal wavelength of 656.285 nm, but in Vega's spectrum this line is located at 656.255 nm. The wavelength shift is calculated thus:

$$\Delta\lambda = \lambda - \lambda_0 = 656.255 \text{ nm} - 656.285 \text{ nm} = -0.030 \text{ nm}$$

and the star is approaching with a speed of

$$v = c\left(\frac{\Delta\lambda}{\lambda_0}\right) = 3 \times 10^5 \text{ km/s}\left(\frac{-0.030 \text{ nm}}{656.285 \text{ nm}}\right) = -13.7 \text{ km/s}$$

(The minus sign indicates motion toward the observer.)

Speed determined in this fashion is called **radial velocity**, because v is the component of the star's motion parallel to our line of sight, or along the "radius" drawn from Earth to the star. Of course, a sizable fraction of a star's motion may be perpendicular to our line of sight. The speed of this transverse movement, called **tangential velocity**, does not affect wavelengths.

Key words

absolute zero	blueshift	electromagnetism	gamma rays
absorption line spectrum	chemical element	electron	infrared radiation
atom	continuous spectrum	*electron volt	ionization
Balmer line	*degrees Celsius	emission line spectrum	joule
Balmer series	*degrees Fahrenheit	energy flux	kelvin
blackbody	Doppler effect	energy level	Kirchhoff's laws
blackbody curves	electromagnetic radiation	energy-level diagram	*luminosity
blackbody radiation	electromagnetic spectrum	frequency	Lyman series

microwaves	Planck's law	spectral analysis	visible light
neutron	proton	spectral line	wavelength
nucleus	radial velocity	spectrum (*plural* spectra)	Wien's law
Paschen series	radio waves	Stefan–Boltzmann law	X rays
photoelectric effect	redshift	tangential velocity	
photon	*solar constant	ultraviolet radiation	

Key ideas

- Electromagnetic radiation (including visible light) has wavelike properties and travels through empty space at the constant speed $c = 3 \times 10^8$ m/s.

- A blackbody is a hypothetical object that is a perfect absorber of electromagnetic radiation at all wavelengths. Stars closely approximate the behavior of blackbodies.

- The Stefan–Boltzmann law states that a blackbody radiates electromagnetic waves with a total energy flux (F) directly proportional to the fourth power of the temperature (T) of the object: $F = \sigma T^4$.

- Wien's law states that the dominant wavelength at which a blackbody emits electromagnetic radiation is inversely proportional to the temperature of the object: λ_{max}(in meters) $= 2.9 \times 10^{-3}/T$.

- The intensities of radiation emitted at various wavelengths by a blackbody at a given temperature are shown by a blackbody curve.

   An explanation of blackbody curves led to the discovery that light has particlelike properties; the particles of light are called photons.

- Planck's law relates the energy (E) of a photon to its frequency (ν) or wavelength (λ): $E = h\nu = hc/\lambda$, where h is Planck's constant.

- Kirchhoff's three laws of spectral analysis state that (1) a hot, glowing object emits a continuous spectrum containing all wavelengths; (2) when the continuous spectrum is viewed through a cool gas, dark spectral lines (absorption lines) characteristic of the elements present in the gas are observed; (3) when heated, or otherwise supplied with energy, the gas emits light that consists of a pattern of bright spectral lines (emission lines) corresponding exactly to the pattern of absorption lines.

- An atom consists of a small, dense nucleus (composed of protons and neutrons), orbited by electrons that occupy only certain orbits or energy levels.

   When an electron jumps from one energy level to another, it emits or absorbs a photon of appropriate energy (and hence of a specific wavelength).

The spectral lines of a particular element correspond to the various electron transitions between allowed energy levels in atoms of that element.

The spectral lines of hydrogen's Balmer series are produced by jumps between the second energy level of the atom and higher levels; the Balmer series is located in the visible portion of the electromagnetic spectrum. Other series of spectral lines involve transitions between other energy levels and are observed at nonvisible wavelengths.

The process by which an atom absorbs an energetic photon and completely loses an electron is called ionization.

- The spectral lines of an approaching light source are shifted toward short wavelengths (a blueshift); the spectral lines of a receding light source are shifted toward long wavelengths (a redshift).

   The Doppler effect explains that the size of a wavelength shift is proportional to the radial velocity between the light source and the observer.

Review questions

1 Describe an experiment in which light behaves like a wave.

2 Describe an experiment in which light behaves like a particle.

*3 (*Basic*) Approximately how many times around the world could a beam of light travel in one second?

*4 (*Basic*) A light source emits infrared radiation at a wavelength of 825 nm. What is the frequency of this radiation?

*5 Announcers at a certain radio station say that they are at "103 FM on your dial," meaning that they transmit at a frequency of 103 MHz. What is the wavelength of the radio waves from this station?

*6   The bright star Regulus in the constellation of Leo, the lion, has a surface temperature of 12,200 K. What is the dominant wavelength (λ_{max}) of its energy emission?

*7   The bright star Procyon in the constellation of Canis Minor, the small dog, emits the greatest intensity of radiation at a wavelength (λ_{max}) of 445 nm. What is the surface temperature of the star?

8   Using Wien's law and the Stefan–Boltzmann law, explain the color changes that are observed as the temperature of a hot, glowing object increases.

*9   What wavelength of electromagnetic radiation is emitted with greatest intensity by this book? To what region of the electromagnetic spectrum does this wavelength correspond?

10   What is a blackbody? In what way is a blackbody black? If a blackbody is black, how can it emit light?

11   Explain why astronomers are interested in blackbody radiation.

12   Describe the experimental evidence that supports our current views of the structure of an atom.

13   Describe the spectrum of hydrogen at visible wavelengths and show how Bohr's model of the atom explains the Balmer lines.

14   Why do different elements have different patterns of lines in their spectra?

15   What is the Doppler effect, and why is it important to astronomers?

*16   The wavelength of H_β in the spectrum of the star Megrez (δ Ursa Majoris) in the Big Dipper is 486.112 nm. Laboratory measurements demonstrate that the normal wavelength of this spectral line is 486.133 nm. Is the star coming toward us or moving away from us? At what speed?

17   Why do you suppose that ultraviolet light can cause skin cancer but ordinary visible light does not?

*18   (Basic)   What is the temperature of the Sun's surface in degrees Fahrenheit?

*19   Your body temperature is 98.6°F. What kind of radiation do you emit? At what wavelength do you emit the most radiation?

*20   The bright star Sirius in the constellation of Canis Major, the large dog, has a radius of 1.67 $R_\odot$ and a luminosity 25 $L_\odot$. What is the star's surface temperature?

*21   During the 1970s and 1980s, balloon-borne instruments have detected 511 KeV photons coming from the direction of the center of our galaxy. What is the wavelength of these photons? To what part of the electromagnetic spectrum do these photons belong?

22   Calculate the wavelength of H_δ. Draw a schematic diagram of the hydrogen atom and indicate the electron transition that gives rise to this spectral line.

*23   The bright star Sirius has a surface temperature of about 10,000 K. Compared to the Sun, how much more energy is emitted each second from each square meter of Sirius' surface?

*24   Imagine a star the same size as the Sun, but with a surface temperature twice that of the Sun. At what wavelength would that star emit most of its radiation? How many times brighter than the Sun would that star be?

*25   (Challenging)   Imagine a star whose diameter is 10 times larger than the Sun's and whose surface temperature is 2900 K. Suppose that both the Sun and this star were located at the same distance from you. Which would be brighter? By what factor?

*26   In the spectrum of the bright star Rigel, H_α has a wavelength of 656.331 nm. Is the star coming toward us or moving away from us? How fast?

*27   (Challenging)   Imagine driving down a street toward a traffic light. How fast would you have to go so that the red light would appear green?

Advanced questions

Tips and tools . . .

Formulas for converting between temperature scales are found in Box 5-1. Relationships between a star's radius, its luminosity, and its surface temperature are discussed in Box 5-2. A simple example using Planck's law to calculate the energy of a photon is given in Box 5-3.

Discussion questions

28   Compare chemical identification based on spectral line patterns with the identification of people by their fingerprints. In what ways is this a good or poor analogy?

29   Suppose you look up at the night sky and observe some of the brightest stars with your naked eye. Is there any way of telling which stars are hot and which are cool? Explain.

30   Why do you think the human eye is most sensitive over the same wavelength range at which the Sun emits the greatest intensity of radiation? Suppose creatures were to evolve

on a planet orbiting a star somewhat hotter than the Sun. To what wavelengths would their "eyes" most likely be sensitive?

Observing projects

31 As soon as weather conditions permit, observe a rainbow. Confirm that the colors are in the same order as shown in Figure 5-3.

32 Turn on an electric stove or toaster oven and carefully observe the heating elements as they warm up. Relate your observations to Wien's law and the Stefan–Boltzmann law.

33 Obtain a glass prism and look through it at various light sources such as an ordinary incandescent light, a neon sign, and a mercury vapor street lamp. **DO NOT LOOK AT THE SUN! Looking directly at the Sun can cause blindness.** Do you have any trouble seeing spectra? What do you have to do to see a spectrum? Describe the differences in the spectra of the various light sources you observed.

For further reading

Bova, B. *The Beauty of Light*. John Wiley, 1988 • A fascinating introduction to all aspects of light. Includes sections on the eye, color, and the technology of producing and detecting light.

Cline, B. *Men Who Made a New Physics*. Signet, 1965 • This book describes the history of the discoveries that led to our modern understanding of light and atoms.

Feynman, R. *QED*. Princeton University Press, 1985 • Using many carefully thought-out analogies, this slim volume discusses how photons and particles interact. The author won a Nobel prize for his work in this field.

Gingerich, O. "Unlocking the Chemical Secrets of the Cosmos." *Sky & Telescope*, July 1981 • This brief article offers some fascinating insights on the work of Kirchhoff and Bunsen.

Gribbin, J. *In Search of Schroedinger's Cat*. Bantam, 1984 • This excellent introduction to quantum mechanics provides both background and historical material along with a fine review of current thinking on the subject.

van Heel, A., and Velzel, C. *What Is Light?* McGraw-Hill, 1968 • This book surveys our modern understanding of the phenomenon of light.

Kuhn, T. *Black-Body Theory and the Quantum Discontinuity*. University of Chicago Press, 1987 • A somewhat technical history of the path to quantum mechanics with emphasis on the work of Max Planck.

Sobel, M. *Light*. University of Chicago Press, 1987 • An excellent nontechnical introduction to all aspects of light.

Snow, C. *The Physicists*. Macmillan, 1981 • This book describes many of the people who contributed to our modern understanding of light and matter.

Weinberg, S. *Subatomic Particles*. Scientific American Books, 1983 • This book beautifully describes the various developments, both theoretical and experimental, leading up to our modern understanding of light and matter. The author won a Nobel prize for his work in this field.

The Summit of Mauna Kea on Hawaii Astronomers prefer to build observatories on mountaintops far from city lights, where the air is dry, stable, and cloudfree. The summit of Mauna Kea, shown here in winter, offers the world's best observing site. The dome in the foreground houses the 10-m Keck reflector, the largest telescope in the world. On the ridge in the background are the 3.6-m Canada-France-Hawaii telescope, the 2.2-m University of Hawaii telescope, and the U.K. Royal Observatory 3.8-m infrared telescope. (California Association for Research in Astronomy)

CHAPTER

6

Optics and Telescopes

The telescope is the astronomer's most important tool. Telescopes are essential because they give us bigger, brighter, sharper images of distant astronomical objects than our eyes do. Refracting telescopes, which use large lenses to collect incoming starlight, were popular in the nineteenth century, but modern astronomers prefer reflecting telescopes, which gather light with large concave mirrors. Astronomers attach a variety of equipment to telescopes with which to record and analyze incoming starlight. Light-sensitive silicon chips, called CCDs, mounted at a telescope's focus are the latest device used to their study distant stars and galaxies. The Earth's atmosphere is transparent to both visible light and radio waves. Ground-based radio telescopes can therefore be used to view the universe at wavelengths much longer than those of visible light. At most other wavelengths, the Earth's atmosphere is opaque, and so observations must be performed from space. Sophisticated orbiting observatories are now giving us unprecedented views of the cosmos all across the electromagnetic spectrum.

The telescope is the single most important tool of astronomy. Using a telescope, we can see extremely faint objects in space far more clearly than we can with the naked eye. Telescopes have played a major role in revealing the universe since Galileo first used one and saw craters on the Moon four centuries ago.

Traditionally, telescopes have been used to detect visible light. Light from a distant object is brought, either by lenses or by mirrors, to a focus where the resulting image is viewed or photographed. Recently, however, astronomers have built telescopes that detect nonvisible forms of light such as X rays and radio waves. Although all the different types of electromagnetic radiation share many basic properties—for example, they all travel at the speed of light—they interact quite differently with matter. For instance, your body is transparent to X rays but not to visible light; your eyes respond to visible light but not to gamma rays; your radio detects radio waves but not ultraviolet light. Consequently, astronomers use fundamentally different kinds of telescopes for various wavelength ranges. For example, a radio telescope that detects radio waves from space differs from either an X-ray telescope or an ordinary optical telescope. Because optical telescopes are the most common and familiar astronomical tool, we shall discuss them in detail before turning to the more exotic instruments capable of revealing the nonvisible sky.

6-1 A refracting telescope uses a lens to concentrate incoming starlight at a focus

Although light travels at 3×10^{10} m/s in a vacuum, it travels at a slower speed through a dense substance like glass. The abrupt slowing of light as it enters a piece of glass is analogous to a person's walking from a boardwalk onto a sandy beach: Her pace suddenly slows as she steps from the smooth pavement into the sand. Similarly, upon its exiting a piece of glass, light resumes its original speed, just as a person stepping back onto a boardwalk easily resumes her original pace.

Furthermore, as it passes from one transparent medium into another, a light ray is bent at an angle oblique to the surface between the media. This phenomenon, which is called **refraction,** is caused by the change in the speed of light. Imagine driving a car from a smooth pavement onto a sandy beach. If the car approaches the beach at an angle, one of the front wheels will be slowed by the sand before the other is, causing the car to veer from its original direction.

To describe the refraction of a light ray entering a piece of glass, imagine drawing a perpendicular to the surface of the glass at the point where the light strikes the glass, as shown in Figure 6-1. As the light ray goes from the less dense medium (such as air or a vacuum) into the more dense medium

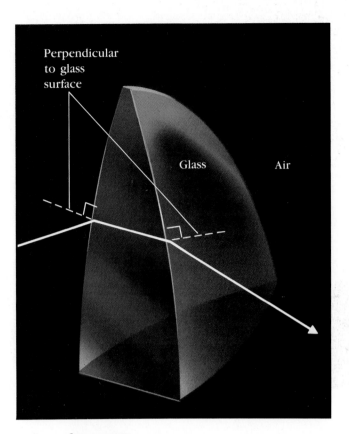

Figure 6-1 Refraction *A light ray entering a piece of glass is bent toward the perpendicular. As the light ray leaves the piece of glass, it is bent away from the perpendicular.*

(like glass), it will be bent toward the perpendicular. This happens in the same way as a car driving obliquely onto sand veers toward the direction perpendicular to the pavement–beach boundary. Upon emerging from the far side of a piece of glass, the light ray resumes its original high speed and is bent away from the perpendicular. The exact amount of refraction depends on the speed of light in the glass, which in turn depends on the chemical composition of the glass. Different kinds of glass produce slightly different amounts of refraction.

Because of the refracting property of glass, a convex lens (one that is fatter in the middle than at the edges) causes incoming light rays to converge at a point called the **focus,** as shown in Figure 6-2. If the light source is extremely far away, the incoming light rays will be parallel and will come to a focus at a specific distance from the lens known as the **focal length** of the lens.

The stars and planets are so far away that light rays from them are essentially parallel. Consequently, a lens always focuses light from an astronomical object as shown in Figure 6-2. An image of the astronomical object is formed at the focus. A second lens can then be used to magnify and examine this image. Such an arrangement of two lenses is called a **refracting telescope, or refractor** (see Figure 6-3). The large-diameter, long-focal-length lens at the front of the telescope is called the **objective lens**, while the smaller, shorter-focal-length lens at the rear of the telescope is called the **eyepiece lens**. Galileo used a small refracting telescope for astronomical observations very soon after the device was invented in Holland for viewing distant objects on Earth.

The **magnification, or magnifying power**, of a refracting telescope is equal to the focal length of the objective lens divided by the focal length of the eyepiece lens. For example, if the objective of a telescope has a focal length of 100 cm and the eyepiece has a focal length of $\frac{1}{2}$ cm, the magnifying power of the telescope is 200 (usually written as 200×).

If you were to build a telescope using only the instructions given so far, you would probably be disappointed with the results. You would see stars surrounded by fuzzy, rainbow-colored halos. This optical defect, called **chromatic aberration**, exists because a lens bends different colors of light through different angles, just as a prism does (recall Figure 5-3).

By adding small amounts of various chemicals to a vat of molten glass, an optician can manufacture different kinds of glass. The speed of light varies slightly from one kind of glass to another, a fact that opticians use to correct for chromatic aberration. Specifically, a thin lens can be mounted just behind the main objective lens of a telescope, as shown in Figure 6-4. By carefully choosing two different kinds of glass

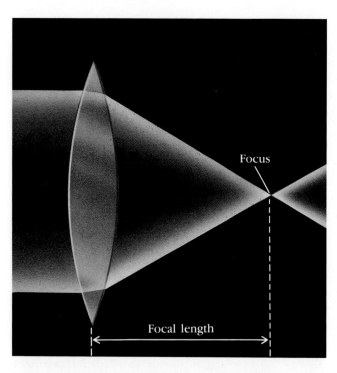

Figure 6-2 A convex lens *A convex lens causes parallel light rays to converge to a focus. The distance from the lens to the focus is the focal length of the lens.*

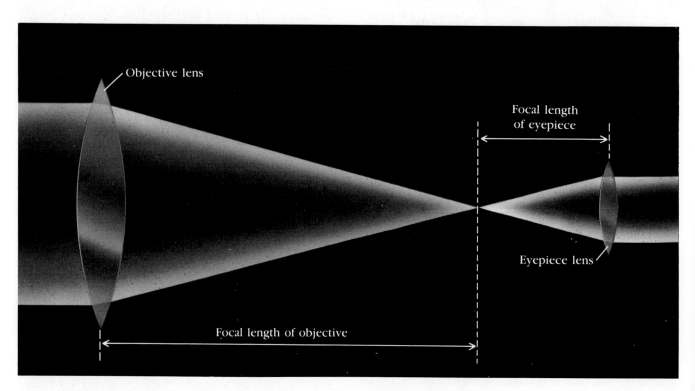

Figure 6-3 A refracting telescope *A refracting telescope consists of a large, long-focal-length objective lens and a small, short-focal-length eyepiece lens. The eyepiece lens magnifies the image formed at the focus of the objective lens.*

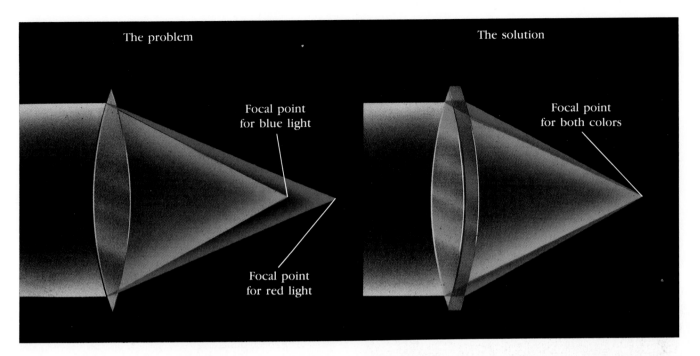

The problem

Focal point
for blue light

Focal point
for red light

The solution

Focal point
for both colors

Figure 6-4 *Chromatic aberration* *A single lens suffers from a defect called chromatic aberration, in which different colors of light have* *slightly different focal lengths. This problem is corrected by adding a second lens made from a different kind of glass than the first lens.*

for these two lenses, the optician can ensure that different colors of light will come to a focus at the same point.

Chromatic aberration is only the most severe of a host of optical problems that must be solved in designing a high-quality refracting telescope. During the nineteenth century, master opticians devoted their lives to overcoming these problems, and several magnificent refractors were constructed in the late 1800s. The largest refracting telescope, completed in 1897, is located at the Yerkes Observatory, not far from Chicago (see Figure 6-5). The objective lens, which has a diameter of 102 cm (40 in.), was built under the expert guidance of Alvan Clark. This was not Clark's first attempt at a large telescope. In 1888, he had completed a similar instrument with an objective lens 91 cm (36 in.) in diameter for Lick Observatory near San Jose, California. Other major refractors are listed in Box 6-1. They all have extremely long focal lengths. For example, the Yerkes refractor has a focal length of 19.35 m (63½ ft).

Few major refracting telescopes have been constructed in the twentieth century. There are many reasons for the modern astronomer's lack of interest in this type of telescope. First, because faint light must pass readily through the objective lens, the glass from which the lens is made must be totally free of defects, such as the bubbles that frequently form when molten glass is poured into a mold. Consequently, the glass for the objective lens is extremely expensive. Second, glass is opaque to certain kinds of light. Even visible light is dimmed substantially as it passes through the thick slab of glass at the front of a refractor, and ultraviolet radiation is largely absorbed by the glass lens. Third, it is impossible to

Figure 6-5 *A large refracting telescope* *This giant refractor was built in the late 1800s and is housed at Yerkes Observatory near Chicago. The objective lens is 102 cm (40 in.) in diameter, and the telescope tube is 19½ m (64 ft) long. (Yerkes Observatory)*

Box 6-1   Major refracting telescopes

The fourteen refractors at the observatories listed in the table are the only ones in the world with objective lenses larger than 65 cm (26 in.) in diameter.

Observatory	Location	Year completed	Objective diameter (cm)	Focal length (cm)
Yerkes Observatory	Williams Bay, Wisconsin	1897	102	1936
Lick Observatory	Mount Hamilton, California	1888	90	1763
Observatoire de Paris	Meudon, France	1889	83	1616
Zentralinstitut für Astrophysik	Potsdam, East Germany	1899	80	1200
Allegheny Observatory	Pittsburgh, Pennsylvania	1914	76	1412
Observatoire de Nice	Mont Gros, France	1886	74	1790
Old Royal Observatory	Greenwich, England	1894	71	848
Archenhold-Sternwarte	Berlin, East Germany	1896	68	2100
Institut für Astronomie	Vienna, Austria	1880	67	1050
Republic Observatory	Johannesburg, South Africa	1925	67	1092
Leander McCormick Observatory	Charlottesville, Virginia	1883	67	991
United States Naval Observatory	Washington, D.C.	1873	66	987
Royal Greenwich Observatory	Herstmonceux, England	1899	66	686
Mount Stromlo Observatory	Canberra, Australia	1956*	66	1080

*First used at Johannesburg, South Africa, in 1925.

produce a large lens that is free of chromatic aberration. Fourth, it is difficult to support these heavy lenses without blocking the path of light into the telescope. All these problems can be avoided by using mirrors instead of lenses.

6-2   A reflecting telescope uses a mirror to concentrate incoming starlight at a focus

To understand **reflection**, imagine drawing a perpendicular to a mirror's surface at the point where a light ray strikes the mirror, as shown in Figure 6-6. The angle between an arriving (incident) light ray and the perpendicular is always equal to the angle between the reflected ray and the perpendicular. Knowing this, Isaac Newton realized that a concave mirror would cause parallel light rays to converge to a focus, as

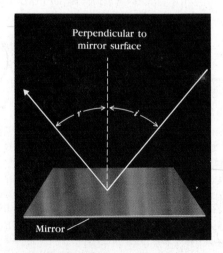

Figure 6-6   *Reflection*   *The angle at which a beam of light approaches a mirror (called the angle of incidence i) is always equal to the angle at which the beam is reflected from the mirror (called the angle of reflection r). Reflection is accurately described by the equation* i = r.

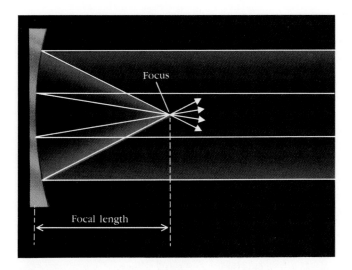

Figure 6-7 *A concave mirror A concave mirror causes parallel light rays to converge to a focus. The distance between the mirror and the focus is the focal length of the mirror.*

shown in Figure 6-7. The distance between the reflecting surface and the focus is the focal length of the mirror.

An image of a distant object is formed at the focus of a concave mirror. In order to view the image, Newton simply placed a small, flat mirror at a 45° angle in front of the focal point, as sketched in Figure 6-8a. This secondary mirror de-

flected the light rays to one side of the **reflecting telescope**, or **reflector**, where the astronomer could place an eyepiece lens to magnify the image. A telescope having this optical design is appropriately called a **Newtonian reflector**. The magnifying power of such a reflecting telescope is calculated in the same way as for a refractor: The focal length of the primary mirror is divided by the focal length of the eyepiece.

Useful modifications of Newton's original design have since been made. The primary mirrors of many major reflectors are so large that the astronomer can actually sit at the undeflected focal point, directly in front of the primary mirror. The "observing cage" in which the astronomer rides blocks only a small fraction of the incoming starlight. This arrangement is called a **prime focus** (see Figure 6-8b).

Another popular optical design, called a **Cassegrain focus,** has the advantage of placing the focal point at a convenient and accessible location. A hole is drilled directly through the center of the primary mirror. A convex secondary mirror placed in front of the original focal point is then used to reflect the light rays back through the hole (see Figure 6-8c). Alternatively, a series of mirrors can be used to channel the light rays away from the telescope to a remote focal point. Heavy optical equipment that could not be mounted directly on the telescope is located at the resulting **coudé focus**, named after a French word meaning bent like an elbow (see Figure 6-8d).

To make a reflector, an optician grinds and polishes a large slab of glass into the appropriate concave shape. The

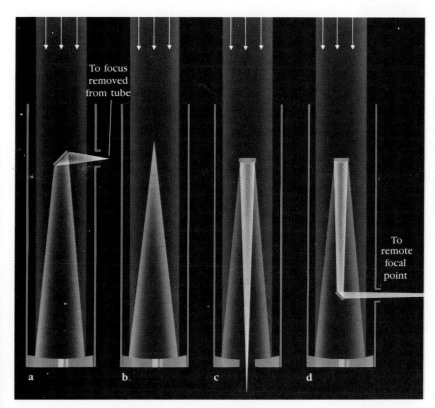

Figure 6-8 *Reflecting telescopes Four of the most popular optical designs for reflecting telescopes: (a) Newtonian focus, (b) prime focus, (c) Cassegrain focus, and (d) coudé focus.*

glass is then coated with silver, aluminum, or a similar highly reflective substance. Defects inside the glass of a reflector, such as bubbles or flecks of dirt, do not detract from such a telescope's effectiveness, because light reflects off the surface of the glass rather than passing through it, as it would with the objective lens of a refracting telescope.

Furthermore, reflection is not affected by the wavelengths of the incoming light. The light of all wavelengths is reflected to the same focus, without losing any wavelengths through absorption. (Some problems of absorption and chromatic aberration may arise with the smaller lens used to magnify the image, but the major difficulties caused by the nature of the large objective lens in the refractor do not exist in the reflector.) Finally, the mirror can be fully supported by braces on its back, so that a large, heavy mirror can be mounted without much danger of breakage or surface distortion.

There are fifteen reflectors around the world with primary mirrors measuring at least 3 m (9.8 ft) in diameter (see

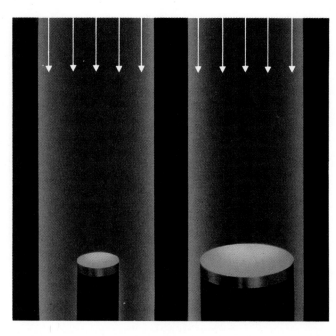

Figure 6-10 Light-gathering power *A large mirror intercepts more starlight than does a small one. Large mirrors therefore produce brighter images than do small mirrors.*

Figure 6-9 The 4-m telescope at Cerro Tololo *This telescope is located on a mountaintop near Santiago, Chile. Its twin is at the Kitt Peak Observatory in Arizona. Both telescopes have been in operation since the early 1970s. (Cerro Tololo Inter-American Observatory)*

Box 6-2). One of the largest, a 6-meter telescope, is located in the Caucasus Mountains in the U.S.S.R., and the famous 200-inch telescope is at the Palomar Observatory in southern California. In the early 1970s, a matching pair of telescopes was built in Arizona and Chile. These twins (see Figure 6-9) allow astronomers to observe the entire sky with essentially the same instrument.

Astronomers strongly prefer large telescopes. A large mirror intercepts and focuses more starlight than does a small mirror (see Figure 6-10). A large mirror therefore produces brighter images and detects fainter stars than does a small one. The **light-gathering power** of a telescope is directly related to the area of the telescope's primary mirror. The area of a mirror is proportional to the square of its diameter, so, for example, the 200-inch mirror at Palomar Observatory has four times the area of the 100-inch mirror at Mount Wilson Observatory. Therefore, the Palomar telescope has four times the light-gathering power of the Mount Wilson telescope.

A large telescope also increases the sharpness of the image and the extent to which fine details can be distinguished. This property is called **resolving power**. With low resolving power, star images are fuzzy and blurred together. With high resolving power, images are sharp and crisp.

ℚ   The resolving power of a telescope is described in terms of the smallest angle between two adjacent stars that can just barely be distinguished as separate objects. That angle is proportional to the wavelength being observed and inversely proportional to the diameter of the telescope. Large modern instruments, like the 4-m telescopes at Kitt Peak and Cerro

Box 6-2 Major reflecting telescopes

There are fifteen reflectors in the world with primary mirrors equal to or larger than 3 m (9.8 ft) in diameter. The observatories where these telescopes reside are listed in the table below. Two of the telescopes in this table have segmented primary mirrors. The Keck Telescope in Hawaii has 36 hexagonal mirrors whose total reflective area equals one 10-m mirror (see Figure 6-12). The Multiple Mirror Telescope at the Whipple Observatory in Arizona has six 1.8-m mirrors whose total reflective area equals that of one 4.5-m mirror (see Figure 6-11).

Observatory	Location	Year completed	Mirror diameter (m)
Keck Observatory	Mauna Kea, Hawaii	1991	10.0
Special Astrophysical Observatory	Zelenchukskaya, U.S.S.R.	1976	6.0
Palomar Observatory	Palomar Mountain, California	1948	5.1
Fred L. Whipple Observatory	Mount Hopkins, Arizona	1979	4.5
La Palma Observatory	Canary Islands	1987	4.2
Cerro Tololo Inter-American Observatory	Cerro Tololo, Chile	1974	4.0
Kitt Peak National Observatory	Kitt Peak, Arizona	1973	4.0
Anglo-Australian Observatory	Siding Spring, Australia	1975	3.9
Royal Observatory, Edinburgh*	Mauna Kea, Hawaii	1979	3.8
Canada–France–Hawaii Observatory	Mauna Kea, Hawaii	1979	3.6
European Southern Observatory	Cerro La Silla, Chile	1976	3.6
European Southern Observatory	Cerro La Silla, Chile	1989	3.6
Astrophysical Research Consortium	Apache Point, New Mexico	1990	3.5
Lick Observatory	Mount Hamilton, California	1959	3.0
Mauna Kea Observatory*	Mauna Kea, Hawaii	1979	3.0

*These two telescopes are used primarily for infrared observations.

Tololo, are calculated to have resolving powers of roughly 0.03 arc sec. In practice, however, this exceptionally high resolving power is never achieved because turbulence and impurities in the air cause star images to jiggle around and twinkle. Even through the largest telescopes, a star still looks like a tiny circular blob rather than a pinpoint of light.

The angular diameter of a star's image, broadened by turbulence, is called the seeing disk. This disk is a realistic measure of the best possible resolution that can be achieved. The size of the seeing disk varies from one observatory site to another and from one night to another. At Palomar and Kitt Peak, the seeing disk is roughly 1 arc sec. The best conditions in the world, where the seeing disk is $\frac{1}{4}$ arc sec, have been reported at the observatories on the top of Mauna Kea, a 14,000-ft volcano on the island of Hawaii. Only in the vacuum of outer space could the theoretical resolving power of a large telescope be achieved.

Significant engineering problems come with building large reflectors. Very large mirrors (more than about 4 m in diameter) are slabs of glass so heavy that the mirror actually sags under its own weight. The mirror's shape changes slightly as the telescope is turned toward different parts of the sky, thereby detracting from the sharpness of the focus and the quality of the resulting image. New techniques for building thin, lightweight mirrors should help alleviate this problem.

Another approach is to mount together several smaller mirrors aimed at the same focal point. For example, the Multiple Mirror Telescope atop Mount Hopkins in Arizona has six mirrors, each measuring 1.8 m (5.9 ft) in diameter, mounted together as shown in Figure 6-11. The total light-gathering power of this arrangement is equivalent to that of one 4.5-m mirror. Its design has proven so successful that astronomers around the world are now planning even larger multiple-mirror telescopes.

The first of these giant multiple-mirror instruments is the 10-meter Keck telescope currently under construction on the summit of Mauna Kea in Hawaii. Thirty-six hexagonal mirrors are mounted side by side to give a primary that is 10 m (400 inches) in diameter, as shown in Figure 6-12. Each individual mirror is 6 ft across, 3 in. thick, and weighs 1400 lb. Astronomers eagerly look forward to the completion of this telescope in 1991.

Just as chromatic aberration plagues refracting telescopes, a defect called **spherical aberration** must be minimized when building reflecting telescopes. At issue is the precise shape of a mirror's concave surface. A spherical surface is easy to grind and polish, but different parts of a spherical mirror have slightly different focal lengths (see Figure 6-13a), which results in a fuzzy image.

One way to eliminate spherical aberration is to polish the mirror's surface to a parabolic shape. A parabola reflects

Figure 6-12 The Keck Telescope This model shows the design of a giant 10-m telescope under construction on the summit of Mauna Kea in Hawaii. Thirty-six hexagonal mirrors, each measuring 1.8 m (5.9 ft) across, will together have the same effect as one mirror 10 m in diameter. (California Institute of Technology)

Figure 6-11 The Multiple Mirror Telescope This aerial photograph shows the six 1.8-m (5.9-ft) mirrors that together constitute the first multiple-mirror telescope. The total reflective area of the six mirrors is equal to that of one 4.5-m mirror. (Multiple Mirror Telescope)

parallel light rays to a common focus (see Figure 6-13b), so many reflecting telescopes have parabolic mirrors.

The only problem with this solution is that the astronomer no longer has a wide-angle view. Unlike spherical mirrors, parabolic mirrors suffer from a defect called **coma**, wherein star images far from the center of the field of view are elongated to look like tiny teardrops.

A telescope with a high-quality, wide-angle field of view uses a spherical mirror to minimize coma and a thin correcting lens at the front of the telescope to eliminate spherical aberration (see Figure 6-13c). The unique shape of this lens is specifically designed to ensure that all light rays have the same focal point.

This optical arrangement is the basic idea of the **Schmidt telescope**, named after its inventor, Bernhard Schmidt, who built the first prototype in the 1930s. Today there are more than a dozen large Schmidt telescopes at major observatories around the world. One of the largest (see Figure 6-14) is

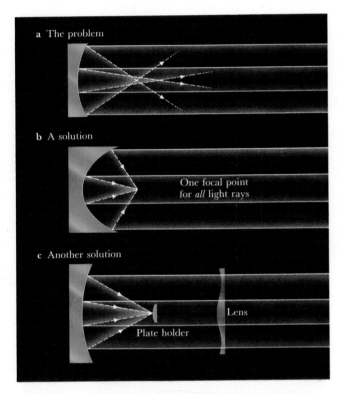

Figure 6-13 *Spherical aberration* (a) *Different parts of a spherically concave mirror reflect light to slightly different focal points. This difficulty can be corrected by either (b) using a parabolic mirror, or (c) using a "correcting lens" in front of the mirror.*

Figure 6-14 *The Schmidt Telescope at Palomar This is one of the largest Schmidt telescopes in the world. Its mirror has a diameter of 1.8 m (5.9 ft), and the correcting lens at the front of the telescope measures 1.2 m (3.9 ft) across. An astronomer is shown guiding the telescope as it takes a wide-angle photograph of the sky. (Palomar Observatory)*

located on Palomar Mountain, a short walk from the giant 5-m reflector. The 5-m telescope has a field of view only 2 arc min across, but the Schmidt telescope produces photographs covering a field 7° in diameter. Schmidt telescopes are designed to work as cameras: the astronomer using the telescope does not see the view until the photograph is developed.

During the early 1950s, a team of astronomers spent several years photographing the sky with the Palomar Schmidt telescope. This work culminated in the famous National Geographic Society–Palomar Observatory Sky Survey. The entire northern hemisphere and the southern hemisphere down to a declination of −33° were covered in 879 pairs of photographs. Each 6° × 6° segment of the sky was photographed on both blue-sensitive and red-sensitive photographic plates. The blue-sensitive plates recorded primarily blue light, whereas the red-sensitive plates responded mostly to red light. The views of celestial objects in these two wavelength ranges were often strikingly different from each other (see Figure 6-15).

The Sky Survey was repeated in the 1980s so that astronomers could search for changes in the sky over the preceding 30 years. In addition, Schmidt telescopes in Chile and Australia have extended this wide-angle coverage to those portions of the southern sky not accessible from Palomar. These magnificent photographs constitute a permanent record of the sky. Many of them reveal large-scale structures that had previously been overlooked by astronomers using bigger telescopes with small fields of view.

6-3 An electronic device is often used to record the image at a telescope's focus

The invention of photography during the nineteenth century was a boon for astronomy. By taking long exposures with a camera mounted at the focus of a telescope, an astronomer could record features too faint to be seen by simply looking through the telescope. Even today, astronomical photography performs an important role by revealing details in galaxies, star clusters, and nebulae.

Astronomers have long realized, however, that a photographic plate is not a very efficient light detector. Only 1 out

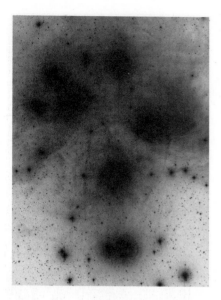
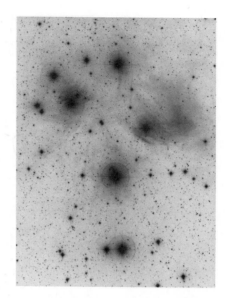

Figure 6-15 The Pleiades on the Sky Survey *Each of these photographs shows a cluster of stars called the Pleiades that is visible to the naked eye in the constellation of Taurus. The blue-sensitive plate is on the left, the red-sensitive plate in the center. Both pictures are printed exactly as they appear on the original plates, with black stars and white sky. Notice that very different features and details are seen in the two contrasting views. For comparison, a full-color photograph is included on the right. The area in each of these views is approximately $1° \times 1\frac{1}{2}°$. (Palomar Observatory; Anglo-Australian Observatory)*

of every 20 photons striking a photographic plate succeeds in triggering the requisite chemical reaction in the photographic emulsion to produce an image. Thus, roughly 95 percent of the light falling onto a photographic plate is wasted.

The most sensitive and efficient light detector currently available to astronomers is the newly invented **charge-coupled device** (CCD). A CCD is a thin piece of silicon (see Figure 6-16), not unlike the silicon chips used in pocket calculators and programmable appliances. The silicon wafer of a CCD is divided into an array of small light-sensitive squares called picture elements or, more commonly, **pixels**. Some of the latest CCDs have 640,000 pixels arranged in 800 columns by 800 rows in an area roughly the size of a postage stamp. When an image from a telescope is focused on the CCD, an electric charge builds up in each pixel in proportion to the number of photons falling on that pixel. When the exposure is finished, the amount of the charge on each pixel is read into a computer. From the computer, the image can be transferred onto ordinary photographic film or to a television monitor. Over certain wavelength ranges, more than 75 percent of the photons falling on a CCD can be recorded.

Figure 6-17 shows one photograph and two CCD images of the same region of the sky, both taken with the same telescope. Notice that many details visible in the CCD images are totally absent in the ordinary photograph. Because of their extraordinary sensitivity and their ability to be used in conjunction with computers, CCDs are playing an increasingly important role in astronomy.

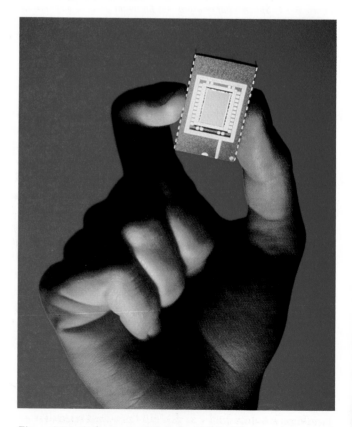

Figure 6-16 A charge-coupled device (CCD) *This tiny silicon rectangle contains 163,840 light-sensitive electric circuits that store images. At the end of each exposure, additional circuits in the silicon chip control the transfer and readout of the data to a waiting computer. (Smithsonian Astrophysical Observatory)*

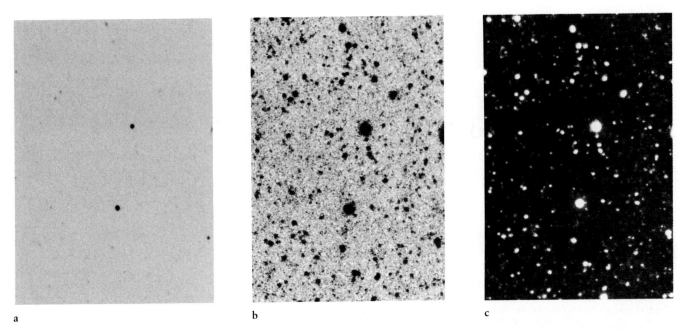

a   b   c

Figure 6-17 Ordinary photography versus a CCD image *Three views of the same part of the sky, each taken with the 4-m Cerro Tololo telescope, demonstrate the superiority of CCDs to ordinary photographic plates. (a) A negative print (black stars and white sky) of a 45-minute exposure on a photographic plate. (b) The sum of fifteen* *500-second CCD images. Notice that many faint stars and galaxies virtually invisible in the ordinary photograph can be clearly seen in this CCD image. (c) This color view was produced by combining a series of CCD images taken through colored filters. The total exposure time was 6 hours. (Courtesy of P. Seitzer, NOAO)*

6-4 Spectrographs record the spectra of astronomical objects

The spectrograph is one of the astronomer's most important tools, perhaps second only to the telescope itself. In its most basic form, a **spectrograph** consists of a slit, two lenses, and a prism, arranged so as to focus the spectrum of an astronomical object onto a small photographic plate as shown in Figure 6-18. This optical device is typically mounted at the focal point of a telescope, with the image of the object to be examined focused on the slit. After the spectrum of a star or galaxy has been photographed, the exposed portion of the photographic plate is covered and light from a known source (usually the emission spectrum of a preselected element like iron or argon) is focused on the spectrograph slit. This process produces a "comparison spectrum" above and below the spectrum of the astronomical object, as shown in Figure 6-19. The wavelengths of the bright spectral lines of the comparison spectrum are already known from laboratory experiments and can therefore serve as reference markers for measuring the wavelengths of the lines in the spectrum of the star or galaxy.

There are drawbacks to this old-fashioned spectrograph. For one, a prism does not disperse the colors of the rainbow evenly. The red part of the spectrum is compressed, while the blue and violet colors are spread out. In addition, the blue and violet wavelengths must pass through a fairly thick sec-

tion of the prism, which absorbs much of the starlight. Indeed, a glass prism is opaque to the near-ultraviolet part of the spectrum. A better device for breaking starlight into the colors of the rainbow is a diffraction grating.

A **diffraction grating** is a piece of glass onto which thousands of closely spaced parallel lines have been cut. Some of

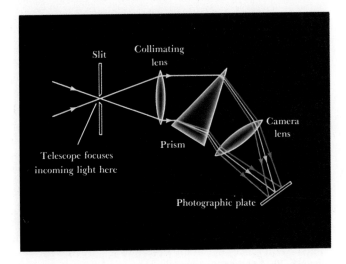

Figure 6-18 A prism spectrograph *This optical device uses a prism to break up the light from a source into a spectrum. The collimating lens directs incoming light rays so they enter the prism parallel to each other. The camera lens then focuses the spectrum onto a photographic plate.*

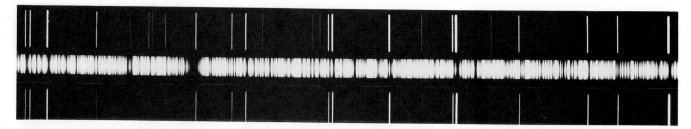

Figure 6-19 A spectrogram *The photographic record of a spectrum is called a spectrogram. This spectrogram shows the spectrum of a star. Numerous dark absorption lines can be seen against the brighter back-ground of the spectrum. Above and below the star's spectrum are emission lines produced by an iron arc. These bright lines are the comparison spectrum. (Palomar Observatory)*

the finest diffraction gratings have more than 10,000 lines per centimeter. These lines are usually cut by drawing a diamond back and forth across the piece of glass. The spacing of the lines must be quite regular, and the best results are obtained when the grooves are beveled.

A diffraction grating can be used in either of two ways. Incoming starlight can either be reflected off the grating (it is then called a **reflection grating**), or the starlight can be passed through the grating (it is then called a **transmission grating**). In either case, light rays of various wavelengths leaving the different parts of the grating interfere with each other so as to produce a spectrum. Figure 6-20 shows the design of a modern grating spectrograph.

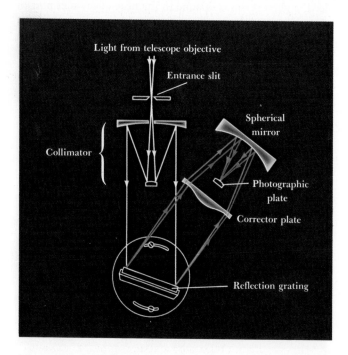

Figure 6-20 A grating spectrograph *This optical device uses a reflection grating to break up the light from a source into a spectrum. The collimator ensures that light rays striking the grating are parallel. A corrector lens and mirror then focus the spectrum onto a photographic plate or a CCD.*

Most major observatories record spectra with charge-coupled devices. A CCD, instead of a photographic plate, is placed at the focus of the spectrograph. When the exposure is finished, electronic equipment is used to measure the amount of charge that has accumulated in each pixel. The final result is a graph (intensity versus wavelength) on which spectral lines appear as peaks or valleys, as on Figure 6-21.

6-5 A radio telescope uses a large concave dish to reflect radio waves to a focus

Until recently, all the information that astronomers could gather about the universe was based on ordinary visible light. But with the discovery of nonvisible electromagnetic radiation, scientists began to wonder if objects in the universe might also emit radio waves, X rays, and infrared and ultraviolet radiation. Surely, views of the universe at these nonvisible wavelengths would enhance our understanding of the cosmos.

The first evidence of radio radiation coming from outer space was provided by the work of a young engineer, Karl Jansky, of Bell Telephone Laboratories. Using long antennas, Jansky began investigating the sources of the radio static that affects short-wavelength radiotelephone communication. By 1932, he realized that one particular kind of radio noise is strongest when the constellation of Sagittarius is high in the sky. The center of our Galaxy is located in the direction of Sagittarius, so Jansky concluded that he was detecting radio waves from an astronomical source.

Astronomers were not quick to pursue this line of research. Only one person, Grote Reber (an electronics engineer living in Illinois), pursued the matter. In 1936, Reber built in his backyard the first **radio telescope** for the purpose of mapping radio emission from the Milky Way. His design was modeled after an ordinary reflecting telescope, with a parabolic "dish" (reflecting antenna) measuring 9.1 m in diameter. The radio receiver at the focal point of the metal dish was tuned to a wavelength of 1.85 m.

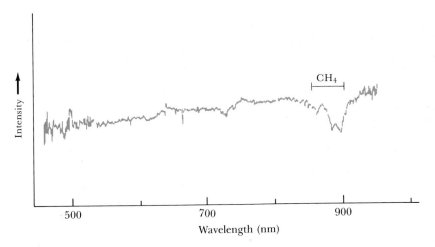

Intensity

CH₄

500   700   900

Wavelength (nm)

Figure 6-21 *A CCD spectrum of Pluto* This graph shows the spectrum of Pluto as recorded by a CCD. The prominent absorption feature near 900 nm is caused by methane (CH₄) and is conclusive evidence for the existence of that substance on the planet. (From observations by J. Apt, N. P. Carleton, and C. D. Mackay)

By 1944, when Reber completed his map of the Milky Way, astronomers had begun to take notice of his discoveries. Shortly after World War II, radio telescopes began to spring up around the world. Radio observatories are today as common as major optical observatories.

Like Reber's prototype, the standard radio telescope has a large parabolic dish. A small antenna tuned to the desired frequency is located at the focus, and the incoming signal is relayed to amplifiers and recording instruments typically located in a room at the base of the telescope's pier. Figure 6-22 shows a modern radio telescope.

At first, astronomers were not enthusiastic about trying to detect radio noise from space, in part because of the poor resolving power of radio telescopes. Recall that resolving power is described in terms of the smallest angular separation between two stars that can be seen as separate objects. That angle is proportional to the wavelength being observed and inversely proportional to the diameter of the telescope. Thus resolving power becomes worse with increasing wavelength; the longer the wavelength, the fuzzier the picture. Because radio radiation has very long wavelengths, astronomers thought that radio telescopes could produce only blurry, indistinct views.

Very large radio telescopes can produce somewhat sharper radio images, because the bigger the dish, the better the resolving power. For this reason, most modern radio telescopes have dishes more than 100 ft in diameter. Nevertheless, even the largest radio dish in existence cannot come close to the resolution of the best optical instruments.

A very clever technique was devised to circumvent this problem and produce high resolution radio images. Unlike ordinary light, radio signals can be carried over electrical wires; which means that two radio telescopes separated by many kilometers can be hooked together. This technique is called **interferometry,** because the incoming radio signals are made to "interfere," or blend together, making the combined signal sharp and clear. The result is most impressive: The effective resolving power of two such radio telescopes is equivalent to that of one gigantic dish with a diameter equal to the distance between the two telescopes.

Interferometry techniques were exploited for the first time in the late 1940s, when astronomers began receiving their first detailed views of radio objects in the sky. Radio telescopes separated by thousands of kilometers were linked together using radio signals to give resolving power much higher than that of optical telescopes. This technique is called **very-long-baseline interferometry (VLBI).** The best possible resolution under such a system would be obtained by two telescopes on opposite sides of the Earth. In that case, features as small as 0.00001 arc sec could be distinguished at radio wavelengths—a resolution 100,000 times better than the sharpest pictures from ordinary optical telescopes.

Figure 6-22 *A radio telescope* The dish of this radio telescope is 45.2 m (148 ft) in diameter. It is one of several large instruments at the National Radio Astronomy Observatory near Green Bank, West Virginia. (NRAO)

Figure 6-23 The Very Large Array (VLA) *The 27 radio telescopes of the VLA system are arranged along the arms of a Y in central New Mexico. The north arm of the array is 19 km long; the southwest and southeast arms are each 21 km long. (NRAO)*

One of the finest arrangements of radio telescopes began operating in 1980 in the desert near Socorro, New Mexico. This system, called the Very Large Array (VLA), consists of 27 parabolic dishes, each 26 m (85 ft) in diameter. The 27 telescopes are arranged along the arms of a gigantic Y that covers an area 27 km (17 miles) in diameter. Only a portion of the VLA is shown in Figure 6-23. This system produces radio views of the sky with a resolution comparable to that of the very best optical telescopes.

Radio astronomers often use "false color" to display their radio views of astronomical objects. An example is shown in Figure 6-24. The most intense radio emission is shown in red, the least intense in blue. Intermediate colors of the rainbow represent intermediate levels of radio intensity. Black indicates that there is no detectable radio emission. Astrono-

mers working at other nonvisible wavelength ranges also frequently use false-color techniques to display images obtained from their instruments.

6-6 Telescopes in orbit around the Earth detect radiation that does not penetrate the atmosphere

As the success of radio astronomy began to mount, astronomers started exploring the possibility of making observations at other nonvisible wavelengths. This was a difficult task because the Earth's atmosphere is opaque to many wavelengths. Very little radiation, other than visible light

a

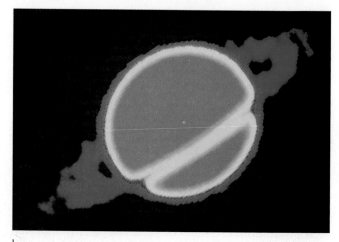

b

Figure 6-24 Optical and radio views of Saturn (a) *This picture was taken by a spacecraft 18 million km from Saturn. Sunlight reflecting from the planet's cloudtops and rings is responsible for this view.*

(b) *This picture, taken by the VLA, shows radio emission from Saturn at a wavelength of 2 cm. Color is used to represent the intensity of radio emission. (NASA; NRAO)*

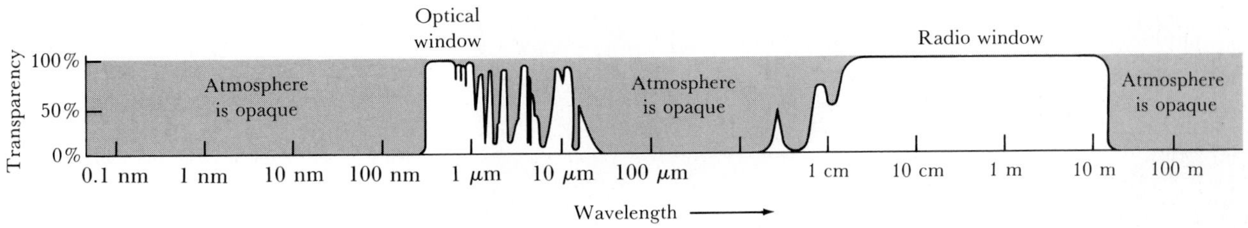

Transparency

100%
50%
0%

Optical window

Radio window

Atmosphere is opaque

Atmosphere is opaque

Atmosphere is opaque

0.1 nm   1 nm   10 nm   100 nm   1 μm   10 μm   100 μm   1 cm   10 cm   1 m   10 m   100 m

Wavelength ⟶

Figure 6-25 The transparency of the Earth's atmosphere This graph shows the percentage of radiation that can penetrate the Earth's atmosphere as a function of wavelength. Oxygen and nitrogen completely absorb all radiation with wavelengths shorter than 290 nm. Water vapor and carbon dioxide effectively block out all radiation from about 10 μm to 1 cm.

and radio waves, manages to penetrate the air we breathe.

The transparency of the Earth's atmosphere is graphed in Figure 6-25. Notice the **optical window**, through which we see visible light from space, and the **radio window**, which permits Earth-based radio astronomy. Also notice the various transparent regions at infrared wavelengths between 1 and 10 μm. Infrared radiation within these wavelength intervals does penetrate the Earth's atmosphere and can be detected with ground-based equipment. This wavelength range is called the near-infrared, because it lies just beyond the red end of the visible spectrum.

Water vapor is the main absorber of infrared radiation from space. Infrared observatories are therefore located at sites with low humidity. For example, the 14,000-ft summit of Mauna Kea on Hawaii is exceptionally dry. Making infrared observations is the primary function of NASA's 3-m telescope there.

Another way of avoiding the water-vapor problem is to take the telescope up in an airplane. That is the basic idea behind the Kuiper Airborne Observatory (KAO) shown in Figure 6-26. This NASA airplane carries a 0.9-m reflecting telescope to an altitude of 12 km (40,000 ft), thus placing the observatory above 99 percent of the atmosphere's water vapor.

A telescope in Earth orbit offers the best arrangement. In 1983, NASA launched the Infrared Astronomical Satellite (IRAS) into a polar orbit 900 km high. This satellite was designed around a small reflecting telescope that surveyed the entire sky at infrared wavelengths. With this instrument, astronomers saw what the sky looks like in the "far infrared," the wavelength range from about 100 μm up to 1 mm.

The European Space Agency (ESO) plans to launch an infrared telescope in the early 1990s. Later that same decade, NASA plans to launch the Space Infrared Telescope Facility (SIRTF), which will have a 0.85-m primary mirror (see Figure 6-27). Both telescopes will be superior to IRAS in making observations at infrared wavelengths that never penetrate the Earth's atmosphere.

The Earth's atmosphere is also transparent to the longest-wavelength ultraviolet light. This wavelength range, extending from about 400 nm down to 290 nm, is called the near-ultraviolet, because it lies just beyond the violet end of the visible spectrum. Astronomers can easily make ground-based observations in this wavelength range—if they do not use glass lenses in their telescopes. Glass is opaque to the near-ultraviolet (which is a big drawback of refracting telescopes), and all lenses must therefore be made of quartz or some other nonabsorbing substance.

Figure 6-26 The Kuiper Airborne Observatory This C-141 jet airplane carries a 1-m reflecting telescope specifically designed for infrared observations. The observing portal through which the telescope is aimed can be seen on the top of the fuselage just in front of the wing. (NASA)

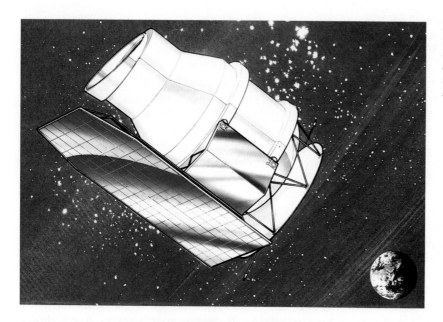

Figure 6-27   The Space Infrared Telescope Facility (SIRTF)   *If all goes as planned, a rocket will carry this 5-ton infrared telescope into a high, circular orbit around the Earth in 1998. The telescope will have a mirror 0.95 m (37 in.) in diameter and will be designed to operate for at least five years. (NASA)*

To see the far-ultraviolet, astronomers must make their observations from space. During the early 1970s, Apollo and Skylab astronauts carried small ultraviolet telescopes above the Earth's atmosphere to give us our first views of the ultraviolet sky. Small rockets have also been used to place ultraviolet cameras briefly above the Earth's atmosphere. A typical view is shown in Figure 6-28, along with a corresponding infrared view from IRAS, a view in visible light, and a star chart.

Some of the best ultraviolet astronomy has been accomplished by the International Ultraviolet Explorer (IUE),

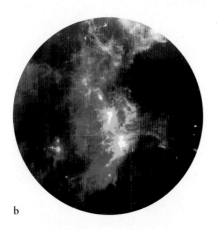

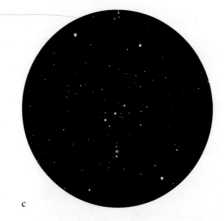

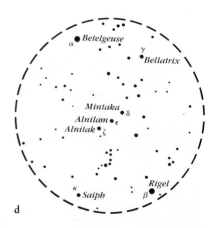

Figure 6-28   Orion seen at ultraviolet, infrared, and visible wavelengths   *An ultraviolet view (**a**) of the constellation of Orion was obtained during a brief rocket flight on December 5, 1975. The 100-second exposure covers the wavelength range 125–200 nm. The false color view (**b**) from IRAS displays infrared intensity according to* color: red for strong 100-μm radiation; green for strong 60-μm radiation; and blue for strong 12-μm radiation. For comparison, an ordinary optical photograph (**c**) and a star chart (**d**) are included. (Courtesy of G. R. Carruthers, NRL; NASA; R. C. Mitchell, Central Washington University)

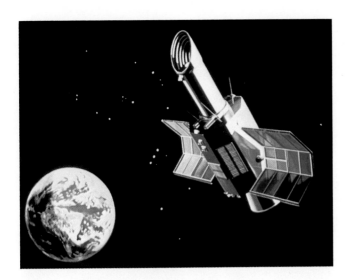

Figure 6-29 The International Ultraviolet Explorer (IUE) Since its launch in 1978, this 671-kg satellite has produced superb observations in the far-ultraviolet. The blue panels extending from either side of the midsection of the satellite are solar-cell arrays that provide electrical power for the radio transmitters and other electronic equipment. (NASA)

launched in 1978. This satellite (see Figure 6-29) is built around a Cassegrain telescope with a 45-cm (18-in.) mirror and a total focal length of 6.74 m (22 ft). Observations cover the range from 116 to 320 nm.

For decades, astronomers have dreamed of having a major observatory in space. Although satellites like the IRAS and IUE gave excellent views of selected wavelength ranges, astronomers were still tantalized by the prospect of a very large telescope that could be operated at any wavelength, from the infrared through the visible range and out into the far-ultraviolet. This dream is the mission of the Hubble Space Telescope (HST), which was carried aloft by the Space Shuttle in 1990 (see Figure 6-30).

Soon after HST was placed in orbit, astronomers discovered that the telescope's 2.4-m primary mirror suffers from spherical aberration. The mirror should have been able to concentrate 70 percent of a star's light into an image 0.1 arc sec in diameter. Instead, only 20 percent of a star's light is focused into the desired 0.1 arc sec spot; the remaining 80 percent is smeared out over an area 1 arc sec in diameter. A star image therefore consists of a central spot of modest brightness surrounded by a hazy glow. If pictures are taken using 100 percent of the incoming starlight, the resulting star images are about 1 arc sec in diameter, which is about the same as the image quality achieved at major ground-based observatories.

One way astronomers can cope with this problem is to use only the 20 percent of incoming starlight that is properly focused and, with computer processing, discard the remaining poorly focused 80 percent. This technique produces sharp, crisp images, but is practical only on brighter objects

with which astronomers can afford to waste light. Unfortunately, many of the observing projects scheduled on the HST involve extremely dim galaxies and nebulae. These observations will be postponed until a Shuttle mission, perhaps in 1993, installs new cameras equipped with correcting lenses to compensate for the spherical aberration in the primary mirror. Meanwhile, many observing projects—particularly those that can succeed with only 20 percent of the incoming starlight or, like spectroscopy, do not require a pinpoint focus—will go forward as planned.

Perhaps the greatest surprises will be the discovery of totally new objects in space. Ever since Galileo turned his telescope toward the skies and first saw four moons orbiting Jupiter, each new generation of astronomical instruments has disclosed unimagined objects and processes that were often more bizarre than the strangest science fiction. This happened twice in the 1960s, for instance, when radio telescopes found quasars and pulsars. And it happened once more in the 1970s, when X-ray telescopes detected bursters. With serendipity so commonplace, perhaps it will happen again in the 1990s.

Because neither X rays nor gamma rays penetrate the Earth's atmosphere, observations at these extremely short wavelengths also must be made from space. Astronomers got their first quick look at the X-ray sky from brief rocket flights during the late 1940s. During the early 1970s, several small satellites viewed the entire X-ray and gamma-ray sky, revealing hundreds of previously unknown sources, including several good black-hole candidates.

Figure 6-30 The Hubble Space Telescope (HST) The Space Shuttle placed this 2.4-m telescope into Earth orbit in 1990. The HST is designed to study the heavens over a wavelength range from 110 nm in the ultraviolet to 1.1 μm in the infrared. As new technology becomes available, the Space Shuttle will return to the HST to repair equipment and replace old components. Astronomers therefore anticipate using this superb telescope well into the twenty-first century. (NASA)

Although heroic in their day, these preliminary efforts pale in comparison to the detailed views and results from three huge satellites launched between 1977 and 1979. Called High Energy Astrophysical Observatories (HEAO), these satellites each carried an array of X-ray and gamma-ray detectors. Thousands of sources were discovered all across the sky. The second satellite in this series was especially successful in producing high-quality X-ray images of a wide range of exotic objects. This satellite was called the Einstein Observatory and was launched near the hundredth anniversary of Albert Einstein's birth. X-ray views from this observatory appear throughout this book, illustrating the discussions of the extraordinary objects and incredibly hot environments that produce these high-energy photons.

A logical successor to the Einstein Observatory is AXAF, the Advanced X-Ray Astrophysics Facility, which is in the planning stages at NASA. Like the Hubble Space Telescope, AXAF will be long-lived, adaptable, and will be controlled by astronomers on the ground. If this project receives funding in the near future, AXAF could be launched in the mid-1990s (see Figure 6-31).

Ordinary optical equipment cannot help astronomers who want to observe the X-ray or gamma-ray sky. The energy carried by electromagnetic radiation varies inversely with wavelength, as mentioned in Chapter 5 (see Box 5-3). In other words, the shorter the wavelength, the higher the energy of a photon. The wavelengths of X rays and gamma rays are so short that these high-energy photons would simply bury themselves in the mirror of an ordinary reflecting telescope. Astronomers therefore use electronic detectors (similar to Geiger counters) that respond to the effects of

Figure 6-32   The Gamma Ray Observatory (GRO)   *The best views of the high-energy gamma-ray sky will come from this satellite, which was carried aloft by the Space Shuttle in 1990. The 15-ton satellite has four gamma-ray detectors that will examine individual sources as well as measure gamma-ray background radiation. (NASA)*

high-energy radiation as it passes through sealed containers of gas or electrically charged metal plates carried on satellites.

Electromagnetic radiation with the shortest wavelengths and therefore the most energy consists of gamma rays. Astronomers have had only their first tantalizing glimpses of this exotic region of the electromagnetic spectrum, and many hopes and expectations lie with the Gamma Ray Observatory (GRO), which was launched in 1990 (see Figure 6-32).

The advantages and benefits of these Earth-orbiting observatories cannot be overemphasized. We are no longer limited to the narrow ranges of whatever wavelengths manage to leak through our shimmering, hazy atmosphere (see Figure 6-33). For the first time we are really seeing the universe.

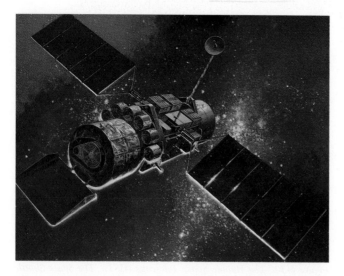

Figure 6-31   The Advanced X-Ray Astrophysics Facility (AXAF)   *If all goes as planned, this X-ray telescope will be placed into Earth orbit in the mid-1990s. Twelve cylindrical mirrors—the largest being about a meter long and 1¼ meters in diameter—will focus incoming X rays to produce images and observations of a wide variety of objects. (Courtesy of TRW)*

Figure 6-33 [right]   The entire sky at five wavelength ranges   *These five views show the entire sky at (a) radio, (b) infrared, (c) visible, (d) ultraviolet, and (e) X-ray wavelengths. The Milky Way stretches horizontally across each picture. The radio view (a) shows the sky at a wavelength of 73 cm (411 MHz), with the brightest regions in red and the dimmest in blue. The infrared view (b) from IRAS shows several infrared wavelengths by color: 100 μm is yellow, 60 μm is red, and 12 μm is blue. In the visible view (c) Orion is at the right, Sagittarius in the middle, and Cygnus toward the left. The ultraviolet view (d) covers 1350 to 2550 nm and shows numerous stars around Orion and Cygnus. The X-ray view (e) from HEAO-1 covers 0.2 to 6 nm, corresponding to a range of photon energies from 6 to 0.2 keV. (Max Planck Institut für Radioastronomie; Jet Propulsion Laboratory; Griffith Observatory; Royal Observatory, Edinburgh; E. Boldt, NASA)*

a Radio

b Infrared

c Visible

d Ultraviolet

e X ray

Key words

Cassegrain focus	interferometry	radio telescope	resolving power
charge-coupled device (CCD)	light-gathering power	radio window (in Earth's atmosphere)	Schmidt telescope
chromatic aberration	magnification	reflecting telescope	seeing disk
coma	magnifying power	reflection	spectrograph
coudé focus	Newtonian reflector	reflection grating	spherical aberration
diffraction grating	objective lens	reflector	transmission grating
eyepiece lens	optical window (in Earth's atmosphere)	refracting telescope	very-long-baseline interferometry (VLBI)
focal length	pixel	refraction	
focus (of a lens or mirror)	prime focus	refractor	

Key ideas

- Refracting telescopes, or refractors, produce images by bending light rays as they pass through glass lenses.

    Chromatic aberration is an optical defect whereby light of different wavelengths is bent in different amounts by a lens.

    Limitations of glass purity, chromatic aberration, opacity to certain wavelengths, and structural difficulties have made it inadvisable to build extremely large refractors.

- Reflecting telescopes, or reflectors, produce images by reflecting light rays to a focus point from curved mirrors.

    Reflectors are not subject to most of the problems that limit the useful size of refractors, although they are plagued by other problems, such as spherical aberration and weight.

- Charge-coupled devices (CCDs) are often used at a telescope's focus to record faint images.

- A spectrograph is a device that uses a diffraction grating and lenses to record the spectrum of an astronomical object.

- A photometer is a device that uses a photoelectric detector to measure the brightness of an astronomical object.

- Radio telescopes have large reflecting antennas or dishes that are used to focus radio waves.

    Very large dishes are needed to produce reasonably sharp radio images. High resolution is achieved with interferometry techniques that link smaller dishes together.

- The Earth's atmosphere absorbs much of the radiation that arrives from space.

    The atmosphere is transparent chiefly in two wavelength ranges known as the optical window and the radio window. A few wavelengths in the near-infrared also reach the ground.

    For observations at other wavelengths, astronomers depend upon telescopes carried above the atmosphere by high-altitude airplanes, rockets, or satellites.

    Satellite-based observatories are giving us a wealth of new information about the universe and are permitting coordinated observation of the sky at all wavelengths.

Review questions

1 Give everyday examples of the phenomena of refraction and reflection.

2 With the aid of a diagram, describe a refracting telescope.

3 What is chromatic aberration, and how can it be corrected?

4 With the aid of a diagram, describe a reflecting telescope. Describe four different ways in which an astronomer can have access to the focal point.

5 Explain some of the advantages of reflecting telescopes over refracting telescopes.

6 What is spherical aberration, and how can it be corrected?

7 Quite often, advertisements appear for telescopes that extol their magnifying abilities. Is this a good criterion for evaluating telescopes? Explain your answer.

8 What kind of telescope would you use if you wanted to take a color photograph entirely free of chromatic aberration? Why?

*9 The observing cage in which an astronomer sits at the prime focus of the 5-m telescope on Palomar Mountain is about 1 meter in diameter. What fraction of the incoming starlight is blocked by the cage?

*10 Compare the light-gathering power of the Palomar 5-meter telescope with that of the fully dark-adapted human eye, which has a pupil diameter of about 5 mm.

11 Why can radio astronomers make observations at any time during the day, whereas optical astronomers are mostly limited to nighttime observing at night?

12 What is meant by the "optical window" and the "radio window"? Why isn't there an "X-ray window" or an "ultraviolet window"?

13 Describe three advantages that the Hubble Space Telescope has over ground-based telescopes.

Advanced questions

> **Tips and tools . . .**
> You may find it useful to review the small-angle formula discussed in Box 1-1. Data concerning the planets are found in the appendixes at the end of this book. An example of how to calculate magnifying power was given in the body of this chapter.

*14 The four largest moons of Jupiter are roughly the same size as our Moon and are about 628 million km from Earth at opposition. What is the size of the smallest surface features that the Hubble Space Telescope (resolution = 0.1 arc sec) will be able to detect? How does this compare with the smallest features that can be seen on the Moon with the unaided human eye (resolution = 1 arc min)?

*15 (*Basic*) Suppose your Newtonian reflector has a mirror 20 cm (8 in.) in diameter and a focal length of 2 m. What magnification do you get with eyepieces whose focal lengths are (**a**) 9 mm, (**b**) 20 mm, and (**c**) 55 mm? What is the telescope's resolving power?

*16 (*Challenging*) Will the HST be able to distinguish any features on Pluto?

17 Show by means of a diagram why the image formed by a simple refracting telescope is "upside down."

18 Optical problems with the newly launched Hubble Space Telescope were announced at about the same time this book went to press. Consult such magazines as *Sky & Telescope* and *Science News* to find out how things are going. Has a congressional committee pinpointed the cause of the problems? Have any pictures been sent back? How do they differ from pictures of the same objects taken by ground-based telescopes? What plans exist for a Shuttle mission to repair HST?

19 The Gamma Ray Observatory was launched at about the same time this book went to press. Consult such magazines as *Sky & Telescope* and *Science News* to find out how the GRO is doing. Have any noteworthy or surprising discoveries been announced?

Discussion questions

20 Discuss the advantages and disadvantages of using a small telescope in Earth orbit versus a large telescope on a mountaintop.

21 If you were in charge of selecting a site for a new observatory, what factors would you consider important?

Observing projects

22 Obtain a telescope during the daytime along with several eyepieces of various focal lengths. If you can determine the telescope's focal length, calculate the magnifying powers of the eyepieces. Focus the telescope on some familiar object, such as a distant lamppost or tree. **DO NOT FOCUS ON THE SUN! Looking directly at the Sun can cause blindness.** Describe the image you see through the telescope. Is it "up-

side down?" How does the image move as you slowly and gently shift the telescope left and right, up and down? Examine the eyepieces, noting their focal lengths. By changing the eyepieces, examine the distant object under different magnifications. How does the field of view and the quality of the image change as you go from low power to high power?

23 On a clear night, view the Moon, a planet, and a star through a telescope using eyepieces of various focal lengths and known magnifying powers. (You may have to consult such magazines as *Sky & Telescope* or *Astronomy* to determine the phase of the Moon and the locations of the planets.) In what way does the image seem to degrade as you view with increasingly higher magnification? Do you see any chromatic aberration? If so, with which object and which eyepiece is it most noticeable?

24 Many towns and cities have amateur astronomy clubs. If you are so inclined, attend a "star party" hosted by your local club. People who bring their telescopes to such gatherings are delighted to show you their instruments and take you on a telescopic tour of the heavens. Such an experience can lead to a very enjoyable lifelong hobby.

For further reading

Bahcall, J., and Spitzer, L. "The Space Telescope." *Scientific American*, July 1982 • This article covers the history, design, and construction of the Hubble Space Telescope.

Bok, B. "The Promise of the Space Telescope." *Mercury*, May/June 1983 • An excellent summary of the questions and issues to be examined by the HST.

Chaisson, E., and Villard, R. "Hubble Space Telescope: The Mission." *Sky & Telescope*, April 1990 • This article, written by two astronomers at the Space Telescope Science Institute, describes the capabilities of the HST and gives an excellent overview of the main projects that it is working on.

Cohen, M. *In Quest of Telescopes.* Sky Publishing and Cambridge University Press, 1980 • A fascinating account of the author's development as a professional astronomer, interwoven with tours of some of the world's most interesting centers of astronomical research.

Davies, J. *Satellite Astronomy.* Ellis Horwood and John Wiley, 1988 • A comprehensive introduction to astronomy done from satellites.

Field, G. "The Future of Astronomy in Space." *Mercury*, July/August, 1984 • In this article, the chairman of the National Academy of Science's Astronomy Survey Committee describes some of the space telescopes and observing programs proposed by that committee.

Fienberg, R. "HST: Astronomy's Discovery Machine." *Sky & Telescope*, April 1990 • Written just prior to launch, this article is a superb "nuts and bolts" guide to the most ambitious astronomical satellite ever built.

―――. "The New, Improved Space Telescope." *Sky & Telescope*, February 1989 • This brief article discusses last-minute improvements to the Hubble Space Telescope prior to its launch in 1990.

―――, and Sinnott, R. "HST: Picking Up the Pieces," *Sky & Telescope*, October 1990 • A well-written article that summarizes the problems with the Hubble Space Telescope.

Giacconi, R. "The Einstein X-Ray Observatory." *Scientific American*, February 1980 • This article briefly discusses many of the discoveries made by the Einstein Observatory during its historic mission.

Harrington, S. "Selecting Your First Telescope." *Mercury*, July/August 1982 • An excellent guide for the person who is thinking about buying a telescope.

Henbest, N., and Marten, M. *The New Astronomy.* Cambridge University Press, 1983 • This coffee-table book contains an impressive collection of photographs showing astronomical objects at a variety of wavelengths.

Janesick, J., and Blouke, M. "Sky on a Chip: The Fabulous CCD." *Sky & Telescope*, September 1987 • This superb article describes how CCDs work and why they have become an indispensable asset at well equipped observatories.

Krisciunas, K. *Astronomical Centers of the World.* Cambridge University Press, 1988 • This book tells of the development of major centers of astronomy, including the observatory complexes on Mauna Kea and Kitt Peak.

Kristian, J., and Blouke, M. "Charge-Coupled Devices in Astronomy." *Scientific American*, October 1982 • This article explains how and why microelectronics is changing the way astronomers record their observations.

Learner, R. *Astronomy Through the Telescope.* Van Nostrand Reinhold, 1981 • A beautifully illustrated history of telescopes.

Preston, R. *First Light: The Search for the Edge of the Universe.* Atlantic Monthly Press, 1987; paperback, 1988 • This eloquent book gives behind-the-scenes vignettes of astronomers working at the Palomar Observatory.

Readhead, A. "Radio Astronomy by Very-Long-Baseline Interferometry." *Scientific American*, June 1982 • This article explains how radio telescopes thousands of miles apart can combine their observations to produce extraordinary views of the radio sky.

Robinson, L. "The Frigid World of IRAS." *Sky & Telescope*, January 1984 • An excellent summary of the historic IRAS mission.

Schorn, R. "Astronomy in the Next Decade." *Sky & Telescope*, April 1982 • This article discusses the planning, politics, and financing of astronomy in the United States.

Sullivan, W. "Radio Astronomy's Golden Anniversary." *Sky & Telescope*, December 1982 • This article consists primarily of a remarkable collection of historical photographs illustrating the early years of radio astronomy.

Tucker, W., and Tucker, K. *The Cosmic Inquirers*. Harvard University Press, 1986 • This delightful book takes the reader behind the scenes in the design, construction, and operation of five major astronomical instruments.

Box 6-3   Early results from the Hubble Space Telescope

During the summer of 1990, astronomers at the Space Telescope Science Institute trained the HST on several star clusters to study the effects of the telescope's spherical aberration. One of the first objects to be examined was 30 Doradus, a vast cloud of glowing hydrogen gas surrounding a cluster of young, massive, hot stars. It is located in a nearby galaxy called the Large Magellanic Cloud (see page 507), which is about 160,000 light years from Earth.

The HST's Wide Field/Planetary Camera produced the image in panel A below, which shows stars clustered around the core of 30 Doradus (compare this with the photograph of 30 Doradus on page 439). Panel B is an enlargement of the inner regions, showing the centrally located star cluster called R136.

This cluster is believed to contain some of the hottest, brightest, most massive stars known to astronomers. Note that the telescope's spherical aberration causes individual star images to be surrounded by a fuzzy glow.

Panel C is a ground-based image of R136 showing details as small as 0.6 arc sec in size, which is typical of the best that can be done from the Earth's surface. Panel D is a computer-enhanced version of panel B, demonstrating how the halos around the stars can be reduced to achieve a resolution of about 0.1 arc sec. At least 60 individual stars are easily discernible. This dramatic enhancement suggests that computer processing will be an important tool in overcoming HST's optical defects.

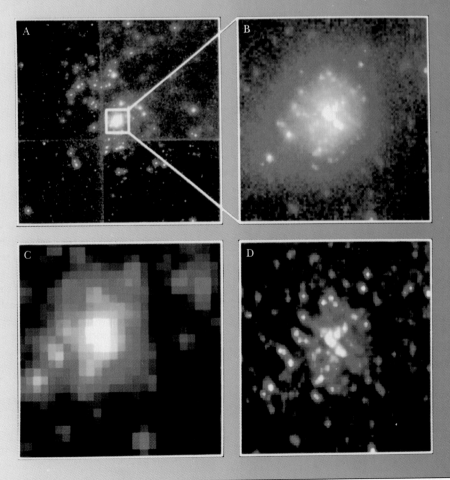

Richard West is a Danish astronomer born in Copenhagen in 1941. He received his doctorate from Copenhagen University in 1964 and joined the European Southern Observatory in 1970. During his work for the ESO Atlases of the Southern Sky, Dr. West discovered comet West in late 1975. While his early research interests concentrated on close binary stars and automatic spectral classification, he is now mostly involved in solar system research, particularly very distant comets and minor planets, of which he has discovered several dozen.

From 1982 to 1985, Dr. West was general secretary of the International Astronomical Union (IAU)—the most authoritative world body in astronomy—and is now president of IAU Commission 20. He is the editor of the ESO in-house journal The Messenger, and, in addition to his professional work, he has written many popular articles and three books.

Richard M. West NEW ESO TELESCOPES

Unlike most other natural sciences, astronomy almost exclusively relies on *remote sensing*. Astronomers perform "observations" rather than "experiments." Most astronomical objects are way beyond our reach, and unlike our physicist colleagues, we cannot change the experimental conditions.

In compensation, astronomy rewards its people by being a science of apparently unlimited opportunities and, above all, of *extreme conditions*. Within its realm, we find the largest distances, the longest periods of time, the most massive objects, the highest temperatures, the strongest electrical and magnetic fields, the highest and lowest densities, and the most extreme energies known. Indeed, the universe is a fantastic laboratory where nature invites us to witness incredible events impossible to stage here on Earth.

Consequently, astronomers traditionally employ the most sophisticated instruments and methods conceived by humans at any given epoch. Examples are megalithic observatories like Stonehenge, and pre-Columbian structures in the Americas, the great sighting instruments of Ulug Begh (of medieval Samarkand) and Tycho Brahe, the giant mirrors of William Herschel, and today's incredibly complex astronomical satellites. Progress in astronomy is intimately related to the application of the best available technology.

The complexity of astronomical projects has grown to the extent that most are now carried out by groups of scientists and engineers, and frequently within an international framework. In a world of scarce financial and human resources, astronomers appreciate the value of close collaboration across borders. It is in this spirit that eight countries in Western Europe (Belgium, Denmark, Germany, France, Italy, the Netherlands, Sweden, and Switzerland) have jointly created an international astronomical organization, the European Southern Observatory (ESO).

ESO operates one of the world's largest observatories on La Silla, an isolated mountaintop in Chile's dry Atacama desert. The revolutionary 3.5-m New Technology Telescope (NTT) was recently installed there. In anticipation of the needs of observational astronomy long into the next century, ESO's flagship project is currently the 16-m Very Large Telescope (VLT), which will become the world's largest optical telescope when it is ready in 1998.

Confronted with the need to push the observational horizon further back in space and time, astronomers realized in the late 1970s that in order to be cost-effective, future giant telescopes must be radically different from the existing ones in the 4- to 6-m class. Some key words of new telescope technology are *active* and *adaptive optics*, *thermal control*, and *remote observing*.

Until now, reflecting telescopes have been *passive* in that the mirrors are stiff and keep their form when the telescope moves. To retain this principle for larger mirrors would lead to excessive mirror masses and consequently to enormous, very expensive telescopes. The concept of *active optics* was therefore pursued at ESO, first by means of extensive tests with a very thin 1-m mirror during 1986–1987. Made of a zero-expansion ceramic material and having a thickness of only 19 mm, this mirror was kept in shape by computer-activated supports. A similar system has now been successfully incorporated into the ESO NTT.

Astronomical observations have fully confirmed the performance of the NTT active optics. Since the NTT "tunes" itself continuously, it always collects the incoming light as efficiently as possible. This reduces the necessary observing time and allows observations of fainter objects than is possible with any other ground-based telescope. Indeed, the $14-million NTT is able to record the light from stars as faint as those which can be seen with the much more expensive Hubble Space Telescope.

To ensure sharper images than have hitherto been possible, the NTT incorporates a new design that virtually eliminates atmospheric turbulence within the telescope's dome. This is achieved by a wind-tunnel tested octagonal dome shape and active thermostatic control of all sources of heat, including the hydrostatic oil pads on which the telescope moves. The NTT also makes use of *adaptive optics*, which effectively eliminate the adverse influence of atmospheric turbulence, to yielding images of astronomical objects almost as sharp as if the telescope were situated in space.

The adaptive optics technique developed by ESO in collaboration with a number of scientific institutes and industrial firms in France is based on a feedback loop. The optical system contains a deformable mirror that can change its surface profile to compensate exactly for the distortion of light by the atmosphere. The information about how to deform the mirror comes from a wavefront sensor, which measures the shape of the distorted light wavefront. It requires a powerful computer to calculate how the actuators (located behind the mirror) have to push and pull the mirror surface.

The prototype system has a mirror with 19 actuators. The mirror is deformed, hence the wavefront is corrected, 100 times a second. It was successfully tested at the ESO

3.6-m telescope at La Silla in 1990. It is now possible to achieve the theoretical limits for optical imaging in the infrared wavelengths by means of a ground-based medium-sized telescope. Our next aim is to perfect the technique for larger telescopes.

A permanent communication link was established in 1987 between the ESO headquarters in Garching, West Germany, and La Silla, via an Intelsat satellite in geosynchronous orbit and a microwave link. During nighttime this link is used for fully interactive remote control of three telescopes at La Silla, 12,000 km away. The remote observer receives the digitized spectra or direct (CCD) images back in real time. These data can be immediately analyzed with on-line image processing software, so that the observer is always kept informed about the quality of the observations and can take appropriate action if something "interesting" turns up.

ESO's next telescope project is the $225-million VLT. The aim is to construct an array of four 8.2-m telescopes with a total area equal to that of a 16-m telescope. It is likely to be placed at Cerro Paranal, in the Atacama desert, where even better observing conditions exist than at La Silla. The first 8-m unit telescope will become available in 1995; all four telescopes should be ready in 1998.

The VLT will have a total of 17 foci, most of which will have stationary instruments with the possibility of rapid switching between operation modes and foci. When the four telescopes are interferometrically coupled, the VLT will achieve angular resolutions of the order of 0.5 milliarc second, unequalled by any other ground-based optical facility.

The NTT is in fact a prestudy for the Very Large Telescope. The scaling factor from the 3.5-m NTT mirror to the 8.2-m unit telescopes is 2.3, a reasonably conservative step in engineering science. Still, the VLT project represents a major technological challenge, especially in regard to the manufacture and maintenance of the very large mirrors.

Optical astronomy is in a phase of rapid development, and with our new telescopes, on the ground as well as in space, we may look forward to a wealth of exciting new information about the universe during the next years.

VLT

7

Our Solar System

With telescopes and space probes, astronomers have amassed a rich body of information about our solar system. Perhaps the most obvious fact is that the planets fall into two distinct classes. Small, rocky planets like the Earth are found near the Sun, while huge, gaseous planets, like Jupiter, are located far from the Sun. This dichotomy is a direct result of conditions under which the planets formed. Seven large moons or satellites, which can be classified as Earthlike planets, and numerous smaller satellites, asteroids, and comets also populate the solar system. Spectroscopy provides crucial information about the chemical composition of the planets. The material from which the solar system formed is debris left over from generations of stars that lived out their lives and shed their matter into space long before the Sun was born. Our solar system was probably created from a huge cloud of interstellar gas and dust that contracted under the force of its own gravity. Most of the matter fell toward the center of that contracting cloud to form the Sun; outlying material accumulated in clumps to form the planets. Although many details of this process remain to be elucidated, we seem to have a basic understanding of the creation of the solar system.

Nebulae in Sagittarius   Planets are probably forming along with new stars in these nebulae (NGC 6559 and IC 1274-5) in Sagittarius. The type of planet to form at a particular distance from a star depends on conditions such as the temperature and substances (rock fragments, ice crystals, and gases) existing at that distance. In our solar system, planets composed primarily of rock formed near the Sun, whose heat drove off ices and gases. Far from the Sun, where temperatures are low, planets were able to retain volatile substances, resulting in worlds composed primarily of gases and ices. (Royal Observatory, Edinburgh)

Looking up at the heavens and wondering about the nature and origin of the Sun, Moon, stars, and planets is a common human experience. Unlike our ancestors, however, we possess a wealth of information about the universe. Especially within the past few decades, telescopic observations and interplanetary spacecraft have given us vast quantities of data from which to draw ideas and to formulate and test theories. The answers are out there, within our grasp. We see the Sun, planets, moons, asteroids, comets, and meteoroids that make up our niche in the universe. Many of these objects are exceedingly ancient, old enough to contain records of the cosmic events that created our solar system.

In addition to gleaning information from the objects that orbit the Sun, we can observe active star formation occurring elsewhere in our Galaxy. Stars and planetary systems are now being formed in many beautiful nebulae scattered across the heavens, as shown at the beginning of this chapter. A general understanding of star creation, coupled with knowledge of the Sun and its satellites, gives us a fairly comprehensive picture of how our solar system was created. Many details still need to be worked out, but the overall scenario seems remarkably sound. For the first time, we can truly appreciate what is unique and what is commonplace about our world. We have begun to fathom our connection with the rest of the cosmos and our place in the universe.

7-1 The planets are classified as either terrestrial or Jovian by their physical attributes

A brief overview of the solar system distinguishes two classes of planets. Notice the striking dichotomy in the orbits of the planets, as shown in Figure 7-1. The orbits of the four inner planets (Mercury, Venus, Earth, and Mars) are crowded in close to the Sun. In contrast, the orbits of the next four planets (Jupiter, Saturn, Uranus, and Neptune) are widely spaced at great distances from the Sun.

As you might expect, the range of surface temperatures that each planet experiences is generally related to its distance from the Sun. (Review Box 5-1 with its discussion of temperature measurements.) The four inner planets are quite warm. For example, noontime temperatures on Mercury climb to 600 K (= 327°C = 621°F), and at noon in the middle of summer on Mars, it is sometimes as warm as 300 K (= 27°C = 81°F). Of course, the outer planets, which receive much less solar radiation, are cooler. Typical temperatures range from about 150 K (= −123°C = −189°F) in Jupiter's cloudtops to 63 K (= −210°C = −346°F) on Neptune. Temperature plays a major role in determining whether various substances exist as solids, liquids, or gases, thereby profoundly affecting the appearance of the planets.

Most of the planets' orbits are nearly circular. As discussed in Chapter 4, Kepler discovered that these orbits are actually ellipses. Astronomers denote the elongation of an ellipse by its **eccentricity** (review Box 4-3 for details). The eccentricity of a circle is zero. Most planets have orbital eccentricities that are very close to zero. The exceptions are Mercury and Pluto. Pluto's noncircular orbit sometimes takes it nearer the Sun than its neighbor, Neptune.

The planetary orbits all lie in nearly the same plane. In other words, the orbits of the planets are inclined at only slight angles to the plane of the ecliptic. Again, however, Pluto is a notable exception. The plane of Pluto's orbit is tilted at about 17° to the plane of the Earth's orbit (see Table 7-1).

As we compare the physical properties of the planets, we again find that they fall naturally into two classes—the four inner planets and the four outer ones—with Pluto as an exception. Some of the most important properties of the planets are their diameters, masses, and densities.

Table 7-1 Orbital characteristics of the planets

	Mean distance from Sun		Orbital period (years)	Eccentricity	Inclination to the ecliptic (°)
	(AU)	(10^6 km)			
Mercury	0.39	58	0.24	0.206	7.0
Venus	0.72	108	0.62	0.007	3.4
Earth	1.00	150	1.00	0.017	0.0
Mars	1.52	228	1.88	0.093	1.9
Jupiter	5.20	778	11.86	0.048	1.3
Saturn	9.53	1426	29.41	0.056	2.5
Uranus	19.19	2870	84.04	0.046	0.8
Neptune	30.06	4497	164.8	0.010	1.8
Pluto	39.53	5914	248.6	0.248	17.1

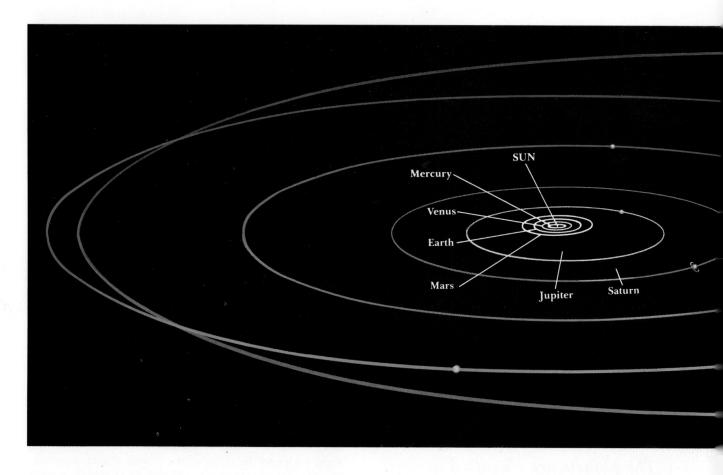

Figure 7-1 The solar system   *This scale drawing shows the distribution of planetary orbits around the Sun. The four inner planets are crowded in close to the Sun, while the five outer planets orbit it at much greater distances.*

The diameter of a planet can be computed from its apparent angular diameter and distance (recall Box 1-1). For example, at its greatest western elongation in 1990, Venus was 1.0158×10^8 km from Earth and had an angular diameter of 24.58 arc sec. Using the small-angle formula, we find that the diameter of Venus is 12,104 km (7523 mi). Similar calculations demonstrate that the other three inner planets are also quite small. Indeed, the Earth, with its diameter of about 12,760 km (7930 mi), is the largest of the four. In sharp contrast, the four outer planets are much larger. First place goes to Jupiter, whose equatorial diameter is about 143,800 km (89,370 mi). Pluto, however, is even smaller than the inner planets, despite its position as the outermost

Table 7-2   Physical characteristics of the planets

	Equatorial diameter		Mass		Average density	
	(km)	(Earth = 1)	(kg)	(Earth = 1)	(kg/m³)	(g/cm³)
Mercury	4,878	0.38	3.3×10^{23}	0.06	5400	5.4
Venus	12,104	0.95	4.9×10^{24}	0.82	5200	5.2
Earth	12,756	1.00	6.0×10^{24}	1.00	5500	5.5
Mars	6,794	0.53	6.4×10^{23}	0.11	3900	3.9
Jupiter	142,800	11.19	1.9×10^{27}	317.8	1300	1.3
Saturn	120,000	9.26	5.7×10^{26}	95.3	700	0.7
Uranus	51,120	4.01	8.7×10^{25}	14.6	1300	1.3
Neptune	49,528	3.88	1.0×10^{26}	17.2	1700	1.7
Pluto	2,300	0.18	1.3×10^{22}	0.002	2000	2.0

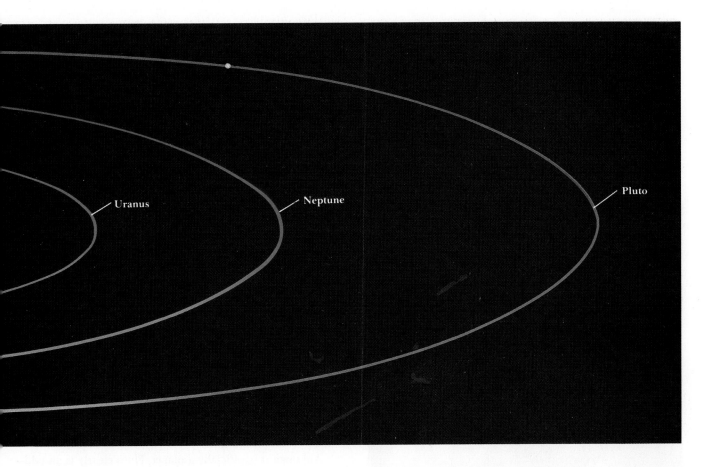

planet. Its diameter is only about 2290 km (1420 mi). Figure 7-2 shows the Sun and the planets drawn to the same scale. The diameters of the planets are given in Table 7-2.

Determining the mass of a planet is most easily accomplished if the planet has a satellite. Since the satellite obeys Kepler's third law, astronomers can use formulas like those in Box 4-3 to calculate the planet's mass, if the satellite's period and semimajor axis are known. If the planet does not have a satellite, astronomers must rely on observations of a comet or spacecraft that passes near the planet. The planet's gravity (which is directly related to the planet's mass) will produce a deflection in the path of the comet or spacecraft. By measuring the size of this deflection and using Newtonian mechanics, astronomers can determine the planet's mass.

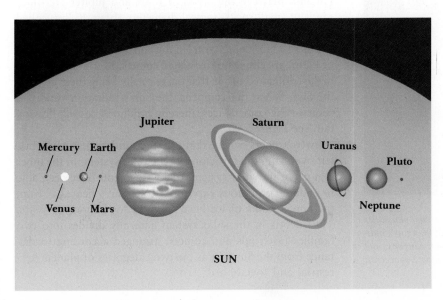

Figure 7-2 The Sun and the planets *This drawing shows the nine planets in front of the disk of the Sun, with all ten bodies drawn to the same scale. The four planets that have orbits nearest the Sun (Mercury, Venus, Earth, and Mars) are small and are made of rock. The next four planets from the Sun (Jupiter, Saturn, Uranus, and Neptune) are large and are composed primarily of gas.*

The four inner planets have low masses, but the next four planets have substantially greater masses (see Table 7-2). Again, first place goes to Jupiter, whose mass is 318 times greater than Earth's.

Average density (mass divided by volume) is a physical property that can often be used to deduce important information about the composition of an object. For convenience's sake, scientists commonly measure average density in grams per cubic centimeter but, by international agreement, express it in kilograms per cubic meter. The four inner planets have very large average densities (see Table 7-2); the average density of the Earth is 5500 kg/m³. For comparison, the average density of a typical rock is about 3000 kg/m³ and the average density of water is 1000 kg/m³. The Earth must therefore contain a large amount of material that is denser than rock. This information provides our first clue that Earthlike planets have iron cores.

In sharp contrast, the outer planets have quite low densities. Saturn has an average density less than that of water. This information strongly suggests that the giant outer planets are composed primarily of such light elements as hydrogen and helium. Pluto is again an exception. Although Pluto is even smaller than the dense inner planets, its average density seems to be much like those of the giant outer planets.

These differences in size, mass, and density lead us to consider the four inner and the four outer planets as two distinct groups. The four inner planets are called **terrestrial planets** because they resemble the Earth (in Latin, "terra"). They are small and dense. Craters, canyons, and volcanoes are common on their hard, rocky surfaces (see Figure 7-3). The outer four planets are called **Jovian planets** because they resemble Jupiter (Jove was another name for the Roman god Jupiter). Vast swirling cloud formations dominate the appearance of these enormous gaseous spheres (see Figure 7-4). These planets probably have solid cores roughly the same size as the Earth, buried beneath atmospheres that are tens of thousands of kilometers thick.

Although it is officially a planet, Pluto clearly is an oddity. Its physical properties are not typical of either the terrestrial or the Jovian planets. Some astronomers suggest that Pluto is a moon that had escaped from Neptune. Many of the satellites of outer planets are composed of ice, which has a density of about 1000 kg/m³. Pluto's density is about 2100 kg/m³, so a significant fraction of Pluto is probably composed of ice.

In addition to the nine planets, many smaller objects orbit the Sun. Between the orbits of Mars and Jupiter are thousands of small rocks (typical diameters about 40 km) called **asteroids**. The largest asteroid, Ceres, has a diameter of about 750 km. Quite far from the Sun, well beyond the orbit of Pluto, are chunks of ice called **comets**. Many comets have highly elongated orbits that occasionally bring them close to the Sun. When this happens, the Sun's radiation vaporizes some of the comet's ices, thereby producing a long flowing tail (see Figure 7-5).

Astronomers believe that asteroids and comets are debris left over from the formation of the solar system. In the inner regions of the solar system, rocky fragments have been able to endure continuous exposure to the Sun's heat. Far from the Sun, chunks of ice have survived for billions of years. Thus debris in the solar system naturally divides into two families (asteroids and comets) arranged according to distance from the Sun just as the two categories of planets (terrestrial and Jovian) are.

Figure 7-3 *A terrestrial planet* Mars is a typical terrestrial planet. Its diameter is only 6800 km, and its average density is 3900 kg/m³, indicating that the planet is composed of rock. Volcanoes, canyons, and craters can be seen in this photograph taken by the Viking 2 spacecraft in August 1976. (NASA)

Figure 7-4 *A Jovian planet* *Jupiter is the largest of the Jovian planets. Its equatorial diameter is 143,800 km and its average density is only 1300 kg/m³, indicating that the planet is composed primarily of light elements. Two of Jupiter's moons, Io and Europa, are seen in this view taken by the* Voyager 1 *spacecraft in February 1979. Each of these satellites is approximately the same size as Earth's moon. (NASA)*

7-2 Seven large satellites can also be classified as terrestrial planets

All the planets except Mercury and Venus have satellites. More than fifty satellites are known (Jupiter, Saturn, and Uranus each have at least fifteen). Certainly, dozens of other satellites remain to be discovered. The known satellites fall into two distinct categories. Seven are giant satellites that are each roughly comparable to Mercury in size. They are listed in Table 7-3 with relevant data. All the other satellites are much smaller, with diameters less than 2000 km.

Table 7-3   The seven giant satellites

Satellite	Parent planet	Diameter (km)	Average density (kg/m³)	(g/cm³)
Moon	Earth	3476	3300	3.3
Io	Jupiter	3630	3600	3.6
Europa	Jupiter	3138	3000	3.0
Ganymede	Jupiter	5262	1900	1.9
Callisto	Jupiter	4800	1800	1.8
Titan	Saturn	5150	1900	1.9
Triton	Neptune	2720	2000	2.0

Figure 7-5   A comet *The solid part of a comet is a chunk of ice roughly 10 km in diameter. When a comet passes near the Sun, solar radiation vaporizes some of the comet's ices and the resulting gases form a tail millions of kilometers long. This photograph shows a comet that was seen in January 1974. (NASA)*

The interplanetary missions of *Voyager 1* and *Voyager 2* revealed many fascinating characteristics of these giant satellites (see Figure 7-6). For example, Jupiter's satellite Io is one of the most geologically active worlds in the solar system, with numerous volcanoes belching forth sulfur-rich compounds. Saturn's largest satellite, Titan, has an atmosphere nearly twice as dense as Earth's atmosphere. Figure 7-7 shows all seven giant satellites along with Mercury, all to the same scale.

Because of their sizes, it is reasonable to include these seven giant satellites in a listing of the planets, despite their low densities and even though they happen to orbit other planets instead of orbiting the Sun independently. The hard surfaces (of either rock or ice) of these satellites would place them in the terrestrial category. Thus expanding our definition of a terrestrial planet gives us ten Earthlike worlds that we can compare with our own planet.

Figure 7-6 A giant satellite *This view, taken from* Voyager 2 *in July 1979, shows Callisto, the second largest moon of Jupiter. Its diameter is almost exactly the same as that of Mercury. Numerous craters pockmark Callisto's icebound surface. (NASA)*

Figure 7-7 The smaller terrestrial worlds *Mercury, our Moon, and the largest satellites of Jupiter, Saturn, and Neptune are shown here to the same scale. Each of these worlds has its own unique characteristics, and all are large enough to be classified as terrestrial planets. Only Titan possesses a substantial atmosphere. (NASA)*

7-3 Spectroscopy reveals the chemical composition of the Sun, planets, and stars

Astronomers have learned a great deal about the composition of the planets. The most accurate determinations have come from spacecraft that have landed on planets and made direct chemical analyses of the atmosphere or soil. Unfortunately, we have thus far obtained such direct information for only four worlds: Venus, the Moon, Mars, and the Earth. In all other cases, astronomers must rely on their ability to analyze the sunlight reflected from the distant planets and their satellites. Under these circumstances, astronomers bring to bear one of their most powerful tools, spectroscopy.

In Chapter 5, we saw that white light can be separated into a spectrum of colors (review Figure 5-3). We also learned that each chemical element produces its own unique pattern of spectral lines (Figure 5-12). **Spectroscopy** is the systematic study of spectra and spectral lines.

Spectral lines provide extremely reliable evidence about the chemical composition of distant objects. If the spectral lines of a certain chemical are found in a star's spectrum, it is safe to conclude that this chemical is present in that star's atmosphere. For example, a portion of the Sun's spectrum is shown in Figure 7-8. Also shown is the spectrum of iron over the same wavelength range. This pattern of spectral lines is iron's own distinctive "fingerprint," which no other chemical can imitate. The fact that some absorption lines in the Sun's spectrum coincide with the iron lines means that some vaporized iron must exist in the Sun's atmosphere.

Figure 7-8 Iron in the Sun *The upper spectrum is a portion of the Sun's spectrum from 420 to 430 nm. Numerous dark spectral lines are visible. The lower spectrum is a corresponding portion of the spectrum of vaporized iron. Several bright spectral lines are seen against a black background. The fact that the iron lines coincide with some of the solar lines proves that there is some iron (albeit a very tiny amount) in the Sun's atmosphere. (Mount Wilson and Las Campanas Observatories)*

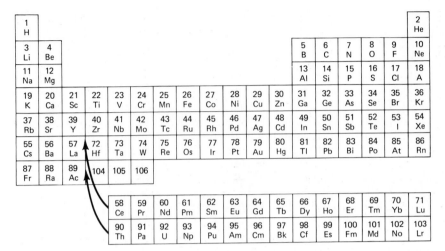

Figure 7-9 The periodic table of the elements The *periodic table is a convenient listing of the elements, arranged according to their atomic weights and chemical properties.*

Over the years, astronomers have used spectrographs (review Figures 6-18 and 6-20) to examine and record the spectra of stars and galaxies. As data accumulated, an important fact emerged. No matter where we look—no matter how far we peer into space—we always find the same chemical elements. These naturally occurring elements of which the Earth is made are the same building blocks for all matter everywhere in the universe. The elements are fundamental because they cannot be broken into more basic chemicals.

A listing of the chemical elements is most conveniently displayed in the form of a **periodic table** (see Figure 7-9). Each element is assigned a unique **atomic number**. Elements are arranged in the periodic table in order of increasing atomic numbers, and with only a few exceptions, this sequence also corresponds to increasing average mass of the atoms of the elements. Thus hydrogen (symbol H), with atomic number 1, is the lightest element. As another example, iron (symbol Fe) has atomic number 26 and is a relatively heavy element. All the elements that appear in a single vertical column of the periodic table have similar chemical properties. For example, the elements listed in the far right column are all gases under Earth-surface conditions of temperature and pressure, and they are all very reluctant to react chemically with other elements.

In addition to listing nearly 100 naturally occurring elements, Figure 7-9 gives several artificially produced elements. All these elements are heavier than uranium

(symbol U) and are highly radioactive, which means that they decay into lighter elements within a short time of being created in laboratory experiments.

An **atom** is the smallest possible piece of an element. Typical atomic diameters are about 0.1 nm. Atoms can be broken down into three basic types of subatomic particles: protons, neutrons, and electrons. Box 7-1 discusses some important details of atomic structure.

Although the number of elements is limited, a wide variety of substances exist in the universe, because atoms can combine to form various **molecules**. For example, two hydrogen atoms can combine with an atom of oxygen to form a molecule of water. Water's chemical formula, H_2O, describes the atoms in its molecule. In a similar way, two oxygen atoms can bond with a carbon atom to produce a molecule of carbon dioxide, whose formula is CO_2. The chemical formula for common table salt is $NaCl$, which means that salt contains one sodium atom combined with each chlorine atom.

Molecules, like atoms, also produce unique patterns of lines in the spectra of astronomical objects. For example, Figure 7-10, showing a portion of the spectrum of Venus, has a series of spectral lines caused by carbon dioxide. It is therefore reasonable to conclude that Venus's atmosphere contains this gas. Indeed, direct measurements by Soviet and American spacecraft have confirmed that carbon dioxide gas forms 97 percent of the Venusian atmosphere.

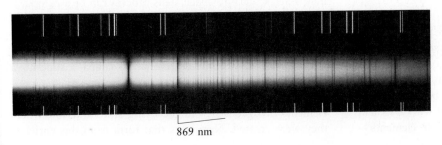

869 nm

Figure 7-10 Carbon dioxide on Venus This near-infrared spectrum of Venus shows a series of spectral lines beginning at 869 nm that is caused by carbon dioxide, the primary constituent of Venus's atmosphere. (Lick Observatory)

Box 7-1 Atoms and isotopes

Laboratory experiments during the first part of the twentieth century revealed that the structure of an atom superficially resembles a tiny solar system, as shown in the diagram. Most of the mass of an atom is concentrated in a dense **nucleus** that is composed of **protons** and **neutrons**. Orbiting this massive nucleus like small planets are **electrons**. Whereas the solar system is held together by gravitational forces, atoms are held together by electrical forces. Each proton carries a positive charge, and each electron carries an equal but opposite negative charge. The electric forces attracting the positively charged protons and the negatively charged electrons keep the atom from coming apart.

Although electrons and protons have equal but opposite charges, they have unequal masses. The electron is one of the lightest subatomic particles, with a mass of only 9.1×10^{-31} kg. The proton is nearly 2000 times heavier: Its mass is 1.7×10^{-27} kg.

A neutron has almost exactly the same mass as a proton. As its name suggests, a neutron has no electric charge—it is electrically neutral. Neutrons serve as buffers between the positively charged protons that are crowded together inside the nucleus. Typically, there are more neutrons than protons in a nucleus, especially in the case of the heaviest elements.

In the elemental state, the number of electrons orbiting an atom is equal to the number of protons in the nucleus, making the atom electrically neutral. The number of protons in an atom's nucleus equals the atomic number for that particular element. A hydrogen nucleus has one proton, a helium nucleus has two protons, and so forth—up to uranium, with 92 protons in its nucleus.

The number of protons in the nucleus of an atom uniquely determines what element that atom is. Nevertheless, the same element may have slightly different numbers of neutrons in its nuclei. For example, consider oxygen, the eighth element on the periodic table. Its atomic number is 8, and every oxygen nucleus

has exactly 8 protons, but it can have 8, 9, or 10 neutrons. Thus, three slightly different kinds of oxygen, called **isotopes**, exist. The isotope with 8 neutrons is by far the most abundant variety. It is written as ^{16}O, or oxygen-16. The rarer isotopes with 9 and 10 neutrons are designated as ^{17}O and ^{18}O, respectively.

The superscript that precedes the chemical symbol for an element is equal to the total number of protons and neutrons in a nucleus of that particular isotope. For example, the common isotope of iron is ^{56}Fe, or iron-56, which means that its nucleus contains a total of 56 protons and neutrons. From the periodic table, however, we see that the atomic number of iron is 26. This means that every iron atom has 26 protons in its nucleus. Therefore, the number of neutrons in an iron-56 nucleus is $56 - 26 = 30$.

It is extremely difficult to distinguish chemically between the various isotopes of a particular element. Ordinary chemical reactions involve only the electrons that orbit the atom, never the neutrons buried in its nucleus.

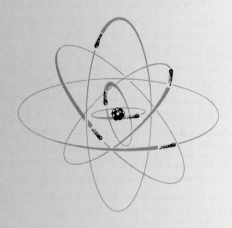

7-4 The relative abundances of the elements are the result of cosmic processes

Some elements are very common, but others are very rare. Hydrogen is by far the most abundant substance in the universe—it makes up nearly three-quarters of the mass of all the stars and galaxies in the universe.

Helium is the second most abundant element. Together, hydrogen and helium account for about 98 percent of the mass of all the material in the universe. In other words, only 2 percent is left for all the other elements combined.

There is a good reason for this overwhelming abundance of hydrogen and helium. Most astronomers believe that the universe began roughly 20 billion years ago with a violent event called the Big Bang. Only the lightest elements—

hydrogen, helium, and a tiny amount of lithium—emerged from the enormously high temperatures following this cosmic event. All the other elements were manufactured by stars later, either by nuclear fusion reactions deep in their interiors or by the violent explosions that mark the end of massive stars' lives. Were it not for stars, there would be no heavy elements in the universe today.

Near the ends of their lives, some stars cast much of their matter back out into space. This process can be a comparatively gentle one in which a star's outer layers are gradually expelled. Figure 7-11 shows the star HD 65750, which is losing material in this fashion. Alternatively, a star may end its life with a spectacular detonation called a **supernova explosion**, which blows the star apart. Either way, the interstellar gases in the galaxy become enriched with heavy elements dredged up from the dying star's interior, where they were created. New stars that form from this enriched

Figure 7-11   The mass-loss star HD 65750   This star is shedding material rapidly and is surrounded by nebulosity (called IC 2220) made visible by light reflecting from dust grains. These dust grains may have condensed from material shed by the star. (Anglo-Australian Observatory)

material thus have an ample supply of heavy elements from which to develop a system of planets, satellites, comets, and asteroids.

Our solar system is made of matter created in stars that existed billions of years ago. The Sun is a fairly young star, only 5 billion years old. All the heavy elements in our solar system were created and cast off by ancient stars during the first 10 billion years of our Galaxy's existence. We are literally made of star dust (see Figure 7-12).

Stars create different elements in different amounts. For example, the elements carbon, oxygen, silicon, and iron are readily produced in the interiors of massive stars, whereas

gold is created only under special circumstances. Therefore, gold is rare in our solar system but carbon is not.

A convenient way to express the relative abundances of the various elements is to say how many atoms of a particular element are found for every trillion (that is, 10^{12}) hydrogen atoms. For example, for every trillion hydrogen atoms in space, there are about 60 billion (6×10^{10}) helium atoms. From spectral analysis of stars and chemical analysis of Earth rocks, Moon rocks, and meteorites, scientists have been able to determine the relative abundance of the elements existing in our part of the Galaxy. The ten most abundant elements are listed in Table 7-4.

In addition to these ten very common elements, there are five elements that are moderately abundant: sodium, alumi-

Figure 7-12   A dusty region of star formation   These young stars in the constellation of Orion are still surrounded by much of the gas and dust from which they formed. The bluish wispy nebulosity is caused by starlight reflecting off abundant interstellar dust grains. These grains are made of heavy elements produced by earlier generations of stars. (Anglo-Australian Observatory)

Table 7-4   Abundances of the most common elements

Atomic number	Element	Symbol	Relative abundance
1	Hydrogen	H	10^{12}
2	Helium	He	6×10^{10}
6	Carbon	C	4×10^{8}
7	Nitrogen	N	9×10^{7}
8	Oxygen	O	7×10^{8}
10	Neon	Ne	4×10^{7}
12	Magnesium	Mg	4×10^{7}
11	Silicon	Si	5×10^{7}
16	Sulfur	S	2×10^{7}
26	Iron	Fe	3×10^{7}

Box 7-2   Thermal motion and the retention of an atmosphere

A moving object possesses energy. The faster it moves, the more energy it has. Energy of this type is called **kinetic energy**. If an object of mass m is moving with a speed v, its kinetic energy is given by

$$\frac{1}{2} mv^2$$

This expression for kinetic energy is valid for all objects, both big and small, from atoms and molecules to planets and stars, as long as their speed is small in relation to the speed of light. If the mass is expressed in kilograms and the speed in meters per second, the energy is expressed in joules (J).

Example: A 2-kg object moving at 300 m/s has a kinetic energy of

$$\frac{1}{2}(2 \times 300^2) = 90{,}000 \text{ J}$$

Any substance with a temperature above absolute zero possesses thermal energy. **Thermal energy** is the kinetic energy of the moving atoms or molecules in a substance. In a cold substance, the atoms move slowly; in a hot substance, they move much more rapidly.

Consider a gas, such as the atmosphere of a star or planet. If the gas is hot, it probably consists of individual atoms moving at high speeds. For example, the Sun's atmosphere consists primarily of hydrogen and helium atoms. If the gas is cool, the atoms typically combine to form molecules. For example, the Earth's atmosphere consists primarily of nitrogen (N_2) and oxygen (O_2) molecules, and the Martian atmosphere is mostly carbon dioxide (CO_2) molecules.

According to the kinetic theory of gases developed during the nineteenth century, the average amount of thermal energy per atom or molecule in a gas is given by

$$\frac{3}{2} kT$$

where T is the temperature of the gas in kelvins and k is the Boltzmann constant, which has the value k = 1.38×10^{-23} J/K.

The thermal energy of an atom or molecule is just kinetic energy, so we can write the equality

$$\frac{1}{2} mv^2 = \frac{3}{2} kT$$

where v represents the average speed of an atom or molecule in a gas with temperature T. Rearranging this equation, we obtain

$$v = \sqrt{\frac{3kT}{m}}$$

(This value is actually slightly higher than the average speed of the atoms or molecules in the gas, but it is close enough for our purposes here. If you are studying physics, you may know that v is actually the root-mean-square average speed.)

Example: Suppose you want to know the average speed of the oxygen molecules that you breathe at a room temperature of 72°F (= 22°C = 295 K). From a reference book, you can find that the mass of an oxygen atom is 2.66×10^{-26} kg. The mass of an oxygen molecule (O_2) is twice the mass of an oxygen atom, or $2(2.66 \times 10^{-26}$ kg) = 5.3×10^{-26} kg. Thus the average speed is

---

num, argon, calcium, and nickel. These elements have abundances in the range of 10^6 to 10^7 relative to the standard trillion hydrogen atoms (see Figure 7-13). Most of the other elements are much rarer. For example, for every trillion hydrogen atoms in the solar system, there are only six atoms of gold.

7-5 The planets were formed by the accumulation of material in the solar nebula during the birth of the Sun

Hydrogen and helium are the most common elements in the universe, but they are not plentiful on the four inner terrestrial planets. Because of warmth from the Sun, temperatures on the four inner planets are comparatively high, and the

higher the temperature of a substance, the greater the speed of its atoms. In the warm inner solar system, the lightweight atoms of hydrogen and helium move swiftly enough to escape from the relatively weak gravity of the comparatively small terrestrial planets.

A different situation prevails on the four Jovian planets. Far from the Sun, temperatures are low, and the atoms of hydrogen and helium move slowly. The relatively strong gravity of the massive Jovian planets easily prevents the lightest gases from escaping into space. In fact, the Jovian planets are composed primarily of hydrogen and helium. (See Box 7-2 for a discussion of the thermal motion of atoms and the ability of a planet's gravity to retain an atmosphere.)

Temperature must also have been an important factor in determining the specific conditions inside the vast cloud of gas and dust called the **solar nebula**, from which the solar system formed. For this reason, it is extremely useful to categorize the more abundant elements and their common

$$v = \left(\frac{3(1.38 \times 10^{-23})(295)}{5.3 \times 10^{-26}} \right)^{1/2} = 4.8 \times 10^2 \text{ m/s} = 0.48 \text{ km/s}$$

or almost exactly 1000 miles per hour.

Obviously, the atoms and molecules in a gas are moving rapidly, even at moderate temperatures. The speeds may even be so great that a planet's gravity is not strong enough to retain an atmosphere.

Astronomers find it useful to speak of the **escape velocity** (or speed) from a planet in trying to decide if an object can permanently leave the planet and escape into interplanetary space. According to Newtonian mechanics, the escape velocity from a planet of mass M and radius R is given by

$$V_{\text{escape}} = \sqrt{\frac{2GM}{R}}$$

where G is the universal constant of gravitation (G = 6.67 × 10^{-11} N m²/kg²).

The table gives the escape velocity for various objects in the solar system. For example, astronauts must leave the Earth with a speed greater than 11.2 km/s (25,100 mi/hr).

In a planet's atmosphere, some molecules are moving slowly, others more swiftly. Nevertheless, the average speed of a particular kind of molecule can be calculated if you know the planet's atmospheric temperature. A good rule of thumb is that a planet can retain a gas if the escape velocity is at least six times greater than the average speed of the molecules in the gas. In such a case, very few molecules will be moving fast enough to escape from the planet's gravity.

Example: Consider the Earth's atmosphere. We saw that the average speed of oxygen molecules is 0.48 km/s at room temperature. The escape velocity from the Earth (11.2 km/s) is much more than six times the average speed of the oxygen molecules, so the Earth has no trouble keeping oxygen in its atmosphere.

A similar calculation for hydrogen molecules (H_2) gives a different result, however. At 295 K, the average speed of a hydrogen molecule is 1.9 km/s. Six times this speed is 11.4 km/s, which is slightly higher than the escape velocity from the Earth. Thus, the Earth does not retain hydrogen in its atmosphere. Any hydrogen released into the air we breathe slowly leaks away into space.

	Escape velocity	
	(km/s)	(mi/hr)
Mercury	4.3	9,600
Venus	10.3	23,000
Earth	11.2	25,100
Moon	2.4	5,400
Mars	5.0	11,000
Jupiter	59.5	133,000
Saturn	35.6	79,600
Uranus	21.2	47,400
Neptune	23.6	52,800

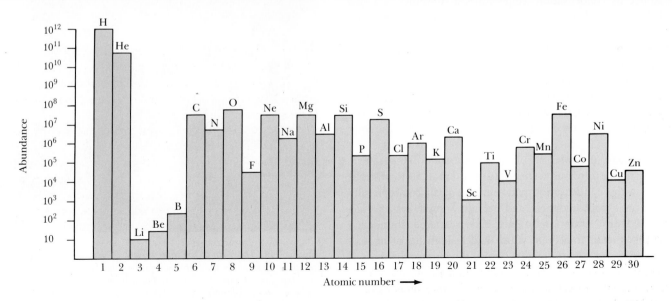

Figure 7-13 *Abundances of the lighter elements This graph shows the abundances of the thirty lightest elements (hydrogen through zinc) compared to a value of 10^{12} for hydrogen. All elements heavier than zinc have abundances less than 1000 atoms per trillion atoms of hydrogen.*

Table 7-5 Common planet-forming substances

Gas	Ice	Rock
Hydrogen (H)	Water (H_2O)	Iron (Fe)
Helium (He)	Methane (CH_4)	Iron sulfide (FeS)
Neon (Ne)	Ammonia (NH_3)	Olivine ($(Mg,Fe)SiO_4$)
——	Carbon dioxide (CO_2)	Pyroxene ($CaMgSi_2O_6$)

compounds according to their behaviors at various temperatures.

First, hydrogen and helium are gaseous, except at extremely low temperatures and extraordinarily high pressures. Second, the rock-forming compounds of iron and silicon are solids except at temperatures exceeding 1000 K. Finally, there is an intermediate class of substances such as water, carbon dioxide, methane, and ammonia. At low temperatures (typically below 200 to 300 K), these common chemicals solidify into the solids called ices. At somewhat higher temperatures, they can exist as liquids or gases. In Table 7-5, some of the common planet-forming substances are listed according to these three broad classifications.

By observing the process of star formation elsewhere in our Galaxy, astronomers can deduce the conditions that led to the formation of our own solar system. Just before the birth of the Sun, atoms in the solar nebula were so widely spaced that no substance could exist as a liquid. Matter in this vast cloud existed either as a gas or as tiny grains of dust and ice. Under such conditions of extremely low pressure, astronomers find it useful to speak of a substance's **condensation temperature** to specify whether it is a solid or a gas. Above its condensation temperature, a substance is a gas; below its condensation temperature, the substance solidifies into tiny specks of dust or snowflakes.

Rock-forming substances have high condensation temperatures, typically in the range of 1300 to 1600 K. Ices have much lower condensation temperatures, ranging from 100 to 300 K. The condensation temperatures of hydrogen and helium are so near absolute zero that these common elements always existed as gases during the creation of the solar system.

Initially, the solar nebula was quite cold. Temperatures throughout the cloud were probably less than 50 K, which is below the condensation temperatures of all common substances except hydrogen and helium. Snowflakes and ice-coated dust grains must have been scattered abundantly across the solar nebula, which had a diameter of at least 100 AU and a mass roughly two to three times the mass of the Sun.

The gravitational pull of particles on one another caused them to begin a general drift toward the center of the solar nebula. Density and pressure at the center of the solar nebula

thus began to increase, producing a concentration of matter called the **protosun**. As the solar nebula continued to contract, atoms in the cloud collided with one another with increasing speed and frequency, thereby causing temperatures deep inside the solar nebula to climb. This process, whereby the gravitational energy of a contracting gas cloud is converted into thermal energy, is called **Helmholtz contraction**, after the nineteenth-century German physicist who first described it.

The solar nebula also must have had an overall slight amount of rotation—or **angular momentum**, as it is properly called (recall Box 4-4). Otherwise, everything would have fallen straight into the protosun, leaving nothing behind to form the planets. Mathematical studies show that the combined effects of gravity and angular momentum can transform a shapeless cloud into a rotating flattened disk that is warm at the center and cold at its edges, as sketched in Figure 7-14. The transformation of the solar nebula into a flattened disk explains why the orbits of the planets are all

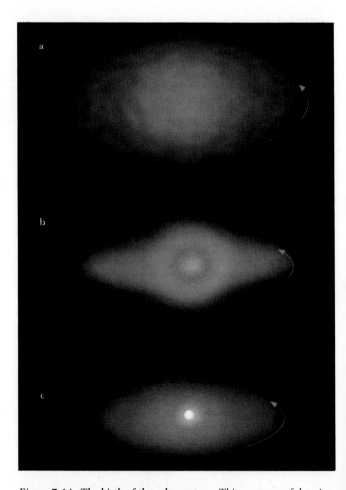

Figure 7-14 The birth of the solar system *This sequence of drawings shows three stages in the formation of the solar system. In (a) a slowly rotating cloud of interstellar gas and dust begins to contract because of its own gravity. A central condensation forms in (b) as the cloud flattens and rotates faster. In (c) a flattened disk of gas and dust surrounds the young Sun, which has begun to shine.*

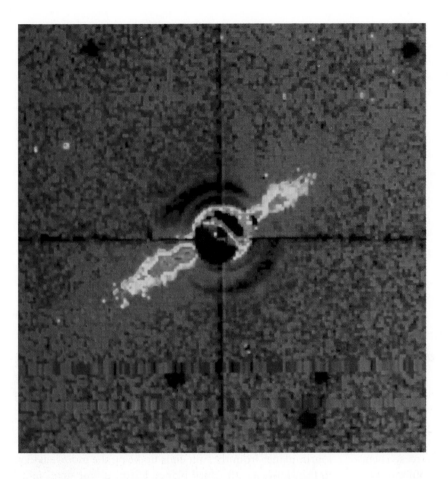

Figure 7-15 A circumstellar disk of matter This computer-enhanced photograph made with a CCD shows a disk of matter orbiting the star β Pictoris. The star is located at the center of the picture, behind a small circular mask that was needed to block the star's light, which would otherwise have overwhelmed the light from the disk. The disk, seen nearly edge-on, is believed to be very young, possibly no more than a few hundred million years old. (University of Arizona and JPL)

nearly in the same plane. Furthermore, astronomers find disks of material surrounding other stars, like the one shown in Figure 7-15. Planets may still be forming in this disk of debris left over from the birth of that star.

Temperatures around the newly created protosun soon climbed to 2000 K. Meanwhile, temperatures in the outer-most regions of the solar nebula remained at less than 50 K. Figure 7-16 shows the probable temperature distribution throughout the solar nebula at the end of this preliminary stage in the formation of our solar system. Obviously, all the common icy substances in the inner regions of the solar nebula were vaporized by the high temperatures. Only the rocky

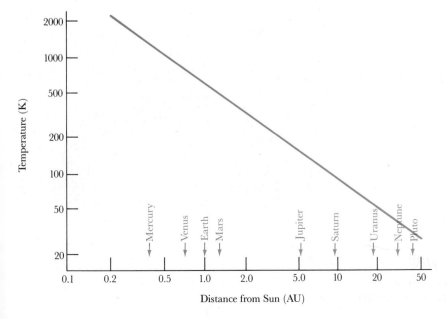

Figure 7-16 Temperature distribution in the solar nebula This graph shows how temperatures probably varied across the solar nebula as the planets were forming. For the inner planets, temperatures ranged from roughly 1200 K at Mercury to 500 K at the orbit of Mars. Beyond Jupiter, temperatures were everywhere less than 200 K.

substances remained solid. That is why the four inner planets are today composed primarily of dense, rocky material. In contrast, snowflakes and ice-coated dust grains were able to survive in the cooler, outer portions of the solar nebula. That is why the Jovian planets have such low average densities. In fact, spectral lines of methane and ammonia have been identified in the spectra of some of the outer planets. In addition, many of the satellites of the Jovian planets are either partially or almost entirely composed of ices.

As we shall see in Chapter 17, a similar dichotomy exists in the debris left over from the formation of the solar system. Near the Sun this debris consists of rocks called asteroids and meteoroids. Far from the Sun the debris is covered with substantial amounts of frozen gases. If one of these icy chunks passes near the Sun, some of its gases are vaporized, resulting in a comet.

7-6 Computer simulations help us understand details of the formation of the planets

The formation of the four inner planets was dominated by the fusing of solid, rocky particles. Initially, neighboring dust grains and pebbles in the solar nebula collided and clung together by electrostatic or gravitational forces. Then, over a period of a few million years, these accumulations of dust and pebbles coalesced into objects called **planetesimals**, with diameters of about 100 km.

During the next stage, the gravitational attraction between the planetesimals caused them to collide and coalesce into still larger objects called **protoplanets**. This accumula-

tion of material through the action of gravity is called **accretion**. In recent years, astronomers have used detailed computer simulations to improve our understanding of this accretion process. A computer is typically programmed to simulate a large number of planetesimals circling a hypothetical newborn sun along orbits dictated by Newtonian mechanics. Such studies have shown that accretion would continue for roughly 100 million years and characteristically form about half a dozen planets.

A particularly successful computer simulation by George W. Wetherill is summarized in Figure 7-17. The calculations begin with 100 planetesimals, each having a mass of 1.2×10^{23} kg. This ensures that the total mass (1.2×10^{25} kg) equals the mass of the four terrestrial planets (Mercury through Mars) plus their satellites. The initial orbits of these planetesimals are inclined to each other by angles of less than 5°, to simulate a thin layer of asteroidlike objects orbiting the protosun.

After an elapsed time simulating 30 million years, the 100 original planetesimals have coalesced into 22 protoplanets. After 79 million years, 11 larger protoplanets remain. Nearly another 100 million years elapse before the total number of growing protoplanets is reduced to six: Figure 7-17c shows four planets following nearly circular orbits after a total elapsed time of 441 million years. It should be noted, however, that this model simulation does not exactly reproduce the inner solar system. In the simulation, the fourth planet from the Sun ends up being the most massive. In fact, however, the third planet, Earth, is nine times more massive than the fourth, Mars. Nevertheless, the agreement of this simulation with most characteristics of the inner solar system is striking.

The material from which the inner protoplanets accreted was rich in mineral-forming elements with high condensa-

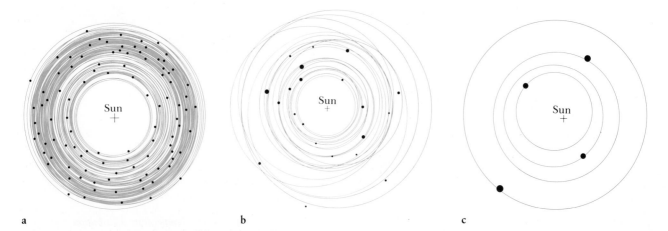

a   b   c

Figure 7-17 *Accretion of the terrestrial planets These three drawings show the results of a computer simulation of the formation of the inner planets. (a) The simulation begins with 100 planetesimals. (b) After 30 million years, these planetesimals have coalesced into 22 protoplanets.*

(c) *This final view is for an elapsed time of 441 million years, but the formation of the inner planets was essentially complete after 150 million years. (Adapted from G. W. Wetherill)*

tion temperatures. Iron, silicon, magnesium, and sulfur were particularly abundant, followed closely by aluminum, calcium, and nickel. The energy released by violent impacts of the planetesimals upon the growing protoplanets, as well as the decay of radioactive elements, melted much of this rocky material. The terrestrial planets therefore may have begun their existence as spheres of molten rock.

Many details of the formation of the planets are topics of current research and debate among scientists. For instance, there are two competing theories to explain how planets like the Earth came to have dense, iron-rich cores surrounded by mantles of less-dense rock. According to the **homogeneous accretion theory**, the inner planets grew by the accretion of planetesimals of the same general composition—a mixture of rock and iron. Initially, therefore, the inner planets were quite homogeneous. As the decay of short-lived radioactive elements melted this matter, gravity caused the denser iron-rich minerals to sink to the centers of the planets, while the less dense silicon-rich minerals floated to their surfaces. This process is called **chemical differentiation**.

An alternative explanation, called the **heterogeneous accretion theory**, argues that planets formed as material condensed out of the solar nebula as it cooled off by radiating its thermal energy into space. Iron and iron oxides were among the first substances to condense, and so the iron-rich cores of the terrestrial planets formed quite early. Then, as the solar nebula continued to cool, silicon-rich minerals condensed and were accreted onto the iron-rich planetary cores.

Neither theory may be entirely correct; an accurate account of the birth of the inner planets may involve ideas from both. Supercomputer simulations offer one of the best methods of examining and evaluating various scenarios that might describe the creation of the solar system. Scientists program a supercomputer with the relevant laws of physics to portray a huge cloud of gas and dust contracting under the action of its own gravity. The more details they include, the more realistic the simulation. Figure 7-18 shows one such simulation. Colors, in the order of the rainbow, are used to display density, from blue for the most dense regions to red for the least dense. In this picture, which shows a region 20 AU across, most of the mass is still in the disk and has not yet fallen into the protosun.

The formation of the outer planets probably also began with the accretion of planetesimals. In this case, the process resulted in a rocky core for each of the Jovian planets that would serve as a "seed" around which the rest of the planet would soon grow. According to calculations by Peter Bodenheimer of Lick Observatory and James Pollack of NASA Ames, the core of a Jovian protoplanet began to capture an envelope of gas as it continued to grow by accretion. For about a million years, this accumulation of both rock and gas proceeded in a leisurely fashion until a critical point at which the masses of the core and the envelope were equal. From that moment on, the rate of accumulation of gas in-

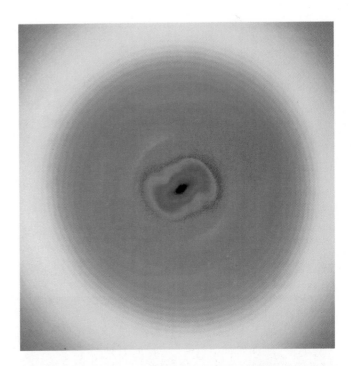

Figure 7-18 A supercomputer simulation of the formation of the solar system *This computer-generated picture shows the density of matter in the equatorial plane of supercomputer simulation of the formation of the solar system. Density is displayed according to the colors of the rainbow, from blue for the most dense regions to red for the least dense. The area shown is 20 AU across. Note the barlike structure in the inner nebula. Such temporary asymmetries are common in rotating disks of matter. (Courtesy of Alan Boss)*

creased dramatically, with the envelope pulling in all the gas it could get. This runaway growth of the protoplanet continued until all the available gas was used up. The final result was a huge planet with an enormously thick, hydrogen-rich atmosphere surrounding an Earth-sized core of rocky material. This scenario occurred at four different distances from the Sun, thus creating the four Jovian planets. Figure 7-19 summarizes this story of the formation of the solar system.

During the millions of years while the planets were forming, temperatures and pressures at the center of the contracting protosun continued to climb. Finally, temperatures at the center of the protosun reached 8 million degrees, hot enough to ignite thermonuclear reactions, and the Sun was born. Sunlike stars take approximately 100 million years to form from a prestellar nebula, which means that the Sun must have become a full-fledged star at roughly the same time the accretion of the inner protoplanets was complete.

The preceding description presents conclusions drawn from many different kinds of evidence, and some of the details of this story will probably be revised as more evidence becomes available from spacecraft and orbiting observatories and better theoretical models are developed and tested.

Figure 7-19 Formation of the planets *This series of sketches shows the major stages in the birth of the solar system spanning 100 million years. Terrestrial planets accrete from rocky material in the warm inner regions of the solar nebula. Meanwhile, the huge, gaseous Jovian planets form in the cold outer regions. (a) The solar nebula in its initial stages. (b) The early solar system after 50 million years. (c) Planetary formation is nearly complete after 100 million years.*

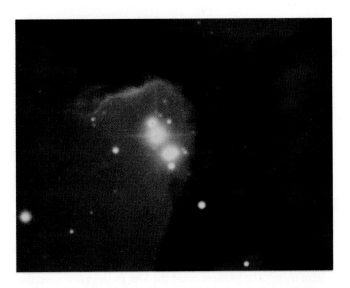

Figure 7-20 Newly formed stars A full-fledged star is born when thermonuclear reactions ignite at its center. This ignition is often accompanied by an outpouring of particles and radiation from the star's surface that sweeps the surrounding space clean of dust and gas from which the star had formed. This photograph shows dusty material being blown away from newly created stars at the center of the so-called Trifid Nebula in the constellation of Sagittarius. (Anglo-Australian Observatory)

However, most of the new discoveries of the past few decades have confirmed the broad outline of this history. Astronomers are thus fairly confident that the solar system did indeed form in the fashion outlined here.

A newborn star adjusts fairly violently to the onset of thermonuclear reactions in its core, and often the star's tenuous outermost layers are vigorously expelled into space (see Figure 7-20). This brief burst of mass loss is observed in many young stars across the sky and is called a **T Tauri wind**, after the star in Taurus (the bull), where it was first identified. The T Tauri wind that heralded the birth of the Sun swept the solar system clean of excess gases, thereby preventing further accretion. Many small rocks were left behind to pelt the planets over the next half billion years. For all practical purposes, however, the formation of the solar system had been completed by the time the T Tauri wind began to blow.

The Sun today continues to lose matter gradually in a mild fashion. This ongoing, gentle mass loss, called the **solar wind**, consists of high-speed protons and electrons leaking away from the Sun's outer layers. In later chapters, we shall see how each of the planets carves out its own distinctive cavity in this solar wind.

Key words

accretion	eccentricity (of an ellipse)	Jovian planet	protosun
angular momentum	*electron	*kinetic energy	solar nebula
asteroid	*escape velocity	molecule	solar wind
atom	Helmholtz contraction	*neutron	spectroscopy
atomic number	heterogeneous accretion theory	*nucleus	supernova explosion
average density		periodic table	terrestrial planet
chemical differentiation	homogeneous accretion theory	planetesimal	*thermal energy
comet		*proton	T Tauri wind
condensation temperature	*isotope	protoplanet	

Key ideas

- The four inner planets of the solar system share many characteristics and are distinctly different from the four giant outer planets.

   The four inner (terrestrial) planets are relatively small (with diameters of 5000 to 13,000 km), have high average densities (4000 to 5500 kg/m³), and are composed primarily of rock.

   The giant outer (Jovian) planets have large diameters (50,000 to 144,000 km), low densities (1000 to 2000 kg/m³), and are composed primarily of hydrogen and helium.

- Spectroscopy, the study of spectra, provides information about the chemical composition of distant objects.

  Spectral lines can be used to identify the elements and chemical compounds in objects from which light is emitted.

- Hydrogen and helium, the two lightest elements, were formed shortly after the creation of the universe. The heavier elements were produced much later, in the centers of stars, and were cast into space when the stars died.

  By mass, 98 percent of the matter in the universe is hydrogen and helium.

- The atoms of various elements can combine to form many different kinds of molecules.

- The basic planet-forming substances can be classified as gases, ices, or rock, depending on their condensation temperatures. The terrestrial planets are composed primarily of rock, whereas the Jovian planets are composed largely of gas.

  The solar system formed from a disk-shaped cloud of hydrogen and helium that also contained ice and dust particles.

  The four inner planets formed through the accretion of dust particles into planetesimals, then into larger protoplanets.

  The four outer planets probably formed through the runaway accretion of gas onto rocky protoplanetary cores.

  The Sun formed by accretion at the center of the nebula. After about 100 million years, temperatures at the protosun's center became high enough to ignite thermonuclear reactions.

  When the protosun became a star, excess gas was vigorously blown away from the Sun, thereby ending the process of planet formation.

Review questions

1   What are the characteristics of a terrestrial planet?

2   What are the characteristics of a Jovian planet?

3   In what ways does Pluto not fit into the usual classifications of either terrestrial or Jovian planets?

4   Why is it reasonable to classify certain large satellites as planets? Which satellites are these?

5   Why do astronomers almost always use the Kelvin temperature scale in their work rather than the Celsius or Fahrenheit scales?

6   What is meant by the "average density" of a planet? What does the average density of a planet tell us?

7   If hydrogen and helium account for 98 percent of the mass of all the material in the universe, why aren't the Earth and Moon composed primarily of these two gases?

8   Why are water (H_2O), methane (CH_4), and ammonia (NH_3) comparatively abundant substances?

9   What occurred during the formation of our solar system to ensure that the terrestrial planets formed close to the Sun and the Jovian planets far from the Sun?

10   Explain how our current understanding of the formation of the solar system can account for the following characteristics of the solar system: (a) All planetary orbits lie in nearly the same plane. (b) All planetary orbits are nearly circular. (c) The planets orbit the Sun in the same direction that the Sun itself rotates.

11   Why is it reasonable to suppose that our solar system was formed over a relatively short period?

12   Would you expect very old stars to possess planetary systems? If so, what types of planets would they have? Explain.

Advanced questions

Tips and tools . . .
The volume of a sphere of radius r is $\frac{4}{3}\pi r^3$. The density of an object is its mass divided by its volume. To compute the mass of Mars, you may have to review Box 4-3 for the appropriate form of Kepler's third law. Be sure to use the same system of units (e.g., meters, seconds, kilograms) in all your calculations. Formulas relating various temperature scales are found in Box 5-1.

*13   A spherical asteroid 1 km in diameter composed of rock and iron with an average density of 5 g/cm^3 strikes the Earth with a speed of 25 km/s. What is the kinetic energy of the asteroid at the moment of impact? How does this energy compare with that released by a 20-kiloton nuclear weapon, like the device that destroyed Hiroshima? (*Hint:* 1 kiloton of TNT equals 4.2×10^{12} J.)

*14   A hydrogen atom has a mass of 1.673×10^{-27} kg, and the temperature of the Sun's surface is 5800 K. What is the average speed of hydrogen atoms at the Sun's surface?

15   (*Basic*) The Sun's mass is 1.989×10^{30} kg and its radius is 6.96×10^8 m. What is the escape velocity from

the Sun's surface? Using your answer to the previous question concerning the average speed of hydrogen atoms at the Sun's surface, explain why hydrogen does not escape from the Sun.

*16 Suppose a spacecraft landed on Jupiter's moon Ganymede, whose diameter is 5262 km and whose mass is 1.48×10^{23} kg. After collecting samples from the satellite's surface, the spacecraft prepares to return to Earth. With what speed must the spacecraft leave Ganymede's surface in order to begin its homeward journey?

17 Suppose you are trying to determine the chemical composition of the atmosphere of a planet by observing its spectrum. Chemicals in the Earth's atmosphere also produce spectral lines (often called *telluric lines*) in the spectra you observe. Can you think of a way of distinguishing between telluric lines and the spectral lines of the distant planet?

*18 Mars has two small satellites, Phobos and Deimos. Phobos circles Mars once every 0.31891 days at an average altitude of 5980 km above the planet's surface. The diameter of Mars is 6794 km. Using this information, calculate the mass and average density of Mars.

*19 At what temperature will the reading on a Celsius thermometer equal that on a Fahrenheit thermometer? At what temperature will the reading on a Kelvin thermometer equal that on a Fahrenheit thermometer?

*20 (Challenging) What would the mass of the Earth be if it had retained hydrogen and helium in the same proportion to the heavier elements as exists elsewhere in the universe? How does your answer compare with the masses of the Jovian planets? What does your answer imply about the cores of the Jovian planets?

*21 Suppose there were a planet having roughly the same mass as the Earth but located at 50 AU from the Sun. What do you think this planet would be made of? On the basis of this speculation, assume a reasonable density for this planet and calculate its diameter. How many times bigger or smaller than the Earth would it be?

Discussion questions

22 Propose an explanation for the fact that the Jovian planets are orbited by terrestrial-like satellites.

23 Suppose that a planetary system is now forming around some protostar in the sky. In what ways might this planetary system turn out to be similar to or different from our own solar system?

24 Suppose astronomers discovered a planetary system in which the planets orbit a star along randomly inclined orbits. How might a theory for the formation of that planetary system differ from that for our own?

Observing projects

25 There are many examples of young stars still embedded in the clouds of gas and dust from which they recently formed. In the winter evening sky, for instance, is the famous Orion Nebula. In the summer night sky, the Lagoon, Omega, and Trifid nebulae are found in the Milky Way. Examine some of these nebulae with the aid of a telescope. Describe their appearance. Can you guess which stars in your field of view are actually associated with the nebulosity? To help you find some of these nebulae, the table below gives their coordinates (right ascension and declination) for the year 2000.

Nebula	Right ascension (h min)	Declination (° ′)
Lagoon	18 03.8	−24 23
Omega	18 20.8	−16 11
Trifid	18 02.3	−23 02
Orion	5 35.4	−5 27

26 Consult such magazines as *Sky & Telescope* or *Astronomy* to determine which planets are visible in your evening sky. Examine these planets through a telescope. Describe their appearance. From what you observe, is there any way of knowing whether you are looking at a planet's surface or cloud cover?

For further reading

Beatty, J., and Chaikin, A., eds. *The New Solar System*, 3rd ed. Sky Publishing and Cambridge University Press, 1990 • This excellent introduction to the solar system consists of twenty-three lavishly illustrated chapters, each written by noted scientists.

Briggs, G., and Taylor, F. *The Cambridge Photographic Atlas of the Planets*. Cambridge University Press, 1982 • A clear, nontechnical survey of the solar system that includes a superb collection of spacecraft photographs from various missions.

Cameron, A. G. W. "The Origin and Evolution of the Solar System." *Scientific American,* September 1975 • This article describes conditions in the solar nebula and the accretion of the planets.

Chapman, C. *Planets of Rock and Ice*. Scribner's, 1982 • The author eloquently conveys the excitement of planetary exploration while taking the reader on a nontechnical tour of the solar system.

Fisher, D. *The Birth of the Earth*. Columbia University Press, 1987 • This book gives a nontechnical survey of the origins of the universe, the solar system, and the Earth.

Frazier, K. *Solar System*. Time-Life Books, 1985 • This lavishly illustrated book gives an overview of our understanding of the solar system.

Lewis, J. "The Chemistry of the Solar System." *Scientific American*, March 1974 • This well-written survey of the solar system compares the chemical composition of the planets.

Moore, P., and Hunt, G. *Atlas of the Solar System*. Rand McNally, 1983 • This excellent reference includes maps, photographs, and tables summarizing information about the Sun and planets.

Morrison, D., and Owen, T. *The Planetary System*. Addison Wesley, 1988 • This superb textbook covers the full scope of planetary science.

Murray, B., ed. *The Planets*. W. H. Freeman and Company, 1983 • This book is a collection of ten articles about the solar system published in *Scientific American* magazine.

Preiss, B., ed. *The Planets*. Bantam, 1985 • This book is an intriguing collection of essays and science-fiction stories about the solar system, written by well-known scientists and science-fiction authors.

Reeves, H. "The Origin of the Solar System." *Mercury*, March/April 1977 • This excellent article by a noted French astronomer describes many intriguing details of the formation of the solar system.

Smoluchowski, R. *The Solar System*. Scientific American Books, 1983 • This book presents a brief, nicely illustrated survey of the solar system.

Wetherill, G. "The Formation of the Earth from Planetesimals." *Scientific American*, June 1981 • This article describes the accretion of the terrestrial planets from the colliding planetesimals that originally orbited the Sun.

The Earth Astronauts often report that the Earth is the most invitingly beautiful object visible from their spacecraft. Our blue-and-white world is the largest of the terrestrial planets. This photograph was taken in 1969 by Apollo 11 astronauts on the way to the first manned landing on the Moon. Most of Africa and portions of Europe can be seen. (NASA)

C H A P T E R

Our Living Earth

The Earth is a geologically active world mostly by water and living organisms. Rocks found in Earth's crust give clues about its history. The Earth's surface is divided into huge plates that jostle and rub against one another, producing mountain ranges, volcanoes, earthquakes, and oceanic trenches. Neither Venus nor Mars, our neighbor planets, seem to exhibit features indicative of such plate activity. From seismic waves of earthquakes, we know that the Earth's interior consists of an iron-rich core surrounded by a thick mantle of ferromagnesian minerals.

Electric currents in the molten portions of the core produce a planetwide magnetic field, which extends far into space and shields us from the solar wind. Earth's atmosphere, rich in nitrogen and oxygen, is quite unlike the carbon dioxide atmospheres of either Venus or Mars. Biological activity of living organisms over billions of years is responsible for the air we breathe. Unfortunately, humans are rapidly altering the chemical composition and thermal balance of Earth's atmosphere, thereby threatening the survival of many species, even our own.

Figure 8-1 *The three largest terrestrial planets* *The Earth and Venus have nearly the same mass, size, and surface gravity. However, Venus's surface is perpetually shrouded by a dense cloud cover containing poisonous, corrosive gases. Mars is only about half the size of Earth and has only one-tenth its mass. The thin, dry Martian atmosphere does not protect the planet from the Sun's ultraviolet radiation. Both the Venusian and the Martian environments are hostile to life forms that thrive on Earth. (NASA)*

Of all the planets that orbit the Sun, we are most familiar with the Earth. We walk on its surface, drink its water, and breathe its air. We therefore know more about the Earth than about any other object in the universe.

Imagine an alien spacecraft approaching the inner solar system. Its occupants would see Mars with its thin atmosphere and barren desert landscapes. Closer to the Sun, they would see Venus with its shroud of corrosive clouds hiding a forbiddingly hot surface. Between these two relatively unpromising planets, they would see the Earth, with an ever-changing ballet of delicate white cloud formations contrasting against the darker browns and blues of its continents and oceans (see Figure 8-1). Perhaps it is only our own bias that leads us to guess that aliens would focus on the Earth as the most interesting and inviting of the terrestrial planets (a summary of Earth data appears in Box 8-1).

8-1 Earth rocks contain clues about the history of our planet's surface

Were aliens to land on Earth, they would find it radically different from either of its neighbors: The Earth is very wet. Nearly 71 percent of the Earth's surface is covered with water. An alien space probe to Earth might send back thousands of photographs such as Figure 8-2, representing a random close-up view of the Earth's surface. In contrast, no place on Earth is as dry as the extremely arid surfaces of Venus and Mars. By Venusian or Martian standards, our Sahara Desert is a veritable swamp.

In those locations where land protrudes above the oceans, you find rocks, which are typical specimens of the Earth's outermost layer, or **crust**. Of course, rocks are composed of chemical elements (recall the periodic table in Figure 7-9), but individual chemical elements are rarely found in a pure state. The exceptions include gold nuggets and the carbon crystals called diamonds. A specimen of a less valuable example, native sulfur, appears in Figure 8-3.

In describing specimens of the Earth's crust, it is useful to distinguish between minerals and rocks. A **mineral** is a naturally occurring solid composed of a single element or chemical combination of elements, often in the form of crystals. A **rock** is a solid part of the Earth's crust that is composed of one or more minerals.

Figure 8-2 *A typical close-up view of Earth's surface* *Nearly three-quarters of the Earth's surface is covered with water. In contrast, there is no liquid water at all on Mercury, Venus, Mars, or the Moon.*

Box 8-1 Earth data

Mean distance from the Sun:	1.000 AU = 1.496×10^8 km
Maximum distance from the Sun:	1.017 AU = 1.521×10^8 km
Minimum distance from the Sun:	0.983 AU = 1.471×10^8 km
Mean orbital velocity:	29.8 km/s
Sidereal period:	365.256 days
Rotation period:	23.9345 hours
Inclination of equator to orbit:	23° 27′
Inclination of orbit to ecliptic:	0°
Orbital eccentricity:	0.0167
Diameter (equatorial):	12,756 km
Mass:	5.976×10^{24} kg
Mean density:	5500 kg/m³
Escape velocity:	11.2 km/s
*Oblateness:	0.0034
Surface temperature range:	maximum = 60°C = 140°F = 333 K
	mean = 20°C = 70°F = 293 K
	minimum = −90°C = −130°F = 183 K
Albedo:	0.39
Mean diameter of Sun as seen from Earth:	0° 32′

*Oblateness measures how flattened a planet or star is, usually as a result of rapid rotation. Oblateness is defined as the ratio of the difference between the equatorial and polar radii to the equatorial radius. The oblateness of a perfect sphere is zero.

Quartz, one of the most common rock-forming minerals in the Earth's crust, is one member of an abundant class of minerals called **silicates** that are based on the element silicon. A quartz crystal consists of an orderly arrangement of silicon and oxygen atoms.

To identify various minerals, geologists use a number of criteria, such as the shape, hardness, and color of the crystals. Three common rock-forming minerals are shown in Figure 8-4. Each has its own unique, easily identifiable characteristics. These particular minerals are found mixed to-

Figure 8-3 Native sulfur Sulfur is one of the very few chemical elements found in a pure state. Carbon crystals, called diamonds, and gold nuggets are also examples of pure elements found in nature. Most minerals are chemical combinations of various elements.

Figure 8-4 Three common minerals A mineral is a specific chemical combination of elements whose atoms are typically arranged in an orderly fashion to produce a crystal. A quartz (SiO_2) crystal is at the left. Feldspar (center) is another silicate. Mica (right) is also a common rock-forming mineral.

gether in a common rock called granite, which is composed of four minerals: quartz, feldspar, mica, and hornblende.

Geologists cannot classify rocks the same way they classify minerals, because rocks contain a mixture of mineral characteristics. For example, the tiny quartz crystals in a piece of granite have a different color and hardness from the shiny flecks of mica in the granite. Therefore, geologists classify rocks according to how they were created. There are three major categories of rocks, corresponding to three rock-forming processes.

Igneous rocks result when minerals cool from a molten state. The formation of igneous rock can be observed directly during a volcanic eruption. Molten rock is called **magma** when it is buried below the surface and **lava** when it flows out upon the surface. Two common igneous rocks, basalt and granite, are shown in Figure 8-5.

Sedimentary rocks are those produced by the action of wind, water, or ice. For example, as winds pile up layer after layer of sand, the grains gradually become cemented together to produce sandstone. And minerals that precipitate out of the oceans can cover the ocean floor with layers of rock such as limestone. These two common sedimentary rocks are shown in Figure 8-6.

Sometimes igneous or sedimentary rocks become buried deep beneath the Earth's surface, where they are subjected to enormous pressures and temperatures. These severe conditions change the rocks' structure, producing the third variety, called **metamorphic rock**. For example, when fine-grained igneous rock is subjected to pressure and heat, it becomes schist. When limestone is metamorphosed, it becomes marble. These two common metamorphic rocks are shown in Figure 8-7.

By studying the kinds of rocks that are found at a particular location, geologists can deduce the history of that site.

For instance, whether an area was once covered by an ancient sea or was flooded by lava from volcanoes is readily apparent from the kinds of rocks that are present.

8-2 Studies of earthquakes reveal the interior structure of the Earth

Although the specimens shown in Figures 8-5, 8-6, and 8-7 are typical samples of the Earth's crust, they are not representative of what makes up our planet's interior. The densities of crustal rocks are typically 3000 to 4000 kg/m^3, but the average density of the Earth as a whole is 5500 kg/m^3. The Earth's interior must therefore be composed of a substance much denser than the crust.

Iron is a good candidate for this substance because it is the most abundant of the heavier elements (recall Table 7-4), and its presence in the Earth's interior is indicated by the existence of the Earth's magnetic field. Furthermore, iron is common in meteoroids that strike the Earth, suggesting that it was in the planetesimals from which the Earth formed.

Geologists strongly suspect that the Earth was entirely molten soon after its formation, about 4.5 billion years ago. Energy released by the violent impacts of numerous meteoroids and asteroids and by the decay of radioactive isotopes likely melted the solid material collected from the earlier planetesimals. Gravity caused abundant, dense iron to sink toward the Earth's center, forcing less dense material to the surface. This process of chemical differentiation then produced a layered structure within the Earth: a central core composed of almost pure iron, surrounded by a mantle of dense, iron-rich minerals, which in turn is surrounded by a thin crust of relatively light silicon-rich minerals.

Figure 8-5 Igneous rocks (basalt and granite) Igneous rocks are created when molten minerals solidify. The sample on the left is basalt, which is a fine-grained mixture of feldspar with ferromagnesian minerals that give the rock its dark color. A granite specimen is on the right. The fine-grained basalt typically forms when molten minerals cool near the Earth's surface; the coarser-grained granite typically forms from slower cooling deeper within the Earth's crust.

Figure 8-6 Sedimentary rocks (limestone and sandstone) Sedimentary rocks are produced, often layer by layer, by the action of wind, water, or ice. The sample on the left is limestone, which is primarily calcite with some impurities that give it a dark color. Sandstone is on the right. A sedimentary rock is typically formed when loose particles of soil or sand are buried a short distance below the Earth's surface and are cemented into rock by chemical changes.

Figure 8-7 Metamorphic rocks (marble and schist) When igneous or sedimentary rocks are subjected to high temperatures and pressures deep in the Earth's crust, they are changed into metamorphic rock. Marble (left) is produced from limestone; schist (right) is formed from fine-grained igneous rock.

The Earth's interior is as inaccessible as the most distant galaxies in space. The deepest wells go down only a few kilometers, barely penetrating the surface of our planet. However, geologists have deduced the basic properties of the Earth's interior by studying earthquakes.

Over the centuries, stresses build up in the Earth's crust. Occasionally, these stresses are relieved with a sudden vibratory motion called an **earthquake**. The exact origin of an earthquake, called its **focus**, is usually deep within the Earth's crust. The point on the Earth's surface directly above the focus is called the earthquake's **epicenter**.

Earthquakes produce three different kinds of **seismic waves** that travel around or through the Earth in different ways and at different speeds. Geologists use sensitive **seismographs** to detect and record these vibratory motions. The rolling motion that people feel near an epicenter is caused by **L waves**, which travel only over the Earth's surface and are analogous to water waves on the surface of the ocean. The two remaining kinds of waves, called **primary** or **P waves** and **secondary** or **S waves**, travel through the Earth. P waves are said to be longitudinal waves because their oscillations are parallel to the direction of wave motion, like a spring that is alternately pushed and pulled. In contrast, S waves

are said to be transverse waves because their vibrations are perpendicular to the direction in which the waves are moving. S waves are analogous to waves produced by a person shaking a rope up and down (see Figure 8-8). P waves travel almost twice as fast as S waves. Consequently, P waves from an earthquake always arrive at a seismographic station before S waves do. By measuring the time delay between the arrival of these two kinds of waves, geologists can deduce the distance to the earthquake's epicenter.

Seismic waves do not travel along straight lines through the Earth. Because of the varying density and composition of the Earth's interior, both S waves and P waves are bent, or refracted. By studying how these waves are bent, geologists can map out the general structure of the Earth's interior.

When an earthquake occurs in the Earth's crust, seismographs within a few thousand kilometers of the epicenter are able to record both S waves and P waves. However, on the opposite side of the Earth, only P waves can be recorded at seismographic stations. This absence of S waves was first explained in 1906 by the geologist R. D. Oldham, who noted that transverse vibrations such as S waves cannot travel far through liquids. Oldham therefore concluded that our planet has a molten core. Furthermore, there is a region called the **shadow zone** in which neither S waves nor P waves from an earthquake can be detected (see Figure 8-9). This shadow zone results from the specific way in which P waves are refracted at the boundary between the solid mantle and the molten core. By measuring the size of the shadow zone,

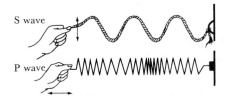

Figure 8-8 Seismic waves Earthquakes produce two kinds of waves that travel through our planet. P (or primary) waves are longitudinal waves analogous to those produced by pushing a spring in and out. S (or secondary) waves are transverse waves analogous to waves produced by shaking a rope up and down.

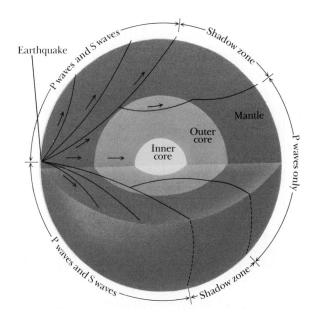

Figure 8-9 The paths of seismic waves Seismic waves are refracted as they pass through the Earth. Furthermore, only the P waves can pass through the Earth's liquid outer core. From tracing the paths followed by seismic waves, geologists have deduced the dimensions of the Earth's mantle and core.

geologists have concluded that the core's diameter is about 7000 km (4300 miles). For comparison, the overall diameter of our planet is 12,700 km (7900 miles).

As the quality and sensitivity of seismographs improved, geologists soon realized that faint traces of P waves could be detected within the shadow zone of an earthquake. This finding led seismologist Inge Lehmann to conclude in 1936 that a small, solid **inner core** exists at the center of our planet. The diameter of this inner core is about 2500 km (1550 miles).

The interior of our planet therefore has a curious structure: a liquid **outer core** sandwiched between a solid inner core and a solid mantle. The explanation for this arrangement lies in how temperature and pressure inside the Earth affect the melting point of rock.

Both temperature and pressure increase with increasing depth below the Earth's surface. The temperature of the Earth's interior rises steadily from about 20°C on the surface to nearly 5000°C at our planet's center (see Figure 8-10).

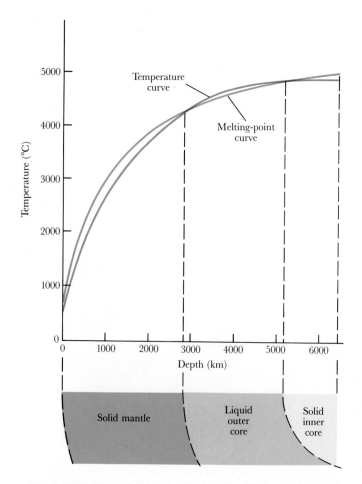

Figure 8-10 Temperature and melting point of rock inside the Earth
The temperature (red curve) rises steadily from the Earth's surface to its center. By plotting the melting point of rock (blue curve) on this graph, we can deduce which portions of the Earth's interior are solid or liquid. (Adapted from F. Press and R. Siever)

The Earth's crust is only about 30 km thick. It is composed of rocks whose melting points are far higher than typical temperatures in the crust. Hence the crust is solid.

The Earth's **mantle**, which extends to a depth of about 2900 km (1800 mi), is largely composed of minerals rich in iron (called *ferrum* in Latin) and magnesium. On the Earth's surface, specimens of these ferromagnesian minerals have melting points slightly over 1000°C. However, the melting point of a substance depends on the pressure to which it is subjected: the higher the pressure, the higher the melting point. As shown in Figure 8-10, the melting point of all the mantle's minerals is everywhere higher than the actual temperature, so the mantle is primarily solid.

At the boundary between the mantle and the outer core, there is an abrupt change in chemical composition, from ferromagnesian minerals to almost-pure iron with a small admixture of nickel. This iron–nickel material has a lower melting point than ferromagnesian minerals, and so the melting-point curve in Figure 8-10 dips below the temperature curve as it crosses from the mantle to the outer core. The melting-point curve remains below the temperature curve down to a depth of about 5100 km (3200 mi). Hence, from depths of about 2900 to 5100 km, the core is liquid.

At depths greater than about 5100 km, the pressure is more than 10^{11} N/m^2 (about 10^4 tons/in.^{2}). Beyond 5100 km, the pressure is so great that the melting point of the iron–nickel mixture exceeds the actual temperature (see Figure 8-10). As a result, the Earth's inner core is solid.

8-3 *The process of plate tectonics is responsible for many phenomena and features of the Earth's surface, including earthquakes, mountain ranges, and volcanoes*

One of the most important geological discoveries of the twentieth century is the realization that our planet has an active, constantly changing crust. We have learned that the Earth's crust is divided into huge **plates** that constantly jostle each other, producing earthquakes, volcanoes, and oceanic trenches.

The idea of huge, moving plates might occur to anyone carefully examining a map of the Earth. South America, for example, would fit snugly against Africa, were it not for the Atlantic Ocean. Indeed, the fit between land masses on either side of the Atlantic Ocean is quite remarkable (see Figure 8-11). This observation inspired people such as Alfred Wegener, a meteorologist, to propose the idea of "continental drift," which suggests that the continents on either side of the Atlantic Ocean have simply drifted apart. After much research, Wegener published in 1924 the theory that there

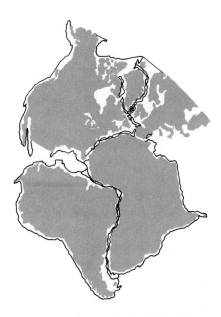

Figure 8-11 **Comparing the continents** *Africa, Europe, Greenland, and North and South America fit together as though they had once been joined. The fit is especially convincing if the edges of the continental shelves (rather than today's shoreline) are used. (Adapted from P. M. Hurley)*

had originally been a single gigantic supercontinent he called Pangaea, which began to break up and drift apart some 200 million years ago. Some geologists refined this theory, arguing that Pangaea must have first split into two smaller supercontinents, which they called Laurasia and Gondwanaland, separated by what they called the Tethys Sea. Gondwanaland later split into Africa and South America, with Laurasia dividing to become North America and Eurasia. According to this theory, the Mediterranean Sea is a surviving remnant of the ancient Tethys Sea.

Initially, however, most geologists greeted Wegener's ideas with ridicule. Although it was generally accepted that the continents do "float" in the denser, somewhat plastic mantle beneath them, few geologists could accept the idea that entire continents could move around the Earth at speeds that must be as great as several centimeters per year. The "continental drifters" could not explain what forces could be shoving the massive continents around.

Then, in the mid-1950s, Bruce C. Heezen of Columbia University and his colleagues began discovering long mountain ranges on the ocean floors, such as the Mid-Atlantic Ridge (see Figure 8-12), which stretches all the way from Iceland to Antarctica. During the 1960s, careful examina-

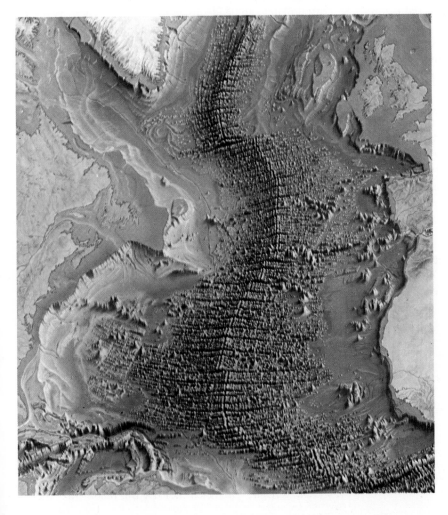

Figure 8-12 **The Mid-Atlantic Ridge** *This artist's rendition shows the floor of the North Atlantic Ocean. The unusual mountain range in the middle of the ocean floor is called the Mid-Atlantic Ridge. It is caused by lava seeping up from the Earth's interior along a rift that extends from Iceland to Antarctica. (Courtesy of M. Tharp and B. C. Heezen)*

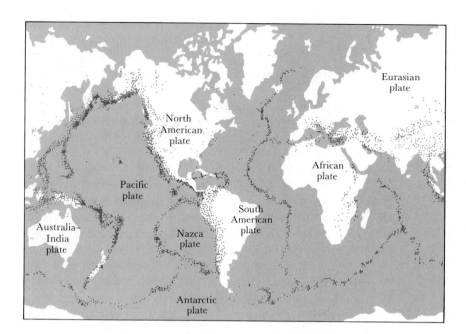

Figure 8-13 **The Earth's major plates** *The boundaries of Earth's major plates are the scenes of violent seismic and geologic activity. Most earthquakes occur where plates separate or collide. Plate boundaries are therefore easily identified by simply plotting earthquake epicenters on a map.*

tion of the ocean floor revealed that rock from the Earth's mantle is being melted and then forced upward along this Mid-Atlantic Ridge, which is therefore a long chain of underwater volcanoes. This upwelling of new material is forcing the ocean floor to spread. The eastern floor of the Atlantic Ocean is moving eastward, the western floor is moving westward. This **sea-floor spreading** is in fact pushing South America and Africa apart, at a speed of roughly 3 cm per year.

Sea-floor spreading provided just the physical mechanism that had been missing from the theories of continental drift. With some embarrassment, geologists began around 1964 to review the old theories and found a great deal of new evidence to support them. This time, however, the emphasis was upon the motions of large plates of the crust (not simply

the continents), and so was born the modern theory of crustal motion that came to be known as **plate tectonics**. Geologists today realize that the boundaries of the Earth's crustal plates can be identified by the seismic activity occurring where these plates are either colliding or separating. The plate's boundaries stand out clearly when the epicenters of earthquakes are plotted on a map (see Figure 8-13).

Geologists believe that convective currents in the Earth's mantle are responsible for the movement of the plates. **Convection** is the transport of heat energy that takes place as lighter, hotter material rises upward while heavier, cooler material sinks downward. Although the mantle is rather solid, it is hot enough to permit an oozing, plastic flow throughout a soft region of the upper mantle called the **asthenosphere**. As sketched in Figure 8-14, hot material

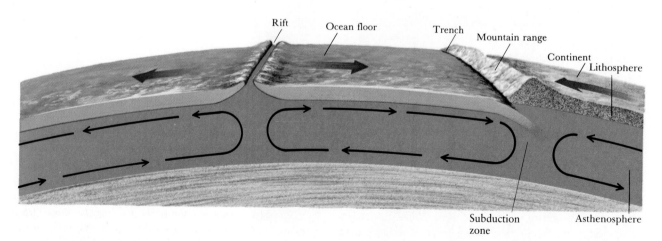

Figure 8-14 **The mechanism of plate tectonics** *Convection currents in the soft upper layer of the mantle (called the asthenosphere) are responsible for pushing around rigid, low-density plates on the crust (called the lithosphere). New crust is formed in oceanic rifts, where lava oozes upward between separating plates. Mountain ranges and deep oceanic trenches are formed where plates collide.*

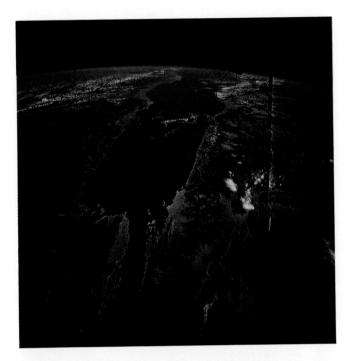

Figure 8-15 *The separation of two plates* *The plates that carry Egypt and Saudi Arabia are moving apart, leaving the trench that contains the Red Sea. In this view taken by* Gemini 12 *astronauts in 1966, Saudi Arabia is on the left and Egypt (with the Nile River) on the right.* *(NASA)*

Figure 8-16 *The collision of two plates* *The plates that carry India and China are colliding. As a result, the Himalaya Mountains have been thrust upward. In this photograph taken by Apollo 7 astronauts in 1968, India is on the left and Tibet on the right; Mt. Everest is one of the snow-covered peaks near the center. (NASA)*

(magma) seeps upward along **oceanic rifts** where plates are separating. Cool crustal material sinks back down into the magma along **subduction zones** where plates are colliding. Above the plastic asthenosphere is a low-density rigid layer called the **lithosphere**, divided into plates that simply ride on the convection currents of the asthenosphere. The boundary between crust and mantle—a boundary between different chemical compositions—is within the lithosphere.

The boundaries between plates are the sites of some of the most impressive geological activity on our planet. Great mountain ranges, such as those along the western coasts of North and South America, are thrust up by ongoing collisions with the plates of the ocean floor. Subduction zones, where old crust is pushed back down into the mantle, are typically the locations of deep oceanic trenches, such as the ones off the coasts of Japan and Chile. Figures 8-15 and 8-16 show two well-known geographic features that resulted from tectonic activity at plate boundaries.

The convection process associated with sea-floor spreading is the primary way in which heat is transported outward from the Earth's interior to the crust. Here our planet differs radically from its neighbors. Venus does not have long mountain chains resembling Earth's Mid-Atlantic Ridge (see Figure 8-17). As we shall see in Chapter 12, Mars also lacks tectonic features. Instead, both Venus and Mars have huge

volcanoes. Olympus Mons, the largest volcano on Mars, rises to an altitude of 24 km (15 mi) above the surrounding volcano-studded plains—nearly three times taller than Mount Everest. Similarly, the summit of Maxwell Montes on Venus is at an altitude of 11 km (7 mi).

Venus and Mars apparently transport heat outward to their surfaces predominantly by a mechanism called **hot-spot volcanism**. Hot spots deep in the mantles of Venus and Mars squirt molten lava up through the crust. In the absence of any tectonic activity to move the crust around, millions of years of eruptions eventually built the enormous volcanoes we observe.

Hot-spot volcanism also occurs on Earth. The Hawaiian Islands, which are in the middle of the Pacific plate, are a fine example. There is a hot spot in the Earth's mantle beneath Hawaii that continuously pumps lava up through the crust. However, the Pacific plate is moving northwest, at the rate of several centimeters per year. New volcanoes are thus created over the hot spot as older volcanoes move away from the magma source, become extinct, and eventually erode and disappear beneath the ocean. The Hawaiian Islands are the most recent additions to a long chain of extinct volcanoes that stretches all the way up to Japan. This chain is a permanent record of the movements of the Pacific plate over the past 70 million years.

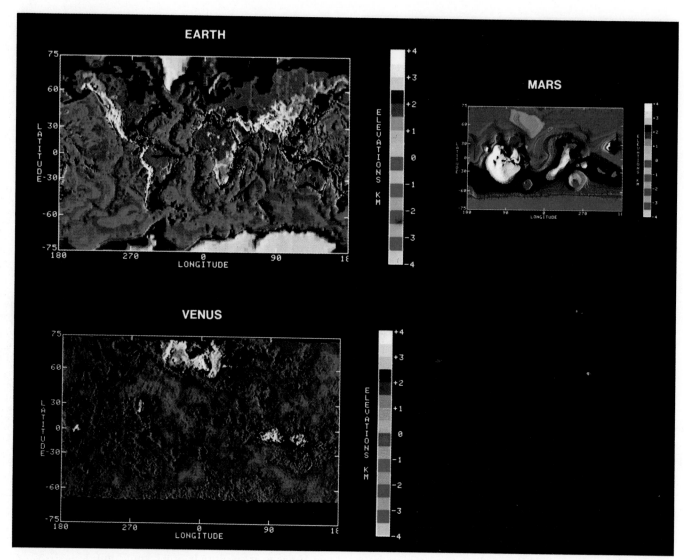

Figure 8-17 The surfaces of Earth, Venus, and Mars *These maps of Earth, Venus, and Mars are reproduced to the same scale. Each map shows surface elevation (Earth's oceans are empty) on a color-coded scale. Elevations up to 4 km above the planet's average radius (sea level for Earth) appear in shades of tan, olive green, and white. Elevations* *down to 4 km below the planet's average radius appear in various shades of blue. Venus's two continents are prominent, as are Earth's seven continents. The large, high-elevation protrusion on Mars is centered on a major volcanic region. (NASA)*

8-4 Absorption of sunlight and the Earth's rotation govern the behavior of our nitrogen–oxygen atmosphere

The Earth differs from its neighbors in yet another way: the chemical composition of its atmosphere. The atmospheres of both Venus and Mars consist almost entirely of carbon dioxide, whereas carbon dioxide accounts for only 0.03 percent of Earth's atmosphere. The Earth's atmosphere is predominantly a 4-to-1 mixture of nitrogen and oxygen, two gases that are found only in small amounts on Venus and Mars (see Table 8-1).

Why is Earth, with its nitrogen–oxygen atmosphere, bracketed by two planets possessing atmospheres of nearly pure carbon dioxide? On Earth, a vast amount of carbon

Table 8-1 The chemical composition of three planetary atmospheres

	Venus	Earth	Mars
Nitrogen	4%	77%	3%
Oxygen	Almost zero	21%	Almost zero
Carbon dioxide	96%	Almost zero	95%
Other gases	Almost zero	2%	2%

Box 8-2   Atmospheric pressure

At sea level on Earth, the air weighs down with a pressure of 14.7 pounds per square inch. By definition, this pressure is called one **atmosphere** (1 atm). Because we are familiar with this pressure, the **atmospheric pressure** on other planets can usefully be expressed in atmospheres also. For example, the atmospheric pressure on Venus is 90 atm, while that on Mars is only 0.01 atm.

There are, however, other ways of expressing pressure. In SI, for example, pressure is measured in newtons per square meter (N/m^2) rather than pounds per square inch. One newton per square meter is often called a **pascal** (Pa), in honor of the seventeenth-century French mathematician Blaise Pascal. These various units are related to each other as follows:

$$1 \text{ lb/in.}^2 = 6.89 \times 10^3 \text{ N/m}^2$$
$$1 \text{ N/m}^2 = 1 \text{ Pa}$$
$$1 \text{ N/m}^2 = 1.45 \times 10^{-4} \text{ lb/in.}^2$$
$$1 \text{ atm} = 14.7 \text{ lb/in.}^2 = 1.01 \times 10^5 \text{ N/m}^2$$

The fact that 1 atm is equal to almost exactly 100,000 N/m^2 has prompted the definition of another unit of pressure called a **bar**, where

$$1 \text{ bar} = 10^5 \text{ N/m}^2 = 0.987 \text{ atm}$$
$$1 \text{ atm} = 1.013 \text{ bar}$$

In other words, a bar is just slightly less than an atmosphere.

In expressing low atmospheric pressures, it is often convenient to use the **millibar**, which is simply 0.001 bar. Thus, we say that the atmospheric pressure on Mars is about 10 millibars, approximately the same as the atmospheric pressure at an altitude of 30 km above the Earth's surface (see Figure 8-18).

Finally, there is a way of expressing atmospheric pressure based on the behavior of a barometer. In its simplest form, a barometer can be made by filling a long tube (sealed at one end) with mercury and inverting the tube in a bowl of mercury, as shown in the diagram. Mercury is a very dense liquid, and gravity pulls strongly downward on the mercury in the tube. However, atmospheric pressure is also pushing down on the mercury in the dish, which tends to force the mercury back up into the tube, because atmospheric pressure cannot act downward on the top of the column inside the sealed tube. Thus the mercury in the tube falls, leaving a vacuum above the mercury column until a point when the downward tug of gravity (the weight of mercury in the tube) is just balanced by the atmospheric pressure that pushes the mercury up into the tube.

Atmospheric pressure is commonly expressed as the height of the column of mercury that it supports in a barometer. For example, at sea level on Earth, normal atmospheric pressure supports a column of mercury (Hg) that is 760 mm (29.9 in.) tall. Thus,

$$1 \text{ atm} = 760 \text{ mm Hg} = 29.9 \text{ in. Hg}$$

When the weather report says that the barometer reads 30.2 inches, you now know that the atmosphere is weighing down on the Earth's surface with a pressure sufficient to support a column of mercury 30.2 in. tall.

---

dioxide is dissolved in the oceans and chemically bound into carbonate rocks, such as limestone and marble. Furthermore, from study of the Earth's geological records, we know that living organisms have been active here for at least the past 3 billion years and that biological processes such as photosynthesis are largely responsible for the chemical composition of the Earth's atmosphere today. As far as we know, large amounts of oxygen in a planetary atmosphere can arise only as a direct result of biological activity.

The average atmospheric pressure at the Earth's surface is 14.7 pounds per square inch (1.01×10^5 Pascals), or 1 atm (see Box 8-2). The pressure decreases smoothly with increasing altitude, falling by roughly half with every $5\frac{1}{2}$ km. Thus, at an elevation of about $5\frac{1}{2}$ km (18,000 ft), the atmospheric pressure is down to about half of what it is at sea level, which means that half of the Earth's atmosphere lies below this elevation. With every breath, a mountain climber at 18,000 feet takes in only half the air molecules she does at sea level.

Seventy-five percent of the mass of the Earth's atmosphere lies below an altitude of 11 km (roughly 7 mi or 36,000 ft) in a region called the **troposphere**. All Earth's weather—clouds, rain, sleet, and snow—occurs in this lowest layer. Commercial jets generally fly at the top of the troposphere to minimize buffeting and jostling.

Whereas atmospheric pressure decreases smoothly with altitude, our atmosphere has a complicated temperature structure, with minima and maxima, as graphed in Figure 8-18. These temperature variations result from the differing absorption rates for sunlight at various altitudes.

Atmospheric temperature decreases upward to −55°C (= −67°F = 218 K) at the top of the troposphere. Above this level is the region called the **stratosphere**, which extends from 11 to 50 km (about 7 to 31 mi) above the Earth's surface. Ozone (O_3) molecules in the stratosphere efficiently absorb solar ultraviolet rays, thereby heating the air in this layer. The temperature increases upward through the stratosphere to about 0°C (= 32°F = 273 K) at its top. International concerns exist that certain chemicals, such as refrigerants and solvents, released into the air are destroying the ozone layer. If it were not absorbed by the ozone in the stratosphere, the ultraviolet radiation from the Sun would

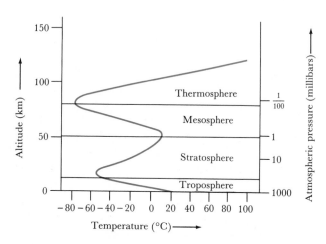

Figure 8-18 Temperature profile of Earth's atmosphere The atmospheric temperature varies with altitude because of the way sunlight interacts with various atoms and ions at different elevations.

beat down on the Earth's surface. Because ultraviolet radiation breaks apart most of the delicate molecules that form living tissue, loss of the ozone layer could lead to literal sterilization of the planet.

Above the stratosphere lies the **mesosphere**. Atmospheric temperature again declines as one moves up through the mesosphere, reaching a minimum of about $-75°C$ ($= -103°F = 198$ K) at an altitude of about 80 km (50 mi). This minimum marks the bottom of the **thermosphere**, above which the Sun's ultraviolet light strips atoms of one or

more electrons. Ionized atoms and molecules of oxygen and nitrogen are the most prevalent species there. The stripped electrons reflect radio waves over a wide range of frequencies. The reason you can tune in a distant AM radio station is that its transmissions bounce off this radio-reflective region that surrounds our planet.

Earth's atmosphere has intricate circulation patterns. If the Earth did not rotate, heated gases would rise in the equatorial regions, flow toward the polar regions, where they would cool and sink to lower altitudes, and then flow back toward the equator. This cycle would produce a single equator-to-pole convection cell in each hemisphere. Indeed, as we shall see in Chapter 11, Venus's atmosphere exhibits this simple convection pattern because Venus rotates very slowly. However, the Earth rotates much more rapidly than does Venus. The Earth's relatively fast rotation breaks up the convection pattern into three cells per hemisphere, as shown in Figure 8-19.

8-5 The Earth's magnetic field produces a magnetosphere that captures particles from the solar wind

As almost anyone with a compass knows, the Earth has a magnetic field. Magnetism arises whenever electrically charged particles are in motion. Many geologists suspect

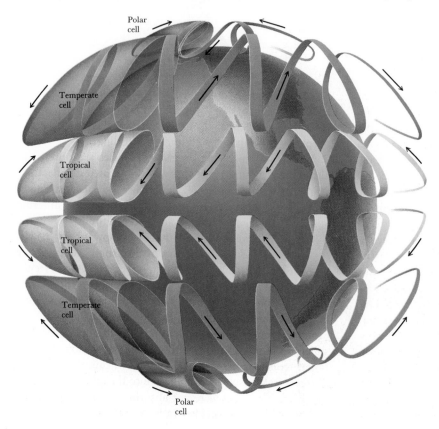

Figure 8-19 Circulation patterns in Earth's atmosphere The dominant circulation in the Earth's atmosphere consists of three convection cells in the northern hemisphere and three cells in the southern hemisphere. The Earth's rotation deflects winds in the convection cells from a purely north–south flow. Thus, at ground level in the northern temperate region (e.g., the continental United States), for instance, the prevailing winds are from the southwest toward the northeast.

that the Earth's magnetic field is caused by electric currents flowing in the liquid portions of the Earth's iron core. As our planet rotates, these currents produce a magnetic field in the same way that a loop of wire carrying an electric current generates a magnetic field. This process, in which the rotation of a planet with an iron core produces a magnetic field, is called the **dynamo effect**. The Earth rotates fast enough to produce a magnetic field that dominates space for tens of thousands of kilometers in all directions and dramatically affects Earth's interaction with the solar wind.

As mentioned in the previous chapter, the solar wind is a constant flow of charged particles (mostly protons and electrons) away from the outer layers of the Sun's upper atmosphere. In the vicinity of the Earth, the particles in the solar wind are moving at supersonic speeds of roughly 400 km/s, or nearly a million miles per hour. If a planet possesses a magnetic field, this field repels and deflects the impinging particles, thereby forming an elongated "cavity" in the solar wind. This cavity is called the planet's **magnetosphere**.

Figure 8-20 is a scale drawing of Earth's magnetosphere. When the supersonic particles in the solar wind first encounter the Earth's magnetic field, they are abruptly slowed to subsonic speeds, producing a bow-shaped **shock wave** that marks the boundary where this sudden decrease in velocity occurs. Still closer to the Earth, there is another well-defined boundary, called the **magnetopause**, where the outward magnetic pressure of the Earth's field is exactly counterbalanced by the impinging pressure of the solar wind. Most of the particles of the solar wind are deflected around the mag-

netopause through a turbulent region called the **magnetosheath**, just as water is deflected to either side of the bow of a ship. The region inside the magnetopause is the true magnetic domain of the Earth.

Our planet's magnetic field is strong enough to trap charged particles that manage to leak through the magnetopause. These particles are trapped in two huge doughnut-shaped rings called the **Van Allen radiation belts**. These belts were discovered in 1958 during the flight of the United States's first successful Earth-orbiting satellite. They are named after the physicist James Van Allen, who insisted that the satellite should carry a Geiger counter to detect charged particles. The inner Van Allen belt, which extends over altitudes of about 2000 to 5000 km, contains mostly protons. The outer Van Allen belt, which is about 6000 km thick, is centered at an altitude of about 16,000 km above the Earth's surface and contains mostly electrons. The source of these charged particles is the solar wind.

The solar wind is an excellent electrical conductor. As it sweeps past the Earth's magnetic field, the entire magnetosphere acts like a gigantic natural generator that produces enormous electrical currents. These currents flow along magnetic field lines down onto the ionosphere surrounding the north and south magnetic poles. As high-speed electrons and protons collide with gases in the upper atmosphere, atoms of oxygen and nitrogen gain energy, boosting the atoms' electrons to upper energy levels. This energy is released as the atoms' electrons cascade back down toward the lower energy level, emitting photons along the way. In this

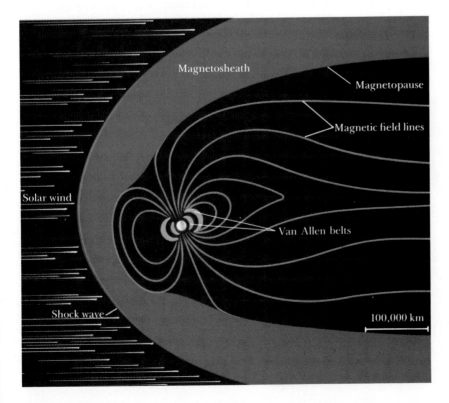

Figure 8-20 **Earth's magnetosphere** *Earth's magnetic field carves out a cavity in the solar wind. Earth's magnetosphere consists of a shock wave, a magnetopause, and a magnetosheath. Because of the strength of the Earth's magnetic field, our planet is able to trap charged particles in two huge doughnut-shaped rings called the Van Allen belts.*

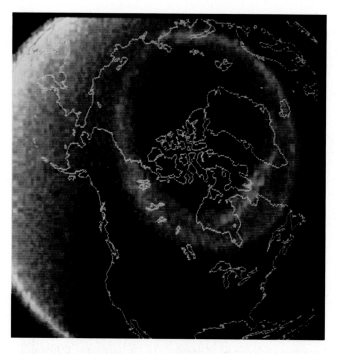

Figure 8-21 *Aurorae around the north magnetic pole* This false-color image was taken at ultraviolet wavelengths by the Dynamics Explorer 1 spacecraft. Coastlines are overlaid in green. The glowing oval of auroral emission, about 4500 km in diameter, is predominantly from oxygen atoms and molecular nitrogen. These emissions rival those from the sunlit hemisphere of the Earth. (Courtesy of L. A. Frank and J. D. Craven, The University of Iowa)

around the Earth's north magnetic pole where charged particles streaming down onto the atmosphere continually produce aurorae.

Occasionally, a violent event on the Sun's surface called a **solar flare** sends a burst of protons and electrons toward the Earth. The resulting auroral display can be exceptionally bright (see Figure 8-22) and can often be seen over a wide range of latitudes.

It is remarkable that the Earth's magnetosphere, dominated as it is by vast radiation belts completely encircling the Earth, was unknown until a few decades ago. Such discoveries remind us of how little we truly understand and how much remains to be learned, even about our own planet.

8-6 A burgeoning human population is profoundly altering Earth's biosphere

One of the extraordinary characteristics of Earth is that it is covered with life. Living organisms thrive in a wide range of environments, from ocean floors to mountain tops, from frigid polar caps to blistering deserts.

Living organisms subsist in a relatively thin layer enveloping the Earth, called the **biosphere**, which includes the oceans, a meter of topsoil covering the Earth's crust, and the lowest few kilometers of the troposphere. A portrait of Earth's biosphere, based on data from NASA's *Nimbus 7* and *NOAA 7* satellites, is shown in Figure 8-23. Ocean colors correspond to concentrations of microscopic free-floating plants called phytoplankton. Reds and oranges show regions of high fecundity; greens and yellows indicate moderate productivity. Land colors portray vegetation mass: dark green for the rain forests, light green and gold for savannas and farmland, and yellow for the deserts. The bio-

way the gases are made to fluoresce like the gases in a fluorescent tube. The result is a beautiful, shimmering display called the **northern lights** (**aurora borealis**) or **southern lights** (**aurora australis**), depending on the hemisphere in which the phenomenon is observed. Figure 8-21 shows a glowing ring

Figure 8-22 *The northern lights (aurora borealis)* A deluge of protons and electrons from a solar flare can produce aurorae that can be seen over a wide range of latitudes. Aurorae typically occur at altitudes between 100 and 400 km above the Earth's surface. (Courtesy of S.-I. Akasofu, Geophysical Institute, University of Alaska)

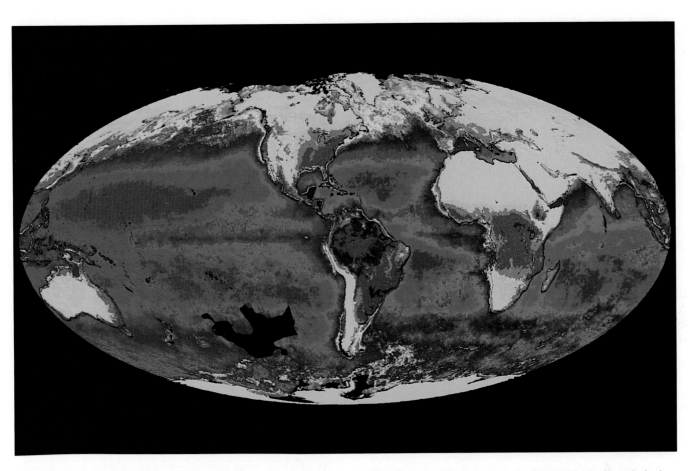

Figure 8-23 The Earth's biosphere This computer-generated picture shows the plant distribution over the Earth's surface. Ocean colors in the order of the rainbow correspond to phytoplankton concentrations, with red and orange for high productivity to blue and purple for low. Land colors designate vegetation, ranging from dark green for the rain forests to yellow for deserts. (Courtesy of G. Feldman, NASA)

sphere, which has taken hundreds of millions of years to evolve to its present state, is a delicate, interactive, highly complex system in which plants and animals depend on each other for their mutual survival.

In recent years, humanity has begun to alter the biosphere in a dramatic and irreversible fashion. This activity is driven by a skyrocketing human population. Figure 8-24 shows the human population over the past 2500 years. Note that the curve began to steepen in the late 1700s with the onset of the Industrial Revolution. In the mid-twentieth century, medical and technological advances ranging from antibiotics to high-yield grains (the so-called "green revolution") permitted the human population to increase at unprecedented rates. In 1960 there were 3 billion people on the Earth. In 1975, 4 billion. The 5-billion mark was passed in 1986. According to the United Nations Population Fund, there will be 6 billion people on Earth by the year 2000, and 10 billion by 2025.

Every human being has basic requirements in terms of food, clothing, and housing. We all need fuel for cooking and heating. To meet these demands, we cut down forests, cultivate grasslands, and build sprawling cities, replete with

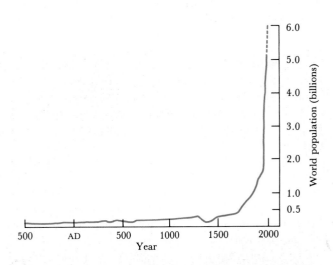

Figure 8-24 The human population Data and estimates from the U.S. Bureau of Census, the Population Reference Bureau, and the United Nations Population Fund were combined to produce this graph showing the human population over the past 2500 years. Note how the population began to rise in the eighteenth century, and the astonishing population increase in the present century.

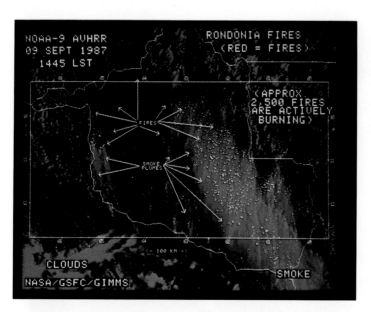

Figure 8-25 *The deforestation of Amazonia* *This picture from NOAA 9 looks down onto the Brazilian state of Rondonia (outlined in yellow), which is located in the western Amazon rain forests where Brazil borders on Bolivia. The red dots are fires and the bluish streaks are smoke plumes. Rondonia is about the same size as the state of Oregon. (Courtesy of J. Tucker, NASA)*

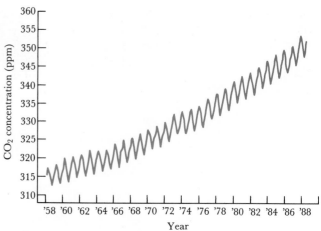

Figure 8-26 *The carbon-dioxide content of the atmosphere* *Measurements of carbon dioxide (in parts per million, ppm) taken at the National Oceanic and Atmospheric Administration's station on Hawaii show that concentrations of this gas have increased by 10 percent since 1958, when the observations started. The saw-toothed pattern is probably due to the absorption of carbon dioxide by plants during the spring and summer.*

roads and factories. A striking example of this activity is occurring in the Amazon rain forest. Tropical rain forests are vital to our planet's ecology because they absorb significant amounts of carbon dioxide and release oxygen through the process of photosynthesis. Although rain forests occupy only 7 percent of the world's land areas, they contain at least 50 percent of all of the species of plants and animals on our planet. Nevertheless, to make way for farms and grazing land, people simply cut down the trees and set them on fire— a process called slash-and-burn. Figure 8-25 shows the Brazilian state of Rondonia photographed by the *NOAA 9* satellite. More than 2500 fires and numerous smoke plumes can be identified. Similar deforestation, along with extensive lumbering operations, is occurring in Malaysia, Indonesia, Papua, and New Guinea. The rain forests that once thrived in Central America, India, and the western coast of Africa are almost gone.

A direct result of this activity is that we are adding carbon dioxide to the Earth's atmosphere faster than plants can extract it. For instance, in 1989 alone we put 18 billion tons of CO_2 into the air we breathe. Figure 8-26 shows how the carbon-dioxide content of the atmosphere has increased over the past thirty years.

Carbon dioxide has the curious property that it is transparent to visible light but not to infrared radiation. Consequently, sunlight has no trouble entering our atmosphere and warming the ground, which then radiates at infrared wavelengths (recall Wien's law). The infrared radiation cannot escape back out into space because it is absorbed by the

carbon dioxide in the air, and so the Earth's atmosphere warms up. This phenomenon, whereby energy from sunlight is trapped in the atmosphere, is called the **greenhouse effect.**

The greenhouse effect is not a new phenomenon. A variety of gases, most importantly water vapor, have warmed the Earth's surface for billions of years. Indeed, without these infrared-absorbing molecules in the air, our planet would be about 30°C (54°F) cooler than it is today. So although humans did not create the greenhouse effect, they are responsible for strengthening it. As a consequence, the average temperature of the Earth's surface is expected to rise, leading to such occurrences as a melting of the polar caps, a rise in sea level, and the flooding of low-lying terrain. Figure 8-27 shows how the global air temperature has changed dur-

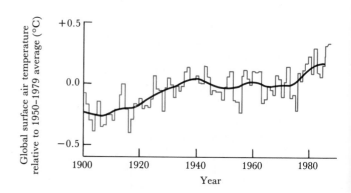

Figure 8-27 *Global temperature change* *Both land and sea surface temperatures were combined to produce this graph, which seems to show a gradual warming trend over the past ninety years. The vertical scale is measured relative to the 1950–1979 average temperature.*

ing the twentieth century. Some scientists view this data with alarm, while others argue that further research is needed to see if the Earth's average temperature is really rising.

As we continue to populate our planet, it is clear that we face many challenges. Age-old modes of behavior that helped ensure the survival of our ancestors may not be the wisest courses of action in the future. Contemplating our behavior and our relationship to the Earth brings to mind the Chinese proverb:

If we don't change the direction in which we are going, we are likely to end up where we are headed.

Key words

asthenosphere	greenhouse effect	*millibar	secondary wave
*atmosphere	hot-spot volcanism	mineral	sedimentary rock
*atmospheric pressure	igneous rock	northern lights	seismic wave
aurora australis (*plural* aurorae)	inner core (of the Earth)	*oblateness	seismograph
	L wave	oceanic rift	shadow zone
aurora borealis	lava	outer core (of the Earth)	shock wave
*bar	lithosphere	P wave	silicate
biosphere	magma	*pascal	solar flare
convection	magnetopause	plate (lithospheric)	southern lights
crust (of the Earth)	magnetosheath	plate tectonics	stratosphere
dynamo effect	magnetosphere	primary wave	subduction zone
earthquake	mantle	rock	thermosphere
epicenter	mesosphere	S wave	troposphere
focus (of an earthquake)	metamorphic rock	sea-floor spreading	Van Allen radiation belt

Key ideas

- The outermost layer, or crust, of the Earth consists of rocks that contain clues about the history of the planet.

    Rocks are composed of minerals, naturally occurring elements, or chemical combinations of elements.

    There are three major categories of rocks: igneous rocks (cooled from molten material); sedimentary rocks (formed by action of wind, water, or ice); and metamorphic rocks (altered in the solid state by extreme heat and pressure).

- Study of seismic waves (vibrations produced by earthquakes) shows that the Earth has a small, solid inner core surrounded by a liquid outer core; the outer core is sur-

rounded by the dense mantle, which in turn is surrounded by the thin low-density crust.

    The Earth's inner and outer cores are composed of almost-pure iron with some nickel mixed in. The mantle is composed of iron–magnesium (ferromagnesian) minerals. The crust is largely composed of silicate minerals.

    There are three types of seismic waves. L waves travel along the planet's surface. Longitudinal P waves and transverse S waves travel through the planet's interior.

    Both temperature and pressure increase steadily with depth inside the Earth.

- The Earth's crust and a small part of its upper mantle form a rigid layer called the lithosphere; this layer is divided into huge plates that move about over the plastic layer called the asthenosphere in the upper mantle.

   Movements of these plates (the process called plate tectonics) are caused by the upwelling of molten material at oceanic rifts, which produces sea-floor spreading.

   Plate tectonics is responsible for most of the major features of the Earth's surface, including mountain ranges, volcanoes, and the shapes of the continents and oceans.

- The Earth's atmosphere is unlike those of the other terrestrial planets in its chemical composition, circulation pattern, and temperature profile.

   Earth's atmosphere is four-fifths nitrogen and one-fifth oxygen, with only a small amount of carbon dioxide. This abundance of oxygen is largely due to the biological activities of life forms on the planet.

   The Earth's atmosphere is divided into layers called the troposphere, stratosphere, mesosphere, and thermosphere; ozone molecules in the stratosphere absorb ultraviolet light.

   Because of the Earth's rapid rotation, the circulation in its atmosphere is complex, with three circulation cells in each hemisphere.

- The Earth's magnetic field produces a magnetosphere that surrounds the planet and blocks the solar wind from hitting the atmosphere.

   A bow-shaped shock wave, where the supersonic solar wind is abruptly slowed to subsonic speeds, marks the outer boundary of the magnetosphere.

   The magnetopause is the location where the magnetic pressure of the Earth's field is exactly counterbalanced by the pressure of the solar wind.

   Most of the particles of the solar wind are deflected around the magnetopause through a turbulent region called the magnetosheath.

   Charged particles from the solar wind are trapped in two huge doughnut-shaped rings called the Van Allen radiation belts.

   A deluge of particles from a solar flare can initiate an auroral display.

- Human activity is changing the Earth's biosphere, on which all living organisms depend.

   Deforestation and the burning of fossil fuels is increasing the carbon-dioxide content of our atmosphere, thereby stimulating the greenhouse effect and perhaps already warming the planet.

Review questions

1 Describe the various ways in which the Earth is unique among the planets of our solar system.

2 Why do you suppose the Earth's surface is not riddled with craters like the surfaces of Mercury, the Moon, and Mars?

3 What is the difference between a rock and a mineral?

4 What is the difference between igneous, sedimentary, and metamorphic rocks? What do these rocks tell us about the sites at which they are found?

5 Look up the chemical compositions of quartz, feldspar, mica, and hornblende, then compare the elements you find with those listed in Table 7-4. Are your findings consistent with the fact that these minerals are very common on the Earth's surface?

6 Describe the interior structure of the Earth. How do we know about the Earth's interior, given that the deepest wells and mines go down only a few kilometers?

7 Describe the process of plate tectonics. Give specific examples of geographic features created by plate tectonics.

8 Explain how the outward flow of energy from the Earth's interior drives the process of plate tectonics.

9 Why do you suppose that active volcanoes, such as Mount St. Helens, are usually located in mountain ranges that border on subduction zones? In what way is Hawaii an exception?

10 Describe the structure of the Earth's atmosphere. Explain how biological activity has affected the chemical composition of our atmosphere. Explain how heat from the Sun and the Earth's rotation affect the circulation of the Earth's atmosphere.

11 Describe the Earth's magnetosphere. If the Earth did not have a magnetic field, do you think aurorae would be more or less common than they are today?

12 What are the Van Allen belts?

Advanced questions

Tips and tools . . .
You'll need to know that the volume of a sphere is $\frac{4}{3}\pi r^3$, where r is the sphere's radius. You may have to consult an atlas to examine the geography of the South Atlantic. Also, remember that the average density of an object is its mass divided by its volume.

13 In what way are the Earth's oceans like an atmosphere? In what way could they be considered part of the Earth's crust? How would your answers be different if the Earth were as close to the Sun as Venus is or as far from the Sun as Jupiter is?

14 The Earth's primordial atmosphere probably contained an abundance of methane (CH_4) and ammonia (NH_3), whose molecules were broken apart by ultraviolet light from the Sun and particles in the solar wind. What happened to the atoms of carbon, hydrogen, and nitrogen that were liberated by this dissociation?

∗15 *Basic:* Africa and South America are separating at a rate of about 3 centimeters per year, as explained in the text. Assuming that this rate has been constant, calculate when these two continents must have been in contact.

16 Why do you suppose that worldwide television transmissions require the use of relay satellites, yet radio programs can be transmitted around the world without the use of such satellites?

∗17 What fractions of the Earth's total volume are occupied by the core, the mantle, and the crust?

∗18 Using data for the mass and size of the Earth listed in Box 8-1, verify that the average density of the Earth is 5500 kg/m^3. Assuming that the average density of material in the Earth's mantle is about 3500 kg/m^3, what must the average density of the core be?

∗19 The Earth's atmospheric pressure decreases by a factor of one-half for every $5\frac{1}{2}$-km increase in altitude above sea level. Construct a plot of pressure versus altitude, assuming the pressure at sea level is 1 bar. Discuss the characteristics of your graph. At what altitude is the atmospheric pressure 1 millibar?

20 On the average, temperature beneath the Earth's crust increases at a rate of 20°C per kilometer. At what depth would water boil? (Assume the surface temperature is 20°C and neglect the effect of the pressure of overlaying rock on the boiling point of water.)

Discussion questions

21 The human population on Earth is currently doubling about every 30 years. Describe the various pressures placed on the Earth by uncontrolled human population growth. Can such growth continue indefinitely? If not, what natural and human controls might arise to curb this growth? It has been suggested that overpopulation problems could be solved by colonizing the Moon or Mars. Do you think this is a reasonable solution? Explain your answer.

22 A recent scientific study suggests that the continued burning of fossil and organic fuels by humans is releasing enough CO_2 to stimulate the greenhouse effect that will melt the polar ice caps. Estimate the volume of the polar ice caps. Assuming that water and ice have roughly the same density, estimate the amount by which the water level of the world's oceans would rise if the polar caps were to melt completely. What portions of the Earth's surface would be inundated by such a deluge?

23 In 1986, scientists discovered a large hole in the Earth's ozone layer over Antarctica. Go to a library and investigate the current status of this ominous development. Is the situation getting better or worse?

Observing project

24 When was the last earthquake in the vicinity of your hometown? How far is your hometown from a plate boundary? What kinds of topography (e.g., mountains, plains, seashore) dominate the geography of your hometown area? Do the topography and the frequency of earthquakes seem to be consistent with your hometown's proximity to a plate boundary?

For further reading

Akasofu, S.-I. "The Aurora: New Light on an Old Subject." *Sky & Telescope*, December 1982 • This brief article, which includes some beautiful photographs, describes the physical processes involved in aurorae.

———. "The Dynamic Aurorae." *Scientific American*, May 1989 • This detailed article by a renowned expert on aurorae describes how interactions of the Earth's magnetic field and the solar wind give rise to a vast generator that powers the aurorae.

Burchfiel, B. C. "The Continental Crust." *Scientific American*, September 1983 • This article describes how the Earth's surface is constantly being reworked by cycles of tectonics, volcanism, erosion, and sedimentation.

Carrigan, C., and Gubbins, D. "The Source of the Earth's Magnetic Field." *Scientific American*, February 1979 • This article discusses the dynamo effect and describes how the ponderous flow of matter in the Earth's core is responsible for our planet's magnetic field.

Cattermole, P., and Moore, P. *The Story of the Earth.* Cambridge University Press, 1985 • This well illustrated book gives a fine introduction to geology and the chronology of our planet.

Clark, W. C. "Managing Planet Earth." *Scientific American*, September 1989 • This lead-off article to single-topic issue of *Scientific American* explores such questions as: What kind of a planet do we want? What kind of a planet can we get?

Cloud, P. "The Biosphere." *Scientific American.* September 1983 • The author explains how the totality of microbial, animal, and plant life on Earth not only is sustained by the lithosphere, the hydrosphere, and the atmosphere, but also has powerfully shaped their evolution.

Frohlich, C. "Deep Earthquakes." *Scientific American,* January 1989 • This article explores the mysteries of deep earthquakes, which occur so far beneath the Earth's surface that high pressures and temperatures should prevent rock from fracturing suddenly.

Graedel, T. E., and Crutzen, P. J. "The Changing Atmosphere." *Scientific American,* September 1989 • Acid rain and smog are among the topics covered in this alarming article about the effects of human activity on the atmosphere.

Ingersoll, A. P. "The Atmosphere." *Scientific American,* September 1983 • This article describes how climates, both past and future, are the result of the dynamic activity of the atmosphere that distributes the energy of solar radiation received by the Earth.

Jeanloz, R., "The Earth's Core." *Scientific American,* September 1983 • This article summarizes our understanding of the Earth's core and shows that turbulent flow in the liquid portions generates the Earth's magnetic field.

Keyfitz, N. "The Growing Human Population." *Scientific American,* September 1989 • This article broaches the difficult question of how poor nations can progress when population growth not only hastens degradation of the environment but also threatens development itself.

Kelley, K., ed. *The Home Planet.* Addison Wesley and Mir, 1988. • This beautiful album of photographs of Earth from space includes moving quotations from U.S. astronauts and Soviet cosmonauts.

Marvin, U. "The Rediscovery of Earth." In Cornell, J., and Gorenstein, P., eds. *Astronomy from Space.* MIT Press, 1983 • This excellent book contains an outstanding chapter on what we can learn about the Earth by observing it from space.

McKenzie, D. P. "The Earth's Mantle." *Scientific American,* September 1983 • This article discusses the processes that occur in the mantle with particular attention to the convection currents in the upper 700 km of ductile rock.

Newell, R., Reichle, H., and Seiler, W. "Carbon Monoxide and the Burning Earth." *Scientific American,* October 1989 • Although industry is usually blamed for carbon monoxide, the authors show that this poisonous gas is also abundant in the tropics, where it comes from burning the rain forests and savannas.

Repetto, R. "Deforestation in the Tropics." *Scientific American,* April 1990 • The article examines the impact of government policies that promote the destruction of tens of thousands of square miles of rain forests each year.

Ruckelshaus, W. D. "Toward a Sustainable World." *Scientific American,* September 1989 • This visionary article argues that moving people and nations toward a sustainable world will involve changes in society comparable in magnitude to the agricultural and industrial revolutions.

Schneider, S. H. "The Changing Climate." *Scientific American,* September 1989 • This article argues that global warming caused by the greenhouse effect should be noticeable by the turn of the century, and significant measures must be taken now to avert the buildup of heat-trapping gases.

Sheffield, C. *Man On Earth.* Macmillan, 1983 • This beautiful coffee-table book includes large, colorful *Landsat* photographs showing evidence of human activity from an altitude of 900 km.

Van Andel, T. *New Views on an Old Planet.* Cambridge University Press, 1985 • This well written summary of the Earth's evolution includes good discussions of continental drift, long-term climatic changes, and the evolution of life.

C H A P T E R

Our Barren Moon

Unlike the Earth, the Moon is a desolate and barren world whose surface has remained essentially unaltered for billions of years. Long ago, debris left over from the formation of the planets rained down on the Moon, extensively cratering its rocky surface. Broad plains, called maria, bear silent witness to vast lava flows that flooded the largest impact basins. Lunar exploration gave many insights about the Moon's history. The Moon is geologically dead and lacks both an atmosphere and a magnetic field. Nevertheless, gravitational interaction with the Earth and Moon produces tides that flex the Moon, causing moonquakes. Radioactive age dating of lunar rocks brought back to Earth demonstrates that the Moon is 4.6 billion years old. The Moon was probably formed when a huge asteroidlike object struck the Earth. That impact ejected a vast amount of molten rock into space, where it cooled and coalesced to form the Moon. This plausible scenario is our first indication that catastrophic collisions between protoplanets and large planetesimals played an important role in shaping the solar system.

One of the most conspicuous features of the Earth is its large satellite, the Moon. We are in fact living on one member of a unique double-planet system. Neither Mercury nor Venus has a natural satellite. Mars does have two moons, but they are very small. The Jovian planets possess several large satellites but, like the moons of Mars, each is quite small compared to its parent planet. In all these cases, each planet is thousands of times more massive than any of its satellites. However, the Earth is only 81 times more massive than its Moon. In other words, the masses of the Earth and Moon are roughly comparable. Why did these two worlds turn out so vastly different? The Earth is a dynamic, living planet; its moon is biologically and geologically dead. The only noteworthy events to occur on the lunar surface over the past 3.5 billion years are occasional impacts by wayward meteoroids—and, of course, the arrival of humanity.

9-1 The Moon's early history can be deduced from the craters, maria, and mountains visible on its surface

Our satellite consistently provides one of the most dramatic sights in the nighttime sky. The Moon is so large and so near the Earth that some of its surface features are readily visible to the naked eye. Observation reveals that the Moon perpetually keeps the same side facing the Earth. This **synchronous rotation** is a stable situation resulting from the gravitational interaction between the Earth and the Moon. Nevertheless, with patience, you can see slightly more than half the lunar surface, because the Moon wobbles slightly as it moves along its orbit. This periodic wobbling, called **libration**, permits us to view 59 percent of our satellite's surface.

With even a small telescope or binoculars, you can see several major different types of lunar terrain (see Figure 9-1). Most prominent are the large, dark, flat areas called **maria** (pronounced MAR-ee-uh). The singular form of this term, **mare** (pronounced MAR-ee), means *sea* in Latin and was introduced in the seventeenth century when observers using early telescopes thought these features were large bodies of water on the Moon. In fact, bodies of liquid water could not possibly exist on our airless satellite. Because there is no atmospheric pressure, a lake or ocean would boil furiously and evaporate rapidly into the vacuum of space. Actually, the maria were formed by huge lava flows that inundated low-lying regions of the lunar surface 3.5 billion years ago. Nevertheless, we have kept their fanciful names, such as Mare Tranquillitatis (Sea of Tranquillity), Mare Nubium (Sea of Clouds), Mare Nectaris (Sea of Nectar), and Mare Serenitatis (Sea of Serenity).

The largest of the 14 maria is Mare Imbrium (Sea of Showers). It is roughly circular and measures 1100 km (700 mi) in diameter (see Figure 9-2). Although the maria seem

Figure 9-1 The Moon *Our Moon is one of seven large satellites in our solar system. The Moon's diameter (3476 km = 2160 mi) is slightly less than the distance from New York to San Francisco. This photograph is a composite of first-quarter and last-quarter views, which causes elongated shadows to enhance all the surface features. (Lick Observatory)*

Figure 9-2 Mare Imbrium from the Earth *Mare Imbrium (Sea of Showers) is the largest of 14 dark plains that dominate the Earth-facing side of the Moon. Because few craters are seen on the maria, they must have been formed by lava flows occurring late in the Moon's geologic history. (Mount Wilson and Las Campanas Observatories)*

Box 9-1 Moon data

Distance from Earth (center to center):	mean = 384,400 km = 238,860 mi
	minimum (perigee) = 356,410 km
	maximum (apogee) = 406,700 km
Distance from Earth (surface to surface):	mean = 376,280 km = 233,810 mi
	minimum (perigee) = 348,290 km
	maximum (apogee) = 398,580 km
Mean orbital velocity:	3680 km/hr
Sidereal period (fixed stars):	27.321661 days
Synodic period (new moon to new moon):	29.530588 days
Inclination of lunar equator to orbit:	6° 41′
Inclination of orbit to ecliptic:	5° 09′
Inclination of lunar equator to ecliptic:	1° 33′
Orbital eccentricity:	0.055
Diameter:	3476 km = 2160 mi
Diameter (Earth = 1):	0.2673
Apparent diameter as seen from Earth:	mean = 31′ 5″
	maximum = 33′ 31″
	minimum = 29′ 22″
Mass:	7.35×10^{22} kg
Mass (Earth = 1):	0.0123
Mean density:	3340 kg/m^3
Surface gravity (Earth = 1):	0.165
Escape velocity:	2.38 km/s
Mean surface temperatures:	day = +130°C = 266°F = 403 K
	night = −180°C = −292°F = 93 K
Albedo:	0.07
Brightest magnitude (when full):	−12.7

quite smooth in telescopic views from the Earth, close-up photographs from lunar orbit reveal small craters and occasional cracks called **rilles** (see Figure 9-3).

Perhaps the most familiar and characteristic features on the Moon are its **craters**. With an Earth-based telescope, some 30,000 of them are visible, in sizes from 1 km to more than 100 km across (see Figure 9-4). Following a tradition established in the seventeenth century, the most prominent craters are named after famous philosophers and scientists, such as Plato, Aristotle, Copernicus, and Kepler.

Craters smaller than about 1 km in diameter cannot be seen from Earth, simply because of optical limitations, such as the resolving power of telescopes. However, photographs from lunar orbit have revealed millions of smaller craters that had previously escaped the scrutiny of Earth-based ob-

servers. Virtually all craters, both large and small, are the result of bombardment by meteoritic material. Nearly all of these craters are circular, indicating that they were not merely gouged out by high-speed rocks. Instead, the violent collisions with the Moon's surface vaporized the rapidly moving meteoroids, and the resulting explosions of hot gas produced the round craters observed today.

Many of the youngest craters are surrounded by light-colored streaks called **rays** that were formed by the violent ejection of material during impact with the surface. Rays emanating from the crater Copernicus, just south of Mare Imbrium, are visible in the lower portion of Figure 9-2. In addition, many large craters have a pronounced **central peak** that forms during a high-speed impact by a sizable meteoroid (see Figure 9-5).

Figure 9-3 Details of Mare Tranquillitatis *Close-up views of the lunar surface reveal numerous tiny craters and cracks on the maria. This photograph was taken in 1969 by astronauts in lunar orbit during a final photographic reconnaissance of potential landing sites for the first manned landing. (NASA)*

Figure 9-4 The crater Clavius *This photograph, taken through the 200-in. Palomar telescope, shows one of the largest craters on the Moon. Clavius has a diameter of 232 km (144 mi) and a depth of 4.9 km (16,000 ft), measured from the crater's floor to the top of the surrounding rim. (Palomar Observatory)*

The flat, low-lying, dark maria cover only 15 percent of the lunar surface. The remaining 85 percent of the Moon's surface is light-colored, heavily cratered terrain at elevations generally higher than those of the maria. This second, more common terrain is called the **terrae**, or **highlands**. (*Terra* means *land* in Latin; in this fanciful terminology, the entire lunar surface is covered by either "land" or "sea.")

One of the surprises stemming from the early days of lunar exploration is that there is only one small mare on the Moon's far side. The lunar far side consists almost entirely of heavily cratered highlands (see Figure 9-6). Detailed observations by astronauts in lunar orbit demonstrated that the maria on the Moon's Earth-facing side are 2 to 5 km below the average lunar elevation. In contrast, the cratered terrae on the lunar far side are typically at elevations up to 5 km above the average lunar elevation.

There are other features on the Moon besides the flat maria and the cratered terrae. For example, there are **mountain ranges**. Although they take their names from famous terrestrial ranges such as the Alps, the Apennines, and the Carpathians, the lunar mountains were not formed in the same way as were mountains on Earth. The Earth's major mountain ranges were created by collisions between lithospheric plates. The lunar mountain ranges, on the other hand, were thrust up by violent impacts of the huge meteoroids that created the maria. Lunar mountains form the rims of vast **basins** that contain the maria. An example of such a

Figure 9-5 Details of a lunar crater *This photograph, taken from lunar orbit by astronauts in 1969, shows a typical view of the Moon's heavily cratered far side. The large crater near the middle of the picture is approximately 80 km (50 mi) in diameter. Note the crater's central peak and the numerous tiny craters that pockmark the lunar surface. (NASA)*

Figure 9-6 Terrae versus maria This photograph by astronauts in 1972 shows the main difference between the near and far sides of the Moon. The Earth-facing side of the Moon (toward the upper left) is characterized by dark, flat maria. The far side of the Moon (toward the lower right) has heavily cratered terrain, with virtually no maria. (NASA)

mountain range can be seen around the edges of Mare Imbrium in Figure 9-2.

From the features we see on the Moon, we can piece together a probable history of the lunar surface. Like the terrestrial planets, the Moon was molten during the later stages of its formation. The heat energy that caused the melting came from the impact of planetesimals and the decay of radioactive elements. Shortly after the Moon's crust cooled, a period of intense meteoritic bombardment peppering the Moon created the numerous craters that form the terrae. Near the end of this cratering period, several large, asteroid-sized objects struck the lunar surface and broke through the thin crust. Lava welled up, flooding low-lying areas and producing maria. Large meteoroids must have struck the Moon from all sides. However, we theorize that on the side of the Moon opposite the maria, the crust must have been thick enough to prevent meteoroid penetration to the molten interior. Not much has happened since those ancient days, as indicated by the fact that the mare surfaces have remained almost unchanged since their formation.

Although this scenario seems quite reasonable, we need supporting evidence. When did most of the cratering occur? When did the lava flows flood the mare basins? Are the light-colored highland rocks really older than the darker mare rock, as we would expect from a telescopic examination of the lunar surface? There was only one way to find out: journey to the Moon and obtain lunar samples for direct examination and analysis.

Lunar samples helped answer many questions about the Moon and also shed light on the origin of the Earth. Because Earth is a geologically active planet, all traces of its origins have been erased. Typical terrestrial surface rocks are only a few hundred million years old, which is only a small fraction of the Earth's age. Moon rocks, however, have been undisturbed for billions of years. Lunar exploration provided a valuable perspective on the creation of the Earth and the birth of the entire solar system.

9-2 Manned exploration of the lunar surface was one of the greatest adventures of human history

The 1960s will long be remembered as the decade when humanity reached out and touched another world (see Box 9-2). During a few historic moments, people walked on the Moon. An incredible dream became a reality.

Lunar missions began in 1959 when the Soviet Union sent three small spacecraft toward the Moon. *Luna 1* missed the Moon altogether, while *Luna 2* impacted as planned on the

Box 9-2 Successful missions to the Moon

Between 1959 and 1976, the Soviets and the Americans landed dozens of probes on the Moon. The most historic lunar missions are listed below, along with a few brief comments. The photo-graph shows the *Apollo 11* liftoff on July 16, 1969, which carried American astronauts to the first manned lunar landing.

Ranger program	Launch date		Comments
Ranger 7	1964	July 28	Hit the Moon in Mare Nubium
Ranger 8	1965	February 17	Hit the Moon in Mare Tranquillitatis
Ranger 9	1965	March 21	Hit the Moon in the crater Alphonsus
Surveyor program			
Surveyor 1	1966	May 30	Landed on Oceanus Procellarum
Surveyor 3	1967	April 17	Landed on Oceanus Procellarum
Surveyor 5	1967	September 8	Landed on Mare Tranquillitatis

Surveyor program	Launch date		Comments
Surveyor 6	1967	November 7	Landed on Sinus Medii
Surveyor 7	1968	January 17	Landed on rim of crater Tycho
Orbiter program			
Orbiter 1	1966	August 10	Successful lunar orbiter
Orbiter 2	1966	November 7	Successful lunar orbiter
Orbiter 3	1967	February 25	Successful lunar orbiter
Orbiter 4	1967	May 4	Successful lunar orbiter
Orbiter 5	1967	August 1	Successful lunar orbiter
Luna and Zond probes			
Luna 2	1959	September 12	Hit the Moon
Luna 3	1959	October 4	Went by the Moon; photographed lunar far side
Zond 3	1965	July 18	Went by the Moon
Luna 9	1966	January 31	Landed on Oceanus Procellarum
Luna 10	1966	March 31	Successful lunar orbiter
Luna 11	1966	August 24	Successful lunar orbiter
Luna 12	1966	October 22	Successful lunar orbiter
Luna 13	1966	December 21	Landed on Oceanus Procellarum
Luna 14	1967	April 7	Successful lunar orbiter
Zond 5	1968	September 14	Went around the Moon
Zond 6	1968	November 10	Went around the Moon and returned to Earth
Zond 7	1969	August 7	Went around the Moon and returned to Earth
Luna 16	1970	September 12	Landed on Mare Fecunditatis and returned lunar rocks to Earth
Luna 17	1970	November 10	Landed on Mare Imbrium; carried vehicle *Lunokhod 1*, which travelled 10.5 km across lunar surface
Luna 19	1971	September 28	Successful lunar orbiter
Luna 20	1972	February 14	Landed on Mare Fecunditatis and returned lunar rocks to Earth
Luna 21	1973	January 8	Landed on Mare Serenitatis; carried vehicle *Lunokhod 2*, which travelled 37 km across lunar surface
Luna 22	1974	May 29	Successful lunar orbiter
Luna 24	1976	August 9	Landed on Mare Crisium and returned lunar rocks to Earth
Apollo program	**Landing date**		
Apollo 11	1969	July 20	Landed on Mare Tranquillitatis
Apollo 12	1969	November 19	Landed on Oceanus Procellarum
Apollo 14	1971	January 31	Landed at Fra Mauro
Apollo 15	1971	July 30	Landed at Hadley-Apennines
Apollo 16	1972	April 21	Landed at crater Descartes
Apollo 17	1972	December 11	Landed at Taurus-Littrow

a

b

Figure 9-7 The Moon from Ranger 9 *and the Earth* (a) *Seconds be-fore it crashed onto the Moon,* Ranger 9 *sent back this picture of the crater Alphonsus (white circle indicates impact target). Note that the* resolution *is far better than that of the Earth-based telescopic view shown in* (b). *The diameter of Alphonsus is about 129 km (80 mi). (NASA)*

lunar surface. Neither spacecraft sent back significant data. *Luna 3*, which was larger and more complex than its predecessors, went behind the Moon in October 1959 and sent back the first pictures—albeit of low resolution—of the lunar far side.

American attempts to reach the Moon began in the early 1960s with Project Ranger. Equipped with six television cameras, the Ranger spacecraft were designed to transmit close-up views of the lunar surface during the final few minutes before crashing onto the Moon. All the early attempts failed, but in 1964 and 1965, *Rangers 7, 8,* and *9* did succeed in sending back more than 19,000 close-up views of the Moon. Figure 9-7 shows a view from *Ranger 9* just before it crashed onto the floor of the crater Alphonsus. A photograph of the same area taken through an Earth-based telescope demonstrates the dramatic increase in resolution that was achieved by the spacecraft.

A complete high-resolution photographic survey of the lunar surface was the mission of the Orbiter program, which involved five spacecraft launched between August 1966 and August 1967. Each spacecraft carried two cameras and 260 feet of black-and-white photographic film. An onboard "processor" developed the film and a "scanner" scanned each picture in thin strips. After being transmitted back to Earth, the strips were placed side by side to reconstruct the photograph. Consequently, all the Orbiter photographs have a characteristic striped appearance (see Figure 9-8). In

all, the five Orbiters sent back a total of 1950 high-resolution pictures covering $99\frac{1}{2}$ percent of the Moon. In some of these photographs, features as small as 1 meter across could be seen. Because these pictures revealed the lunar surface in

Figure 9-8 The Moon's far side from Orbiter 5 *In 1967, the last of the lunar orbiters sent back this picture of the Moon's far side. The large, flat-bottomed crater to the left is Mare Moscoviense, measuring about 250 km (155 mi) in diameter. Mare Moscoviense is one of the few marelike features on the Moon's densely cratered far side. (NASA)*

Figure 9-9 *An astronaut visits* **Surveyor 3** *Surveyor 3 successfully soft-landed in Oceanus Procellarum (The Ocean of Storms) on February 8, 1967. Like the four other Surveyors, it sampled the lunar soil and sent back pictures to Earth. Surveyor 3 was visited 2½ years later by an Apollo 12 astronaut who took pieces of the spacecraft back to Earth for analysis. (NASA)*

close detail, they were essential in helping NASA scientists choose landing sites for the astronauts.

An important step in preparing for a manned landing on the Moon was to soft-land a spacecraft on the lunar surface. This was the purpose of the Surveyor program. Between June 1966 and January 1968, five Surveyors successfully touched down on the Moon, sending back pictures and data directly from the lunar surface. These missions dispelled a variety of fears, such as the worry that a spacecraft or an astronaut stepping onto the lunar surface might simply sink out of sight into a deep layer of dust. Figure 9-9 shows an astronaut visiting *Surveyor 3*, 2½ years after it had landed on the Moon.

The decision to transport humans to the lunar surface was primarily political. American technology of the early 1960s was suffering from an inferiority complex. The Soviet Union had been first to launch an Earth-orbiting satellite (*Sputnik 1* on October 4, 1957), first to orbit a man around the Earth (Yuri Gagarin on April 12, 1961), and first to send pictures back from the Moon (*Luna 3* on October 6, 1959). In addition to the military implications of these successes, the possibility of seeing an American place and American flag on the Moon excited politicians and the public into embracing the manned space program with gusto.

In view of the value placed on human life, the steps to fulfill President Kennedy's 1961 pledge to land a man on the

Moon before 1970 were tedious and expensive. Developmental work included six one-man Earth-orbiting flights (the Mercury project from May 1961 through May 1963) and ten two-man Earth-orbiting flights (the Gemini series from March 1965 through November 1966). The Apollo program, which finally took people to the Moon, used three-man crews and a spacecraft consisting of two detachable sections. Upon reaching the Moon, one astronaut would stay behind in lunar orbit in the command and service module while the other two astronauts would descend to the lunar surface in the lunar module. After completing their activities, the two astronauts would return to lunar orbit in their lunar module and "dock" with the command module. Once they, their equipment, and moon rocks were transferred to the command module, they would "jettison" the lander for the trip back to Earth.

There were six manned lunar landings between 1969 and 1972. The first two, *Apollo 11* and *Apollo 12*, set down in maria. The remaining four landings (*Apollo 14* through *Apollo 17*) were made in progressively more challenging terrain, culminating in rugged mountains east of Mare Serenitatis. Major factors in the choice of landing sites were the astronauts' safety and the opportunity to explore a wide variety of geologically interesting features. Figure 9-10 shows one of the Apollo bases.

Almost everything that American astronauts did on the Moon could have been accomplished years earlier at greatly

Figure 9-10 *The* **Apollo 15** *base   The last three Apollo missions used surface vehicles called Lunar Rovers that greatly enhanced the astronauts' mobility. This photograph shows the Apollo 15 landing site at the foot of the Apennine Mountains near the eastern edge of Mare Imbrium. The hill in the background, called Hadley Delta, is about 5 km behind the lunar module and rises about 3 km above the surrounding plain. (NASA)*

reduced cost using robots, the tactic used by the Soviet Union. Between 1966 and 1976, a series of Luna and Zond spacecraft orbited the Moon and soft-landed on its surface. The first soft-landing was achieved by *Luna 9*, four months before the arrival of the United States' *Surveyor 1*. The first lunar satellite was *Luna 10*, which orbited the Moon four months before *Orbiter 1*. *Luna 16, 20,* and *24* picked up samples of rock and returned them to Earth. *Luna 17* and *21* included vehicles that roamed the lunar surface taking numerous pictures. No spacecraft has returned to the Moon since the successful mission of *Luna 24* in August 1976.

9-3 Measurements on the lunar surface show that the Moon has no magnetic field but may have a small, solid core

The Apollo Moon landings gave scientists an unprecedented opportunity to investigate and explore an alien world. The lunar landing modules were therefore packed with an assortment of apparatus and equipment that the astronauts deployed around the landing sites (see Figure 9-11). For example, all the missions carried magnetometers, which confirmed that the Moon has no magnetic field. This suggests that the Moon does not have a molten core. Careful mag-

netic measurements of lunar rocks indicated, however, that the Moon did have a weak magnetic field when the rocks solidified, billions of years ago. The implication is that the Moon originally had a small, molten, iron-rich core, but that the core solidified as the Moon cooled, and the lunar magnetic field disappeared. Geologists estimate that, if the Moon does in fact have an iron-rich core, it is probably less than 700 km (435 mi) in diameter.

Seismometers set up by the astronauts indicated that the Moon exhibits far less seismic activity than does the Earth. Roughly 3000 **moonquakes** were detected per year, whereas a similar seismometer on Earth would record hundreds of thousands of earthquakes per year. Furthermore, typical moonquakes are far weaker than typical earthquakes. A major moonquake, which typically measures 0.5 to 1.5 on the Richter scale, would go unnoticed by a person standing near the epicenter. For comparison, a major earthquake is in the range of 6 to 8 on that scale.

Analysis of the feeble moonquakes reveals that most of them originate at depths of 600 to 800 km below the surface, deeper than most earthquakes. The depth of the deepest earthquakes is regarded as an indicator of the boundary between the solid lithosphere and the plastic asthenosphere. The lithosphere is brittle enough to fracture and produce seismic waves, whereas the deeper rock of the asthenosphere oozes and flows rather than cracking. Applying the same reasoning to the Moon, we may conclude that the Moon's lithosphere is about 800 km thick (see Figure 9-12).

Figure 9-11 Seismic equipment on the Moon This view of the Apollo 11 *base shows astronaut Edwin Aldrin standing alongside a seismometer that detected moonquakes and transmitted data back to Earth. By combining data from seismometers left on the Moon by Apollo astronauts, scientists have been able to deduce the structure of the Moon's interior. The* Apollo 11 *lunar module and some additional equipment are in the background. (NASA)*

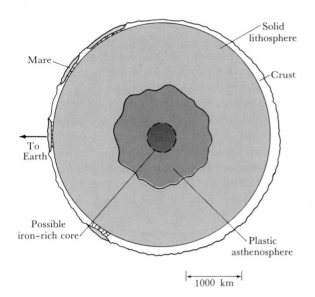

Figure 9-12 The internal structure of the Moon *Like the Earth, the Moon probably has a crust, a mantle, and a core. The lunar crust has an average thickness of about 60 km on the Earth-facing side but about 100 km on the far side. The crust and solid upper mantle form a lithosphere about 800 km thick. The plastic (nonrigid) asthenosphere probably extends all the way to the base of the mantle. If the Moon has an iron-rich core, it is solid and is less than 700 km in diameter. Although the main features of the Moon's interior are analogous to those of the Earth's interior, the proportions are quite different (see Figure 8-9).*

The lunar asthenosphere probably extends down to the iron-rich core at a depth of more than 1400 km below the lunar surface. The asthenosphere and the lower part of the lithosphere presumably are composed of relatively dense ferromagnesian rocks. The upper part of the lithosphere is a less dense, silicate-rich crust, with an average thickness of about 60 km on the Earth-facing side and up to 100 km on the far side. For comparison, the thickness of the Earth's crust ranges from 5 km under the oceans to about 30 km under major mountain ranges on the continents.

Heat-flow experiments were set up at the *Apollo 15* and *Apollo 17* bases on the Moon. Sensitive heat detectors measured a surface heat loss about one-third as great as that measured on the surface of the Earth. The existence of this outward heat flow indicates that the Moon is not totally cold and dead, which supports the idea that there is enough heat inside the Moon to create a nonrigid asthenosphere.

Seismometers left on the Moon continued to transmit data for nearly eight years, so geologists were able to monitor meteoritic impacts. The Apollo seismometers were even sensitive enough to detect a hit by a grapefruit-sized meteoroid anywhere on the lunar surface. Long-term monitoring showed that the Moon is struck by 80 to 150 meteoroids per year with masses in the range of 100 g to 1000 kg (roughly $\frac{1}{4}$ lb to 1 ton).

9-4 Gravitational interactions produce significant tidal effects in the Earth–Moon system

Long-term monitoring of the seismometers left at the Apollo landing sites on the Moon revealed that the frequency of moonquakes reaches a maximum at new moon and at full moon. Geologists concluded from this pattern that the frequency of moonquakes is influenced by tidal forces.

Tidal forces are caused by the differences in gravitational pull across an object by a second object. For instance, consider the gravitational pull of the Moon on the Earth, as sketched in Figure 9-13*a*. The strength and direction of the gravitational force of the Moon at several locations on the Earth are indicated with arrows. Notice that the side of the Earth closest to the Moon feels a greater gravitational pull than the Earth's center does. Because of this difference, the side of the Earth facing the Moon feels an overall pull, called a tidal force, toward the Moon. Similarly, the Earth's center is subjected to a greater gravitational pull toward the Moon than is any place on the Earth's hemisphere facing away

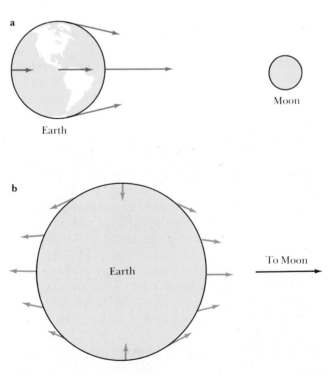

Figure 9-13 Tidal forces *The Moon induces tidal forces on the Earth because the strength of the Moon's gravity differs from one place on Earth to another. (a) Red arrows indicate the strength and direction of the Moon's gravitational pull at selected points on the Earth. (b) Blue arrows indicate the strength and direction of the tidal forces acting on the Earth. At any location, the tidal force equals the Moon's gravitational pull at that point—minus the gravitational pull of the Moon at the center of the Earth.*

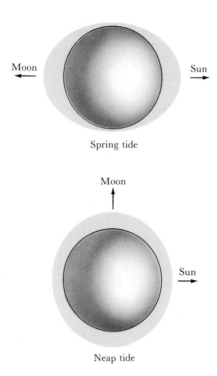

Figure 9-14 *Tides on the Earth* *The gravitational forces of the Moon and Sun deform the oceans. The greatest deformation (spring tides) occurs when the Sun, Earth, and Moon are aligned. The least deformation (neap tides) occurs when the Sun, Earth, and Moon form a right angle.*

from the Moon. Over this hemisphere, the tidal forces are directed away from the Moon. The net effect, as sketched in Figure 9-13*b*, is that tidal forces tend to deform the Earth, stretching it into the shape of a football.

The tides in the oceans are the result of tidal forces caused by both the Moon and the Sun trying to distort the shape of the Earth. This tidal distortion is greatest when the Sun, Moon, and Earth are aligned (at either new moon or full moon), producing large shifts in water level that are called **spring tides** (see Figure 9-14). At first quarter and last quarter, when the Sun and Moon are at right angles from the Earth, the tidal distortion of the oceans is the least pronounced, producing smaller tidal shifts called **neap tides.**

It is easy to notice ocean tides, because the water level goes up and down. Harder to detect are similar motions of the solid Earth as it is deformed and stretched by the gravitational forces of the Sun and Moon, though to a much lesser extent than the oceans.

Just as the gravitational forces of the Sun and Moon deform the Earth, so also the Sun and Earth deform the Moon. The Moon is alternately squeezed and stretched as it orbits our planet. Because the Earth is 81 times as massive as the Moon, the Earth produces large tidal deformations of the lunar surface. The average tidal deformation on the Earth is about 1 foot, but about 60 feet on the Moon. The strongest gravitational stresses occur when the Sun, Moon, and Earth are aligned (at new moon and full moon)—the time when

the Apollo seismometers reported the highest frequency of moonquakes.

Tidal effects also explain why the Moon keeps the same side facing the Earth. Under the Earth's gravitational pull, the Moon is slightly elongated—bulging in much the same way as the Moon's gravity causes the oceans to bulge out on opposite sides of the Earth. Under the gravitational pull of the Earth, the most stable configuration exists when the tidally induced bulge on the Moon points at the Earth. Tidal forces thus induce the common phenomenon of synchronous rotation. We shall see in later chapters that the tiny Martian moons and many of the satellites of the Jovian planets also exhibit synchronous rotation, always keeping the same side facing the parent planet.

The tidal interactions between the Earth and the Moon have had other major effects on the relationship between these two bodies over the ages. The Earth rotates about its axis faster than the Moon revolves around the Earth. Therefore, as the Moon produces a tidal bulge in the oceans, the Earth's rotation carries that bulge forward so that high tide is slightly ahead of the Moon rather than staying just below it (see Figure 9-15). As viewed from the Moon, the bulge on the Earth is always aimed slightly ahead of the Moon's own position. The gravitational pull of this bulge produces a small but constant force tugging the Moon forward and gradually lifting it into a more distant orbit around the Earth. In other words, the Moon is very slowly but relentlessly spiraling away from the Earth at a rate of about 4 cm per year.

As the Moon moves farther and farther away from the Earth, its sidereal period becomes longer and longer (recall Kepler's third law). Furthermore, constant friction between

Figure 9-15 *The Moon's tidal recession* *Because of the Earth's rapid rotation, the tidal bulge on the oceans is dragged ahead of the Moon's position in the sky. This bulge on the leading side of the Earth produces a small forward force on the Moon that causes it to spiral slowly away from the Earth.*

the oceans and the Earth's surface is gradually causing the Earth's rate of rotation to decrease. The Earth's day is therefore becoming longer and longer, by approximately 0.002 seconds per century. At some point in the distant future, the Earth will be rotating so slowly that a solar day will equal a lunar month. At that time, the Earth's tidal bulge will be aimed directly at the Moon, which will cause the Moon to stop spiraling away from the Earth. Calculations tell us that this stable situation will be reached when the solar day and the lunar month both equal 47 of our present solar days. The Earth will then keep its same side facing the Moon just as the Moon presently keeps the same side facing the Earth.

9-5 Lunar rocks were formed 3 to 4.5 billion years ago

The Apollo astronauts brought back 382 kg (842 lb) of lunar rocks, which proved to be a very important source of information about the early history of our double-planet system. All of the lunar samples turned out to be igneous rocks, and all of them appear to have formed through cooling of molten lava. The samples are almost completely composed of the same minerals that are found in terrestrial volcanic rocks: pyroxene, olivine, feldspar, and so forth. No sedimentary rocks were found by the Apollo astronauts, indicating that the Moon has never had an atmosphere or oceans to create such rocks. They found no true metamorphic rocks either, although many of the lunar samples have been modi-

fied by meteoritic impacts. In fact, the entire lunar surface is covered with a layer of fine powder and rock fragments produced by 4.5 billion years of relentless meteoritic bombardment. This layer, which ranges in thickness from 1 to 20 m, is called the **regolith** (from the Greek meaning *blanket of stone*); the term *soil* suggests the presence of decayed biological matter, and so does not apply to the Moon's surface.

Astronauts who visited the maria discovered that these dark regions of the Moon are covered with basaltic rock quite similar to the dark-colored rocks formed by lava from volcanoes on Hawaii and Iceland. The rock of these low-lying lunar plains is called **mare basalt** (see Figure 9-16).

In contrast to the dark maria, the lunar highlands are covered with a light-colored rock called **anorthosite** (see Figure 9-17). On Earth, anorthositic rock is found only in such very old mountain ranges as the Adirondacks in the eastern United States. In comparison to the mare basalts, which have more of the heavier elements like iron, magnesium, and titanium, anorthosite is rich in calcium and aluminum. Anorthosite therefore has a lower density than basalt. The anorthositic magma apparently floated to the lunar surface when the Moon was molten, solidifying as it cooled to form the lunar crust. The denser mare basalts formed later from lava that oozed out from the lunar interior and filled the mare basins. The action of tidal forces acting on the heavy maria helped establish the Moon's synchronous rotation with the maria facing the Earth.

The Apollo astronauts brought back many specimens of **impact breccias**, which are made of various rock fragments that have been cemented together by meteoritic impact (see

Figure 9-16 Mare basalt *This 1.531-kg (3½-lb) specimen of mare basalt was brought back by Apollo 15. It is a vesicular basalt, so called because of the holes, or vesicles, covering 30 percent of its surface. Gas must have dissolved under pressure in the lava from which this rock solidified. When the lava reached the airless lunar surface, bubbles formed as the pressure dropped. Some of the bubbles were frozen in place as the rock cooled. (NASA)*

Figure 9-17 Anorthosite *The light-colored lunar terrae are covered with an ancient type of rock called anorthosite, which is believed to be the material of the original lunar crust. This sample, called the Genesis rock by the Apollo 15 astronauts who picked it up at the base of the Apennine Mountains, has an age of approximately 4.1 billion years. (NASA)*

Figure 9-18 *A lunar breccia Meteoritic impacts can cement rock fragments together to form breccias. This particular impact breccia was collected by Apollo 16 astronauts from the rim of a crater near their landing site. (NASA)*

Figure 9-19 *A glass-coated breccia This glass-coated lunar rock was picked up near the rim of a crater by the Apollo 15 astronauts. Natural glass forms whenever molten rock cools very rapidly. Notice the dark color of the shiny glass coating. (NASA)*

Figure 9-18). On Earth, breccias formed by sedimentary or volcanic action are fairly common, but impact breccias are quite rare.

Meteoritic bombardment is the only source of "weathering" to which rocks on the lunar surface are subjected. In addition to making breccias and churning up the regolith, meteoritic impacts also melt rocks to produce glass. Many lunar samples are coated with a thin layer of smooth, dark glass created when the surface of the rock was suddenly melted and then rapidly solidified (see Figure 9-19). Moreover, small black glass beads are common in the lunar regolith. Presumably, these glass spheres were formed from droplets of molten rock hurled skyward by the impact of a meteoroid. Finally, many lunar samples bear the scars of high-speed meteoritic dust. Dust grains traveling at thousands of kilometers per hour produce tiny **zap craters** on the exposed surfaces of moon rocks (see Figure 9-20). Because of these tiny, glass-lined craters, Moon rocks often seem to sparkle when held in the sunshine.

Although evidence of meteoritic impacts dominates our impressions of the lunar surface, the rate of this weathering is actually quite slow. Geologists estimate that it takes tens of millions of years to wear away a layer of rock only 1 mm thick. Consequently, features formed 3 billion years ago are still well preserved today. The astronauts' footprints themselves will remain sharply imprinted on the lunar surface for millions of years to come.

Although lunar rocks bear a strong resemblance to terrestrial rocks, there are some important differences. For one, every terrestrial rock contains some water, but lunar rocks are totally dry. There is absolutely no evidence that water has ever existed on the Moon. In the absence of both an atmosphere and water, it is not surprising that the astro-

nauts found no traces of life. After the first few Apollo missions, the astronauts were quarantined for a few days to make sure they had not brought back any lunar microorganisms—germs that might cause some terrible disease. However, in view of the total sterility of the lunar samples, this precaution was soon dropped. The lunar rocks are, however,

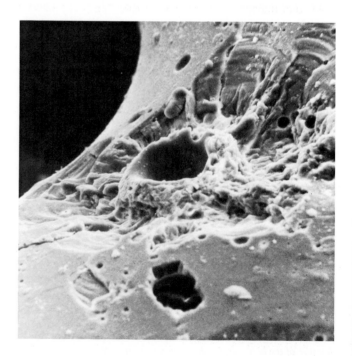

Figure 9-20 *A zap crater The upper surfaces of many moon rocks are covered with nearly microscopic craters produced by the impact of high-speed meteoritic dust grains. These glass-lined craters are typically less than 1 mm in diameter. (NASA)*

stored in rooms filled with dry nitrogen. Geologists justifiably fear that the rocks would rust if exposed to the oxygen and water vapor in the Earth's air.

By carefully measuring the abundances of trace amounts of radioactive elements in lunar samples (see Box 9-3), geologists confirmed that anorthosite is more ancient than the mare basalts. This result had been expected, because the lunar highlands are densely cratered, whereas the basaltic surfaces of the mare show relatively few craters. Typical anorthositic specimens from the highlands are between 4.0 and 4.3 billion years old, and one rock brought back by the *Apollo 17* astronauts is nearly 4.6 billion years old. All these extremely ancient specimens represent material from the Moon's original crust. In sharp contrast, all the mare basalts are between 3.1 and 3.8 billion years old. Apparently, the mare basalts solidified from ferromagnesian-rich lavas that gushed up from the Moon's mantle and flooded the mare basins between 3.1 and 3.8 billion years ago, just about the time when the oldest rocks on the Earth's present surface layers were being formed.

9-6 The Moon probably formed from debris cast into space when a huge asteroid struck the proto-Earth

Prior to the Apollo program, there were three different theories about the origin of the Moon. First, the **fission theory** holds that the Moon was pulled out from a rapidly rotating proto-Earth. Second, the **capture theory** suggests that the Moon was formed elsewhere in the solar system and drawn into orbit about the Earth by gravitational forces. Third, the **cocreation theory** proposes that the Earth and the Moon were formed at the same time, but separately. A fourth theory, called the **collisional ejection theory**, was proposed in 1984 and suggests that the Earth was struck by an object perhaps as large as Mars and the collision ejected debris from which the Moon formed. Clues provided by the lunar rocks along with what we know about the formation of the planets seems to favor the last of those hypotheses.

One fact used to support the fission theory is that the Moon's average density (3340 kg/m³) is comparable to that of the Earth's outer layers, as would be expected if the Moon had been formed from material flung out by a proto-Earth's rapid rotation. However, most geologists counter this argument by pointing to the significant differences between lunar and terrestrial rocks—particularly the differences in water content and in the relative abundances of volatile and refractory elements.

Volatile elements (such as potassium and sodium) melt and boil at relatively low temperatures, whereas **refractory elements** (such as titanium, calcium, and aluminum) melt and boil at much higher temperatures. Compared to terrestrial rocks, the lunar rocks have slightly greater proportions of the refractory elements and slightly lesser proportions of the volatile elements. The implication is that the Moon formed from material somewhat hotter than the material from which the Earth was created. Some of the volatile elements boiled away, leaving the young moon relatively enriched in the refractory elements.

Proponents of the capture theory use these differences to argue that the Moon formed elsewhere and later was captured by the Earth. After all, neither Mercury nor Venus nor Mars has a large satellite—why should the Earth have been unique in forming as a double-planet system? Jupiter and Saturn do have satellites even larger than our Moon, but those satellites (and the smaller moons of Uranus) all follow orbits that lie in the equatorial planes of their parent planets. The rotation of each Jovian planet and the revolution of its satellites share the same orientation in space, as would be expected if the planet and satellites had been formed as part of a single process. The plane of the Moon's orbit, on the other hand, is much closer to the plane of the ecliptic than it is to the plane of the Earth's equator. This pattern would be expected if the Moon had originally been in orbit about the Sun but later was captured by the Earth.

There are difficulties with the capture theory, however. If the Earth did capture the Moon intact, then a host of conditions must have been met, entirely by chance. The Moon must have coasted to within 50,000 km of the Earth at exactly the right speed to leave a solar orbit and adopt an Earth orbit without ever actually hitting our planet. Of course, just because something is highly improbable is not proof that it did not happen. Most geologists agree that the Moon must have formed in the same general part of our solar system as the Earth. An analysis of meteorites shows that the abundances of the isotopes of oxygen (^{16}O, ^{17}O, and ^{18}O) differ measurably among rocks from different parts of the solar system. The abundances in terrestrial and lunar rocks are sufficiently similar to indicate that both worlds formed at nearly the same distance from the Sun.

Although special conditions must be met for a planet to capture a giant rock from solar orbit, much less stringent conditions exist for the capture of swarms of tiny rocks. Great numbers of rock fragments must have swarmed about the protosun during the formation of the solar system. Most of this rocky debris probably orbited our infant star in the plane of the ecliptic along with the protoplanets. Energy radiated from the contracting protosun would have baked the water and volatile elements out of these smaller rock fragments, many of which were presumably captured into orbit about the proto-Earth. Then, just as planetesimals accreted to form the proto-Earth in orbit about the Sun, the fragments in orbit about the Earth accreted to form the Moon—or so says the cocreation theory. This theory fails to explain why our nearest neighbors—Venus and Mars—were not similarly bestowed with comparatively large moons.

In the mid-1980s, several teams of scientists began exploring a new theory inspired by the realization that the

Box 9-3 Radioactive age dating

The Apollo program gave geologists the opportunity to get their hands on extremely ancient rocks. By examining these lunar samples, scientists began to piece together a history of important events that happened shortly after the creation of our solar system. To get the story right, however, geologists had to accurately determine the ages of the lunar rocks. Fortunately, most rocks contain trace amounts of radioactive elements such as uranium. The relative abundances of various radioactive isotopes and their decay products provide the key that geologists use to determine the ages of rocks. As we saw in Box 7-1, every atom of a particular element has the same number of protons in its nucleus. However, different isotopes of the same element have different numbers of neutrons in their nuclei. For example, the common isotopes of uranium are ^{235}U and ^{238}U. Each isotope of uranium has 92 protons in its nucleus (correspondingly, uranium is the element 92 on the periodic chart; see Figure 7-8). However, a ^{235}U nucleus contains 143 neutrons, whereas a ^{238}U nucleus has 146 neutrons.

In lightweight elements, the number of protons in the nucleus is typically equal to or nearly equal to the number of neutrons. For example, the common isotope of oxygen is ^{16}O, which has eight protons and eight neutrons in its nucleus. In heavy elements, however, the number of neutrons exceeds the number of protons. These extra neutrons serve as buffers between the many positively charged protons crowded together in the nucleus. Without the shielding effect of these extra neutrons, the repulsive electric forces between the closely packed protons would break apart the nucleus. A nucleus is said to be stable if it effectively resists the natural forces that tend to break it apart.

A nucleus is unstable or radioactive if it contains too many protons or too many neutrons. In either case, the unstable nucleus ejects particles to achieve a stable nucleus. For example, ^{235}U, which is radioactive, naturally casts off helium (^{4}He) nuclei. This process, which removes two neutrons and two protons from an unstable nucleus, is a common mode of radioactive decay called **alpha decay**. In particular, we find

$$^{235}_{92}U \rightarrow {}^{231}_{90}Th + {}^{4}_{2}He$$

where the subscripts refer to the number of protons that each nucleus contains. Because uranium decays into thorium, ^{235}U

is called the **parent isotope** and ^{231}Th is called the **daughter isotope**.

Some radioactive isotopes decay rapidly, others decay slowly. Physicists find it convenient to talk about the decay rate in terms of an isotope's half-life. The **half-life** of an isotope is the time interval in which one-half of the parent nuclei decay into daughter nuclei. For example, the half-life of ^{235}U is 710 million years. This means that, if you start out with 1 kg of ^{235}U, after 710 million years you will have only $\frac{1}{2}$ kg of ^{235}U left; the other $\frac{1}{2}$ kg of your sample will have turned into other elements.

When ^{235}U decays into ^{231}Th, the process doesn't stop there. The thorium isotope is extremely unstable and has a half-life of only 25.6 hours. At a rapid rate, ^{231}Th decays into an isotope of protactinium, ^{231}Pa. The protactinium in turn decays into actinium. The process continues on and on, until a stable isotope of lead (^{207}Pb), which does not decay, is finally formed. The entire **decay series** of ^{235}U is shown in the table below:

Parent isotope	Half-life	Daughter isotope
^{235}U	710,000,000 years	^{231}Th
^{231}Th	25.6 hours	^{231}Pa
^{231}Pa	34,000 years	^{227}Ac
^{227}Ac	21.6 years	^{227}Th
^{227}Th	18 days	^{223}Ra
^{223}Ra	11.7 hours	^{219}Rn
^{219}Rn	3.9 seconds	^{215}Po
^{215}Po	0.002 second	^{211}Pb
^{211}Pb	36 minutes	^{211}Bi
^{211}Bi	2.2 minutes	^{207}T1
^{207}T1	4.8 minutes	^{207}Pb

Notice that only the first step in the decay of ^{235}U takes a long time. Compared to this first step, all the other steps occur quickly. In other words, we can describe the decay series simply

early history of the solar system was dominated by collisions between objects. As we have seen (recall Figure 7-17), the formation of the inner planets involved smaller objects colliding and fusing together to build larger objects. Some of these collisions must have been quite spectacular, especially near the end of the planet-forming episode when most of the mass of the inner solar system was contained in the protoplanets and a few dozen large asteroidlike objects. According to the collisional ejection theory, one such asteroidlike object struck the proto-Earth 4.6 billion years ago. A

supercomputer simulation of this cataclysm is shown in Figure 9-21. Energy released during the collision produces a huge plume of vaporized rock that squirts out from the point of impact. As this ejected material cools, it coalesces to form the Moon, as shown in Figure 9-22.

The collisional ejection theory is in agreement with many of the known facts about the Moon. For example, the vaporization of ejected rock during impact would have driven off volatile elements and water, leaving material that strongly resembles moon rocks. If the collision took place after chemas the conversion of uranium (^{235}U) into lead (^{207}Pb) with a half-life of 710 million years. If the relative abundances of ^{235}U and ^{207}Pb in a rock are known, its age can be calculated.

There are several parent–daughter isotope combinations that are useful for "age dating" rocks. Four are listed in the table below:

Radioactive parent isotope	Half-life (billions of years)	Stable daughter isotope
Potassium (^{40}K)	1.3	Argon (^{40}Ar)
Rubidium (^{87}Rb)	47.0	Strontium (^{87}Sr)
Uranium (^{235}U)	0.7	Lead (^{207}Pb)
Uranium (^{238}U)	4.5	Lead (^{206}Pb)

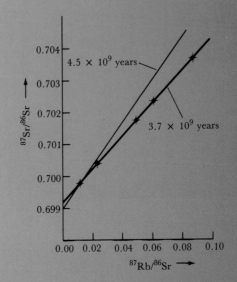

To see how geologists date rocks, consider the slow conversion of rubidium (^{87}Rb) into strontium (^{87}Sr). Over the years, the amount of ^{87}Rb in a rock decreases, while the amount of ^{87}Sr experiences a corresponding increase. Dating the rock is not simply a matter of measuring its ratio of rubidium to strontium, however, because the rock already had some strontium in it when it was formed. Geologists must therefore determine how much fresh strontium came from the decay of rubidium after the rock's formation. To do this, geologists use as a reference a third isotope whose concentration has remained constant. In this case, they use ^{86}Sr, which is stable and is not created by radioactive decay; it is said to be **nonradiogenic**. Specifically, dating a rock entails comparing the ratio of radiogenic and nonradiogenic strontium (^{87}Sr/^{86}Sr) in it to the ratio of radioactive rubidium to nonradiogenic strontium (^{87}Rb/^{86}Sr). Because the half-life for the conversion of ^{87}Rb into ^{87}Sr is known, the rock's age can then be calculated from these ratios.

This particular isotope combination has turned out to be highly reliable in dating the lunar rocks. The graph shows data for lunar lava sample 10044 brought back from Mare Tranquillitatis by the *Apollo 11* astronauts. The rock is a typical mare basalt. At various locations around the sample, geologists D. A. Papanastassiou and G. J. Wasserburg measured the abundance ratios ^{87}Sr/^{86}Sr and ^{87}Rb/^{86}Sr. The data were then plotted on a graph of ^{87}Sr/^{86}Sr versus ^{87}Rb/^{86}Sr. Each dot on the graph represents a pair of measurements from one tiny part of rock 10044.

Notice that the data points fall along a straight line. This means that, although different amounts of rubidium and strontium were found in various parts of the rock, everywhere the same percentage of ^{87}Rb had decayed into ^{87}Sr. Because the half-life of the ^{87}Rb/^{87}Sr decay is known, lines can be drawn on this graph corresponding to the abundance ratios that would be found in rocks of various ages. These lines, called **isochrons**, are steeper for greater ages because ^{87}Rb/^{86}Sr decreases as rock ages, whereas ^{87}Sr/^{86}Sr increases.

The isochron for 4.5 billion years, the age of the oldest lunar rocks, is shown on the graph. The data for rock 10044, however, lie along a shallower line that corresponds to an age of 3.7 billion years. That is the age of the rock. We can therefore conclude that igneous rock sample 10044 solidified when the Moon was slightly less than a billion years old. This rock is thought to be a sample of lava that flowed out onto the lunar surface after the impact of a large meteoroid that penetrated the young Moon's fragile crust.

ical differentiation had occurred on Earth, then relatively little iron would have been ejected, which would account for the Moon's small iron-rich core. Furthermore, most of the debris from the collision would lie in or near the plane of the ecliptic, if the orbit of the impacting asteroid had been in that plane. Indeed, the impact of an asteroid large enough to create the Moon could have also tipped the Earth's rotation axis relative to the Moon's orbit.

The surface of the newborn Moon was probably kept molten for many years, both from heat released during the impact of rock fragments falling into the young satellite and from the decay of such radioactive isotopes as ^{26}Al (aluminum-26). After a few hundred million years, the rain of rock fragments tapered off, and the most radioactive isotopes had largely decayed into stable isotopes. The Moon gradually cooled, and the low-density lava floating on the Moon's surface began to solidify into the anorthositic crust that exists today. The heavy barrage of large rock fragments ended about 4 billion years ago, with the final impacts producing the craters that cover the lunar highlands.

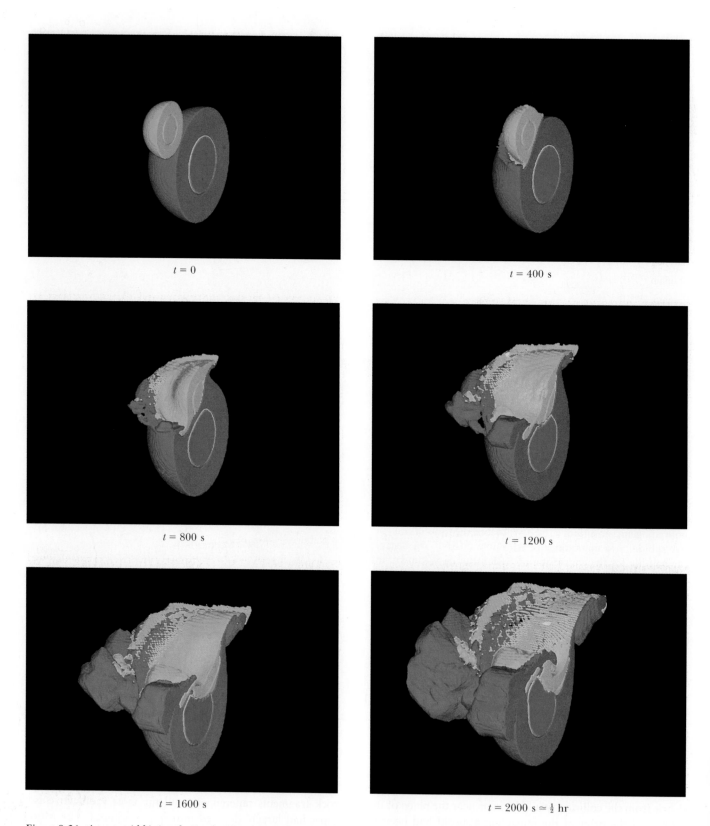

$t = 0$

$t = 400$ s

$t = 800$ s

$t = 1200$ s

$t = 1600$ s

$t = 2000$ s $\simeq \frac{1}{2}$ hr

Figure 9-21 An asteroid hitting the Earth *This supercomputer simulation shows the effects of a Mars-sized asteroid hitting the Earth. The mass of the asteroid is one-tenth of the Earth's mass, and its speed at impact is 8 km/s (18,000 mi/hr). Colors identify core and mantle material of the Earth and the asteroid. Cut-away views reveal the location and movement of this material at six intervals during the first half hour after the impact. According to the collisional ejection theory, the Moon formed from debris ejected by such a collision. (Courtesy of M. Kipp and J. Meloch; Sandia National Laboratories)*

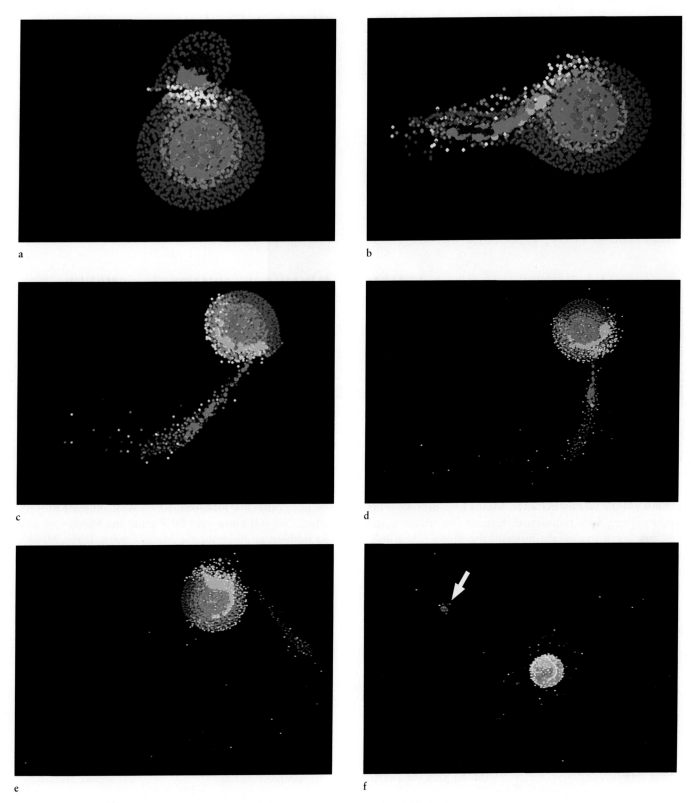

a

b

c

d

e

f

Figure 9-22 *The Moon's creation* *This supercomputer simulation shows the creation of the Moon from material ejected by the impact of a large asteroid on the Earth. To follow the ejected material as it moves away from the Earth, successive views show increasingly larger volumes of space. Blue and green indicate iron from the cores of the Earth and* *the asteroid; red, orange, and yellow indicate rocky mantle material. In this simulation, the impact ejects both mantle and core material, but most of the iron falls back onto the Earth. The surviving ejected rocky matter coalesces to form the Moon (indicated by arrow) during its first orbit of the Earth. (Courtesy of W. Benz)*

Figure 9-23 **Eratosthenes** *Eratosthenes is a young crater 61 km in diameter on the southern edge of Mare Imbrium. Another young crater, Copernicus, is near the horizon in this photograph taken by astronauts in 1972. Both craters also appear near the lower edge of the Earth-based view in Figure 9-2. (NASA)*

Recorded among the final scars at the end of this crater-making era are the impacts of more than a dozen asteroid-sized objects, each measuring at least 100 km across. As these huge rocks rained down on the young Moon, they blasted out the vast mare basins. Meanwhile, heat from the decay of long-lived radioactive elements like uranium and thorium began to melt the inside of the Moon. Then, over the period from 3.8 to 3.1 billion years ago, great floods of molten rock gushed up from the lunar interior, filling the impact basins and creating the maria we see today.

Very little has happened on the Moon since those ancient times. A few fresh craters have been formed (see Figure 9-23), but the astronauts visited a world that has remained largely unchanged for over 3 billion years. The history of our own planet was just beginning 3 billion years ago, when the first organisms began populating the new oceans.

Many questions and mysteries still remain. The six American and three Soviet lunar landings have brought back samples from only nine locations, barely scratching the Moon's surface. We still know very little about the Moon's far side and nothing at all about the Moon's poles. Is the Moon's interior molten? Does the Moon really have an iron core? How old are the youngest lunar rocks? Did lava flows occur over western Oceanus Procellarum only 2 billion years ago, as crater densities there suggest? Is the Moon really geologically dead, or does it just look dead, simply because our examination of the lunar surface has been so cursory? Such questions can only be answered by returning to the Moon.

Key words

*alpha decay	central peak (of a crater)	*daughter isotope	impact breccia
anorthosite	cocreation theory (of Moon's formation)	fission theory (of Moon's formation)	*isochrons
basin (lunar)	collisional ejection theory (of Moon's formation)	*half-life	libration
capture theory (of Moon's formation)	crater	highlands	mare (*plural* maria)
			mare basalt

moonquake

mountain range (lunar)

neap tide

*nonradiogenic

*parent isotope

ray (lunar)

refractory element

regolith

rille

spring tide

synchronous rotation

terra (*plural* terrae)

tidal force

volatile element

zap crater

Key ideas

- The Earth-facing side of the Moon displays light-colored, heavily cratered highlands and dark-colored, smooth-surfaced maria; the Moon's far side has no maria.

- Much of our knowledge about the Moon came from lunar exploration in the 1960s and early 1970s.

  The Moon has no magnetic field today, although it may have had a weak magnetic field billions of years ago.

  Analysis of seismic waves from moonquakes and meteoroid impacts indicates that the Moon has a crust thicker than the Earth's (and thickest on the far side of the Moon), a mantle with a thickness equal to about 75 percent of the Moon's radius, and perhaps a small, solid iron core.

  The Moon's lithosphere is far thicker than that of the Earth, and the Moon's asthenosphere probably extends from the base of the lithosphere to the core.

- Gravitational interactions between the Earth and the Moon produce tides in the oceans of the Earth and in the solid bodies of both worlds.

  Tidal interactions have locked the Moon into synchronous rotation with the Earth and have given rise to a small constant acceleration of the Moon in its orbit, thereby causing it to spiral slowly outward away from the Earth.

- The anorthositic crust exposed in the highlands was formed between 4.0 and 4.3 billion years ago, whereas the mare basalts solidified between 3.1 and 3.8 billion years ago.

  The Moon's surface has undergone very little change over the past 3 billion years.

  Meteoroid impacts have been the only significant "weathering" agent on the Moon; the Moon's regolith ("soil" layer) was formed by meteoritic action, and glasses and breccias of meteoritic origin are very common.

  All of the lunar rock samples are igneous rocks formed largely of minerals found in terrestrial rocks; the lunar rocks contain no water and also differ from terrestrial rocks in being relatively enriched in the refractory elements and depleted in the volatile elements.

- The collisional ejection theory of the Moon's origin holds that the proto-Earth was struck by a huge asteroid, and debris from this collision coalesced to form the Moon.

  The Moon was molten in its early stages, and the anorthositic crust solidified from low-density magma that floated to the lunar surface; the mare basins were created later, by the impact of planetesimals, and filled with lava from the lunar interior.

Review questions

1 Why is more lunar detail visible through a telescope when the Moon is near quarter phase than when it is at full phase?

2 Why are temperature variations between day and night on the Moon much more severe than on the Earth?

3 Why are Moon rocks so much older than Earth rocks, even though both worlds formed at nearly the same time?

4 On the basis of moon rocks brought back by the astronauts, explain why the maria are dark colored but the lunar highlands are light colored.

5 Why are there so few craters on the maria?

6 Briefly describe the main differences and similarities between Moon rocks and Earth rocks.

7 How would you prove to someone that the Moon has no atmosphere?

8 Why do you suppose that no Apollo mission landed on the far side of the Moon?

9 What is the difference between spring tides and neap tides?

10 How would our theories of the Moon's history have been affected if astronauts had discovered sedimentary rock on the Moon?

11 Some people who support the fission theory have proposed that the Pacific Ocean basin is the scar left when the Moon pulled away from the Earth. Explain why this idea is probably wrong.

12 Imagine that you are planning a lunar landing mission. What type of landing site would you select in order to obtain bedrock? Where might you land to search for evidence of recent volcanic activity?

Advanced questions

> **Tips and tools . . .**
> Recall that the average density of an object is its mass divided by its volume, and that the volume of a sphere is $\frac{4}{3}\pi r^3$, where r is the sphere's radius. Recall also that the acceleration of gravity on the Earth's surface is 9.8 m/s². You may find it useful to know that a 1-pound weight (1 lb) presses down on the Earth's surface with a force of 4.448 newtons. You might want to review Newton's law of gravitation in Chapter 4 and consult Box 9-1 and the appendices for any additional data.

*13 *(Basic)* Using the diameter and mass of the Moon given in Box 9-1, verify that the Moon's average density is 3340 kg/m³. Explain why this average density implies that the Moon's interior contains much less iron than the Earth's.

*14 How much would an 80-kg person weigh on the Moon? How much does that person weigh on the Earth?

*15 Estimate the rate at which the Moon is gaining mass from meteoritic impacts. On the basis of your calculation, how long will it be before the Moon doubles its mass? Compare your answer with the age of the universe (about 20 billion years) and explain why your extrapolation into the distant future is probably unreasonable.

*16 *(Challenging)* Find the average distance between the Earth and the Moon when the length of the day and the lunar month will both be equal to 47 of our present days.

17 Can you think of any other weathering processes that might occur on the Moon besides those related to meteoritic impacts?

Discussion questions

18 Comment on the idea that without the presence of the Moon in our sky, astronomy would have developed far more slowly.

19 Why are nearly all of the lunar maria located on the side of the Moon facing the Earth? Why are there more craters on the far side of the Moon than on the near side?

20 Compare the advantages and disadvantages of exploring the Moon with astronauts as opposed to using mobile unmanned instrument packages.

21 Describe how you would empirically test the idea that human behavior is related to the phases of the Moon. What problems are inherent in such testing?

Observing projects

22 Use a telescope to observe the Moon. Compare the texture of the lunar surface you see on the maria with that of the lunar highlands. How does the visibility of details vary with distance from the terminator (the boundary between day and night on the Moon)?

23 Observe the Moon through a telescope every few nights over a period of two weeks between new moon and full moon. Make sketches of various surface features such as craters, mountain ranges, and maria. How does the appearance of these features change with the Moon's phase? Which features are most easily seen at a low angle of illumination? Which features show up best with the Sun nearly overhead?

For further reading

Beatty, J. K. "The Making of a Better Moon." *Sky & Telescope*, December 1986 • This brief but excellent article summarizes the collisional ejection hypothesis and includes a "report card" that evaluates the various theories of the Moon's origin.

Cadogan, P. "The Moon's Origin." *Mercury*, March/April 1983 • This brief article, written before the collisional ejection hypothesis was formulated, describes what we have learned from the Moon rocks.

———. *The Moon—Our Sister Planet*. Cambridge University Press, 1981 • This thoughtful and thorough introduction to lunar geology was written by a scientist who participated in the analyses of Moon rocks brought back by both Apollo astronauts and Soviet automated spacecraft.

Cooper, H. *Apollo on the Moon* and *Moon Rocks*. Dial, 1970 • These two books by a respected journalist give accounts of the Apollo Moon landings and the analysis of the Moon rocks the astronauts brought back.

Cortright, E., ed. *Apollo Expeditions to the Moon*. NASA SP-350, 1975 • The story of the Apollo program is told by astronauts and other NASA personnel.

French, B. *The Moon Book*. Penguin, 1977 • This layperson's introduction to the Moon was written by a noted lunar scientist.

_____. "What's New On the Moon?" *Sky & Telescope*, March and April, 1977 • These two articles give an excellent summary of what we have learned from the Moon rocks and the Apollo program.

Goldreich, P. "Tides and the Earth–Moon System." *Scientific American*, April 1972 • This article explores the consequences of the gravitational interaction between the Earth and the Moon.

Hartmann, W. "The Moon's Early History." *Astronomy*, September 1976 • This article, written by a noted planetary scientist, describes the events that shaped the Moon.

_____, Phillips, R., and Taylor, G., eds. *Origin of the Moon*. Lunar & Planetary Institute, 1986 • This voluminous collection of papers presented at a conference in 1984 offers an excellent in-depth survey of our current understanding of the Moon's formation.

Lewis, R. *The Voyages of Apollo: The Exploration of the Moon*. Quadrangle, 1975 • This book tells the story of the Apollo program.

Mason, B. "The Lunar Rocks." *Scientific American*, October 1971 • This article, written while the Apollo program was still in progress, presents an excellent first look at the Moon rocks.

Moore, P. *The Moon*. Rand McNally, 1981 • This comprehensive yet brief atlas covers the gamut with many attractive illustrations and an excellent collection of maps and photographs.

Pellegrino, C., and Stoff, J. *Chariots for Apollo: The Making of the Lunar Module*. Atheneum, 1985 • This well researched history of the construction of the lunar module focuses on both the technology and the personalities involved.

Schmitt, H. "Exploring Taurus-Littrow: *Apollo 17*." *National Geographic*, September 1973 • This superb description of the final manned lunar landing was written by the astronaut-geologist who was there.

Weaver, K. "First Explorers on the Moon: The Incredible Story of *Apollo 11*." *National Geographic*, December 1969 • This beautifully illustrated article tells the story of the first manned lunar landing.

Wood, J. "The Moon." *Scientific American*, September 1975 • This brief article gives an excellent summary of what we learned from the Moon rocks.

10

Sun-Scorched Mercury

Because it is quite small and orbits near the Sun, Mercury was a planet of many mysteries until recently. Earth-based visual observations fail to reveal any surface details, although radio and radar observations have provided information about the planet's temperature and rotation. When the *Mariner 10* spacecraft flew by Mercury in 1974, sending us dramatic pictures of its surface and crucial information about its magnetic field, much of our ignorance was dispelled. Like the Moon, Mercury is heavily cratered and retains the scars of countless impacts that occurred soon after the planets were formed. Unlike the Moon, Mercury has a large iron core and a magnetic field. Mercury's interior is therefore similar to Earth's.

Mercury and the Moon   Mercury, like our Moon, has a heavily cratered surface and virtually no atmosphere. Mercury's diameter is 4878 km, the Moon's is 3476 km (both worlds are shown here to the same scale). For comparison, the distance from New York to Los Angeles is 3944 km (2451 mi). Daytime temperatures at the equator on Mercury reach 430°C (800°F), which is hot enough to melt lead and tin. One-half of Mercury was photographed at close range by the Mariner 10 spacecraft in the mid-1970s. (NASA)

Until 1974, we knew little about the smallest planet to form in the warm inner regions of the solar nebula. Information about Mercury was difficult to obtain for two simple reasons: Mercury's small size and its nearness to the Sun. In fact, Mercury is so close to the Sun that most people—including many astronomers—have never seen it. Finally, in 1974, an unmanned mission to the inner planets revealed that Mercury has a dual personality: a Moonlike surface but an Earthlike interior.

10-1 Earth-based optical observations of Mercury are difficult to make and often prove disappointing

Mercury circles the Sun at an average distance of 0.387 AU (57.9 million kilometers, or 36 million miles). Mercury's orbit is, however, more eccentric than the orbit of any other planet except Pluto. The distance between Mercury and the Sun varies by nearly 24 million kilometers, from 0.306 AU at perihelion to 0.467 AU at aphelion. Figure 10-1 is a scale drawing of the orbits of Mercury and the Earth. Box 10-1 summarizes data about the planet.

The best opportunities to see Mercury occur when the planet is positioned as far from the Sun in the sky as it can be, at its greatest eastern or western elongation. For a few days near the time of greatest eastern elongation, Mercury appears as an "evening star," hovering low over the western horizon for a short time after sunset. Alternatively, near the time of greatest western elongation, Mercury can be glimpsed as a "morning star," heralding the rising Sun in the brightening eastern sky.

Because its orbit is so close to the Sun (see Figure 10-1), Mercury's maximum elongation is only 28°. The celestial sphere rotates at 15° per hour (360° divided by 24 hours), so Mercury never rises more than two hours before sunrise nor sets more than two hours after sunset. Unfortunately, Mercury's elliptical orbit and its inclination to the ecliptic often place Mercury much less than 28° from the horizon at the moment of sunset or sunrise. Some of the elongations are thus "favorable" for viewing Mercury and others are not, as sketched in Figure 10-2. Mercury's synodic period is 115.9 days (approximately $\frac{1}{3}$ year), so a total of six or seven greatest elongations (both eastern and western) occur each year (see Box 10-2). Typically, only two of these elongations will be favorable for viewing the planet.

Although it is usually inconveniently placed, Mercury is not necessarily dim. Indeed, Mercury is often one of the brightest objects in the sky. Sometimes it shines with a magnitude of −1.9 at greatest brilliancy, which is brighter than the brightest stars.

Mercury shines by reflected sunlight, as do all the planets. The fraction of incoming sunlight that a planet reflects is called its **albedo** (from the Latin for *white*). Mercury reflects about 10 percent of the sunlight that falls on its rocky surface, making its albedo about 0.1, roughly comparable to that of weathered asphalt.

Mercury travels around the Sun faster than any other object in the solar system, taking only 88 days to complete a

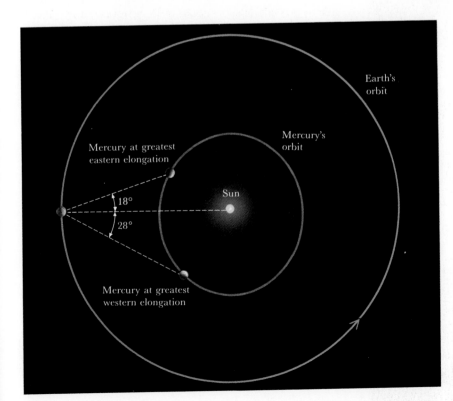

Figure 10-1 Mercury's orbit *Mercury orbits the Sun every 88 days. The distance between Mercury and the Sun varies from 70 million km (44 million mi) at aphelion to 46 million km (29 million mi) at perihelion. As seen from the Earth, the angle between Mercury and the Sun at greatest eastern or western elongation can be as large as 28° when Mercury is near aphelion, or as small as 18° near perihelion.*

Box 10-1 Mercury data

Mean distance from Sun:	$0.387 \text{ AU} = 5.79 \times 10^7 \text{ km}$
Maximum distance from Sun:	$0.467 \text{ AU} = 6.98 \times 10^7 \text{ km}$
Minimum distance from Sun:	$0.307 \text{ AU} = 4.60 \times 10^7 \text{ km}$
Mean orbital velocity:	47.9 km/s
Sidereal period:	87.969 days
Rotation period:	58.646 days
Inclination of equator to orbit:	0°
Inclination of orbit to ecliptic:	7° 00′ 16″
Orbital eccentricity:	0.206
Diameter (equatorial):	4878 km
Diameter (Earth = 1):	0.382
Apparent diameter as seen from Earth:	maximum = 12.9″
	minimum = 4.5″
Mass:	$3.3 \times 10^{23} \text{ kg}$
Mass (Earth = 1):	0.055
Mean density:	5420 kg/m³
Surface gravity (Earth = 1):	0.38
Escape velocity:	4.3 km/s
Oblateness:	0
Mean surface temperatures:	day = +350°C = 662°F = 623 K
	night = −170°C = −274°F = 103 K
Albedo:	0.1
Brightest magnitude:	−1.9
Mean diameter of Sun as seen from Mercury:	1° 22′ 40″

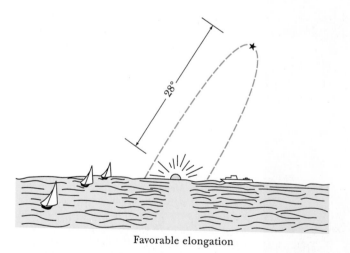

Favorable elongation

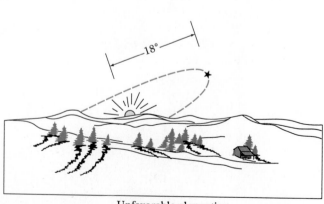

Unfavorable elongation

Figure 10-2 Favorable versus unfavorable elongations Geometric factors such as the tilt of the Earth's axis, the inclination and ellipticity of Mercury's orbit, and the latitude of the observer on the Earth com- bine to make an elongation either "favorable" or "unfavorable" for viewing Mercury.

Box 10-2   Elongations of Mercury, 1991–1999

Year	Western elongations	Eastern elongations
1991	January 14, May 12, September 7, December 27	March 27, July 25, November 19
1992	April 23, August 21, December 9	March 9, July 6, October 31
1993	April 5, August 4, November 22	February 21, June 17, October 14
1994	March 19, July 17, November 6	February 4, May 30, September 26
1995	March 1, June 29, October 20	January 19, May 12, September 9
1996	February 11, June 10, October 3	January 2, April 23, August 21, December 15
1997	January 24, May 22, September 16	April 6, August 4, November 28
1998	January 6, May 4, August 31, December 20	March 20, July 17, November 11
1999	April 16, August 14, December 2	March 3, June 28, October 24

full orbit. Its synodic period is 116 days, and thus Mercury passes through inferior conjunction at least three times a year. You might therefore expect to see Mercury occasionally silhouetted against the Sun. Such a passage in front of the Sun is called a **solar transit**. In fact, transits of Mercury across the Sun are not very common, because Mercury's orbit is tilted 7° to the plane of the ecliptic (see Figure 10-3). Mercury therefore usually lies well above or below the solar disk at the moment of inferior conjunction.

In order for a solar transit to be seen, the Sun, Mercury, and Earth must be in nearly perfect alignment. This arrangement occurs only in May and November when the Earth is located near the line of nodes of Mercury's orbit. Only two transits will occur during the remainder of this century (on November 6, 1993, and November 15, 1999).

The closer Mercury is to the Sun, the farther it can be from the line between the Earth and the Sun's center and still appear in transit from the Earth. November transits, which occur when Mercury is near perihelion, are roughly twice as common as May transits. However, the longest transits occur in May when Mercury is near aphelion and therefore

is traveling comparatively slowly along its orbit. The maximum duration of a solar transit is nine hours. Figure 10-4 is a photograph of one such transit. Note how tiny the planet seems in comparison to the Sun.

Naked-eye observations of Mercury are best made at dusk or dawn, but the best telescopic views are obtained at midday when the planet is high in the sky, far above the degrading atmospheric effects near the horizon. A yellow filter can eliminate much of the scattered blue light from the sky. The photographs in Figure 10-5, among the finest Earth-based views of Mercury, were taken at midday.

Because of its small size and its nearness to the Sun, you cannot see much surface detail on Mercury through an Earth-based telescope. At best only a few faint, hazy markings can be identified. During the 1880s, the Italian astronomer Giovanni Schiaparelli used these markings in an attempt to produce the first map of Mercury. Unfortunately, his telescopic views of Mercury were so vague and indistinct that he made a major error, which went uncorrected for over half a century. Schiaparelli erroneously concluded that Mercury keeps the same side always facing the Sun.

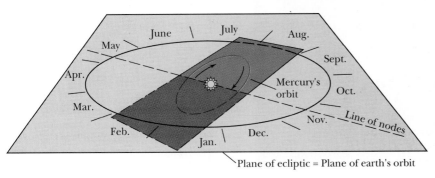

Figure 10-3 *The inclination of Mercury's orbit Mercury's orbit is inclined 7° to the plane of the ecliptic. The angular diameter of the Sun as seen from Earth is about ½°, so Mercury is usually located several degrees north or south of the Sun at inferior conjunction. Solar transits are seen only in May and November, when both Mercury and the Earth are near the line of nodes of Mercury's orbit.*

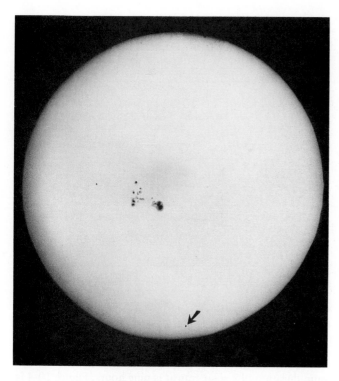

Figure 10-4 *A solar transit of Mercury Roughly a dozen solar transits of Mercury occur in each century. This photograph shows the tiny planet (arrow) silhouetted against the Sun during the transit of November 14, 1907. (Yerkes Observatory)*

10-2 Radio and radar observations of Mercury revealed its rotation rate

Synchronous rotation, in which the rotation period of an object equals its period of revolution, is a common phenomenon in our solar system. For example, the Moon exhibits synchronous rotation, keeping the same side exposed to our Earth-based view. The two moons of Mars and many of the satellites of Jupiter and Saturn also keep the same side facing their parent planet. Newtonian mechanics demonstrates that this situation is a stable one, usually called a **1-to-1 spin–orbit coupling**, but this is not how Mercury rotates. The first clue about Mercury's true rotation period came in 1962 when astronomers detected radio radiation coming from it.

As we learned in Chapter 5, every object emits blackbody radiation. The dominant wavelength of the radiation emitted by an object depends on its temperature. For example, from Wien's law we know that the wavelength (λ_{max}) at which the most intense radiation is emitted depends on the object's temperature (T), according to the following formula:

$$\lambda_{max} = \frac{2.9 \times 10^{-3}}{T}$$

where wavelength is measured in meters and temperature is measured in kelvins. For instance, an object at 100 K emits radiation primarily at a wavelength of 2.9×10^{-5} m (about 30 μm). Only at absolute zero does an object emit no radiation at all.

If Schiaparelli had been correct, one side of Mercury would remain in perpetual, frigid darkness, never exposed to the warming rays of the nearby Sun. Thus, radio astronomers at the University of Michigan were surprised in 1962 to discover that the wavelengths of radio radiation coming from Mercury indicated that the temperature on the planet's nighttime side was not nearly as cold as expected. Schiaparelli's belief in synchronous rotation was so well accepted that some astronomers speculated about an atmosphere on Mercury whose winds could carry warmth from the daytime side of the planet around to the nighttime side. As discussed in detail in Box 7-2, however, the temperatures on Mercury are too high and the planet's gravity too weak to retain any substantial atmosphere.

A breakthrough came in 1965 when Rolf B. Dyce and Gordon H. Pettengill used the giant 1000-ft radio telescope

Figure 10-5 *Earth-based views of Mercury These two views are among the finest photographs of Mercury ever produced with an Earth-based telescope. Hazy markings are faintly visible on the tiny planet. (New Mexico State University Observatory)*

Figure 10-6 *The Arecibo radio telescope* This telescope is the largest one on Earth. The dish, which is mounted permanently in a bowl-shaped valley in Puerto Rico, measures 305 m (1000 ft) in diameter. The dish is not movable, so this telescope can be used only to examine objects that are near the zenith. (Arecibo Observatory)

at the Arecibo Observatory in Puerto Rico (see Figure 10-6) to bounce powerful radar pulses off Mercury. The outgoing radiation consisted of microwaves of a very specific wavelength. In the reflected signal echoed back from the planet, the microwaves were shifted from their original wavelength. This wavelength shift, which occurred because of the Doppler effect (review Figure 5-20), provided detailed information about Mercury's rotation. Microwaves reflected from the planet's approaching side were shortened, whereas those from its receding side were lengthened. The radar pulse went out at one specific wavelength, but it came back spread over a small wavelength range. From the width of the spread, Dyce and Pettengill deduced Mercury's speed of rotation and from that they computed its period of rotation. They found that Mercury's rotation period is approximately 59 days.

Giuseppe Colombo, an Italian physicist with a long-standing interest in Mercury, found this number intriguing. Colombo noted that Mercury's sidereal period is 87.969 days and that

$$\frac{2}{3}(87.969 \text{ days}) = 58.65 \text{ days}$$

Colombo therefore boldly speculated that Mercury's true rotation period is exactly 58 days and $15\frac{1}{2}$ hours.

This was not an idle guess. Colombo realized that a rotation period of 58.64 days would mean that Mercury is locked into a **3-to-2 spin–orbit coupling**, meaning that the planet makes three complete rotations on its axis for every two complete orbits around the Sun. This is a much rarer dynamic stability than the 1-to-1 spin–orbit coupling our Moon and many other satellites exhibit. Both types of spin–orbit coupling are shown in Figure 10-7.

Colombo's enlightened guess was dramatically confirmed in the mid-1970s by several flybys of the *Mariner 10* space-

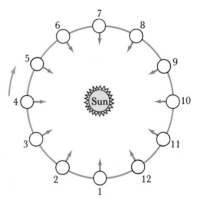

a 1-to-1 spin–orbit coupling

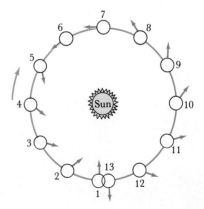

b 3-to-2 spin–orbit coupling

Figure 10-7 *Spin–orbit coupling* (a) The simplest kind of spin–orbit coupling is 1-to-1 coupling, where the rotation period equals the orbital period. The planet always keeps the same side facing the Sun. (b) Mercury exhibits a 3-to-2 coupling: during each revolution around the Sun, the planet rotates $1\frac{1}{2}$ times on its axis. Consequently, Mercury rotates three times about its axis during two complete orbits of the Sun.

craft. After leaving the Earth in 1973, *Mariner 10* first flew past Venus. The trajectory of this flyby was designed so that Venus's gravitational pull would redirect the spacecraft toward Mercury. This flight was the first of several historic gravity-assisted missions in which a near encounter with one planet was used to catapult a spacecraft gravitationally on toward another planet. The flight path of *Mariner 10* is shown in Figure 10-8.

After its first encounter with Mercury, in the spring of 1974, *Mariner 10* was placed in a 176-day orbit about the Sun that brings the spacecraft back to the planet every two Mercurian years. During the second and third flybys, the spacecraft's cameras sent back pictures showing that the same side of Mercury was facing the Sun as on the first flyby. Thus it was proved that Mercury does indeed rotate three times about its axis for every two orbits it makes of the Sun. Because two Mercurian years equal one Mercurian solar day (see Figure 10-7), wherever the spacecraft returns to Mercury, the same side of the planet is facing the Sun as on the original visit.

Mercury's very slow 58.65-day rotation period is responsible for an unusual phenomenon. Mercury's speed along its orbit varies in accordance with Kepler's second law. Its orbital velocity is greatest (59 km/s) at perihelion and least (39 km/s) at aphelion, as calculated from the formulas in Box 4-3. As seen from Mercury's surface, the Sun rises in the east and sets in the west, just as it does on Earth. When Mercury is near perihelion, however, the planet's rapid motion along its orbit outpaces its leisurely rotation about its axis. The usual east–west movement of the Sun across Mercury's sky is interrupted. The Sun actually stops and moves backward (from west to east) for a few Earth days. If you were standing on Mercury watching a sunset near the time of passing the perihelion, the Sun would not simply set. It would dip below the western horizon—and then come back up, only to set a second time a day or two later.

10-3 Photographs from **Mariner 10** revealed Mercury's heavily cratered, lunarlike surface

Our first detailed knowledge about Mercury's surface was acquired in the spring of 1974, when *Mariner 10* coasted over the planet's surface. *Mariner 10* (see Figure 10-9) is fairly typical of the spacecraft that were involved in missions to the planets during the 1970s. Two television cameras mounted on a steerable platform obtained pictures of the planet. In addition, a device called an infrared radiometer measured the planet's surface temperature. An ultraviolet

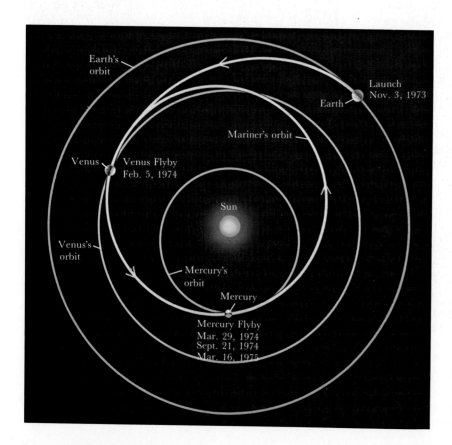

Figure 10-8 **The orbit of Mariner 10** *After coasting past Mercury, on March 29, 1974, Mariner 10 went into an orbit whose period is exactly two Mercurian years. Consequently, Mercury and the spacecraft pass close to each other once every 176 days. However, fuel to stabilize the spacecraft lasted only for the first three encounters.*

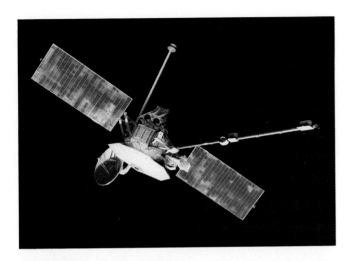

*Figure 10-9 **The** Mariner 10 **spacecraft*** Mariner 10 *weighed 500 kg (1100 lb) and measured 7.6 m (25 ft) from one end of its extended solar panels to the other. A total of 8122 solar cells provided electricity for its television cameras, magnetometers, spectrometers, and associated equipment. (NASA)*

spectrometer searched for a Mercurian atmosphere, and magnetometers measured Mercury's magnetic field. Throughout the mission, charged particle and plasma detectors examined the solar wind.

The flight path of *Mariner 10* carried the spacecraft through Mercury's shadow, past the darkened nighttime side of the planet, on March 29, 1974. Naturally, no television pictures could be taken of Mercury's unilluminated hemisphere. The photography was therefore divided into two parts: "incoming views" taken prior to the closest approach, and "outgoing views" taken after *Mariner 10* had emerged from Mercury's shadow. The best incoming and outgoing photomosaics of the entire planet can be seen in Figures 10-10 and 10-11.

As *Mariner 10* closed in on Mercury, scientists were surprised by the Moonlike pictures appearing on their television monitors. It was obvious that Mercury is a barren, desolate, heavily cratered world. Figure 10-12 shows a typical closeup view sent back from *Mariner 10*.

*Figure 10-10 **Mariner 10's incoming view*** *This view of Mercury was sent back from Mariner 10 as the spacecraft sped toward the planet. Eighteen pictures taken at 42-second intervals were assembled into this photomosaic. The spacecraft was 200,000 km (124,000 mi) from the planet at the time these pictures were taken. (NASA)*

*Figure 10-11 **Mariner 10's outgoing view*** *This view of Mercury was sent back from Mariner 10 as the spacecraft coasted away from the planet, shortly after emerging from Mercury's shadow. Eighteen pictures taken at 42-second intervals were assembled into this photomosaic. These pictures were taken from a distance of 210,000 km (130,000 mi). (NASA)*

Figure 10-12 **Mercurian craters and intercrater plains** *This view of Mercury's northern hemisphere was taken by* Mariner 10 *at a range of 55,000 km (34,000 mi) from the planet's surface. Numerous craters and extensive intercrater plains appear in this photograph, which covers an area 480 km (300 mi) wide. (NASA)*

Figure 10-13 **Lunar craters** *This Earth-based photograph shows a portion of the Moon's southern hemisphere during last quarter moon. Densely packed craters fill the view, which covers an area approximately 600 km (370 mi) wide. (Mount Wilson and Las Campanas Observatories)*

Although the first impression of Mercury was that it had a lunar landscape, closer scrutiny of Mercury's surface revealed some significant nonlunar characteristics. For comparison, Figure 10-13 shows the Moon's southern hemisphere. Notice how the lunar craters are densely packed, with one overlapping the next. In sharp contrast, Mercury's surface has extensive **intercrater plains**.

Astronomers believe that most of the craters on both Mercury and the Moon were produced during the 700 million years after the planets formed. The strongest evidence comes from analysis and dating of the Moon rocks brought back by the *Apollo* astronauts. Debris remaining after planet formation rained down on these young worlds, gouging out most of the craters we see today.

Astronomers agree that at first the Moon and the terrestrial planets must have been completely molten spheres of liquid rock. After a few hundred million years, their surfaces solidified as the rock cooled. Nevertheless, large meteoroids easily punctured the planets' thin, cooling crusts, allowing molten lava to well up from their interiors. Older craters were obliterated as seas of molten rock flooded across portions of the planets' surfaces. As we saw in Chapter 9, the lunar maria are the result of extensive lava flooding.

The planets did not cool at the same rate. The volume of a planet is proportional to the cube of its radius, whereas its surface area is proportional to the square of the radius. Con-

sequently, a small planet has a higher ratio of surface area to volume than does a big planet. A small planet can thus more easily radiate its internal heat into space and can cool off more rapidly than a big planet.

Because Mercury is larger than the Moon, it took longer for a thick protective crust to form on Mercury than it did on the Moon. Throughout Mercury's early history, molten rock seeped up through cracks and fissures in its young, frail crust, and volcanic activity was probably pervasive. The resulting lava flows certainly inundated many older craters, leaving behind the broad, smooth intercrater plains seen by *Mariner 10*.

Mariner 10 also revealed numerous long cliffs, called **scarps**, meandering across Mercury's surface (see Figure 10-14). These scarps probably formed while the planet was cooling from its initial molten state. As the interior solidified, the planet contracted, causing the crust to wrinkle. Most geologists agree that the scarps are the ridges and wrinkles thrust up during this period of compression. The Mercurian scarps are named after famous sailing ships. The Victoria scarp shown in Figure 10-14 takes its name from Magellan's ship, the first to sail around the world. Some scarps rise as much as 3 kilometers (2 miles) above the surrounding plains.

While examining the incoming views of Mercury sent back by *Mariner 10*, scientists noticed unusual, closely

Figure 10-14 The Victoria scarp This view of Mercury's northern hemisphere was taken from a range of 77,800 km (48,300 mi). A long scarp extends southward for several hundred kilometers. This photograph covers an area measuring roughly 550 km (340 mi) across. (NASA)

packed hills jumbled over a region slightly south of the planet's equator. Figure 10-15 is a wide-angle view of this puzzling area; the hills appear as tiny wrinkles that cover most of the photograph. The hills are about 5 to 10 km wide and have elevations between 100 and 1800 m (300–5900 ft). The total area covered by this chaotic terrain is about 500,000 km² (193,000 mi²).

This peculiar, hilly terrain was explained by what scientists saw in the outgoing views of Mercury. Looking back at the receding planet, *Mariner 10* photographed a huge **impact basin** just north of the equator, along the line dividing day from night. (This dividing line is called the **terminator**.) Slightly more than half of the impact basin is hidden on the night side of the terminator in the *Mariner 10* views (see Figure 10-16).

This huge impact feature is called the Caloris Basin, from the Latin word for *hot*, because the Sun is directly over the Caloris Basin during alternating perihelion passages. It is therefore the hottest place on the planet once every 176 days.

Figure 10-15 Unusual, hilly terrain The tiny, fine-grained wrinkles on this picture are actually closely spaced hills, part of a jumbled, hilly terrain that covers nearly half a million square kilometers of Mercury's "incoming hemisphere." The large smooth-floored crater near the center of this photograph has a diameter of 170 km (106 mi). (NASA)

Figure 10-16 The Caloris Basin Mariner 10 sent back this view of Mercury's "outgoing hemisphere," which shows a huge impact basin on the terminator. Although the center of the impact basin is hidden in the shadows (just beyond the left side of the picture), several semicircular rings of mountains reveal its extent. The outer rim of the basin is defined by a ring of mountains up to 2 km (6500 ft) high. The diameter of the basin is 1300 km (810 mi). (NASA)

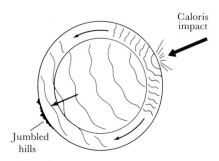

Figure 10-17 The Caloris impact *Seismic waves from the Caloris impact became slightly focused as they traveled through the planet. These focused waves pushed up the numerous jumbled hills on the opposite side of Mercury.*

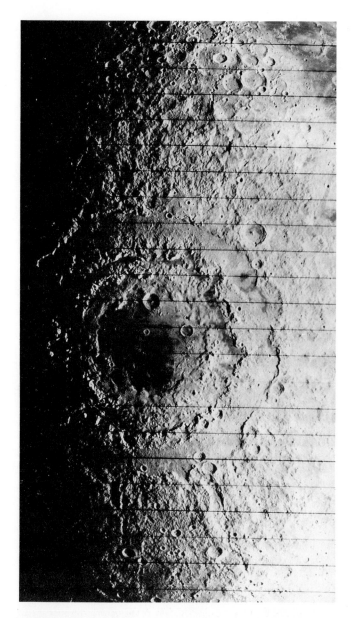

Figure 10-18 The Orientale Basin on the Moon *The Orientale Basin is one of several large impact basins on the Moon that resemble the Caloris Basin on Mercury. The outermost ring of peaks, called the Cordillera Mountains, forms a circle 900 km (560 mi) in diameter. This photograph was sent back in 1967 from the* Lunar Orbiter 4 *spacecraft. The black lines across the picture exist because the photograph was transmitted to Earth in thin strips. (NASA)*

The Caloris Basin is exactly on the opposite side of the planet from the jumbled terrain that *Mariner 10* saw on the "incoming hemisphere." This basin is 1300 km (810 mi) in diameter. It was formed by the impact of a large meteoroid near the end of the crater-making period of bombardment that dominated the first billion years of our solar system. We know that the Caloris impact occurred late in this period because there are relatively few fresh craters on the lava flow that filled up the basin.

The Caloris impact must have been a violent event that shook the planet with earthquakelike disturbances. Geologists contend that seismic waves from the Caloris impact became focused as they passed through Mercury (see Figure 10-17). As this concentrated seismic energy reached the far surface of the planet, jumbled hills were pushed up.

Features similar to those of the Caloris Basin exist on our Moon. The best example is the Orientale Basin, shown in Figure 10-18. Its diameter is 900 km (560 mi). The scarcity of fresh craters in this basin attests to its late formation. Furthermore, the astronomers' theory about the seismic upthrusting of jumbled hills on Mercury is supported by the fact that similar (though less extensive) chaotic hills are also found on the side of the Moon opposite the Orientale Basin.

10-4 Mercury has an iron core and a magnetic field, like Earth

Mercury's average density is 5420 kg/m³. As mentioned in Chapter 7, the average density of the Earth is 5520 kg/m³. Typical rocks from Earth's surface (as well as from the Moon and Mars) are composed primarily of silicon and other lightweight, mineral-forming elements. These surface rocks have a density of only about 3000 kg/m³. The higher average density of our planet is caused by the Earth's iron core. By studying how the Earth vibrates during earth-

quakes, geologists have deduced that our planet's iron core is approximately 7000 km (4300 mi) in diameter. For comparison, the Earth's overall diameter is 12,700 km (7900 mi).

Mercury's average density is just slightly less than the Earth's, so you might suspect Mercury to have a proportionately smaller iron core than Earth's, but this is not the case. The Earth is 18 times more massive than Mercury.

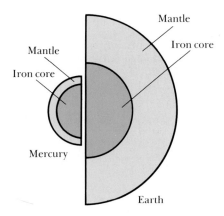

Figure 10-19 The internal structures of Mercury and Earth Mercury is the most iron-rich planet in our solar system. Mercury's iron core occupies an exceptionally large proportion of the planet's interior.

The weight of this larger mass pushing down on the Earth's interior compresses the Earth's core much more than Mercury's core is compressed. Calculations have shown that the uncompressed density of the Earth would be about 4000 kg/m^3, whereas Mercury's uncompressed density would be 5300 kg/m^3. In fact, Mercury is the most iron-rich planet in our solar system, with iron accounting for 65–70 percent of the planet's mass. The iron core of Mercury has a diameter equal to three-fourths of the planet's diameter, whereas the diameter of the Earth's core is only slightly more than one-half of the Earth's diameter. Figure 10-19 is a scale drawing of the interior structures of Mercury and Earth.

Mercury's high abundance of iron is the result of the high condensation temperature of ferromagnesian minerals. In the warm inner regions of the primordial solar nebula, the temperature was sufficiently high that iron-rich minerals were primarily the ones that readily condensed into solids. At greater distances from the protosun, larger percentages of other minerals were included. Thus the iron content of the terrestrial planets decreases with increasing distance from the Sun.

Independent evidence of Mercury's large iron core came from *Mariner 10*'s magnetometers, which discovered that Mercury has a magnetic field. Iron is the only common element that could account for the existence of a magnetic field on a terrestrial planet.

As we saw in Chapter 8, the magnetic field around the Earth is similar to the magnetism that surrounds an electromagnet (a coil of wire in which electricity is flowing) and probably originates with electric currents flowing in the liquid portions of our planet's iron core. In a process called the dynamo effect these currents create a planetwide magnetic field as they are carried around by the Earth's rotation.

Scientists were surprised to find that Mercury also has a magnetic field. Mercury rotates much more slowly than the Earth (59 days versus 24 hours), and most scientists believed this leisurely rotation would be too slow to induce a magnetic field by the dynamo effect. Nevertheless, Mercury does have a weak field, 100 times weaker than the Earth's field. The source of Mercury's magnetic field is still a topic of debate among both astronomers and geologists.

10-5 Mercury's magnetic field shields the planet from the solar wind

Mariner 10's other instruments measured the surface temperature on Mercury and searched for traces of an atmosphere there. Temperatures on Mercury vary from 700 K (= 427°C = 800°F) at local noon on the equator at perihelion to 100 K (= −173°C = −280°F) at local midnight. This 600 K temperature range is greater than that on any other planet or satellite in our solar system. Earth-based observers have recently discovered a tenuous atmosphere of vaporized sodium and potassium around Mercury. The only atmosphere *Mariner 10* detected was a thin scattering of particles captured from the solar wind. As was explained in Box 7-2, Mercury's high surface temperature and low surface gravity preclude the planet retaining a substantial atmosphere.

The solar wind is a constant flow of charged particles (mostly protons and electrons) away from the outer layers of the Sun's upper atmosphere. If a planet possesses a magnetic field, this field will repel and deflect the impinging particles, thereby forming an elongated "cavity" in the solar wind called a **magnetosphere**.

Charged particle detectors on *Mariner 10* mapped the structure of Mercury's magnetosphere. The results are shown in Figure 10-20. When the particles in the solar wind first encounter Mercury's magnetic field, they are abruptly slowed, producing a bow-shaped shock wave that marks the boundary where this sudden decrease in velocity occurs and the magnetosheath begins. The **magnetosheath** is a turbulent region where most of the subsonic particles from the solar wind are deflected around the planet, just as water is deflected to either side of the bow of a ship. Still closer to Mercury, the **magnetopause**—within which lies the true magnetic domain of the planet—marks the boundary where the outward pressure of the planet's magnetic field is exactly counterbalanced by the impinging gas pressure of the solar wind. Thus Mercury's magnetic field produces a magnetosphere that blocks the solar wind from reaching the surface of the planet.

These features of Mercury's magnetic environment were not entirely obvious when *Mariner 10* first flew past the planet on March 29, 1974. Fortunately, the spacecraft came back to Mercury for a second and then a third look. Important features of Mercury's magnetosphere were determined

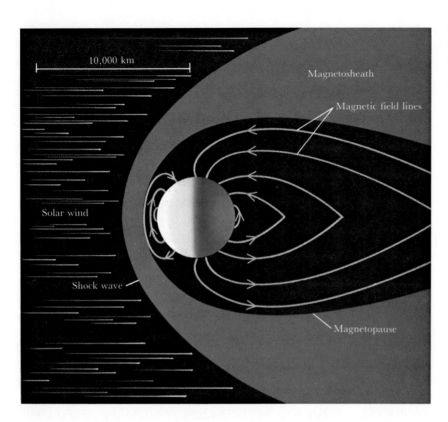

Figure 10-20 *Mercury's magnetosphere Mercury's weak magnetic field is just strong enough to carve out a cavity in the solar wind, preventing the impinging particles from striking the planet's surface directly.*

during these additional flybys, but *Mariner 10*'s cameras did not see any new portions of Mercury's surface. Because two Mercurian years equals one Mercurian solar day and the orbital period of *Mariner 10* was exactly two Mercurian years, the spacecraft's second and third visits to Mercury sent back pictures of only the same craters seen during the first flyby.

Mariner 10 still coasts over Mercury's sun-scorched surface every 176 days, but its stabilizing rockets had only enough fuel to last through the third encounter. Today the spacecraft tumbles aimlessly, its eyes blind and its radio voice silent. There are currently no plans to return to Mercury, and the side of the planet hidden from *Mariner 10* will no doubt remain unexplored for the rest of this century.

Key words

albedo	magnetopause	spin–orbit coupling
impact basin	magnetosheath	solar transit
intercrater plains (on Mercury)	magnetosphere	synchronous rotation
	scarps	terminator

Key ideas

- At its greatest eastern and western elongations, Mercury is only 28° from the Sun, so it can be seen only briefly after sunset or before sunrise.

Solar transits of Mercury occur only in May and November and only about a dozen times per century.

Poor telescopic views of Mercury's surface led to the mistaken impression that the planet keeps the same side always toward the Sun, a process called 1-to-1 spin–orbit coupling.

- Radio observations and radar observations in the 1960s revealed that Mercury in fact has 3-to-2 spin–orbit coupling. For an observer on Mercury, one solar day would last for two Mercurian years.

- The *Mariner 10* spacecraft made three useful passes near Mercury in the mid-1970s, providing pictures of its surface.

The Mercurian surface is pocked with craters like those of the Moon, but extensive, smooth inter-crater plains appear between these craters; these features appear to have formed as the crust of the planet solidified.

Long, high cliffs called scarps meander across the surface of Mercury. These scarps probably formed as the planet cooled, solidified, and shrank.

The impact of a large object long ago formed the huge Caloris Basin and shoved up jumbled hills on the opposite side of the planet.

- Instruments on *Mariner 10* provided information about the magnetic field, temperature, and thin atmosphere of Mercury, and indicated that the planet has an iron core much like that of the Earth.

The iron core of Mercury has a diameter equal to three-fourths of the planet's diameter, whereas the diameter of the Earth's core is only slightly more than one-half of the Earth's diameter.

The surface temperatures on Mercury range from 100 to 700 K, the greatest range of temperature on any of the planets.

Mercury's atmosphere consists of only a thin scattering of sodium and potassium atoms and particles captured from the solar wind.

Mercury's magnetic field produces a magnetosphere surrounding the planet that blocks the solar wind from the surface of the planet.

Review questions

1 Why are naked-eye observations of Mercury best made at dusk or dawn, whereas telescopic observations are best made around noon?

2 Why can't you see any surface features on Mercury when it is closest to the Earth?

3 Explain why Mercury does not have a substantial atmosphere.

4 Explain why November solar transits of Mercury, which occur near the time of perihelion passage, are more common than May transits.

5 Why do astronomers believe Mercury to be the most iron-rich planet in our solar system?

6 How is Mercury's rotation rate related to its orbital period?

7 Explain why *Mariner 10* was able to photograph only one side of Mercury even though the spacecraft returned to the planet on three separate occasions.

8 Compare the surfaces of Mercury and our Moon. How are they similar? How are they different?

9 What is "chemical differentiation," and what role did it play in shaping the interiors of planets such as Mercury and the Earth?

10 With the aid of a drawing, describe Mercury's magnetosphere.

11 If the albedo of Mercury were increased, would the planet's surface temperature go up or down? Explain your answer.

12 How can you tell an old crater from a new one?

13 Explain why the Sun is directly over the Caloris Basin on Mercury only during every other perihelion passage.

14 Why is it reasonable to presume that none of the large satellites in our solar system, including our Moon, possess a substantial magnetic field?

Advanced questions

Tips and tools . . .
You may need to refresh your memory about the small-angle formula, found in Box 1-1, and about Wien's law and the Doppler effect, which are both discussed in Chapter 5. The criteria for a planet to be able to retain an atmosphere are discussed in Box 7-2.

*15 Suppose you have a superb telescope that can resolve features as small as 1 arc sec across. What is the size of the smallest surface features you should be able to see on Mercury? How does your answer compare with the size of the Caloris Basin? (*Hint:* Assume that you choose to observe Mercury when it is at greatest elongation, about 25° from the Sun.)

*16 (*Basic*) Find the value of λ_{max} for radiation coming from the sunlit side of Mercury.

*17 In view of Mercury's 58.6 day rotation period, what difference in wavelength is observed for a spectral line at 500 nm reflected from either the approaching or receding edge of the planet?

*18 (*Challenging*) Calculate the minimum molecular weight of a gas that could in theory be retained as an atmosphere by Mercury if the average daytime temperature were 620 K. Are there any abundant gases that meet this minimum criterion? Why doesn't Mercury have an atmosphere of these gases? (*Hint:* The most convenient way to express the mass of a molecule is by its molecular weight, μ, times the mass of a hydrogen atom, $m_H = 1.67 \times 10^{-27}$ kg. The molecular weight of a molecule equals the sum of the atomic weights of its atoms, which can be looked up on a periodic table of the elements. Thus, for instance, the molecular weight of CO_2 is $12.0 + 16.0 + 16.0 = 44.0$, and so the molecule's mass is $44 \times m_H$.)

*19 How much would an 80-kg person weigh on Mercury? How does that compare with that person's weight on the Moon? How much does that person weigh on Earth?

*20 Using the fact that the orbital period of *Mariner 10* is twice that of Mercury, calculate the length of the semimajor axis of the spacecraft's orbit.

Discussion questions

21 If you were planning a return mission to Mercury, what features and observations would be of particular interest to you?

22 What evidence do we have that the surface features on Mercury were not formed during recent geological history?

Observing projects

23 Refer to Box 10-2 to determine the dates of the next two or three greatest elongations of Mercury. Consult such magazines as *Sky & Telescope* and *Astronomy* to see if any of these greatest elongations is going to be a favorable one. If so, make plans to be one of those rare individuals who has

actually seen the innermost planet of the solar system. Set aside several evenings (or mornings) around the date of the favorable elongation to reduce the chances of being "clouded out." Select an observing site that has a clear, unobstructed view of the horizon where the Sun sets (or rises). Make arrangements to have a telescope at your disposal. Search for the planet on the dates you have selected and make a drawing of its appearance through your telescope.

24 **This observing project should be performed only under the direct supervision of an astronomer who knows how to point a telescope safely at Mercury.** Make arrangements to view Mercury during broad daylight. This is best done by visiting an observatory where the coordinates (right ascension and declination) of Mercury's position can be used to point the telescope. **DO NOT LOOK AT THE SUN! Looking directly at the Sun can cause blindness.**

For further reading

Davies, M., et al., eds. *Atlas of Mercury*. NASA SP-423, 1978 • This oversized book from NASA includes many high-quality enlargements of the *Mariner 10* pictures of Mercury.

Dunne, J., and Burgess, E. *The Voyage of* Mariner 10: *Mission to Venus and Mercury*. NASA SP-424, 1978 • This NASA publication gives a fine overview of the entire *Mariner 10* mission.

Hartmann, W. "The Significance of the Planet Mercury." *Sky & Telescope*, May 1976 • This excellent article by a noted planetary scientist discusses the importance of what we have learned from the *Mariner 10* mission to Mercury.

Murray, B. "Mercury." *Scientific American*, May 1976 • This article is one of the best summaries of what we now know about the innermost planet.

———, and Burgess, E. *Flight to Mercury*. Columbia University Press, 1977 • This exciting book, which includes an excellent selection of photographs, interweaves the history of the *Mariner 10* mission with newsworthy events that often overshadowed the mission.

Strom, R. *Mercury: The Elusive Planet*. Smithsonian Institution Press, 1987 • This clear, well written overview of the *Mariner* mission to Mercury is a superb nontechnical introduction to the innermost planet.

———. "Mercury: The Forgotten Planet." *Sky & Telescope*, September 1990 • This fascinating article explains some of the latest ideas about Mercury, including the theory that Mercury's large iron core resulted from a devastating impact 4.5 billion years ago.

Cloud-Covered Venus

Venus and the Earth have many similar properties, including size, mass, and average density. Venus's thick covering of clouds had long kept Earth-based astronomers from learning much more about the planet until radar was used to successfully penetrate the Venusian clouds. Radar observations not only demonstrated that Venus rotates backward but also gave us remarkably good images of the planet's surface, which includes two large continents and is mostly covered with gently rolling hills. The most surprising body of information came from space probes to Venus: Soviet and American missions discovered that Venus has an extremely hot, dense atmosphere of carbon dioxide with clouds of sulfuric acid droplets. Several active volcanoes are probably responsible for the high sulfur content of the Venusian clouds. The greenhouse effect is responsible for Venus's high temperature, and so an understanding of our sister planet may give important insights about the history and future of our own world.

Venus   Venus and Earth have nearly the same size, mass, and surface gravity. However, Venus's thick cloud cover efficiently traps ultraviolet radiation from the Sun, resulting in a surface temperature on Venus of 750 K (= 480 °C = 900°F), even hotter than Mercury's. Unlike Earth's clouds, which are made of water droplets, Venus's clouds are very dry and are made of droplets of concentrated sulfuric acid, along with sulfur dust. Active volcanoes may be responsible for maintaining these sulfur-rich clouds. This photograph was obtained by the Pioneer Venus Orbiter *in 1979. (NASA)*

At first glance, Venus looks like Earth's twin. The two planets have almost the same mass, the same diameter, the same average density, and the same surface gravity. (Box 11-1 lists basic data about Venus.) However, Venus is closer to the Sun than the Earth is, so it is exposed to more intense sunlight, transforming this potentially Earthlike planet into a world that is extremely hostile to living organisms. Venus is an inferno whose crushing, poisonous atmosphere is drenched in sulfuric acid.

11-1 The surface of Venus is hidden beneath a thick, highly reflective cloud cover

Venus's orbit is almost twice as large as Mercury's. Consequently, at its greatest elongation, Venus appears about 47° away from the Sun (see Figure 11-1). This distance is a comfortable one for viewing the planet without interference from the Sun's glare. At its greatest eastern elongation, Venus is easily spotted high above the western horizon after sunset, when it is called an "evening star." Alternatively, at greatest western elongation the planet rises nearly three hours before the Sun. With the arrival of dawn, Venus is positioned high in the eastern sky and is often called the "morning star." Box 11-2 lists the dates of greatest eastern and western elongation between 1991 and 1999.

Venus is easy to identify, because it is often one of the brightest objects in the night sky. Venus's cloud cover reflects 76 percent of the sunlight that falls on it (its albedo is 0.76), and Venus sometimes reaches a magnitude of −4.4 as seen from the Earth, which is 16 times brighter than the brightest star. Only the Sun and the Moon outshine Venus at its greatest brilliancy.

Earth-based telescopic views of Venus reveal that it is enveloped by a thick, nearly featureless, unbroken layer of clouds. Evidence for Venus's thick atmosphere comes from observations near the time of inferior conjunction when

Box 11-1 Venus data

Mean distance from the Sun:	0.723 AU = 1.082×10^8 km
Maximum distance from the Sun:	0.728 AU = 1.089×10^8 km
Minimum distance from the Sun:	0.718 AU = 1.075×10^8 km
Mean orbital velocity:	35.0 km/s
Sidereal period:	224.70 days
Rotation period:	243.01 days (retrograde)
Inclination of equator to orbit:	177.4°
Inclination of orbit to ecliptic:	3° 23′ 40″
Orbital eccentricity:	0.007
Diameter:	12,104 km
Diameter (Earth = 1):	0.949
Apparent diameter as seen from Earth:	maximum = 65.2″
	minimum = 9.5″
Mass:	4.87×10^{24} kg
Mass (Earth = 1):	0.8150
Mean density:	5240 kg/m³
Surface gravity (Earth = 1):	0.903
Escape velocity:	10.4 km/s
Oblateness:	0
Mean surface temperature:	480°C = 900°F = 750 K
Albedo:	0.65
Brightest magnitude:	−4.4
Mean diameter of Sun as seen from Venus:	44′ 15″

Box 11-2 Configurations of Venus, 1991–1999

Greatest eastern elongation		Inferior conjunction		Greatest western elongation		Superior conjunction	
1991	June 13	1991	August 22	1991	November 2	1992	June 13
1993	January 19	1993	April 1	1993	June 10	1994	January 17
1994	August 25	1994	November 2	1995	January 13	1995	August 20
1996	April 1	1996	June 10	1996	August 19	1997	April 2
1997	November 6	1998	January 16	1998	March 27	1998	October 30
1999	June 11	1999	August 20	1999	October 30	2000	June 11

clouds illuminated by scattered sunlight produce a luminescent ring encircling the planet (see Figure 11-2). Although this highly reflective cloud cover makes Venus dazzlingly bright, it is also responsible for our long-standing ignorance of the planet. Because of this perpetual shroud, we did not even know until recently how fast Venus rotates.

Until the twentieth century, very little was gleaned from telescopic observations of Venus. Nineteenth-century observers did realize that it has an atmosphere. Otherwise, the only noteworthy viewing of Venus was largely confined to phenomena such as solar transits, when Venus passes directly in front of the Sun at inferior conjunction.

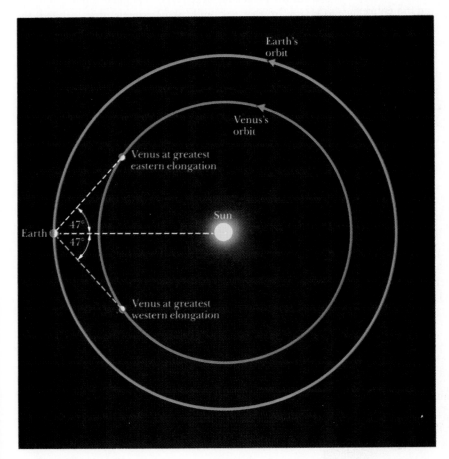

Figure 11-1 Venus's orbit *Venus travels around the Sun along a nearly circular orbit with a period of 224.7 days. The average distance between Venus and the Sun is 108 million kilometers (the average distance between the Earth and Sun is 150 million kilometers). At its greatest eastern elongation, Venus appears as a prominent evening star; at its greatest western elongation, it is a prominent morning star.*

Figure 11-2 Venus at inferior conjunction This photograph of Venus at inferior conjunction shows an illuminated atmosphere surrounding the planet. If Venus did not have an atmosphere to scatter sunlight, a thin crescent rather than a ring would be seen. (Lowell Observatory)

Solar transits of Venus are much rarer than those of Mercury. Venus is farther from the Sun than Mercury is, and the required Sun–Venus–Earth alignment occurs much less frequently. The last Venusian transits occurred in 1874 and 1882. Not one transit of Venus will occur during the entire twentieth century. Venusian transits always occur in pairs separated by eight years. The next pair will occur on June 8, 2004, and June 6, 2012.

The transits of 1761 and 1769 allowed the size of the solar system to be measured accurately for the first time, using the common surveying method of parallax (recall Figure 4-6). Prior to these transits, astronomers knew only the *relative* sizes of the planetary orbits, which were measured in terms of an "astronomical unit" of unknown length. During the eighteenth-century transits, observers could simultaneously measure Venus's position from such widely separated locations as Great Britain and Tahiti (where an expedition was led by Captain James Cook). These measurements, when combined with the distances between the observation sites, revealed the actual distance between the Earth and Venus at the time of inferior conjunction. The true dimensions of the solar system had finally been determined.

11-2 Venus's rotation is slow and retrograde

In the mid-1950s, astronomers began to realize that Venus has some unusual characteristics, very unlike those of the Earth. Clues about Venus's rotation came in 1956 when Robert S. Richardson of the Mount Wilson Observatory

measured the Doppler shift of spectral lines in sunlight reflected from the planet. As discussed in Chapter 5 (review Figure 5-20), the wavelength of a spectral line is affected by relative motion between the source and the observer. As Venus rotates, one side of the planet is approaching us and its spectral lines are thus blueshifted; the other side of the planet, which is receding from us, exhibits redshifted spectral lines (see Figure 11-3).

To everyone's surprise, Richardson's observations indicated that Venus's rotation is **retrograde**, or backward. In other words, sunrise on Venus occurs in the west.

This retrograde rotation is surprising because the other planets (except for Uranus) rotate about their axes in the same direction that they revolve about the Sun. If you could view our solar system from a great distance above the Earth's north pole, you would see all the planets orbiting the Sun counterclockwise. Closer examination would reveal that most of the planets also rotate counterclockwise on their axes. Furthermore, with only a very few exceptions, even the satellites of the planets move counterclockwise along their orbits.

This situation, whereby most of the planets and their satellites rotate in the same direction, is the result of the overall rotation of the primordial solar nebula. As we saw in Chapter 7, the solar nebula must have possessed angular momentum, otherwise everything would have fallen directly inward toward the protosun. Much of this angular momentum was tied up in the gas and planetesimals that orbited the protosun, and so the planets that formed from this material inherited their angular momentum, thus preserving their general direction of rotation. It is difficult to imagine how a planet could have bucked this trend, although one theory suggests that primordial Venus could have been hit by a huge planetesimal in such a way that the impact actually reversed Venus's direction of rotation. There is no direct evidence, however, that such an impact ever took place.

Light reflected from
receding side of
Venus is redshifted

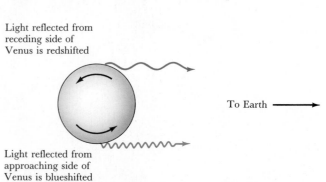

To Earth

Light reflected from
approaching side of
Venus is blueshifted

Figure 11-3 The Doppler effect and Venus's rotation Light reflected from Venus's approaching hemisphere is blueshifted whereas light reflected from the planet's receding hemisphere is redshifted. By measuring the wavelength shift across the planet, astronomers were able to determine Venus's rate of rotation. All forms of electromagnetic radiation—sunlight reflecting off of Venus's cloud cover as well as microwaves reflecting from its surface—are subject to the Doppler effect and therefore exhibit wavelength shifts caused by the planet's rotation.

Like the Earth's atmosphere, the Venusian clouds are transparent to radio waves and microwaves. In the early 1960s, improved equipment made it possible to send microwave radiation to Venus and detect the waves reflected from the surface of that planet. The process of bouncing microwave radiation from an object is known as **radar**, an acronym from "radio detection and ranging."

The first radar observations of Venus were made in 1961 by a team of scientists headed by William B. Smith at the Lincoln Laboratory of the Massachusetts Institute of Technology. Their goal was to use the Doppler effect to measure Venus's rotation rate. The wavelengths of radar waves reflected from the approaching side of the planet are shortened, whereas those from the receding side are lengthened. However, radar waves are reflected back from the planet's surface rather than from the clouds. Although the microwaves leave Earth at one precise frequency, their echoes come back spread over a small range of frequencies. By measuring the frequency spread, Smith and his colleagues concluded that Venus's rotation is slow and probably retrograde.

These observations were repeated in 1962 by Roland L. Carpenter and Richard M. Goldstein at the Jet Propulsion Laboratory of the California Institute of Technology. They confirmed that Venus has a retrograde rotation with a sidereal period of approximately 240 days. Over the next several years, the accuracy of this figure was refined with NASA's deep-space tracking antenna at Goldstone, California (see Figure 11-4). Today we know that the sidereal period of Venus's retrograde rotation is 243.01 days. This value is an interesting one. To see why, we must compare the length of a solar day on Venus to Venus's synodic period.

Sidereal period is measured with respect to the stars, not the Sun. In other words, if you were standing on Venus and could see through the clouds, you would have to wait 243.01 Earth days to see the same star pass again overhead. Meanwhile, of course, Venus is moving around the Sun, with an orbital period of 224.70 Earth days. Thus, Venus's sidereal rotation period is not the same as a solar day, which is the time from one local noon to the next. The length of a solar day on Venus can be calculated from the planet's sidereal rotation and orbital periods as follows:

$$\frac{1}{\text{solar day}} = \frac{1}{243.01} + \frac{1}{224.70} = 0.008565 = \frac{1}{116.8}$$

Figure 11-4 A deep-space tracking antenna Although this antenna was built by NASA to track interplanetary spacecraft, it can also be used to beam powerful pulses of microwaves toward Venus. The antenna's dish measures 64 m (210 ft) in diameter. (Jet Propulsion Laboratory)

In other words, an astronaut on Venus would have to wait 116.8 days from one local noon to the next.

Remember that a planet's synodic period is the time interval between successive occurrences of the same orbital configuration, such as the interval from one inferior conjunction to the next. From the data in Table 4-1, we see that Venus's synodic period is 584 days. Note that

$$5 \times 116.8 = 584.0$$

Thus, Venus's synodic period is equal to five Venusian solar days.

These numbers reveal an interesting circumstance. Earth-based astronomers must wait 584 days between successive inferior conjunctions of Venus. Meanwhile, five solar days elapse on Venus. Consequently, at each inferior conjunction the same side of Venus is turned toward the Earth. No one has been able to explain how and why this relationship was established. It may simply be a coincidence.

11-3 The surface of Venus is very warm because of the greenhouse effect

As we have noted, the Venusian clouds are transparent to radio waves. Astronomers in the 1950s tried to deduce the surface temperature of Venus by measuring radio waves emitted from the planet's surface. At any temperature, an object emits radiation at all wavelengths (review the blackbody curves shown in Figure 5-8). By measuring the relative strengths of radiation at different wavelengths and comparing the results with blackbody curves like those in Figure 5-8, you can estimate the temperature of the object emitting the radiation. Therefore, an analysis of the relative strengths of radio waves from Venus at different wavelengths should provide a measurement of the planet's surface temperature.

At first, no one could believe the results of such studies, but the observations were carefully repeated again and again, always with the same surprising result: The temperature at Venus's surface is above the melting point of lead. Measurements by spacecraft that have landed on the planet in fact indicate a surface temperature of 750 K (= 480°C = 900°F). After their initial skepticism, astronomers quickly realized that there is a straightforward explanation for Venus's high surface temperature.

Perhaps you have had the experience of parking your car in the sunshine on a warm summer day. You roll up the windows, lock the car, and go on an errand. After a few hours, you return to discover that the interior of your automobile has become stiflingly hot, typically at least 20°C warmer than the outside air temperature.

What happened to make your car so warm? First, sunlight entered your car through the windows. The radiation

was absorbed by the dashboard, the steering wheel, and the upholstery, raising their temperatures. Every object emits blackbody radiation appropriate to its temperature. Wien's law relates the Kelvin-scale temperature (T) to the dominant wavelength (λ_{max}) at which the most intense radiation is emitted. For example, the Sun's surface temperature is about 5800 K. Thus sunlight has its greatest intensity at wavelength λ_{max} of 500 nm. As you may have expected from our biological evolution, 500 nm is in the middle of the visible spectrum. This sunlight typically raises the temperature of a car's upholstery to around 330 K (= 57°C = 135°F). At that temperature, the seats in your car reradiate the energy primarily with a λ_{max} of 8.8 μm (= 8800 nm). Radiation of wavelength 8.8 μm is in the infrared portion of the electromagnetic spectrum. Your car windows, which are transparent to incoming sunlight, are opaque to infrared wavelengths. This energy therefore is trapped inside your car and absorbed by the air and interior surfaces. As more sunlight comes through the windows and is trapped, the temperature continues to rise. This phenomenon is called the **greenhouse effect** (see Figure 11-5).

Carbon dioxide is responsible for a similar phenomenon in Venus's atmosphere, and to a lesser extent in the Earth's. Like your car windows, carbon dioxide is transparent to visible light but is opaque to infrared radiation. In Chapter 8, we found that the small percentage of carbon dioxide in the Earth's atmosphere produces a comparatively gentle greenhouse effect that warms the Earth. Warmed by sunlight, the ground emits infrared radiation, some of which is blocked by the carbon dioxide, causing the air temperature to rise.

In contrast to Earth, 96 percent of Venus's thick atmosphere is carbon dioxide (recall Table 8-1). Although most of the visible sunlight striking the Venusian clouds is reflected back into space, enough reaches the Venusian surface to heat

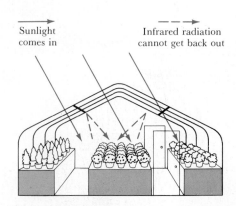

Figure 11-5 The greenhouse effect *Incoming sunlight easily penetrates the windows of a greenhouse and is absorbed by the objects inside. These objects then reradiate this energy, but at infrared wavelengths, to which the glass windows are opaque. The trapped infrared radiation is absorbed by the air and objects there, causing the temperature inside the greenhouse to rise. A similar phenomenon occurs in Venus's atmosphere to produce a high temperature on that planet's surface.*

it. The warmed surface in turn emits infrared radiation, which cannot penetrate Venus's CO_2-rich atmosphere. This trapped radiation produces the high temperatures found on Venus.

11-4 Space flights to Venus discovered neither a magnetic field nor a magnetosphere

In the 1960s, both the United States and the Soviet Union began sending probes to Venus. The Americans sent fragile, lightweight spacecraft past the planet and studied the Venusian environment with remote sensing devices. The Soviets,

who had more powerful rockets, sent massive vehicles that plunged directly into the Venusian clouds.

The first successful mission to Venus was the flight of *Mariner 2* in 1962. During its three-month interplanetary voyage, this spacecraft discovered the solar wind. The solar wind, you will recall, consists of charged particles escaping from the Sun at supersonic speeds. Our first measurement of the speed (roughly 400 km/s) and density (1 particle/cm³) of the solar wind came from *Mariner 2*.

As the spacecraft passed Venus, the magnetometer on *Mariner 2* failed to detect any magnetic field whatsoever. This lack of a magnetic field was confirmed during subsequent missions. As discussed in Chapter 8, the Earth's magnetic field is explained by the dynamo effect: Currents in the iron core of the rotating Earth produce our planet's magnetism. Venus has nearly the same density and size as Earth, so Venus probably has an iron core very similar to Earth's. Therefore, we conclude that the leisurely rotation rate of Venus is too slow to induce a planetwide magnetic field.

With virtually no magnetic field, Venus is incapable of producing a magnetosphere to protect itself from the solar wind. Therefore the solar wind impinges directly on Venus's upper atmosphere, where it strips many of the atoms of one or more electrons. An atom that carries an excess charge because it has gained or lost one or more electrons is called an **ion**. The ions in Venus's atmosphere have a positive electric charge, because they have lost negatively charged electrons. The electromagnetic interaction of these ions and the supersonic charged particles in the solar wind produces a well-defined **shock wave** (see Figure 11-6). A shock wave, you will recall, occurs where a supersonic flow of particles abruptly becomes subsonic. Along a boundary called the **ionopause**, inside the shock wave, the pressure of the ions just counterbalances the pressure from the solar wind. The ionopause is analogous to the magnetopause that surrounds a planet with a magnetic field.

11-5 Spacecraft descending into the clouds provided detailed information about cloud layers in Venus's dense atmosphere

While the United States was busy landing astronauts on the Moon, Soviet scientists concentrated on building spacecraft that could survive a descent into the Venusian cloud cover. The task proved to be more frustrating than anyone had expected. Finally, in 1970, a probe called *Venera 7* managed to transmit data for a few seconds directly from the Venusian surface. Soviet missions during the early 1970s measured a surface temperature of 750 K (= 480°C = 900°F) and a pressure of 90 atmospheres. Recall from Box 8-2 that one atmosphere (1 atm) is simply the average air pressure at

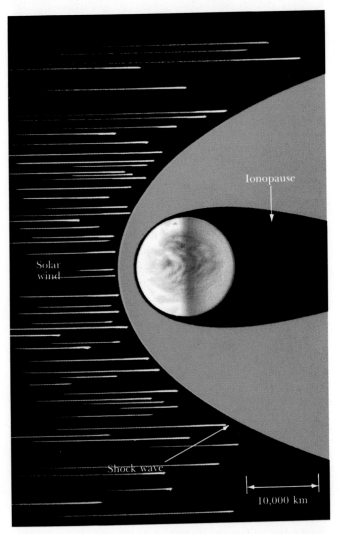

Figure 11-6 *Venus's interaction with the solar wind* Venus has no magnetic field, so the solar wind strikes the uppermost layers of the planet's atmosphere. Interaction with ions in the upper atmosphere produces a shock wave and an ionopause, as shown in this scale drawing.

Figure 11-7 **Pioneer Venus Orbiter** *This spacecraft carried twelve instruments to explore the Venusian environment. These instruments included three spectrometers, a magnetometer, an electron temperature probe, an electric field detector, and a gamma-ray burst detector. A radar altimeter was used to map 93 percent of Venus's surface. The spacecraft weighed 590 kg (1300 lb) and measured $2\frac{1}{2}$ m ($8\frac{1}{3}$ ft) in diameter. (NASA)*

Figure 11-8 **Pioneer Venus Multiprobe** *This entire spacecraft (with probes attached) was almost exactly the same size as the Orbiter, but this one weighed 910 kg (nearly a ton). The bus and four probes carried a total of eighteen scientific instruments. Fifteen ports and windows in the large probe provided access to seven instruments. The small probes were identical to each other and less complicated. Each contained three basic scientific instruments. (NASA)*

sea level on Earth (14.7 pounds per square inch). Thus, the Venusian atmosphere weighs down on the planet with a crushing pressure of $\frac{2}{3}$ ton per square inch.

During the 1970s, the Soviets and Americans used a variety of spacecraft to explore the Venusian atmosphere. A particularly successful American effort involved the *Pioneer Venus Orbiter* and the *Pioneer Venus Multiprobe* (see Figures 11-7 and 11-8), which arrived at Venus in December 1978. Explosive charges on the *Multiprobe* separated the four blunt-nosed probes from the cylindrical "bus," and all five objects then hurtled into the Venusian cloud cover. As expected, the bus burned up in the clouds, but it did send back data about the upper atmosphere to Earth. The four aerodynamically shaped probes made numerous measurements at lower altitudes, and one even continued to send data from the surface for over an hour.

All the Soviet and American atmospheric probes carried instruments that measured pressure and temperature as they descended to the Venusian surface. The results of all the missions are summarized in Figures 11-9 and 11-10. Graphs like these are important, because they show how pressure and temperature are related to the altitude above a planet's surface—fundamental information from which the detailed

structure of a planet's atmosphere can be deduced. The pressure and temperature profiles of the Venusian atmosphere are simple: Both pressure and temperature decrease smoothly with increasing altitude. As we saw in Chapter 8 (recall Figure 8-18), Earth's atmosphere has a much more complicated relationship between temperature and altitude. In the chapters on Jupiter and Saturn, we shall see how pressure and temperature variation with altitude can profoundly affect a planet's appearance.

Soviet spacecraft discovered that Venus's clouds are confined to a 20-km-thick layer located between 48 and 70 km above the planet's surface. Below that, between the altitudes of 30 and 50 km, is a 20-km-thick haze layer. Beneath the haze layer, the Venusian atmosphere is remarkably clear, all the way down to the surface.

The *Multiprobe* discovered distinct cloud layers within Venus's cloud cover. An **upper cloud layer** extends from altitudes of 68 km down to 58 km. A denser and more opaque **middle cloud layer** extends from 58 km down to 52 km. Finally, the **lower cloud layer**, from 52 to 48 km, contains the densest and most opaque of the Venusian clouds, even though it is only 4 km thick. These layers are indicated on Figures 11-9 and 11-10.

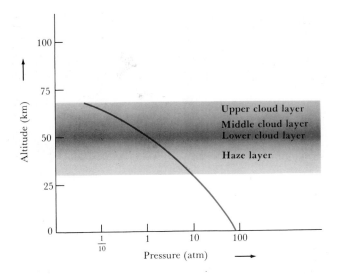

Figure 11-9 *Temperature in the Venusian atmosphere* *The tempera-ture in Venus's atmosphere rises smoothly from a minimum of about 170 K (about −100°C, or −150°F) at an altitude of 100 km to a maxi-mum of nearly 750 K (about 480°C, or 900°F) on the ground.*

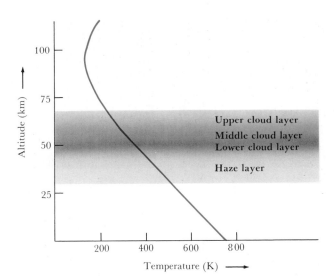

Figure 11-10 *Pressure in the Venusian atmosphere* *The pressure at the Venusian surface is a crushing 90 atm (1300 pounds per square inch). Above the surface, atmospheric pressure decreases smoothly with increasing altitude.*

11-6 Space probes provided detailed information about the chemical composition and weather in Venus's corrosive atmosphere

In 1932, Walter S. Adams and Theodore Dunham of the Mount Wilson Observatory identified carbon dioxide (CO_2) in the planet's spectrum. (A portion of Venus's spectrum showing carbon dioxide lines appears in Figure 7-10.) As-tronomers could not reliably estimate the percentage of this gas in the planet's atmosphere from these early observations, but the lack of other spectral patterns suggested that CO_2 composed most of the Venusian atmosphere. Direct mea-surements by the Soviet probes showed that 96 percent of Venus's dense atmosphere is indeed carbon dioxide. Nitro-gen accounts for most of the remaining 4 percent.

Venus appears yellowish or yellow-orange to the human eye, and data from the *Multiprobe* indicated why: The upper clouds contain substantial amounts of sulfur dust. Over the temperature range of these upper clouds, sulfur is distinctly yellow or yellow-orange. At lower elevations, large concen-trations of sulfur compounds such as sulfur dioxide and hy-drogen sulfide were found, along with droplets of sulfuric acid. Just as Earth's clouds are composed of water droplets, Venusian clouds are composed of droplets of concentrated sulfuric acid. Because of the tremendous atmospheric pres-sure on Venus, the droplets do not fall as a rain; instead, they are suspended in the clouds like an aerosol.

The sulfuric acid in Venus's atmosphere causes a number of chemical reactions. For example, reactions with fluorides and chlorides in surface rocks give rise to hydrofluoric acid (HF) and hydrochloric acid (HCl). Further reactions pro-duce fluorosulfuric acid (HSO_3F), which is one of the most corrosive substances known to chemists, capable of dissolv-ing lead, tin, and most rocks. The Venusian clouds are a cauldron of chemical reactions hostile to metals and other solid materials.

As the *Multiprobe* hurtled into the Venusian atmosphere, the *Orbiter* began a lengthy series of observations and mea-surements that would keep scientists occupied for many years. For example, the *Orbiter* sent back numerous photo-graphs of Venus in the ultraviolet wavelengths at which its atmospheric markings stand out best (see Figure 11-11). By following individual cloud markings, scientists determined that Venus's atmosphere rotates in a retrograde direction around the planet in only four days. This rapid atmospheric motion is in sharp contrast to the slow rotation of the solid planet itself.

The *Multiprobe* also determined the dominant circulation patterns in the Venusian atmosphere. Warmed by the Sun, hot gases in the equatorial regions rise upward and travel in the upper cloud layer toward the cooler polar regions. At the polar latitudes, the cooled gases sink downward to the lower cloud layer, in which they are transported back toward the equator. Recall from Chapter 8 that this process of heat transfer, whereby hot gases rise while cooler gases sink, is called **convection**.

Figure 11-11 Venus from the Orbiter *These four ultraviolet views of Venus were taken in May 1980 at a distance of roughly 50,000 km. All the views show variations of the so-called Y feature caused by the rapid retrograde motion of the clouds around the planet. (NASA)*

The circulation of Venus's atmosphere is dominated by two huge **convection cells** (one in the northern hemisphere and another in the southern hemisphere) that circulate gases between the equatorial and polar regions of the planet (see Figure 11-12). These convection cells, which are almost entirely contained within main cloud layers, are called **driving cells**, because they propel similar circulation cells above and below the main cloud deck somewhat like a meshed set of gears.

Because Venus's entire atmosphere rotates around the planet in only four days, strong prevailing winds blow from east to west. These winds stretch out the driving cells and produce the characteristic V-shaped, chevronlike patterns that dominate the planet's appearance.

11-7 Volcanoes probably created the atmospheres of the terrestrial planets

Before space probes analyzed the atmospheres of Venus and Mars, there were four main theories, or hypotheses, about the origin of the atmospheres of the terrestrial planets. The first, called the **accretion hypothesis**, held that the "ices" (carbon dioxide, water, and so on) had been trapped in the planets when they were formed and later had been "outgased" by volcanic activity. The other three hypotheses assumed that the planets had originally been free of ices and that their atmospheres had come from some outside source. The **solar nebular hypothesis** held that the ices had been gathered by the planets from the remnants of the solar nebula. The **solar wind hypothesis** argued that the ices had been captured from the solar wind. The **comet–asteroid hypothesis** argued that the ices arrived on the planets in the comets, meteoroids, and asteroids that collided with them during the early stages of the solar system when such collisions were relatively common.

The best way to test these competing hypotheses is to study the nonreactive (or inert) gases such as neon, argon, and krypton in the planetary atmospheres. These gases are not incorporated into surface rocks, because they are chemically unreactive, and they do not escape into space, either, because they are composed of heavy atoms that move too slowly to escape from a planet's atmosphere (recall Box 7-2). Thus any inert gases ever added to an atmosphere should still be there today, and therefore the relative abundances of the inert gases in planetary atmospheres should provide important clues about the origins of planetary atmospheres.

The atmospheres of Venus, Earth, and Mars contain similar proportions of neon and argon isotopes (especially the ratio of ^{20}Ne to ^{36}Ar), but the Sun's atmosphere has quite

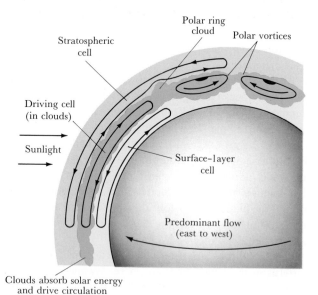

Figure 11-12 The circulation in Venus's atmosphere The main convection mechanism in Venus's atmosphere occurs in the clouds. This convection cell drives similar circulation patterns above and below the cloud layer. Circular wind patterns (vortices) in the polar regions force the cloud layer to bulge upward at polar latitudes, producing a polar ring cloud that is clearly visible in many Orbiter *photographs. (Adapted from A. Seiff)*

different ratios of these isotopes. It is reasonable to assume that the Sun's atmosphere has essentially the same composition as the solar wind and the original solar nebula. We can therefore rule out the solar wind and the solar nebula as likely sources of the planetary atmospheres. In this way, chemical analyses by Mars and Venus probes helped limit the likely answer to a choice between the accretion hypothesis and the comet–asteroid hypothesis.

Venus, Earth, and Mars should have received roughly equal bombardment by comets and asteroids, so they should have similar amounts of inert gases in their atmospheres if the comet–asteroid hypothesis is valid. But the *Pioneer* probes found that argon is 60 times more abundant in Venus's atmosphere than in the Earth's. This evidence would seem to eliminate the comet–asteroid hypothesis, leaving the accretion hypothesis as the only likely explanation for the origin of atmospheres of the terrestrial planets. The bulk of these planetary atmospheres were thus produced by volcanic outgasing, and the resulting primitive atmospheres were probably composed primarily of water vapor (H_2O), carbon dioxide (CO_2), carbon monoxide (CO), hydrogen (H_2), and nitrogen (N_2). Some researchers believe that methane (CH_4) and ammonia (NH_3) were also present at first, but were later transformed into carbon dioxide and nitrogen by atmospheric chemistry.

Although comets and asteroids did not contribute significant quantities of gas to the primordial atmospheres of the terrestrial planets, major impacts on these planets probably did have profound consequences. For instance, we have seen that our Moon was probably formed by a catastrophic impact of a Mars-sized asteroid. That impact could have stripped the Earth of much of its primordial atmosphere. Since Venus does not have a moon, it probably was not subjected to a comparable cataclysm near the end of the planet-forming era 4.5 billion years ago.

Certainly, planetary atmospheres have been modified over the ages by gases from outside sources. For example, the Venusian atmosphere has a very high ratio of argon to krypton, similar to the ratio that exists in the solar atmosphere. We therefore conclude that the atmosphere of Venus, nearer to the Sun, has been more affected by the solar wind than have the atmospheres of the other terrestrial planets. Nonetheless, the preponderance of evidence suggests that outgasing by volcanoes (see Figure 11-13) has been the primary process in the formation of atmospheres on the terrestrial planets.

11-8 Active volcanoes are probably responsible for Venus's clouds

There is compelling evidence for significant volcanic activity occurring today on Venus. All of the major volcanic effluents observed in Earth's volcanoes have been detected in Venus's

Figure 11-13 Mount St. Helens eruption of May 1980 A preponderance of data suggests that the terrestrial planets obtained their atmospheres from volcanic outgasing. Geologists study this outgasing process in eruptions of volcanoes on Earth. It is possible that active volcanoes are responsible for maintaining Venus's sulfur-rich cloud cover. (USGS)

atmosphere. Because many of these chemicals are highly reactive and very short-lived, they must be constantly replenished. In addition, an antenna on *Pioneer Venus Orbiter* often picked up low-frequency radio bursts thought to be strokes of lightning. Lightning discharges have often been seen in the plumes of erupting volcanoes on Earth. Long-term observations by *Pioneer Venus Orbiter* suggest that sulfurous gases are being injected into the Venusian atmosphere by processes that could only be volcanic. When the spacecraft arrived at Venus in 1978, its ultraviolet spectrometer recorded unexpectedly high levels of sulfur dioxide and sulfuric acid, which steadily declined over the next several years. A similar anomalously high abundance of haze particles may have occurred in the late 1950s. Larry W. Esposito of the University of Colorado has proposed that in both the late 1950s and the late 1970s, energetic volcanic eruptions injected sulfur dioxide into the upper atmosphere. It therefore seems that Venus's sulfur-rich clouds are regularly replenished by active volcanoes.

Clear evidence for volcanoes comes from Earth-based radar observations of Venus's surface. Astronomers have often used the giant 305-m (1000-ft) dish at the Arecibo Observatory in Puerto Rico (see Figure 10-6) to map Venus because microwaves easily penetrate the Venusian clouds and reflect off its surface. The Arecibo dish is used to transmit a powerful, brief burst of microwaves toward Venus. These microwaves strike mountain tops on Venus slightly sooner than deep valleys, and so echoes from high elevations arrive back at the Arecibo dish slightly ahead of those reflected from lower elevations. In addition, Venus's rotation causes the microwave burst to become spread out in wavelength according to the Doppler effect (recall Figure 11-3). By analyzing the radar echo, which is spread out in both arrival time and wavelength, astronomers are able to construct a map of the Venusian surface. This method is successful only when Venus is near inferior conjunction. At all other times the Venus–Earth distance is too great and the radar echo too weak to produce good data.

Figure 11-14 is a radar view of Theia Mons, one of several huge volcanoes that comprise a region called Beta Regio, roughly 1300 kilometers north of Venus's equator. Theia Mons rises to an altitude of 6 km and has gently sloping sides that extend over an area 1000 km in diameter. A volcano having this characteristic shape is called a **shield volcano**, because in profile it resembles an ancient Greek war-

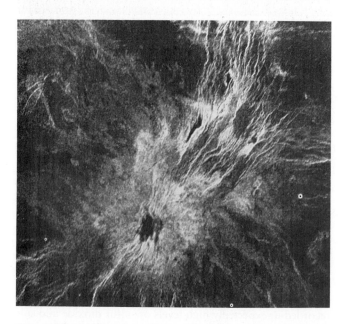

Figure 11-14 A Venusian volcano This high-resolution radar image of Theia Mons shows details of the lava flows and rifts that surround the volcano. The volcano's summit is the darkish spot toward the bottom center of the picture, which covers an area 1700 km by 1500 km—roughly three times the size of the state of Texas. (Courtesy of D. B. Campbell, Arecibo Observatory)

rior's shield lying on the ground. Shield volcanoes exist on both Earth and Mars, the best-known example being the Hawaiian Islands. Shield volcanoes on Venus and Mars, as well as the Hawaiian Islands on Earth, are thought to be the result of **hot-spot volcanism**, whereby a hot region beneath the planet's surface extrudes molten rock at a particular location over a long period of time.

Carbon dioxide and water vapor are among the most abundant gases in volcanic vapors on Earth. Venus's atmosphere contains abundant carbon dioxide, but very little water. In contrast, the Earth has abundant water in its oceans, but there is very little carbon dioxide in its atmosphere. If Venus and the Earth got their atmospheres mainly through the same outgasing process, what happened to all the water on Venus? And where is all the carbon dioxide on Earth?

In the upper Venusian atmosphere, intense ultraviolet radiation from the Sun breaks down water molecules into separate hydrogen and oxygen atoms. The light hydrogen atoms escape into space. Oxygen, which is one of the most chemically active elements, readily recombines with other substances in the atmosphere. As a result, Venus is left with almost no water.

The carbon dioxide on Earth is dissolved in the oceans and chemically bound into carbonate rocks, such as limestone and marble, that formed in those oceans. If the Earth became as hot as Venus, so much carbon dioxide would be boiled out of the oceans and baked out of the crust that our planet would soon develop a thick, oppressive carbon dioxide atmosphere much like that of Venus.

Thomas A. Donahue of the University of Michigan has proposed an interesting outline of Venus's early history. Venus and Earth are so similar in size and mass that it is reasonable to suppose that Venusian volcanoes outgased an amount of water vapor roughly comparable to the total content of the Earth's oceans. Although some of this water on Venus might originally have collected in oceans, heat from the Sun soon vaporized the liquid to create a thick cover of water-vapor clouds. Calculations demonstrate that this water vapor would have added 300 atm of pressure to Venus's existing 90 atm of carbon dioxide. Thus, the early Venusian atmosphere must have weighed down on the planet's surface with a pressure of 3 tons per square inch.

This thick, humid atmosphere efficiently trapped heat from the Sun, creating a greenhouse effect far more extreme than the greenhouse effect that operates today on Venus. Calculations demonstrate that the ground temperature would have increased to 1800 K (2700°F), which is hot enough to melt rock. The Venusian surface was therefore probably molten down to a depth of 450 km (280 mi). Soon, however, the dissociation of the water molecules and the subsequent loss of hydrogen to space left behind the carbon-dioxide atmosphere we find today. Hostile as it may seem, the modern environment on Venus is probably quite mild compared to that of earlier times.

11-9 Radar maps of Venus reveal gently rolling hills, two continents, and large volcanoes

The *Pioneer Venus Orbiter* mapped 93 percent of the planet's surface, using a radar altimeter that bounced microwaves off the ground directly below the spacecraft. By measuring the time delay of the radar echo, scientists could determine the heights and depths of Venus's hills and valleys to an accuracy of ±200 m. The results are shown in Figure 11-15, which uses a color code to denote altitude.

Venus is remarkably flat. About 60 percent of the planet's surface is covered with gently rolling hills that vary by less than 1 km from the planet's average radius of 6051.4 km. (This value for the planet's radius is commonly used as a reference level, as we use average sea level here on Earth.)

There are two large "continents" rising well above Venus's generally level surface. In the northern hemisphere, there is **Ishtar Terra**, named after the Babylonian goddess of love. Ishtar Terra, which is approximately the same size as Australia, consists of a high plateau called Lakshmi Planum that is ringed by high mountains. The second Venusian continent, **Aphrodite Terra**, named after the Greek equivalent of

Venus, lies just south of the equator. Aphrodite, slightly bigger than Ishtar, has an area about half that of Africa.

In 1983, Soviet scientists sent two spacecraft called *Venera 15* and *Venera 16* to Venus. Using sophisticated onboard radar, these spacecraft mapped 120 million square kilometers of the planet's northern hemisphere. The Venera data were combined with *Pioneer Venus* altimetry and Arecibo data to produce the extraordinary map shown in Figure 11-16. Color is used to indicate surface roughness: Purples and blues represent smooth areas, green is rougher, and red indicates the roughest terrain. Figure 11-17 is a computer-synthesized oblique view of Ishtar that looks northward across the planet.

The broad, smooth area near the center of the map is Lakshmi Planum, the huge plane that dominates the continent of Ishtar. The central craters of two large volcanoes are seen near the center of the plane. The huge mountain (colored red because of its extreme roughness) to the east of Lakshmi Planum is Maxwell Montes, named after the physicist James Clerk Maxwell. The peak of Maxwell is nearly 12 km above Venus's reference level, making it the tallest protrusion on the planet. For comparison, Mount Everest on Earth rises 9 km above sea level. The roughness of Maxwell's surface suggests that it is very young. Older

a

b

Figure 11-15 **Topographic globes of Venus's surface** *These computer-generated globes show the vertical relief on Venus according to a color code. The average elevation (like sea level on Earth) is shaded light blue. Dark blue and violet denote lower elevations. Green and yellow indicate higher elevations, with the highest peaks shown in red. View (a) has the north pole tilted toward the observer to show Ishtar Terra near the top* *of the globe. Maxwell Montes, on the eastern edge of Ishtar Terra, is the highest region feature on the planet and is colored red. View (b) has the south pole tilted toward the observer to show Aphrodite Terra. The blank circles cover the polar regions, which were not observed by the spacecraft. (NASA)*

Figure 11-16   Venus's shaded-relief color-coded radar map   This map is based on Venera 15 and 16 radar images and Pioneer Venus altimetry. Color represents surface roughness: purple and blue indicate smooth areas, green is rougher, and red is the roughest terrain. Lakshmi Planum is the flat, purple area near the center of the map. Maxwell Montes is the red patch of mountains toward the right. Several large circular features surrounded by concentric rings are seen toward the upper left. The origin of these circular structures is unknown, but they may be the result of hot-spot volcanism. (Courtesy of A. S. McEwen, USGS)

Figure 11-17   Oblique view of Lakshmi Planum   This northward-looking oblique view of Ishtar is oriented as if the observer is located 25° above the horizon. Vertical relief was exaggerated by a factor of 50 to show the planet's topography. Lakshmi Planum is left of center and Maxwell Montes is toward the right. Color indicates surface roughness, as in Figure 11-16. (Courtesy of A. S. McEwen, USGS)

features are quickly worn smooth by the harsh Venusian environment, where a 1-mph wind has the impact of a 90-mph wind on Earth.

The main part of Maxwell is a vast mountain whose rectangular base measures nearly 700 km by 400 km, an area roughly the size of the state of Colorado. Steep slopes, perhaps with cliffs up to 6 km high, dominate the western edge of Maxwell. The banded features that cover central Maxwell are parallel ridges 10 to 20 km wide. The smooth circular area on the eastern side of Maxwell, which is about 100 km in diameter, could be a lava-filled volcanic crater. Many astronomers suspect that Maxwell is an enormous volcano. The fact that there are no long chains of volcanoes on Venus suggests that Venus's surface does not move around the way the Earth's surface does.

Several Soviet spacecraft, like the one in Figure 11-18, have landed on the surface of Venus. Figure 11-19 is a panoramic view taken by one of these landers that set down at a site covered with broken slabs of rock. Soviet scientists suggest that this region was once covered with a thin layer of lava that fractured upon cooling to create the rounded, interlocking shapes seen in the photograph. This hypothesis agrees with the analysis by the spacecraft's instruments, which indicates that the soil composition is similar to the lava rocks called basalt that are common on Earth and on the Moon.

Soviet and American scientists have discovered several large circular features on Venus. One of the largest, called Artemis Chasma, some 1800 km in diameter, is located along the southern edge of Aphrodite Terra (see Figure 11-13b). Some scientists believe that these features are impact craters. Others suggest that they are huge volcanic domes that have collapsed, leaving folded crust around their periphery. This issue is one of many to be explored by the

Figure 11-18 A Venera lander Many of the Venera missions involved two spacecraft (a bus and a lander) with a total weight of roughly 10 tons. Upon arrival at Venus, the lander separated from the bus and descended through the Venusian clouds with parachutes. A shock-absorbing ring mounted at the base of the spacecraft cushioned the touchdown. (TASS)

a

b

Figure 11-19 A Venusian landscape (a) This color photograph from Venera 13 shows that the rocks on the Venusian surface appear orange because the thick, cloudy atmosphere absorbs the blue component of sunlight. (b) When computer processing is used to remove the effects of orange illumination, the true grayish color of the rocks is seen. In this panorama, the rocky plates covering the ground may be fractured segments of a thin layer of lava, or they may be crusty layers of sediment that have been cemented together by chemical and wind erosion. (Courtesy of C. M. Pieters and the U.S.S.R. Academy of Sciences)

Figure 11-20 *The* **Magellan** *spacecraft Using sophisticated radar techniques, Magellan will map nearly 90 percent of Venus's surface. This spacecraft was launched from Earth in May 1989 and arrived at Venus in August 1990. (NASA)*

Magellan spacecraft (see Figure 11-20) launched in 1989. This spacecraft will map the Venusian surface with unprecedented resolution, answering many questions about the extent of volcanism and impact cratering.

After examining all the available data on Venus, many astronomers believe they are looking at a planet that has evolved geologically much as Earth did, but to a lesser extent. For instance, Venus's crust may simply be too thin and fragile to support plate tectonics. Perhaps Venus's atmosphere has slowed the rate of heat loss from the planet, thus delaying the formation of a crust as thick and sturdy as that on Earth. The Venusian surface may, as a result, contain clues about the appearance of our planet prior to the onset of plate tectonics.

In a certain sense, Venus is the Earth's twin. Its surface tells us of our geologic past, and its atmosphere suggests that Earth may have an awesome future. Five billion years from now, as our Sun swells to become a red giant star, the oceans will boil and shroud the Earth in a thick cloud cover. As the greenhouse effect drives temperatures toward 1000 K, vast amounts of carbon dioxide will be baked out of the Earth's rocks. In looking at Venus's atmosphere, perhaps we see the distant fate of our own world.

Key words

accretion hypothesis	driving cell	Ishtar Terra	shield volcano
Aphrodite Terra	greenhouse effect	lower cloud layer	shock wave
comet–asteroid hypothesis	hot-spot volcanism	middle cloud layer	solar nebular hypothesis
convection	ion	radar	solar wind hypothesis
convection cell	ionopause	retrograde rotation	upper cloud layer

Key ideas

- Venus is similar to the Earth in its size, mass, average density, and surface gravity, but it is covered by unbroken, highly reflective clouds that conceal its other features from Earth-based observers.

- Space probes reveal that 96 percent of the Venusian atmosphere is carbon dioxide; most of the balance of the atmosphere is nitrogen.

   Venus's clouds consist of droplets of concentrated sulfuric acid, along with substantial amounts of yellowish sulfur dust. Active volcanoes on Venus may be a continual source of this sulfurous material.

   Venus's clouds are confined to a range of altitudes between 48 and 70 km above the planet's surface. A haze layer extends down to an elevation of 30 km, beneath which the atmosphere is clear.

   The surface pressure and temperature on Venus are 90 atm and 750 K, respectively. Both temperature and pressure decrease as altitude above the planet increases.

   The circulation of the Venusian atmosphere is dominated by two huge convection currents in the cloud layers, one in the northern hemisphere and one in the southern hemisphere.

   Venus's high temperature is caused by the greenhouse effect, as the dense carbon-dioxide atmosphere traps and retains energy from sunlight.

- Venus rotates slowly in a retrograde direction with a solar day of 117 Earth days and a sidereal rotation period of 243 Earth days. Thus there are approximately two Venusian solar days in a Venusian year.

   The Venusian atmosphere rotates rapidly in a retrograde direction around the planet with a period of only about four Earth days.

- Venus has no detectable magnetic field or magnetosphere.

- The atmospheres of the terrestrial planets were probably formed from "ices" trapped in the original protoplanets and later released in volcanic activity.

- Water was lost from Venus by the action of ultraviolet radiation on the upper atmosphere. Carbon dioxide on Earth has been dissolved in the oceans and chemically bound into rocks.

- The surface of Venus is surprisingly flat, mostly covered with gently rolling hills. There are two major "continents" and several large volcanoes.

   The surface of Venus shows little evidence of the motion of large crustal plates, of the type that played a major role in shaping the Earth's surface.

Review questions

1 Venus takes 440 days to move from greatest western elongation to greatest eastern elongation, but it needs only 144 days to go from greatest eastern elongation to greatest western elongation. With the aid of a diagram, explain why.

2 As seen from Earth, the magnitude of Venus changes as it moves along its orbit. Describe the main factors that determine Venus's variations in brightness as seen from Earth.

3 Why is it hotter on Venus than on Mercury?

4 In earlier astronomy books, Venus is often referred to as the Earth's twin. What physical properties do the two planets have in common? In what physical ways are the two planets dissimilar?

5 What is the greenhouse effect, and what role does it play in the atmospheres of Venus and the Earth?

6 Why are there no oceans on Venus? Where has all Venus's water gone?

7 Why is there so much carbon dioxide in Venus's atmosphere while very little of this gas is present in Earth's atmosphere?

8 Compare the ways in which Mercury, Venus, and Earth interact with the solar wind.

9 What evidence exists for active volcanoes on Venus?

10 How might Venus's cloud cover change if all of Venus's volcanic activity suddenly stopped? How might these changes affect the overall Venusian environment?

11 Suppose that Venus had no atmosphere at all. How would the albedo of Venus then compare with that of Mercury or the Moon? Explain your answer.

Advanced questions

Tips and tools . . .
You should recall that Wien's law relates the temperature of a blackbody to λ_{max}. The circumference of a circle of radius r is $2\pi r$. The linear speed of a point on a planet's equator is the planet's circumference divided by its rotation period. The albedos of Earth and Venus are found in Boxes 8-1 and 11-1, respectively.

*12 (Basic) At what wavelength does Venus's surface emit the most radiation? Do astronomers have telescopes that can detect this radiation? Why can't we use such telescopes to view the planet's surface?

*13 What is the maximum wavelength shift of yellow light ($\lambda = 550$ nm) across the face of Venus because of its rotation?

14 Why do you suppose that Venus does not have a magnetic field but Mercury does, even though both planets rotate rather slowly?

*15 Which planet looks brighter: the sunlit side of Venus seen from 1 AU, or the sunlit side of Earth seen from 1 AU? By what factor?

16 Suppose that a planet's atmosphere was opaque to visible light but transparent to infrared radiation. How would this affect the planet's surface temperature? Contrast and compare this hypothetical planet's atmosphere with the greenhouse effect in Venus's atmosphere.

*17 (Challenging) Water has a density of 1000 kg/m³, so a column of water n meters tall produces a pressure equal to $n \times 1000$ kg/m². On either Earth or Venus, which have nearly the same surface gravity, a mass of 1 kg weighs about 2.2 lb. Calculate how deep you would have to descend into the ocean to be subject to a pressure equal to the atmospheric pressure on Venus's surface.

Discussion questions

18 Describe the apparent motion of the Sun during a "day" on Venus relative to (a) the horizon and (b) the background stars.

19 If you were designing a space vehicle to land on Venus, what special features would you think necessary? In what ways would this mission and landing craft differ from a spacecraft designed for a similar mission to Mercury?

20 Some scientists believe that the Magellan spacecraft will reveal numerous craters on the Venusian surface. Other scientists disagree. What do you suppose are the pros and cons of this argument?

Observing projects

21 Refer to Box 11-2 to see if Venus is near a greatest elongation. If so, view the planet through a telescope. Make a sketch of the planet's appearance. From your sketch can you determine if Venus is closer to us or further from us than the Sun is?

22 Observe Venus once a week for a month and make a sketch of the planet's appearance on each occasion. From your sketches, can you determine if Venus is approaching us or moving away from us?

23 **This observing project should be performed only under the direct supervision of an astronomer who knows how to** point a telescope safely at Venus. Make arrangements to view Venus during broad daylight. This is best done by visiting an observatory where the coordinates (right ascension and declination) of Venus's position can be used to point the telescope. **DO NOT LOOK AT THE SUN! Looking directly at the Sun can cause blindness.**

For further reading

Bazilevskiy, A. T. "The Planet Next Door." Sky & Telescope, April 1989 • In this article, a noted Soviet scientist describes the results and implications of the most recent Venera missions to Venus.

Beatty, J. "Report from a Torrid Planet." Sky & Telescope, May 1982 • This discussion of the results from the Venera 13 and 14 missions includes four panoramic views of the Venusian surface.

———. "A Soviet Space Odyssey." Sky & Telescope, October 1985 • This article describes the Soviet Vega missions which landed two spacecraft on Venus and floated balloon-borne instrument packages in its atmosphere.

Burgess, E. Venus: An Arrant Twin. Columbia University Press, 1985 • This brief, nontechnical survey of our modern understanding of Venus includes results from the Pioneer Venus and Venera missions.

Chapman, C. "The Vapors of Venus and Other Gassy Envelopes." Mercury, September/October 1983 • The atmospheres of Venus and Earth are compared in this nontechnical article.

Fimmel, R., et al. Pioneer Venus. NASA SP-461, 1983 • This generously illustrated book from NASA tells the complete story of the Pioneer Venus mission, including many engineering details about the spacecraft.

Hunt, G., and Moore, P. The Planet Venus. Faber & Faber, 1982 • This excellent introduction to our sister planet covers early ideas and observation as well as the recent U.S. and Soviet mission.

Pettengill, G., et al. "The Surface of Venus." Scientific American, August 1980 • This article gives an excellent overview of how Earth-based radar observations are used to map the surface of Venus.

Prinn, R. C. "The Volcanoes and Clouds of Venus." Scientific American, March 1985 • This excellent article discusses evidence for active volcanoes on Venus and describes how they affect the planet's atmosphere.

Schubert, G., and Covey, C. "The Atmosphere of Venus." Scientific American, July 1981 • This fine article describes the dynamics of Venus's atmosphere and compares it with Earth's atmosphere.

CHAPTER 12

The Martian Invasions

People have long speculated about the possibility of Martian life, because Mars has many Earthlike characteristics. Around 1900, some astronomers claimed to have seen networks of linear features ("canals") on the Martian surface, leading to speculation that an advanced alien civilization there had built them. Nearly a century later the Mariner and Viking spacecraft reported many surprising facts about Mars—including the discovery of an enormous volcano and a huge canyon—but they found no canals, and no signs of life. The Viking spacecraft have sent back an enormous amount of data about the Martian surface, interior, atmosphere, and satellites that have given us a basic understanding of Mars. Nevertheless, many questions about the Martian environment remain, questions that could be answered by another mission to this fascinating world.

Mars Many of the major features of the Martian surface are seen in this view from Viking 2 as the spacecraft approached the dawn side of the planet in August 1976. Clouds flank the western slopes of the huge volcano Olympus Mons near the top of the picture. In the middle is a vast canyon called Valles Marineris. This canyon stretches nearly 4000 km along the planet's equator. At the bottom of the photograph, carbon-dioxide snow lines the floor of the Argyre Basin and surrounding craters. (NASA)

Mars is the only planet whose surface features can be seen through Earth-based telescopes. Early telescopic observers reported seeing seasonal variations that seemed to indicate vegetation on the Martian surface, and some reported geometric patterns that might represent a planetwide system of artificial canals. Scientists speculated about the possibility of life on Mars, and science-fiction writers wove popular stories about invasions of Earth by hostile Martians. Ironically, the first actual invasion came in 1976 when automated spacecraft from the Earth landed on the Martian surface. These probes sent back pictures and data indicating that Mars is a barren, desolate world. The possibility of some form of Martian life has not been completely ruled out, but it now seems quite likely that Mars is as sterile and lifeless as the Moon.

12-1 Earth-based observations originally suggested that Mars might have some form of extraterrestrial life

The first reliable record of surface features on Mars is found in observations recorded by the Dutch physicist Christian Huygens in November 1659. Using a refracting telescope of his own design, Huygens identified a prominent, dark, triangular feature that we now call Syrtis Major. After observing this feature for several weeks, Huygens concluded that the rotation period of Mars is approximately 24 hours. This observation was the first in a series that would soon lead to speculations about life on Mars, a planet which seemed re-

markably similar to Earth. (Basic data about Mars are listed in Box 12-1.)

The first accurate measurements of Mars's rotation period were made in 1666 by the Italian astronomer Giovanni Domenico Cassini. He determined that a Martian day is nearly 37½ minutes longer than an Earth solar day. Cassini was also the first to see the Martian polar caps, which bear a striking superficial resemblance to the Arctic and Antarctic polar caps on Earth (see Figure 12-1). More than a century elapsed, however, before the famous German-born English astronomer William Herschel first suggested that the Martian polar caps might be made of ice or snow.

Herschel was also the first to determine the inclination of Mars's axis of rotation. Just as Earth's equatorial plane is tilted 23½° from the plane of its orbit, Mars's equator makes an angle of over 25° with its orbit. This striking coincidence means that Mars experiences Earthlike seasons. However, the Martian seasons last nearly twice as long as Earth's, because Mars takes nearly two years to orbit the Sun.

The Martian surface exhibits interesting seasonal variations. During spring and summer in a Martian hemisphere, the polar cap shrinks and the dark markings (which often look greenish) become very distinct. Half a Martian year later, with the approach of fall and winter, the dark markings fade and the polar cap grows. The implication that there is Martian vegetation was so strong that in 1802 the German mathematician Karl Friedrich Gauss proposed that we signal the Martian inhabitants by drawing huge geometric patterns in the Siberian snow. His plan was never carried out.

There are favorable—and not-so-favorable—times to observe Mars. Because Mars's orbit is noticeably elongated (see Figure 12-2), the best views are obtained when Mars is simultaneously at opposition and near perihelion. This configuration is called a **favorable opposition** because the Earth–Mars distance can be as small as 56 million kilometers (35 million miles). At such times, Mars appears brilliant and red in the sky, outshining all the stars and gleaming with a magnitude as bright as −2.8. Through a telescope, the ruddy planet presents a disk nearly 26 arc sec in diameter. That is the same angular diameter as a moderate-sized (50-km) lunar crater viewed from the Earth. The Earth-based photograph in Figure 12-1 was taken during the favorable opposition of 1971. Data concerning oppositions during the 1990s are given in Box 12-2.

At unfavorable oppositions, such as those when Mars is near aphelion, the Earth–Mars distance can be as large as 101 million kilometers (63 million miles). And when Mars is not at opposition, the Earth–Mars distance can be much greater than 100 million kilometers. Mars then appears as a tiny reddish dot, virtually devoid of features even when viewed through a large telescope.

Speculations about life on Mars were furthered by observations of the Italian astronomer Giovanni Virginio Schiaparelli during a favorable opposition in 1877. Schiaparelli reported seeing 40 straight-line features crisscrossing the Martian surface. He called these dark linear fea-

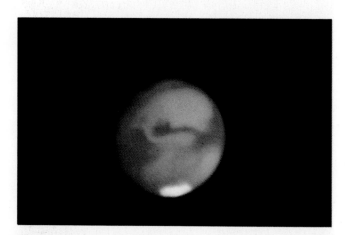

Figure 12-1 Mars viewed from the Earth This high-quality Earth-based photograph of Mars was taken a few days after the opposition of 1971 when the Earth–Mars distance was only 56¼ million km (35 million miles). At that time, Mars presented a disk nearly 25 arc sec in angular diameter. Circumstances as favorable as these will not be repeated for the rest of the century. This photograph portrays what you might typically see through a moderate-sized telescope under excellent observing conditions. Notice the prominent southern polar cap. (Courtesy of S. M. Larson)

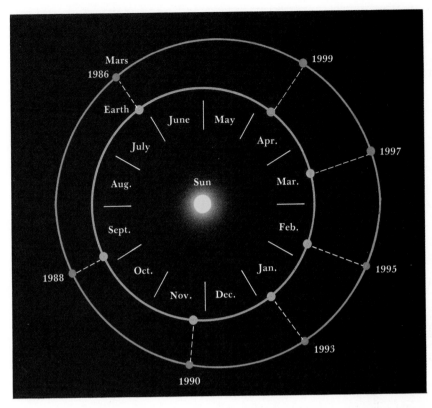

Figure 12-2 The orbits of Earth and Mars The Earth catches up with Mars every 780 days (nearly 2⅓ years). These close encounters, properly called oppositions, are the best times to observe the red planet. Because Mars's orbit is noticeably elongated, however, some oppositions afford better views than others. The last three favorable oppositions of the twentieth century occurred in 1971, 1986, and 1988. The next favorable oppositions will occur in 2003 and 2005.

tures **canali**, an Italian term meaning *natural water channels*, which was soon mistranslated into English as *canals*. The alleged discovery of canals seemed to imply the existence on Mars of intelligent creatures capable of substantial engineering feats. This speculation led Percival Lowell, who came

from a wealthy Boston family, to finance a major new observatory near Flagstaff, Arizona, primarily for additional studies of Mars. By the end of the nineteenth century, Lowell had reported observations of 160 Martian canals (see Figure 12-3).

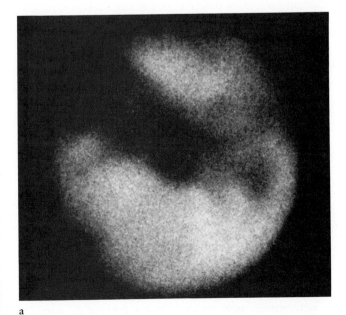

a

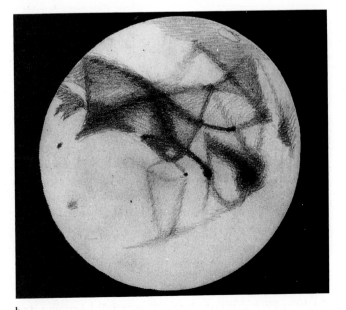

b

Figure 12-3 Martian canals During moments of "good seeing," when the air is exceptionally calm and clear, some Earth-based observers reported seeing networks of lines crisscrossing the Martian surface. Some people suggested that these lines, called canals, are irrigation ditches constructed by intelligent creatures. This photograph (**a**) and the corresponding drawing (**b**) were made during the opposition of 1926. (Lick Observatory)

Box 12-1 Mars data

Mean distance from the Sun:	$1.524 \text{ AU} = 2.279 \times 10^8$ km
Maximum distance from the Sun:	$1.666 \text{ AU} = 2.491 \times 10^8$ km
Minimum distance from the Sun:	$1.382 \text{ AU} = 2.067 \times 10^8$ km
Mean orbital velocity:	24.1 km/s
Sidereal period:	686.98 days = 1.88 years
Rotation period:	$24^h\ 37^m\ 23^s$
Inclination of equator to orbit:	25° 11′
Inclination of orbit to ecliptic:	1° 50′ 59″
Orbital eccentricity:	0.093
Diameter:	6794 km
Diameter (Earth = 1):	0.532
Apparent diameter as seen from Earth:	maximum = 25.7″
	minimum = 3.5″
Mass:	6.42×10^{23} kg
Mass (Earth = 1):	0.107
Mean density:	3940 kg/m³
Surface gravity (Earth = 1):	0.380
Escape velocity:	5.0 km/s
Oblateness:	0.006
Surface temperatures:	maximum = 20°C = 70°F = 293 K
	minimum = −140°C = −220°F = 133 K
Albedo:	0.15
Brightest magnitude:	−2.0
Mean diameter of Sun as seen from Mars:	21′

Not all astronomers saw the canals. In 1894, the American astronomer Edward Emerson Barnard, working at the Lick Observatory, complained that "to save my soul I can't believe in the canals as Schiaparelli draws them." Incidentally, Barnard was apparently the first to report seeing craters on Mars. However, he did not publish these observations, for fear of ridicule.

The skepticism of cautious observers like Barnard was drowned out by the more fanciful pronouncements of Lowell and his colleagues. It soon became fashionable to speculate that the Martian canals formed an enormous planetwide irrigation network to transport water from melting polar caps to vegetation near the equator. In view of the planet's reddish, desertlike appearance, Mars was thought to be a dying planet whose inhabitants must have to go to great lengths to irrigate their farmlands. No doubt the Martians would readily abandon their arid ancestral homeland and invade the Earth for its abundant resources. Hundreds of science-fiction stories and dozens of monster movies owe their existence to the canali of Schiaparelli.

12-2 Space probes to Mars found craters, volcanoes, and canyons— but no canals

Three American spacecraft journeyed past Mars in the 1960s and sent back pictures that clearly showed numerous flat-bottomed craters (see Figure 12-4), but not one canal. Schiaparelli's canali had been an optical illusion.

From studies of the relatively crater-free lunar maria, scientists knew that the major crater-making era in our solar system had ended 3 billion years ago. Like the Moon, the Martian surface must therefore be extremely ancient and have apparently experienced few changes over the past 3 billion years. The prospect of abundant Martian life suddenly seemed extremely remote.

Astronomers were quick to point out that the partially eroded, flat-bottomed appearance of the Martian craters is probably a result of dust storms that rage across the planet's surface. From time to time over the past two centuries,

Box 12-2 Martian oppositions, 1990–1999

Five Martian oppositions occur between 1990 and 1999, as listed below. Only the 1990 opposition afforded good Earth-based views of Mars. None of the remaining four will be particularly favorable. The previous favorable Martian opposition occurred in September 1988, when Mars and Earth were separated by only 58 million km. The next favorable oppositions will occur in 2003 and 2005. A star chart showing the path of Mars during the 1993 opposition is given in Figure 4-1.

Date of opposition		Earth–Mars distance (millions of km)	Apparent diameter (arc sec)	Magnitude
1990	November 27	77	18	−1.7
1993	January 7	93	15	−1.2
1995	February 12	100	14	−1.0
1997	March 17	97	14	−1.1
1999	April 24	85	16	−1.5

Figure 12-4 Martian craters Numerous flat-bottomed craters are seen in this mosaic of about 100 images taken by a Viking orbiter in 1980. North is at the top. The large, heavily cratered circular area in the upper half of this view is known as Arabia. The dark area to the right of Arabia, called Syrtis Major, can be seen from Earth during favorable oppositions. Carbon-dioxide snow covers the floors of craters near the Martian south pole. (NASA, USGS)

Figure 12-5 **Mariner 9** *The 1-ton Mariner 9 spacecraft went into orbit about Mars in November 1971. It operated flawlessly for nearly a full year, sending back a total of 7329 close-up pictures of the Martian surface and the two tiny Martian moons. (NASA)*

astronomers have watched the faint surface markings disappear under a reddish-orange haze as thin Martian winds stirred up finely powdered dust. Some of these storms actually obscure the entire planet. Over the ages, deposits of dust from these storms in the craters have given the craters their characteristic flat-bottomed shape.

The early missions also confirmed that the Martian atmosphere is primarily composed of carbon dioxide. In 1947 the American astronomer Gerard P. Kuiper had detected spectral lines of this gas in the spectrum of sunlight reflected from Mars, but data from spacecraft were required before it could be demonstrated that carbon dioxide forms 95 percent of the Martian atmosphere. These measurements also indicated that the Martian atmosphere is extremely thin. Atmospheric pressure on the Martian surface is typically in the range of 5 to 10 millibars, which is roughly the same as the air pressure at an altitude of 30 km (100,000 ft) above the Earth's surface. This extremely low atmospheric pressure was unexpected, especially because of the huge dust storms

visible from Earth. The material whipped up by the rarified Martian winds must be extremely fine-grained powder. This assumption is consistent with the fact that, over 3 billion years, numerous dust storms have not completely obliterated the Martian craters.

The flybys of the 1960s gave us only fleeting glimpses of a small fraction of the Martian surface; the best was yet to come. After traveling 400 million kilometers, *Mariner 9* went into orbit about Mars on November 13, 1971 (see Figure 12-5). To everyone's dismay, Mars happened to be embroiled in a planetwide dust storm, which completely obscured the surface. After a few weeks, however, as the dust started to subside, scientists realized that they were looking down on several volcanoes far taller and broader than any comparable protrusions on Earth.

The largest Martian volcano, Olympus Mons, rises 24 km (15 miles) above the surrounding plains—nearly three times the height of Mount Everest (see Figure 12-6). The highest volcano on Earth, Mauna Loa in the Hawaiian

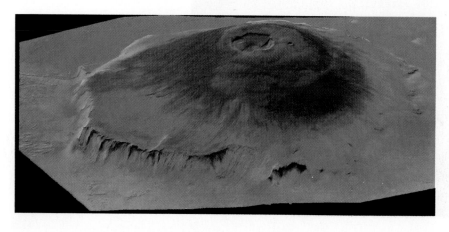

Figure 12-6 **Olympus Mons** *This computer-enhanced view looks eastward across the largest volcano on Mars. Olympus Mons is nearly three times as tall as Mount Everest. Its cliff-ringed base measures 600 km (370 mi) in diameter. (NASA, USGS)*

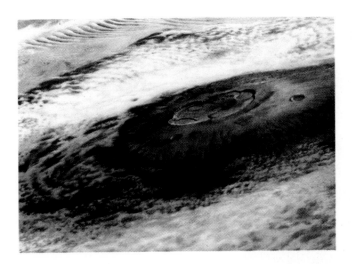

Figure 12-7 *The Olympus caldera* *This view of the summit of Olympus Mons is based on a mosaic of six pictures taken by one of the Viking orbiters. The caldera consists of overlapping, volcanic craters. It measures roughly 70 km across. The volcano is wreathed in mid-morning clouds formed from ice (H_2O) brought upslope by cool air currents. The cloudtops are about 8 km below the volcano's peak. (NASA)*

Islands, has a summit only 8 km above the ocean floor. Both the Martian volcanoes and the Hawaiian Islands were formed by hot-spot volcanism. Recall, however, that the process of plate tectonics also affects the Hawaiian Islands. The huge size of Olympus Mons strongly suggests an absence of plate tectonics on Mars. Instead, a hot spot in Mars's mantle probably kept pumping lava upward through the same vent for millions of years, producing one giant volcano rather than a long chain of smaller ones.

The base of Olympus Mons is ringed with cliffs, has a diameter of nearly 600 km (370 mi), and covers an area as big as the state of Missouri. The volcano's summit has collapsed to form a volcanic crater, called a **caldera**, large enough to contain the state of Rhode Island (Figure 12-7).

Olympus Mons is one of several large volcanoes clustered together on a huge bulge called the Tharsis Rise, which covers an area 2500 kilometers in diameter centered just north of the Martian equator (see Figure 12-8). Ground levels over this vast dome-shaped region are typically 5 to 6 kilometers higher than the average ground level for the rest of the planet. Another major, though less dramatic, grouping of

Figure 12-8 *The Valles Marineris hemisphere of Mars* *This mosaic composed of 102 Viking orbiter images covers nearly a full hemisphere of the planet. North is at the top. Three large volcanoes that dominate the Tharsis Rise seen at the left. The center of this view shows the entire Valles Marineris canyon system, over 3000 km long and up to 8 km deep. Note the vast riverbed that begins in the north-central canyons and meanders northward to a darkish area called Acidalia Planitia. (NASA, USGS)*

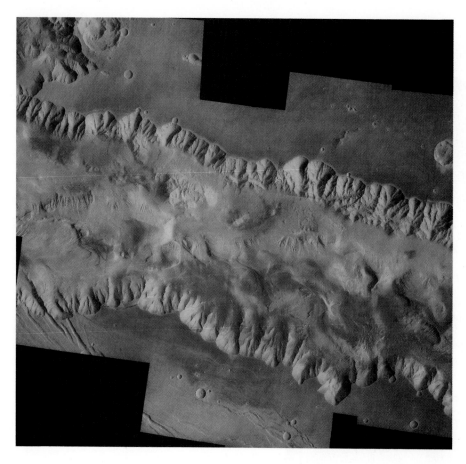

Figure 12-9 *A segment of Valles Marineris This mosaic of Viking orbiter photographs shows a segment of Valles Marineris called Candor Chasm. The canyon is about 100 km wide in this region. The canyon floor has two major levels. The northern (upper) canyon floor is 8 km beneath the surrounding plateau, whereas the southern canyon floor is only 5 km below the plateau. (NASA, USGS)*

volcanoes is located nearly on the opposite side of the planet from Tharsis, in a region called Elysium. None of these volcanoes is active today.

For reasons we do not yet understand, most of the volcanoes on Mars are in the northern hemisphere, but most of the impact craters are in the southern hemisphere. Between the two hemispheres, *Mariner 9* discovered a vast canyon running roughly parallel to the Martian equator (see Figures 12-8 and 12-9). If this canyon were located on Earth, it could stretch all the way from New York to Los Angeles. In honor of the Mariner spacecraft that revealed so much of the Martian surface, this enormous chasm has been designated Valles Marineris.

Valles Marineris stretches over 3000 km, beginning with heavily fractured terrain in the west and ending with ancient cratered terrain in the east. Many geologists suspect Valles Marineris to be a fracture in the Martian crust caused by internal stresses. It may be like the rift valleys on Earth, such as the Red Sea (recall Figure 8-15), that result from plate tectonics.

Plate tectonics did not significantly affect the surface of Mars, because the planet cooled rapidly during its early history. In Chapter 10, we learned that small bodies lose heat much more rapidly than large bodies. Mars's diameter is only half that of the Earth, and its mass is just one-tenth of the Earth's mass (only about twice the mass of Mercury). Plate tectonics on Mars bogged down early in the planet's history as the rapidly cooling lithosphere became thick and sluggish.

12-3 Surface features indicate that water once flowed on Mars

Valles Marineris was not formed by water erosion, but *Mariner 9* photographs did reveal many features that look like dried-up riverbeds (see Figure 12-10). Because craters overlap many of the riverbeds, scientists estimate these erosional features to be about 4 billion years old.

The Martian riverbeds were totally unexpected, because liquid water cannot exist today on Mars. Water is liquid over only a certain range of temperatures and pressures: At low temperatures, water becomes ice; at high temperatures, it becomes steam. The atmospheric pressure above a body of water also affects the state of the water. If the pressure is very low, molecules escape easily from the liquid's surface,

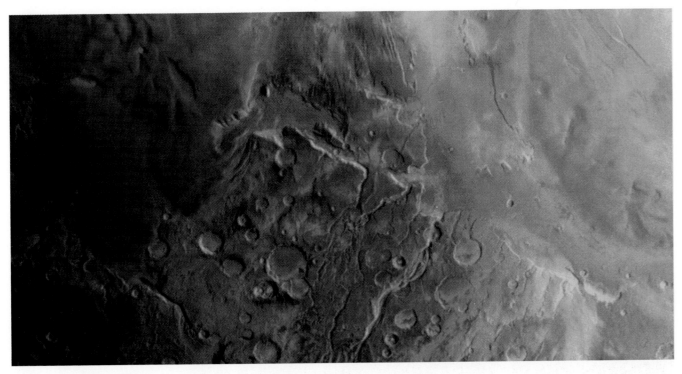

Figure 12-10 *Ancient river channels on Mars* *Many dried riverbeds cut across this cratered terrain, located only 5° south of the Martian equator. The existence of riverbeds on Mars suggests that the planet's atmosphere was thicker and its climate more Earthlike long ago, so that* *liquid water could flow across its surface. Any water exposed to the sparse Martian atmosphere today would rapidly boil away. (NASA, USGS)*

causing the water to vaporize. Any liquid water on Mars would boil furiously and evaporate immediately into the thin Martian air. Data from *Mariner 9* furthermore demonstrate that the Martian atmosphere is very dry; it never rains on Mars. If all the water vapor could somehow be squeezed out of the Martian atmosphere, it would not fill one of the five Great Lakes in North America.

If water once flowed on Mars, where might that water be today? Scientists immediately looked to the Martian polar caps as a possible frozen reservoir. The winter temperature at the Martian poles is low enough ($-140°C = -220°F$) that carbon dioxide freezes. It was not clear, therefore, how much of the polar caps consists of frozen carbon dioxide (dry ice) and how much is water ice.

In 1972, *Mariner 9* sent back photographs from orbit as summer came to the northern hemisphere of Mars. During the Martian spring, the polar cap had receded rapidly. This rapid shrinkage strongly suggested that a thin layer of carbon dioxide frost might be evaporating quickly in the sunlight. However, with the arrival of summer, the rate of recession slowed abruptly, suggesting that a thicker layer of water ice had been exposed. Scientists concluded that the **residual polar caps** that survive through the Martian summers contain a large quantity of frozen water (see Figure 12-11). Calculating the volume of water ice in the residual

Figure 12-11 *The north polar ice cap* *This mosaic of Viking images shows the northern polar cap of Mars in midsummer. Most of the carbon-dioxide frost that covers northern latitudes during the winter has evaporated away. The residual polar cap shown here is mostly water ice. The reddish streaks consist of dust and dark bluish areas are sand dunes forming a huge "collar" that surrounds the polar cap. (NASA, USGS)*

Figure 12-12 An ancient dried lake? *The layered terrain visible in the scene may be the residue of a huge ancient lake. Note the erosional patterns that emanate from the cliffsides at the lower left and flow across the canyon floor. The area shown in this mosaic of Viking orbiter images is about the same size as the state of Ohio. (NASA, USGS)*

caps is difficult, however, because we do not know how thick the layer of ice is. Scientists are thus still debating the exact amount of frozen water stored in the Martian polar caps.

Another important clue about water on Mars came in 1976 from a pair of spacecraft called the Viking orbiters. Many Viking photographs show erosional features that look like the results of flash floods, suggesting that water once raged across the Martian surface in great torrents. Close-up views show that these flash-flood features usually emerge from craters or from collapsed, jumbled terrain. It therefore seems likely that frozen water exists in a layer of **permafrost**

under the Martian surface, similar to that beneath the tundra in the far northern regions on Earth. Apparently, heat from a meteoroid impact or from volcanic activity in the distant past occasionally melted this subsurface ice. The ground then collapsed and millions of tons of rock pushed the water to the surface (see Figure 12-12).

From the widths and depths of the Martian flash-flood channels, scientists estimate peak flood discharges of 10^7 to 10^9 m³/s. For comparison, the average discharge of the Amazon is 10^5 m³/s. The largest known flash flood on the Earth occurred about 2 million years ago in eastern Washington, when a natural dam gave way. Its peak discharge is estimated to have been 10^7 m^3/s. Most of the Martian flash floods therefore must have been greater than anything known on Earth.

Viking photographs also showed evidence of gradual water flow on Mars. Intricate branched patterns of dry stream beds and delicate channels meandering among flat-bottomed craters strongly suggest that water once flowed leisurely across the Martian surface. All these erosional features, from flash-flood channels to meandering stream beds, clearly indicate that the Martian climate must have been much warmer and more Earthlike 4 billion years ago.

12-4 Earth and Mars began with similar atmospheres that evolved very differently

How could the Martian surface have been warm enough to permit substantial amounts of flowing water? The likeliest mechanism is the greenhouse effect (recall Figure 11-5). The atmospheres of both Earth and Mars are transparent to sunlight, so they are not heated directly by the Sun. Instead, these atmospheres are heated from below, by energy reradiated from the ground at longer, infrared wavelengths. Water vapor and carbon dioxide absorb these wavelengths, trapping the heat and warming the atmosphere.

The average worldwide temperature of the Earth's atmosphere is increased by about 33°C through the greenhouse effect. But this effect is today not very efficient in warming the sparse Martian atmosphere, which permits infrared radiation to escape into space. If the greenhouse effect was once important on Mars, the carbon-dioxide content of the Martian atmosphere must then have been much higher than it is today. Calculations indicate that 1000 millibars of carbon dioxide would have been needed to sustain a greenhouse effect capable of permitting free-flowing water. Thus, the primordial atmosphere of Mars would have been comparable in density and pressure to the Earth's atmosphere today.

We have seen that the terrestrial planets obtained their atmospheres primarily through volcanic outgasing (recall Figure 11-13). Water vapor, carbon dioxide, and nitrogen are among the most common gases in volcanic vapors. These three gases probably constituted the bulk of the primordial atmospheres of both Earth and Mars 4 billion years ago.

On Earth, most of the water is in the oceans. Nitrogen, which is not chemically very reactive, is still in the atmosphere. Carbon dioxide dissolves in water and thus rain washed carbon dioxide out of the air. The dissolved carbon dioxide reacted with rocks in rivers and streams, and the combined residue was ultimately deposited on the ocean floors, where it turned into carbonate rocks such as limestone. Thus, most of the Earth's carbon dioxide is tied up in the Earth's crust. Plate tectonics cycles carbonate rocks through volcanoes, which then continually reintroduce car-

bon dioxide back into the atmosphere. Without volcanoes, rain would wash all the carbon dioxide from the Earth's air in only a few thousand years. Of course, the oxygen in Earth's atmosphere is the result of photosynthesis by plant life (see Figure 12-13a).

On Mars, rainfall 4 billion years ago washed much of the planet's carbon dioxide from its atmosphere and probably led to the creation of carbonate rocks in which the carbon dioxide is today chemically bound. Thus, on both Earth and Mars, carbonate rocks are significant reservoirs of the primordial carbon dioxide outgased from ancient volcanoes. But plate tectonics did not occur on Mars, and its carbonate rocks were not recycled through volcanoes. Martian rainfall therefore permanently removed carbon dioxide from the atmosphere. According to calculations by James B. Pollack of NASA's Ames Research Center, a 1000-millibar carbon-dioxide Martian atmosphere could have survived for only 10 to 100 million years. However, volcanic activity early in Mars's history must have recycled some carbon dioxide, thus prolonging the greenhouse effect for perhaps half a billion years. Ultimately, rainfall did succeed in destroying the greenhouse effect that had allowed water to exist in the first place. In effect, water on Mars led to its own disappearance.

As both the water vapor and the carbon dioxide became depleted, ultraviolet light from the Sun could penetrate the thinning Martian atmosphere to strip it of nitrogen. Nitrogen molecules (N_2) normally do not have enough thermal energy to escape from Mars, but they can acquire the needed energy through a process called **dissociative recombination**. First, an electron is knocked out of a nitrogen molecule by an energetic ultraviolet photon. Then the electron reunites with the molecule and gives up the energy it acquired from the ultraviolet photon, thereby breaking the molecule into two atoms and giving each of these atoms enough speed to escape from the planet's gravity. Carbon dioxide, which is also affected by this photochemical reaction, experiences "nonthermal escape" from the Martian atmosphere (see Figure 12-13b).

12-5 The Viking landers sent back close-up views of the Martian surface

The discovery that water once existed on the surface of Mars rekindled speculation about Martian life forms. By the mid-1960s, although it was clear that Mars has neither civilizations nor fields of plants, some sort of microbial life forms still seemed possible. Searching for Martian microbes was one of the main objectives of the ambitious and highly successful Viking missions.

The two Viking spacecraft, which were launched from Earth during the summer of 1975, arrived at Mars almost a year later. Each spacecraft consisted of two modules: an orbiter and a lander. After landing sites had been chosen, each

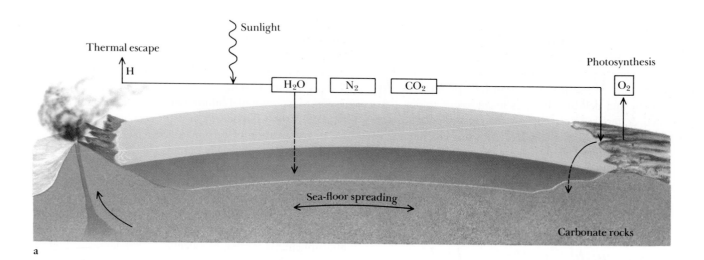

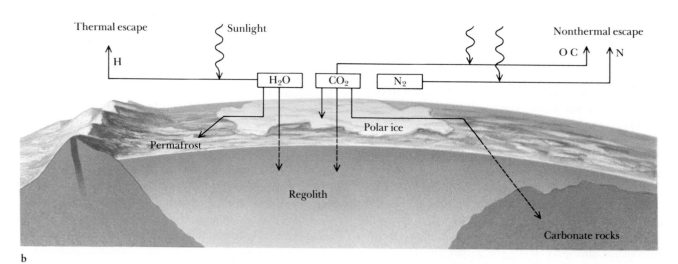

Figure 12-13 The atmospheres of Earth and Mars The primitive atmospheres of Earth and Mars were probably similar, but they evolved differently. (a) On Earth, water condensed into oceans, and nitrogen remained in the air, while carbon dioxide was removed from the atmosphere by rainfall and then stored in carbonate rocks. (b) On Mars, rainfall destroyed the thick carbon-dioxide atmosphere that kept the planet warm enough (through the greenhouse effect) for liquid water to exist. Today water ice is found in the polar caps and the Martian permafrost. Some water, along with carbon dioxide, may also be chemically bound in the fine-grained Martian regolith. Ultraviolet light from the Sun is responsible for breaking apart both carbon dioxide and nitrogen molecules into their constituent atoms, which then escape into space. (Adapted from R. M. Haberle)

Viking lander separated from its orbiter and began a descent to the planet's surface (see Figure 12-14). A heat shield, a parachute, and retrorockets slowed the landers for a gentle touchdown. As shown in Figure 12-15, each lander is roughly the size of a small automobile.

The *Viking 1* landing site was a rocky field called Chryse Planitia (the Golden Plains) at latitude 22° north of the Martian equator. The *Viking 2* lander set down in Utopia Planitia (the Utopian Plains) at latitude 48° north of the Martian equator, nearly on the opposite side of the planet from Chryse.

The *Viking 1* site at Chryse is a moderately cratered plain near the mouth of a large flash-flood channel. The upraised rim of a crater is visible on the horizon in Figure 12-16. Such craters are probably the source of the jagged rocks that litter the scene. Some light-colored bedrock is exposed near the horizon.

Some small sand drifts were seen at the *Viking 1* site. Careful scrutiny of pictures taken over several months reveals that these drifts are pushed around slightly by the Martian winds. New rocks are occasionally exposed and others are covered up, providing what may be the best explanation

Figure 12-14 The Viking descent This drawing schematically shows the sequence of events during the descent of one of the Viking spacecraft on Mars. A heat shield, a parachute, and retrorockets were used to ensure a gentle landing. (NASA)

for the seasonal changes observed from Earth for the past two centuries. Seasonal winds transporting fine red dust alternately expose, then cover dark rocks, mimicking vegetational color changes over the long Martian year.

A quirk of human color vision is probably responsible for the greenish hues reported for the dark surface markings by many Earth-based observers. When a neutral grey area is viewed next to a bright red-orange area, the eye perceives the grey to have a blue-green cast, the so-called complementary colors of red-orange. Perhaps the wishful thinking of astronomers envisioning Martian vegetation also contributed to observations of greenish surface patterns on Mars.

Figure 12-15 The Viking lander From the base of its foot pads to the top of its antenna, each Viking lander is about 2 m (6 ft) tall and weighs 576 kg (1270 lb) without fuel. An extendable arm for retrieving soil samples is visible in the foreground. The two cylinders (one with an American flag) protruding from the top of the lander contain the two television cameras. The dish antenna was used to relay data to the Viking orbiter overhead, which in turn sent the information back to Earth. (NASA)

Figure 12-16 The Viking 1 site This view looks southward over the Chryse Planitia (the Golden Plains). The horizon is about 3 km (2 mi) from the spacecraft. Part of the upraised rim of a crater on the horizon is visible at the upper left. A few light-colored patches of exposed bedrock also appear in this picture. (NASA)

Figure 12-17 *A panorama of the* **Viking 2 site** *This mosaic of three summertime views shows about 180° of the Utopian plains. Northwest is at the left; southeast on the right. The flat, featureless horizon is approximately 3 km (2 mi) away from the spacecraft. (NASA)*

The *Viking 2* site at Utopia is situated approximately 1500 kilometers (930 miles) closer to the north pole than is Chryse. As shown in Figure 12-17, Utopia is remarkably crater-free, although many of the rocks in this view may have been ejected from a large crater about 200 kilometers east of the spacecraft.

12-6 Instruments on the Viking landers sent back detailed information about the Martian climate

In all the pictures sent back from Mars, the sky has a distinctly pinkish-orange tint. This coloration is thought to be caused by extremely fine-grained dust suspended in the Martian atmosphere.

Both landers carried equipment to measure meteorological conditions on Mars. These instruments promptly confirmed that the surface atmospheric pressure is in the range of 6 to 8 millibars, as had been expected from previous missions. The atmospheric pressure at sea level on Earth is about 1000 millibars, so the density of the Martian atmosphere is less than 1 percent that of the Earth's atmosphere. Direct chemical analysis of the Martian atmosphere also confirmed its high carbon-dioxide content: The Viking instruments reported a carbon-dioxide abundance of 95 percent, nitrogen 2.7 percent, and argon 1.6 percent. The remaining fraction of a percent is mostly oxygen and carbon monoxide, with a small amount of water vapor.

Above the Martian surface, atmospheric density decreases with increasing altitude, much as it does on Earth. The atmospheric temperature also declines with increasing altitude, but the temperature profile of the Martian atmosphere does not have the pronounced swings in temperature that characterize the Earth's atmosphere (see Figure 12-18).

Atmospheric temperature on Earth (recall Figure 8-18) exhibits a maximum at an altitude of 50 km, because of the absorption of ultraviolet radiation by our ozone layer. We may thus conclude that Mars has no ozone layer to protect its surface from the Sun's ultraviolet rays.

After only a few weeks on the Martian surface, the Viking landers provided data showing clearly that the atmospheric pressure at both landing sites was dropping steadily. Mars seemed to be rapidly losing its atmosphere, leading some scientists to joke that soon all the air would be gone. A straightforward explanation was, however, readily available: Winter was coming to the southern hemisphere. At the Martian south pole, it became so cold that large amounts of carbon dioxide began solidifying out of the atmosphere, covering the ground with dry-ice snow (see Figure 12-19). The condensation of dry-ice snow was removing gas from the

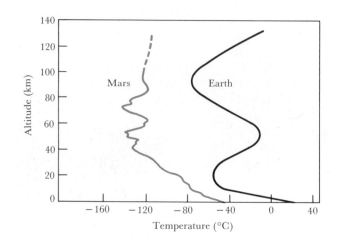

Figure 12-18 *Temperature of the Martian atmosphere* Unlike Earth, *Mars lacks an ozone layer in its atmosphere. Consequently, the temperature profile of the Martian atmosphere does not show a maximum at 50 km altitude, as does the Earth's temperature profile.*

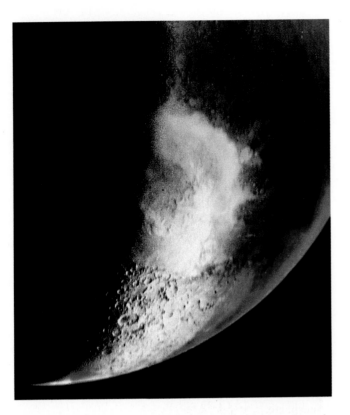

Figure 12-19 The south pole of Mars *During the Martian winter, the temperature drops so low that carbon dioxide freezes out of the Martian atmosphere. A thin coating of carbon-dioxide frost covers a heavily cratered region around Mars's south pole. Seasonal freezing and evaporation of carbon dioxide at both poles causes planetwide variations in atmospheric pressure. (NASA)*

Martian atmosphere, thereby causing the atmospheric pressure to decline.

Several months later, when spring came to the southern hemisphere, the dry-ice snow rapidly evaporated and the atmospheric pressure returned to its pre-winter levels. When winter arrived in the northern hemisphere, another decrease in atmospheric pressure was observed as dry-ice snow blanketed the northern latitudes (see Figure 12-20). Thus it is clear that the Martian surface experiences extreme seasonal variations in temperature and atmospheric pressure.

Summer comes to the southern hemisphere of Mars at the same time that the planet makes its closest approach to the Sun. Along the retreating edge of the polar cap, the sharp temperature contrast between the evaporating dry ice and the soil warmed by the sunshine gives rise to strong winds. These winds often produce swirling dust storms.

Aside from an occasional dust storm, the weather on Mars may seem boring to someone accustomed to the temperate zones of Earth. Mars's atmospheric pressure varies with the seasons in a regular and predictable fashion. The atmospheric temperature varies with the time of day, again in a monotonously repetitive way. Actually, the daily temperature variation on Mars is quite similar to the temperature changes observed in a desert on Earth. The warmest time of day occurs about two hours after local noon, with the lowest temperatures being recorded just before sunrise. The thin, dry Martian atmosphere is not capable of retaining much of the day's heat, however, so the range of daily temperatures on Mars is larger than that on Earth (see Figure 12-21).

Figure 12-20 Winter in Mars's northern hemisphere *This picture, taken during midwinter, shows a thin frost layer that lasted for about a hundred days at the Viking 2 lander site. Freezing carbon dioxide adheres to water-ice crystals and dust grains in the atmosphere, causing them to fall to the ground. The sky is therefore not as pink as it was in the summertime view of Figure 12-17. (NASA)*

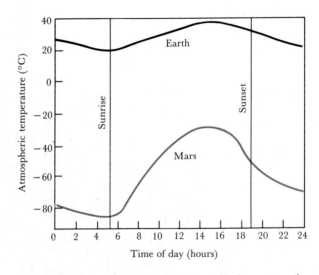

Figure 12-21 Diurnal temperature variation *The lower curve shows the daily temperature variation on Mars at the Viking 1 site. The upper curve shows the daily temperature variation at a desert site in California. The daily temperature range on Mars is about three times greater than that on Earth. The thin, dry Martian air does not retain heat as well as the Earth's atmosphere does.*

12-7 Geological analysis by the Viking landers showed that the Martian regolith contains abundant iron

Just as meteorologists were delighted with the success of their atmospheric measurements, geologists were pleased with the chemical analysis that the Viking landers performed on the Martian regolith. Some of the most important data came from a device called an X-ray fluorescence spectrometer, which measured the chemical composition of the rocks at both lander sites and showed them to be rich in iron. Specifically, iron and silicon comprise about two-thirds of a Martian rock's content. Surprisingly high concentrations of sulfur were also found. Sulfur is 100 times more abundant in Martian rocks than in terrestrial rocks.

Each Viking lander had a scoop at the end of a mechanical arm to obtain regolith and rock samples for analysis (see Figure 12-22). Bits of the regolith were observed to cling to a magnet mounted on the scoop, indicating that a significant fraction of the regolith (3 to 7 percent by weight) is magnetic material, possibly the iron-rich mineral maghemite. In general, the Martian regolith at both lander sites can best be described as an iron-rich clay. The familiar reddish color of Mars may in fact be caused by rust (iron oxides) in the regolith.

The high iron content of the Martian crust is interesting, because Mars has a lower average density (3940 kg/m^3) than the other terrestrial planets (more than 5000 kg/m^3 for Mercury, Venus, and Earth). It appears that Mars's total iron content is less than that of the other terrestrial planets and what iron Mars does have is distributed throughout the body of the planet rather than being concentrated in a dense core. In other words, Mars apparently did not undergo chemical differentiation. The reasons for this unusual his-

tory are poorly understood, and geologists hoped to obtain important information about the interior of Mars from seismometers on the Viking landers. Unfortunately, one of the seismometers failed to deploy. With only one properly functioning seismometer, geologists were unable to pinpoint the locations and depths of marsquakes, so the structure of the Martian interior remains largely unknown.

The absence of an Earth-like iron-rich core in Mars seems consistent with its almost total absence of a planetwide magnetic field. Soviet missions demonstrated that Mars's magnetic field is less than 0.004 times as strong as the Earth's field. Mars rotates nearly as fast as the Earth, but an Earth-like magnetic dynamo definitely is not operating inside Mars. If Mars does have an iron-rich core, it is reasonable to assume that the core is not molten nor, as indicated above, very large.

Mars's magnetic field is so weak that it is largely ineffectual in warding off the solar wind. In fact, the solar wind may actually impinge directly on the outermost layers of the Martian atmosphere, just as it does on Venus's (recall Figure 11-6). Unfortunately, our knowledge of this aspect of the Martian environment is woefully incomplete. An alternative possibility is that Mars's weak field just manages to carve out a magnetosphere that barely encloses the planet's atmosphere.

12-8 Biological experiments failed to detect any conclusive evidence of life on Mars

The meteorological and geological equipment on the Viking landers dramatically broadened our understanding of the Martian environment, but most of the anticipation and ex-

Figure 12-22 Digging in the Martian soil *Viking's mechanical arm with its small scoop protrudes from the right side of this view on the Chryse Plains. Figure 12-14 shows the fully extended arm. Several small trenches dug by the scoop in the Martian regolith appear near the left side of the picture. (NASA)*

citement centered on the Viking biological experiments. Each lander carried a compact biological laboratory designed to perform three different tests for microorganisms in the Martian soil. Additional equipment was expected to shed light on the possibility of Martian life.

For example, the television cameras might reveal some life forms. Biologists joked about "titanium mushrooms" or other strange creatures that the biological tests might fail to detect yet that should be classified as living because they grew and multiplied. However, careful scrutiny of the thousands of pictures sent back over many months failed to reveal objects or changes that could be attributed to biological processes.

The three biological experiments carried out by the Viking landers were based on the general proposition that living things must alter their environment. They eat, they breathe, and they give off waste products. In each of the three experiments, a sample of the Martian regolith was placed in a closed container, with or without a certain nutrient substance. The container was then examined for any changes in its contents.

The **gas-exchange experiment** was designed to detect any processes that might be broadly classifiable as respiration. A small sample of regolith was placed in a sealed container along with a controlled amount of gas and nutrients. The gases in the container were then monitored to see if their chemical composition changed.

The **labeled-release experiment** was designed to detect processes resembling metabolism. A small sample of regolith was moistened with nutrients containing radioactive carbon atoms. Any organisms in the regolith should eat the food, then emit gases containing the telltale radioactive carbon.

The **pyrolytic-release experiment** was designed to detect photosynthesis, the biological process by which plants on Earth synthesize organic compounds from carbon dioxide, using sunlight as an energy source. In the Viking experiments, a regolith sample was placed in a container along with radioactive carbon dioxide and exposed to artificial sunlight. If plantlike photosynthesis were to occur, some of the radioactive carbon from the gas should become incorporated into the microorganisms in the regolith.

The first data returned by the Viking biological experiments caused great excitement: In almost every case, rapid and extensive changes were detected inside the sealed containers. However, further analysis of the data led to the conclusion that these changes were due solely to nonbiological chemical processes. It appears that the Martian regolith is rich in chemicals that effervesce (fizz) when moistened. A large amount of oxygen is apparently tied up in the regolith in the form of unstable chemicals called peroxides and superoxides that break down in the presence of water to release oxygen gas.

A machine impressively called a gas-chromatograph mass spectrometer performed detailed chemical analyses of the Martian soil. At both landing sites, surprising results were obtained: Absolutely no organic compounds were detected in the soil. This finding is strong evidence that life has never existed on Mars, because this equipment should at least have detected the corpses of Martian microorganisms or their waste products. This negative result is even more surprising because meteorites falling on Mars should have contaminated the planet's surface with an easily detectable amount of organic material. Yet the equipment, which was undoubtedly in perfect working order, failed to find any organic compounds whatsoever. This discovery was our first hint that natural chemical processes on Mars may have literally sterilized the planet's surface.

The chemical reactivity of the Martian regolith may be a direct result of the absence of a protective ozone layer in the Martian atmosphere. Ultraviolet radiation from the Sun beats down on the Martian surface, easily dissociating molecules of carbon dioxide (CO_2) and water vapor (H_2O) by knocking off oxygen atoms, which then become loosely attached to chemicals in the regolith. Alternatively, this process can produce ozone (O_3) and hydrogen peroxide (H_2O_2), which also become incorporated in the regolith. In all these cases, the loosely attached oxygen atom makes the regolith extremely reactive in the presence of water.

Here on Earth, hydrogen peroxide is commonly used as an antiseptic. When you pour this liquid on a wound, it fizzes and froths as the loosely attached oxygen atoms chemically combine with organic material, thereby destroying germs. The Viking landers may have failed to detect any organic compounds on Mars because the superoxides and peroxides in the regolith make it literally antiseptic.

12-9 The two Martian moons resemble asteroids

Two tiny moons move around Mars in orbits close to the planet's surface. These satellites are so small that they were not discovered until the favorable opposition of 1877. While Schiaparelli was seeing canals, the American astronomer Asaph Hall spotted these two moons, which orbit very close to the planet and almost directly above its equator. He named them Phobos ("fear") and Deimos ("panic"), after the mythical horses that drew the chariot of the Greek god of war, Mars.

Phobos is the inner and larger of the Martian moons. It circles Mars in only 7 hours 39 minutes, moving over the Martian surface at an average distance of only 6000 kilometers (3700 miles). (Recall that our Moon orbits at an average distance of 376,280 km above the Earth's surface.) This orbital period is much shorter than the Martian day; Phobos rises in the west and gallops across the sky in only $5\frac{1}{2}$ hours, as viewed by an observer near the Martian equator. During this time, Phobos appears several times brighter in the Martian sky than Venus does from Earth.

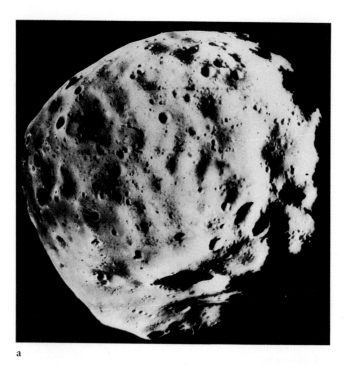

a

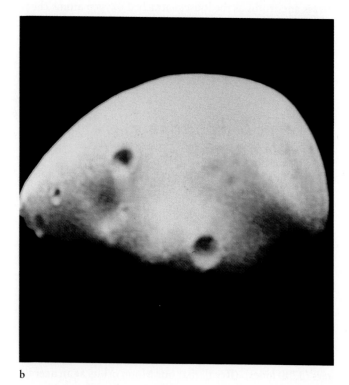

b

Figure 12-23 *Phobos and Deimos* *Viking orbiters provided these images of Mars's tiny moons.* (a) *Phobos, the larger of Mars's two moons, is potato-shaped and measures roughly 28 by 23 by 20 km (about 17 by 14 by 12 mi).* (b) *Deimos seems to be less cratered than Phobos and measures roughly 16 by 12 by 10 km (about 10 by 7 by 6 mi).* (NASA)

Deimos, which is farther from Mars and somewhat smaller than Phobos, appears about as bright from Mars as Venus does from Earth. Deimos's orbit, 20,000 kilometers (12,400 miles) above the Martian surface, is at almost the right distance for a synchronous orbit—an orbit in which a satellite seems to hover above a single location on the planet's equator. As seen from the Martian surface, Deimos rises in the east and takes about three full days to creep slowly from one horizon to the other.

The *Mariner 9* mission treated astronomers to their first close-up views of the Martian satellites. Numerous additional views provided by the Viking orbiters revealed Phobos and Deimos to be jagged, heavily cratered, football-shaped rocks (see Figure 12-23). Phobos was found to be roughly 28 by 23 by 20 km, with Deimos being slightly smaller, at roughly 16 by 12 by 10 km. It was further revealed that both Phobos and Deimos rotate synchronously about Mars, always keeping their same sides facing the red planet.

The origin of the Martian moons is unknown, but it is possible that they are captured asteroids. Mars is quite near the asteroid belt, where thousands of large rocks orbit the Sun. Like Phobos and Deimos, many asteroids have low albedos (typically less than 10 percent) because of a high carbon content. Perhaps two of these asteroids wandered close enough to Mars to become permanently trapped by the planet's gravitational field. Future measurements of the surface composition of the Martian moons may shed light on their origin.

12-10 More Martian probes are needed to answer important questions about Mars and about Earth

Both the Soviet Union and the United States are planning additional missions to Mars. The next American mission will use a small satellite called the *Mars Observer*. The launch is anticipated in the 1990s. Using remote-sensing techniques, the *Mars Observer* will analyze the mineral composition of the Martian surface and also determine the role of water in the Martian climate. The results could lead to a deeper understanding of the Earth's weather and geology.

Although the possibility of Martian organisms seems quite remote, some scientists point out that the polar regions of Mars may have conditions more suitable for life. These regions could be explored by another mission to the Martian surface that might also send back data about the Martian interior. Ideally, such an ambitious undertaking would utilize a rover vehicle to send back data and pictures from a wide range of sites. Another possibility is to land a spacecraft that would scoop up some Martian rocks and regolith, blast off, and bring the samples back to Earth for laboratory analysis. The Soviet Union has been particularly successful with missions of this type to the Moon. Both Soviet and American scientists are also planning manned missions to Mars, with the ultimate goal of establishing a manned base there in the twenty-first century. NASA's long range plans call for sending four people to Mars in 2011 for a 600-day stay on the red planet, with their return to Earth scheduled in 2018. A courageous mission of this magnitude would realize within our lifetimes some of the great dreams of science-fiction writers, as well as scientists.

Key words

caldera

canali

dissociative recombination

favorable opposition

gas-exchange experiment

labeled-release experiment

permafrost

pyrolytic-release experiment

residual polar cap

Key ideas

- Earth-based observers found that the Martian solar day is nearly the same as that on the Earth, that Mars has polar caps that expand and shrink with the seasons, and that the Martian surface undergoes seasonal color changes. A few observers reported a network of linear features called canals; these observations led to many speculations about Martian life.

  The best Earth-based views of Mars are obtained at favorable oppositions, when Mars is simultaneously at opposition and near perihelion.

- Spacecraft have returned close-up views showing that the Martian surface has numerous flat-bottomed craters, several huge volcanoes, a vast canyon, and dried-up riverbeds—but no canals.

- The Martian atmosphere is composed mostly of carbon dioxide; the atmospheric pressure on the surface there is about 1 percent that produced by the Earth's atmosphere and shows seasonal variations.

  Great dust storms sometimes blanket Mars. Fine-grained dust in its atmosphere gives the Martian sky a pinkish-orange tint.

  Mars has no ozone layer, so its surface is exposed to ultraviolet radiation.

- The flash-flood features and dried riverbeds on the Martian surface indicate that water once flowed on Mars.

  Liquid water would quickly boil away in Mars's thin atmosphere, but its polar caps contain frozen water, and a layer of permafrost may exist beneath the Martian regolith.

  Mars's primordial atmosphere probably contained enough carbon dioxide to support a greenhouse effect that permitted liquid water to exist on the planet's surface.

- The winter expansion of Mars's polar caps is caused by deposition of a thin layer of frozen carbon dioxide (dry ice).

- Mars apparently never experienced chemical differentiation; the planet might not have an iron-rich core.

- Mars has two small, football-shaped satellites that move in orbits close to the surface of the planet; they may be captured asteroids.

- Chemical reactions in the Martian regolith, together with ultraviolet radiation from the Sun, apparently act to sterilize the Martian surface.

- Little is known about the interior of Mars; even the absence of a dense core has not been confirmed by seismic studies.

Review questions

1 Why is Mars red?

2 Explain why Mars has the longest synodic period of all the planets, although its sidereal period is only 687 days.

3 Compare the cratered regions of Mercury, the Moon, and Mars. Assuming that the craters on all three worlds

originally had equally sharp rims, what can you conclude about the environmental histories of these worlds?

4   Explain why the Earth's sky is blue, that of Venus is yellow, and that of Mars is pink.

5   How would you tell which craters on Mars had been formed by meteoritic impacts and which by volcanic activity?

6   Compare the volcanoes of Venus, Earth, and Mars. Cite evidence that hot-spot volcanism is or was active on all three worlds.

7   What geologic features (or lack thereof) on Mars have convinced scientists that plate tectonics did not shape significantly the Martian surface?

8   What two types of Martian terrain suggest that Mars may have had a denser atmosphere in the past than it does today? What third type of terrain suggests that this episode of a denser atmosphere must have been comparatively short-lived?

9   Why is it reasonable to assume that the primordial atmospheres of Venus, Earth, and Mars were roughly the same?

10   Compare the evolution of the atmospheres of Venus, Earth, and Mars. Explain how the atmospheres of these three planets turned out so differently from each other, even though they all started with roughly the same chemical composition.

11   With carbon dioxide being just about as abundant in the Martian atmosphere as in the Venusian atmosphere, why do you suppose there is no greenhouse effect on Mars today?

12   Why do you suppose that Phobos and Deimos are not round like our Moon?

13   What is the current status of the issue of life on Mars? Do you think Mars might be as barren and sterile as the Moon?

14   Imagine you are an astronaut living at a base on Mars. Describe your day's activities, what you see, the weather, the spacesuit you are wearing, and so on.

Advanced questions

> **Tips and tools . . .**
> You will need to remember the small angle formula, given in Box 1-1. Also, you may need to review the form of Kepler's third law, stated in Box 4-3, that explicitly includes mass.

*15   (Basic)   Suppose you have a telescope with a resolving power of 1 arc sec. What is the smallest feature you could see on the Martian surface during the opposition of 1993? Suppose you had access to the Hubble Space Telescope, which has a resolving power of 0.1 arc sec. What is the smallest feature you could see on Mars with the HST during the 1993 opposition?

16   Imagine you are on Mars, looking back at the Earth. Assuming the human eye has a resolving power of 50 arc sec, would you see the Earth and Moon as a "double star"?

*17   (Challenging)   Use the orbital data for Phobos to calculate the mass of Mars. How does your answer compare with the mass of Mars given in Box 12-1?

18   The synodic period of Mars is 780 days, but the dates given in Box 12-2 show that the intervals between successive oppositions vary. Explain this apparent contradiction.

19   Is it reasonable to suppose that the polar regions of Mars might harbor life forms, even though the Martian regolith is sterile at the Viking lander sites?

Discussion questions

20   Suppose someone told you that the Viking mission failed to detect life on Mars simply because the tests were designed to detect terrestrial life forms, not Martian life forms. How would you respond?

21   Do you think sedimentary rocks might exist on Mars? Is it possible that sedimentary rocks might contain fossils of early Martian life forms? What sort of adaptations would these early life forms have had to make in order to survive to the present time?

22   Compare the scientific opportunities for long-term exploration offered by the Moon and Mars. What difficulties would be inherent in establishing a permanent base or colony on each of these two worlds?

Observing project

23   Consult such magazines as Sky & Telescope or Astronomy to determine Mars's location among the constellations. If Mars is suitably placed for observation, arrange to view the planet through a telescope. Draw a picture of what you see. What magnifying power seems to give you the best image? Can you distinguish any surface features? Can you see a polar cap or dark markings? If not, can you offer an explanation for Mars's bland appearance?

For further reading

Batson, R., et al., eds. *The Atlas of Mars*. NASA SP-438, 1979 • This NASA atlas contains photographs and maps of Mars produced by the U.S. Geological Survey.

Boston, P., ed. *The Case for Mars*. Univelt/American Astronautical Society, 1984 • These proceedings of a 1981 conference discuss the prospects and technology for a manned mission to Mars.

Burgess, E. *To the Red Planet*. Columbia University Press, 1978 • This book offers a fine summary of the Viking missions to Mars.

Carr, M. "The Surface of Mars: A Post-Viking View." *Mercury*, January/February 1983 • This article by the leader of the Viking Orbiter Imaging Team gives an excellent summary of what we have learned about the Martian surface.

Cooper, H. *The Search for Life on Mars*. Holt, Rinehart, & Winston, 1980 • This book by a noted journalist profiles the Viking missions to Mars.

Ezell, E., and Ezell, L. *On Mars: Exploration of the Red Planet, 1958–1978*. NASA SP-4212, 1984 • This nontechnical history of Mars exploration provides a detailed summary of the engineering, science, politics, and people involved in the Mariner and Viking projects.

Gore, R. "Sifting for Life in the Sands of Mars." *National Geographic*, January 1977 • This beautifully illustrated article summarizes the findings of the Viking missions.

Haberle, R. M. "The Climate of Mars." *Scientific American*, May 1986 • This article argues that Mars's climate was once like Earth's early climate, but it evolved differently.

Hartmann, W. K. "What's New on Mars?" *Sky & Telescope*, May 1989 • This fine article by a noted planetary scientist discusses the current state of our knowledge about Mars.

Horowitz, N. H. *To Utopia and Back*. W. H. Freeman and Company, 1986 • This excellent book discusses the search for life on Mars.

Parker, D., et al. "Mars' Grand Finale," *Sky & Telescope*, April 1989 • This article by members of the Association of Lunar and Planetary Observers contains many fine photographs of Mars taken during the 1988 opposition, whose favorable circumstances will not be repeated until 2003.

Pollack, J. "The Atmospheres of the Terrestrial Planets." In Beatty, J., et al., eds. *The New Solar System*, 2nd ed. Cambridge University Press, 1982 • This chapter compares the chemical composition, circulation, and evolution of the atmospheres of Earth, Venus, and Mars.

Schiltz, P. "Polar Wandering on Mars." *Scientific American*, December 1985 • The author of this intriguing article argues that Mars's entire lithosphere has shifted, moving polar regions toward the equator.

Spitzer, C., ed. *Viking Orbiter Views of Mars*. NASA SP-441 • This magnificent album from NASA contains both color and black and white photographs of the Martian surface.

Toon, O. B. "How Climate Evolved on the Terrestrial Planets." *Scientific American*, February 1988 • This superb article compares the evolution of the atmospheres of Venus, Earth, and Mars.

Jupiter with Io and Europa *Jupiter is the largest and most massive planet in the solar system. Jupiter's diameter equals 11¼ Earth diameters, and its mass equals 318 Earth masses. However, Jupiter is composed primarily of hydrogen and helium in abundances similar to the Sun's chemical composition. This photograph from Voyager 1 in 1979 shows Europa (right) and Io (in front of Jupiter). Both satellites are nearly the same size as our Moon. The Great Red Spot, a persistent storm in the Jovian atmosphere about three times the size of the Earth, is on the left. (NASA)*

13

Jupiter:
Lord of the Planets

Jupiter is an active, vibrant, multicolored world, more massive than all the other planets combined. Jupiter has a distinctly striped appearance when seen through a telescope, because its rapid rotation stretches its weather systems completely around the planet. Some features in the Jovian cloudcover are fleeting, whereas others, like the Great Red Spot, seem to endure year after year. Four spacecraft have given us brief but revealing close-up views of Jupiter's colorful, turbulent clouds and probed its enormous magnetosphere. Jupiter's hydrogen-rich interior is so highly compressed by the planet's gravity that the gas has become a metal, capable of conducting electricity and supporting the planet's vast magnetic field. The Galileo spacecraft, now on its way to Jupiter, will send a probe into the Jovian atmosphere to give us crucial information about the largest planet in the solar system.

More than any other single factor, it was the temperature distribution throughout the young solar nebula that dictated the final forms and natures of the planets that orbit the Sun. In the warm inner regions of this ancient nebula, the dust grains that survived the heat consisted primarily of metals, silicates, and oxides. The temperature was simply too high to allow substantial condensation of such volatile substances as water, methane, and ammonia. The four planets that formed close to the Sun were therefore composed almost entirely of rocky material. Their surface gravities were too low and their surface temperatures too high to retain any of the abundant—but lightweight—hydrogen and helium gases that made up most of the solar nebula.

The orbit of Jupiter, however, is about five times as large as the orbit of the Earth, and sunlight reaching Jupiter is only $\frac{1}{25}$ as bright as that reaching the Earth. In the past, as now, it was cold at this distance of $\frac{1}{2}$ billion miles from the Sun (recall Figure 7-16). In the young solar nebula, the dust grains at this distance from the protosun were covered with thick, frosty coatings of frozen water, methane, and ammonia. These volatile substances thus became important constituents of the planets that accreted in the outer reaches of our solar system.

Many astronomers think that the Jovian planets were formed in a two-step process. First, accretion of ice-coated dust grains led fairly quickly to formation of a large protoplanet having at least several times the mass of the Earth. Then the strong gravitational pull of this protoplanet attracted and retained substantial quantities of the hydrogen and helium existing in the cool outer reaches of the solar nebula. The Canadian-American astrophysicist Alastair Cameron has carried out calculations to show that this gathering of lightweight gases would have become very efficient and rapid after the protoplanet had grown beyond a certain mass.

This process created the largest planet in our solar system—a planet whose bulk is composed of approximately 82 percent hydrogen, 17 percent helium, and only 1 percent all other elements (by weight). This composition is probably close to the original composition of the solar nebula and is comparable to the composition of the Sun's atmosphere today (recall Table 7-4). Saturn—and, to a lesser extent, Uranus and Neptune—are thought to have formed in much the same way as Jupiter.

13-1 Huge, massive Jupiter is composed largely of lightweight gases

Jupiter is huge: Its mass is 318 times greater than the mass of the Earth. Indeed, the mass of Jupiter is $2\frac{1}{2}$ times the combined masses of all the other planets, satellites, asteroids, meteoroids, and comets in the solar system. Jupiter's equatorial diameter is $11\frac{1}{4}$ times as large as the Earth's. Jupiter's

volume is about 1430 times larger than the Earth's. (Box 13-1 lists basic data about Jupiter.)

Jupiter's average density can be computed from its mass and size: It is only 1330 kg/m^3. This low average density is entirely consistent with our picture of a huge sphere of hydrogen and helium compressed by its own gravity. In fact, the Swiss-American astronomer Rupert Wildt first suggested back in the 1930s that Jupiter is mostly hydrogen, basing this hypothesis largely upon the calculated average density of the planet. Unfortunately, it is difficult to detect hydrogen and helium in Jupiter's atmosphere, because these molecules do not produce prominent spectral lines at wavelengths visible in the light reflected from the planet. Consequently, evidence to confirm the composition of Jupiter was slow in coming.

Wildt did find prominent spectral lines of methane (CH_4) and ammonia (NH_3) in the spectrum of sunlight reflected from Jupiter. That each of these molecules contains three or four hydrogen atoms was one bit of evidence leading Wildt to suspect a great abundance of hydrogen in Jupiter's atmosphere. However, at the low temperatures in Jupiter's upper atmosphere, hydrogen exists only as H_2 molecules, which produce only weak spectral lines (they were finally detected in 1960). Conclusive confirmation of the existence of helium in the Jovian atmosphere was not obtained until the 1970s, when infrared instruments on space probes passing near the planet detected evidence of collisions there between hydrogen molecules and helium atoms. This indirect evidence finally confirmed Wildt's brilliant hypothesis about the composition of Jupiter's atmosphere, which is today known to consist of approximately 86 percent hydrogen, 14 percent helium, with very small amounts of methane (0.09 percent), ammonia (0.02 percent), water vapor (0.008 percent), and other gases.

Through an Earth-based telescope, Jupiter appears as a colorful, intricately banded sphere, as shown in Figure 13-1. Figure 13-2 is a close-up view from the spacecraft *Voyager 1*. The most prominent features are the alternating dark and light bands parallel to Jupiter's equator that are shaded in subtle tones of red, orange, brown, and yellow. The dark, reddish bands are called **belts**, the light-colored bands **zones**. These features are not the only conspicuous markings. A large, red-orange oval called the **Great Red Spot** is often visible in Jupiter's southern hemisphere. This remarkable feature, which has been observed since the mid-1600s, appears to be a long-lived storm in the planet's dynamic atmosphere. Many careful observers have also reported smaller spots and blemishes that last for only a few weeks or months in Jupiter's turbulent clouds.

The best time to observe Jupiter is, of course, when the planet is at opposition. At such times, Jupiter outshines all the stars in the sky (its brightest magnitude is −2.6). Through a telescope, Jupiter presents a disk nearly 50 arc sec in diameter—roughly twice the angular diameter of Mars under the most favorable conditions. Jupiter takes almost a

Figure 13-1 *Jupiter from Earth* *Various belts and zones are easily identified in this Earth-based view of the largest planet in our solar system. The Great Red Spot was exceptionally prominent when this photograph was taken. (Courtesy of S. Larson)*

Figure 13-2 *Jupiter from* **Voyager 1** *This view was sent back from Voyager 1 in 1979, when the spacecraft was only 30 million kilometers from Jupiter. Features as small as 600 km across can be seen in the turbulent cloudtops of this giant planet. Complex cloud motions surround the Great Red Spot. (NASA)*

dozen years to orbit the Sun, so this planet appears to meander across the 12 constellations of the zodiac at the rate of approximately one per year. Successive oppositions occur at intervals of about 13 months (see Box 13-2).

13-2 Details of Jupiter's rotation give clues about the planet's internal structure

Although it is the largest and most massive planet in our solar system, Jupiter has the fastest rate of rotation. At its equatorial latitudes, Jupiter completes a full rotation in only 9 hours 50 minutes 30 seconds. However, Jupiter is not a solid, rigid object. By observing features in the belts and zones, we see that the polar regions of the planet rotate a little more slowly than do the equatorial regions. Near the poles, the rotation period of Jupiter's atmosphere is about 9 hours 55 minutes 41 seconds. The first person to notice this **differential rotation** of Jupiter was the Italian astronomer Giovanni Domenico Cassini in 1690. You may recall that Cassini is the same gifted observer who gave us the first accurate determination of Mars's rotation rate.

Jupiter's colorful cloudtops are the turbulent, uppermost layer of the planet's thick atmosphere. Are the observed ro-

tation rates of these clouds anything like the rotation rates of deeper levels, or that of a solid central core? Intriguing clues come from the fact that Jupiter emits radio waves. In particular, radio emissions from Jupiter with wavelengths in the range of 3 to 75 cm vary slightly in intensity with a period of 9 hours 55 minutes 30 seconds. As we shall see later, this radio emission is directly associated with Jupiter's magnetic field, which is anchored deep inside the planet. Thus, radio observations reveal Jupiter's **internal rotation period**, which is slightly different from the atmospheric rotation we can observe through a telescope.

Jupiter's rapid rotation profoundly affects the overall shape of the planet. Even a casual glance through a small telescope shows that Jupiter is slightly flattened, or **oblate**. The diameter across Jupiter's equator (142,800 km) is 6.5 percent larger than its diameter from pole to pole (133,500 km). Thus Jupiter is said to have an **oblateness** of 6.5 percent, or 0.065.

If Jupiter were not rotating, it would be a perfect sphere. A massive, nonrotating object naturally settles into a spherical shape in which every atom on its surface experiences the same intensity of gravity aimed directly at the object's center. However, because Jupiter is rotating, every part of the planet also experiences an outward-directed "centrifugal force" that is proportional to each part's distance from the axis of rotation. Equatorial regions are farther from the planet's axis of rotation than are polar regions; hence the

Box 13-1 Jupiter data

Mean distance from the Sun:	5.203 AU = 7.783×10^8 km
Maximum distance from the Sun:	5.455 AU = 8.157×10^8 km
Minimum distance from the Sun:	4.951 AU = 7.409×10^8 km
Mean orbital velocity:	13.1 km/s
Sidereal period:	11.86 years
Rotation period:	equatorial = $9^h \, 50^m \, 30^s$
	internal = $9^h \, 55^m \, 30^s$
Inclination of equator to orbit:	3° 04′
Inclination of orbit to ecliptic:	1° 18′
Orbital eccentricity:	0.048
Diameter:	equatorial = 142,800 km
	polar = 133, 500 km
Diameter (Earth = 1):	equatorial = 11.19
	polar = 10.47
Apparent diameter as seen from Earth:	maximum = 50.1″
	minimum = 30.4″
Mass:	1.90×10^{27} kg
Mass (Earth = 1):	317.8
Mean density:	1314 kg/m^3
Surface gravity (Earth = 1):	2.53
Escape velocity:	60 km/s
Oblateness:	0.065
Mean surface temperature (at cloudtops):	−110°C = −166°F = 163 K
Albedo:	0.52
Brightest magnitude:	−2.7
Mean diameter of Sun as seen from Jupiter:	6′ 09″

planet's equatorial diameter is slightly larger than its polar diameter. This centrifugal stretching of the equatorial dimensions gives Jupiter its characteristic oblate shape.

At every point throughout Jupiter, the inward force of gravity is exactly balanced by the outward pressure of the compressed material plus the outward centrifugal effects of the planet's rotation. This balance is called **hydrostatic equilibrium**. The shape and density of a planet adjust themselves to ensure that this balance is always maintained.

Jupiter's shape is an excellent indicator of its internal structure. Two planets with the same mass, same average density, and same rotation will have a slightly different degree of oblateness if one planet has a compact core but the other does not. The less-dense material surrounding a compact core necessarily extends to a greater distance from the planet's center and therefore experiences more deformation

by "centrifugal forces" than does the material of a planet with a uniform distribution of matter. All other things being equal, therefore, a planet with a dense core will be more oblate than will a planet without a compact central core.

Detailed calculations by William Hubbard of the University of Arizona and Soviet scientists V. Zharkov and V. Trubitsyn strongly suggest that 4 percent of Jupiter's mass is concentrated in a dense rocky core. In other words, Jupiter's oblateness is consistent with its having a rocky core nearly 13 times as massive as the entire Earth. Some of this core was probably the original "seed" around which the proto-Jupiter accreted. However, additional amounts of rocky material were added later by meteoritic material falling onto the planet.

Although Jupiter's rocky core alone is 13 times as massive as the entire Earth, it is probably not much bigger than our

Box 13-2   Oppositions of Jupiter, 1991–1999

The synodic period of Jupiter is 398.9 days, or approximately a year and a month. This interval between successive oppositions is obvious in the following table, which lists the oppositions during the last decade of this century.

Date		Angular diameter (arc sec)	Magnitude
1991	January 28	45.7	−2.1
1992	February 28	44.6	−2.0
1993	March 30	44.2	−2.0
1994	April 30	44.5	−2.0
1995	June 1	45.6	−2.1
1996	July 4	47.0	−2.2
1997	August 9	48.6	−2.4
1998	September 16	49.7	−2.5
1999	October 23	49.8	−2.5
2000	November 28	48.5	−2.4

planet. The tremendous crushing weight of the remaining bulk of Jupiter—equal to the mass of 305 Earths—compresses Jupiter's core down to a sphere 20,000 km in diameter (Earth's diameter is 12,800 km). The pressure at Jupiter's center is about 80 million atmospheres, meaning that the rocky material of Jupiter's core is squeezed to a density of about 20,000 kg/m^3. The corresponding temperature at the planet's center is probably about 25,000 K. In contrast, the temperature at Jupiter's cloudtops is only 165 K.

13-3 Radio observations suggested Jupiter's magnetic field and metallic hydrogen interior

In the 1950s, astronomers began discovering radio emissions coming from Jupiter. Most of this radiation is **thermal**; that is, it comes from the planet's surface and is exactly what would be expected from a warm object. However, some of the radiation is **nonthermal**. It is found in two broad wavelength ranges. At wavelengths of a few meters are sporadic bursts of **decametric** ("ten meter") **radiation** that probably is caused by electrical discharges associated with powerful electric currents in Jupiter's ionosphere. As we shall see in Chapter 14, these discharges may be the result of complex electromagnetic interactions between Jupiter and its large satellite, Io. At wavelengths of a few tenths of a meter (par-

ticularly 3 to 75 cm) is the **decimetric** ("tenth meter") **radiation** that revealed Jupiter's internal rotation period. This decimetric radiation has also provided important clues about the interior of Jupiter's enormous bulk between its rocky core and its colorful cloudtops. A radio view of Jupiter is seen in Figure 13-3.

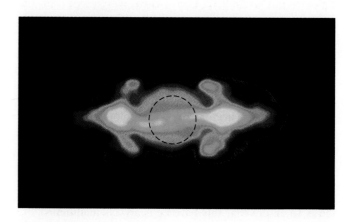

Figure 13-3 A radio view of Jupiter The Very Large Array (VLA) was used to produce this map of synchrotron emission from Jupiter at a wavelength of 21 cm. The image is about five Jupiter diameters wide and is elongated parallel to the planet's magnetic equator. Jupiter is at the center of the image, indicated by the dashed circle. The strongest emission, shown in white, comes from a ring of trapped electrons encircling Jupiter similar to the Van Allen belts that surround the Earth. (NRAO)

The first person to explain the source of decimetric radiation was the Soviet astrophysicist Iosif Shklovskii. In the 1950s, Shklovskii argued convincingly that decimetric radiation is characteristic of radio waves emitted by high-speed electrons moving through a magnetic field like the one surrounding Jupiter. Astronomers now realize that this process is very important in a wide range of circumstances throughout the universe, from pulsars and quasars to the radio emissions of entire galaxies. Whenever electrons traveling near the speed of light encounter a magnetic field, they spiral around that field and emit copious radio waves. Energy produced in this fashion is called **synchrotron radiation.**

If some of Jupiter's radiation is indeed synchrotron radiation, then Jupiter must have a magnetic field. From the intensity of this type of radiation, it was soon apparent that Jupiter's magnetic field must be extremely strong, roughly 19,000 times stronger than Earth's. Additional calculations showed that a molten iron center inside Jupiter's rocky core could not possibly have created such an enormous planetwide magnetic field. What then could be the source of Jupiter's magnetism?

The answer lies in the fact that Jupiter is largely composed of hydrogen. A hydrogen atom consists of a single proton orbited by a single electron. Deep inside Jupiter, pressures are so great that the electrons are stripped from their protons. The result is a mixture of protons and electrons. The electrons, no longer bound to their protons, are free to wander around on their own. Their motion, directed by Jupiter's rotation, creates an electric current, just as the ordered movement of electrons in a copper wire constitutes an electric current. In other words, the highly compressed hydrogen deep inside Jupiter behaves like a metal; thus it is called **liquid metallic hydrogen.**

Detailed calculations by the American scientists Edwin Salpeter and David Stevenson strongly suggest that molecular hydrogen is transformed into liquid metallic hydrogen when the pressure exceeds 3 million atmospheres. This transition occurs at a depth of about 20,000 km below Jupiter's cloudtops. Thus, the internal structure of Jupiter consists of three distinct regions (see Figure 13-4): a rocky core, surrounded by a 40,000-km-thick layer of liquid metallic hydrogen, surrounded in turn by a 20,000-km-thick layer of ordinary molecular hydrogen. The colorful cloud patterns that we can see through telescopes are located in the outermost 100 km of the exterior layer of molecular hydrogen.

It is clear from Figure 13-4 that the vast majority of Jupiter's enormous bulk is electrically conductive liquid metal. Because of Jupiter's rapid rotation, electric currents in this thick layer of liquid metallic hydrogen generate a powerful magnetic field, in much the same way that liquid portions of the Earth's core produce the Earth's magnetic field. Thus, the Jovian magnetic field is so much stronger than ours partly because Jupiter's liquid metallic region is so much larger than the Earth's core and partly because Jupiter rotates so much faster than the Earth.

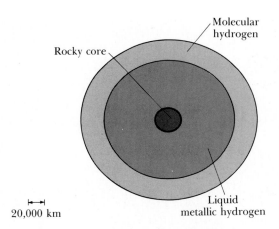

Figure 13-4 The structure of Jupiter *Most scientists believe that Jupiter probably has a rocky core approximately 20,000 km in diameter. This core is surrounded by a 40,000-km-thick layer of liquid metallic hydrogen, which is responsible for Jupiter's powerful magnetic field. The outer, 20,000-km-thick, layer is composed primarily of ordinary hydrogen.*

13-4 Spacecraft mapped details of Jupiter's enormous magnetosphere

Jupiter's powerful magnetic field surrounds the planet with an enormous magnetosphere, one large enough to envelop the orbits of many of its moons. From the Earth, however, our evidence of this magnetosphere is limited to the faint hiss of radio static. For many years astronomers realized that our understanding and appreciation of this extraordinary planet would be vastly increased if we could send spacecraft past Jupiter to probe its magnetosphere and view its dynamic atmosphere at close range. This program was carried out by four historic missions to Jupiter during the 1970s.

The first Jupiter flyby occurred in December 1973, when *Pioneer 10* coasted over the planet's cloudtops (see Figure 13-5). An identical spacecraft, *Pioneer 11*, followed in December 1974. Each spacecraft carried an array of instruments to detect charged particles, associated radiation, and magnetic fields. The Pioneer missions therefore gave us our first look at the Jovian magnetosphere and revealed its awesome dimensions. The shock wave that surrounds Jupiter's magnetosphere is nearly 30 million kilometers across. In other words, if you could see Jupiter's magnetosphere from Earth, it would cover an area in the sky 16 times larger than the full Moon.

Pioneer 10 and *11* crossed Jupiter's magnetopause several times, reporting the boundary at distances ranging from 3 to 7 million kilometers above the planet. Evidently, Jupiter's magnetosphere is sensitive to fluctuations in the solar wind. It expands and contracts by a factor of two, rapidly and frequently. In contrast, dramatic changes in the size of the Earth's magnetosphere are extremely rare, in spite of occasional "gusts" in the solar wind.

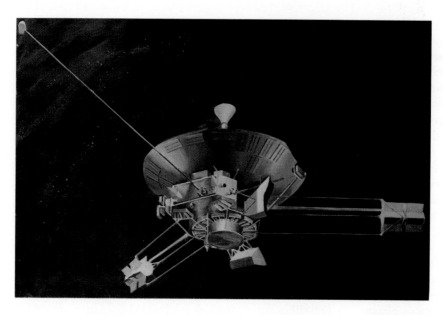

Figure 13-5   The Pioneer spacecraft   This drawing shows one of two identical spacecraft that flew past Jupiter in the 1970s. Instead of using solar panels, the spacecraft drew its power from two electric generators containing radioactive isotopes. Numerous instruments on board made measurements of Jupiter's magnetic environment and sent back pictures of Jupiter's clouds. (NASA)

The inner regions of our magnetosphere are dominated by two huge Van Allen belts (recall Figure 8-20) that are filled with charged particles. A similar entrapment of charged particles occurs around Jupiter. In addition, however, Jupiter's rapid rotation spews the particles out into a huge electrically charged **current sheet** (see Figure 13-6). This current sheet lies in the plane of Jupiter's magnetic equator. Jupiter's magnetic axis is inclined 11° from the planet's axis of rotation. The orientation of Jupiter's magnetic field is the reverse of Earth's—a compass would point toward the south on Jupiter.

With planets such as the Earth and Mercury, the magnetopause occurs where the pressure of the planet's magnetic field counterbalances the pressure of the solar wind. However, in 1979 the flybys of spacecraft *Voyager 1* and *2* demonstrated that this is not the case with Jupiter.

The Voyager detectors indicated that the inner regions of Jupiter's magnetosphere contain a hot, gaslike mixture of

charged particles called a **plasma.** A plasma is formed when a gas is heated to such extremely high temperatures that electrons are torn off the atoms of the gas. As a result, a plasma is an electrically neutral mixture of positively charged ions and negatively charged electrons. The plasma that envelops Jupiter consists primarily of electrons and protons, with some ions of helium, sulfur, and oxygen. These charged particles are caught up in Jupiter's rapidly rotating magnetic field and accelerated to extremely high speeds.

The Voyager instruments relayed a startling discovery: The plasma surrounding Jupiter has an astoundingly high temperature of 300 million to 400 million K. The Jovian plasma is the hottest place in the solar system! Even the center of our Sun is only about 20 million K. However, the Jovian plasma is very thin: The average density in the hottest regions is about one charged particle per 100 cm³.

Because of its high temperature, Jupiter's huge plasma has a great deal of energy and exerts pressure that holds off

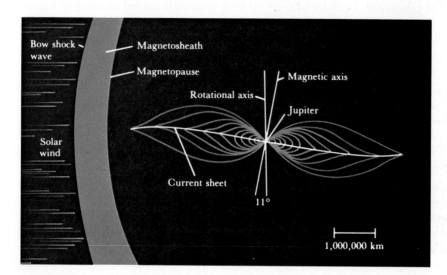

Figure 13-6   Jupiter's magnetosphere   Like other planetary magnetospheres, Jupiter's has a bow shock wave, a magnetosheath, and a magnetopause. In Jupiter's case, gas pressure from a hot plasma keeps the magnetosphere inflated, holding off the solar wind. Particles trapped inside Jupiter's magnetosphere are spewed out into a vast current sheet by the planet's rapid rotation. Jupiter's axis of rotation is inclined to its magnetic axis by about 11°.

the solar wind. The Voyager data suggest that the pressure balance between the solar wind and the hot plasma inside the Jovian magnetosphere is precarious. A gust in the solar wind can blow away some of the plasma, at which point the magnetosphere deflates rapidly, to as little as one-half its original size. Charged-particle detectors carried by the Voyager spacecraft recorded several bursts of hot plasma that may have been associated with magnetospheric deflations. However, additional electrons and ions accelerated by Jupiter's rotating magnetic field soon replenish the plasma and the magnetosphere expands again.

13-5 Pictures taken during flybys show many details in Jupiter's clouds

The most dramatic data from the Pioneer and Voyager flybys were spectacular close-up color pictures of Jupiter's dynamic atmosphere. The two projects carried very different imaging systems. The Pioneer spacecraft rotated rapidly for stability, so its pictures of Jupiter had to be assembled from a series of thin strips produced as the cameras of the whirling spacecraft scanned across the planet's face. The Voyager spacecraft, designed several years later, had more sophisticated technology and employed television systems. As expected, the Voyager pictures were superior to the Pioneer views in resolution, color discrimination, and detail.

Even after allowing for differences in the photographic equipment used, it was obvious that Jupiter's atmosphere underwent some fundamental changes during the four years between the Pioneer and Voyager flybys (see Figure 13-7).

The changes in Jupiter's cloud cover are most apparent in the area surrounding the Great Red Spot. Over the past three centuries, Earth-based observers have reported many long-term variations in the spot's size and color. At its largest, the Great Red Spot measures 40,000 by 14,000 km, so large that three Earths could fit side by side across it. At other times (as in 1976–1977), the spot almost faded from view. During the Voyager flybys of 1979, the Great Red Spot was only slightly larger than the Earth.

Figure 13-8 shows two contrasting views of the Great Red Spot. During the Pioneer flybys, the Great Red Spot was embedded in a broad white zone that dominated the planet's southern hemisphere. By the time of the Voyager missions a

Pioneer 10, December 1973

Pioneer 11, December 1974

Voyager 1, March 1979

Voyager 2, July 1979

Figure 13-7 *Four views of Jupiter These four close-up pictures of Jupiter span 5½ years and show major changes in the planet's upper atmosphere. Noticeable changes are apparent even in the Voyager views, which were separated by only four months. All four pictures show the Great Red Spot. The black dot on the* Pioneer 10 *view is the shadow of Jupiter's moon Io. (NASA)*

a December 1974   **b** July 1979

few years later, however, the cloud structure had changed dramatically. A dark belt had broadened and encroached on the Great Red Spot from the north, and the entire region was now apparently embroiled in much greater turbulence.

Careful examination of cloud motions in and around the Great Red Spot reveal that the spot rotates counterclockwise with a period of about six days. Furthermore, winds to the north of the spot are blowing to the west, but winds south of the spot are moving toward the east. The circulation around the Great Red Spot is thus like a wheel spinning between two oppositely moving surfaces (see Figure 13-9). This surprisingly stable wind pattern has survived for at least three centuries.

13-6 Infrared observations probed the vertical structure of Jupiter's atmosphere

Infrared data from the Pioneer and Voyager spacecraft confirmed an extraordinary fact that had earlier been deduced from Earth-based observations: Jupiter emits more energy

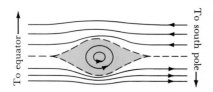

Figure 13-9 *Circulation around the Great Red Spot* *The Great Red Spot spins counterclockwise, completing a full revolution in about six days. Meanwhile, the winds to the north and south of the spot blow in opposite directions. Consequently, circulation associated with the spot resembles a wheel spinning between two oppositely moving surfaces. (Adapted from A. P. Ingersoll)*

than it receives from the Sun—in fact, nearly twice as much. Many scientists believe that this excess heat escaping from Jupiter is energy left over from the formation of the planet 4.5 billion years ago. As gases from the solar nebula fell into the protoplanet, vast amounts of gravitational energy were converted into the thermal energy that still radiates into space.

In order for heat to flow upward through the Jovian atmosphere and radiate out into space, the temperature must be warmer deep inside the atmosphere than it is at the cloudtops. Infrared measurements have confirmed that temperature rises with increasing depth in the Jovian cloud cover. This pattern is analogous to what happens in the Earth's atmosphere, which is cool at the highest cloudtops (at the top of the troposphere) but is warmer near the ground.

Over the range of temperatures in the Jovian atmosphere, its gases emit energy primarily as infrared radiation. Figure 13-10 shows nearly simultaneous photographs of Jupiter at infrared and visible wavelengths. In the infrared picture, the brighter parts of the image correspond to higher temperatures. There is a striking correlation between brightness in the infrared image and color in the visible-light image. In other words, the range of colors in Jupiter's clouds corresponds to differing temperatures and hence to different depths in the atmosphere. (It is customary to discuss these features in terms of their depth as measured from the cloudtops down, rather than from their altitude above a surface, because the exact location of any solid surface on Jupiter is unknown.) Brown clouds in visible-light images correspond to the brightest parts of the infrared picture, so these clouds must be the warmest—hence the deepest—layers that we can see in the Jovian atmosphere. Whitish clouds form the next layer up, followed by red clouds in the highest layer.

The Jovian atmosphere has a minimum temperature of about 110 K (= −160°C = −260°F) at an altitude above the cloudtops where the atmospheric pressure is about 100

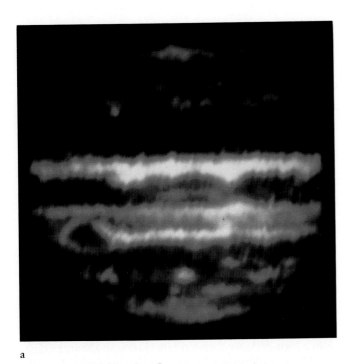

a

b

Figure 13-10 Infrared and visible views of Jupiter (a) *This infrared photograph was taken through the 200-inch Palomar telescope. The brightest parts of the image correspond to holes in the clouds where deeper and warmer regions of the Jovian atmosphere are visible. Dark parts of the image correspond to the cool cloudtops.* (b) *This image, in* visible light, was taken by Voyager 1 *at almost the same time as the infrared picture. Comparisons between the two photographs show that cloud color is correlated with depth in the Jovian atmosphere. The brown clouds are roughly 100°C warmer and 100 km deeper than the red and whitish clouds.* (NASA)

millibars. By analogy with the Earth's atmosphere (recall Figure 8-18), we can call this level the boundary between the stratosphere and the troposphere. As on Earth, all the weather on Jupiter takes place below the stratosphere. At 100 km below the top of the Jovian troposphere, conditions are fairly hospitable, by human standards (see Figure 13-11).

From spectroscopic observations and calculations of temperature and pressure in the Jovian clouds, scientists conclude that the cloud layers have different chemical compositions. The uppermost cloud layer is composed of crystals of frozen ammonia. About 25 km deeper in the troposphere, ammonia (NH_3) and hydrogen sulfide (H_2S) com-

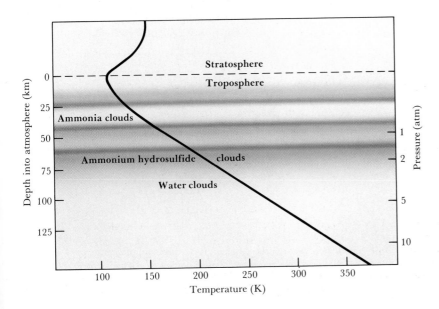

Figure 13-11 The vertical structure of Jupiter's upper atmosphere *This graph displays a temperature and pressure profile of Jupiter's upper atmosphere, as deduced from measurements at infrared and radio wavelengths. Three major cloud layers are shown, along with the colors that predominate at various depths.*

bine to produce ammonium hydrosulfide (NH$_4$SH) crystals. Farther down, about 100 km below the top of the troposphere, the layer of clouds is composed of water droplets and snowflakes of frozen water.

13-7 Studies using computers explain some of the phenomena we see in the Jovian clouds

Because the Great Red Spot appears dark in infrared photographs, we know that its cool clouds are located near the top of the Jovian cloud layers. The Great Red Spot must therefore be a high-pressure system.

Anyone familiar with weather forecasting knows that the Earth's weather is dramatically affected by both high-pressure and low-pressure systems. A **high-pressure system** (commonly called a high) is simply a place where a more-than-usual amount of air happens to be located. This excess air weighs down on the Earth's surface to produce high barometric readings (recall Box 8-2). On the other hand, a **low-pressure system** (commonly called a low) is a place where there is a less-than-normal amount of air, thus producing low barometric readings. If we could see air, highs would look like bulges in the atmosphere (where extra amounts of air are piled up), and lows would look like depressions, or troughs.

Gravity has a strong influence on the basic dynamics of the Earth's weather. Gravity causes air to flow "downhill"

from high-pressure bulges into low-pressure troughs. Because the Earth is rotating, however, the wind flow from the highs toward the lows is not along straight lines. The Earth's rotation causes a deflection of the winds, producing either a clockwise or a counterclockwise flow about the high- and low-pressure regions. This deflection is the result of the **Coriolis effect,** which occurs whenever motion is superimposed on rotation.

Imagine riding on a merry-go-round. If you throw a ball across the merry-go-round, you will find that the ball is deflected to one side. The Coriolis effect explains that this deflection occurs because the platform from which you are observing the ball's motion is rotating. In the same way, winds flowing into a low or out of a high are deflected from their straight-line paths, resulting in either clockwise or counterclockwise motion.

Figure 13-12a shows wind flow about highs and lows in the Earth's atmosphere. In the northern hemisphere, winds blowing toward a low-pressure region rotate counterclockwise about the low, forming a **cyclone.** Winds blowing away from a high-pressure region rotate clockwise about the high, forming an **anticyclone.** In the southern hemisphere, the directions of rotation are reversed: Cyclonic winds about a low rotate clockwise, and anticyclonic winds about a high rotate counterclockwise.

Highs and lows also dominate the weather on Jupiter, but the planet's rapid rotation draws them into bands that completely encircle Jupiter, as shown in Figure 13-12b. The light-colored zones, where we see high-elevation clouds, are high-pressure systems stretched all the way around the planet. Similarly, dark-colored belts, where we see deep into

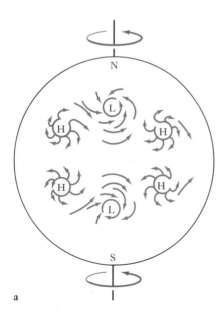

a

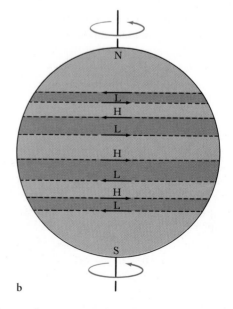

b

Figure 13-12 Wind-flow patterns on Earth and Jupiter (a) *Because of the Earth's rotation, cyclonic winds flowing into a low-pressure region or anticyclonic winds flowing out of a high-pressure region rotate either clockwise or counterclockwise, depending on the hemisphere in*

which the weather system is located. (**b**) *Jupiter's rapid rotation stretches low- and high-pressure systems all the way around the planet. Light-colored zones are high-pressure systems, and the dark-colored belts are regions of low-pressure.*

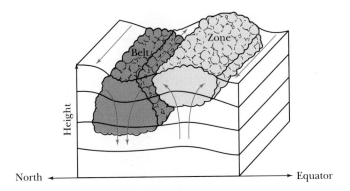

Figure 13-13 *The belt–zone structure of Jupiter's clouds* *Gases warmed by heat from Jupiter's interior rise upward in the light-colored zones. Cool gases descend in the dark-colored belts. The horizontal surfaces indicate contours of constant pressure.*

Note the changes that occurred during the four months between the flybys of *Voyager 1* and *Voyager 2*. The regular spacing of light-colored plumes is apparent in the equatorial regions of both pictures. During the interval between the two flybys, the Great Red Spot moved westward and the white ovals moved eastward.

These beautiful Voyager photographs might suggest Jupiter's clouds to be the result of incomprehensible turmoil. You might wonder if there is anything constant in the Jovian atmosphere. Surprisingly, there is. Telescopic observations over the past 80 years, along with the Voyager data, demonstrate that wind speeds in the Jovian atmosphere are remarkably stable. Although Jupiter's colorful bands do change quite rapidly, its underlying wind patterns do not.

Jupiter's persistent wind patterns consist of counterflowing eastward and westward winds. The speeds of these **zonal jets** are plotted in Figure 13-17. Although the winds near Jupiter's equator blow eastward with speeds slightly greater than 100 miles per second (224 miles per hour), reversals of wind direction occur in a regular pattern from the equator toward the poles. Many scientists suspect that this stationary pattern of alternating wind-flow directions is caused by currents deep inside Jupiter.

Computer simulations to show whirlpools and eddies caught between counterflowing streams help us understand

Jupiter's cloud cover, are low-pressure systems encircling the planet. Gases warmed by Jupiter's internal heat rise upward in the zones, while cool gases from the upper atmosphere descend in the belts (see Figure 13-13). High-speed winds mark the boundaries between the cyclonic flow of the belts and the anticyclonic flow in the zones.

Superimposed on Jupiter's regular pattern of belts and zones are low- and high-pressure features that also exhibit cyclonic or anticyclonic flow. The six-day counterclockwise rotation of the Great Red Spot in Jupiter's southern hemisphere is consistent with infrared observations that the reddish cloudtops are cool and at a high elevation. The Great Red Spot is thus a long-lasting, high-pressure bulge in Jupiter's southern hemisphere.

The Voyager pictures also showed other anticyclones in Jupiter's southern hemisphere. These features appear as **white ovals** (two are visible in Figure 13-8*b*), indicating that their cloudtops are not quite as high as the Great Red Spot. Nevertheless, the wind flow in these ovals clearly is counterclockwise.

Most of the white ovals are observed in Jupiter's southern hemisphere. In contrast, **brown ovals** are more common in Jupiter's northern hemisphere (see Figure 13-14). White ovals are the high-altitude cloudtops of high-pressure systems, but brown ovals result from holes in Jupiter's cloud cover that permit us to see down into warmer regions of the Jovian atmosphere. Like the Great Red Spot, white ovals are apparently long-lived; Earth-based observers have reported seeing them in the same locations since 1938. However, a brown oval lasts for only a year or two. Computer-generated Figure 13-15 shows how Jupiter would look if you were located directly over either the planet's north pole or its south pole. The regular spacing of such cloud features as ripples, plumes, and light-colored wisps is also obvious.

Computer processing was also used to "unwrap" Jupiter to produce maplike views of the planet such as Figure 13-16.

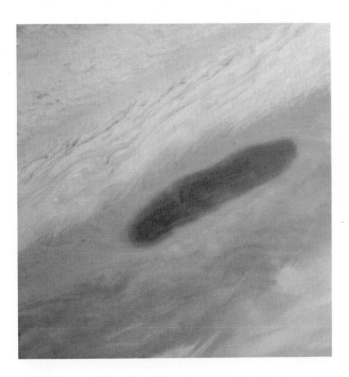

Figure 13-14 *A brown oval* *Large brown ovals in Jupiter's northern hemisphere are caused by openings in the main cloud layer that reveal warm, dark-colored gases below. The length of this oval is roughly equal to the Earth's diameter. Voyager 1 was 4 million km from Jupiter when this picture was taken. (NASA)*

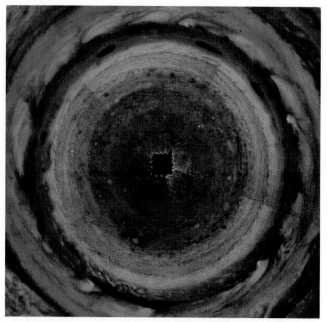

a North pole

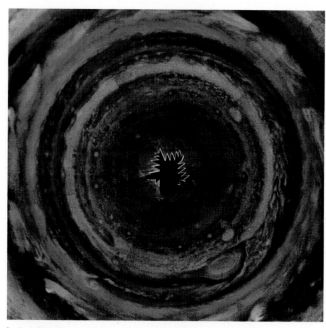

b South pole

Figure 13-15 Jupiter's northern and southern hemispheres Computer processing was used to construct these views that look straight down onto Jupiter's north and south poles. (a) In this northern view, light-colored plumes are evenly spaced around the equatorial regions. Several brown ovals are visible. (b) In this southern view, the three biggest white ovals are separated by almost exactly 90° of longitude. In both views, the banded belt–zone structure is absent near the poles. The black spots are areas not photographed by the spacecraft. *(NASA)*

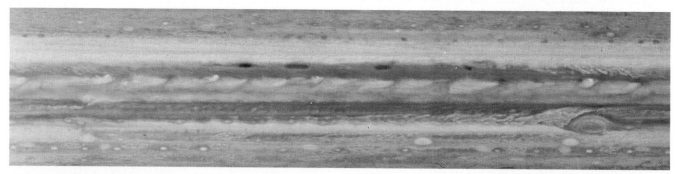

a *Voyager 1* view

b *Voyager 2* view

Figure 13-16 A comparison of Voyager 1 and Voyager 2 views Computer processing was used to produce these two "unwrapped" views of Jupiter from (a) Voyager 1 and (b) Voyager 2. Each view was aligned with respect to Jupiter's magnetic axis so that displacements to the right or left would represent real cloud motions. Notice that the Great Red Spot moved westward, but the white ovals moved eastward. *(NASA)*

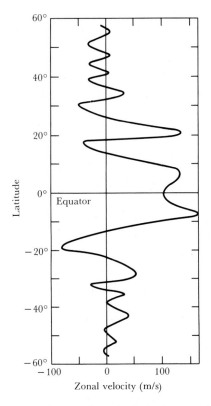

Figure 13-17 Prevailing zonal winds Jupiter's atmosphere is dominated by a persistent pattern of counterflowing eastward and westward winds. This graph shows wind velocities over latitudes ranging from 60°N to 60°S of the Jovian equator.

both the long-term and the short-term features in Jupiter's clouds. Andrew Ingersoll and his colleagues at the California Institute of Technology have pioneered these calculations. For example, Figure 13-18 shows the behavior of a small, unstable whirlpool, technically called a **vortex**. This whirlpool is spinning too slowly to remain intact and so is torn apart by the counterflowing winds. However, larger, rapidly rotating vortices do survive in these simulations. The white ovals and the Great Red Spot endure by simply rolling along with the wind currents. Figure 13-19 shows a simulation in which two stable vortices merge to form a larger, stable vortex. The long-lived white ovals apparently maintain themselves in this fashion.

13-8 Many aspects of Jupiter's clouds, such as their colors, are still poorly understood

The 1970s saw a dramatic increase in our understanding and appreciation of the largest planet in our solar system, but many mysteries still remain. For example, we still do not know what gives Jupiter's clouds their colors. The crystals of ammonia, ammonium hydrosulfide, and frozen water in Jupiter's three main cloud layers are all white. Thus the browns, blues, reds, and oranges must be caused by other

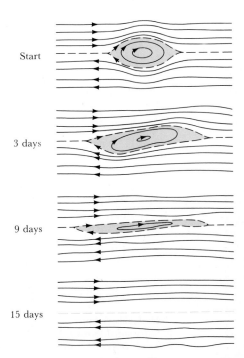

Figure 13-18 The demise of an unstable vortex In this computer simulation, a small vortex is rotating too slowly to remain intact. After slightly more than a week, the vortex is pulled apart by counterflowing zonal jets. (Adapted from A. P. Ingersoll)

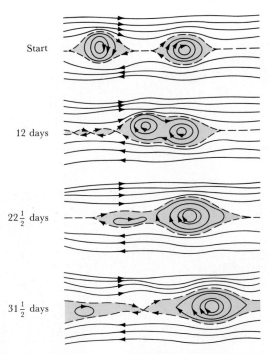

Figure 13-19 The merging and maintenance of stable vortices This computer simulation shows the collision and merger of two rapidly spinning, stable vortices. The result is a larger vortex and the ejection of material. Such mergers occur near the Great Red Spot. (Adapted from A. P. Ingersoll)

chemicals. Some scientists believe that sulfur, which can assume many different colors depending on its temperature, plays an important role in Jupiter's clouds. Others think that phosphorus might be involved, especially in the Great Red Spot. The Sun's ultraviolet radiation, which is capable of inducing chemical reactions, may also play a role in coloring the Jovian clouds.

Perhaps the most intriguing proposal is that Jupiter's colors might be due to organic compounds, possibly resulting from biological activity. Many of Jupiter's colors are similar to the colors of organic material created in laboratory experiments that attempt to simulate the first steps in the creation of life. These experiments, discussed in detail in the Afterword, involve passing electrical discharges (to simulate lightning) through a mixture of methane, ammonia, hydrogen, and water vapor. These gases are relatively abundant in Jupiter's atmosphere, and there is ample evidence of lightning there (see Figure 13-20). Many of the organic compounds produced in such experiments may have been brewing for 4 billion years in Jupiter's clouds. Is it possible that those clouds might harbor extraterrestrial life?

To explore these issues, NASA launched *Galileo* in a roundabout path toward Jupiter (see Figure 13-21). The spacecraft carries an atmospheric probe that will be released into Jupiter's atmosphere just before *Galileo* goes into orbit around the planet in 1995. Using a heat shield and a para-

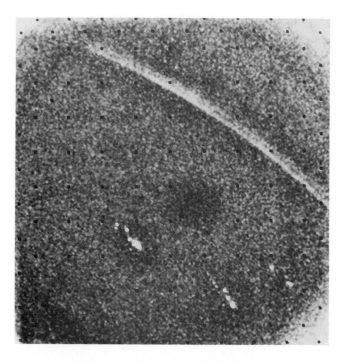

Figure 13-20 Aurorae and lightning in the Jovian night *This photograph of Jupiter's dark side is a 3-minute exposure from* Voyager 1. *A huge aurora arches across the northern horizon. In the lower half of the picture, flashes from about 20 enormous lightning bolts illuminate the clouds. (NASA)*

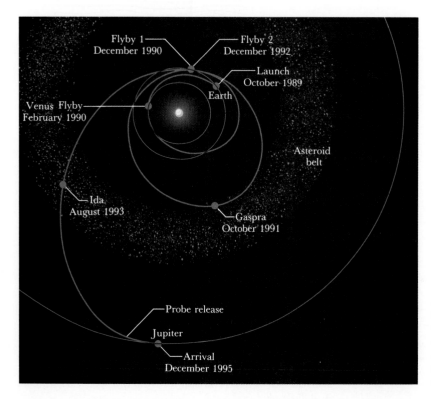

Figure 13-21 Galileo's *trajectory to Jupiter* *To build up the necessary speed to reach the outer solar system,* Galileo *will pass by Venus once and the Earth twice, relying on their gravity to accelerate and redirect the spacecraft toward Jupiter.* Galileo *will also encounter two asteroids, Gaspra and Ida, as it passes through the asteroid belt.*

Figure 13-22 Galileo's atmospheric probe *Just before arriving at Jupiter, Galileo will release an atmospheric probe that will make direct measurements and chemical analyses as it descends through the Jovian cloud cover. Scientists look forward to this historic event, which will occur in December 1995. (NASA)*

chute, the probe will descend through the Jovian clouds, making a wide range of measurements and chemical analyses (see Figure 13-22). Just as the Voyager flybys revealed phenomena that stagger the imagination, the Galileo mission undoubtedly will provide a few answers and many new mysteries.

Key words

anticyclone	decimetric radiation	low-pressure system	vortex (*plural* vortices)
belts	differential rotation	nonthermal radiation	white ovals
brown ovals	Great Red Spot	oblate	zonal jets
Coriolis effect	high-pressure system	oblateness	zones
current sheet	hydrostatic equilibrium	plasma	
cyclone	internal rotation period	synchrotron radiation	
decametric radiation	liquid metallic hydrogen	thermal radiation	

Key ideas

- Jupiter, whose mass is equivalent to the masses of 318 Earths, is composed of 82 percent hydrogen, 17 percent helium, and only 1 percent all other elements. This chemical composition is very similar to that of the Sun.

  Jupiter may have a rocky core with a mass of about 13 Earth masses. This core is surrounded by a 40,000-km-thick layer of liquid metallic hydrogen and an outer layer of molecular hydrogen about 20,000 km thick.

  The visible features of Jupiter—belts, zones, the Great Red Spot, ovals, and colored clouds—exist in the outermost 100 km of the molecular hydrogen layer.

The outer layers of the Jovian atmosphere show differential rotation, with the equatorial regions rotating slightly faster than the polar regions. The internal rotation rate, determined from the study of radio waves, is nearly the same as the polar rotation rate.

Because of its rapid rotation, Jupiter is noticeably oblate.

- Jupiter has a strong magnetic field created by currents in the metallic hydrogen layer. Its huge magnetosphere contains a vast current sheet of electrically charged particles.

  Charged particles in the densest portions of Jupiter's magnetosphere emit synchrotron radiation at radio wavelengths.

  The Jovian magnetosphere encloses a plasma of charged particles that is hotter than the center of the Sun but has very low density. The magnetosphere exists in a delicate balance between pressures from the plasma and from the solar wind, and its size fluctuates drastically.

- The colored ovals visible in the Jovian atmosphere represent gigantic storms in the troposphere; some, such as the Great Red Spot, are quite stable and persist for many years.

  The Jovian ovals are cyclonic or anticyclonic storms created at the boundaries between zonal jets (wind streams) moving in opposite directions around the planet.

- Jupiter emits more heat than it receives from the Sun; presumably the planet is still cooling.

- There are three cloud layers in Jupiter's troposphere. The reasons for the distinctive colors of these different layers are not yet known.

- Some scientists speculate that life may exist in the layers of the Jovian atmosphere where pressures and temperatures are not too different from those on Earth; organic molecules almost certainly do exist there.

Review questions

1   Why do the magnitudes of Jupiter at opposition vary so little, compared to the oppositions of Mars (see Boxes 12-2 and 13-2)?

2   Which planet, Mars or Jupiter, is easier to observe with an Earth-based telescope? Explain your answer.

3   Compare Jupiter's chemical composition with that of the Sun. What does this tell us about Jupiter's formation?

4   If Jupiter does not have any observable solid surface and its atmosphere rotates differentially, how are astronomers able to determine the planet's rotation rate?

5   Why do astronomers believe that Jupiter does not have a large iron-rich core, even though the planet possesses a strong magnetic field?

6   What is liquid metallic hydrogen?

7   What is thought to be the source of Jupiter's excess internal heat?

8   What are the "belts" and "zones" in Jupiter's atmosphere? Is the Great Red Spot more like a belt or a zone?

9   What is the difference between a cyclone and an anticyclone?

10   Explain the statement, "Since the Great Red Spot appears dark in infrared photographs, we know that its cool clouds are located at a high elevation."

11   What data and techniques have been used to determine the internal structure of Jupiter?

12   Compare and contrast Jupiter's magnetosphere with the magnetosphere of a terrestrial planet like Earth. Why is the size of the Jovian magnetosphere variable, whereas the Earth's is not?

Advanced questions

> **Tips and tools . . .**
> For a discussion of center of mass, see Box 4-3, which also contains a very useful form of Kepler's third law. Newton's universal law of gravitation is the basic equation from which a planet's surface gravity can be calculated.

*13   (Basic)   Find the location of the center of mass of the Sun–Jupiter system. Is the center of mass inside or outside the Sun?

*14   Using orbital data for a Jovian satellite of your choice (see Appendix 3), calculate the mass of Jupiter. How does your answer compare with the mass quoted in Box 13-1?

*15   How much would a 150-lb person weigh on Jupiter's "surface"?

*16   (Challenging)   Estimate the wind velocities in the Great Red Spot, which rotates with a period of about six days.

17   What sort of experiment would you design in order to establish whether Jupiter has a rocky core?

Discussion questions

18 Describe some of the semipermanent features in Jupiter's atmosphere. What factors influence their longevity? Compare and contrast these long-lived features with some of the transient phenomena seen in Jupiter's clouds.

19 Suppose you were designing a mission to Jupiter involving an airplanelike vehicle that would spend many days (months?) flying through the Jovian clouds. What observations, measurements, and analyses should this aircraft be prepared to make? What dangers might the aircraft encounter, and what design problems would you have to overcome?

20 How is *Galileo* doing? Consult such magazines as *Sky & Telescope* or *Science News* to see how the mission is proceeding. Where is the spacecraft right now? How many planets has it passed? Are all its instruments functioning properly?

Observing projects

21 Consult such magazines as *Sky & Telescope* or *Astronomy* to determine the visibility of Jupiter. If Jupiter is visible in the night sky, make arrangements to view the planet through a telescope. What magnifying power seems to give you the best view? Draw a picture of what you see. Can you seen any belts and zones? How many? Can you see the Great Red Spot?

22 Make arrangements to view Jupiter's Great Red Spot through a telescope. The Calendar Notes section of *Sky & Telescope* lists the universal dates and times when the center of the Great Red Spot should pass across Jupiter's central meridian when the planet is observable from North America. The Great Red Spot is well placed for viewing for 50 minutes before and after meridian transit.

For further reading

Burgess, E. *By Jupiter*. Columbia University Press, 1982 • This nontechnical book reviews our understanding of Jupiter in light of the Pioneer and Voyager flybys.

Carroll, M. "Project Galileo: The Phoenix Rises." *Sky & Telescope*, April 1987 • This article describes the long-delayed Galileo mission to Jupiter.

Hunt, G., and Moore, P. *Jupiter*. Rand McNally, 1981 • This occasionally technical book is a useful reference atlas for Jupiter and its satellites.

Ingersoll, A. "The Meteorology of Jupiter." *Scientific American*, March 1976 • This article gives an exceptionally clear description of the dominant weather patterns on Jupiter.

Johnson, T., and Yeates, C. "Return to Jupiter: Project Galileo." *Sky & Telescope*, August 1983 • This article by Jet Propulsion Laboratory scientists gives an excellent description of the Galileo spacecraft, but the dates quoted are wrong because the mission was delayed for many years.

Morrison, D., and Samz, J. *Voyage to Jupiter*. NASA SP-439, 1980 • This superb book, which describes the Voyager missions to Jupiter, includes an excellent selection of color photographs.

Washburn, M. *Distant Encounters: The Exploration of Jupiter and Saturn*. Harcourt Brace Jovanovich, 1983 • This entertaining book describes the Pioneer and Voyager missions to the outer solar system.

Wolfe, R. "Jupiter." *Scientific American*, 1975 • Written shortly after the Pioneer flybys, this article describes Jupiter's interior with special emphasis on the role of liquid metallic hydrogen.

Moon

Io

Europa

Mercury

Ganymede

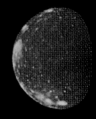
Callisto

The Galilean satellites with our Moon and Mercury Jupiter's four large satellites are shown here along with our Moon and Mercury. All six worlds are reproduced to the same scale. Io has numerous active volcanoes and Europa has a smooth, icy surface; both are roughly the same size as our Moon. Ganymede and Callisto are each covered with a 1000-km-thick layer of ice; both are roughly the size of Mercury. Saturn's largest moon, Titan, is also about the size of Mercury. Neptune's largest moon, Triton, is slightly smaller than our Moon. (NASA)

C H A P T E R

14

The Galilean Satellites of Jupiter

In 1610, Galileo discovered four moons circling Jupiter that are each about the same size as our Moon or Mercury. In 1979, two Voyager spacecraft flew by the Galilean satellites and found extraordinary panoramas. Io, the innermost of the four, is covered with colorful sulfur compounds deposited by numerous active volcanoes. This geologically active satellite orbits Jupiter deep within its magnetosphere and modulates the radio emissions from the planet. The layer of ice that envelopes Europa is crisscrossed with a profusion of fine cracks. Ganymede and Callisto are encased in thick "oceans" of ice whose frozen surfaces bear the scars of meteoroid impacts. In 1995, the *Galileo* spacecraft will go into orbit about Jupiter and provide long-term, high-resolution coverage of the four satellites. Our understanding of the physical processes on terrestrial planets, including the Earth, is being shaped by the ongoing exploration of Jupiter's fascinating moons.

Galileo Galilei called them the Medicean Stars to attract the attention of a wealthy Florentine patron of the arts and sciences. Since their discovery in 1610, the four giant moons of Jupiter have played an important role in our understanding of the universe. To Galileo, they were observational evidence to support the heretical Copernican cosmology. To the modern astronomer, the Voyager flybys of 1979 revealed four extraordinary terrestrial worlds, different from anything astronomers had ever seen or even imagined. We call them the **Galilean satellites.** They are named after the mythical lovers and companions of the Greek god Zeus: Io, Europa, Ganymede, and Callisto.

14-1 The Galilean satellites are easily seen with Earth-based telescopes

When viewed through an Earth-based telescope, the Galilean satellites look like pinpoints of light. With patience, you can follow these four worlds as they orbit Jupiter. Because their orbital periods are fairly short (ranging from 1.8 days for Io to 16.7 days for Callisto), major changes in the positions of the satellites are easily noticed from one night to the next. By consulting a current issue of the *Astronomical Almanac*, you can schedule your observations to include transits, eclipses, and occultations. When one of the four moons passes between Jupiter and the Sun in a transit, the satellite's shadow is seen as a black dot against the planet's colorful cloudtops (examine Figure 13-7). In an eclipse, one of the moons seems suddenly to disappear, then reappear, as it moves behind Jupiter, passing into and out of the planet's enormous shadow. In an **occultation**, a satellite passes in front of a star, blocking its light for Earth-based viewers. Because these moons' apparent magnitudes range from about 4.7 for Ganymede to nearly 6 for Callisto, you might think that all four satellites should be visible to the naked eye. Binoculars or a telescope are necessary, however, because of the overwhelming glare of Jupiter (see Figure 14-1).

Observations of eclipses have been used to measure the diameters of the Galilean satellites. When a satellite emerges from Jupiter's shadow, it does not blink on instantly. Instead, there is a brief interval during which the satellite gets brighter and brighter as more and more of its surface becomes exposed to sunlight. The duration of this interval depends on the orbital speed of the satellite (which is known from Kepler's laws) and the diameter of the satellite. Consequently, by accurately measuring the time it takes for a satellite to move into or out of Jupiter's umbra, we can calculate the satellite's diameter.

Stellar occultations have been used in a similar fashion. By measuring the time it takes for one of the four satellites to pass in front of a background star we can calculate the satellite's diameter. In addition, the Galilean satellites occasionally occult each other. All four satellites orbit Jupiter in the plane of the planet's equator. Every six years, the Earth

Figure 14-1 The Galilean satellites *This photograph, taken by an amateur astronomer with a small telescope, shows the four Galilean satellites alongside an overexposed image of Jupiter. Each Galilean satellite is bright enough to be seen with the unaided eye, were it not overwhelmed by the glare of nearby Jupiter. (Courtesy of J. Jenkins)*

passes through this equatorial plane and mutual occultations of the satellites occur for a few days. Once again, timing the occultations enables us to calculate the satellites' diameters.

The brightness of each of the moons varies slightly as it moves along its orbit. These variations can logically be attributed to dark and light areas on the moons' surfaces that are alternately exposed to or hidden from our view as the satellites rotate. Careful measurements have shown that the apparent magnitude of each satellite varies with a period equal to the satellite's orbital period. In other words, each Galilean satellite rotates exactly once on its axis during each trip around its orbit. Hence each Galilean satellite has synchronous rotation that keeps the same hemisphere perpetually facing Jupiter, just as our Moon keeps the same side facing the Earth. Tidal forces between a planet and its satellite are responsible for such 1-to-1 spin–orbit coupling.

14-2 Data from spacecraft have greatly improved our knowledge about the Galilean satellites

Accurate measurements of the diameters of the Galilean satellites came from the Voyager flybys, because direct photography revealed measurable disks (see Figure 14-2). These measurements confirmed and refined the data obtained from Earth. The two inner Galilean satellites, Io and Europa, are approximately the same size as our Moon. The two outer satellites, Ganymede and Callisto, are comparable in size to

Figure 14-2 Io and Europa *The Galilean satellites look only like pinpoints of light when viewed through an Earth-based telescope, but the Earth-orbiting Hubble Space Telescope produces views of Jupiter comparable to this photograph from Voyager 1. Surface features as small as 400 km (240 mi) across are visible. Io (left) and Europa (right) are each approximately the same size as our Moon. (NASA)*

Mercury. The Voyager data (accurate to better than ±10 km) are listed in Box 14-1.

As early as the 1920s, fairly accurate determinations of the masses of the Galilean satellites had been made from Earth. The satellites often pass near each other as they orbit Jupiter. During these encounters, gravitational interactions between the satellites cause slight perturbations of their orbits. By measuring these tiny orbital deflections and using Newtonian mechanics, astronomers were able to calculate the masses of the satellites. Europa was discovered to be the least massive (its mass is two-thirds that of our Moon). Ganymede is by far the most massive of the four, with at least twice the mass of our Moon. As you might expect, the Pioneer and Voyager flybys produced the best determinations of the satellites' masses. Data based on deflections of the spacecrafts' trajectories are given in Box 14-1.

As soon as reliable mass and diameter measurements were available, it became apparent that the average densities of the four satellites are related to their distances from Jupiter. The innermost satellite, Io, has the highest average density (3550 kg/m³), which is slightly greater than the density of our Moon (3340 kg/m³). The next satellite out, Europa, has an average density of 3040 kg/m³. Recalling that typical rocks in the Earth's crust have densities around 3000 kg/m³, it is reasonable to suppose that both Io and Europa are made primarily of rocky material.

The two outer satellites also exhibit decreasing density with increasing distance from Jupiter. Both Ganymede and Callisto have an average density of less than 2000 kg/m³, indicating that a sizable fraction of each world is made of something less dense than rock. Figure 14-3 shows the Galilean satellites to the same scale.

| Io | Europa | Ganymede | Callisto |

Figure 14-3 The Galilean satellites *The four Galilean satellites are shown here to the same scale. Io and Europa have diameters and densities comparable to our Moon and are composed primarily of rocky ma-* *terial. Although Ganymede and Callisto are roughly as big as Mercury, their average densities are low. Each of these two outer satellites is covered with a thick layer of water and ice. (NASA)*

Box 14-1   The Galilean satellites

Four of the sixteen known satellites that orbit Jupiter are large enough to be classified as terrestrial worlds in their own right. The table below lists basic data about these four worlds, with data about Mercury and our Moon for comparison.

	Mean distance from Jupiter (km)	Sidereal period (days)	Diameter (km)	Mass (kg)	Mass (Moon = 1)	Mean density (kg/m^3)
Io	421,600	1.77	3630	8.92×10^{22}	1.21	3550
Europa	670,900	3.55	3138	4.87×10^{22}	0.66	3040
Ganymede	1,070,000	7.16	5262	1.49×10^{23}	2.03	1940
Callisto	1,880,000	16.69	4800	1.06×10^{23}	1.44	1810
Mercury	——	——	4878	3.30×10^{23}	4.49	5420
Moon	——	——	3476	7.35×10^{22}	1.00	3340

14-3 The formation of the Galilean satellites probably mimicked the formation of our solar system

The arrangement of the Galilean satellites parallels the arrangement of the planets' characteristics when they are grouped according to their distance from the Sun. Moving outward from the Sun, the planets' average density steadily declines, from more than 5000 kg/m^3 for Mercury to less than 1000 kg/m^3 for Saturn (recall Table 7-2). Scientists therefore suspect that the same general processes that formed our solar system were also at work during the formation of the Galilean satellites, though on a much smaller scale.

The low densities of Ganymede and Callisto suggest that these two satellites are composed of roughly equal amounts of rock and ice. In fact, in the early 1970s, the American astronomer John Lewis pointed out that these low densities are exactly what one would expect for relatively small objects formed by accretion of ice-covered dust grains. More recently, NASA scientists James Pollack and Fraser Fanale constructed theoretical models for the formation of the Galilean satellites, including in their simulations the fact that Jupiter emits twice as much energy as it receives from the Sun. They calculated that frozen water could be retained and incorporated into satellites at the distances of Ganymede and Callisto, but concluded that only rocky material would condense at the orbital distances of Io and Europa, because of Jupiter's warmth. Thus, Jupiter's gravity and heat produced two distinct classes of Galilean satellites, just as warmth from the proto-Sun caused a dichotomy between small, dense, rocky, inner planets and huge, gaseous, low-density outer planets.

Confirmation of water ice on the Jovian satellites came in the early 1970s from Earth-based spectroscopic observations. Various teams of astronomers measured the intensity of reflected sunlight at wavelengths from 0.3 to 5.3 μm. The spectra of Europa and Ganymede exhibited strong absorptions at 1.5 and 2.0 μm (see Figure 14-4). Scientists soon

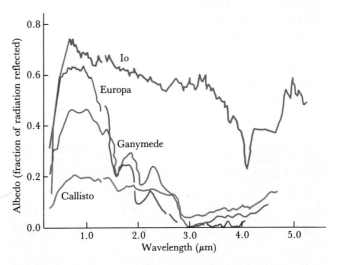

Figure 14-4   Infrared spectra of the Galilean satellites *Reflected sunlight from Europa and Ganymede shows prominent absorption at infrared wavelengths, which is caused by water ice. Notice that Io's spectrum differs radically from those of the other three, strongly suggesting that the innermost Galilean satellite is unusual. The strong absorption at 4.1 μm in Io's spectrum is probably produced by frozen SO_2. The depressed reflectivity of Callisto is caused by a dark coating, perhaps meteoritic dust. (Adapted from R. Clark and T. McCord)*

recognized these spectral lines as the characteristic pattern of water–ice molecules. Since Europa is composed mostly of rock, its ice content must be limited to a relatively thin surface layer. Although Callisto also was expected to have an icy surface, the spectral lines of ice were much weaker in its spectrum. Callisto is, in fact, the dimmest of the Galilean satellites, reflecting much less light over a wide range of wavelengths than do any of its companions. As a result, scientists began to suspect Callisto to be covered with a dark, dusty coating of meteoritic material that subdues the reflective properties of the underlying layer of ice.

14-4 The Voyager spacecraft discovered several small moons and a ring around Jupiter

The first moon to be photographed at close range by *Voyager 1* was a tiny, reddish satellite called Amalthea, one of Jupiter's many small moons. Amalthea circles the planet in only 11.7 hours, along an orbit much closer to Jupiter than that of Io (see the diagram in Box 14-2). Amalthea proved to be an irregularly shaped object similar to a large asteroid (see Figure 14-5). It measures about 270 kilometers along its longest axis and 155 kilometers in the shortest direction. Thus Amalthea is about ten times larger than the moons of

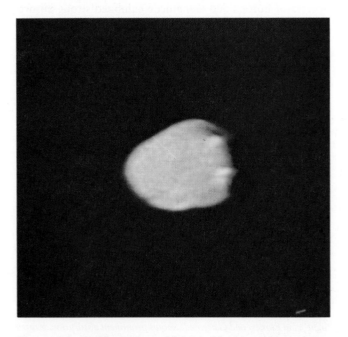

Figure 14-5 Amalthea *Tiny, reddish Amalthea is one of Jupiter's four innermost satellites. It is an irregularly shaped, asteroidlike object measuring 270 km along its longest dimension. This photograph, taken by Voyager 1, has a resolution of 8 km. Amalthea was discovered in 1892 by the American astronomer E. E. Barnard at the Lick Observatory. (NASA)*

Mars but ten times smaller than the smallest Galilean satellites. Like those satellites, Amalthea orbits Jupiter with synchronous rotation, keeping its longest axis pointed toward the planet.

A total of 16 satellites are now known to orbit Jupiter (see Box 14-3), three of which were discovered during the Voyager flybys. These newly identified moons, along with Amalthea and the Galilean satellites, all orbit Jupiter in the plane of that planet's equator. In contrast, the remaining eight moons are all extremely tiny, with estimated diameters typically of less than 50 km, and circle Jupiter along large orbits that are inclined at steep angles to the planet's equatorial plane. Furthermore, the four outermost satellites all have retrograde orbits. Because of the oddity of these orbits, it is reasonable to suppose that the eight outer moons may be wayward asteroids captured by Jupiter's powerful gravitational field. Indeed, Newtonian mechanics shows that it is easier for Jupiter to capture an asteroid into a retrograde orbit than a direct one.

In addition to discovering several Jovian moons, the Voyager cameras revealed a faint ring around Jupiter (see Figure 14-6). This ring, which is probably composed primarily of dust along with some small rock fragments, lies in Jupiter's equatorial plane, closer to the planet than even the orbit of Amalthea. The sharp outer boundary of the ring is only 1.81 Jupiter radii from the planet's center. This ring, the small inner moons, and the Galilean satellites together form a system of orbiting material that apparently is a common feature of Jovian planets. As we shall see in the next two chapters, Saturn, Uranus, and Neptune also have rings and families of satellites whose orbits lie in their planets' equatorial planes.

14-5 Calculations and observations before the Voyager flybys suggested that Io might be volcanically active

Infrared spectra provided the first indication that Io is a most unusual world (examine Figure 14-4). Only three scientists had the insight, however, to recognize the unusual circumstances that profoundly affect Io.

Three days before *Voyager 1* flew past Io, Stanton Peale of the University of California at Santa Barbara and Patrick Cassen and Ray Reynolds of NASA's Ames Research Center published calculations indicating that Io is acted upon by tremendous tidal forces. As it orbits Jupiter, Io is repeatedly caught in a gravitational tug-of-war between the huge planet on one side and the remaining Galilean satellites on the other. This gravitational battle distorts Io's orbit, causing this satellite to vary its distance from Jupiter. As the distance between Io and Jupiter varies, tidal stresses on Io alternately squeeze and flex it. This constant tidal stressing in turn causes frictional heating of Io's interior.

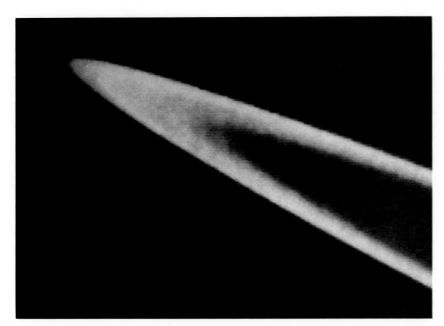

Figure 14-6 Jupiter's ring *A portion of Jupiter's faint ring is seen in this photograph from Voyager 2. The ring, which is closer to Jupiter than any of the planet's satellites, is probably composed of tiny rock fragments. The brightest portion of the ring is about 6000 km in width. The outer edge of the ring is sharply defined, but the inner edge is somewhat fuzzy. A tenuous sheet of material extends from the ring's inner edge all the way down to the planet's cloudtops. (NASA)*

Calculations show that the heat pumped into Io in this fashion could be as great as 10^{13} watts, which is equivalent to 2400 tons of TNT exploding every second. This energy must eventually make its way to the satellite's surface, so Peale, Cassen, and Reynolds predicted that "widespread and recurrent volcanism" should exist on Io. Although the paths of the Voyager spacecraft were designed to give maximal photographic coverage of the surfaces of the Galilean satellites, no one expected to obtain photographs of erupting volcanoes on Io. After all, a spacecraft making a single trip past the Earth would be highly unlikely to catch a large volcano actually erupting. Yet that is exactly what happened at Io.

14-6 Io is covered with colorful deposits of sulfur compounds ejected from numerous active volcanoes

Within a few hours after *Voyager 1* passed Amalthea, Io loomed into view and the spacecraft began sending back a series of strange and unexpected pictures, like the one shown in Figure 14-7. Baffled by what they were seeing, scientists at the Jet Propulsion Laboratory (JPL) in Pasadena (where the pictures were being received and enhanced by computer processing) jokingly compared Io to pizzas and rotten oranges.

Figure 14-7 Io *This close-up view of Io was taken by Voyager 1. Notice the extraordinary range of colors from white, yellow, and orange to black. Scientists believe that these brilliant colors are caused by surface deposits of sulfur ejected from Io's numerous volcanoes. (NASA)*

Box 14-2   The Voyager flybys and the *Galileo* orbiter

From Earth-based observations, all four Galilean satellites were known to keep one side constantly facing Jupiter, just as the same side of our Moon constantly faces Earth. With this fact in mind, the trajectories of the two Voyager spacecraft were planned to maximize the pictorial coverage of the satellites' surfaces. For example, during its flyby on March 5, 1979, *Voyager 1* photographed the Jupiter-facing hemispheres of Ganymede and Callisto. As shown in the diagram on the far right, *Voyager 2* encountered Callisto and Ganymede before reaching Jupiter on July 9, 1979, and thus photographed the outward-facing hemispheres of these two worlds. Consequently, nearly 80 percent of the surfaces of Ganymede and Callisto were photographed at close range.

Europa, whose Jupiter-facing side had been seen from afar by *Voyager 1*, was photographed at a closer distance by *Voyager 2*. Io and Amalthea revolve about Jupiter so rapidly that they were successfully photographed by both Voyagers.

In December, 1995, *Galileo* will arrive at Jupiter (see artist's rendition below). After releasing a probe into Jupiter's atmosphere, the spacecraft will make a single, inbound pass by Io, which lies deep in Jupiter's magnetosphere. Scientists fear that the spacecraft's electronics might suffer significant radiation damage if more than one Io flyby is attempted. *Galileo* will then begin long-term surveillance of the remaining Galilean satellites, using their gravity to redirect the spacecraft's orbit. For instance, the bottom right diagram below shows how Ganymede's gravity can either increase or decrease the size of *Galileo*'s orbit, depending on Ganymede's position at the time of the flyby. Such maneuvers will permit the spacecraft to go from one Galilean satellite to another. Seen from above Jupiter's orbital plane, the succession of distorted elliptical paths will resemble the petals of a flower. The resulting flybys will give scientists the opportunity to study three icy worlds that were glimpsed only briefly by the Voyagers.

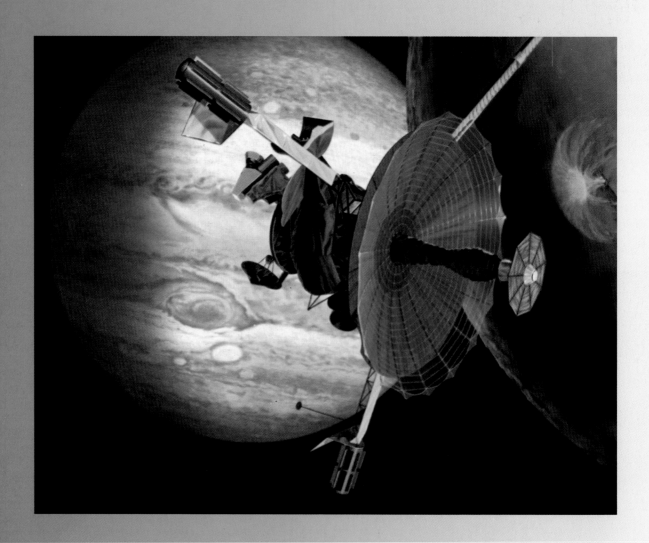

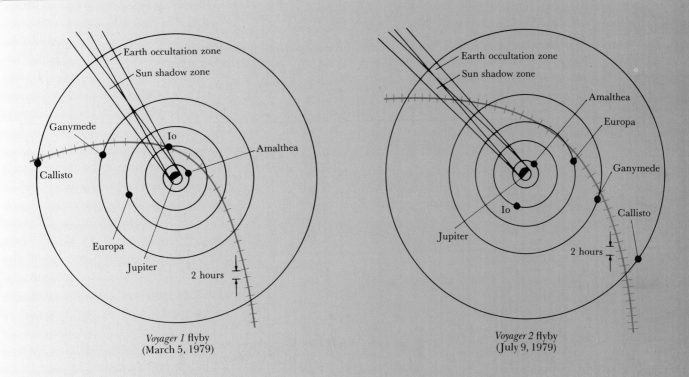

Voyager 1 flyby
(March 5, 1979)

Voyager 2 flyby
(July 9, 1979)

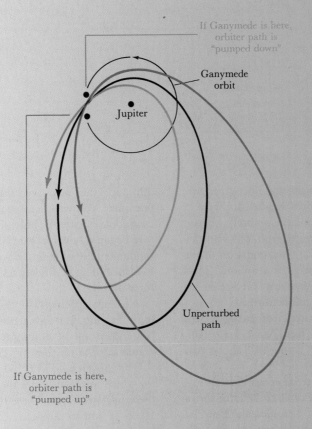

Box 14-3 Jupiter's family of moons

Sixteen confirmed satellites orbit Jupiter. Nearest Jupiter are four small moons. Next come the four giant Galilean satellites. The remaining eight moons, all very tiny, are located at great distances from Jupiter. The sixteen satellites are listed in the table with data about their orbits.

	Average radius of orbit		Orbital period	Year of
	(km)	(Jupiter radii)	(days)	discovery
Metis	128,000	1.79	0.30	1981
Adrastea	129,000	1.81	0.30	1979
Amalthea	181,000	2.54	0.50	1892
Thebe	222,000	3.11	0.67	1980
Io	421,000	5.91	1.77	1610
Europa	670,900	9.40	3.55	1610
Ganymede	1,070,000	15.0	7.15	1610
Callisto	1,883,000	26.4	16.69	1610
Leda	11,094,000	155	239	1974
Himalia	11,480,000	161	251	1904
Lysithea	11,720,000	164	259	1938
Elara	11,740,000	164	260	1905
Ananke*	21,200,000	297	631R	1951
Carme*	22,600,000	317	692R	1938
Pasiphae*	23,500,000	329	735R	1908
Sinope*	23,700,000	332	758R	1914

*The outermost four satellites move in a retrograde direction about Jupiter, indicated by R with the orbital period.

A major clue to these puzzling vistas was uncovered several days after the Jupiter flyby when Linda Morabito, a navigation engineer at JPL, noticed a large umbrella-shaped cloud protruding from Io in one photograph. She had discovered an erupting volcano. Careful reexamination of the close-up photographs revealed eight giant eruptions. Thus was the brilliant Peale–Cassen–Reynolds prediction of "widespread and recurrent volcanism" confirmed.

The volcanoes on Io are named after gods and goddesses traditionally associated with fire in Greek, Norse, Hawaiian, and other mythologies. For instance, Figure 14-8 shows two views of the symmetric plume of Prometheus.

The plumes and fountains of material spewing from Io's volcanoes rise to astonishing heights of 70 to 280 km above the satellite's surface. To reach these altitudes, the material must emerge from the volcanoes' vents with speeds between 300 and 1000 m/s. Even the most violent terrestrial volcanoes, like Vesuvius, Krakatoa, and Mount St. Helens, have eruption velocities of only around 100 m/s. Scientists therefore began to suspect that Io's volcanoes must operate in a fundamentally different way from volcanoes here on Earth. Evidence of such differences was found in the Voyager pictures and data.

No impact craters like those on our Moon were seen on Io, indicating that the satellite's surface is extremely young. Material from the volcanoes apparently obliterates impact craters soon after they are created.

The Voyager cameras did reveal numerous black dots on Io, which apparently are the volcanic vents through which the eruptions occur. These black spots, which are typically 10 to 50 km in diameter, cover 5 percent of Io's surface. Lava flows (see Figure 14-9) radiate from many of these

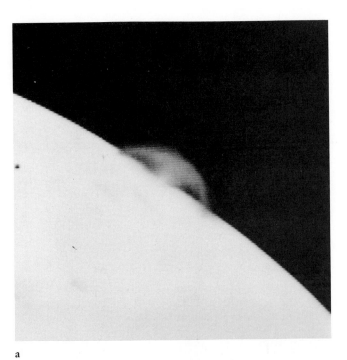

a

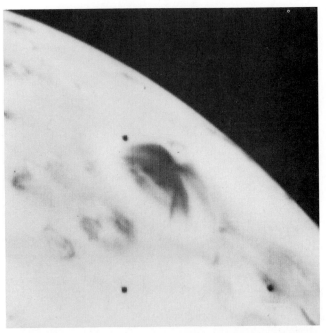

b

Figure 14-8 Prometheus on Io *These two views from Voyager 1, taken two hours apart, show details of the plume of the volcano called Prometheus.* (a) *The plume's characteristic umbrella shape is silhouetted against the blackness of space.* (b) *When viewed against the light background of Io's surface, jets of material give the plume a spiderlike appearance. The plume of Prometheus rises to an altitude of 100 km above Io's surface. (NASA)*

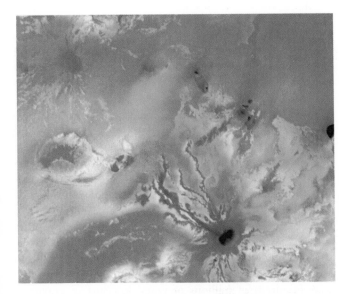

Figure 14-9 A volcanic center on Io *No impact craters are seen in close-up pictures of Io such as this one taken by Voyager 1. Long, meandering lava flows radiate from many of the black dots that apparently are the sites of intense volcanic activity. This photograph covers an area measuring 1000 by 800 km, approximately twice the size of California. (NASA)*

black dots, some of which are located at the origins of volcanic plumes.

Evidence supporting the volcanic nature of Io's black spots came from infrared interferometer spectrometers carried on each of the Voyager spacecraft. By measuring the intensity of infrared radiation across Io's surface, scientists discovered that some of the black spots have temperatures as high as 20°C, in sharp contrast to the surface temperature surrounding them, which is only −146°C. In fact, the Voyager cameras captured the actual onset of a minor eruption (see Figure 14-10).

After the discovery of widespread volcanic activity on Io, scientists soon concluded that it is sulfur (S) ejected from the volcanoes that is responsible for the satellite's brilliant colors. Sulfur is normally bright yellow. If heated and suddenly cooled, however, it can assume a range of colors, from orange and red to black.

The infrared spectrometer on *Voyager 1* detected primarily sulfur and sulfur dioxide in the plumes of material erupting from Io's volcanoes. Sulfur dioxide (SO_2) is an acrid gas commonly discharged from volcanic vents here on Earth. When this gas is released into the cold vacuum of space from eruptions on Io, it crystallizes into white snowflakes. It is

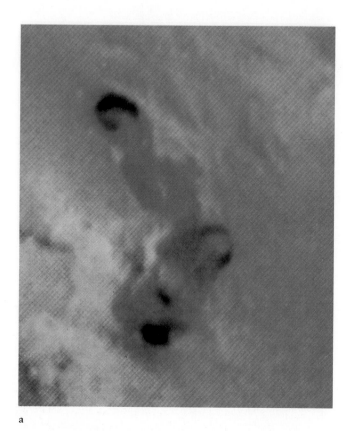

a

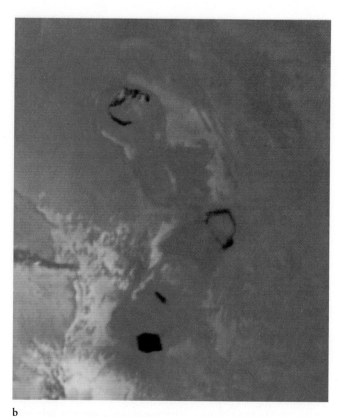

b

Figure 14-10 The venting of gases on Io *These two photographs were taken six hours apart by Voyager 1. In the second photograph, bright bluish patches appeared over a black crescent-shaped feature. These blue-white spots are probably caused by sulfur-dioxide gas escaping* *from Io's interior.* (a) *This view was taken at a distance of 374,000 km.* (b) *This clearer photograph was taken at a distance of 130,000 km.* *(NASA)*

likely that the whitish deposits on Io (see Figure 14-7) are frozen sulfur dioxide (SO_2 frost or snow).

The abundant sulfur and sulfur dioxide on Io suggest an explanation for the mechanism of Io's volcanoes. The term *volcano* may be the wrong word altogether.

We have seen that material is ejected from volcanic vents on Io at much higher velocities than is observed in even the most explosive volcanic eruptions here on Earth. Another major difference is that Io's volcanic vents are not located at the tops of volcanic mountains. Many volcanoes on Earth, and Mars as well, have an easily recognized conical shape with a caldera at the summit. Few of Io's calderas (the black spots) are associated with major changes in elevation.

In all these respects, Io's volcanoes are more similar to terrestrial geysers than to terrestrial volcanoes. In a geyser, such as those in Yellowstone Park in Wyoming, water seeps down to volcanically heated rocks, is suddenly changed to steam, then erupts explosively through a vent. Geologist Susan Kieffer calculates that if the Old Faithful geyser were to erupt under the low gravity and vacuum that surround Io, this geyser would send a plume of water and ice to an altitude of 40 kilometers.

Both sulfur and sulfur dioxide are molten at depths of only a few kilometers below Io's surface, because of the heat generated by the tidal flexing of the satellite. The planetary geologists Eugene Shoemaker and Bradford Smith have pointed out that sulfur dioxide could be the principal propulsive agent driving Io's eruptions. Just as the explosive conversion of water into steam produces a geyser on Earth, a sudden conversion of liquid sulfur dioxide into a high-pressure gas could produce an eruption on Io. Kieffer's calculations indicate that this explosive expansion of sulfur dioxide could result in eruption velocities of up to 1000 m/s.

The Voyager's instruments failed to detect any other abundant gases, such as the water vapor and carbon dioxide that are emitted from terrestrial volcanoes. Apparently, Io has been completely outgased by volcanic activity extending over hundreds of millions of years. Io's surface gravity is comparable to that of our Moon, so Io has not been able to retain its volatile gases; this satellite has almost no atmosphere. The atmospheric pressure on Io, which is primarily caused by sulfur dioxide, is a scant 10^{-4} millibars.

Each volcano on Io ejects an estimated 10,000 tons of material per second. This material is a hot mixture of molten sulfur and sulfur-dioxide gas under high pressure as it first gushes from a volcanic vent, but the gas–liquid mixture rapidly cools and solidifies in the cold, nearly perfect vacuum around the satellite. It then takes about half an hour for the fine particles of sulfur dust and sulfur-dioxide snow to fall back down onto Io's surface.

Altogether, Io's volcanoes and vents eject roughly 1 trillion tons of matter each year. This amount is sufficient to cover the entire surface of Io to a depth of 10 m in a thousand years, or to alter the color of an area of a thousand square kilometers in a few weeks. Io's surface is thus constantly changing, so it is probably safe to say that there are no long-lived—or even semipermanent—features on this satellite. This rate of deposition of matter is easily sufficient to obliterate any impact craters rapidly.

Nearly 100,000 tons of sulfur and sulfur dioxide erupt from Io each second, but not all of this matter returns to the satellite's surface. Apparently, a small fraction—perhaps 1 ton per second—manages to escape Io's gravity and become part of Jupiter's magnetosphere. High temperatures in the magnetosphere easily strip one or more electrons from each sulfur or oxygen atom. The result is a huge doughnut-shaped ring called the **Io torus**, circling Jupiter at the distance of Io's orbit, that contains a plasma composed primarily of electrons and ions of sulfur and oxygen. Although these ions are not plentiful, they produce radiation that is detected by Earth-based astronomers (see Figure 14-11).

The Io torus lies in the plane of Jupiter's magnetic equator, inclined by 11° to the plane of Io's orbit.

The Voyager flybys also illuminated Io's role in the bursts of the decametric radio radiation (radiation with wavelengths of tens of meters) that come from Jupiter. Jupiter rotates once every ten hours, whereas Io takes 1.77 days to complete an orbit. Thus, Jupiter's magnetic field constantly sweeps past Io at high speed. Because of this rapid motion of the Jovian magnetic field, a complex, powerful electromagnetic interaction is set up between Io and Jupiter. Io develops a strong electric charge, and an electric current flows between it and Jupiter. Although details are not yet fully understood, this electric current is believed to be responsible for the decametric radiation bursts.

14-7 Europa is covered with a smooth layer of ice that is crisscrossed with numerous cracks

Voyager 1 did not pass near Europa, but *Voyager 2* got close enough to capture the excellent view shown in Figure 14-12. Europa is a smooth-surfaced world with no mountains and very few craters, crisscrossed by a spectacular series of streaks and cracks. Most of the cracks appear to be filled

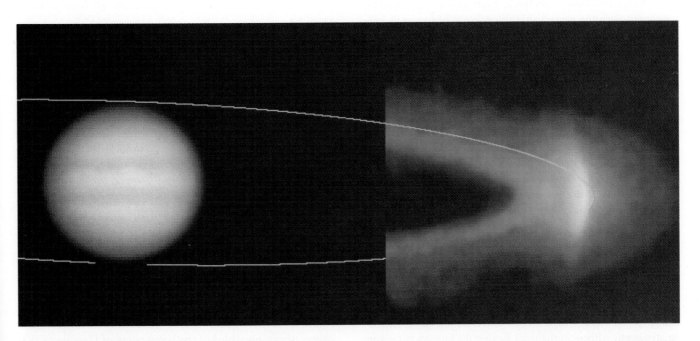

Figure 14-11 Io's plasma torus *This view of Io's plasma torus was obtained with an Earth-based telescope. Regions populated with S^{2+} ions appear in purple hues; regions dominated by S^+ appear in green. These ions come from sulfurous material ejected by Io's volcanoes. Because of the overwhelming glare of Jupiter, only the outer edge of the torus could be photographed. An artist added the drawing of Jupiter. (Courtesy of J. Trauger)*

Figure 14-12 Europa Europa's icy surface is covered by numerous streaks and cracklike features that give this satellite a fractured appearance. The streaks are typically 20 to 40 km wide. Surface features as small as 5 km across can be seen in this picture taken by Voyager 2. (NASA)

with dark-colored material, but some have a light-colored substance in them.

We have seen that Europa is only slightly less dense than our Moon. Spectroscopic observations from Earth indicate that Europa has frozen water on its surface. These two facts together suggest that Europa's surface may be covered with an ice layer 100 kilometers thick. Such a covering would be consistent with the remarkable smoothness of Europa's surface. An "ocean" of ice 100 km deep would certainly hide mountain ranges and other topographic features. But what has caused the network of cracks? And why are impact craters so rare on Europa?

With Jupiter on one side and the two largest Galilean moons periodically passing on its opposite side, Europa is caught in a tidal tug-of-war similar to Io's. However, the tidal effects of Jupiter on Europa are considerably less than those on Io. The gravitational force of Jupiter is inversely proportional to the square of the distance from the planet's center. However, the corresponding tidal force is inversely proportional to the cube of the distance. In other words, because Europa is about 1.6 times farther from Jupiter than Io is, the tidal effects of Jupiter on Europa are $1/(1.6)^3$, or only one-quarter as great as on Io.

NASA scientist Patrick Cassen has suggested that tidal flexing of Europa is responsible for the network of cracks covering its surface. Some of the darkest streaks do in fact follow paths along which the tidal stresses are calculated to be strongest. Although the tidal flexing on Europa is far too weak to produce volcanoes there, it is thought to supply enough energy to jostle and churn the satellite's icy coating. This movement would explain why only a few small impact

craters have survived, though Europa's surface is much older than Io's.

There are various arguments against Cassen's theory. Presumably, the streaks on Europa were caused by cracks in the icy coating through which water gushed up, then froze. These cracks are a few tens of kilometers wide. To accommodate this many cracks, Europa's surface area would have had to increase by 10 to 15 percent since the cracks first began to appear. Yet it seems unreasonable to suppose that Europa is expanding like an inflating balloon. Apparently, more is happening on Europa than meets the eye. Perhaps old surface material is somehow being pulled back down into the mushy layer below the ice coating and recycled, just as the Earth's crust is pulled back down into the Earth's mantle in subduction zones (recall Figure 10-19). Europa's surface may thus represent a water-and-ice version of plate tectonics.

14-8 Ganymede and Callisto have heavily cratered, icy surfaces

Only 100 km of ice and water on top of an otherwise rocky world suffices to explain Europa's density, but a much thicker layer of water and ice must surround Ganymede and Callisto. To be consistent with their average densities, which are slightly less than 2000 kg/m³, the rocky cores of these two outer satellites must be enveloped in mantles of water and ice nearly 1000 km thick. The diagrams in Figure 14-13 show the probable interior structure for all four Galilean satellites.

The two outer Galilean satellites exhibit the kind of ancient, cratered surface normally associated with Moonlike landscapes. Ganymede even looks somewhat like our Moon (see Figure 14-14). Of course, the craters on both Ganymede and Callisto are made of ice rather than rock.

Ganymede is the largest satellite in our solar system. Its diameter is 5276 km, which is slightly greater than the diameter of Mercury. Ganymede looks vaguely like our Moon, but there are significant differences. For one, Ganymede has two very different kinds of terrain, which are distinguished by both their appearance and their age (see Figure 14-15). Certain dark, polygon-shaped regions are presumed to be the oldest surface features on Ganymede, because they exhibit a high density of craters. Light-colored, heavily grooved terrain found between the dark, angular islands is much less cratered and is therefore younger.

The largest single feature on Ganymede is the vast, dark, circular island of ancient crust called Galileo Regio seen in Figure 14-14. It measures 4000 kilometers in diameter and covers nearly one-third of the hemisphere of Ganymede that faces away from Jupiter. It is the only surface feature on the Galilean satellites that can be detected with Earth-based telescopes.

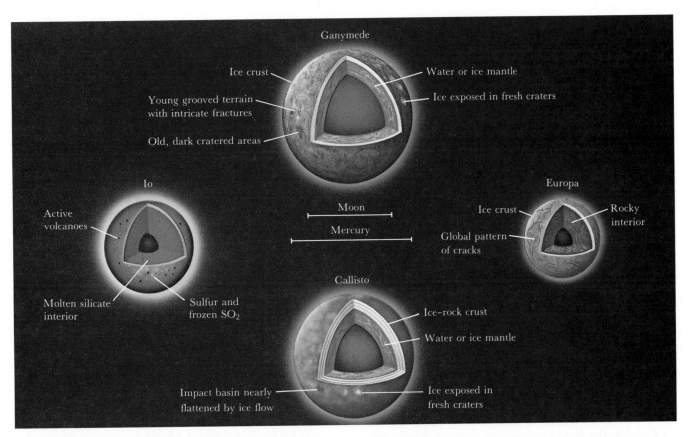

Figure 14-13 Interiors of the Galilean satellites *These cross-sectional diagrams show the probable internal structure of the four Galilean sat-* *ellites, based on their average densities and on information from the Voyager flybys.*

Figure 14-14 Ganymede *This view from Voyager 2, taken at a distance of 1.2 million km, shows the hemisphere that always faces away from Jupiter. This hemisphere is dominated by a huge, dark circular area called Galileo Regio, which is the largest remnant of Ganymede's ancient crust. (NASA)*

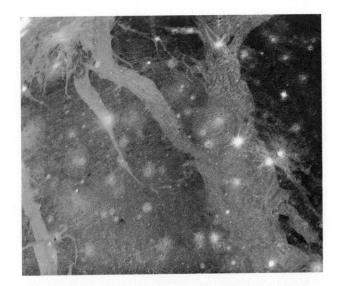

Figure 14-15 Young and old terrain on Ganymede *This close-up view of Ganymede was taken by Voyager 2 at a range of only 312,000 km. Features as small as 5 km across can be seen. Dark, angular islands from Ganymede's ancient crust are separated by younger, light-colored, grooved terrain. The southwest edge of Galileo Regio appears at the right side of the picture. (NASA)*

It is easy to distinguish between the young and the old craters on Ganymede. The youngest craters are surrounded by bright rays of freshly exposed ice (see Figure 14-15). Older craters clearly have been covered with deposits of dark meteoritic dust. The most ancient craters—sometimes called ghost craters or palimpsests, an archeological term for a parchment that was scraped clean and reused—are barely visible on Galileo Regio and other dark islands made of old crust. The degradation and near obliteration of the oldest craters probably also involved a slow plastic flow of Ganymede's icy surface.

The Voyager photographs also show that the small craters on Ganymede are better preserved than the large ones. Eugene Shoemaker of the U.S. Geological Survey points out that the preservation of Ganymede's craters is linked to the thermal and structural history of the satellite's icy crust. When Ganymede formed 4.5 billion years ago, it may have been completely covered with an ocean of water roughly 1000 km deep. During perhaps the next 200 million years, the water cooled and a thick coating of ice developed. Today this layer of solid ice is probably about 100 km thick. Beneath it lies a 900-km-thick slushy mantle of water and ice. The dark, angular islands are remnants of the ancient original crust.

Craters also tell us about the history of the younger, light-colored terrain, which is covered with numerous grooves. Cratering on this grooved topography varies in density, from about the same as on the ancient crust down to one-tenth that amount. We thus suspect that this grooved terrain

formed over a long time. The process probably began quite early in Ganymede's history and continued through the period of intense meteoritic bombardment. The age of the grooved terrain thus ranges from about 4.5 to 3.5 billion years.

High-resolution photographs, such as Figure 14-16, show that this grooved terrain actually consists of parallel mountain ridges up to 1 km high, spaced 10 to 15 km apart. These features suggest that the process of plate tectonics may have dominated Ganymede's early history. Water seeping upward through cracks in the satellite's original crust would freeze and force apart fragments of that crust, producing jagged, dark islands of old crust separated by bands of younger, light-colored, heavily grooved ice. The cracks thus play a role in Ganymedean plate tectonics that is analogous to the role of the oceanic rifts on Earth. However, unlike thinner-crusted Europa, where tectoniclike activity may still occur today, tectonics on Ganymede bogged down 3 billion years ago, when the satellite's cooling crust gradually froze to unprecedented depths.

Callisto, Jupiter's outermost Galilean satellite, looks very much like Ganymede: Numerous impact craters are scattered over a dark, ancient, icy crust (see Figure 14-17). There is one obvious difference, however: Callisto has no younger, grooved terrain. From this fact we can infer that tectonic activity never began on Callisto. Perhaps because of Callisto's greater distance from Jupiter, the ocean that enveloped the young satellite 4.5 billion years ago froze more rapidly and to a greater depth than on Ganymede, forever preventing tectonic processes. Callisto's icy crust may in fact

Figure 14-16 Grooved terrain on Ganymede *This picture, taken by Voyager 1 at a range of 145,000 km, shows an area roughly as large as Pennsylvania. The smallest visible features are about 3 km across. The numerous parallel mountain ridges visible here are spaced 10 to 15 km apart and have heights up to 1000 m. (NASA)*

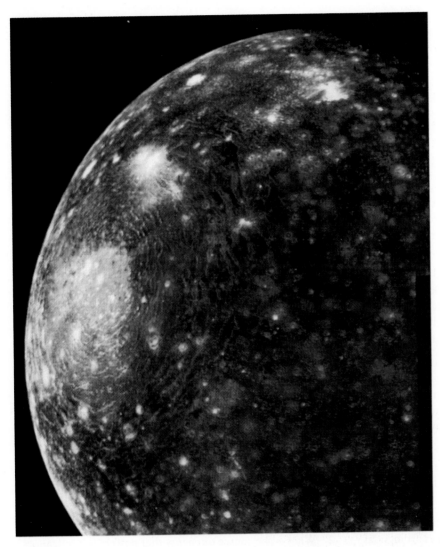

Figure 14-17 Callisto *Callisto, the outermost Galilean satellite, is almost exactly the same size as Mercury. Numerous craters pockmark Callisto's icy surface, as seen in this mosaic of views from* Voyager 1 *taken at a distance of about 400,000 km. Note the series of faint, concentric rings that cover the left third of the image. These rings outline a huge impact basin called Valhalla that dominates the Jupiter-facing hemisphere of this frozen, geologically inactive world. (NASA)*

be several times thicker than Ganymede's and extend to depths of several hundred kilometers. It is bitterly cold on Callisto: Voyager instruments measured a noontime temperature of −118°C (−180°F), and the nighttime temperature plunges to −193°C (−315°F).

Voyager 1 photographed a huge impact basin on Callisto seen faintly in Figure 14-17. Located on Callisto's Jupiter-facing hemisphere, this so-called Valhalla Basin was produced by the impact of an asteroid-sized object early in the satellite's history. Valhalla consists of a large number of concentric rings, 50 to 200 km apart and having diameters up to 3000 km.

The great age of Valhalla can be inferred both from the presence of impact craters there and from the absence of substantial vertical relief in the concentric rings. The Valhalla impact occurred probably around 4 billion years ago,

when the satellite's young, relatively thin crust still let plastic flow reduce the height of the upraised rings in the ice.

The Voyager pictures also showed traces of a Valhallalike impact on Ganymede's Galileo Regio. Segments of a system of concentric rings cover a large portion of this dark island of ancient crust. However, no obvious impact feature exists at the geometric center of this ring system. Apparently, as grooved terrain developed, it completely obliterated the impact basin.

In looking at the Galilean satellites, we see a neat, orderly progression from a high-density, geologically active world (Io) nearest Jupiter to a low-density, geologically dead world (Callisto) farthest from Jupiter. As the Voyager spacecraft sped on to rendezvous with Saturn, many scientists expected to find there another orderly system neatly laid out for their examination. They were in for a rude surprise.

Key words

Key ideas

- The four Galilean satellites of Jupiter orbit the planet in the plane of its equator, with synchronous rotation. Their periods of rotation and revolution about the planet are relatively short (from about 2 to 17 days).

  The two inner Galilean moons, Io and Europa, have roughly the same size and density as our Moon. The two outer Galilean moons, Ganymede and Callisto, are roughly the size of Mercury and are lower in density than either the Moon or Mercury.

  Several small moons and a faint ring of mostly dust also orbit in the plane of Jupiter's equator, inside the orbit of Io. Eight small moons move in much larger orbits that are noticeably inclined to the plane of Jupiter's equator; some have retrograde orbits.

- Io is covered with a colorful layer of sulfur compounds deposited by frequent explosive eruptions from volcanic vents.

  Io's volcanic eruptions resemble terrestrial geysers. The energy heating Io's interior comes from tidal forces that flex the moon.

  The Io torus is a ring of electrically charged particles circling Jupiter at the distance of Io's orbit. Interactions between this ring and Jupiter's magnetic field produce the strong decametric radio emissions associated with Jupiter.

- Europa is covered with a smooth layer of frozen water crisscrossed by an intricate pattern of long cracks that are probably produced by tidal flexing of the moon.

- The heavily cratered surface of Ganymede is composed of frozen water with large polygons of dark, ancient surface separated by regions of heavily grooved, lighter-colored, younger terrain.

  Plate tectonics apparently operated during the early history of Ganymede.

- Callisto has a heavily cratered crust of frozen water, but plate tectonics apparently never operated on this moon, presumably because it quickly developed a thick, solid crust.

  The impact basin called Valhalla on Callisto provides evidence for plastic flow in the icy crust of this moon.

- The inner moons, Galilean satellites, and ring of Jupiter probably all formed through a process of accretion, similar to that which formed our solar system but on a smaller scale; the outer moons were possibly asteroids captured later by Jupiter's gravity.

- The Galilean satellites help us understand physical processes on the terrestrial planets, because they provide examples of how these processes operate under conditions quite different from those prevalent on the inner planets of our solar system.

Review questions

1 How does the Galilean satellite system resemble our solar system? How is it different?

2 What is the source of energy that powers Io's volcanoes?

3 With all its volcanic activity, why doesn't Io possess a thick atmosphere?

4 Why are numerous impact craters found on Ganymede and Callisto—but not on Io or Europa?

5 How would you account for the existence of the non-Galilean satellites of Jupiter as well as for Jupiter's ring?

6 Long before the Voyager flybys, Earth-based astronomers reported that Io appeared brighter than usual for the few hours after it emerged from Jupiter's shadow. From what we know about the material ejected from Io's volcanoes, explain this brief brightening of Io.

7 Compare and contrast the surface features of the four Galilean satellites, discussing their relative geological activity and the evolution of these four satellites.

8 What is the Io torus and what is its source?

9 Compare and contrast Valhalla (see Figure 14-17) with Mare Orientale on our Moon and the Caloris Basin on Mercury.

10 Describe the plans for the *Galileo* orbiter to conduct long-term surveillance of the Galilean satellites. Why will this spacecraft make many close approaches to Europa, Ganymede, and Callisto, but not to Io?

Advanced questions

> **Tips and tools . . .**
> Kepler's third law tells us that P^2 divided by a^3 is a constant. The small-angle formula is discussed in Box 1-1. The best seeing conditions on Earth give a seeing disk $\frac{1}{4}$ arc second in diameter. The orbits of the Galilean satellites are almost perfect circles, and so the orbital speeds of these satellites can be easily calculated from the data listed in Box 14-1.

11 (*Basic*) Using orbital data given in Box 14-1, demonstrate that the Galilean satellites obey Kepler's third law.

12 Invent a system of units in which $P^2 = a^3$ for Jupiter's satellites.

*13 (*Basic*) What is the size of the smallest feature you should be able to see on a Galilean satellite though a large telescope under conditions of excellent seeing when Jupiter is near opposition?

*14 Using the diameter of Io (3630 km) as a scale, estimate the height to which the plume of Prometheus rises above the surface of Io in Figure 14-8.

*15 (*Challenging*) If material is ejected into space from Io's volcanoes at the rate of 1 ton/s, how long will it be before Io loses 10 percent of its mass? How does your answer compare with the age of the solar system?

*16 (*Challenging*) How long does it take for Ganymede to enter or leave Jupiter's shadow?

17 Consult magazines like *Science News* or *Sky & Telescope* to learn about the current status of the Galileo mission. Is all well with the spacecraft? When is it scheduled to arrive at Jupiter?

Discussion questions

18 Speculate on the possibility that Europa, Ganymede, or Callisto might harbor some sort of marine life.

19 Suppose you were planning four missions to the moons of Jupiter that would land a spacecraft on each of the Galilean satellites. What kinds of questions would you want these missions to answer, and what kinds of data would you want your spacecraft to send back? In view of the different environments on the four satellites, how would the designs of the four spacecraft differ? Be specific about the possible hazards and problems each spacecraft might encounter in landing on the four satellites.

Observing projects

20 Observe Jupiter through a pair of binoculars. Can you see all four Galilean satellites? Make a drawing of what you observe.

21 Observe Jupiter through a small telescope on three or four consecutive nights. Make a drawing each night showing the positions of the satellites relative to Jupiter. Record the time and date of each observation. Consult the section called "Satellites of Jupiter" in the *Astronomical Almanac* for the current year to see if you can identify the satellites by name. This section of the *Astronomical Almanac* shows the apparent paths of the Galilean satellites as they move back and forth, from one side of Jupiter to the other. Note that universal time (UT) is used in the *Astronomical Almanac*, and so you must convert your local time to UT. Universal time, which is the same as Greenwich mean time, is counted from 0 hours beginning at midnight in Greenwich, England. Thus, for instance, to convert Eastern Standard Time (EST) to universal time you must add five hours. For example, 10:30 PM EST on January 3 is the same as 3:30 UT on January 4.

22 Make arrangements to observe an eclipse, a transit, or an occultation of one of the Galilean satellites. Consult a listing of such phenomena in the section called "Satellites of Jupiter" in the *Astronomical Almanac* for the current year. Choose the phenomenon you would like to see and calculate its scheduled time by converting the universal time given in the *Astronomical Almanac* to your local time. Arrange to be at a telescope at the scheduled time to observe the phenomenon you have selected.

23 If you are fortunate enough to have access to a large telescope with a primary mirror 1 meter or more in diameter, make arrangements to view Jupiter through that telescope. Describe what you see. Under conditions of excellent seeing when Jupiter is near opposition, the Galilean satellites should look like tiny discs rather than starlike pinpoints of light.

For further reading

Carroll, M. "Project Galileo: The Phoenix Rises." *Sky & Telescope*, April 1987 • This brief article discusses the Galileo mission.

Croswell, K. "Io: Jupiter's Fiery Satellite." *Space World*, July 1988 • This well written article surveys current understanding of Io.

Griffin, R. "Barnard and His Observations of Io." *Sky & Telescope*, November 1982 • This brief article discusses pre-Voyager visual observation of Io, including some attempts to map its surface.

Johnson, T. "The Galilean Satellites." In Beatty, J., et al., eds. *The New Solar System.* 3rd ed. Sky Publishing and Cambridge University Press, 1990 • A well-written, up-to-date summary of our current understanding of the Galilean satellites.

_____, and Soderblom, L. "Io." *Scientific American*, December 1983 • This article, which is devoted to volcanism on Io, includes an enlightening discussion of the properties of sulfur and sulfur dioxide.

Morrison, D. "Four New Worlds: The Voyager Exploration of Jupiter's Satellites." *Mercury*, May/June 1980 • This article by a noted planetary scientist discusses what we have learned from the Voyager flybys.

Soderblom, L. "The Galilean Moons of Jupiter." *Scientific American*, January 1980 • An excellent selection of photographs accompany this fine article, which summarizes the discoveries made during the Voyager flybys.

CHAPTER 15

The Spectacular Saturnian System

Saturn is a fascinating and beautiful world primarily because of the system of thin, flat rings that encircles it. In the early 1980s, two Voyager flybys led to many important discoveries about Saturn, its rings and many satellites. These explorations allowed a close comparison of the atmospheric structures of Saturn and Jupiter, suggesting differences in how these two planets evolved. The Voyager spacecraft glimpsed surface features on many of Saturn's satellites and showed that some of these satellites profoundly affect the structure and appearance of the rings. Most interesting is Titan, Saturn's largest moon, which is enveloped by a thick nitrogen-rich atmosphere where methane snow and rain may fall beneath a smog of carbon–hydrogen compounds. These worlds pose many new puzzles for planetary scientists that could be answered by the *Cassini* spacecraft, tentatively scheduled to orbit the ringed planet early in the twenty-first century.

Jupiter, Saturn, and Earth This montage shows Jupiter, Saturn, and Earth reproduced to the same scale. Saturn, like Jupiter and the Sun, is composed primarily of hydrogen and helium. Note that the cloud features on Saturn are much less distinct than those on Jupiter. Saturn has a lower surface gravity than does Jupiter, which causes the Saturnian cloud layers to be at greater depths in Saturn's atmosphere than in Jupiter's. The sunlight reflected from these deeper cloud layers thus suffers a greater amount of absorption, giving Saturn a faded appearance compared to Jupiter. (S. P. Meszaros; NASA)

The magnificent rings of Saturn make this planet one of the most spectacular objects visible in the nighttime sky to an amateur astronomer using even a small telescope. Saturn is so far away, however, that our Earth-based telescopes can reveal only the coarsest, large-scale features. Astronomers were totally unprepared for the breathtaking vistas that unfolded as the Voyager spacecraft sped past Saturn's shimmering cloudtops in the early 1980s. (Box 15-1 lists basic data about Saturn.)

15-1 Earth-based observations reveal gaps in Saturn's rings as well as faint markings on the planet

When Galileo focused his telescope on Saturn in the early 1600s, he saw few details, but he did notice two puzzling lumps protruding from opposite edges of the planet's disk. In 1655, the Dutch astronomer Christian Huygens observed Saturn with a better telescope and noted no protrusions. Having faith in Galileo's observations, Huygens suggested that Saturn might be surrounded by a flattened ring that just happened to be edge-on as viewed from Earth in 1655, making it temporarily almost impossible to see. This brilliant deduction was confirmed ten years later, by which time Saturn had moved far enough along its orbit for the ring to be seen again.

As the quality of telescopes improved, details of Saturn's ring and its cloud cover became visible. In 1675, G. D. Cassini discovered a dark division in the ring, an apparent gap about 5000 kilometers wide. Astronomers also discovered

Figure 15-1 Saturn from the Earth *This view of Saturn is one of the best ever produced by an Earth-based observatory. Sixteen original color images taken on the same night with the 1.5-m telescope at the Catalina Observatory were combined to make this photograph. Note the prominent Cassini division in the rings, and the belts and zones in the Saturnian atmosphere. (NASA)*

stripes in Saturn's clouds similar to the belts and zones on Jupiter (see Figure 15-1). The contrast between Saturn's belts and zones is less dramatic than the colorful patterns in Jupiter's atmosphere.

After Cassini's discovery of the gap in Saturn's ring, astronomers began to view this ring as a system of rings. The **Cassini division** separates the outer **A ring** from the brighter **B ring** closer to the planet. By the mid-1800s, astronomers using improved telescopes could detect a faint **C ring**, or crepe ring, that lies just inside the B ring (see Figure 15-2).

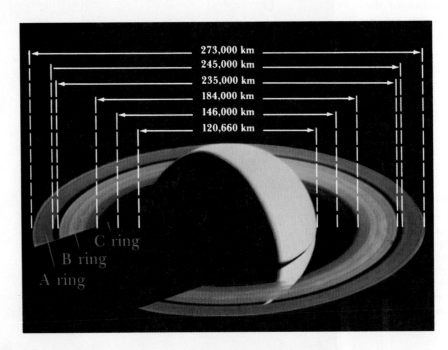

Figure 15-2 Saturn's classic rings *Details of Saturn's rings are visible in this photograph sent back by Voyager 1. The equatorial diameter of Saturn as well as the diameters of the inner and outer edges of the rings are given in the overlay. Closest to the planet is the 19,000-km-wide C ring, so faint that it is almost invisible in this view. Next outward from Saturn is the broad, bright B ring, whose width is 25,500 km. The outermost ring that can be seen from Earth is the A ring, which is 14,000 km wide. The 5000-km-wide Cassini division lies between the B ring and the A ring. (NASA)*

Box 15-1 Saturn data

Mean distance from the Sun:	9.529 AU = 1.426×10^9 km
Maximum distance from the Sun:	10.044 AU = 1.503×10^9 km
Minimum distance from the Sun:	9.014 AU = 1.348×10^9 km
Mean orbital velocity:	9.6 km/s
Sidereal period:	29.41 years
Rotation period:	equatorial = $10^h\ 13^m\ 59^s$
	internal = $10^h\ 39^m\ 25^s$
Inclination of equator to orbit:	27°
Inclination of orbit to ecliptic:	2° 29′
Orbital eccentricity:	0.054
Diameter:	equatorial = 120,000 km
	polar = 107,100 km
Diameter (Earth = 1):	equatorial = 9.26
	polar = 8.27
Apparent diameter as seen from Earth:	maximum = 20.9″
	minimum = 15.0″
Mass:	5.69×10^{26} kg
Mass (Earth = 1):	95.3
Mean density:	700 kg/m³
Surface gravity (Earth = 1):	1.07
Escape velocity:	35.6 km/s
Oblateness:	0.108
Mean surface temperature (at cloudtops):	−180°C = −292°F = 93 K
Albedo:	0.76
Brightest magnitude:	−0.3
Mean diameter of Sun as seen from Saturn:	3′ 22″

Earth-based views of the Saturnian ring system change dramatically as Saturn orbits slowly about the Sun (a Saturnian year is equal to 29½ Earth years; see Box 15-2 for a listing of Saturn oppositions during the 1990s). This change is observed because the rings, which lie in the plane of Saturn's equator, are tilted 27° from the plane of Saturn's orbit. Thus, over the course of a Saturnian year, the rings are viewed from various angles by an Earth-based observer (see Figure 15-3). At one time, the observer looks "down" on the rings, but half a Saturnian year later the "underside" of the rings is exposed to our Earth-based view. In between, the rings are viewed edge-on, when they seem to disappear entirely. This disappearance indicates that the rings are very thin—less than 2 km thick, according to recent estimates. The last edge-on presentation of Saturn's rings was in 1980, and the next will occur in 1995.

15-2 Saturn's rings are composed of numerous fragments of ice and ice-coated rock

Astronomers have long known that Saturn's rings could not possibly be solid, rigid, thin sheets of matter. In 1857 the Scottish physicist James Clerk Maxwell proved mathematically that such a broad, thin sheet would break apart. He therefore concluded that Saturn's rings were composed of "an indefinite number of unconnected particles."

Supporting observational evidence came four decades later when James Keeler at the Lick Observatory observed Doppler shifts in sunlight reflected from Saturn's rings. By measuring the displacement of spectral lines of sunlight reflected from Saturn's rings, Keeler determined that the inner

Figure 15-3 **The changing appearance of Saturn's rings** *Saturn's rings are tilted 27° from the plane of Saturn's orbit. Earth-based observers therefore see the rings at various angles as Saturn moves around its orbit. Note that the rings seem to disappear entirely when viewed edge-on. (Lowell Observatory)*

Box 15-2 Oppositions of Saturn, 1991–1999

The synodic period of Saturn is 378 days, or approximately 1 year and 2 weeks. This interval between successive oppositions is obvious in the following table, which lists oppositions during the last decade of this century.

Date of opposition	Magnitude
1991 July 26	+ 0.3
1992 August 7	+ 0.4
1993 August 19	+ 0.5
1994 September 1	+ 0.7
1995 September 14	+ 0.8
1996 September 26	+ 0.7
1997 October 10	+ 0.4
1998 October 23	+ 0.2
1999 November 6	0.0

portions of Saturn's rings are moving more rapidly about the planet than are the outer portions, which would be expected if the rings were formed of numerous tiny moonlets circling Saturn along their own individual orbits. The orbital speeds measured for various parts of the rings are in fact in complete agreement with the orbital velocities predicted from Kepler's third law.

Saturn's rings are very bright, with an albedo of 0.8, so the particles that form the rings must be highly reflective. Astronomers long suspected the rings were made of ice and ice-coated rocks, but confirming evidence was not obtained until the 1970s, when American astronomers Gerard Kuiper and Carl Pilcher identified the spectral characteristics of frozen water in the near-infrared spectrum of the rings. Additional measurements from Earth-based observatories and the Voyager spacecraft tell us that the temperature of the rings ranges from −180°C (−290°F) in the sunshine to less than −200°C (−300°F) in Saturn's shadow. Water ice is in no danger of melting or evaporating at these temperatures.

Changes in the radio signals received from *Voyager* as it passed behind the ring system enabled astronomers to estimate the particle sizes as ranging from snowflakes less than 1 mm in diameter up to icy boulders tens of meters across. It seems reasonable to suppose that all this material is ancient debris that failed to accrete (fall together) into satellites. As a matter of fact, the ring particles are so close to Saturn that they will never be able to form moons.

Imagine a collection of small fragments of rock orbiting a planet. The gravitational attraction of neighboring rocks tends to pull the rock fragments together to produce a larger satellite. However, because the various rock fragments are at differing distances from the parent planet, they experience different amounts of gravitational pull from the planet. The difference in strength of these gravitational forces collectively produces an effect called a **tidal force**, which tends to keep the fragments separated. At a certain distance from the planet's center called the **Roche limit**, these attractive and disruptive effects are exactly balanced. Inside the Roche limit, fragments will not accrete to form a larger body; instead, they will tend to spread out into a ring around the planet. All large satellites are found only outside a planet's Roche limit. If any large moon were to come inside a planet's Roche limit, the planet's tidal forces would cause the satellite to break up into fragments. The Roche limit for Saturn lies just outside the outer edge of the A ring. As we shall see in the next chapter, such a tidal disruption may be the catastrophic fate of Neptune's large satellite, Triton, whose orbit is gradually decaying.

The Roche limit applies only to objects held together by gravity. In a rock fragment, chemical bonds (electromagnetic forces) between atoms and molecules hold the rock together. These chemical forces are much stronger than the disruptive tidal force of a nearby planet, so the rock does not break apart. In the same way, people walking around on the Earth's surface (which is inside the Earth's Roche limit) are in no danger of coming apart, because we are held together by comparatively strong intermolecular forces rather than gravity.

15-3 The internal structure of Saturn was deduced from measurements of its flattened shape

Although Earth-based observations of Saturn indicated that the composition and structure of its atmosphere are similar to those of Jupiter, there are significant differences between the two planets. The average density of Saturn is 700 kg/m³, only about half that of Jupiter. Such a low density means that Saturn must be composed largely of very light elements. The spectral lines of methane (CH_4) and ammonia (NH_3) that appear in sunlight reflected from Saturn are a strong indication of a high abundance of hydrogen.

Saturn is even more oblate than Jupiter (examine Figure 15-1). Saturn's equatorial diameter is about 10 percent larger than its polar diameter—that is, its oblateness is about 0.11. As explained in Chapter 13, the oblateness of a planet is directly related to its speed of rotation and to the degree to which its mass is concentrated near its center. Saturn rotates a little more slowly than Jupiter does: A solar day near Saturn's equator is about 24 minutes longer than a solar day near Jupiter's equator. The greater oblateness of Saturn thus cannot be caused by faster rotation, so it must result from a greater concentration of mass at Saturn's center. This concentration cannot be caused by greater compression, because Saturn is smaller and less dense than Jupiter. We may conclude, therefore, that Saturn's rocky core must be larger and more massive than Jupiter's. Detailed calculations suggest that about 26 percent of Saturn's mass is contained in its rocky core, whereas Jupiter's core contains only about 4 percent of its entire mass.

Despite Saturn's more massive core, its average density (700 kg/m³) is much less than that of Jupiter (1310 kg/m³), so Saturn's density must decrease more rapidly than Jupiter's, proceeding outward from the core toward the surface. From information about Saturn's average density, oblateness, and probable chemical composition, astronomers have constructed a model of the planet's internal structure. This model suggests that, in a general way, Saturn resembles Jupiter in structure: a solid, rocky core surrounded by a mantle of liquid metallic hydrogen, which in turn is surrounded by a layer of liquid and gaseous molecular hydrogen (see Figure 15-4).

Saturn's mantle of liquid metallic hydrogen, like Jupiter's, is thought to be the source of the planet's magnetic field. Information about Saturn's magnetosphere was sent back during the *Pioneer 11* flyby in August 1979 (see Figure 15-5). Saturn's magnetic field turns out to be somewhat weaker than Jupiter's, as expected, because of Saturn's slightly slower rotation and much smaller volume of liquid metallic hydrogen.

Saturn's magnetosphere is intermediate between those of Jupiter and Earth. It has a bow shock, a magnetosheath, a magnetopause, and an elongated tail pointing away from the

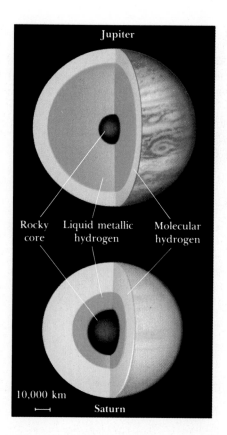

Figure 15-4 The internal structures of Saturn and Jupiter There are three distinct layers in Saturn's interior, as in Jupiter's. Each planet's rocky core is surrounded by a layer of liquid metallic hydrogen, which in turn is enveloped in a thick layer of liquid and gaseous molecular hydrogen. Both diagrams are drawn to the same scale.

Sun. The position of the bow shock varies between 1 and 1.5 million kilometers from the planet because of variations in the solar wind pressure.

Although Saturn's magnetic field is almost as strong as Jupiter's, it contains far fewer charged particles. There are two reasons for this deficiency of charged particles in Saturn's magnetosphere. First, Saturn lacks a continuous source of particles, like the volcanoes of Io that dump one ton of material into Jupiter's magnetosphere every second. Second, many charged particles are absorbed by the fragments in Saturn's rings, thus depleting the particle concentration in the inner magnetosphere. The charged particles that do exist in Saturn's magnetosphere are concentrated in radiation belts similar to the Van Allen belts in the Earth's magnetosphere.

15-4 Like Jupiter, Saturn emits more radiation than it receives from the Sun

Both Jupiter and Saturn have internal sources of energy. Each planet radiates more energy than it receives as sunlight. Jupiter's internal heat is thought to be the residual energy trapped as the planet accreted from the original solar nebula

Figure 15-5 The best view of Saturn from Pioneer 11 Pioneer 11 *was the first of three spacecraft to visit Saturn. Although the Pioneer instruments were designed primarily to measure physical conditions around the planet, the spacecraft did send back pictures of modest quality. This view was taken at a range of 2½ million km. (NASA)*

4.5 billion years ago. Jupiter has been slowly cooling off ever since, as this energy escapes in the form of infrared radiation.

Saturn is both smaller and less massive than Jupiter. One would thus expect Saturn to have cooled more rapidly than Jupiter and hence to emit less energy today. But in fact Saturn radiates about 2½ times as much energy as it receives from the Sun, whereas Jupiter emits only about 1½ times the heat it absorbs from sunlight. What might explain why Saturn emits so much more heat than Jupiter?

For some time before the space probes to the Jovian planets, astronomers had suspected both Jupiter and Saturn to have compositions similar to that of the original solar nebula, and to that of the Sun's atmosphere today. Each of these giant planets is both massive enough and cool enough to have retained all the gases that originally accreted from the solar nebula. The Voyager flybys (see Figure 15-6) did confirm that Jupiter has an abundance of elements (82 percent hydrogen, 17 percent helium, and 1 percent all other elements, by weight) that is similar to the solar abundance. Surprisingly, however, the Voyager spacecraft reported that Saturn's atmosphere has less helium than expected. The chemical composition of Saturn's atmosphere is, by weight, 88 percent hydrogen, 11 percent helium, and 1 percent all other elements.

Edwin E. Salpeter of Cornell University and David J. Stevenson of the California Institute of Technology (Caltech) offered a brilliant hypothesis that links Saturn's apparent deficiency of helium to the excess heat radiated by the planet. According to this theory, Saturn did indeed cool more rapidly than Jupiter, which triggered a process analogous to the development of a rainstorm here on Earth. When the air is cool enough, humidity in the Earth's atmosphere condenses into raindrops, which fall to the ground. On

Figure 15-6 A Voyager flyby After passing by Jupiter, both Voyager spacecraft continued on to Saturn. Voyager 1 passed near Saturn in November 1980, then headed out of the solar system. Voyager 2 coasted past Saturn in August 1981 and then continued on to encounter Uranus in January 1986 and Neptune in August 1989. (NASA)*

Figure 15-7 Saturn from Voyager 2 *Voyager 2 sent back this picture when the spacecraft was still 2 months and 34 million km away from its closest approach to the planet. Faint belts and zones are clearly visible. (NASA)*

Figure 15-8 Saturn's clouds from Voyager 1 *This view of Saturn's cloudtops was taken by Voyager 1 at a range of 1¾ million km. Note that there is substantially less contrast between the belts and zones on Saturn than there is on Jupiter. (NASA)*

Saturn, though, it is helium droplets that rain downward through the planet's atmosphere toward its core. Helium is deficient in Saturn's upper atmosphere simply because it has fallen farther down into the planet. Furthermore, as the helium droplets descend toward the planet's core, their gravitational energy is converted into thermal energy (heat) that eventually escapes from Saturn's surface.

The precipitation of helium from Saturn's clouds is calculated to have begun 2 billion years ago. The energy released adequately accounts for the extra heat radiated by Saturn since that time. Similar calculations for Jupiter indicate that it is only now reaching the stage where a significant amount of helium precipitation can begin in its outer layers. Saturn has therefore given us important clues about the probable course of Jupiter's future evolution.

15-5 Saturn's atmosphere extends to a greater depth and has higher wind speeds than Jupiter's atmosphere

Saturn's atmosphere, like Jupiter's, contains trace amounts of methane (CH_4), ammonia (NH_3), and water vapor (H_2O). These compounds are the simplest combinations of hydrogen with carbon, nitrogen, and oxygen. Also like Jupiter, Saturn has three distinct cloud layers: an upper layer of frozen ammonia crystals; a middle layer of crystals of ammonia hydrosulfide (NH_4SH), and a lower layer of frozen water crystals.

Although their atmospheres have similar structures and compositions, Saturn and Jupiter are by no means identical in appearance. Saturn's clouds lack the colorful contrast of Jupiter's (see Figure 15-7), although some of the Voyager photographs do show faint hints of belts and zones, as

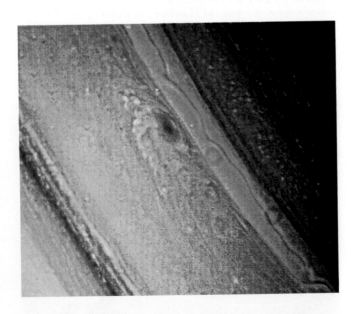

Figure 15-9 Eddy currents in Saturn's atmosphere Computer processing exaggerates the colors in this Voyager 2 picture of Saturn's northern middle latitudes. The wavy line in the light blue ribbon is a pattern moving eastward at 150 m/s (340 mi/hr). The dark oval and the two puffy, blue-white spots below it are eddies drifting westward at roughly 20 m/s (45 mi/hr). (NASA)

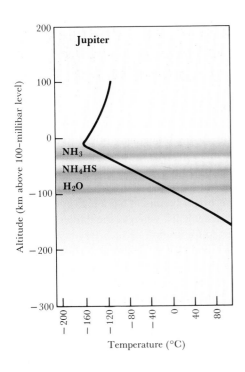

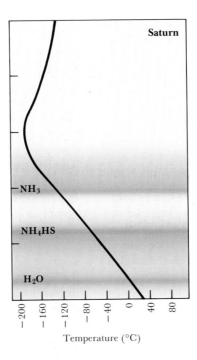

Figure 15-10 **The temperature profiles of Jupiter and Saturn** *The structures of the upper atmospheres of Jupiter and Saturn are displayed on these graphs of temperature versus depth. Note that Saturn's atmosphere is more "spread out" than Jupiter's, as a direct result of Saturn's weaker surface gravity. (Adapted from A. P. Ingersoll)*

shown in Figure 15-8. After special computer enhancement to exaggerate greatly the colors in the Voyager photographs, details such as storm systems and ovals became visible (see Figure 15-9).

The different appearances of Saturn and Jupiter are related to the different masses of the two planets. Jupiter's strong surface gravity compresses its atmosphere. Its three cloud layers are compressed into a range of 75 km in the planet's upper atmosphere. Saturn's somewhat weaker surface gravity subjects its atmosphere to less compression, so

that the same three cloud layers are spread out over a range of nearly 300 km (see Figure 15-10). The colors of Saturn's clouds are less dramatic than Jupiter's because deeper layers are partly obscured by the thick atmosphere above them.

By following features in the Saturnian clouds, scientists have determined the wind speeds in the planet's upper atmosphere. There are counterflowing easterly and westerly currents in Saturn's upper atmosphere, as in Jupiter's. However, Saturn's equatorial jet is much broader—and much faster—than Jupiter's (see Figure 15-11). In fact, wind speeds near

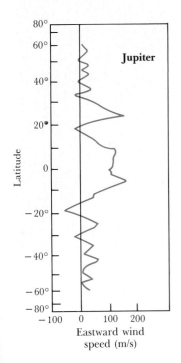

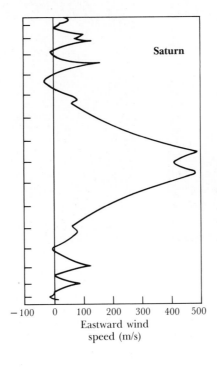

Figure 15-11 **Wind speeds on Jupiter and Saturn** *The average wind speeds on Jupiter and Saturn are plotted here for latitudes ranging from 80° north to 80° south. The positive numbers are eastward velocities; the negative numbers are westward velocities. Although both planets exhibit counterflowing currents, Saturn's equatorial zonal jet is much broader and faster than Jupiter's.*

Saturn's equator approach 500 m/s (1100 miles per hour), which is approximately two-thirds the speed of sound in Saturn's atmosphere.

15-6 Saturn's rings consist of thousands of narrow, closely spaced ringlets

During the Voyager flybys of 1980 and 1981, the spacecrafts' cameras sent back pictures showing unexpected details in the ring structures of Saturn. The broad rings were seen to consist of hundreds upon hundreds of closely spaced bands, or **ringlets**, of particles (see Figure 15-12). Although intriguing explanations have been proposed, scientists still do not understand just why Saturn's A, B, and C rings are divided into thousands of thin ringlets.

The Voyager cameras also sent back high-quality pictures of the **F ring**, which was first detected during the 1979 flyby of *Pioneer 11*. The F ring is visible just beyond the outer edge of the A ring in Figure 15-12. Close-up views revealed a startling and mysterious fact: The F ring is kinky and braided; it actually consists of several intertwined strands (see Figure 15-13). One *Voyager 2* image shows a total of five strands, each about 10 km across. Voyager scientists were at a loss to explain this complex structure. The braids, kinks, knots, and twists in the F ring pose a challenging puzzle.

15-7 The sizes and concentration of particles in the rings were deduced from studies of scattered sunlight

Through Earth-based telescopes, we can see only the sunlit side of Saturn's rings. From this perspective, the B ring appears very bright, the A ring moderately bright, the C ring dim, and the Cassini division is dark. The proportion of sunlight reflected back toward the Sun—the albedo—is directly related to the concentration of the fragments or particles in the ring. The B ring is bright because it has a high concentration of ice and rock fragments, whereas the darker Cassini division has a lower concentration of such fragments.

Eight hours after it had crossed from the northern to the southern side of the rings, *Voyager 1* took the photograph that appears as Figure 15-14. The Sun was shining down on the northern side of the rings at the time of the Voyager flybys, so Figure 15-14 shows sunlight that has passed through the rings. As expected, the B ring looks darkest, because little sunlight gets through its dense collection of fragments, whereas the Cassini division looks bright, because sunlight passes relatively freely through its more widely spaced fragments. However, the very fact that the Cassini division does appear bright is clear evidence that it contains some fragments. If it contained no fragments at all to scatter the sunlight, we would see only the blackness of space through it.

Another division is visible in Figure 15-14 in the outer half of the A ring. This 270-km-wide gap is named the **Encke**

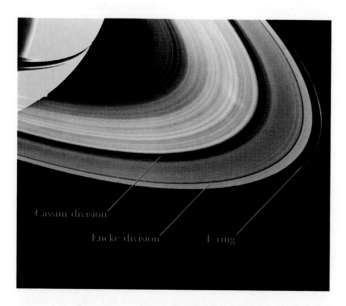

Figure 15-12 Saturn's rings from Voyager 1 *Voyager 1 took this photograph from a distance of about 1½ million km. Because the C ring scatters light differently from the way the A or B rings do, it has a bluer color. The broad Cassini division is clearly visible, as is the narrow Encke division in the outer A ring. The very thin F ring is visible just beyond the outer edge of the A ring. (NASA)*

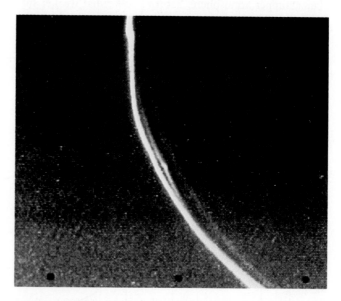

Figure 15-13 Details of the F ring *This Voyager 1 photograph of the F ring was taken from a distance of 750,000 km. The total width of the F ring is about 100 km. Within this span are several discontinuous strands, each measuring roughly 10 km across. (NASA)*

Figure 15-14 *Details of Saturn's A ring* *This view of the underside of Saturn's rings was taken by* Voyager 1. *Both the Encke division and the thin F ring are clearly visible toward the right side of the picture. The Cassini division and the outer edge of the B ring are on the left side of the picture. The Cassini division appears bright in this view of the shaded side of the rings. (NASA)*

Figure 15-15 *A false-color view of ring details* *Computer processing has severely exaggerated the subtle color variations in this view of the sunlit side of the rings from* Voyager 2. *Note that the C ring and Cassini's division appear bluish. Also note the distinct color variations across both the A and the B rings. (NASA)*

division, after the German astronomer Johann Franz Encke, who reported seeing it in 1838. Many astronomers have argued that Encke's report was erroneous, however, because his telescope did not have sufficient resolving power to produce an image of this narrow gap in the rings. The first undoubted observation of the Encke division was made in the late 1880s by the American astronomer James Keeler, using the newly constructed 36-in. refractor at the Lick Observatory. For this reason, the Encke division has been called the Keeler gap by some astronomers.

The process by which light bounces off particles or fragments in its path is called **scattering**. The proportion of light scattered in various directions depends upon both the size of the particles and the wavelength of the light. By measuring the albedo at various wavelengths and at various angles to the incident (incoming) sunlight, scientists can estimate the size of the particles or fragments that are scattering the sunlight. Similar measurements made with radiation at other wavelengths give more accurate estimates of the sizes of fragments that are larger than the wavelengths of visible light.

The Voyager scientists carefully measured the brightness of Saturn's rings from many angles as the spacecraft flew past the planet. They also measured how radio waves emitted by the spacecrafts' transmitters were scattered by the particles in the rings. These data indicate that the largest particles in Saturn's rings are roughly 10 m across. More abundant are snowball-sized particles about 10 cm in diameter. The smallest particles are just a few micrometers wide, smaller than snowflakes.

The F ring consists primarily of tiny, micrometer-sized particles. The A and B rings have a mixture of particles of various sizes. The Cassini division and the C ring contain relatively few tiny particles.

Subtle differences in color from one ring to the next also give important clues about the nature of the particles in the different rings. These variations are clearly visible in Figure 15-15, in which the colors have been exaggerated by computer processing. Scientists believe that these color variations correspond to slight differences in the chemical compositions of the rings. Although the main chemical constituent is frozen water, trace amounts of other chemicals—perhaps coating the surfaces of the ice particles—probably cause the colors seen in the computer-enhanced view. These trace chemicals have not yet been identified. Nevertheless, the existence of color variations and their probable persistence over millions of years suggests that the icy particles do not migrate substantially from one ringlet to another.

The Voyager scientists also noted radial "spokes" in the B ring, which were particularly evident in black-and-white photographs like Figure 15-16. Studies of how light is scattered from these spokes suggest that they are made of fine, dustlike particles suspended a few tens of meters above the main plane of the rings. A buildup of electric charges on the ring's particles (through the absorption of protons and electrons from Saturn's magnetosphere) may produce electrostatic forces that suspend fine dust particles at some distance from the larger ring fragments, producing the spokes we see in pictures like Figure 15-16.

Figure 15-16 Spokes in Saturn's B ring *Radial, wedge-shaped spokes are visible here in the central, most opaque portions of Saturn's B ring. The micrometer-sized dustlike particles do not scatter sunlight back toward the Sun. Thus, the spokes appear dark in this picture from Voyager 2, taken with the Sun behind the camera. (NASA)*

15-8 Saturn's innermost satellites affect the appearance and structure of its rings

Astronomers have long known that one of Saturn's moons, Mimas, has an effect on its ring system. Mimas is a moderate-sized satellite that orbits Saturn every 22.6 hours. According to Kepler's third law, particles in Cassini's division should orbit Saturn approximately every 11.3 hours. Consequently, on every second orbit, the particles in Cassini's division line up between Saturn and Mimas. During these repeated alignments, the combined gravitational forces of Saturn and Mimas cause small fragments to deviate from their original orbits. This 2-to-1 resonance with Mimas depletes Cassini's division of the particles and fragments that would otherwise scatter sunlight back toward Earth. Earth-based astronomers therefore see the Cassini division as a relatively dark band between neighboring rings.

In addition to revealing new details about the A, B, C, and F rings, the Voyager cameras also discovered three new ring systems: the D, E, and G rings. The **D ring** is Saturn's innermost ring system. It consists of a series of extremely faint ringlets located between the inner edge of the C ring and the Saturnian cloudtops. The **E ring** and the **G ring** both lie far from the planet, well beyond the outer edge of the A ring (see Box 15-3). Both of these outer ring systems are extremely faint, fuzzy, and tenuous. Each lacks the ringlet structure so prominent in the main ring systems. The E ring lies along the orbit of Enceladus, one of Saturn's icy satellites. Some scientists suspect that water geysers on Enceladus are the source of ice particles in the E ring, much as Io's volcanoes produce a torus of material along its orbit around Jupiter.

The Voyager cameras also discovered two tiny satellites that follow orbits on either side of the F ring (see Figure 15-17). Peter Goldreich of Caltech and Scott Tremaine at Princeton pointed out that it is the gravitational forces of these two satellites that keep the F ring particles in place. The outer satellite moves around Saturn at a slightly slower speed than that of the ice particles in the ring. As the ring particles pass the outer satellite, they experience a tiny gravitational tug, which tends to slow them down. As a result, these particles lose a little energy, causing them to fall into orbits a bit closer to Saturn. Meanwhile, the inner satellite is orbiting the planet somewhat faster than are the F ring particles. As the satellite moves past the particles, its gravitational pull tends to speed them up, nudging them into a slightly higher orbit. The combined effect of these two satellites therefore focuses the icy particles into a well-defined narrow band about 100 km wide. Because of their confining influence, these two moons, Prometheus and Pandora, are called **shepherd satellites**.

A shepherd satellite also is responsible for the sharp outer edge of the A ring. A tiny moon circles Saturn just beyond the outer edge of the A ring. As particles near the edge of the A ring pass by this slowly moving shepherd satellite, named Atlas, they feel a gravitational drag that slows them down slightly, preventing them from wandering into orbits farther from Saturn.

Figure 15-17 The F ring and its two shepherds *Two tiny satellites (Prometheus and Pandora), each measuring only about 50 km across, orbit Saturn on either side of its F ring. The gravitational effects of these two shepherd satellites focus and confine the particles in the F ring to a band about 100 km wide. This Voyager 2 picture was taken from a range of 10½ million km. (NASA)*

Box 15-3 Saturn's rings

Several broad ring systems surround Saturn. The A, B, and C rings are visible through Earth-based telescopes. Although observers had often reported seeing additional faint ring systems, the true locations of the D, E, F, and G rings were not established until the Voyager flybys in the early 1980s. Information about each of the seven ring systems is given here in order outward from the planet. The accompanying diagram shows the locations of the rings along with the orbits of the inner satellites.

· **The D ring** extends from Saturn's cloudtops out to the inner edge of the C ring at 1.21 Saturn radii (73,000 km) from the planet's center. The D ring, discovered by the Voyager spacecraft, consists of a series of very faint ringlets.

· **The C ring**, discovered in 1850, can be seen through Earth-based telescopes under good observing conditions. The C ring extends from 1.21 Saturn radii out to the inner edge of the B ring at 1.53 Saturn radii (92,000 km) from the planet's center.

· **The B ring**, the brightest of Saturn's ring systems, extends from 1.53 Saturn radii out to the inner edge of the Cassini division at 1.95 Saturn radii (117,600 km) from the planet's center.

· **The A ring** extends from the outer edge of the Cassini division at approximately 2.03 Saturn radii (122,500 km) out to 2.26 Saturn radii (136,600 km) from the planet's center. The sharp outer edge of the A ring is controlled by the "shepherd satellite" called Atlas.

· **The F ring** is very thin. It is located at 2.33 Saturn radii (140,600 km) from the planet's center. The F ring, which was discovered during the *Pioneer 11* flyby, is only 100 km wide. The F ring is confined by the two "shepherd satellites," Pandora and Prometheus.

· **The G ring** is extremely faint and tenuous. It is located at about 2.8 Saturn radii (169,000 km) from the planet's center and lacks the detailed ringlet structure that dominates the appearance of the inner rings.

· **The E ring**, like the G ring, is extremely faint. The E ring extends from about 3.5 to 5.0 Saturn radii (roughly 200,000 to 300,000 km) and is devoid of fine-scale ringlet structure. The material in the E ring may be associated with the satellite Enceladus, whose orbit is 3.95 Saturn radii from Saturn's center.

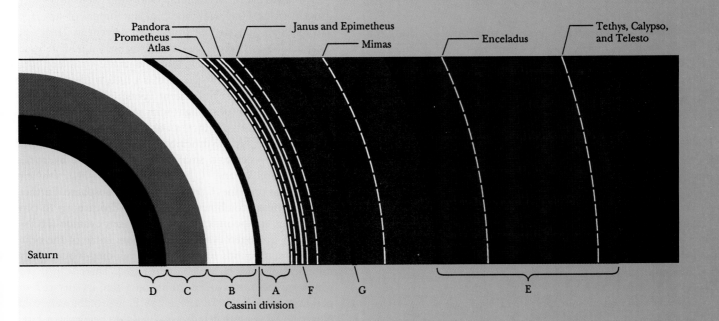

15-9 Titan has a thick, opaque atmosphere rich in methane, nitrogen, and hydrocarbons

Long before the Voyager flybys, astronomers knew that Titan is an extraordinary world. It was discovered by Christian Huygens in 1655, the same year that he proposed that Saturn has rings. By the early 1900s, several scientists had begun to suspect that Titan might have an atmosphere, because it is cool enough and massive enough to retain heavy gases. Confirming evidence came in 1944 when the American astronomer Gerard P. Kuiper discovered the spectral lines of methane in sunlight reflected from Titan. Titan is now known to be the only satellite in the solar system to have an appreciable atmosphere.

Because of its atmosphere, Titan became a primary target for the Voyager missions. To everyone's chagrin, however, the Voyagers spent hour after precious hour sending back featureless images (see Figure 15-18). Titan's cloud cover is so thick that it blocks any view of the surface. The haze surrounding Titan is so dense that little sunlight penetrates down to the ground, and its surface must be a dark, gloomy place.

In size, mass, and average density, Titan is quite similar to the largest Jovian satellites. We might thus expect Titan's

internal structure to resemble that of Ganymede or Callisto—a rocky core surrounded by a mantle of frozen water nearly 1000 kilometers thick.

However, Titan's thick atmosphere distinguishes it from all other satellites. The atmospheric pressure at Titan's surface is 1.6 bars, which is 60 percent greater than the atmospheric pressure at sea level on Earth, even though Titan's surface gravity is lower than the Earth's. Thus, considerably more gas must be weighing down on Titan than on Earth. Calculations show that about ten times more gas lies above each square meter of Titan's surface than lies above each square meter of Earth's surface.

What factors led to the formation of Titan's atmosphere that did not exist for either Ganymede or Callisto? For one, Titan formed in a much cooler part of the solar nebula. In contrast to the warmer conditions near Jupiter, the ices from which Titan accreted probably contained substantial amounts of frozen methane and ammonia. As Titan's interior warmed up (through the decay of naturally occurring radioactive isotopes), these ices vaporized, producing an atmosphere around the young satellite. Methane (CH_4) is stable in sunlight, so it remained in the atmosphere, but ammonia (NH_3) is easily broken down into nitrogen and hydrogen by the Sun's ultraviolet radiation. Titan's gravity is too weak to retain hydrogen, which escapes into space. In fact, the Voyager instruments detected a huge torus of dilute hydrogen gas surrounding Saturn along Titan's orbit. Even today, hydrogen is still escaping from Titan at a substantial rate.

The dissociation of ammonia and subsequent loss of hydrogen leaves Titan with an abundant supply of nitrogen. Voyager data suggest that roughly 90 percent of Titan's atmosphere is nitrogen. The two next-most-abundant gases are methane and argon. On Earth, methane—a major component of "natural gas"—is used as a fuel; while argon—an inert gas—is used in fluorescent lights.

The interaction of sunlight with methane induces chemical reactions that produce a variety of carbon–hydrogen compounds called **hydrocarbons**. The Voyager infrared spectrometers detected small amounts of many hydrocarbons such as ethane (C_2H_6), acetylene (C_2H_2), ethylene (C_2H_4), and propane (C_3H_8) in Titan's atmosphere. Furthermore, nitrogen combines with these hydrocarbons to produce other compounds, such as hydrogen cyanide (HCN). Although hydrogen cyanide is a poison, some of the other nitrogen compounds formed are the building blocks of the organic molecules on which life is based. There is little reason to suspect the existence of life on Titan; its surface temperature of 95 K ($-178°C = -288°F$) is prohibitively cold. Nevertheless, a more detailed study of the chemistry of Titan may shed light on the origins of life on Earth.

The atmospheric pressure and temperature on Earth are near the **triple point** of water, so that water is found on Earth in all three phases: liquid, solid, and gas. At Titan's surface, however, the atmospheric pressure and temperature

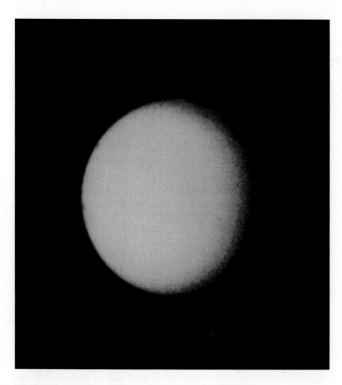

Figure 15-18 **Titan** *This view of Titan was taken by* Voyager 2 *from a distance of 4½ million km. Very few features are visible in the thick, unbroken haze that surrounds this large satellite. The main haze layer is located nearly 300 km above Titan's surface. (NASA)*

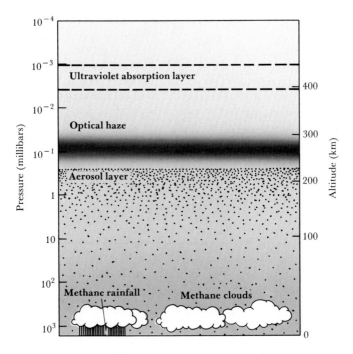

Figure 15-19 Titan's atmosphere *The Voyager data suggest that Titan's atmosphere has three distinct layers. The uppermost layer absorbs ultraviolet radiation from the Sun. The middle layer is opaque to visible light. And the lowest layer is an aerosol of particles suspended in the atmosphere. Methane rain clouds may exist near Titan's surface, which is at a temperature of about 95 K. (Adapted from T. Owen)*

worlds of our solar system. With its diameter of 5150 km, Titan is intermediate in size between Mercury and Mars. Second, there are six moderate-sized satellites, all discovered before 1800. These six satellites have properties and diameters that lead us to group them in pairs: Mimas and Enceladus (400 to 500 km in diameter), Tethys and Dione (roughly 1000 km in diameter), and Rhea and Iapetus (about 1500 km in diameter). Finally, there are several tiny moons that may in fact be either captured asteroids (as in the case of Phoebe, Saturn's outermost satellite) or jagged fragments of ice and rock produced by impacts and collisions. The shepherd satellites, for instance, may be such fragments.

Saturn's six moderate-sized satellites share several common characteristics. All six circle the planet along **regular orbits**, which is to say, in the plane of Saturn's equator and in the same direction the planet itself rotates. These six satellites also exhibit synchronous rotation. Finally, all six have average densities in the range of 1200 to 1400 kg/m^3, a value consistent with a composition largely of ice.

Mimas, the innermost of the six middle-sized moons, is heavily cratered. In fact, one of the craters on Mimas (see Figure 15-20) is so large that the impact forming it must have come close to shattering this moon into fragments.

Although the cratered surface of Mimas indicates that it is a geologically dead world, its neighbor and near-twin Enceladus has extensive surface regions that lack craters (see

are near the triple point of methane. Thus, methane may play a role on Titan similar to that of water on Earth. Methane snowflakes may fall onto frozen methane polar caps, with methane raindrops descending into methane rivers, lakes, and seas at warmer locations.

Some molecules are capable of joining together in long, repeating molecular chains to form substances called **polymers**. Many of the hydrocarbons and carbon–nitrogen compounds in Titan's atmosphere can form such polymers. Droplets of some polymers remain suspended in the atmosphere to form the kind of mixture called an **aerosol** (see Figure 15-19), but the heavier polymer particles settle down onto Titan's surface, probably covering it with a thick layer of sticky, tarlike goo. Darrell F. Strobel of the Naval Research Laboratory estimates that the deposits of hydrocarbon sludge on Titan may be ½ km deep.

15-10 The features of the icy surfaces of Saturn's six moderate-sized moons provide clues to their histories

Nearly two dozen moons are known to orbit Saturn (see Box 15-4). They can be grouped according to size in three categories. First, planet-sized Titan is one of the terrestrial

Figure 15-20 Mimas *Mimas is the smallest and innermost of Saturn's six moderate-sized satellites. This view was taken by Voyager 1 at a range of nearly 500,000 km. The huge impact crater, named Herschel after a famous astronomer, after the legendary English king, is 130 km in diameter; Mimas itself is only 400 km in diameter. (NASA)*

Box 15-4 Saturn's satellites

Nearly two dozen satellites are known to orbit Saturn. The table below lists the most prominent of these, with data about their orbits and physical properties. All the larger satellites are spherical and their diameters are listed in the *Size* column. The smaller satellites are nonspherical. Three dimensions—width, length, and height—are given for each of these satellites. The masses, and hence the densities, of the smaller satellites are not yet known.

The A-ring shepherd, previously called 1980S28, has been renamed Atlas; satellites 1980S27 and 1980S26 are the F-ring shepherds and have been named Prometheus and Pandora, respectively. Janus and Epimetheus (previously named 1980S3 and 1980S1, respectively) are called **coorbital satellites**, because they move in almost the same orbit. Tethys, Calypso, and Telesto are also coorbital satellites, as are Dione and Helene (formerly called 1980S6). In the latter two cases, the tiny satellites occupy specific locations along the orbits of the larger moons, where a balance exists between the gravitational pulls of Saturn and the larger moon. These locations, called the Lagrangian points, are discussed in Box 17-1.

	Distance from center of Saturn		Orbital period (hrs)	Size (km)	Density (kg/m^3)	Albedo
	(km)	(Saturn radii)				
1981S13	133,600	2.22	13.75	20	——	0.4–0.7
Atlas	137,670	2.28	14.45	20 × 20 × 40	——	0.9
Prometheus	139,350	2.31	14.71	140 × 100 × 80	——	0.6
Pandora	141,700	2.35	15.08	110 × 90 × 70	——	0.9
Epimetheus	151,420	2.51	16.66	140 × 120 × 100	——	0.8
Janus	151,470	2.51	16.67	220 × 200 × 160	——	0.8
Mimas	185,540	3.08	22.62	392	1400	0.5
Enceladus	238,040	3.95	32.88	500	1200	1.0
Tethys	294,670	4.88	45.31	1060	1200	0.9
Calypso	294,670	4.88	45.31	34 × 28 × 26	——	0.6
Telesto	294,670	4.88	45.31	24 × 22 × 22	——	0.5
Dione	377,420	6.26	65.69	1120	1400	0.7
Helene	378,060	6.26	65.69	36 × 32 × 30	——	0.7
Rhea	527,100	8.74	108.42	1530	1300	0.7
Titan	1,221,860	20.25	382.69	5150	1880	0.2
Hyperion	1,481,000	24.55	510.64	410 × 260 × 220	——	0.3
Iapetus	3,560,800	59.02	1,903.94	1460	1200	0.5–0.05
Phoebe*	12,954,000	214.7	13,210.8R	220	——	0.06

Saturn's outermost satellite moves in a retrograde direction about the planet, indicated by R with the orbital period.

Figure 15-21). These regions apparently have been "resurfaced" within the past 100 million years by some form of geological activity. Furthermore, Enceladus is the most highly reflective large object in our solar system, with an albedo greater than that of newly fallen snow. Enceladus seems to be covered with extremely pure ice that is totally free of the contaminating rock or dust thought to explain the lower albedos of the other icy satellites.

One hypothesis suggests that Enceladus, like Jupiter's moon Io, is heated by tidal forces. The satellite Dione orbits Saturn in 65.7 hours, almost exactly twice Enceladus's orbital period of 32.9 hours. Enceladus therefore is repeatedly caught in a tidal tug-of-war between Saturn and Dione. The tidal flexing thus produced may melt Enceladus's interior, forming water geysers on its surface in much the same way volcanoes are produced in Io. Such geysers may produce

Figure 15-21 *Enceladus This high-resolution image of Enceladus was obtained by* Voyager 2 *from a distance of 191,000 km. Ice flows and cracks strongly suggest that Enceladus's surface has been subjected to recent geological activity. The youngest, crater-free ice flows are estimated to be less than 100,000 years old. (NASA)*

Figure 15-22 *Tethys Voyager 1 obtained this view of Tethys at a distance of 1.2 million km. Ithaca Chasma is the long canyon near the edge of the sunlit hemisphere. (NASA)*

deposits of fresh ice and snow on the surface and may also be the source of the particles that form the thin E ring. No direct evidence for such geysers, or of a molten interior, was obtained by the Voyager spacecraft, however.

Tethys and Dione are the next-largest pair of Saturn's six moderate-sized satellites. Tethys is heavily cratered, like Mimas (see Figure 15-22). Also like Mimas, Tethys has one exceptionally huge impact crater; called Odysseus, it is 400 km in diameter. Stretching three-fourths of the way around Tethys is a 2000-km-long complex of valleys called Ithaca Chasma that may be a huge crack produced by the impact that created Odysseus.

Dione is slightly larger than Tethys. The surfaces of its leading hemisphere (that is, the hemisphere pointing toward its direction of orbital motion) and trailing hemisphere are quite different. Dione's leading hemisphere is heavily cratered, like the surface of Tethys, but its trailing hemisphere is covered by a network of strange, wispy markings (see Figure 15-23). These wisps may be troughs and valleys in the icy surface, or may be fresh deposits of frozen water formed by outgasings from Dione's interior.

Rhea and Iapetus are the largest, and most distant, of Saturn's moderate-sized moons. Rhea resembles Dione in that its trailing hemisphere is covered by a network of wispy

Figure 15-23 *Dione Bright wispy streaks crisscross Dione's trailing hemisphere. (A similar network of light-colored wisps covers the trailing hemisphere of Rhea.) The nature and cause of these streaks are not known. This view of Dione was obtained by* Voyager 1 *from a distance of 695,000 km. (NASA)*

Figure 15-24 Rhea *The leading hemisphere of Rhea is heavily cratered, like the leading hemisphere of Dione. This view was taken by Voyager 1 at a distance of 128,000 km. Surface features as small as 3 km across are visible. (NASA)*

Figure 15-25 Iapetus *Voyager 2 took this photograph of Iapetus from a distance of 1.1 million km. The leading hemisphere of Iapetus is extremely dark. One theory suggests that this hemisphere may have been coated with material drifting inward from Saturn's outermost satellite, Phoebe. (NASA)*

markings and its leading hemisphere is heavily cratered (see Figure 15-24). Whatever process produced the wisps on the one side of Dione has apparently also been active on the corresponding side of Rhea.

Iapetus is the most distant of Saturn's moderate-sized satellites, orbiting 3½ million kilometers from the planet. Iapetus has been known to be an unusual moon ever since its discovery, because its albedo varies widely but in a regular cycle as it moves about the planet. The Voyager spacecraft confirmed earlier hypotheses about there being extreme differences between the leading and trailing hemispheres of Iapetus (see Figure 15-25). The leading hemisphere is as black as asphalt, but the trailing hemisphere is highly reflective, like the surfaces of the other moderate-sized moons.

Most scientists agree that the darkness of Iapetus's leading hemisphere comes from a relatively thin layer of some light-absorbing substance on the surface, but the nature and source of this dark coating are unknown. One possibility is that the coating was formed by substances outgased from the satellite's interior. Another intriguing hypothesis involves Phoebe, the most distant of Saturn's known satellites.

Phoebe moves in a retrograde direction along an orbit tilted well away from the plane of Saturn's equator. This

motion would lead us to suspect that Phoebe is a captured asteroid. Its dark surface does indeed resemble the surfaces of a particular class of asteroids that are called carbonaceous asteroids, because of their high carbon content. Bits and pieces of this dark charcoal-like material drifting toward Saturn may have been swept up onto the leading hemisphere of Iapetus.

The Voyager spacecraft revealed many new and unsuspected facts about Saturn's complex system of rings and satellites. There is much yet to be learned about the Saturnian system, information that should bring better understanding of the creation and evolution of other planets, including the Earth. The giant moon Titan may even harbor important clues about the origin of life on Earth. Scientists are designing a new spacecraft, called *Cassini*, for a return mission to Saturn and Titan. *Cassini* will carry a cone-shaped probe that will descend through Titan's dense atmosphere and plop onto its surface—be it solid, goo, or liquid. After dispatching the probe, *Cassini* will go into orbit about Saturn to conduct long-term observations. The spacecraft could be launched toward Saturn as early as 1996 and arrive at the planet 6½ years later, but delays in funding may postpone the mission.

Key words

A ring	B ring	Cassini division	D ring
aerosol	C ring	*coorbital satellites	E ring

Encke division	hydrocarbon	ringlets	shepherd satellite
F ring	polymer	Roche limit	tidal force
G ring	regular orbit	scattering (of light)	triple point

Key ideas

• Saturn is circled by a system of thin, broad rings lying in the plane of the planet's equator. This system is tilted away from the plane of Saturn's orbit, which causes the rings to be seen at various angles by an Earth-based observer over the course of a Saturnian year.

> Three major, broad rings can be seen from Earth. The faint C ring lies nearest Saturn. Just outside it is the much brighter B ring, and outside that is a dark gap called the Cassini division, beyond which is the moderately bright A ring. Other, fainter rings were observed by the Voyager spacecraft.

> The principal rings of Saturn are composed of numerous fragments of ice and ice-coated rock ranging in size from a few micrometers to around 10 m.

> The rings exist inside the Roche limit of Saturn, where disruptive tidal forces are stronger than the gravitational forces attracting the fragments to each other.

> Each of the major rings is composed of a great many narrow ringlets.

> The faint, narrow F ring, which is just outside the A ring, has an intricately braided structure of unknown origin.

> Some of the ring boundaries are produced by so-called shepherd satellites, whose gravitational pull restricts the orbits of the ring fragments.

• Saturn's internal structure and atmosphere are similar to those of Jupiter, but Saturn has a larger rocky core and a thinner mantle of liquid metallic hydrogen; therefore Saturn has a more oblate shape and a weaker magnetic field.

> Saturn's atmosphere contains less helium than does Jupiter's. This lower abundance may be the result of precipitation of helium downward into the planet. If helium rain does fall, the resulting conversion of gravitational into thermal energy would account for Saturn's surprisingly strong heat output, which is $2\frac{1}{2}$ times as great as the incoming solar energy.

> Saturn has belts, zones, zonal jets, and ovals like those of Jupiter.

> The cloud layers in Saturn's atmosphere are spread out over a greater altitude range than those of Jupiter.

• The largest Saturnian satellite, Titan, is a terrestrial world with a dense nitrogen atmosphere in which methane may play a role similar to that of water on the Earth. A variety of hydrocarbons are produced there by the interaction of sunlight with methane, which form an aerosol layer in Titan's atmosphere and probably a thick sludge on its surface.

• Six moderate-sized moons circle Saturn in regular orbits: Mimas, Enceladus, Tethys, Dione, Rhea, and Iapetus. They are all probably composed largely of ice, but their surface features and histories vary significantly.

• Saturn has more than a dozen much smaller satellites whose tilted orbits and retrograde revolution suggest that they may be captured asteroids.

Review questions

1 Is there any way we can infer from naked-eye observations that Saturn is the most distant of the planets visible without a telescope? Explain.

2 Why do you suppose there are fewer transits, eclipses, and occultations of the Saturnian satellites than of the Galilean satellites?

3 It has been claimed that Saturn would float if one had a large enough bathtub. Using the mass and size of Saturn given in Box 15-2, confirm that the planet's average density is 700 kg/m³, and comment on this claim.

4 Compare the interiors of Jupiter and Saturn. Why does Saturn have a larger core than Jupiter, even though Saturn is less massive than Jupiter?

5 Explain why Saturn is more oblate than Jupiter, even though Saturn rotates more slowly than Jupiter.

6 Compare the internal sources of energy in Jupiter and Saturn that cause both planets to emit more energy than they receive from the Sun in the form of sunlight.

7   Compare the atmospheres of Jupiter and Saturn. Why does Saturn's atmosphere look "washed out" compared to Jupiter's?

8   Describe the structure of Saturn's rings. What evidence is there that ring particles do not migrate significantly between ringlets?

9   Is "shepherd satellite" an appropriate term for the objects thus named? Explain how they operate.

10   Why do you suppose that Titan is the only satellite in our solar system to possess a substantial atmosphere?

12   Compare the role of methane on Titan with that of water on Earth.

Advanced questions

> **Tips and tools . . .**
> The small-angle formula is given in Box 1-1. Saturn's satellites, whose orbital parameters are given in Box 15-4, obey Kepler's third law. A discussion of the retention of an atmosphere is found in Box 7-2.

*13   (Basic)   After its optical problems are solved, the Hubble Space Telescope will have a resolving power of 0.1 arc sec. Is this sufficient to produce resolved images of any of Saturn's satellites? If so, which ones?

*14   It is well known that Cassini's division involves a 2-to-1 resonance with Mimas. Does the location of the Keeler gap—133,500 km from Saturn's center—correspond to a resonance with one of the other satellites? If so, which one?

*15   (Challenging)   Find the escape velocity on Titan. Assuming an average atmospheric temperature comparable to that of Saturn, what is the limiting molecular weight of gases that could be retained by Titan's gravity?

16   Consult such magazines as Sky & Telescope or Science News to learn the current status of the Cassini mission. Has the U.S. Congress approved funds for the mission? What is the estimated launch date? When will Cassini arrive at Saturn?

*17   (Challenging)   Use Kepler's third law to calculate the orbital speeds of particles at the outer edge of the A ring and at the inner edge of the B ring. Use your answers to calculate the size of the Doppler shift of sunlight (wavelength $\cong$ 500 nm) reflected from Saturn's brightest rings.

18   As seen by Earth-based observers, the intervals between successive edge-on presentations of Saturn's rings alternate between 13 years 9 months and 15 years 9 months. Why do you think these two intervals are not equal?

Discussion questions

19   Compare and contrast the internal structures and atmospheres of Jupiter and Saturn. Wherever possible, describe the reasons for these differences or similarities in terms of such physical parameters as their mass, chemical composition, and surface gravity.

20   Comment on the suggestion that Titan may harbor life forms.

21   NASA, the Jet Propulsion Laboratory, and the European Space Agency are currently planning a mission to Saturn that will place the Cassini spacecraft in orbit about the planet. In your opinion, what issues should be examined, what data should be collected, and what kinds of questions might such a mission answer?

Observing projects

22   View Saturn through a small telescope. Make a sketch of what you see. Estimate the angle at which the rings are tilted to your line of sight. Can you see the Cassini division? Can you see any belts or zones in Saturn's clouds? Is there a faint, starlike object near Saturn that might be Titan? What observations could you perform to test whether or not the starlike object is a Saturnian satellite?

23   If you have access to a moderately large telescope, make arrangements to observe several of Saturn's satellites. Consult the section entitled "Satellites of Saturn" in the current Astronomical Almanac, which includes a diagram showing the orbits of Mimas, Enceladus, Tethys, Dione, Rhea, Titan, and Hyperion. Plan your observing session by looking up the dates and times of the most recent greatest eastern elongations of the various satellites. You will have to convert from universal time (UT), which is the same as Greenwich mean time, to your local time zone. Then, using the tick marks along the orbits in the diagram, estimate the positions of the satellites relative to Saturn at the time you will be at the telescope. At the telescope, you should have no trouble identifying Titan, which reaches a magnitude of +8 at opposition. Tethys, Dione, and Rhea should be the next easiest satellites to find because they each have a magnitude of +10 at opposition. Can you confidently identify any of the other satellites?

For further reading

Beatty, J. "Report on the Voyager Encounters with Saturn." *Sky & Telescope*, January, October, and November 1981 • This series of articles boasts a magnificent collection of the best images from the Voyager flybys of Saturn.

Burnham, R. "The Saturnian Satellites." *Astronomy*, November 1981 • This easily understood article discusses the various icy satellites of Saturn.

Gore, R. "Saturn: Riddle of the Rings." *National Geographic*, July 1981 • A discussion of some of the puzzling aspects of Saturn's rings is lavishly illustrated in the tradition of this famous periodical.

Ingersoll, A. "Jupiter and Saturn." *Scientific American*, December 1981 • Written shortly after the Voyager flybys, this enlightening article compares and contrasts the two largest planets.

Morrison, D. "The New Saturn System." *Mercury*, November/December 1981 • This comprehensive article, written shortly after the Voyager flybys, gives an exciting description of what had just been learned.

_____ . *Voyages to Saturn*. NASA SP-451, 1982 • This excellent book by a noted planetary scientist gives an insider's view of the Saturn flybys.

Osterbrock, D., and Cruikshank, D. "Keeler's Gap in Saturn's A Ring." *Sky & Telescope*, August 1982 • This interesting historical article probes a controversy about who actually discovered Encke's division.

Owen, T. "Titan." *Scientific American*, February 1982 • This article, which discusses many details learned from the Voyager missions, speculates that the chemistry of Titan's atmosphere may resemble that of Earth before life arose.

Pollack, J., and Cuzzi, J. "Rings in the Solar System." *Scientific American*, November 1981 • This excellent article on the structure and evolution of planetary rings focuses primarily on Saturn, but also discusses the rings around Jupiter and Uranus.

Soderbloom, L., and Johnson, T. "The Moons of Saturn." *Scientific American*, January, 1982 • This article presents a coherent, comprehensive survey of Saturn's moons as revealed by the Voyager spacecraft.

Washburn, M. *Distant Encounters:* Voyagers 1 *and* 2 *at Jupiter and Saturn*. Harcourt, Brace, Jovanovich, 1983 • This eloquent summary of the Voyager missions, written by a journalist, is filled with awe and enthusiasm.

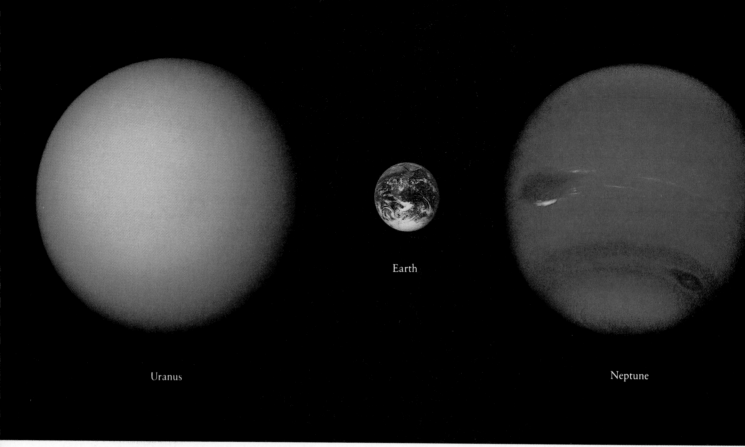

Earth

Uranus

Neptune

Neptune, Earth, and Uranus These images of Uranus, the Earth, and Neptune are to the same scale. Uranus and Neptune are quite similar in mass, size, and chemical composition. Both planets are surrounded by thin, dark rings, quite unlike Saturn's, which are broad and bright. Almost everything we know about the outer two Jovian planets was gleaned from Voyager 2, *which flew past Uranus in 1986 and Neptune in 1989. (NASA)*

16

The Outer Worlds

Shrouded in mystery because of their remoteness, Uranus and Neptune were unveiled by *Voyager 2*. Surrounded by a system of small moons and thin, dark rings, Uranus is tipped on its side so that its axis of rotation lies nearly in its orbital plane. This remarkable orientation suggests that Uranus may have been the victim of a staggering impact by a massive planetesimal. Neptune is a bluish world with markings that give it a Jupiterlike appearance. Neptune even has a huge storm reminiscent of the Great Red Spot. Most intriguing of all is Neptune's giant moon, Triton. Nearly devoid of impact craters, this geologically active world is covered with puzzling features, including frozen lakes and geysers that squirt nitrogen-rich vapors. Triton's retrograde orbit suggests that this strange world may have been gravitationally captured by Neptune. Such an event would have severely disrupted any pre-existing Neptunian satellite system, perhaps catapulting satellites away from the planet altogether. Such may have been the genesis of Pluto, which shares many of Triton's general characteristics.

At first glance, Uranus and Neptune appear to be twins. They have nearly the same size, the same mass, the same density, and the same chemical composition. When *Voyager 2* flew past both planets in the late 1980s, however, astronomers discovered that Uranus and Neptune are really quite different, despite their superficial similarities. Many remarkable characteristics of these two worlds were revealed as *Voyager 2* examined their meteorology, satellites, and magnetospheres. Both planets are surrounded by a system of thin, dark rings that are very difficult to detect from Earth. The surface features of their satellites suggest that they have been subjected to significant modification by impacts and collisions as well as by the tidal effects of their parent planets. The brief glimpses of Uranus and Neptune provided by *Voyager 2* have given us unparalleled insights into these two frigid planets far from the Sun.

16-1 Uranus was discovered by chance, but Neptune's existence was predicted with Newtonian mechanics

Uranus was discovered by chance, on March 13, 1781, by the then-little-known astronomer William Herschel. This German musician had emigrated to England, then became fascinated with astronomy. Using a telescope of his own construction, Herschel was systematically surveying the sky when he noticed a faint, fuzzy object that he first thought to be a distant comet. By the end of 1781, however, it had become clear that the object had a planetlike orbit far out beyond Saturn, beyond where comets can normally be seen. Herschel had not only discovered the seventh planet from the Sun but in doing so had doubled the diameter of the known solar system.

Herschel acquired instant fame. He was appointed court astronomer to King George III and given the financial resources to pursue his scientific interests. During the next few decades, Herschel made discoveries that established him as one of the great astronomers of all time.

Although Herschel received the credit for discovering Uranus, many other astronomers had sighted this planet in earlier times, but had mistaken it for a dim star. At opposition, Uranus reaches a magnitude of +5.6, which means that it can be seen faintly with the naked eye under good observing conditions. Uranus is plotted as a dim star on at least 20 star charts drawn between 1690 and 1781.

By the beginning of the nineteenth century, it had become painfully clear to astronomers that they could not predict the exact orbit of Uranus. The planet's observed positions were slightly different from the positions predicted using Newtonian mechanics. By 1830, the discrepancy between observation and prediction had become so great (2 arc min) that some scientists had begun to suspect a flaw in Newton's laws. The Astronomer Royal of Great Britain, George Airy,

was one of those who thought that Newton's law of gravitation might not be accurate at great distances from the Sun.

In 1843, a 24-year-old student at Cambridge University in England began to explore an earlier suggestion that it might be the gravitational pull of some unknown, distant planet that was causing Uranus to deviate slightly from its predicted orbit. John Couch Adams began trying to calculate the orbit of a planet that could account for the unexplained perturbations in Uranus's orbit. Adams obtained from Airy all the data available about Uranus's exact observed positions. After two years of hard work, Adams reached the conclusion that, in 1822, Uranus had caught up with and passed a more distant planet. Uranus accelerated slightly as it approached the unknown planet, then decelerated slightly after 1822 as it receded from the planet. Adams predicted that the unknown planet would be found at a certain position in the constellation of Aquarius.

In October 1845, Adams submitted his calculations and prediction to Airy, but Airy was unconvinced and the matter was dropped. Meanwhile, in France, the more established astronomer Urbain Leverrier performed the same calculations independently. In June 1846, Leverrier published his prediction that there was an unknown planet beyond Uranus. Leverrier's and Adam's predicted positions for the planet differed by less than 1°. This shocked Airy, who quickly instructed James Challis, the director of the Cambridge Observatory, to begin a thorough search for the unknown planet in the constellation of Aquarius. The search was hampered by the lack of an up-to-date star map at Cambridge Observatory for this region. Many uncharted stars had to be observed, then followed over many nights to see whether any of them showed a planetlike motion with respect to other stars. While the search was going on in Cambridge, Leverrier wrote to Johann Gottfried Galle at the Berlin Observatory, urging a search for the unknown planet. This letter was received in Berlin on September 23, 1846, and the planet was sighted that very night. Because excellent star charts were available in Berlin, Galle was quickly able to locate the only uncharted star that had the expected brightness and was in the predicted location.

Some French astronomers insisted the new planet be named Leverrier, but Leverrier himself proposed that it should be called Neptune. After years of debate between English and French astronomers, the credit for the discovery of Neptune eventually came to be divided equally between Adams and Leverrier.

An interesting postscript to this story is that, in January 1613, Neptune happened to be located quite near Jupiter. At that time, Galileo was regularly observing the motions of the four large satellites of Jupiter. Galileo's drawings of Jupiter and its satellites show a "star" less than 1 arc min from the predicted location of Neptune during those winter nights. Neptune can be as bright as magnitude +7.7, making it visible through small telescopes. Galileo even noted in his observation log that on one night this star seemed to have moved in relation to the other stars. It is therefore quite possible that Galileo may have in fact sighted Neptune over two hundred years before anyone else.

16-2 Earth-based observations suggest that Uranus and Neptune are similar to each other but different from Jupiter and Saturn

Even through a large telescope, both Uranus and Neptune are dim, uninspiring sights. Each planet appears as a small, hazy, featureless disk with a faint greenish-blue tinge (see Figures 16-1 and 16-2). Basic information about these two worlds is found in Boxes 16-1 and 16-2.

Uranus is unique among the planets because its axis of rotation lies very nearly in the plane of its orbit (see Figure 16-3). As a result, when Uranus moves along its 84-year orbit, its north and south poles alternately point toward or away from the Sun, producing highly exaggerated seasonal changes on the planet. For instance, near Uranus's south pole during the summer, the Sun remains above the horizon while northern latitudes are subjected to a continuous, frigid winter night. Half a Uranian year later, the situation is reversed.

The spectra of Uranus and Neptune are remarkably similar to each other, both showing broad absorption caused by methane at infrared wavelengths. Hydrogen has also been detected on both planets, and the presence of helium has been deduced. Nevertheless, the outer two Jovian planets are significantly different from the inner two. The large spheres of Jupiter and Saturn are composed primarily of hydrogen and helium in nearly the same abundance as the Sun. Uranus and Neptune, however, are distinctly smaller and less massive than either Jupiter or Saturn. If Uranus and Neptune also had solar abundances of the elements, their smaller masses would produce less compression, and thus lower average densities, than those of Jupiter and Saturn. In fact, though, Uranus and Neptune have average densities comparable to or greater than those of Jupiter or Saturn. We may therefore conclude that Uranus and Neptune contain greater proportions of the heavier elements than the solar abundances. Each planet probably has a rocky core, composed primarily of iron and silicon, that is roughly the size of the Earth.

Uranus and Neptune have rocky cores similar to those of Jupiter and Saturn, but the remainder of their internal structure differs from that of the two largest planets. Some scien-

Figure 16-1 Uranus *Nearly 3 billion kilometers from the Sun, Uranus receives only $\frac{1}{400}$ the intensity of sunlight that we experience here on Earth. Uranus is therefore a dim, frigid world that never shines brighter than magnitude +5.6 in our skies. Even though its diameter is about four times that of Earth, Uranus always shows us a disk less than 4 arc sec across, which is roughly the size of a golf ball seen from a distance of 1 kilometer. (New Mexico State University Observatory)*

Figure 16-2 Neptune *At $4\frac{1}{2}$ billion km from the Sun, Neptune's surface receives only $\frac{1}{900}$ the intensity of sunlight we receive on Earth. Neptune is therefore dimmer than Uranus; it never shines brighter than magnitude +7.7 in our skies. Although it is about the same size as Uranus, Neptune looks smaller through Earth-based telescopes, simply because it is farther away. The maximum possible angular diameter of Neptune's disk is 2.2 arc sec. That is the same size as a dime seen from a distance of 1 km. (New Mexico State University Observatory)*

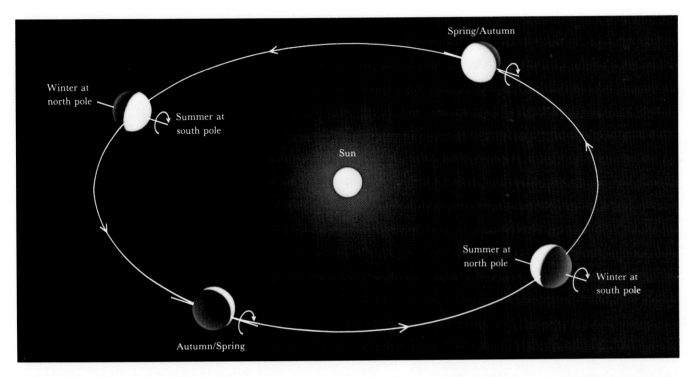

Figure 16-3 Exaggerated seasons on Uranus *Uranus's axis of rotation is tilted so steeply that it lies nearly in the plane of the planet's orbit. The seasonal changes on Uranus are thus severely exaggerated. For instance, during midsummer at Uranus's south pole, the Sun appears nearly overhead for many Earth years, while the planet's northern regions are subjected to a long, continuous winter night.*

tists suspect that Uranus and Neptune have superdense, slushy atmospheres that extend all the way down to their rocky cores. These atmospheres may contain extremely dense water clouds. Indirect evidence for such clouds comes from the apparent deficiency of ammonia on both planets. This gas is easily dissolved in water, which would explain the low abundance of ammonia in the observable outer layers of these planets.

The outer layers of Uranus and Neptune are predominantly composed of hydrogen and helium in a gaseous state and at low density. The temperature in the upper atmospheres of both planets is so low (about 60 K) that methane and water there condense to form clouds of ice crystals. Methane freezes at the lowest temperature and thus forms the highest clouds. This gas efficiently absorbs red light, giving Uranus and Neptune their blue-green color.

The picture that emerges from these observations places Uranus and Neptune in a class somewhere between the giant Jovian planets of Jupiter and Saturn and the terrestrial planets in the inner solar system. This classification is not what we would expect from the most straightforward model of how the solar system formed. According to this simple model, the preponderance of hydrogen and helium should increase progressively with distance from the vaporizing heat of the Sun. Yet Uranus and Neptune contain a greater percentage of heavy elements than do Jupiter and Saturn. If we knew the origin of the heavy material in Uranus and

Neptune, we would have a better understanding of the conditions that prevailed far from the Sun as our solar system was being formed.

16-3 Uranus is nearly featureless and its magnetic field is oriented at an unusual angle

At the time of the 1986 *Voyager* 2 flyby, Uranus's south pole was aimed almost directly at the Sun. Thus the face-on, fully illuminated view of Uranus's daytime hemisphere seen in Figure 16-4 looks nearly straight down onto Uranus's south pole. The Voyager pictures showed that Uranus is remarkably featureless. Severe computer enhancement was required to make the faint cloud markings visible. By following these markings, scientists determined that the winds on Uranus blow in the same direction the planet rotates, just as they do on Venus, Earth, Jupiter, and Saturn. The Uranian winds move at speeds of 40 to 160 meters per second (90 to 360 miles per hour), whereas on Earth the jet streams blow at about 50 meters per second (110 miles per hour).

The Uranian winds arrange the planet's cloud features into east–west bands similar to those on Jupiter and Saturn. The equatorial region receives little sunlight, but its temperature is about the same as that at the sunlit pole. Heat must

Box 16-1 Uranus data

Mean distance from the Sun:	19.19 AU = 2.871×10^9 km
Maximum distance from the Sun:	20.07 AU = 3.003×10^9 km
Minimum distance from the Sun:	18.31 AU = 2.739×10^9 km
Mean orbital velocity:	6.8 km/s
Sidereal period:	84.04 years
Rotation period:	17.2 hours
Inclination of equator to orbit:	97.9°
Inclination of orbit to ecliptic:	0.77°
Orbital eccentricity:	0.046
Diameter:	51,120 km
Diameter (Earth = 1):	4.01
Apparent diameter as seen from Earth:	maximum = 3.7″
	minimum = 3.1″
Mass:	8.70×10^{25} kg
Mass (Earth = 1):	14.6
Mean density:	1300 kg/m^3
Surface gravity (Earth = 1):	0.92
Escape velocity:	21.2 km/s
Oblateness:	0.03
Mean surface temperature (at cloudtops):	−216°C = −357°F = 57 K
Albedo:	0.51
Brightest magnitude:	+ 5.6
Mean diameter of Sun as seen from Uranus:	1′ 41″

therefore be efficiently transported between the poles and the equator. Perhaps this north–south heat transport, which is perpendicular to the wind flow, mixes and homogenizes the atmosphere to make Uranus appear featureless.

Uranus rotates differentially, as do Jupiter and Saturn. A day on Uranus is between 16.2 and 16.9 hours long, depending on the latitude. This rotation period is only for the cloudtops, however. To determine the rotation period for the underlying body of the planet, scientists looked to Uranus's magnetic field, which is presumably anchored in the planet's core, or at least in the deeper and denser layers of its atmosphere.

Figure 16-4 Uranus from Voyager 2 No distinctive cloud patterns are to be seen in any of the Voyager views of Uranus. The blue-green appearance of Uranus comes from methane in the planet's atmosphere that absorbs red wavelengths from the sunlight. Severe computer enhancement of the Voyager pictures does reveal faint cloud features, from which the rotation period of Uranus's atmosphere was determined to be about 16½ hours. (NASA)

Box 16-2 Neptune data

Mean distance from the Sun:	30.06 AU $= 4.497 \times 10^9$ km
Maximum distance from the Sun:	30.36 AU $= 4.546 \times 10^9$ km
Minimum distance from the Sun:	29.76 AU $= 4.456 \times 10^9$ km
Mean orbital velocity:	5.4 km/s
Sidereal period:	164.8 years
Rotation period:	16.11 hours
Inclination of equator to orbit:	$29.6°$
Inclination of orbit to ecliptic:	$1.77°$
Orbital eccentricity:	0.010
Diameter:	49,528 km
Diameter (Earth = 1):	3.88
Apparent diameter as seen from Earth:	maximum = 2.2″
	minimum = 2.0″
Mass:	1.03×10^{26} kg
Mass (Earth = 1):	17.23
Mean density:	1660 kg/m^3
Surface gravity (Earth = 1):	1.12
Escape velocity:	23.6 km/s
Oblateness:	0.026
Mean surface temperature (at cloudtops):	$-216°$C $= -357°$F $= 57$ K
Albedo:	0.35
Brightest magnitude:	$+ 7.7$
Mean diameter of Sun as seen from Neptune:	1′ 04″

The Voyager magnetometer demonstrated that Uranus's magnetic field is about as strong as Earth's. Charged particles from the solar wind emit radio radiation as they encounter Uranus's magnetic field. The intensity of this radio emission changes as the planet rotates. Scientists knew that by observing these changes they could determine Uranus's internal rotation rate.

Scientists were surprised by the results of this straightforward technique. The data indicated that the magnetic axis of Uranus is inclined by nearly 60° from the planet's axis of rotation. Periodic variations in radio emissions from Uranus revealed its internal rotation rate to be about 17.23 hours. Uranus's magnetic axis thus sweeps out a broad cone in space every 17 hours 14 minutes.

Except for Uranus and Neptune, all the other planets have their rotational and magnetic axes nearly aligned. For instance, both Earth and Jupiter have their **magnetic axes** inclined by less than 12° from their rotational axes. What could have caused such a misalignment on Uranus? One possibility is that Uranus's magnetic field might be undergoing a reversal, a phenomenon known from geological data to have occurred on Earth, but never actually observed in progress anywhere. Another possibility is that the misalignment of Uranus's magnetic field came from one or more catastrophic collisions with planet-sized bodies that also tilted the planet's rotation axis. The Uranian moons seem to show evidence of such collisions.

16-4 Uranus is orbited by satellites that bear the scars of many shattering collisions

Prior to the *Voyager 2* flyby, five moderate-sized satellites were known to orbit Uranus. All five are named after sprites and spirits in English literature. The two largest moons, Titania and Oberon, were found by William Herschel in 1789. Another English astronomer, William Lassell, discov-

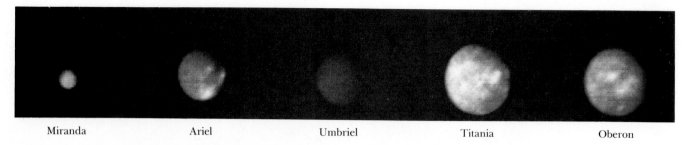

| Miranda | Ariel | Umbriel | Titania | Oberon |

Figure 16-5 The principal Uranian satellites This "family portrait" of Uranus's largest moons shows the five satellites to the same scale and correctly displays their relative reflectivities. All five satellites have only slight color variations on their surfaces, with their average color being nearly gray. Note how dark Umbriel is. *(NASA)*

ered Ariel and Umbriel, in 1851. The smallest of the five, Miranda, was first detected by Gerard P. Kuiper in 1948 at the McDonald Observatory in Texas.

Photographs from *Voyager 2* (see Figure 16-5) gave accurate information about the sizes of the five principal Uranian moons. Titania and Oberon are nearly the same size, with diameters of about 1600 km (1000 miles). Ariel and Umbriel are also about the same size, with diameters of about 1200 km (750 miles). Miranda is about 500 km (300 miles) in diameter. *Voyager 2* also discovered 10 very small satellites orbiting Uranus, most of which have diameters of less than 100 km (60 miles). Box 16-3 contains data about the Uranian moons.

The average density of the Uranian moons is about 1500 kg/m³, which is consistent with all 15 worlds being composed of a mixture of ice and rock. Voyager data suggest that the larger moons are 50 percent water, 20 percent carbon- and nitrogen-based materials, and 30 percent rock. In addition, all the Uranian satellites have very low albedos. Umbriel is one of the darkest moons in our solar system, but recent impact craters on Oberon and Titania appear to have exposed fresh, undarkened ice there. Thus, the low albedo of the Uranian moons is probably caused by a coating of dark, sootlike material. Perhaps the carbon-rich material that is suspected of coating Iapetus (recall Figure 15-25) is significantly more abundant at Uranus than at Saturn. Umbriel's extreme darkness may simply have resulted from that satellite's luck in escaping recent impacts that have exposed fresh ice on Oberon and Titania.

Ariel exhibits mysterious volcanic features that are difficult to explain. For instance, Figure 16-6 shows a region where some sort of thick, viscous fluid flowed down a channel to flood a low-lying area. A mixture of water, methane, and ammonia can have a very low melting point. Could that

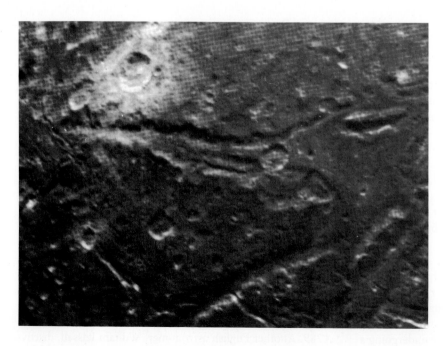

Figure 16-6 Volcanism on Ariel? This close-up view of Ariel shows a channel that once carried a viscous fluid. As the material flowed down the channel (from left to right in this view), it partly buried a crater when it inundated a low-lying area at the right of the picture. Note that the large impact crater near the upper left corner has uncovered fresh, undarkened ice. *(NASA)*

Box 16-3 Uranus's satellites

Ten of Uranus's fifteen known satellites were discovered during the 1986 *Voyager 2* flyby. These ten tiny moons (listed first in the following table) orbit closer to the planet than the five moderate-sized satellites visible to Earth-based observers.

	Distance from center of Uranus (km)	Orbital period (hr)	Diameter or size* (km)	Density (kg/m³)	Albedo
Cordelia	49,700	8.0	40	—	<0.1
Ophelia	53,800	9.0	50	—	<0.1
Bianca	59,200	10.4	50	—	<0.1
Cressida	61,800	11.1	60	—	<0.1
Desdemona	62,700	11.4	60	—	<0.1
Juliet	64,600	11.8	80	—	<0.1
Portia	66,100	12.3	80	—	<0.1
Rosalind	69,900	13.4	60	—	<0.1
Belinda	75,300	14.9	60	—	<0.1
Puck	86,000	18.3	170	—	0.07
Miranda	129,900	33.9	484	1260	0.34
Ariel	190,900	60.5	1160	1650	0.40
Umbriel	266,000	99.5	1190	1440	0.19
Titania	436,300	208.9	1610	1590	0.28
Oberon	538,400	323.1	1550	1500	0.24

The sizes given for the 10 smallest satellites are upper limits.

have been the substance that flowed on Ariel? Is this geological formation evidence for a warmer climate there in the past? No one knows.

Miranda has a landscape unlike that of any other world in our solar system. From afar, Miranda appears to consist of several large, angular, blocklike segments (see Figure 16-7). From close range, Voyager's cameras revealed dramatic topography, including a cliff twice as high as Mount Everest (see Figure 16-8).

Some planetary scientists theorize that Miranda's extraordinary surface may be the result of an impact that shattered the satellite. Prior to that impact, Miranda's core was made up of dense rock and its outer layers of less dense ice. After the impact, blocks of debris drifted back together, because of mutual gravitational attraction, and formed a chaotic mix of rock and ice. The landscape we see today on Miranda is the result of huge, dense rocks trying to settle toward the satellite's center as blocks of less dense ice are forced upward toward the surface.

Detailed examination of the Voyager photographs has led several planetary scientists to suggest that each of the Uranian moons may have had one or more shattering impacts. This suggestion agrees with the speculation that a catastrophic collision between Uranus and an Earth-sized object knocked Uranus on its side and set the planet's core rotating at the oblique angle at which we see it today. Prior to the collision, Uranus's satellites presumably orbited the upright planet in the plane of its equator, just like the regular satellites of the other Jovian planets. After the collision, however, tidal forces exerted by Uranus's equatorial bulge on the satellites disrupted their orbits. Only after many years and numerous collisions did the remaining chunks of rock and ice settle into the highly tilted plane of Uranus's equator, where we find them today.

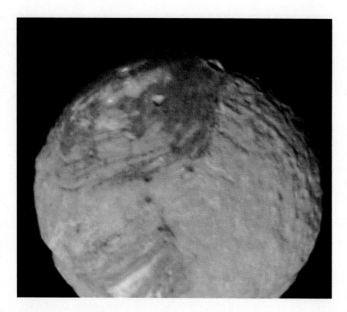

Figure 16-7 Miranda *The patchwork appearance of Miranda suggests that this satellite consists of huge chunks of rock and ice that came back together after an ancient, shattering impact by a large asteroid. The curious banded features that cover much of Miranda are paralleled by numerous closely spaced valleys and ridges, which may have formed as dense, rocky material sank toward the satellite's core. (NASA)*

Figure 16-8 Cliffs on Miranda *This close-up view of Miranda shows impact craters and parallel networks of valleys and ridges. In the upper right corner are enormous cliffs that jut upward 20 km to an elevation twice as high as Mount Everest. These cliffs may consist of low-density ice that was thrust upward as higher-density rock sank toward Miranda's core. (NASA)*

16-5 Uranus is circled by a system of thin, dark rings

Uranus's rings were discovered by accident, as was the planet itself. On March 10, 1977, Uranus was scheduled to occult (move in front of) a faint star, as seen from the Indian Ocean and adjacent areas on Earth. This event afforded an excellent opportunity to measure Uranus's diameter and study its upper atmosphere. Because Uranus's position and speed along its orbit were already known, astronomers needed only to measure how long the star was hidden to deduce Uranus's size. In addition, by carefully measuring how the starlight faded when Uranus passed in front of the star, they planned to deduce important properties of Uranus's upper atmosphere.

With these goals in mind, a team of astronomers headed by James L. Elliot of Cornell University journeyed to the Indian Ocean in NASA's Kuiper Airborne Observatory (recall Figure 6-26). To everyone's surprise, the background star briefly blinked on and off several times just before the occultation and again immediately after it. The astronomers concluded that Uranus must be surrounded by a series of narrow rings—nine in all. During the Voyager flyby, two more rings were discovered (all are listed in Box 16-4).

Uranus's rings are quite different from those of Saturn. Saturn's rings are broad and bright, but Uranus's are narrow

and dark. Most of the Uranian rings are less than 10 km wide. Typical particles in Saturn's rings are chunks of ice with the reflectivity and dimensions of snowballs, but typical particles in Uranus's rings could be compared with large lumps of coal roughly 1 m in size. The Uranian rings reflect only about 1 percent of the sunlight that falls on them.

Two contrasting views of the Uranian rings are shown in Figures 16-9 and 16-10. Figure 16-9, obtained by *Voyager 2* as it approached the planet, shows the sunlit side of the rings. The nine rings that were discovered from Earth are evident. The most prominent ring, which also happens to be the outermost one, is called the ϵ ring; it has segments that are nearly 100 km wide. Perhaps it is kinky and braided like Saturn's F ring (recall Figure 15-13), which also happens to have an overall width of about 100 km.

All of Uranus's rings are located less than 2 Uranian radii from the planet's center, well within the planet's Roche limit. Some sort of mechanism, possibly one involving shepherd satellites, efficiently confines particles to their narrow orbits. *Voyager 2* searched for shepherd satellites, but found only two. The others may be so small and black that they have simply escaped detection. Or perhaps they aren't there, and our ideas need revision.

Figure 16-10 was obtained when *Voyager 2* passed into Uranus's shadow and looked back toward the Sun. This picture demonstrates that the gaps between the rings are not empty. Broad bands of tiny dust grains gleam in the sunlight.

Box 16-4 Uranus's rings

Nine thin, dark rings around Uranus were discovered by Earth-based observers in 1977. Two more rings were discovered during the *Voyager 2* flyby in 1986. All nine Uranian rings are very nearly in the plane of Uranus's equator. They are also quite circular, with eccentricities between 0.001 and 0.1.

Ring	Distance from center of Uranus (km)	Width of ring (km)
1986U2R	37,000–39,500	2500
6	41,850	1–3
5	42,240	2–3
4	42,580	2–3
α	44,730	7–12
β	45,670	7–12
η	47,180	0–2
γ	47,630	1–4
δ	48,310	3–9
1986U1R	50,040	1–2
ε	51,160	22–93

Figure 16-9 *Uranus's rings (back-scattered view)* This view looks down onto the sunlit side of Uranus's rings. The nine principal rings that were discovered by Earth-based observers can be easily seen. Prominent in the upper-right corner is the ε ring, which is about 100 km wide at this location. The ring particles are so dark that they reflect only 1 percent of the light that falls on them. To make the rings visible, the photograph's contrast had to be enhanced so much that the dark sky appears mottled. (NASA)

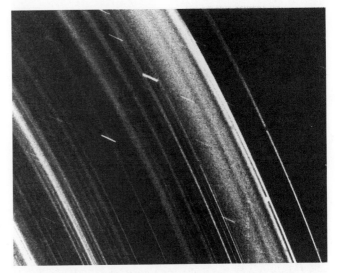

Figure 16-10 *Uranus's rings (forward-scattered view)* This view, taken when Voyager 2 was in Uranus's shadow, looks back toward the Sun. Numerous fine dust particles between the main rings gleam in the sunlight. This dust is probably fine-grained debris from collisions between larger particles in the main rings. The short streaks are star images, blurred because of the spacecraft's motion during the exposure. (NASA)

Collisions among ring particles are probably a continuous source of this dust. Some planetary scientists suggest that the Uranian rings are being eroded before our eyes and will probably be gone in only a few hundred million years.

16-6 Neptune is a cold, bluish world with Jupiterlike atmospheric features and a Uranuslike interior

The arrival of *Voyager 2* at Neptune in August 1989, capped one of NASA's most ambitious and most successful missions. Scientists were overjoyed at the detailed, close-up pictures and wealth of data sent back by the spacecraft about a world so remote that is barely distinguishable from a star through Earth-based telescopes.

At first glance, Neptune looks like a bluish Jupiter (see Figure 16-11). Methane in Neptune's atmosphere absorbs longer visible wavelengths but not the shorter ones. Sunlight reflected from Neptune's clouds is thus depleted of reds and yellows, making the planet appear quite blue.

The most prominent feature in Neptune's atmosphere, called the **Great Dark Spot**, gives the planet its distinctly Jupiterlike appearance. The Great Dark Spot is roughly the same size as the Earth, measuring about 12,000 by 8000 km.

It falls at about the same latitude on Neptune as the Great Red Spot on Jupiter, and occupies an area scaled down by the same factor that Neptune's diameter is smaller than Jupiter's.

The wind flow in the Great Dark Spot is anticyclonic, just as it is in Jupiter's Great Red Spot. Both features are high pressure systems out of which winds flow in a counterclockwise direction. This **anticyclonic flow** on Neptune is difficult to observe, because the only conspicuous features are high-elevation clouds that hover over the Great Dark Spot. These clouds do not move along with the winds in the atmosphere and may be compared with the so-called **lenticular clouds** on Earth, which seem to hang over mountain tops. On Earth, winds blow through these clouds at high speeds, but moisture carried by the winds condenses into visible ice crystals only where flow over the mountain has pushed the gases to cooler regions of the upper atmosphere. As winds descend down the leeward side of the mountain, the ice crystals turn back into vapor and the clouds disappear. Neptune's whitish, cirrus-like clouds are made of methane ice crystals. As prevailing winds on Neptune carry the methane to elevated heights over the Great Dark Spot, the gas condenses into visible clouds. Astronomers were surprised to discover that the Neptunian winds over the Great Dark Spot are near-supersonic. At 700 m/s (about 1500 mi/hr), Neptune's winds are even faster than Saturn's.

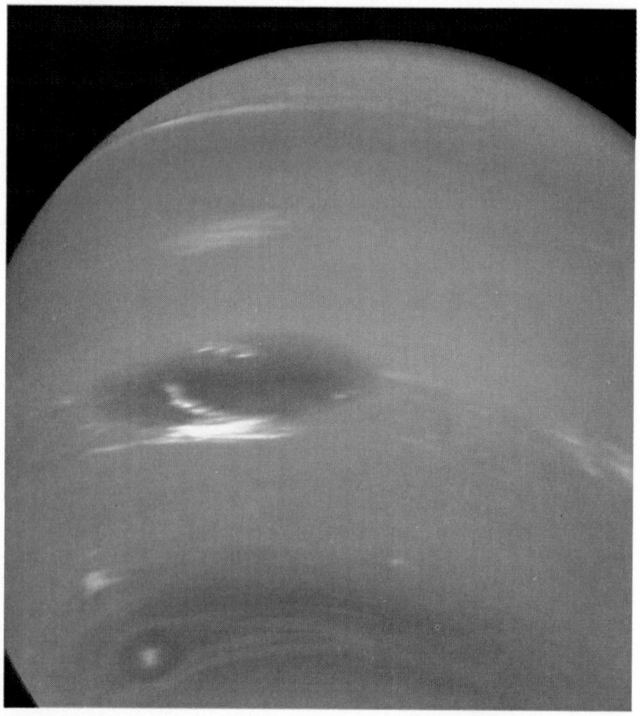

Figure 16-11 Neptune This view from Voyager 2 looks down on the southern hemisphere of Neptune. The Great Dark Spot, which is about the same size as the Earth, is near the center of this picture. Note the white, wispy methane clouds. Toward the lower left is a smaller, dark spot, located 54° south of Neptune's equator. (NASA)

Figure 16-12 Cirrus clouds over Neptune This picture shows vertical relief in Neptune's bright, methane cloud streaks. Voyager 2 photographed these clouds north of Neptune's equator near the terminator (the border between day and night on the planet). Note the shadows cast by the clouds onto the main cloud deck. (NASA)

The high elevation of Neptune's cirrus clouds was confirmed in photographs that showed these clouds casting shadows on the main cloud deck (see Figure 16-12). Knowing the angle of the Sun when the pictures were taken, scientists estimated that these clouds are 50 to 70 km above the cloud deck. Between the cirrus and the cloud deck, the atmosphere is quite clear. For comparison, cirrus clouds on Earth occur at an elevation of 7 to 9 km above sea level.

Neptune's belts and zones are much fainter than those on Jupiter. Most prominent is a broad, darkish band at high southern latitudes. Embedded in this band is a smaller dark spot, about a third the size of the Great Dark Spot. White wispy clouds seem to hover over the smaller dark spot just as they do over the Great Dark Spot.

Neptune radiates more energy than it receives from the Sun, which means that the planet has an internal heat source—yet another resemblance to Jupiter. Computer models for Neptune suggest that it could still be radiating primordial heat, just as Jupiter does. Uranus may have a similar heat source, but it is probably masked by the greater amount of energy that Uranus absorbs from sunlight.

Temperatures in the Neptunian atmosphere range from about 60 K at the equator and the poles to about 50 K at mid-latitudes. This situation, whereby the poles and the equator are both warmer than mid-latitudes, may be caused by the movement of gases in the upper atmosphere. At mid-latitudes, gases cool as they rise upward, but at the poles and equator the gases warm as they descend to lower elevations.

The dominant wind flow in the cloud layer that constitutes Neptune's visible surface is quite unlike that on any other Jovian planet. Voyager 2 detected periodic variations in radio emission from Neptune every 16 hours 3 minutes. It is reasonable to suppose that these emissions reveal the underlying rotation period of Neptune, because they are driven by the planet's magnetic field which is embedded in its interior. Most of Neptune's cloud cover revolves about the planet more slowly than the planet rotates. Some cloud features take as long as 18 hours to go once around the planet. On Jupiter, Saturn, and Uranus, the clouds revolve around the planets more rapidly than the planets themselves rotate. Thus Neptune's atmospheric circulation is retrograde.

While its atmosphere resembles that of Jupiter, Neptune's interior resembles that of Uranus. Both planets probably have a rocky core surrounded by a thick atmosphere primarily composed of hydrogen and helium with a slushy mixture of methane, water, and ammonia.

While probing Neptune's magnetosphere, Voyager 2 made the surprising discovery that the planet's magnetic axis is inclined to its axis of rotation by 47°. This is reminiscent of Uranus, whose magnetic axis is inclined by nearly 60° to its rotational axis. The magnetic fields of all four Jovian planets are oriented opposite to that of Earth (see Figure 16-13). A compass on a Jovian planet points southward.

As we shall see shortly, the orbits of Neptune's satellites suggest that Neptune may have gravitationally captured its largest moon, Triton. Such an event would have significantly perturbed Neptune, but it is not known if that incident was responsible for offsetting Neptune's magnetic axis.

Like Uranus, Neptune is surrounded by a system of thin, dark rings (see Figure 16-14 and Box 16-5). It is so cold at Uranus and Neptune that the ring particles can retain methane ice. Scientists speculate that eons of radiation damage have converted this methane ice into darkish carbon compounds, thus accounting for the low reflectivity of the rings. One of the rings contains clumpy concentrations of particles that produce the appearance of separate arcs of material. These arcs were in fact detected from Earth several years before Voyager 2 arrived at Neptune.

Having completed its mission to the outer planets, Voyager 2 is headed out of the solar system. For many years to come, scientists will continue to monitor the spacecraft's instruments as they probe the solar wind at extreme distances from the Sun. At present there are no plans for further missions to either Uranus or Neptune.

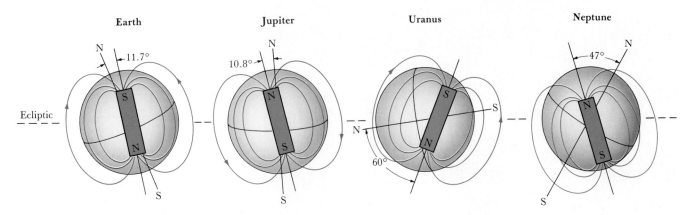

Figure 16-13 *The magnetic fields of four planets* This drawing shows how the magnetic fields of Earth, Jupiter, Uranus, and Neptune are tilted relative to their rotation axes. Note that the magnetic fields of Uranus and Neptune are offset from the center of the planets and steeply inclined to their rotation axes. In contrast, the magnetic fields of Earth and Jupiter are not offset and are tilted only slightly.

Box 16-5 Neptune's rings

Four dark rings around Neptune were discovered during the *Voyager 2* flyby in 1989. The so-called "main ring" (1989N1R) and "inner ring" (1989N2R) are quite thin and are clearly seen in Figure 16-14. The two broad rings (1989N3R and 1989N4R) are diffuse sheets of fine particles.

Ring	Distance from center of Neptune (km)	Width of ring (km)
1989N3R	41,900	1,700
1989N2R	53,200	<15
1989N4R	53,200–59,000	5,800
1989N1R	62,900	<50

16-7 *Triton is a frigid, icy world with a young surface and a tenuous atmosphere*

Triton, the outermost giant satellite of the solar system, was discovered in 1846. Because of its great distance from Earth, however, astronomers were unable to learn much about this remote world until the *Voyager 2* flyby of Neptune in 1989. Even through a large telescope Triton is a faint, starlike pin-point of light. *Voyager 2* revealed an icy satellite whose puzzling surface features present a challenge to planetary scientists (see Figure 16-15).

An important property of Triton, known since its discovery, is its retrograde orbit. Triton goes around Neptune opposite to the direction in which the planet rotates. No other giant satellite has such an orbit, and it is difficult to imagine how any satellite might form near a planet in an orbit opposing the direction of the planet's rotation. Only a few of the outer satellites of Jupiter and Saturn have retrograde orbits

Figure 16-14 *Neptune's rings* Two main rings are easily seen in this view, which blocks out an over-exposed image of Neptune. Careful examination of this picture also reveals a faint, inner ring. A faint sheet of particles, whose outer edge is located between the two main rings, extends inward toward the planet. *(NASA)*

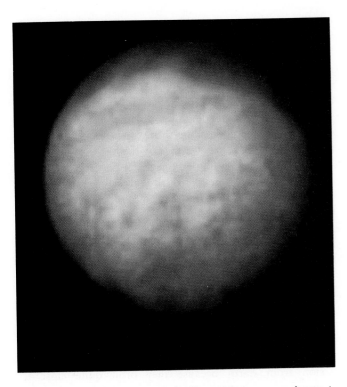

Figure 16-15 Triton Features as small as 100 km across can be seen in this image of Triton, photographed by Voyager 2 from a distance of about 5 million km. The dark areas near the top of this image are part of a belt of dark markings near Triton's equator. (NASA)

and these bodies are probably captured asteroids. Some scientists have therefore hypothesized that Triton may have been captured long ago by Neptune's gravity.

Triton is gradually spiraling in toward Neptune. Triton's orbit is decaying because of tidal interaction between the satellite and the planet. Triton raises a tidal bulge on Neptune just as our Moon distorts the Earth. In the case of the Earth–Moon system, the tidal bulge produces a small forward force on the Moon that causes it to spiral away from the Earth (recall Figure 9-15). However, Triton's orbit is retrograde, which makes the tidal bulge on Neptune lag behind the satellite's position in the sky, producing a small backward force that is causing Triton to spiral gradually in toward Neptune. In roughly 100 million years, Triton will be inside Neptune's Roche limit. The giant satellite will then be torn to pieces by Neptune's tidal forces and the planet will develop a spectacular ring system as rock fragments gradually spread out along the satellite's former orbit.

The orbit of Triton is tilted with respect to the plane of Neptune's equator (by about 20°), which in turn is tilted to the plane of Neptune's orbit about the Sun (by about 29°). Because of its 16-hour rotation, Neptune is slightly oblate and so the planet possesses an equatorial bulge in the plane of its equator. The gravitational pull of this equatorial bulge causes the satellite's orbit to precess, resulting in significant changes in Triton's seasons over the years. There are years in which Triton experiences modest seasonal variations analogous to those on Earth, and others (as when *Voyager 2* passed by in 1989) in which Triton has exaggerated, Uranus-like seasons, with the Sun shining down onto the satellite's south pole.

A high-resolution image of Triton's south polar region is shown in Figure 16-16. Note that very few craters are seen, indicating that some process must have obliterated the scars of numerous ancient impacts left over from the first billion years of the solar system. A major modification of Triton's surface is consistent with the idea that the satellite once orbited the Sun on its own and was captured by Neptune, perhaps 3 or 4 billion years ago. Upon being captured, Triton most likely would have started off in a highly elliptical orbit, but today the satellite's orbit is remarkably circular. Triton's original elliptical orbit would have been "circularized" by tidal forces exerted on the satellite by Neptune's gravity. With each revolution about that original elliptical orbit, the changing distance to Neptune would have stretched and flexed Triton, causing its orbit to become more circular while depositing significant energy into the satellite itself.

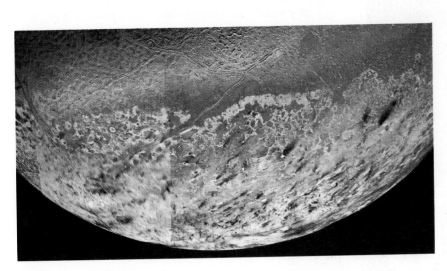

Figure 16-16 Triton's south polar cap Approximately a dozen high-resolution images were combined to produce this view of Triton's southern hemisphere. The pinkish polar cap is probably made of nitrogen frost. A notable scarcity of craters suggests that Triton's surface was either melted or flooded by icy lava after the era of bombardment that characterized the early history of the solar system. (NASA)

This energy would have melted much of the satellite's interior and the resulting volcanism would have obliterated Triton's original surface features, including craters.

The pinkish ice that constitutes Triton's south polar cap is probably a layer of nitrogen frost. With the arrival of summer at Triton's south pole, this frost layer slowly evaporates, and many of the surface markings seen in Figure 16-16 are possibly the result of the northward flow of the resulting nitrogen gas. These winds are very tenuous, however, because Triton's atmosphere is so very thin. Atmospheric pressure on the satellite's surface is only 10^{-5} of that at sea level on Earth.

Triton exhibits some surface features seen on other icy worlds, such as long cracks resembling those on Europa and Ganymede. Other features are unique to Triton and are quite puzzling. For example, near the **terminator** in Figure 16-16, you can see a wrinkled terrain that resembles the skin of a cantaloupe. Triton also has a few frozen lakes like the one shown in Figure 16-17. Some scientists speculate that these

Figure 16-17 A frozen lake on Triton? Some scientists believe that this lakelike feature is the caldera of an ice volcano. The flooded basin is about 200 km wide and 400 km long, and so covers an area about the size of the state of West Virginia. (NASA)

lakelike features are the calderas of extinct ice volcanoes. It is unlikely that any lava is flowing on Triton today, however, because the satellite is so very cold. Voyager instruments measured a surface temperature of 37 K ($-236°C = -395°F$), making Triton the coldest world we have ever visited. Nevertheless, Voyager cameras did glimpse two towering plumes of gas extending up to 8 km above the satellite's surface. Perhaps these plumes are nitrogen gas escaping through vents or fissures warmed by the feeble summer Sun.

Voyager 2 also photographed several small satellites orbiting Neptune. Nereid, which was sighted in 1949 by Gerard Kuiper just after he discovered Miranda, was photographed from afar. Nereid has the most eccentric orbit of any satellite in the solar system. The distance between Nereid and Neptune varies from 1.4 million to 9.7 million km as this tiny moon moves along its highly elliptical orbit. Triton completes its backward orbit in only 6 days, but Nereid takes nearly 360 days to go once around Neptune.

Nereid has a diameter of about 170 km, but it is not the second largest moon of Neptune. *Voyager 2* discovered six moons orbiting Neptune, two of which are slightly bigger than Nereid (see Box 16-6). These small bodies escaped detection from Earth because they orbit quite close to Neptune.

16-8 Pluto was discovered after a laborious search of the heavens

Speculations about a ninth planet date back to the late 1800s, when a few astronomers suggested that Neptune's orbit was being perturbed by an unknown planet. Such statements were more than a bit premature. Nevertheless, encouraged by the fame of Adams and Leverrier, several people set out to become the discoverer of "Planet X." Two Boston gentlemen, William Pickering and Percival Lowell, were prominent in this effort.

Modern calculations show that there were, in fact, no statistically significant perturbations of Neptune's orbit. It is thus not surprising that no planet was found at the positions predicted by Pickering, Lowell, and others, but the search continued.

Before he died in 1916, Percival Lowell urged that a special wide-field camera be constructed to help search for Planet X. After many delays, the camera was finished in 1929 and installed at the Lowell Observatory in Arizona, where a young astronomer, Clyde W. Tombaugh, had joined the staff to carry on the project. On February 18, 1930, Tombaugh finally discovered the long-sought planet. It was disappointingly faint—fifteenth magnitude—and showed no discernible disk. The planet was named for Pluto, the mythological god of the underworld, whose name has Percival Lowell's initials as its first two letters. The discovery was publicly announced on March 13, 1930, the 149th anniver-

Box 16-6 Neptune's satellites

Eight satellites are known to orbit Neptune. Triton and Nereid were discovered from Earth; the rest were discovered during the *Voyager 2* flyby in August 1989. The newly discovered moons are named with the year of their discovery, the letter "N," and a number indicating their order of discovery. Three of these moons are shown alongside an overexposed Neptune in the accompanying photograph. Unlike Triton and Nereid, the orbits of all of the newly discovered moons lie nearly in the plane of Neptune's equator.

	Distance from center of Neptune (km)	Orbital period (hr)	Diameter (km)	Albedo
1989N6	48,000	7.2	50	0.06?
1989N5	50,000	7.4	100	0.06?
1989N3	52,500	7.9	150	0.06?
1989N4	62,000	10.3	180	0.05
1989N2	73,600	13.2	190	0.06
1989N1	117,600	26.9	420	0.06
Triton	354,800	141.1	2760	0.7–0.9
Nereid	5,513,400*	360 days	170	0.14

*The semimajor axis of Nereid's highly elliptical orbit is given.

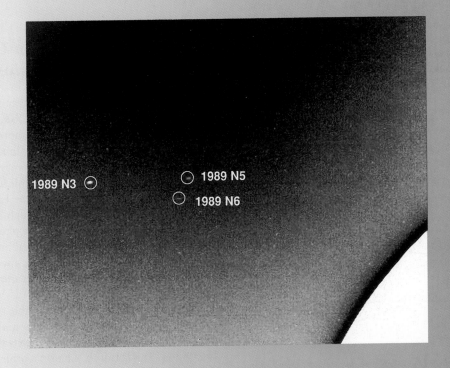

Figure 16-18 Pluto *Pluto was discovered in 1930 by searching for a dim, starlike object that slowly moves in relation to the background stars. These two photographs were taken one day apart. Although Pluto* *does not show a measurable disk, astronomers estimate that its diameter is about two-thirds the diameter of our Moon. (Lick Observatory)*

sary of the discovery of Uranus. Two typical photographs of Pluto appear in Figure 16-18.

Pluto was smaller than Pickering and Lowell had expected, so Tombaugh kept on searching. For the next 13 years, he examined endless pairs of photographs like the pair in Figure 16-18, looking for the telltale displacement of a faint starlike image. None was ever found. If a trans-Pluto planet exists, it must be fainter than eighteenth magnitude. If there is a tenth planet, it must be either very small—more like an asteroid than a planet in size—or very far from the Sun.

Pluto's orbit about the Sun is more elliptical and more steeply inclined to the plane of the ecliptic than is the orbit of any other planet in our solar system (see Figure 16-19). In fact, Pluto's orbit is so eccentric that this planet is sometimes

closer to the Sun than Neptune is. For example, when Pluto was at perihelion in 1989, it was more than 100 million km closer than was Neptune to the Sun.

A solar day on Pluto lasts 6 days 9 hours 18 minutes. Although we cannot see any surface details on Pluto, its magnitude varies with this period. There are features on Pluto—perhaps a large frozen lake—that are periodically exposed to our Earth-based view, causing the variations in brightness that divulge the planet's rotation period.

16-9 Pluto and its moon Charon are roughly comparable in size

In 1978, while examining some photographs of Pluto, James W. Christy of the U.S. Naval Observatory noticed that the image of the planet on a photographic plate was slightly elongated. Pluto seemed to have a lump on one side (see Figure 16-20). Numerous other photographs of Pluto were promptly examined. Some showed this lump, but others did not. It soon became evident that the lump was actually a satellite that is occasionally far enough away from Pluto to show up as a slight elongation in the planet's image. Christy proposed that the newly discovered moon be christened Charon (pronounced KAR-en), after the mythical boatman who ferried souls across the River Styx to Hades, the domain ruled by Pluto.

Charon's orbit is quite remarkable. For one thing, the average distance between Charon and Pluto is a scant 19,700 km—less than $\frac{1}{20}$ the distance between the Earth and our Moon. Furthermore, Charon's orbital period of 6.3874 days is the same as the rotational period of Pluto. In other words, Pluto rotates synchronously with the revolution of its satellite. As seen from the satellite-facing side of Pluto,

Figure 16-19 Pluto's orbit *Pluto's orbit is more elliptical and more steeply inclined to the plane of the ecliptic than is the orbit of any other planet in our solar system. In fact, its orbit is so eccentric that Pluto is occasionally closer to the Sun than Neptune.*

Box 16-7 Pluto data

Mean distance from the Sun:	39.53 AU = 5.914×10^9 km
Maximum distance from the Sun:	49.33 AU = 7.380×10^9 km
Minimum distance from the Sun:	29.73 AU = 4.447×10^9 km
Mean orbital velocity:	4.7 km/s
Sidereal period:	248.6 years
Rotation period:	6.3874 days
Inclination of equator to orbit:	122°
Inclination of orbit to ecliptic:	17.1°
Orbital eccentricity:	0.248
Diameter:	2290 km
Diameter (Earth = 1):	0.18
Apparent diameter as seen from Earth:	0.1″
Mass:	10^{22} kg
Mass (Earth = 1):	0.002
Mean density:	2000 kg/m³
Surface gravity (Earth = 1):	0.06
Escape velocity:	1 km/s
Mean surface temperature:	−223°C = −369°F = 50 K
Albedo:	0.4
Brightest magnitude:	+15
Mean diameter of Sun as seen from Pluto:	50″

Charon neither rises nor sets but instead seems to hover in the sky, as if perpetually suspended above the horizon.

Soon after the discovery of Charon, astronomers realized that they were about to witness an alignment that only oc-curs every 124 years. From 1985 through 1990, the highly inclined plane of Charon's orbit swept across the inner solar system, allowing astronomers to view rare mutual eclipses of Pluto and its moon. As the bodies passed in front of

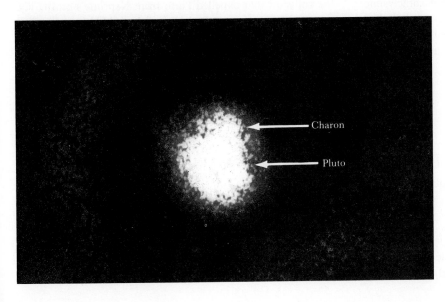

Figure 16-20 Pluto and Charon *Pluto's moon Charon appears as a slight elongation or lump on one side of this greatly enlarged image of the planet. Pluto and Charon are separated by only 19,700 km. The Pluto–Charon system deserves to be called a double planet because these two objects resemble each other in mass and size more closely than does any other planet–satellite pair in the solar system. (Courtesy of J. W. Christy and R. S. Harrington)*

each other, their combined brightness diminished in ways that revealed their sizes and surface characteristics. Data from the eclipses gave Pluto's diameter as 2290 km and Charon's as 1280 km. For comparison, our Moon's diameter (3476 km) is about $1\frac{1}{2}$ times as large as Pluto's.

We know both the separation and the orbital period of the Pluto–Charon system, so we can use Kepler's third law (recall Box 4-3) to determine the total mass of Pluto and Charon. For Pluto and Charon together, the mass is about 0.002 Earth masses. Since we also know the sizes of Pluto and Charon, we can calculate their total volume and, dividing their mass by their volume, arrive at roughly 2000 kg/m³ for their average density. Pluto and Charon are therefore probably made up of a mixture of rock and ice.

The 1985–1990 **eclipse season** gave astronomers the opportunity to observe brightness changes during mutual eclipses of Pluto and Charon. During these events, the reflectivity of Pluto and Charon would change slightly in response to whether dark or light patches on these bodies were exposed to our Earth-based view. Marc Buie at the Space Telescope Science Institute and his colleagues used a supercomputer to go through thousands of photometric observations of Pluto–Charon eclipses to calculate what sort of surface features would give the observed brightness variations. Their simulations suggest that Pluto has bright polar caps and a large, dark equatorial spot (see Figure 16-21). The hemisphere opposite the large dark spot has a narrow, light gray patch extending along the equator. Charon, like Triton, apparently has a bright southern polar cap.

In many respects Pluto resembles the icy satellites of the Jovian planets, especially Triton. Pluto and Triton have about the same size, density, and surface temperature. They both rotate retrograde with periods of about 6 days. Although sparse, they both have atmospheres of nitrogen and methane that can be detected from Earth. These similarities, as well as Pluto's unusual orbit about the Sun, make it reasonable to wonder whether Pluto might be an escaped satellite that once orbited Neptune. Some sort of cataclysmic event might have occurred in the ancient past—perhaps the capture of Triton—that flung Nereid into its highly elliptical orbit and catapulted Pluto away from Neptune altogether.

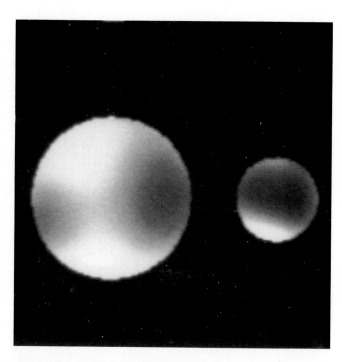

Figure 16-21 *Computer-modeled image of Pluto and Charon* A supercomputer was used to deduce albedo variations on the surfaces of Pluto and Charon that could account for brightness variations that were observed as the two bodies transited and eclipsed each other. Pluto apparently has bright polar caps and a large, dark equatorial spot, turned toward Charon in this view. (Courtesy of M. Buie, K. Horne, and D. Tholen)

The hypothesis that Pluto might be an escaped satellite of Neptune was first proposed by R. A. Lyttleton of Cambridge University in 1936 and vigorously championed by Gerard Kuiper. Problems with this proposal have been long recognized, however. For one, the present-day orbits of Neptune and Pluto do not intersect. The two planets are never closer than 384 million km apart, despite the proximity suggested in Figure 16-15. Perhaps an additional body was involved in the collision that expelled Pluto from Neptune's grasp. The gravitational effects of that body could have supplied enough energy and momentum to tip Pluto's orbit away from Neptune's.

Key words

anticyclonic flow	Great Dark Spot	magnetic axis
eclipse season	lenticular cloud	terminator

Key ideas

- Uranus was discovered by chance, Neptune at a location predicted with Newtonian mechanics, and Pluto after a long search. If another planet exists beyond Pluto, it is either very small or very far away.

- Both Uranus and Neptune have a rocky core surrounded by a liquid mantle of water, ammonia, and methane, with an outer gaseous envelope composed predominantly of hydrogen and helium.

- Uranus is unique among the planets and satellites of our solar system in that its axis of rotation lies nearly in the plane of its orbit, producing greatly exaggerated seasonal changes on the planet.

  Uranus's magnetic axis is inclined by 60° from its axis of rotation, while Neptune's is inclined by 47°. All the other planets have their magnetic and rotational axes roughly parallel.

  Uranus has a system of thin, dark rings and five satellites similar to the moderate-sized moons of Saturn; 10 more small Uranian satellites were discovered during the *Voyager 2* flyby.

  The unusual orientation of Uranus's rotational and magnetic axes, as well as the surface features of some of its satellites (especially Miranda) suggest that Uranus may have been "knocked on its side" by a collision with a planetlike object during the early history of our solar system.

- Neptune's surface features resemble those of a bluish Jupiter, having a Great Dark Spot, and faint belts and zones.

  Neptune's magnetic field is like Uranus's, in that its magnetic axis is inclined steeply to the planet's axis of rotation.

  Like Uranus, Neptune is surrounded by a system of thin, dark rings. The low reflectivity of the ring particles may be due to radiation-damaged methane ice.

- Neptune has an icy, terrestrial-sized satellite, Triton, with a tenuous nitrogen atmosphere.

  The scarcity of craters on Triton suggests that its surface was either molten or flooded with icy lava sometime after the era of bombardment that left numerous craters on such worlds as Mercury and our Moon.

- Pluto and its moon Charon have roughly comparable masses and move in a highly elliptical orbit steeply inclined to the plane of the ecliptic.

  The Pluto–Charon system may be an escaped satellite of Neptune.

Review questions

1 Could astronomers in antiquity see Uranus? If so, why do you suppose it was not recognized as a planet?

2 Why do you suppose that the discovery of Neptune is rated as one of the great triumphs of science, whereas the discoveries of Uranus and Pluto are not?

3 Why are Uranus and Neptune distinctly bluer than Jupiter and Saturn?

4 Describe the seasons on Uranus. Why are the Uranian seasons different from those on any other planet?

5 How does the orientation of Uranus's and Neptune's magnetic axes differ from those of other planets?

6 Briefly describe the evidence supporting the idea that Uranus was struck by a large planetlike object several billion years ago.

7 Briefly describe the evidence supporting the idea that Triton was captured by Neptune.

8 Compare the rings that surround Jupiter, Saturn, Uranus, and Neptune. Briefly discuss their similarities and differences.

9 Why is it reasonable to suppose that Neptune will someday be surrounded by a broad system of rings, perhaps similar to those that surround Saturn?

10 How can astronomers distinguish a faint solar system object like Pluto from background stars within the same field of view?

11 Describe the circumstantial evidence supporting the idea that Pluto is an escaped satellite of Neptune.

Advanced questions

> **Tips and tools . . .**
> The small-angle formula is given in Box 1-1. After its optical problems are fixed, the Hubble Space Telescope will have a resolving power of about 0.1 arc sec. Problem 14 involves some geometry and you need to remember that the circumference of a circle is π times its diameter. Newton's formula for the gravitational force between two objects was given in Chapter 4.

*12 (Basic) If Earth-based telescopes can resolve angles down to 0.25 arc sec, how large could an object be at Pluto's average distance from the Sun and still not present a resolvable disk?

*13 (Basic) Can the Hubble Space Telescope resolve the Great Dark Spot? Is it capable of seeing any features on Pluto?

*14   Using the orbital data given in Box 16-1, estimate the maximum duration of a stellar occultation by Uranus.

*15   (Challenging)   At what planetary configuration is the gravitational force of Neptune on Uranus a maximum? Calculate the fraction by which the sunward gravitational pull on Uranus is reduced by Neptune at that configuration.

16   Suppose you were standing on Pluto. Describe the motions of Charon relative to the Sun, the stars, and your own horizon. Would you ever be able to see a total eclipse of the Sun? Under what circumstances would you never see Charon?

17   Suppose you wanted to search for planets beyond Pluto. Why might it be advantageous to do your observations in infrared rather than visible wavelengths? Use Wien's law to calculate the wavelength range best suited for your search. Could such observations be done at an Earth-based observatory?

*18   Calculate the ratio of the diameters of the Great Red Spot and the Great Dark Spot and compare it to the ratio of the diameters of Jupiter and Neptune. Does this comparison strengthen or weaken the similarities between the Great Red Spot and the Great Dark Spot?

Discussion questions

19   Discuss the evidence presented by the outer planets that catastrophic impacts of planetlike objects occurred during the early history of our solar system.

20   Would you expect the surfaces of Pluto and Charon to be heavily cratered? Explain.

21   At NASA and the Jet Propulsion Laboratory, some scientists are discussing the possibility of placing spacecraft in orbit about Uranus and Neptune early in the twenty-first century. What kinds of data should be collected, and what questions would you like to see answered by these missions?

Observing projects

22   Make arrangements to view Uranus through a telescope. During the early 1990s, Uranus is in Sagittarius, so the planet is best seen in June, July, or August. To help you find the planet, a star chart showing the path of Uranus against the background stars is published each January in Sky & Telescope. Use that star chart at the telescope to find the planet. Are you certain that you have found Uranus? Can you see a disk? What is its color?

23   If you have access to a large telescope, make arrangements to view Neptune. Throughout the 1990s, Neptune is in Sagittarius, so the planet is best seen in June, July, or August. Neptune is most easily found if you know its right ascension and declination, which are given in the Astronomical Almanac. Alternatively, you could consult the star chart published in the January issue of Sky & Telescope. Can you see a disk? What is its color?

For further reading

Bennett, J. "The Discovery of Uranus." Sky & Telescope, March 1981   •   This fascinating historical article describes Herschel's discovery of Uranus.

Berry, R. "Neptune Revealed." Astronomy, December 1989   •   This summary of Voyager results at Neptune includes an outstanding collection of photographs.

Binzel, R. "Pluto." Scientific American, June 1990   •   This superb article summarizes virtually everything known about Pluto.

Brown, R., and Cruikshank, D. "The Moons of Uranus, Neptune, and Pluto." Scientific American, July 1985   •   This article, written before the Voyager flybys, contains many insights gleaned from Earth-based observations.

Cuzzi, J., and Esposito, L. "The Rings of Uranus." Scientific American, July 1987   •   The authors of this article argue that Uranus's rings are actually a fleeting phenomenon.

Elliot, J., et al. "Discovering the Rings of Uranus." Sky & Telescope, June 1977   •   The team of astronomers who discovered Uranus's rings describe their airborne observations.

Harrington, R., and Harrington, B. "The Discovery of Pluto's Moon." Mercury, January/February 1979   •   This article gives authoritative insights into the discovery of Pluto's moon.

Ingersoll, A. P. "Uranus." Scientific American, January 1987   •   This superb article summarizes the results of the Voyager 2 flyby.

Johnson, T., et al. "The Moons of Uranus." Scientific American, April 1987   •   This beautifully illustrated article asserts that Uranus's moons had a geologically vigorous history.

Kinoshita, J. "Neptune." Scientific American, November 1989   •   Pictures of Neptune and Triton grace this article about the Voyager flyby.

Moore, P. "The Discovery of Neptune." Mercury, July/August, 1989   •   A prolific popularizer of astronomy gives a detailed history of how Neptune was found.

Tombaugh, C. "The Search for the Ninth Planet: Pluto." Mercury, January/February 1979   •   The discoverer of Pluto describes his historic observations.

CHAPTER 17

Interplanetary Vagabonds

The planets are not the only objects that orbit the Sun. Asteroids, meteoroids, and comets are small but significant members of our solar system, providing valuable information about the early history of the planets. Some meteoroids are fragments of asteroids that have undergone significant melting and chemical differentiation. Others seem to have escaped such processing and thus contain important chemical clues about conditions in the solar nebula 4.5 billion years ago. Likewise, comets are dusty chunks of ice that contain frozen samples of the solar nebula. When one of these "dirty snowballs" passes near the Sun, solar radiation vaporizes some of its ice, producing a long flowing tail. After many such passages near the Sun, a comet's ice is depleted, leaving only a swarm of dust particles that rain down on Earth in a meteor shower. Meteoroids and asteroids occasionally strike the Earth. Indeed, there is substantial evidence that an asteroid was responsible for the extinction of the dinosaurs some 63 million years ago. Even today, life on Earth is threatened by the devastation that could be wrought by a wayward asteroid.

The head of Comet Halley   This photograph shows the bluish head of Halley's Comet as it approached the Sun in December 1985. Comet Halley orbits the Sun with an average period of 76 years along a highly elliptical path that stretches from just inside the Earth's orbit to slightly beyond the orbit of Neptune. This color photograph was constructed from three black-and-white photographs taken in rapid succession with red, blue, and green filters on the same telescope. (Anglo-Australian Observatory)

When William Herschel discovered a new planet beyond Saturn, in 1781, he proposed that it be named Georgium Sidus, in honor of King George III of Great Britain. Astronomers in other countries rejected this suggestion and adopted the name Uranus, which had been proposed by Johann Elert Bode. This young German astronomer was far more famous, however, for his earlier popularization of a simple rule that describes the average distances of the planets from the Sun. Although Bode did not discover this rule, it is usually known today as Bode's law. Most astronomers now regard this "law" as merely a coincidence, but it did lead directly to the discovery of a large number of previously unknown objects that orbit the Sun.

17-1 Bode's law led astronomers to search for a planet between the orbits of Mars and Jupiter

Bode's rule for remembering the distances of planets from the Sun goes like this:

1 Write down the sequence of numbers 0, 3, 6, 12, 24, 48, 96, ... (Note that each number after the second one is simply twice the preceding number.)

2 Add 4 to each number in the sequence.

3 Divide each of the resulting numbers by 10.

As shown in Table 17-1, the final result is a series of numbers that corresponds remarkably well with the distances of most of the planets from the Sun.

Astronomers regarded Bode's rule as merely a useful trick for remembering the planetary distances—until Herschel's unexpected discovery of Uranus, very near the orbit for a planet beyond Saturn predicted by Bode's scheme. Suddenly it seemed far more likely that Bode's rule might actually represent some physical property of the solar system. The numerical sequence soon became known as **Bode's law,** an unfortunate name, because it is neither a physical law nor was it invented by Bode. Johann Titius, a German physicist and mathematician had first published it in 1766.

Astronomers now looked with new interest at the location of the "missing planet" in the sequence of Bode's law—the gap between the orbits of Mars and Jupiter (see Table 17-1). Six German astronomers who jokingly called themselves the "Celestial Police" organized an international group to begin a careful search for the missing planet. Before their search got under way, however, the anticipated announcement came from Sicily.

The Sicilian astronomer Giuseppe Piazzi had been carefully preparing a map of faint stars in the constellation of Taurus. On January 1, 1801, he noticed a dim, previously uncharted star that shifted its position slightly over the next several nights. Suspecting that he might have found the missing planet suggested by Bode's law, Piazzi excitedly wrote to Bode in Berlin.

Unfortunately, Piazzi's letter did not reach Bode until late March. By that time, Piazzi's object was too near the Sun to be seen. To make matters worse, Piazzi had made only a limited number of observations, and the best mathematical techniques of the day could predict an orbit only if the object were traveling along a circle or parabola. However, an object orbiting the Sun may follow any curve that is a conic section (recall Figure 4-18). An elliptical orbit was suspected for Piazzi's object, and astronomers feared that it might have been lost.

Upon hearing of the plight of these astronomers, the brilliant young mathematician Karl Friedrich Gauss took up the challenge and developed a general method of computing orbits from observations. Gauss's method yields an orbit from only three sets of observations, regardless of the conic section along which an object is traveling. In other words, an object's orbit can be calculated from only three separate measurements of its right ascension and declination.

In November 1801, Gauss finished his computations and predicted that Piazzi's object would be found in the constellation of Virgo. One of the self-styled "Celestial Police," Baron Franz Xavier von Zach, sighted Piazzi's object on December 31, 1801, only a short distance from the position given by Gauss. At Piazzi's request, the object was named Ceres (pronounced SEE-reez), after the Roman goddess of agriculture, popular in Sicily.

Ceres orbits the Sun once every 4.6 years at an average distance of 2.77 AU, which is in remarkable agreement with the distance given by Bode's law for the missing planet. Ceres is very small, however. At opposition, its magnitude is only +7.4, and its diameter is estimated to be only a scant 1000 km. Ceres thus did not qualify as a full-fledged planet, and astronomers continued the search.

Table 17-1 Bode's law

Bode's progression	Planet	Actual distance (AU)
(0 + 4)/10 = 0.4	Mercury	0.39
(3 + 4)/10 = 0.7	Venus	0.72
(6 + 4)/10 = 1.0	Earth	1.00
(12 + 4)/10 = 1.6	Mars	1.52
(24 + 4)/10 = 2.8	?	—
(48 + 4)/10 = 5.2	Jupiter	5.20
(96 + 4)/10 = 10.0	Saturn	9.54
(192 + 4)/10 = 19.6	Uranus	19.18
(384 + 4)/10 = 38.8	Neptune	30.06
(768 + 4)/10 = 77.2	Pluto	39.44

17-2 *Numerous small objects orbit the Sun between the orbits of Mars and Jupiter*

On March 28, 1802, Heinrich Olbers discovered another faint, starlike object that moved against the background stars. He called it Pallas, after the Greek goddess of wisdom. Like Ceres, Pallas orbits the Sun every 4.6 years at an average distance of 2.77 AU. Pallas is even dimmer and smaller than Ceres, reaching a magnitude of only +8.0 at opposition and having an estimated diameter of only 600 km. Obviously, Pallas was not the missing planet either.

The discovery of two small objects with similar orbits at the distance expected for the missing planet led astronomers to suspect that Bode's missing planet might have somehow broken apart or exploded. The search for other small objects went on. Only two more were found, Juno and Vesta, until the mid-1800s, by which time telescopic equipment and techniques had improved. Astronomers then began to stumble across many more such objects orbiting the Sun between the orbits of Mars and Jupiter. These objects are known today as **asteroids** or **minor planets**.

The next major breakthrough came in 1891 when the German astronomer Max Wolf began using photographic techniques to search for asteroids. Some three hundred asteroids had been found up to that time, each painstakingly discovered by scrutinizing the skies for faint, uncharted starlike objects that slowly shifted their positions from one night to the next. With the advent of astrophotography, however, the floodgates were opened. An astronomer simply aims a camera-equipped telescope at the stars and takes a long exposure. If an asteroid happens to be in the field of view, it leaves a distinctive, blurred trail on the photographic plate, because of its movement along its orbit during the long exposure (see Figure 17-1). Using this technique, Wolf alone discovered 228 asteroids.

Although thousands of asteroids have been sighted, only 3000 have well-determined orbits. An additional 6000 asteroids have "passable" orbits, but the orbits of 20,000 more have never been determined. The orbits of all officially discovered asteroids are published annually in the Soviet catalogue *Ephemerides of Minor Planets*.

To become the official discoverer of an asteroid, you must do a lot more than merely produce one photograph with a blurred trail whose path matches no known orbit listed in the *Ephemerides*. You must also track the asteroid long enough to compute an accurate, reliable orbit (important data may be available from colleagues who could have inadvertently photographed your asteroid on earlier occasions). Then you must prove the accuracy of your orbit by locating the asteroid again on at least one succeeding opposition. At that time, an official number will be assigned to your asteroid (Ceres is 1, Pallas is 2, and so forth). You are also given the privilege of selecting a name for your asteroid.

Figure 17-1 *Two asteroids* *Asteroids can be detected by their blurred trails on time-exposure photographs of the stars. The images of two asteroids are seen in this picture. Astronomers sometimes find asteroids accidentally while photographing various portions of the sky for other purposes. (Yerkes Observatory)*

Ceres is unquestionably the largest asteroid. With a diameter of nearly 1000 km, Ceres accounts for about 30 percent of the mass of all the asteroids combined. Only three asteroids (Ceres, Pallas, and Vesta) have diameters greater than 300 km. Thirty other asteroids have diameters between 200 and 300 km, and there are 200 more that are bigger than 100 km across.

Astronomers estimate that roughly 100,000 asteroids exist that are bright enough to appear on Earth-based photographs. The vast majority are less than 1 km across. Like Ceres, Pallas, Vesta, and Juno, most asteroids orbit the Sun at distances between 2 and 3½ AU. This region of our solar system between the orbits of Mars and Jupiter is called the **asteroid belt** (see Figure 17-2). The asteroids whose orbits lie entirely within this region are called **belt asteroids**.

The combined matter of all the asteroids (including an estimate for those not yet officially known) would produce an object barely 1500 km in diameter, considerably smaller than our Moon. Therefore, if the asteroids are fragments of Bode's missing planet, it cannot have been large enough to rank with the terrestrial planets. Furthermore, geologists and physicists have never been able to produce a good theory to explain how a planet could fragment or explode. There is thus little support today for the hypothesis of a shattered planet. It seems more reasonable to suppose instead

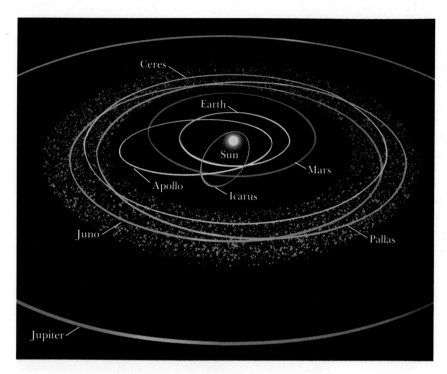

Figure 17-2 *The asteroid belt* *Most asteroids orbit the Sun in a belt 1½ AU wide between the orbits of Mars and Jupiter. The orbits of Ceres, Pallas, and Juno are indicated. The orbits of the asteroids Apollo and Icarus are also shown.*

that the asteroids are debris left over from the formation of the solar system out of the solar nebula.

Constant gravitational perturbations caused by the enormous mass of Jupiter probably kept planetesimals in the region between Mars and Jupiter from ever accreting into larger objects. As a result, the "missing planet" never had a chance to form. The matter that remains today in the gap between the orbits of Jupiter and Mars appears to be simply a remnant of scattered debris from the original solar nebula that elsewhere accreted into planets.

17-3 Jupiter's gravity affects the structure of the asteroid belt and captures asteroids along its orbit

The orbits of the asteroids are affected slightly by the gravitational pulls of the various planets. Most notable are the effects of Jupiter because of its large mass and proximity to the asteroid belt. Mars also perturbs asteroid orbits, but to a much lesser extent than Jupiter. The combined effects of such gravitational perturbations over the ages causes the asteroids to prefer certain orbits while avoiding others. It would be a monumental task to compute the resulting distribution of asteroid orbits because the gravitational forces of many bodies would have to be included. Nevertheless, we can appreciate some basic characteristics of the asteroid belt by considering the effects of Jupiter alone.

Imagine a belt asteroid as it moves along its orbit between the orbits of Jupiter and Mars. Each time the faster-moving

asteroid catches up with and passes the massive Jupiter, it experiences a slight gravitational tug toward Jupiter. This force tends to alter the asteroid's orbit slightly. However, over the ages these close passes occur at many points along the asteroid's orbit, so their effects tend to cancel one another.

Now imagine an asteroid that circles the Sun once every 5.93 years, in exactly half of Jupiter's orbital period. On every second trip around the Sun, the asteroid finds itself lined up again and again between Jupiter and the Sun, always at the same location and with the same orientation. These gravitational perturbations, which add up, deflect the asteroid from its original 5.93-year orbit, leaving a gap in the asteroid belt. According to Kepler's third law, a period of 5.93 years corresponds to a semimajor axis of 3.28 AU. Because of Jupiter, there are no asteroids that orbit the Sun at this average distance.

Similarly, we would expect to find a gap corresponding to an orbital period of one-third Jupiter's period, or 3.95 years. Other gaps should exist for various other simple commensurabilities between the periods of asteroids and Jupiter. The data graphed in Figure 17-3 show that such gaps do exist. They are called **Kirkwood gaps**, in honor of the American astronomer Daniel Kirkwood, who first drew attention to them. The major divisions in the ring structure of Saturn may have a similar origin, created by gravitational perturbations of the Saturnian moons on the icy fragments in the rings.

Although Jupiter's gravitational pull depletes certain orbits in the asteroid belt, it also captures asteroids, at certain locations much farther from the Sun. As explained in Box 17-1, there are two locations along Jupiter's orbit where the

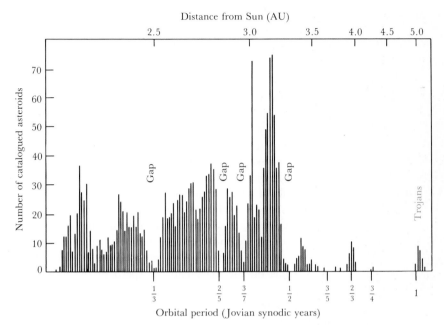

Distance from Sun (AU)

Figure 17-3 *The Kirkwood gaps* *This histogram displays the numbers of asteroids at various distances from the Sun. Notice that very few asteroids have orbits whose orbital periods correspond to such simple fractions as ⅓, ⅖, ⅗, and ½ of Jupiter's orbital period. Gravitational perturbations resulting from repeated alignments with Jupiter have deflected asteroids away from these orbits.*

gravitational forces of the Sun and Jupiter work together to hold asteroids in orbit. These two locations are called the **Lagrangian points** L_4 and L_5. Point L_4 is located one-sixth of the way around Jupiter's orbit ahead of the planet, and point L_5 occupies a similar position behind the planet (see Figure 17-4).

The asteroids trapped at Jupiter's Lagrangian points are called **Trojan asteroids**, named individually after heroes of

the Trojan War. Approximately four dozen Trojan asteroids have been catalogued so far, and some astronomers believe there may several hundred rock fragments orbiting near each Lagrange point.

17-4 Asteroids occasionally collide with each other and with the inner planets

In addition to the belt asteroids and the Trojan asteroids, there are other asteroids distinguished by highly elliptical orbits that bring them into the inner regions of the solar system. This class is usually divided into two groups: asteroids that cross Mars's orbit (called **Amor asteroids**, after their prototype, Amor) and asteroids that cross Earth's orbit (called **Apollo asteroids**, after their prototype, Apollo).

It is probable that the Amor and Apollo asteroids are somehow related. Edward Anders of the University of Chicago argues that gravitational perturbations by Mars deflect Amor asteroids into Earth-crossing orbits, causing Amor asteroids to become Apollo asteroids.

Occasionally an Apollo asteroid passes quite close to Earth. Figure 17-5 shows Eros as it passed within 23 million km of our planet in 1931. On June 14, 1968, Icarus passed Earth at a distance of only 6 million km. One of the closest near-misses in recent history occurred on October 30, 1937, when Hermes passed Earth at a distance of 900,000 km— only a little more than twice the distance to the Moon. A similar close call occurred on March 23, 1989, when an asteroid called 1989FC passed within 800,000 km of the Earth. If this asteroid had struck the Earth, the impact

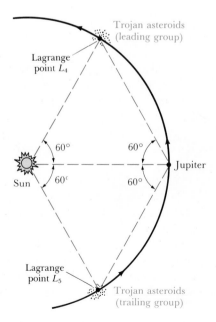

Figure 17-4 *The Trojan asteroids* *The combined gravitational forces of Jupiter and the Sun trap asteroids at the two Lagrangian points along Jupiter's orbit. The asteroids at these locations are named after Homeric heroes of the Trojan War.*

Figure 17-5 Eros Like all known Apollo asteroids, Eros occasionally passes near the Earth. This photograph was taken in February 1931, when the distance between Eros and the Earth was only 23 million km. The dimensions of Eros are roughly 10 by 20 by 30 km. This asteroid rotates with a period of 5.27 hours. (Yerkes Observatory)

would have been equivalent to the explosion of a thousand 20-megaton hydrogen bombs.

During these close encounters, astronomers can examine details about the asteroids. For example, an asteroid's magnitude is often observed to vary in a periodic fashion, presumably because of the different surfaces turned toward us as the asteroid rotates. These periodic variations in magnitude reveal the asteroid's rate of rotation. Typical asteroid rotation periods are in the range of 5 to 20 hours.

Careful scrutiny of an asteroid's variations in magnitude can also reveal the asteroid's shape and dimensions. Only the largest asteroids like Ceres, Pallas, and Vesta are spherical, because only they have enough gravity to pull themselves into a spherical shape. The smaller asteroids permanently retain the odd shapes produced by collisions with other asteroids. A small asteroid looks dim when seen end-on but appears brighter when seen broadside. Therefore, measurements of an asteroid's variations in magnitude can tell astronomers a lot about its shape.

There is ample evidence that interasteroid collisions have fragmented asteroids into small pieces. In 1918, the Japanese astronomer Kiyotsugu Hirayama drew attention to groups of asteroids that share nearly identical orbits. These groupings, now called **Hirayama families**, presumably resulted from fragmentation of parent asteroids. For example, of the three dozen known Amor asteroids, 20 can be grouped into four Hirayama families.

A collision between kilometer-sized asteroids must be an awesome event. Typical collision velocities are estimated to be 1 to 5 km/s (2000 to 11,000 mi/hr), which is more than

sufficient to shatter rock. In collisions at the low end of this velocity range, the resulting fragments may not achieve escape velocity from each other and may then reassemble, because of their mutual gravitational attraction. Alternatively, several large fragments may end up orbiting each other. This is probably what happened to both Pallas and Victoria, which are **binary asteroids**, each consisting of a main asteroid and a large satellite. Only in a high-velocity collision is there enough energy to shatter an asteroid permanently and create a Hirayama family.

Interasteroid collisions produce numerous chunks of rock, many of which eventually rain down on Venus, Earth, and Mars. Fortunately for us, the vast majority of these asteroid fragments, usually called **meteoroids**, are quite small. On rare occasions, however, a large fragment does collide with our planet. The result is an **impact crater**, whose diameter depends on the mass and the speed of the impinging object.

One of the most impressive, best-preserved terrestrial impact craters is the famous Barringer Crater near Winslow, Arizona (see Figure 17-6). The crater measures 1.2 km across and is 200 m deep. The crater was formed 25,000 years ago when an iron-rich object measuring roughly 50 m across struck the ground with a speed estimated at 11 m/s (25,000 mi/hr). The resulting blast was equal to the detonation of a 20-megaton hydrogen bomb.

Iron is one of the more abundant elements in the universe (recall Table 6-4), as well as being one of the most common rock-forming elements (review Table 6-5), so it is not sur-

Figure 17-6 The Barringer Crater An iron meteoroid measuring 50 m across struck the ground in Arizona 25,000 years ago. The result was this beautifully symmetrical impact crater measuring 1.2 km in diameter and 200 m deep at its center. (Meteor Crater Enterprises)

Box 17-1 Lagrangian points and the restricted three-body problem

In the late 1700s, the great French mathematician Joseph Louis Lagrange (pronounced la-GRAHNZH) tackled the famous three-body problem: the task of predicting the motions of three objects moving freely in space under the mutual influences of their gravitational forces. Lagrange succeeded in solving a restricted form of the problem, in which one of the objects is assumed to be small enough that its gravity has no effect on the other two, more massive, objects. (Mathematicians later proved, however, that there is no way to find a precise solution to the general three-body problem.) Among the practical applications of a solution to the restricted three-body problem are plotting a course for a spaceship from the Earth to the Moon, and calculating the path of an asteroid affected chiefly by the gravitational pulls of Jupiter and the Sun.

Lagrange discovered some interesting facts. The easiest way to display the combined gravitational fields of the two massive objects (we shall call them M_1 and M_2, where $M_1 > M_2$) is to draw so-called **equipotential contours**. We can think of an equipotential contour map as a sort of topographic map showing "hills" and "valleys" in the gravitational field (see the diagram). Any small object in this field will feel a force pulling it in the "downhill" direction. The massive objects M_1 and M_2 (along with the entire gravitational field) revolve about the center of mass of the system, which is nearer M_1 than M_2 because $M_1 > M_2$.

Note that there are three locations along a line through M_1 and M_2 where contour lines cross. A tiny object placed at one of these locations would be balanced unsteadily at a local "flat spot" in the gravitational field. These three points are the unstable Lagrangian points L_1, L_2, and L_3 (see the bottom diagram). These points are said to be unstable because, if the small object moves ever so slightly away from the exact Lagrangian point, it will experience a gravitational pull in the "downhill" direction, away from the Lagrangian point. An object such as an asteroid cannot therefore be permanently trapped at one of the unstable Lagrangian points.

Notice that there are two points at the bottoms of the teardrop-shaped "valleys" on either side of the line through M_1 and M_2. These are the stable Lagrangian points L_4 and L_5. If a small object is moved slightly away from one of these points, it will experience a gravitational pull in the "downhill" direction, back toward the Lagrangian point. A small object like an asteroid can thus be permanently trapped at one of the stable Lagrangian points. In fact, the Trojan asteroids do move in small orbits around the two stable Lagrangian points, each of which is located at a corner of an equilateral triangle, with Jupiter and the Sun at the other corners (see Figure 17-4).

The Trojan asteroids at L_4 and L_5 along Jupiter's orbit have been known for many years. (The discovery of the first one, in 1906, provided the first practical proof of Lagrange's theoretical ideas about these points.) Of course, stable Lagrangian points exist at many places in the solar system. While passing Saturn, the Voyager spacecraft discovered tiny satellites at the L_4 and L_5 points of the Saturn–Tethys and Saturn–Dione systems. Small clouds of dust-grain–sized particles have been observed at the L_4 and L_5 points of the Earth–Moon system. A group called the L_5 Society argues that the L_5 point of the Earth–Moon system would be the ideal location for a huge space station with a permanent human population. Despite careful searches, no asteroids have been found at the stable Lagrangian points of the Earth–Sun and the Saturn–Sun systems.

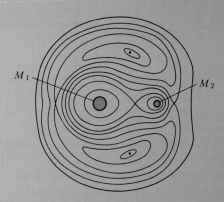

The equipotential contours

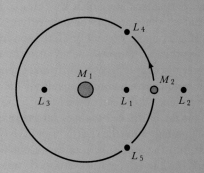

The five Lagrangian points

prising that iron is an important constituent of asteroids and the smaller objects called meteoroids. Another element, iridium—one of several elements that geologists call **siderophiles**, or "iron lovers"—is common in iron-rich minerals but rare in ordinary rocks. Measurements of iridium in

the Earth's crust can thus tell us about the rate at which meteoritic material has been deposited on the Earth over the ages. Geologist Walter Alvarez and his physicist father, Luis Alvarez, from the University of California at Berkeley, were involved in making such measurements in the late 1970s.

Figure 17-7 *Iridium-rich clay* This photograph of strata in the Apennine Mountains of Italy shows a dark-colored layer of iridium-rich clay sandwiched between white limestone (below) from the late Mesozoic era and grayish limestone (above) from the early Cenozoic era. This iridium-rich layer may be the result of an asteroid whose impact caused the extinction of the dinosaurs. The coin is the size of a U.S. quarter. (Courtesy of W. Alvarez)

Working at a site of exposed marine limestone in the Apennine Mountains in Italy, the Alvarez team discovered an exceptionally high abundance of iridium in a dark-colored layer of clay between limestone strata (see Figure 17-7). Since this discovery in 1979, a comparable layer of iridium-rich material has been uncovered at various sites around the world. In every case, geological dating reveals that this apparently worldwide layer of iridium-rich clay was deposited about 63 million years ago.

Paleontologists were quick to realize the profound significance of this particular date. For it was around 63 million years ago that all the dinosaurs became extinct. In fact, at that time a staggering 65 percent of all the species on Earth disappeared within a relatively brief span of time.

The Alvarez discovery suggests a startling explanation for the dramatic extinction of more than half the life forms that inhabited our planet at the end of the Mesozoic era. Perhaps an asteroid hit the Earth at that time. An asteroid 10 km in diameter slamming into the Earth would have thrown enough dust into the atmosphere to block out sunlight for several years. As the temperature dropped drastically and plants died for lack of sunshine, the dinosaurs would have perished, along with many other creatures in the food chain that was based on vegetation. The dust eventually settled, depositing an iridium-rich layer around the world. Tiny, rodentlike creatures capable of ferreting out seeds and nuts became prominent among the animals that managed to survive this holocaust, setting the stage for the rise of mammals in the Cenozoic era.

Some geologists and paleontologists are not yet convinced that a meteoroid impact was responsible for the extinction of the dinosaurs at the end of the Mesozoic era, but many scientists agree that this hypothesis fits the available evidence better than other explanations that have been offered.

17-5 Meteorites are classified as stones, stony irons, or irons, depending on their composition

A meteoroid, like an asteroid, is a chunk of rock in space. There is no official dividing line between meteoroids and asteroids, but the term *asteroid* is generally applied only to objects larger than a few hundred meters across.

A **meteor** is the brief flash of light (sometimes called a *shooting star*) that is visible at night when a meteoroid enters the Earth's atmosphere (see Figure 17-8). Frictional heat is generated as the meteoroid plunges through the atmosphere, the result being a fiery trail across the night sky.

If a piece of rock survives its fiery descent through the atmosphere and reaches the ground, it is called a **meteorite**. People have been finding specimens of meteorites for thousands of years, and descriptions of them appear in ancient Chinese, Greek, and Roman literature. Our ancestors placed special significance on these "rocks from heaven." There have been numerous examples of meteorite veneration, such as the sacred black stone of Kaaba enshrined at Mecca.

The extraterrestrial origin of meteorites was hotly debated until as late as the eighteenth century. Upon hearing a lecture by two Yale professors, President Thomas Jefferson

Figure 17-8 A meteor *A meteor is produced when a piece of interplanetary rock or dust strikes the Earth's atmosphere at high speed. Exceptionally bright meteors, such as the one shown in this long exposure (notice the star trails), are usually called fireballs. (Courtesy of R. A. Oriti)*

is said to have remarked, "I could more easily believe that two Yankee professors could lie than that stones could fall from Heaven." Although several **meteorite falls** had been widely witnessed and specimens from them had been collected (as, for example, in a 1751 fall near Zagreb, Yugoslavia), many scientists were reluctant to accept that rocks could fall to Earth from outer space.

Conclusive evidence for the extraterrestrial origin of meteorites came on April 26, 1803, when fragments pelted the French town of L'Aigle. The austere French Academy, whose members were among the last holdouts, sent the noted physicist J. B. Biot to investigate. His exhaustive report finally convinced the scientific community that meteorites were in fact extraterrestrial.

Meteorites are classified into three broad categories: stones, irons, and stony irons. As their name suggests, **stones,** or **stony meteorites,** look like ordinary rocks at first glance, but they are sometimes covered with a **fusion crust** (see Figure 17-9). This crust is produced by a momentary melting of the meteorite's outer layers during its fiery descent through the atmosphere. When a stony meteorite is cut in two and polished, tiny flecks of iron can sometimes be found in the rock (see Figure 17-10).

Although stony meteorites account for nearly 96 percent of all the meteoritic material that falls to Earth, stones are the most difficult specimens to find. If they lie undiscovered and become exposed to the weather for a few years, they look almost indistinguishable from common terrestrial rocks. Meteorites with a high iron content are much easier to find, because they can be located with a metal detector. Consequently, iron and stony iron meteorites dominate most museum collections.

Stony iron meteorites consist of roughly equal amounts of rock and iron. Olivine is commonly the mineral suspended in the matrix of iron, as in the case of the **pallasite** shown in Figure 17-11. Only about 1 percent of the meteorites that fall to Earth are stony irons.

Iron meteorites (see Figure 17-12), or **irons,** account for about 3 percent of the material that falls on the Earth. Iron

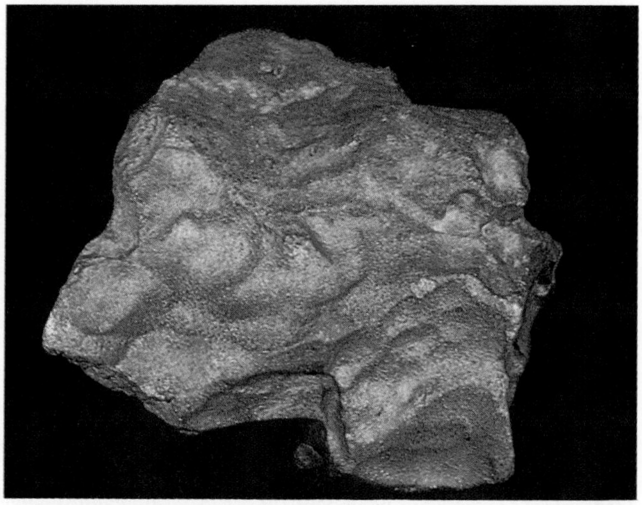

Figure 17-9 A stony meteorite *Of all meteorites that fall on the Earth, 93 percent are stones. Many freshly discovered specimens, like the one shown here, are coated with dark fusion crusts. This particular stone fell near Plainview, Texas. (From the collection of R. A. Oriti)*

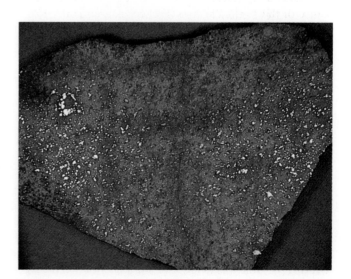

Figure 17-10 A cut and polished stone *When stony meteorites are cut and polished, some are found to contain tiny specks of iron mixed in the rock. This specimen was discovered near Neenach, California. (From the collection of R. A. Oriti)*

Figure 17-11 *A stony iron meteorite* *Stony irons account for slightly less than 2 percent of all the meteorites that fall to Earth. This particular specimen is a variety of stony iron called a pallasite. It fell near Antofagasta, Chile. (From the collection of R. A. Oriti)*

type called **octahedrites**. When an octahedrite is cut, polished, and briefly dipped into a dilute solution of acid, its unique crystalline structure, called **Widmanstätten patterns** (see Figure 17-13), is revealed.

Nickel–iron crystals can grow to lengths of several centimeters only if the molten metal cools slowly over many millions of years. Thus, Widmanstätten patterns constitute conclusive proof of a meteorite's authenticity. Octahedrites cool at rates of 1 to 10 K per million years during the time the crystals are forming. So Widmanstätten patterns are never found in counterfeit meteorites, or "meteorwrongs," as they are humorously called.

17-6 Meteorites provide significant information about the formation of the solar system

The existence of Widmanstätten patterns in meteorites suggests that some asteroids were partly molten for a substantial period after their formation. The size of an octahedrite's parent asteroid can be estimated by calculating how much rock must have insulated the molten nickel–iron interior to produce its long-term cooling rate. Such calculations suggest that typical meteorites are fragments of parent asteroids 200 to 400 km in diameter.

The three main types of meteorites may have come from different parts of a parent asteroid in the following manner. As soon as the asteroid accreted from planetesimals 4.5 billion years ago, rapid decay of short-lived radioactive isotopes heated the asteroid's interior to temperatures above

meteorites usually have no stone inclusions, but many contain from 10 to 20 percent nickel.

In 1808, Count Alois von Widmanstätten, director of the Imperial Porcelain works in Vienna, discovered a conclusive test for the authenticity of the most common type of iron meteorite. About 75 percent of all iron meteorites are of a

Figure 17-12 *An iron meteorite* *Irons are composed almost entirely of nickel–iron minerals. The surface of a typical iron is covered with thumbprintlike depressions caused by ablation (removal by melting) during its high-speed descent through the atmosphere. This specimen was found near Henbury, Australia. (From the collection of R. A. Oriti)*

Figure 17-13 *Widmanstätten patterns* *When cut, polished, and etched with a weak acid solution, most iron meteorites exhibit interlocking crystals in designs called Widmanstätten patterns. These patterns appear only in the type of iron meteorite called octahedrites. This octahedrite was found near Henbury, Australia. (From the collection of R. A. Oriti)*

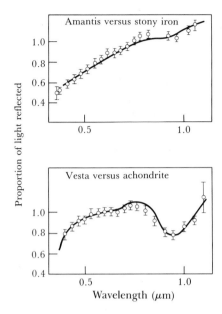

Figure 17-14 *Asteroid versus meteorite spectra These graphs compare the reflectance spectra of two kinds of meteorites (solid lines) with the spectra of two asteroids (circles with probable-error bars). From such comparisons we can determine the composition of an asteroid's surface. (Adapted from C. R. Chapman)*

the melting point of rock. Over the next few million years, chemical differentiation occurred: Iron and associated siderophile elements sank toward the asteroid's center, displacing the lighter silica-associated **lithophile elements** upward toward the asteroid's surface. After the asteroid cooled and its core solidified, interasteroid collisions fragmented the parent body into meteoroids. Iron meteorites are specimens from the asteroid's core, stones being samples of its crust. Stony irons presumably come from regions intermediate between the asteroid's core and its crust.

This theory is supported by analyses of the spectra of meteorites and asteroids. Different kinds of rock reflect or absorb characteristic fractions of incident light over various ranges of wavelength. By comparing the spectrum of an asteroid with those of different kinds of meteorites, it is often possible to deduce the surface composition of the asteroid. Two examples are shown in Figure 17-14. The reflectance spectrum of Vesta strongly resembles the spectrum of a particular kind of stony meteorite called an achondrite. The spectrum of Amantis resembles that of a stony iron meteorite, suggesting that the asteroid's rocky crust may have been torn away, revealing material closer to its core.

Meteorites derived from the fragmentation of large asteroids have been subjected to substantial processing during the first billion years of solar-system formation. These meteoritic specimens are thus not representative of the primordial material from which our solar system was originally created. There is, however, a class of rare meteorites, called **carbonaceous chondrites**, that show no evidence of ever having been

subjected to the metamorphic processes, such as heating and melting, that affect the structure and composition of most other meteorites.

About 6 percent of all the stones that fall on the Earth are carbonaceous chondrites. Their primordial nature is inferred from their high content of volatile compounds, sometimes including as much as 20 percent water. Moreover, carbonaceous chondrites are rich in complex organic compounds. The water and volatile compounds would have been driven out and the large organic molecules been broken down if these meteorites had been subjected to any significant heating.

Shortly after midnight on February 8, 1969, the night sky around Chihuahua, Mexico, was illuminated by a brilliant blue-white light moving across the heavens. The dazzling display was witnessed by hundreds of people, many of whom thought the world was coming to an end. As the light moved across the sky, it disintegrated in a spectacular, noisy explosion that dropped thousands of rocks and pebbles over the terrified onlookers. Within hours, teams of scientists were on their way to collect specimens of a carbonaceous chondrite, collectively called *the Allende meteorite*, after the locality (see Figure 17-15).

Specimens were scattered in an elongated ellipse called a **strewnfield**, approximately 50 km long by 10 km wide. Larger specimens, which had been least decelerated by the atmosphere, were found at one end of the strewnfield. The largest piece (110 kg) was located at the very tip of the ellipse. Most fragments were coated with a fusion crust, but minerals immediately beneath the crust showed no signs of thermal damage. Ablation peels surface material away as fast as it becomes heated during flight through the atmosphere; heat had little chance to penetrate the meteorite's interior.

Figure 17-15 *A piece of the Allende meteorite This carbonaceous chondrite fell near Chihuahua, Mexico, in February 1969. Note the meteorite's dark color, caused by a high abundance of carbon. The amount of argon-40 found in the meteorite corresponds to that which would have been generated by the decay of radioactive potassium-40 in about 4.6 billion years. Geologists therefore believe that this meteorite is a specimen of primitive planetary material. The ruler is 6 inches long. (Courtesy of J. A. Wood)*

One of the most significant discoveries to come from the Allende meteorite was made by Gerald J. Wasserburg and his colleagues at the California Institute of Technology. They found a very small amounts of a short-lived radioactive isotope of aluminum, ^{26}Al, in samples of the meteorite. This isotope, with a half-life of only 720,000 years, rapidly decays into stable isotope of magnesium, ^{26}Mg.

The fact that *any* ^{26}Al was found in the meteorite means that a substantial amount of the radioactive aluminum must have been created shortly before the formation of our solar system. Astronomers therefore had evidence of an energetic nuclear process that occurred in our vicinity roughly 4.5 billion years ago, about the time the Sun was born.

One of nature's most violent and spectacular phenomena, called a **supernova explosion**, occurs during the death of a massive star. The doomed star blows itself apart in a cataclysm that hurls matter outward at tremendous speeds, as we shall see in Chapter 22. During this detonation, violent collisions between nuclei produce a host of radioactive isotopes, including ^{26}Al. It seems possible that a supernova occurred very near the Sun's location some 4.5 billion years ago. In addition to contaminating the interstellar medium with radioactive ^{26}Al, the supernova's shock wave may have compressed the interstellar gas and dust, triggering the birth of our solar system.

Besides telling us about the creation of the solar system, the study of meteorites may shed light on the origin of life on Earth. **Amino acids**, the building blocks of proteins upon which terrestrial life is based, are among the organic compounds occasionally found inside carbonaceous chondrites. Interstellar organic material falling on our planet may have played a role in the appearance of simple organisms on our planet nearly 4 billion years ago.

17-7 A comet is a dusty chunk of ice that becomes partly vaporized as it passes near the Sun

Many rocks and chunks of ice that originally condensed out of the primordial solar nebula still continue to orbit the Sun. Just as heat from the protosun produced two classes of planets, the terrestrial and the Jovian, two main types of interplanetary material were created. Near the Sun, interplanetary debris consists of the rocks called asteroids or meteoroids. Far from the Sun, there are the numerous loose conglomerations of ice and dust called **comets**.

In sharp contrast to asteroids, which travel around the Sun along roughly circular orbits that are largely confined to the asteroid belt and to the plane of the ecliptic, comets travel around the Sun along highly elliptical orbits inclined at random angles to the ecliptic. As a comet approaches the Sun, solar heat begins to vaporize the ices. The liberated gases begin to glow, producing a fuzzy, luminous ball called

a **coma**. Continued action by the solar wind and radiation pressure together blow these luminous gases outward into a long, flowing **tail**. The result is one of the most awesome sights ever visible in the night sky (see Figure 17-16).

The solid part of a comet, called the **nucleus**, is a roughly half-and-half mixture of ice and dust, typically measuring a few kilometers across. Harvard astronomer Fred L. Whipple, a pioneer in comet research, coined the term *dirty snowball* to reflect the fact that in comets bits and pieces of dust and rocky material are mixed in with the ice.

The first pictures of a comet's nucleus were obtained by several spacecraft that flew past Comet Halley in 1986 (see Figure 17-17). Halley's potato-shaped nucleus is darker than a piece of coal and reflects only about 4 percent of the light that strikes it. This dark color is probably caused by a layer of carbon-rich compounds and dust left behind as the comet's ice evaporated.

Figure 17-16 Comet West *A comet is always named after the person who first sees it. Astronomer Richard M. West first noticed this comet, on a photographic plate taken with a telescope, in mid-1975. After passing near the Sun, Comet West became one of the brightest comets in recent years. This photograph shows the comet gracing the predawn sky in the spring of 1976. (Courtesy of H. Vehrenberg)*

Several jets about 15 km long protrude from Halley's nucleus and squirt fountains of dust toward the Sun. The vents from which this material escapes are localized, probably covering only about 10 percent of the surface of the nucleus. These vents seem to be active only when exposed to the Sun. The rotation period of Halley's nucleus is about 53 hours. The vents shut off when the nucleus's rotation brings them into darkness, away from the Sun.

The overall structure of a comet is outlined in Figure 17-18. The nucleus of a comet is never visible to Earth-based astronomers, because it is small, dim, and buried in the glare of the coma. A coma is typically a million km in diameter and a comet's tail can stretch to over 100 million km in length. Also not visible to the human eye is the **hydrogen envelope**, a huge tenuous sphere of gas typically measuring about 20 million km in diameter surrounding the comet's nucleus. This hydrogen comes from water molecules (H_2O) escaping from the comet's evaporating ice. The Sun's radiation breaks up the water molecules and excites the hydrogen atoms so that they emit ultraviolet light at the wavelength of the Lyman alpha (L_α) spectral line. Figure 17-19 shows two views of Comet Kohoutek: as it appeared to Earth-based observers in 1973, and as photographed by an ultraviolet camera from a rocket. The ultraviolet view was the photograph that first gave astronomers evidence of the enormous extent of the hydrogen envelope.

Figure 17-17 The nucleus of Comet Halley In March 1986, five spacecraft passed near Comet Halley. This closeup picture was taken by a camera on board the Giotto spacecraft and shows the potato-shaped nucleus of the comet. The nucleus is darker than coal and measures 15 km in the longest dimension and about 8 km in the shortest. The Sun illuminates the comet from the left. Two bright jets of dust extend 15 km from the nucleus toward the Sun, suggesting that major activity emanates primarily from the sunlit side. (Max Planck Institut für Aeronomie)

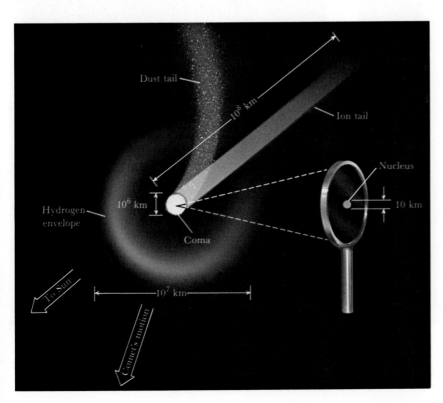

Figure 17-18 The structure of a comet The solid part of a comet, the nucleus, is roughly 10 km in diameter. The coma can be as large as 100,000 to 1 million km across. The hydrogen envelope is typically 10 million km in diameter. A comet's tail can be as long as 1 AU.

Figure 17-19 Comet Kohoutek and its hydrogen envelope These two photographs of Comet Kohoutek are reproduced to the same scale. [Left] The comet in visible light. [Right] This view of the comet in ultraviolet wavelengths reveals a huge hydrogen cloud surrounding the comet's head. (Johns Hopkins University; Naval Research Laboratory)

Figure 17-20 The head of Comet Brooks With an exceptionally large, bright coma, this comet dominated the night skies during October 1911. It was named Comet Brooks after its discoverer. (Lick Observatory)

Figure 17-21 Comet Ikeya-Seki Although the head of Comet Ikeya-Seki was tiny, its tail was 1 AU long and dominated the predawn sky during late October 1965. This comet was named after its two Japanese codiscoverers. (Lick Observatory)

Comets come in a wide range of shapes and sizes. For example, the comet shown in Figure 17-20 had a large, bright coma but a short, stubby tail. In contrast, the comet seen in Figure 17-21 had an inconspicuous coma, but its tail had an astonishing length of 1 AU, long enough to stretch all the way from the Earth to the Sun.

It has long been known that comets' tails always point away from the Sun (see Figure 17-22), regardless of the direction of the comet's motion. The implication that something from the Sun was "blowing" the comet's gases radially outward led Ludwig Biermann to predict the existence of the solar wind a full decade before it was actually discovered in 1962 by instruments on a spacecraft.

In fact, the Sun usually produces two comet tails: an **ion tail** and a **dust tail**. Ionized atoms (that is, atoms missing one or more electrons) are swept directly away from the Sun by the solar wind to form the ion tail. The dust tail is formed when photons of light strike dust particles freed from the evaporating nucleus. Photons carry momentum, and so when light falls on an object, the impact of that momentum exerts a pressure on the object. This pressure, called **radiation pressure**, is quite weak, but fine-grained dust particles in a comet's coma offer little resistance and are blown away from the comet to produce a dust tail. The relatively straight ion tail, sometimes called a **Type I tail**, can exhibit a dramatic structure that changes from night to night (see Figure 17-23). The more amorphous dust tail, sometimes called a **Type II tail**, is typically arched. On rare occasions the geometry of the Earth, the comet, and its arched dust tail is such that the dust tail appears to be sticking out the front of the comet (see Figure 17-24).

Astronomers discover at least a dozen new comets in a typical year. Some are **short-period comets**, which circle the Sun in less than 200 years. Like Halley's Comet, they appear again and again at predictable intervals. However, the majority of comets discovered each year are **long-period comets**, which take roughly 100,000 to 1 million years to complete an orbit of the Sun. These comets travel along extremely elongated orbits and consequently spend most of their time at distances of roughly 40,000 to 50,000 AU from the Sun, or about one-fifth of the way to the nearest star.

Because astronomers discover long-period comets at the rate of roughly one per month, it is reasonable to suppose that there is an enormous population of comets out there 50,000 AU from the Sun. This reservoir of cometary nuclei surrounding the Sun is called the **Oort cloud**, after the Dutch astronomer Jan Oort, who first proposed its existence in the 1950s. Estimates of the number of "dirty snowballs" in the Oort cloud range from 100,000 million to more than a trillion. Only such a large reservoir of cometary nuclei would

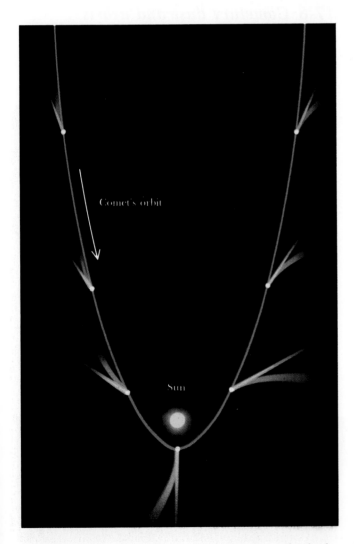

Figure 17-22 The orbit and tail of a comet *The solar wind and radiation pressure from sunlight blow a comet's dust particles and ionized atoms away from the Sun. Consequently, a comet's tail always points away from the Sun.*

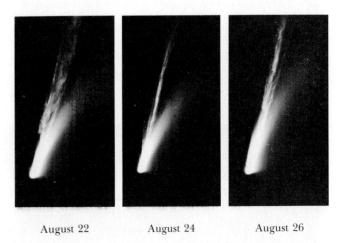

August 22   August 24   August 26

Figure 17-23 The two tails of Comet Mrkos *Comet Mrkos dominated the evening sky during August 1957. These three views, taken at two-day intervals, show dramatic changes in the comet's ion (Type I) tail. In contrast, the slightly curved dust (Type II) tail remained fuzzy and featureless. (Palomar Observatory)*

Figure 17-24 *The antitail of Comet Arend-Roland* *Comet Arend-Roland, seen in April 1957, exhibited an "antitail." Actually, this antitail was merely the end of the dust tail. The Earth and the comet were oriented in such a way that the end of the arched dust tail, as seen past the comet's head, looked like a spike sticking out of the comet's head. (Lick Observatory)*

explain why we see so many long-period comets, even though each one takes up to 1 million years to travel once around its orbit.

The Oort cloud was probably created 4.5 billion years ago from numerous icy planetesimals that orbited the Sun in the vicinity of the newly formed Jovian planets. During near-collisions with the giant planets, many of these chunks of ice and dust were gravitationally catapulted into the highly elliptical orbits that they now occupy, much in the same way that Pioneer and Voyager spacecraft were flung far from the Sun during their flybys of the Jovian planets. During a return trip toward the Sun, a long period comet may again encounter a Jovian planet whose gravity forces the comet into a short-period orbit, where the comet is eventually destroyed by frequent passages near the Sun.

17-8 Cometary dust and debris rain down on the Earth during meteor showers

A comet loses about 0.1 percent of its ice each time it passes near the Sun. The ice content of a typical comet thus becomes completely vaporized after about 1000 perihelion passages, leaving only a swarm of meteoritic dust and pebbles. A comet's nucleus will disintegrate more quickly if it happens to pass very near the Sun as what is called a **Sun-grazing comet**. A comet's nucleus can sometimes be observed to fragment (see Figure 17-25), a good indication of its low cohesive strength.

As a comet's nucleus evaporates, residual dust and rock fragments spread out to form a **meteoritic swarm**, a loose collection of debris that continues to circle the Sun along the comet's orbit (see Figure 17-26). If the Earth's orbit happens to pass through this swarm, a **meteor shower** is seen as the dust particles strike Earth's upper atmosphere.

Nearly a dozen meteor showers can be seen each year (see Box 17-2). Note that the **radiants** for these showers (that is, the places among the stars from which the meteors appear to

| March 8 | March 12 | March 14 | March 18 | March 24 |

Figure 17-25 *The fragmentation of Comet West* *Shortly after passing near the Sun in 1976, the nucleus of Comet West broke into four pieces.* *This series of five photographs clearly shows that disintegration. (New Mexico State University Observatory)*

Box 17-2 Meteor showers

At least ten notable meteor showers occur each year. They are listed with relevant information in the table below. The date of maximum is the best time to observe a particular shower, although good displays can often be seen a day or two before or after the maximum. The radiant is the location among the stars from which meteors seem to be coming. The hourly rate is given for a single observer under excellent conditions. The velocity is the average speed of the meteoritic material as it strikes the atmosphere.

In order to see a fine meteor display, you need a clear, moonless sky. The Moon's presence above the horizon can significantly detract from the number of faint meteors you will be able to see. In addition, the early morning hours (between roughly 2 AM and dawn) are the best times to make your observations. As sketched in the diagram, you are on the leading side of the Earth during the early morning hours, and all meteor-producing particles in the Earth's path are swept into the atmosphere above you. On the other hand, you are on the trailing side of the Earth during the evening hours, where only high-speed particles manage to catch up with our planet to produce meteors.

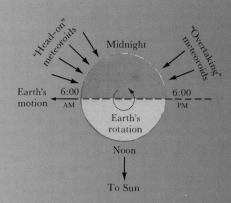

		Radiant			
Shower	Date of maximum	Right ascension (hr min)	Declination (°)	Hourly rate	Velocity (km/s)
Quadrantids	January 3	15 28	+50	40	40
Lyrids	April 22	18 16	+34	15	50
Aquarids	May 4	22 24	0	20	64
Aquarids	July 30	22 36	−17	20	40
Perseids	August 12	3 4	+58	50	60
Orionids	October 21	6 20	+15	20	66
Taurids	November 4	3 32	+14	15	30
Leonids	November 16	10 8	+22	15	70
Geminids	December 13	7 32	+32	50	35
Ursids	December 22	14 28	+76	15	35

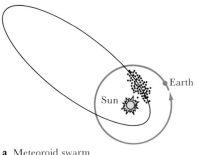

a Meteoroid swarm

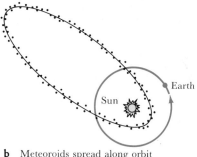

b Meteoroids spread along orbit

Figure 17-26 Meteoritic swarms Rock fragments and dust from "burned out" comets continue to circle the Sun. (a) If the comet is only recently extinct, the particles will still be tightly concentrated in a compact swarm. The most spectacular meteor showers occur when the Earth happens to pass through such a swarm. (b) Over the ages, the particles gradually spread out along the old comet's elliptical orbit. This configuration produces the most predictable meteor showers, because the Earth must pass through the evenly distributed swarm on each trip around the Sun.

Figure 17-27 Aftermath of the Tunguska event In 1908, a piece of a comet's nucleus struck the Earth's atmosphere over the Tunguska region of Siberia. Trees were blown down for many kilometers in all directions around the impact site. (Courtesy of Sovfoto)

come) are not confined to the constellations of the zodiac. Meteor showers can come from various parts of the sky, just as comets are sighted at various locations not related to the plane of the ecliptic.

As incredible as it may seem, an estimated total of 300 tons of extraterrestrial rock and dust fall on the Earth each day. The fluffy, low-density material from comets burns up in the atmosphere, so only the denser matter originating from asteroids typically reaches the ground. There is evidence, however, that a comet struck the Earth in the recent past.

On June 30, 1908, a spectacular explosion occurred over the Tunguska region of Siberia that released about 10^{15} joules of energy, equivalent to a nuclear detonation of several hundred kilotons. The blast knocked a man off his porch some 60 km away, and was audible more than 1000 km away. Millions of tons of dust were injected into the atmosphere, causing a decrease in air transparency detected as far away as California.

Preoccupied with political upheaval and World War I, Russia did not send a scientific expedition to the site until 1927. At that time, Soviet researchers found a location where trees had been seared and felled radially outward in an area about 30 km in diameter (see Figure 17-27). There was no clear evidence of a crater and no significant meteorite samples were found. For many years it was a great mystery why there were no fragments from this enormous body.

The most likely explanation of this event is that a small comet collided with the Earth. No impact crater was formed, and the trees at "ground zero" were left standing upright, although they were completely stripped of branches and leaves. This type of destruction is what would be expected from a loosely consolidated ball of cometary ices vaporizing with explosive force before striking the ground.

Large-sized objects occasionally strike the Earth with a destructive force that could easily be mistaken for the detonation of a nuclear weapon. A comet or small asteroid approaching our planet in the daytime sky from the general direction of the Sun would probably not be discovered by astronomers, so its impact would occur without warning. In spite of safeguards and diplomacy, Armageddon could be triggered by an entirely natural phenomenon.

Key words

amino acid	coma (of a comet)	hydrogen envelope	lithophile element
asteroid	comet	impact crater	long-period comet
asteroid belt	dust tail	ion tail	meteor
binary asteroid	*equipotential contour	iron meteorite	meteorite
Bode's law	fusion crust	Kirkwood gaps	meteoritic swarm
carbonaceous chondrite	Hirayama family	Lagrangian points	meteoroid

meter shower

minor planet

nucleus (of a comet)

Oort cloud

radiant (of a meteor shower)

radiation pressure

short-period comet

siderophile element

stony iron meteorite

stony meteorite

supernova explosion

tail (of a comet)

Trojan asteroid

Widmanstätten patterns

Key ideas

- Bode's law is a numerical sequence that gives the distances of most of the planets (Mercury through Uranus) from the Sun in AU. This "law" inspired nineteenth-century astronomers to search for a planet in a gap between the orbits of Mars and Jupiter.

- Thousands of belt asteroids with diameters ranging from a few kilometers up to 1000 kilometers circle the Sun between the orbits of Mars and Jupiter.

  Gravitational perturbations by Jupiter deplete certain orbits within the asteroid belt. The resulting gaps, called Kirkwood gaps, occur at simple fractions of Jupiter's orbital period.

  Jupiter's gravity also captures asteroids in two locations, called Lagrangian points, along Jupiter's orbit.

  Some asteroids, called Amor and Apollo asteroids, move in elliptical orbits that cross the orbits of Mars and Earth.

- Small rocks in space are called meteoroids. If a meteoroid enters the Earth's atmosphere, it produces a fiery trail called a meteor. If part of the object survives the fall, the fragment that reaches the Earth's surface is called a meteorite.

  Meteorites are grouped into three major classes, according to composition: iron, stony iron, or stony meteorites.

  Rare stony meteorites called carbonaceous chondrites may be relatively unmodified material from the solar nebula; these meteorites often contain organic material and may have played a role in the origin of life on Earth.

  When a large meteoroid strikes the Earth, it forms an impact crater whose diameter depends on both the mass and the speed of the impinging object.

  An asteroid may have struck the Earth 63 million years ago, probably causing the extinction of the dinosaurs and many other species.

- Analysis of isotopes in certain meteorites suggests that a nearby supernova explosion may have triggered the formation of the solar system 4.5 billion years ago.

- A comet is a chunk of ices and rock fragments that generally moves in a highly elliptical orbit about the Sun.

  As a comet approaches the Sun, its icy nucleus develops a luminous coma, surrounded by a vast hydrogen envelope; an ion tail and a dust tail extend from the comet, pushed away from the Sun by the solar wind and radiation pressure.

  Fragments of "burned out" comets produce meteoroid swarms; a meteor shower is seen when the Earth passes through a meteoroid swarm.

  Millions of cometary nuclei probably exist in the Oort cloud some 50,000 AU from the Sun.

Review questions

1 What is Bode's law, and why is it not really a "law"?

2 Why are asteroids, meteorites, and comets of special interest to astronomers who want to understand the early history of the solar system?

3 Describe the asteroid belt.

4 What are Kirkwood's gaps and what causes them?

5 Can you think of another place in our solar system where a phenomenon similar to Kirkwood's gaps occurs? Explain.

6 What are the Trojan asteroids, and where are they located?

7 Why do you suppose there aren't any asteroids at the L_1, L_2, or L_3 Lagrangian points in the Sun–Jupiter system?

8 Are there any examples in the solar system of objects being trapped at the L_4 and L_5 Lagrangian points other than in the Sun–Jupiter system?

9 Where on Earth might you find large numbers of stony meteorites that are not significantly weathered?

10 Suppose you found a rock you suspect might be a meteorite. Describe some of the things you could do to see if it were a meteorite or a "meteorwrong."

11 Why do astronomers believe that meteoroids come from asteroids, whereas meteor showers are related to comets?

12 With the aid of a drawing, describe the structure of a comet.

13 Explain why comets are generally brighter after passing perihelion.

14 What is the Oort cloud, and how might it be related to planetesimals left over from the formation of the solar system?

Advanced questions

Tips and tools . . .
You will need to use Kepler's third law in two of the problems below. A spherical object intercepts an amount of sunlight proportional to the square of its radius. The volume of a sphere of radius r is $\frac{4}{3}\pi r^3$.

---

*15 (*Basic*) Suppose that a planet were indeed 77.2 AU from the Sun, as predicted by Bode's law. How long would it take such an object to orbit the Sun? At what rate (in degrees per year) would this object appear to move in relation to the background stars? Would this motion be detectable with Earth-based telescopes?

*16 (*Challenging*) Suppose that a double asteroid is observed in which one member is 16 times brighter than the other. Suppose that both members have the same albedo and that the larger of the two is 200 km in diameter. What is the diameter of the other member?

*17 Find the orbital periods of Sun-grazing comets whose aphelion distances are (**a**) 100 AU, (**b**) 1000 AU, (**c**) 10,000 AU, and (**d**) 100,000 AU. Assuming that these comets can survive only a hundred perihelion passages, calculate their lifetimes.

*18 Use the percentages of stones, irons, and stony iron meteorites that fall to Earth to estimate what fraction of their parent asteroids originally consisted of an iron core and a stony mantle. How do these percentages compare with that for the Earth? (*Note:* Assume that the percentages of stones and irons that fall to Earth indicate the fractions of parent asteroids occupied by rock and iron, respectively.)

Discussion questions

19 Suppose it were discovered that the asteroid Hermes had been disturbed in such a way as to put it on a collision course with Earth. Describe what humanity could do within the framework of present technology to counter such a catastrophe.

20 From the abundance of craters on the Moon and Mercury, we know that numerous asteroids and meteoroids struck the inner planets early in the history of our solar system. Is it reasonable to suppose that numerous comets also pelted the planets 3 to 4 billion years ago? Speculate about the effects of such a cometary bombardment, especially with regard to the evolution of the primordial atmospheres on the terrestrial planets.

21 Compare the consequences of a global thermonuclear war with that of an asteroid hitting the Earth.

Observing projects

22 Make arrangements to view a meteor shower. Dates of major meteor showers are given in Box 17-2. Choose a shower that will occur near the time of a new moon. Informative details concerning upcoming meteor showers are often published a month ahead of time in such magazines as *Sky & Telescope* and *Astronomy*. Set your alarm clock for the early morning hours (1 to 3 AM). Get comfortable in a reclining chair or lie on your back so that you can view a large portion of the sky. Make note of how long you observe, how many meteors you see, and what location in the sky they seem to come from. How well does your observed hourly rate agree with published estimates such as those in Box 17-2? Is the radiant of the meteor shower apparent from your observations?

23 Make arrangements to view a comet through a telescope. Since astronomers discover roughly a dozen comets each year, there is usually a comet visible somewhere in the sky. Unfortunately, they often are quite dim, and so you will need to have access to a moderately large telescope. Consult recent issues of the IAU *Circular*, published by the International Astronomical Union's Central Bureau for Astronomical Telegrams, which contains the latest word on predicted positions and anticipated brightnesses of comets in the sky. Also, if there is an especially bright comet in the sky, useful information about it might be found in the latest issue of *Sky & Telescope*. Is a comet within reach of a telescope at your disposal? If so, can you distinguish the comet from background stars? Can you see its coma? Can you see a tail?

24 Make arrangements to view an asteroid. At opposition, some of the largest asteroids are as bright as magnitude +5, and so they can be seen through a modest telescope. Check the "Minor Planets" section of current issue of the *Astronomical Almanac* to see if any bright asteroids are near opposition. If so, check the current issue as well as the most recent January issue of *Sky & Telescope* for a star chart showing the asteroid's path among the constellations. You will need such a chart to distinguish the asteroid from background stars. Also, the right ascensions and declinations of Ceres, Pallas, Juno, and Vesta are listed in the *Astronomical Almanac*. Observe the asteroid on at least two occasions separated by a few days. On each night, draw a star chart of the objects in your telescope's field of view. Has the position of one starlike object shifted between observing sessions? Does the position of the moving object agree with the path plotted on published star charts? Do you feel confident that you have in fact seen the asteroid?

For further reading

Beatty, J. K. "An Inside Look at Halley's Comet." *Sky & Telescope*, May 1986 • Published only a few months after the Comet Halley flybys, this article contains some fine photographs of the comet's nucleus as well as a "first look" at the data sent back by the spacecraft.

Chapman, R., and Brandt, J. "Comets and their Origin." *Mercury*, January–February 1985 • This article, written in preparation for the 1986 return of Comet Halley, gives a fine summary of the basic properties of comets in general.

Dodd, R. *Thunderstones and Shooting Stars: The Meaning of Meteorites*. Harvard University Press, 1986 • This clear introduction to the recovery, classification, and study of meteorites was written by a noted geologist.

Gingerich, O. "Newton, Halley, and the Comet." *Sky & Telescope*, March 1986 • This fascinating article by a noted science historian describes the relationship between Newton and Halley.

Hartmann, W. "The Smaller Bodies of the Solar System." *Scientific American*," September 1975 • This article by a noted planetary scientist discusses asteroids and meteorites.

Knacke, R. "Sampling the Stuff of a Comet." *Sky & Telescope*, March 1987 • This article, last of three in a special issue of *Sky & Telescope* describing results from the Comet Halley flybys, discusses the chemical composition of comets.

Kowal, C. *Asteroids*. Ellis Horwood/John Wiley, 1988 • This fascinating book covers the history and science of asteroids and speculates about future space missions to take advantage of asteroid resources.

Olson, R. "Giotto's Portrait of Halley's Comet." *Scientific American*, July 1979 • This article describes historical sightings of Comet Halley, including that by Giotto di Bondone after whom a spacecraft was named.

Russell, D. "The Mass Extinctions of the Late Mesozoic." *Scientific American*, January 1982 • This article discusses evidence that an asteroid hit the Earth 63 million years ago, causing the extinction of the dinosaurs and many other species.

Sagan, C., and Druyan, A. *Comet*. Random House, 1985 • This beautiful volume on comet science and lore is stylishly written, well organized, and contains excellent illustrations.

Weissman, P. "Realm of the Comet." *Sky & Telescope*, March 1987 • This article, the first of three in a special issue of *Sky & Telescope* describing results from the Comet Halley flybys, discusses the Oort cloud.

Whipple, F. "The Nature of Comets." *Scientific American*, February 1974 • The scientist who invented the "dirty snowball" model discusses the structure and behavior of comets.

———. "The Spin of Comets." *Scientific American*, March 1980 • This article, which focuses on Comet Encke, describes the rotation of a cometary nucleus and the thrust of gases evaporating from it.

———. *The Mystery of Comets*. Smithsonian Institution Press, 1985 • This appealing popular-level book includes personal reminiscences by a distinguished astronomer who has spent most of his life studying comets.

———. "The Black Heart of Comet Halley." *Sky & Telescope*, March 1987 • This article, the second of three in a special issue of *Sky & Telescope* describing results from the Comet Halley flybys, discusses the structure and properties of the comet's nucleus.

The solar corona This composite view of the Sun's outer atmosphere combines a white-light view taken during a solar eclipse with an X-ray image captured by a camera on board a rocket. The white-light image shows streamers extending far above the solar surface. The X-ray view shows near-surface features in shades of yellow, orange, and red. By matching the two views, scientists can study coronal structures over a range of wavelengths and construct a three-dimensional picture of solar activity. (High Altitude Observatory, NCAR, University of Colorado)

C H A P T E R

Our Star

The Sun is a typical star. At the Sun's center, thermonuclear reactions convert hydrogen into helium to provide the energy by which the Sun shines. Ionized gases, photons, and magnetic fields interact on a colossal scale as this energy makes its way toward the solar surface, where it is radiated into space. The Sun's surface is the scene of a host of bewildering phenomena including sunspots, which are sites of concentrated magnetic field. The number of sunspots varies with an 11-year period, revealing recurrent processes which are part of a more general 22-year solar cycle that affects the entire solar atmosphere. Because we can view it from close range, studies of the Sun give us important insights into the nature of stars in general. Nevertheless, astronomers are still puzzled by many basic questions. For instance, we do not understand why sunspots exist in the first place. There is also much contention over particles emanating from the Sun's core. For the clues that it can give us about the universe as well as for its importance to life on Earth, solar astronomy is an exciting and rewarding field of research.

Box 18-1 Sun data

Distance from Earth:	mean = 1 AU = 149,598,000 km
	maximum = 152,000,000 km
	minimum = 147,000,000 km
Light travel time to Earth:	8.3 min
Mean angular diameter:	32 arc min
Radius:	696,000 km = 109 Earth radii
Mass:	1.99×10^{30} kg = 3.33×10^{5} Earth masses
Composition (by mass):	74 percent hydrogen
	25 percent helium
	1 percent all other elements
Mean density:	1410 kg/m^3
Mean temperatures:	surface = 5800 K
	center = 15.5×10^{6} K
Spectral type:	G2
Luminosity:	3.90×10^{26} W
Apparent magnitude:	−26.8
Absolute magnitude:	+4.8
Distance from center of Galaxy:	7.7 kpc = 25,000 ly
Orbital period around center of Galaxy:	200,000,000 years
Orbital velocity around center of Galaxy:	230 km/s

The Sun is an average star. Its mass, size, surface temperature, and chemical composition lie roughly midway between the extremes exhibited by other stars. Unlike other stars, however, our Sun is available for detailed, close-up examination. Studying the Sun offers excellent insights into the nature of stars in general.

Understanding the Sun is important because the Sun is our source of heat and light. Life would not be possible on Earth without the energy provided by the Sun. Even a small change in the Sun's size or surface temperature could dramatically alter conditions on the Earth, either melting the polar caps or producing another ice age. (Data about the Sun are listed in Box 18-1.)

Although the Sun is a commonplace star, it is a dramatic arena where we can observe the fascinating and complicated interaction of matter and energy on an immense scale. Beautiful and bewildering phenomena occur as columns of hot gases gush up to the solar surface, interact with the Sun's magnetic field, and dissipate energy into its outer atmosphere. The source of all this energy lies buried at the Sun's center.

18-1 The Sun's energy is produced by thermonuclear reactions in the core of the Sun

During the nineteenth century, geologists and biologists found convincing evidence that the Earth must have existed in more or less its present form for hundreds of millions of years. This certainly posed severe problems for physicists, who found it impossible to explain how the Sun had been shining for so long, radiating immense amounts of energy into space. If the Sun were made of coal, for example, which produces its heat and light by chemical burning, it could last for only 5000 years.

An explanation for how solar energy is created was proposed in the mid-1800s by Lord Kelvin, after whom the temperature scale is named, and Hermann von Helmholtz. They argued that the tremendous weight of the Sun's outer layers pressing inward from all sides should cause the Sun to contract gradually. As the Sun's gravity causes it to contract, its interior gases become compressed. Whenever a gas is compressed, its temperature rises. (Diesel engines operate on this principle, which you can demonstrate experimentally with a bicycle pump.) Gravitational contraction causes the Sun's gases to become hot enough to radiate energy out into space.

This process, called **Kelvin–Helmholtz contraction**, actually does occur during the earliest stages of the birth of a star, as a large ball of interstellar gas shrinks in response to the inward pull of its own gravity. During this contraction process, gravitational energy (the energy associated with gravitational forces) is converted into thermal energy (the energy associated with heat), making the gases glow with a spectrum similar to that of a blackbody.

Kelvin–Helmholtz contraction cannot, however, be the major source of the Sun's energy. Helmholtz's own calculations showed that the Sun must contract so rapidly to produce the energy it emits that it would have had to extend beyond the Earth's orbit only 25 million years ago. Physicists remained unable to explain how the Earth could have existed in its present form for the hundreds of millions of years needed to produce its present surface features and life forms.

An important key to the source of the Sun's energy was provided in 1905 by Albert Einstein's special theory of relativity. One of the implications of Einstein's theory is that matter and energy are mutually interconvertible, according to the simple equation

$$E = mc^2$$

In other words, a mass (m) can be converted into an amount of energy (E) equivalent to mc^2, where c is the speed of light. Because c is a large number and thus c^2 is huge, a small amount of matter can be converted into an awesome amount of energy.

Astronomers began to wonder if the Sun's energy output might come from the conversion of matter into energy. In the 1920s, the British astronomer Arthur Eddington showed that temperatures near the center of the Sun must be much greater than had previously been thought. Another British astronomer, Robert Atkinson, then suggested that under these conditions hydrogen nuclei near the Sun's center might fuse together to produce helium nuclei in a reaction that would transform a tiny amount of mass into a large amount of energy.

Recall that the nucleus of a hydrogen (H) atom consists of a single proton. The nucleus of a helium (He) atom consists of two protons and two neutrons. Neutrons and protons are similar enough particles (see Box 7-1 for details) that they can be interconverted in nuclear reactions. In the nuclear process

$$4\ H \rightarrow He + energy$$

two of the four protons from hydrogen are changed into neutrons to produce a single helium nucleus. This reaction also releases two positively charged electrons, called **posi-**

trons, that carry off the electric charges relinquished by the transmuted protons. In addition, two massless particles, called **neutrinos**, carry off some energy and momentum. Together, these four supplementary particles ensure that the reaction conserves electric charge, energy, and momentum.

In the conversion of hydrogen into helium, matter is lost because the ingredients (four hydrogen nuclei) have a combined mass just slightly more than the product (one helium nucleus). The mass lost during this reaction can be calculated as follows:

$$
\begin{aligned}
4 \text{ hydrogen atoms} &= 6.693 \times 10^{-27} \text{ kg} \\
-1 \text{ helium atom} &= -6.645 \times 10^{-27} \text{ kg} \\
\hline
\text{mass lost} &= 0.048 \times 10^{-27} \text{ kg}
\end{aligned}
$$

Thus, 0.7 percent of the mass of the hydrogen going into the nuclear reaction does not show up in the mass of the helium. This lost mass is converted into energy, in an amount predicted by the equation $E = mc^2$ as follows:

$$E = mc^2 = (0.048 \times 10^{-27} \text{ kg})(3 \times 10^8 \text{ m/s})^2$$
$$= 4.3 \times 10^{-12} \text{ joule}$$

This quantity is a tiny amount of energy, because it results from the creation of only a single helium atom. Consider, however, the conversion of 1 kg of hydrogen into helium. Although a kilogram of hydrogen goes into this reaction, only 0.993 kg of helium comes out. Using Einstein's equation, we find that the missing 0.007 kg of matter has been transformed into 6.3×10^{14} joules of energy. This amount of energy is the same as that released by the chemical burning of 20,000 metric tons of coal.

The Sun's mass, usually designated $M_\odot$, is 2×10^{30} kg ($= 333,000$ Earth masses) and the Sun's total power output, called its **luminosity** ($L_\odot$), is 3.9×10^{26} watts. To produce this luminosity, 6×10^{11} kg of hydrogen must be converted into helium within the Sun each second. This rate is prodigious, but the Sun contains a vast amount of hydrogen. In fact, the Sun contains enough hydrogen to continue to give off energy at the present rate for another 5 billion years.

This process of nuclear fusion, by which hydrogen is converted into helium at the Sun's center, is called **hydrogen burning**, even though nothing is actually burned in the conventional sense of the word. Box 18-2 describes the details of the nuclear reactions involved in hydrogen burning. Normally, the positive electric charge on protons is effective in keeping protons far apart, because like charges repel each other. But in the extreme heat (15 million K) of the Sun's center, the protons are moving so fast that they can penetrate each other's electric repulsion so deeply that a powerful short-range force, called the **strong nuclear force**, takes over and causes the particles to stick together. At the same time, a second short-range force, called the **weak nuclear force**, transforms one of the protons into a neutron, along with the emission of a positron and a neutrino. Hydrogen burning is

Box 18-2   The proton–proton chain and the CNO cycle

There are two main ways in which hydrogen burning proceeds inside a star. In each case, four hydrogen protons combine to form one helium nucleus, with a slight loss of mass, which is converted into energy. The temperature at the star's core determines which sequence occurs.

For stars with masses not greater than the Sun's mass, the central temperature does not exceed 16 million K, and hydrogen burning proceeds via the proton–proton chain. For stars more massive than the Sun, the central temperature is above 16 million K, and hydrogen burning occurs through a series of reactions called the CNO cycle.

The proton–proton chain is essentially the direct fusion of hydrogen into helium. The fusion occurs in three steps.

First, two protons (each denoted by ^{1}H) combine to form an isotope of hydrogen called deuterium (^{2}H), which consists of one proton and one neutron bound together. In this step, one of the two protons turns into a neutron, releasing a positron (e^+), and a neutrino (ν). Thus, the first step is

$$\text{①} \quad {}^1\text{H} + {}^1\text{H} \rightarrow {}^2\text{H} + e^+ + \nu$$

A positron (e^+) is just like an ordinary electron (e^-), except that it has a positive rather than a negative electric charge. A neutrino is a particle to which most matter is virtually transparent. Consequently, neutrinos are very difficult to detect, and their properties are currently a topic of research and debate. Neutrinos have no charge, but they may have a small mass. A neutrino is produced whenever a proton is converted into a neutron. Conversely, an antineutrino ($\bar{\nu}$) is liberated when a neutron turns into a proton.

In the second step of the proton–proton chain, a third proton combines with the deuterium nucleus to produce a low-mass isotope of helium (^{3}He), whose nucleus contains two protons and one neutron. This reaction releases energy (γ stands for a gamma-ray photon):

$$\text{②} \quad {}^1\text{H} + {}^2\text{H} \rightarrow {}^3\text{He} + \gamma$$

In the final step, two ^{3}He nuclei combine to produce an ordinary nucleus of helium (^{4}He = 2 protons + 2 neutrons), with the release of two protons:

$$\text{③} \quad {}^3\text{He} + {}^3\text{He} \rightarrow {}^4\text{He} + {}^1\text{H} + {}^1\text{H}$$

In massive stars, where the central temperatures exceed 16 million K, carbon serves as a catalyst by which hydrogen burning proceeds. Cornell University physicist Hans Bethe was instrumental in discovering the steps through which a carbon nucleus absorbs protons and finally spits out a helium nucleus. Along the way, the carbon is transformed into nitrogen and oxygen; hence this process is called the CNO cycle. There are six steps in the CNO cycle:

$$\text{1)} \quad {}^{12}\text{C} + {}^1\text{H} \rightarrow {}^{13}\text{N} + \gamma$$
$$\text{2)} \quad {}^{13}\text{N} \rightarrow {}^{13}\text{C} + e^+ + \nu$$
$$\text{3)} \quad {}^{13}\text{C} + {}^1\text{H} \rightarrow {}^{14}\text{N} + \gamma$$
$$\text{4)} \quad {}^{14}\text{N} + {}^1\text{H} \rightarrow {}^{15}\text{O} + \gamma$$
$$\text{5)} \quad {}^{15}\text{O} \rightarrow {}^{15}\text{N} + e^+ + \nu$$
$$\text{6)} \quad {}^{15}\text{N} + {}^1\text{H} \rightarrow {}^{12}\text{C} + {}^4\text{He}$$

Both the proton–proton chain and the CNO cycle have the overall effect of converting four hydrogen nuclei (four protons) into one helium nucleus, two positrons, two neutrinos, and some high-energy, short-wavelength gamma radiation. For the CNO cycle to function, the ^{12}C nucleus, which acts as a nuclear catalyst, must be present. Because it is restored at the end of the cycle, however, carbon is not used up by the process.

---

called a thermonuclear reaction, or thermonuclear fusion, because it can occur only at high temperatures. In later chapters, we will see that such other thermonuclear reactions as helium burning, carbon burning, and oxygen burning occur late in the lives of many stars.

18-2 A theoretical model of the Sun shows how energy gets from the Sun's center to its surface

Although the Sun's interior is hidden from our view, we can use the laws of physics to calculate what is going on below the solar surface (see Figure 18-1). The results of such computations constitute a model of the Sun, which tells us physical characteristics at every depth inside the star. A model of the Sun is usually displayed as a table or graph on which such quantities as pressure, temperature, and density are given for various depths beneath the solar surface.

To develop a model of the Sun or any other stable star, we first note that the Sun is not undergoing any dramatic changes. The Sun is not exploding nor collapsing; neither is it significantly heating nor cooling. The Sun is thus in balance both mechanically and thermally.

Mechanical balance, often called hydrostatic equilibrium, means simply that a star is supporting its own weight. Because of gravity, the tremendous weight of the Sun's outer layers pressing inward from all sides tries to make the star contract. However, as gravity compresses the Sun, gas pressure increases inside the star. The greater the compression, the higher the internal pressure. Hydrostatic equilibrium is achieved when the pressure at every depth in the star is exactly sufficient to support the weight of the overlying layers.

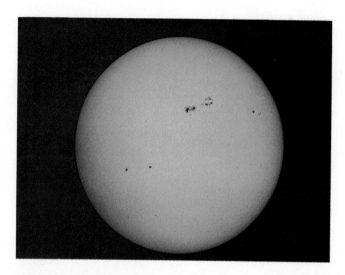

Figure 18-1 *The Sun's surface* *The Sun is the only star whose surface details can be examined through Earth-based telescopes. When viewing the Sun, astronomers always take great care (with extremely dark filters or by projecting the Sun's image on a screen) to avoid severe damage to their eyes. The Sun is so bright that its rays, focused by the lens of an unprotected eyeball, destroy the retina. Therefore,* **never look directly at the Sun.** *(Celestron International)*

Thermal balance, often called **thermal equilibrium**, means simply that a star keeps on shining. Vast amounts of energy escape from the Sun's surface each second. In a state of thermal equilibrium, this energy is constantly resupplied from the Sun's interior at a steady, persistent rate. But exactly how is energy transported from the Sun's center to its surface?

Experience teaches us that energy always flows from hot regions to cooler ones. For example, if you heat one end of a metal bar with a blowtorch, the other end of the bar eventually becomes warm. This method of energy transport is called **conduction.** This process of conduction varies significantly from one substance to another (copper is a good heat conductor, but wood is not), depending on the arrangement and interaction of the atoms. Conditions inside stars like the Sun are not favorable for conduction, so this process is not an efficient means of energy transport. Nevertheless, as we shall see in a later chapter, conduction is important in compact stars such as white dwarfs.

Inside main-sequence stars, energy moves from center to surface by two other means: convection and radiative diffusion. **Convection** involves the circulation of gases between hot and cool regions. Just as a hot-air balloon floats skyward, so hot gases rise toward a star's surface, whereas cool gases sink back down toward the star's center. The net effect of such physical movement of gases is to transfer heat energy from the center toward the surface.

In **radiative diffusion,** photons created in the thermonuclear inferno at a star's center diffuse outward toward the star's surface. The paths of individual photons as they are knocked about between atoms and electrons inside the star

are quite random. The overall photon migration, however, is outward from the hot core, where photons are constantly created, toward the cooler surface, where they escape into space. In all, it takes roughly 100,000 years for energy created at the Sun's center to reach the solar surface and finally escape as sunlight.

The concepts of hydrostatic equilibrium, thermal equilibrium, and energy transport can together be expressed in the form of a set of mathematical equations collectively called the **equations of stellar structure**. These equations can be solved to yield the conditions of pressure, temperature, and density that must exist inside the star if the conditions for equilibrium are to be satisfied.

Astrophysicists use high-speed computers to solve the equations of stellar structure and develop a detailed theoretical model for the structure of a given star. The astrophysicist begins with astronomical data about the star's surface (for example, the Sun's surface temperature is 5800 K, its luminosity is 3.9×10^{26} watts, and the gas pressure and density at the surface are almost zero). Using the equations of stellar structure, the astrophysicist then calculates conditions layer by layer in toward the star's center. In this way, the astrophysicist discovers how temperature, pressure, and density increase with increasing depth below a star's surface. We have learned through this process that the temperature at the Sun's center is 15.5 million K, the pressure is 3.4×10^{11} atm, and the density is 150,000 kg/m^3.

As a result of solving the equations of stellar structure, we know the behavior of various physical quantities such as pressure, density, and temperature throughout a star. This description constitutes the **stellar model**, which can be presented in tables or graphs. Table 18-1 and Figure 18-2 present a theoretical model of the Sun.

In Table 18-1, note that the fraction of luminosity rises to 100 percent at about one-quarter of the way from the Sun's center to its surface. In other words, the Sun's energy production occurs within a volume extending out to $\frac{1}{4}$ solar radius.

Also note that the fraction of mass rises to nearly 100 percent at 0.8 solar radius, which is a distance of only 560,000 km from the Sun's center (recall that 1 solar radius is nearly 700,000 km). Consequently, the outermost 140,000 km of the Sun contain very little matter, which explains why the pressure and density are so low from 0.8 to 1.0 solar radius.

The way in which energy flows from the Sun's center toward its surface depends on how easy it is for photons to move through the gas. If the solar gases are comparatively transparent, photons can travel moderate distances before being scattered or absorbed, and so energy is transported by radiative diffusion. If the gases are rather opaque, photons cannot easily get through the gas. Heat then builds up and the gases start to churn as convection carries hot regions upward while cooler regions sink downward. From the center of the Sun out to 0.8 solar radius, energy is transported by radiative diffusion, and so this region is called the **radia-**

Table 18-1   A theoretical model of the Sun

Fraction of radius	Temperature (× 10⁶ K)	Fraction of central pressure	Density (kg/m³)	Fraction of mass	Fraction of luminosity
0.0 (center)	15.5	1.00	160,000	0.00	0.00
0.1	13.0	0.46	90,000	0.07	0.42
0.2	9.5	0.15	40,000	0.35	0.94
0.3	6.7	0.04	13,000	0.64	1.00
0.4	4.8	0.007	4,000	0.85	1.00
0.5	3.4	0.001	1,000	0.94	1.00
0.6	2.2	0.003	400	0.98	1.00
0.7	1.2	4×10^{-5}	80	0.99	1.00
0.8	0.7	5×10^{-6}	20	1.00	1.00
0.9	0.3	3×10^{-7}	2	1.00	1.00
1.0 (surface)	0.006	4×10^{-13}	0.00030	1.00	1.00

tive zone. Convection dominates the energy flow in the Sun's tenuous outer layer. We thus say that the Sun has a **convective zone**, or **convective envelope**. These aspects of the Sun's internal structure are sketched in Figure 18-3. In all, it takes about 100,000 years for a photon to make its way from the Sun's center to its surface.

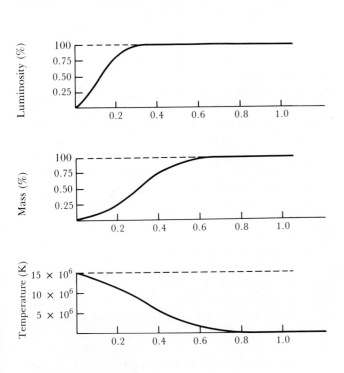

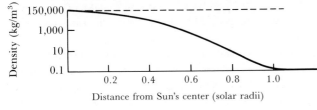

Figure 18-2   A theoretical model of the Sun   The Sun's internal structure is displayed here with graphs that show how the density, temperature, mass, and luminosity vary with the distance from the Sun's center.

18-3 The mystery of the missing neutrinos inspires speculation about the Sun's interior

During hydrogen burning, protons change into neutrons, with each such transformation releasing a neutrino. Neutrinos have no electric charge and are generally believed to have no mass, either, which means they resemble photons and travel through space at the speed of light. Some physicists theorize that neutrinos might have a tiny mass, probably less than one ten-thousandth the mass of an electron. If neutrinos do have mass, they travel at speeds less than that of light.

Neutrinos are difficult to detect, because they do not interact much with ordinary matter. In fact, neutrinos interact so weakly and so infrequently with matter that they easily pass through the Earth as if it were not there. On rare occasions, a neutrino might strike a neutron and convert it into a proton. Such collisions are highly improbable, however, so even objects as big and massive as the Earth are virtually transparent to neutrinos.

The Sun is also largely transparent to neutrinos, allowing these particles to stream outward, unimpeded, from its core. If astronomers could detect these particles, they would have an excellent means of probing the Sun's center. About 10^{38} neutrinos are produced at the Sun's center each second, meaning that the number of neutrinos passing through each square meter of the Earth is surprisingly large: about 10^{14} neutrinos m⁻²s⁻¹. In other words, roughly 100 billion neutrinos pass through every square inch of your body every

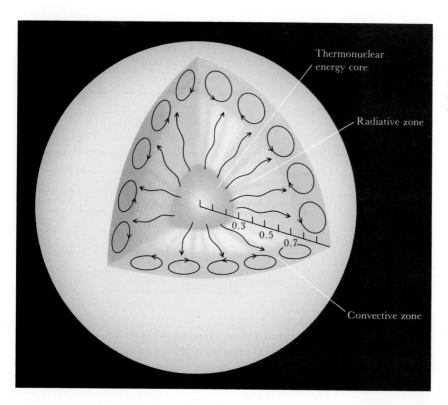

Figure 18-3 *The Sun's internal structure* *Thermonuclear reactions occur in the Sun's core, which extends out to a distance of 0.25 solar radius from the center. Energy is transported outward, via radiative diffusion, to a distance of 0.8 solar radius. Convection is responsible for energy transport in the Sun's outer layers.*

second! If astronomers could detect even a few of these particles, it might be possible to build a "neutrino telescope" that could be used to "see" the thermonuclear inferno that is hidden from ordinary view.

Inspired by such possibilities, Raymond Davis of the Brookhaven National Laboratory designed and built a large neutrino detector. This device uses 100,000 gallons of perchloroethylene cleaning fluid (C_2Cl_4) in a huge tank buried deep in a mine in South Dakota (see Figure 18-4). Because matter is virtually transparent to neutrinos, most of the neutrinos from the Sun pass right through Davis's tank, with no effect whatsoever. On rare occasions, though, a neutrino strikes the nucleus of one of the chlorine atoms in the cleaning fluid and converts one of its neutrons into a proton, creating a radioactive atom of argon. The rate at which this argon is produced tells us the **flux** of solar neutrinos (i.e., the number of neutrinos from the Sun arriving at the Earth per square meter per second).

This experiment, which began in the mid-1960s, has been repeated with extreme care for more than twenty years. On the average, solar neutrinos create one radioactive argon atom every three days in Davis's tank. To the continuing consternation of astronomers, this rate corresponds to only one-third of the neutrinos predicted from the "standard model" of the Sun presented in Table 18-1 and Figure 18-2. Most astronomers now agree that Davis's experiment is probably not at fault, which compels them to reexamine their ideas about the Sun. Three possible solutions to the "mystery of the missing solar neutrinos" involve neutrino

Figure 18-4 *The solar neutrino experiment* *This tank contains 100,000 gallons of perchloroethylene buried 1½ km below ground at the Homestake Gold Mine in South Dakota. The Earth shields the tank from stray particles so that only solar neutrinos can convert chlorine atoms in the fluid into radioactive argon atoms. The number of argon atoms created in the tank is a direct measure of the flux of neutrinos from the Sun. (Courtesy of R. Davis, Brookhaven National Laboratory)*

oscillations, nonstandard solar models, and hypothetical particles whimsically named WIMPS.

If there are three types of neutrinos, then perhaps only one of them is being detected in Davis's experiment. Some theorists predict that if neutrinos have mass, they may continually change from one type to another. This hypothetical phenomenon, called neutrino oscillation, would explain Davis's results, because two-thirds of the time the neutrinos entering his tank would not be of the type he can detect, allowing him to see only one-third of the anticipated neutrino flux.

Another possible solution involves the fact that neutrinos are created in the Sun's core by nuclear reactions that are temperature-sensitive. The Sun's center would have to be only 10 percent lower in temperature to produce a drop in neutrino production that would bring the neutrino flux into agreement with Davis's experiment. Calculations using the equations of stellar structure reveal, however, that if you lower the Sun's central temperature by a million degrees, other obvious features of the Sun, such as its size and surface temperature, would differ from what we observe.

In 1985, American astrophysicists John Faulkner and Ron Gilliland proposed a third and controversial explanation for the missing solar neutrinos. They suggested that the universe may contain numerous "weakly interacting massive particles," or **WIMPS**. Like neutrinos, WIMPS can easily pass through vast amounts of matter. Unlike neutrinos, however, WIMPS are quite massive, perhaps several times more so than a proton or neutron. Over the years, many billions of these hypothetical WIMPS could have been gravitationally captured by the Sun, so that today they are deep inside the Sun's thermonuclear core, freely orbiting its center. As they move along their orbits, these particles could pick up energy at the Sun's center and transport it farther out within the hydrogen-burning core. This simple movement of energy could lower the Sun's central temperature

just enough to bring the neutrino flux into line with the results of Davis's experiments, without significantly altering any other observable aspects of the Sun.

The Faulkner–Gilliland proposal is actually not as outlandish as it might at first seem. As we shall see when we study galaxies, most of the matter in the universe seems to be dark, cold, and quite invisible. Some studies suggest that ordinary matter constitutes only 10 percent of the mass of all the matter in the universe. The remaining 90 percent, which seems to be made of something quite different from ordinary atoms and elements, is very hard to detect. Perhaps solving the mystery of the Sun's center will be our first clue toward discovering what the universe is really made of.

18-4 The photosphere is the lowest of three main layers in the Sun's atmosphere

Although astronomers often speak of the solar surface, the Sun really has no surface at all. As you move in toward the Sun, you encounter ever more dense gases, but there is no sharp boundary such as the surfaces of the Earth or the Moon. The Sun appears to have a surface (for example, see Figure 18-1) simply because there is a specific layer in the Sun's atmosphere from which most of the visible light comes. This layer, which is about 300 to 400 km thick, is appropriately called the Sun's **photosphere** ("sphere of light").

Convection in the Sun's outer layers affects the appearance of the photosphere. Under good observing conditions with a telescope—and special dark filters to protect their eyes from damage—astronomers can often see a blotchy pattern called **granulation** (see Figure 18-5). Each light-

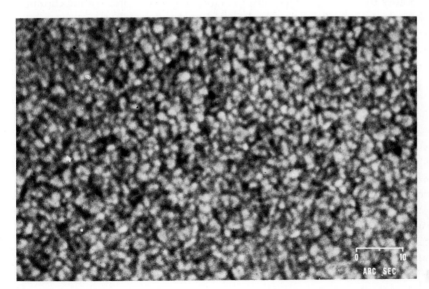

Figure 18-5 Solar granulation *High-resolution photographs of the Sun's surface reveal a blotchy pattern called granulation. The granules, each measuring about 1000 km across, are convection cells in the Sun's outer layers. (San Fernando Observatory)*

colored **granule** measures about 1000 km across and is surrounded by a darkish boundary. High-resolution spectroscopy reveals that there is a blueshifting of spectral lines in the central, bright regions of a granule and a redshifting along the dark, intergranule boundaries. These Doppler shifts show that hot gases rising upward in the granules cool off, spill over the edges of the granules, then plunge back down into the Sun along the intergranule boundaries. The brightness difference between the center and the edge of a granule corresponds to a temperature drop of 300 K.

Granules are individual convection cells in the Sun's outer layers. Time-lapse photography shows that granules form, disappear, and re-form in cycles lasting only a few minutes. At any one time, roughly 4 million granules cover the solar surface. Each granule occupies an area equal to Texas and Oklahoma combined (about 10^6 km^2).

The photosphere appears darker around the edge, or **limb**, of the Sun than it does toward the center of the solar disk. This phenomenon, called **limb darkening,** is caused by photon absorption in the solar atmosphere. The photons that are emitted in warm layers fairly deep within the solar atmosphere are able to escape into space only along the relatively short paths directly away from the Sun's center. The photons that reach the Earth from the edges of the Sun's visible disk come from cooler layers nearer the top of the solar atmosphere, because the photons from the deeper layers are absorbed before they can escape along the much longer paths they must travel toward the Earth. In other words, the Sun's limb appears darker than the middle of the disk because we are looking obliquely at the photosphere near the edge of the disk. We therefore do not see as deeply into the Sun there as we do near the center of the disk. Near the limb, the gas we observe is cooler and thus appears dimmer than the deeper, warmer gas seen near disk center (see Figure 18-6).

The Sun's photosphere shines with a continuous, nearly perfect blackbody spectrum corresponding to its average temperature of about 5800 K. The temperature declines to about 4000 K in the uppermost layers of the photosphere.

All the absorption lines in the Sun's spectrum are produced in this relatively cool layer by atoms extracting energy from the sunlight at various wavelengths (see Figure 18-6).

18-5 The chromosphere is located between the photosphere and the Sun's outermost atmosphere

The relatively cool region immediately above the photosphere is the lowest layer of the **chromosphere** ("sphere of color"), the second of the three major levels in the Sun's atmosphere. During a total solar eclipse, the chromosphere is visible as a 2000-km-thick pinkish layer around the eclipsed Sun, seen just beyond the edge of the dark Moon. The spectrum of the chromosphere is dominated by emission lines, many of which have exactly the same wavelengths as the absorption lines in the spectrum of the photosphere. (Recall Kirchhoff's laws of spectral analysis, summarized in Figure 5-13.) When viewed against the glowing photosphere, the solar spectrum exhibits absorption lines. However, when seen with the dark sky as a background, the spectrum of the chromosphere exhibits emission lines from the atoms in this cooler layer above the photosphere giving up the energy they had extracted from the radiation pouring out of the Sun.

The characteristic reddish-pink color of the chromosphere comes from the Balmer line H$_\alpha$ at 656.3 nm, one of the brightest emission lines in the chromosphere's spectrum. Other bright chromospheric emission lines include the so-called H and K lines of singly ionized calcium, at 396.8 and 393.3 nm, both of which appear as very dark absorption lines in the photosphere's spectrum. The spectrum of the chromosphere also exhibits emission lines of ionized helium and of ionized metals not seen in the photospheric spectrum. These lines must therefore originate in areas of the chromosphere that are hotter than the photosphere.

Figure 18-6 Limb darkening *Light reaching the Earth from the central regions of the solar disk originates in deep, warm layers of the photosphere. However, the light coming to us from the Sun's limb originates in the higher, cooler layers of the photosphere. Consequently, the center of the Sun's disk appears brighter than the limb. (The height of the photosphere is exaggerated in this diagram.)*

Figure 18-7 Spicules and the Sun's chromosphere
Spicules are jets of cool gas that rise up into warmer
regions of the Sun's outer atmosphere. Numerous
spicules are visible in this H_α filtergram, which also
shows many details about the chromosphere. Spicules are located along the irregularly shaped boundaries between large organized cells called supergranules. (NOAO)

The photosphere emits almost no light at the wavelengths of H_α and the calcium H and K lines. The photospheric spectrum has broad, dark absorption lines at these wavelengths. However, the chromosphere is especially bright at these wavelengths, allowing astronomers to study details of the chromosphere by viewing the Sun through special filters transparent to light only at the wavelengths of H_α or the calcium lines.

Figure 18-7 is a photograph of the solar surface taken through an H_α filter. Such photographs, often called **filtergrams**, show what is going on in the chromosphere. Note the numerous dark, brushlike spikes protruding upward. These spikes, called **spicules**, are jets of gas surging up out of the Sun. A typical spicule rises at the rate of 20 km/s, reaches a height of about 7000 km, then collapses and fades away after a few minutes. At any one time, roughly 300,000 spicules exist, covering a few percent of the Sun's surface.

The location of spicules is directly associated with the gas motions of the Sun's outer layers. Detailed observations of the solar surface made in the 1960s revealed the existence of large organized cells called **supergranules**. A typical supergranule is about 30,000 km in diameter and contains many hundreds of ordinary granules. Gases rise upward in the middle of a supergranule, move horizontally outward toward its edge, and descend back into the Sun. Spicules outline the boundaries between supergranules.

18-6 The corona is the outermost layer of the Sun's atmosphere

The outermost region of the Sun's atmosphere is called the **corona**. It extends from the top of the chromosphere out to a distance of several million kilometers, where it gradually becomes the solar wind. (The solar wind, as described earlier, consists of high-speed protons and electrons constantly escaping from the Sun.) We can now see that the **solar atmosphere** has a total of three distinct layers—photosphere, chromosphere, and corona—as sketched in Figure 18-8. Everything below the photosphere is called the **solar interior**.

The total amount of visible light emitted by the solar corona is comparable to the brightness of the Moon at full moon—only about one-millionth as bright as the photosphere. Consequently, the corona can be viewed only when the photosphere is blocked out during a total eclipse or in a specially designed telescope called a **coronagraph**. Figure 18-9 is an exceptionally detailed photograph of the Sun's corona. Numerous **coronal streamers** are visible extending outward to a distance of 6 solar radii.

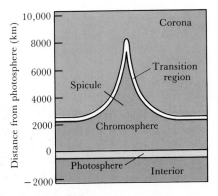

Figure 18-8 The solar atmosphere The Sun's atmosphere has three distinct layers. The lowest, the photosphere, is about 100 km thick. The chromosphere above it extends to an altitude of about 2000 km, with spicules jutting up to 10,000 km above the photosphere. The third, outer layer, the corona, extends many millions of kilometers out into space and merges with the solar wind. (Adapted from J. A. Eddy)

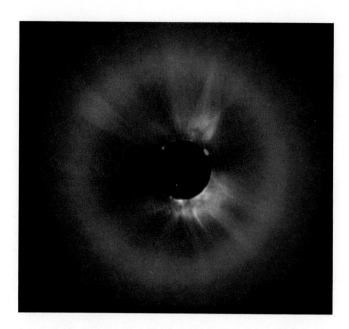

Figure 18-9 *The solar corona This extraordinary photograph of the corona was taken from a jet 40,000 feet above Montana during the total solar eclipse of February 26, 1979. Numerous coronal streamers are visible, extending outward to distances of 4 million km above the solar surface. (Los Alamos Scientific Laboratory)*

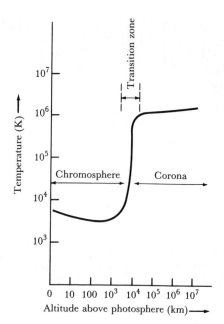

Figure 18-10 *Temperatures in the Sun's upper atmosphere In a narrow region about 10,000 km above the Sun's photosphere, the temperature rises abruptly to about 1 million K. Sound or magnetic waves propagating outward from the Sun's convective zone may provide the energy that heats the Sun's tenuous outer layers to extraordinarily high temperatures.*

Around 1940, astronomers realized that the spectrum of the Sun's corona contains the emission lines of certain highly ionized elements. For example, there is a prominent green line at 530.3 nm that is caused by Fe XIV (iron atoms, each stripped of 13 electrons). Extremely high temperatures are required to strip that many electrons from atoms, so the corona must be very hot. Figure 18-10 shows how temperature varies with distance above the photosphere. Coronal temperatures are in the range of 1 to 2 million kelvins.

The corona is, however, nearly a vacuum. The typical density of coronal gases is only about 10^{11} particles per cubic meter. (The density of the photosphere is about 10^{23} particles per cubic meter, and the density of the air we breathe is about 10^{25} particles per cubic meter.) Despite its high temperature, the corona emits relatively dim light, because of its low density.

Astronomers do not fully understand why the corona is so hot. There is plenty of energy to heat the corona available from the churning gas motions in the Sun's convective envelope, but no one has been able to explain how that energy is transported to the corona. Some researchers suspect that sound waves or waves in the Sun's magnetic field carry energy from the photosphere to the corona, but astronomers do not understand the details of how this happens.

Recent observations from space show that the corona exhibits both complicated structure and activity. For example, in Figure 18-11 there is a huge, bubblelike disturbance called a **solar transient**. These short-lived protuberances erupt suddenly and expand rapidly outward through the

corona. Solar transients probably occur as often as once a day, but they were unknown until the Sun could be examined with a coronagraph carried aloft on board Skylab. They were later identified as the source of the bursts of radio noise long known on the Earth.

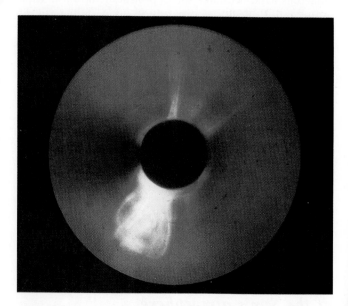

Figure 18-11 *A solar transient During the early 1970s, astronauts on Skylab discovered huge bubbles erupting outward through the Sun's corona. The leading edge of this bubble moved outward from the Sun at a speed of 500 km/s. This solar transient was photographed on June 10, 1973. (NASA; High Altitude Observatory)*

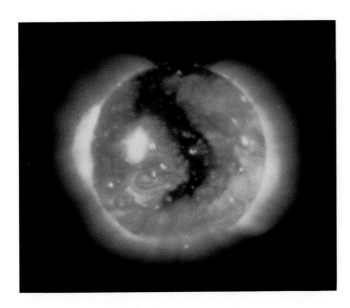

Figure 18-12 *A coronal hole in an X-ray photograph* This X-ray picture of the Sun was taken by Skylab astronauts on August 21, 1973. A huge, dark, boot-shaped coronal hole dominates this view of the inner corona. Numerous bright points are also visible. (NASA; Harvard College Observatory)

The Skylab missions also obtained X-ray photographs of the corona. The corona has temperatures in the million-kelvin range, so from Wien's law we know that the corona should be shining brightly at X-ray wavelengths of roughly 3 nm. The Skylab pictures revealed a blotchy, irregular inner corona (see Figure 18-12). Note the large dark area, called a **coronal hole** because it is devoid of the usual hot, glowing coronal gases. Many astronomers suspect that coronal holes are the main corridors through which particles of the solar wind escape from the Sun.

In addition to dark coronal holes, X-ray photographs also reveal numerous bright spots that are hotter than the surrounding corona. The temperatures in these bright points occasionally reach 4 million K. Many of the bright coronal hot spots seen in X-ray pictures like Figure 18-12 are located over sunspots.

18-7 Sunspots are one of many phenomena associated with a 22-year solar cycle

Superimposed on the basic structure of the solar atmosphere are a host of phenomena that vary with a 22-year period. Irregularly shaped dark regions called **sunspots** are the most easily recognized of these phenomena, because they occur in the photosphere (see Figure 18-13). The dark central core of a sunspot, called the **umbra**, is usually surrounded by a less-

dark border called the **penumbra**. Although sunspots vary greatly in size, typical ones measure a few tens of thousands of kilometers across.

On rare occasions, a sunspot group is large enough to be seen with the naked eye. Chinese astronomers recorded such sightings 2000 years ago, and a huge sunspot group visible to the naked eye was seen in 1989. (ALWAYS USE SPECIAL DARK FILTERS OR OTHER MEANS TO PROTECT YOUR EYES WHEN OBSERVING THE SUN! Looking directly at the Sun can cause blindness.) Of course, a telescope gives a much better view, and so it was not until Galileo that anyone was able to examine sunspots in detail. In fact, Galileo discovered that he could determine the Sun's rotation rate by tracking sunspots as they moved across the solar disk (see Figure 18-14). He found that the Sun rotates once in about four weeks. A typical sunspot group lasts about two months, so a specific one can be followed for two solar rotations.

More careful observations by the British astronomer Richard Carrington in 1859 demonstrated that the Sun does not rotate as a rigid body. Carrington found that the equatorial regions rotate more rapidly than do the polar regions. A sunspot near the solar equator takes only 25 days to go once around the Sun. At 30° north or south of the equator, however, a sunspot takes $27\frac{1}{2}$ days to complete a rotation. The rotation period at 75° north or south of the solar equator is about 33 days, and at the poles it may be as long as 35 days. This phenomenon is known as **differential rotation**, because different parts of the Sun rotate at slightly different rates.

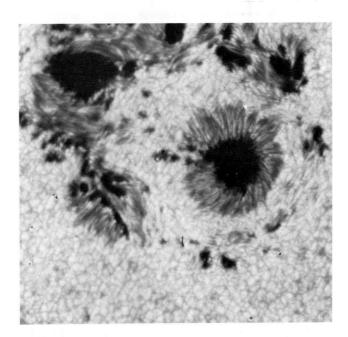

Figure 18-13 *A sunspot group* This photograph of a sunspot group was taken by a balloon-borne telescope at an altitude of roughly 80,000 feet. Note the featherlike appearance of the penumbrae. Also notice the granulation in the undisturbed surrounding photosphere. (Project Stratoscope; Princeton University)

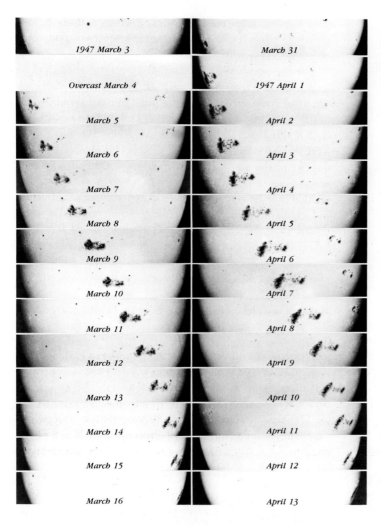

1947 March 3

Overcast March 4

March 5

March 6

March 7

March 8

March 9

March 10

March 11

March 12

March 13

March 14

March 15

March 16

March 31

1947 April 1

April 2

April 3

April 4

April 5

April 6

April 7

April 8

April 9

April 10

April 11

April 12

April 13

Figure 18-14 *The Sun's rotation* By observing the same group of sunspots from one day to the next, Galileo found that the Sun rotates once in about four weeks. The equatorial regions of the Sun actually rotate somewhat faster than the polar regions, a phenomenon called differential rotation. This series of photographs shows the same sunspot group over 1½ solar rotations. (Mount Wilson and Las Campanas Observatories)

Careful observations over many years revealed that the numbers of sunspots change in a periodic fashion. In some years there are many sunspots, in other years almost none. This phenomenon, first noticed by the German astronomer Heinrich Schwabe in 1843, is called the **sunspot cycle**. As shown in Figure 18-15, the average number of sunspots varies with a period of about 11 years. A time of exceptionally many sunspots is a **sunspot maximum**, as occurred in 1959, 1970, late 1980, and 1990. Conversely, the Sun was almost devoid of sunspots at times of **sunspot minimum** in 1965, 1976, and 1987.

The locations of the sunspots also vary in a periodic fashion over the sunspot cycle. At the beginning of a cycle, just after sunspot minimum, sunspots first start to appear at latitudes 30° north and south of the solar equator. Over the succeeding years, the sunspots occur closer and closer to the equator, until at the end of the cycle they are virtually all on the solar equator. Observations showing this equatorial mi-

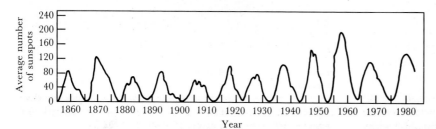

Figure 18-15 *The sunspot cycle* The number of sunspots on the Sun varies in a periodic fashion. Large numbers of sunspots were seen in 1959, 1970, 1980, and 1990, at times of sunspot maximum. Exceptionally few sunspots were observed in 1965, 1976, and 1987.

Year

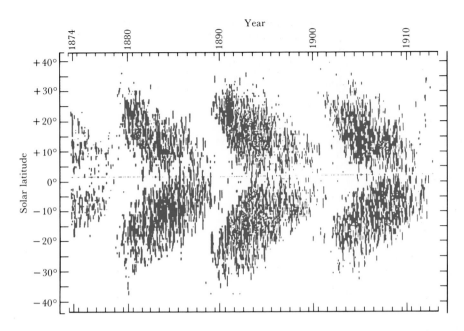

Figure 18-16 The Maunder butterfly diagram
The location of sunspots varies in a regular fashion over the sunspot cycle. The first sunspots in a cycle appear at great distances from the solar equator, but the last spots are formed very near the equator. At sunspot maximum, in the middle of the cycle, most sunspots occur at latitudes of 10° to 15° north and south of the equator.

gration of sunspots are displayed in Figure 18-16. The data on this graph cover areas shaped somewhat like butterflies. The graph is called a **Maunder butterfly diagram**, after E. Walter Maunder, who discovered this phenomenon of sunspot migration in 1904.

In 1908, the American astronomer George Ellery Hale made the important discovery that sunspots are associated with intense magnetic fields on the Sun. When Hale focused a spectroscope on sunlight coming from a sunspot, he found that many spectral lines appeared to consist of several closely spaced spectral lines not usually observed (see Figure 18-17). This "splitting" of a single spectral line into two or more lines is called the **Zeeman effect** after the Dutch physicist Pieter Zeeman, who first observed it in his laboratory in

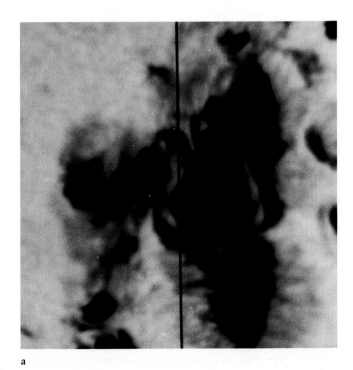

a

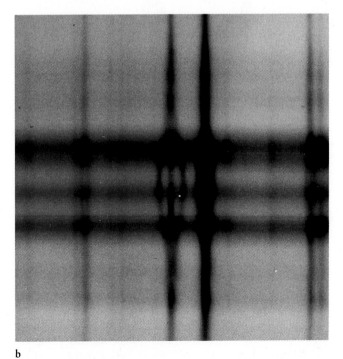

b

Figure 18-17 Zeeman splitting by a sunspot's magnetic field (a) The black line drawn across the sunspot indicates the location toward which the slit of the spectrograph was aimed. (b) In the resulting spectrogram, one line in the middle of the normal solar spectrum is split into three components. The separation between the three lines corresponds to a magnetic field strength in excess of 4000 G. (NOAO)

Box 18-3 The Zeeman effect

In the 1890s, the Dutch physicist Pieter Zeeman placed some hot gases between the poles of a powerful magnet and examined the spectra of the glowing gases through a spectroscope. He found that each spectral line was split into two, three, or more closely spaced lines, depending on the particular gases used. Electrons circle the nuclei of atoms at specific energy levels that are dictated by the laws of quantum mechanics. As we saw in Chapter 5 (recall Figure 5-18), only certain specific energy levels are allowed. Spectral lines are produced when electrons jump from one energy level to another.

An electron moving around a nucleus is actually a miniature electric current, which produces a tiny magnetic field. When an atom is placed in a magnetic field, the tiny magnetic fields caused by the atom's moving electrons try to line up with the overall field. According to the principles of quantum mechanics, this alignment occurs only at specific angles of inclination. Each one of these angles of inclination corresponds to a slightly different energy level for the electron. The number of allowed inclinations increases for higher energy levels.

The diagram shows an example of how the Zeeman effect is produced. When the magnetic field is not present, electrons jumping from a lower energy level up to a higher level absorb photons of a specific wavelength, producing a single absorption line in the spectrum. However, when the magnetic field is turned on around the light source, the lower energy level is replaced by two slightly different energy levels, corresponding to two different alignments of a particular electron orbit in the outside magnetic field. Similarly, the higher energy level is replaced by four slightly different energy levels. Eight different electron jumps are now possible, corresponding to the absorption of photons having eight slightly different wavelengths. In this case, some of the

jumps involve nearly the same energy, so only six distinct spectral lines can be resolved in the spectrum with the magnetic field turned on.

The stronger the magnetic field, the greater the difference in the energy levels. Thus, astronomers can deduce the strength of the magnetic field surrounding the atoms or ions that emit the light by measuring the separations of the lines.

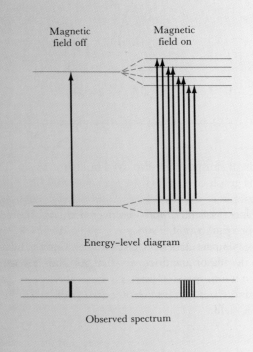

1896. Zeeman showed that the effect is produced when the atoms producing a spectral line are subjected to an intense magnetic field.

The more intense the magnetic field, the wider is the separation of the split lines in the Zeeman effect (see Box 18-3). The strength of a magnetic field is usually expressed in terms of a unit called the **gauss** (G), named after the German mathematician Karl Friedrich Gauss, who experimented with magnets. For example, the strength of the Earth's magnetic field at its north and south magnetic poles is about 0.7 G. The splitting of the Fe I spectral line in Figure 18-17 into three lines corresponds to an intense magnetic field of 4130 G.

Hale's discovery demonstrates that sunspots are places where a powerful, concentrated magnetic field protrudes through the hot gases of the photosphere. Because of the temperature, many of the atoms in the photosphere are ionized, so that the photosphere is a mixture of electrically charged ions and electrons, technically called a **plasma**. A

plasma is an extremely good conductor of electricity that interacts vigorously with magnetic fields. A magnetic field restricts and constrains the motions of a plasma. A sunspot's intense magnetic field greatly inhibits the natural convective motions in it. Energy cannot flow freely upward from the Sun's convective zone, so the gases in this region of the photosphere cool off. In fact, temperatures in a sunspot are typically 4000 to 4500 K, or more than 1000 K cooler than the surrounding, undisturbed photosphere. Because of this lowered temperature, sunspots look dark in contrast to their surroundings.

A host of exotic phenomena occur around and above sunspots as a direct result of their intense magnetic fields. Huge, arching columns of gas called **prominences** often appear above sunspot regions (see Figure 18-18). Some of these formations, called **quiescent prominences**, may hang suspended for days above the solar surface. In contrast, **eruptive prominences** blast material outward from the Sun at speeds of roughly 1000 km/s.

Figure 18-18 A solar prominence *A huge prominence arches above the solar surface in this Skylab photograph taken December 19, 1973. The radiation that exposed this picture is from singly ionized helium (He II) at a wavelength of 30.4 nm that corresponds to a temperature of about 50,000 K. (Naval Research Laboratory)*

The most violent, eruptive events on the Sun, called **solar flares**, occur in complex sunspot groups. Within only a few minutes, temperatures in a compact region may soar to 5 million K. Vast quantities of particles and radiation are blasted out into space. Most astronomers believe that prominences and flares involve concentrated portions of the Sun's magnetic field.

The magnetic effects of a sunspot reach into the corona, where they cause distinctive emission in certain spectral lines of highly ionized atoms. This high-altitude activity allows astronomers to study the solar cycle toward the polar lati-

tudes, where the slanted angle of Earth-based observers prohibits a clear view of the solar surface. In 1987, for instance, Richard C. Altrock reported that his long-term observations of a green emission line of Fe XIV showed that each sunspot cycle begins much earlier than had previously been believed. The magnetic effects of a new cycle actually start at latitudes of 75° north and south of the solar equator several years before the sunspots of the previous cycle have faded away. The locations of these magnetic disturbances gradually migrate toward the equator. Their paths merge with the Maunder butterfly diagram (see Figure 18-16) at about 30° from the equator when the first sunspots of a traditional cycle appear at that latitude. The sunspot cycle can thus be seen as a series of overlapping cycles.

Astronomers construct artificial pictures known as **magnetograms**, which display the magnetic fields in the solar atmosphere, by combining two photographs taken at wavelengths on either side of a magnetically split spectral line. The magnetogram in Figure 18-19 shows the entire solar surface, along with an ordinary photograph of the Sun taken at the same time. Dark blue indicates the areas of the photosphere covered by one magnetic polarity (north), and yellow indicates the area covered by the opposite (south) magnetic polarity.

Many sunspot groups are said to be **bipolar**, meaning that they have roughly comparable areas covered by north and south magnetic polarities (see Figure 18-20). The sunspots on the side of the group toward which the Sun is rotating are called the "preceding members" of the group. The spots that follow behind are referred to as the "following members."

After years of studying the solar magnetic field, George Ellery Hale was able to piece together a remarkable magnetic description of the solar cycle. First of all, Hale discovered that the preceding members of all sunspot groups in one solar hemisphere have the same magnetic polarity. We now

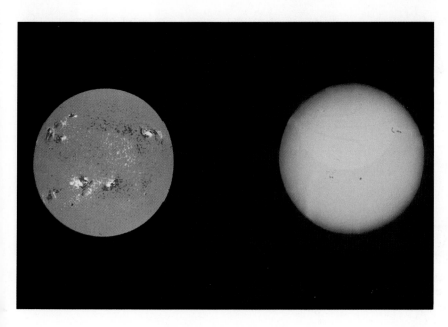

Figure 18-19 Two views of the Sun *The optical view of the Sun (right) was taken at the same time as the magnetogram (left). Notice how regions with strong magnetic fields are related to the locations of sunspots. Careful examination of the magnetogram also reveals that the magnetic polarity of the sunspots in the northern hemisphere is opposite that of those in the southern hemisphere. (NOAO)*

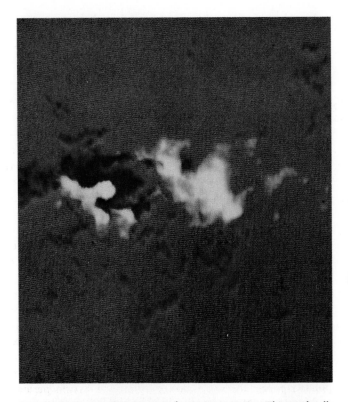

Figure 18-20 *A magnetogram of a sunspot group* This artificially colored picture displays the intensity and polarity of the magnetic field associated with a large sunspot group. One side of a typical sunspot group has one magnetic polarity, and the other side has the opposite magnetic polarity. (NOAO)

know that this polarity is the same as that of the hemisphere in which the group is located. In other words, in the hemisphere of the Sun's north magnetic pole, the preceding members of all sunspot groups have north magnetic polarity. Likewise, in the south magnetic hemisphere all the preceding members have south magnetic polarity.

In addition, Hale found that the polarity pattern completely reverses itself every 11 years. In other words, the hemisphere with a north magnetic polarity at one solar maximum has a south magnetic polarity at the next solar maximum. For this reason astronomers prefer to speak of a **solar cycle** with a period of 22 years, instead of the sunspot cycle of 11 years.

In 1960, the American astronomer Horace Babcock proposed a description that seems to account for many features of the 22-year solar cycle. Babcock's scenario, called a **magnetic-dynamo model**, makes use of two basic properties of the Sun: its differential rotation and its convective envelope. Differential rotation causes the magnetic field in the photosphere to become wrapped around the Sun (see Figure 18-21). The magnetic field then becomes concentrated at latitudes on either side of the solar equator. Convection in the photosphere causes the concentrated magnetic field to become tangled, and kinks erupt through the solar surface. Sunspots appear where the magnetic field protrudes through the photosphere. Careful examination of the orientation of the magnetic field in Figure 18-21 reveals that the preceding members of a sunspot group should indeed have the same

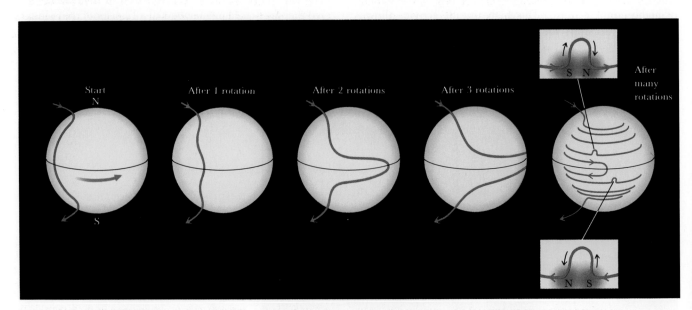

Figure 18-21 *Babcock's magnetic dynamo* A possible partial explanation for the sunspot cycle involves the way the Sun's magnetic field gets wrapped around the Sun by differential rotation. Sunspots appear where the concentrated magnetic field breaks through the solar surface.

magnetic polarity as that of the hemisphere in which they are located.

There are many things that the magnetic-dynamo model fails to explain, however. Most embarrassing is a basic lack of understanding of the physics of sunspots. Astronomers still do not know what holds a sunspot together week after week. Our best calculations predict that a sunspot should break up and disperse almost as soon as it forms.

Our understanding of sunspots is further confounded by irregularities in the solar cycle. For example, the overall reversal of the Sun's magnetic field is often piecemeal and haphazard. One pole may reverse polarity long before the other. For several weeks the Sun's surface may show two north poles and no south pole at all. To make matters worse, there is strong historical evidence that all traces of sunspots and the sunspot cycle have vanished for years at a time. For example, virtually no sunspots were seen from 1645 through 1715, a period called the **Maunder minimum**. Similar sunspot-free periods apparently occurred at irregular intervals in earlier times as well.

The Maunder minimum offers the best evidence for a connection between the Sun and Earth's weather. During this sunspot-free period, Europe experienced years of record low temperatures often referred to as the Little Ice Age, whereas the western United States was subjected to severe drought. Some scientists speculate that solar activity may have an ongoing, though subtle, effect on terrestrial climates.

Not so subtle would be the effects of a varying solar diameter. Jack Eddy of the High Altitude Observatory in Colorado and other astronomers have recently reported controversial observations suggesting that the Sun's diameter may be shrinking slightly. This topic is currently a matter for heated debate, because astronomers are at a loss to explain these reported changes in the Sun's size. Nevertheless, some scientists suspect that the Sun may in fact be undergoing very slow, long-term variations in its physical properties.

18-8 Solar seismology and new satellites are among the latest tools for solar research

There are many reasons why solar astronomers would like to study the Sun's interior. For instance, the Sun's magnetic field that we observe today might have started as a "primordial field" that was compressed and trapped deep within the Sun as the solar nebula contracted 4.5 billion years ago. To search for the roots of the Sun's magnetic field and to study other deeply buried phenomena, astronomers need a way to probe the solar interior.

Geologists are able to determine the Earth's interior structure by using seismographs to record vibrations during earthquakes. Although there are no true sunquakes, the Sun does vibrate at a variety of frequencies, somewhat like a ringing bell. These vibrations were first noticed in 1960 by Robert Leighton at Caltech, who made high-precision Doppler-shift observations of the photosphere. These measurements revealed that portions of the Sun's surface move up and down by about 10 km every 5 minutes (see Figure 18-22). Beginning in the mid-1970s, several astronomers have reported detecting slower vibrations having periods ranging from 20 to 160 minutes. The possibility of detecting extremely slow pulsations has inspired astronomers calling themselves helioseismologists to set up telescopes at the South Pole, where the Sun can be observed continuously for many days. This new field of solar research is called **solar seismology**.

The vibrations of the Sun's surface may be compared with sound waves. If you were within the Sun's atmosphere, you would first notice a deafening roar, somewhat like a jet engine, produced by the turbulence associated with granulation. Superimposed on this noise would be a variety of nearly pure tones. You would be unable to hear these tones,

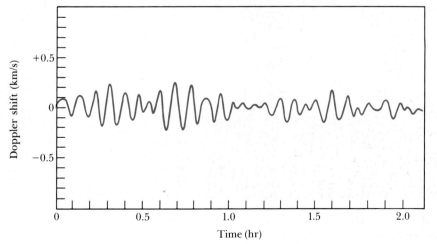

Figure 18-22 Vibrations of the Sun's surface *Doppler-shift measurements give the speed at which the Sun's photosphere bobs up and down. These data extend over 2 hours and show numerous oscillations with a period of about 5 minutes. The gradual variation in the sizes of the oscillations is the result of interference between many vibrations and wavelengths. (Adapted from O. R. White)*

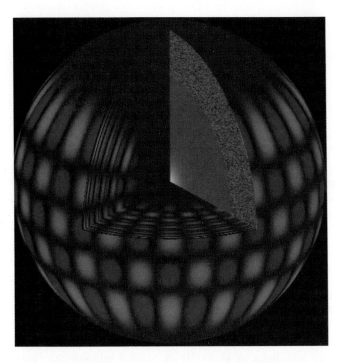

Figure 18-23 *A sound wave resonating in the Sun This computer-generated image shows one of the millions of ways in which the Sun vibrates because of sound waves resonating in its interior. The regions that are moving outward are colored blue, those moving inward, red. The cutaway shows how deep these oscillations are believed to extend. (National Solar Observatory)*

however, for they would be sixteen octaves below the lowest note on the piano keyboard.

In 1970, Roger Ulrich at UCLA pointed out that sound waves moving upward from the solar interior would be reflected back upon reaching the solar surface. However, as a reflected sound wave descended back into the Sun, the increasing density and pressure would bend the wave so severely that it would be turned around and aimed back out again toward the solar surface. In other words, sound waves bounce back and forth between the solar surface and layers deep within the Sun. These sound waves can reinforce each other and resonate if their wavelength is the right size, just as sound waves of a particular wavelength resonate inside an organ pipe.

There are millions of ways in which the Sun might oscillate as a result of waves resonating in its interior. Figure 18-23 is a computer-generated illustration that shows one such vibrational mode. Helioseismologists are able to deduce detailed information about the solar interior from measurements of these oscillations. For instance, it is possible to infer the rotation rate of the Sun's interior. By comparing the speeds of sound waves that travel east to west with those that travel west to east, helioseismologists have attempted to figure out the Sun's rotation rate at different depths and latitudes. As shown in Figure 18-24, the Sun's surface rotation pattern may persist throughout the outer 20 percent of the

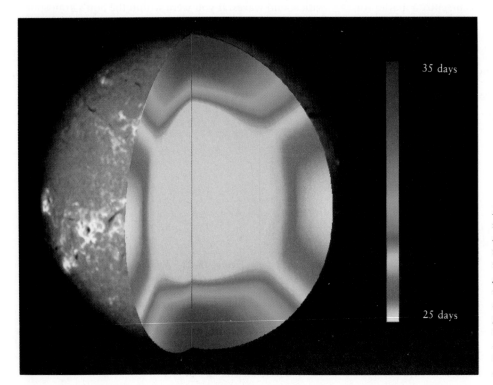

35 days

25 days

Figure 18-24 *Rotation rates in the solar interior This cutaway picture of the Sun shows how the solar rotation rate varies with depth and latitude. Colors represent rotation periods according to the scale given at the right. Note that the surface rotation pattern, which varies from 25 days at the equator to 35 days near the poles, persists throughout the Sun's convective envelope. The Sun's radiative core seems to rotate like a rigid body. (Courtesy of K. Libbrecht; Big Bear Solar Observatory)*

Sun, where it is driven by large-scale convection. Further in, convection ceases and the Sun seems to rotate like a rigid object with a period of about 27 days.

While some solar astronomers have been probing the Sun's core, others have concentrated on making observations of the solar atmosphere from Skylab and Earth-orbiting satellites. During the 1970s, a series of satellites called the Orbiting Solar Observatories provided a wealth of data about the Sun at ultraviolet, X-ray, and gamma-ray wave-

lengths. In 1980, the *Solar Maximum Mission* (SSM) spacecraft was launched to study solar flares. After being repaired by Space Shuttle astronauts who visited it in 1984, the satellite continued to send back spectra and pictures of violent solar phenomena until it fell to Earth in 1989. Meanwhile, astronomers are looking forward to the launching of the *Solar Polar Orbiter*, also called *Ulysses*, which will examine the Sun's polar regions. This mission will help unravel some of the mysteries that surround the Sun.

Key words

bipolar sunspot group	granulation	photosphere	stellar model
chromosphere	granule	plasma	strong nuclear force
*CNO cycle	hydrogen burning	positron	sunspot
conduction	hydrostatic equilibrium	prominence	sunspot cycle
convection	Kelvin–Helmholtz contraction	*proton-proton chain	sunspot maximum
convective zone		radiative diffusion	sunspot minimum
corona	limb darkening	radiative zone	supergranule
coronal hole	luminosity (of the Sun)	solar atmosphere	thermal equilibrium
*deuterium	magnetic-dynamo model	solar cycle	thermonuclear fusion
differential rotation (of the Sun)	magnetogram	solar flare	umbra (of a sunspot)
equations of stellar structure	Maunder butterfly diagram	solar interior	weak nuclear force
	model (of the Sun)	solar seismology	Zeeman effect
flux	neutrino	solar transient	
gauss	penumbra (of a sunspot)	spicule	

Key ideas

- The Sun's energy is produced by a thermonuclear process called hydrogen burning, in which four hydrogen nuclei combine to produce a single helium nucleus, releasing energy in the process.

    The energy released in a nuclear reaction corresponds to a slight reduction of mass, according to Einstein's equation $E = mc^2$.

    A thermonuclear reaction is a nuclear reaction that occurs only at very high temperatures; hydrogen-

burning reactions occur, for instance, only at temperatures of more than about 8 million K.

- A stellar model is a theoretical description of a star's interior derived from calculations based upon the laws of physics.

    The solar model suggests that hydrogen burning occurs in a core extending from the Sun's center to about 0.25 solar radius.

The solar core is surrounded by a radiative zone extending to about 0.8 solar radius, in which energy travels outward through radiative diffusion.

The radiative zone is surrounded by a convective zone of gases at relatively low temperatures and pressures in which energy travels outward primarily through convection.

The visible surface of the Sun is a 100-km-thick layer called the photosphere at the bottom of the solar atmosphere. The gases in this layer shine with almost perfect blackbody radiation corresponding to a temperature of 5800 K; in the photosphere, convection produces features called granules.

Above the photosphere is a layer of cooler, less-dense gases called the chromosphere; spicules extend upward into the chromosphere from the photosphere, along the boundaries of supergranules.

The outermost layer of thin gases in the solar atmosphere, which blends into the solar wind at great distances from the Sun, is called the corona; the gases of the corona are very hot but of a very low density. Solar transients, coronal streamers, and coronal holes are other features associated with the corona.

- The solar surface's features vary periodically in a 22-year solar cycle.

    Sunspots are relatively cool regions produced by local concentrations of the Sun's magnetic field; the average number of sunspots increases and decreases in a regular cycle of approximately 11 years.

    A solar flare is a brief eruption of hot, ionized gases from a sunspot group.

- The magnetic-dynamo model suggests that many features of the solar cycle are caused by the effects of the Sun's differential rotation and convection on the Sun's magnetic field.

- Space missions to observe the Sun from outside the Earth's atmosphere, along with studies of solar seismology, are expected to yield further understanding of the Sun—and hence of stellar processes in general—in coming years.

Review questions

*1   Using the mass and size of the Sun given in Box 18-1, calculate the average density of the Sun. Compare your answer with the average densities of the Jovian planets.

2   Describe the dangers in attempting to observe the Sun. How have astronomers learned to circumvent these observational problems?

3   Give an everyday example of hydrostatic equilibrium. Give an example of thermal equilibrium.

4   Give some everyday examples of conduction, convection, and radiative diffusion.

5   Why do thermonuclear reactions occur only in the Sun's core?

*6   How much energy would be released if each of the following masses were converted *entirely* into its equivalent energy? (a) a carbon atom with a mass of 2×10^{-26} kg (b) one kilogram (c) a planet as massive as the Earth (6×10^{24} kg)

*7   Use the answers to the previous question to calculate how long the Sun must shine in order to release an amount of energy equal to that produced by the complete mass-to-energy conversion of (a) a carbon atom, (b) one kilogram, and (c) the Earth.

*8   The brightest appearing star in the sky, Sirius, has a luminosity of 40 $L_\odot$, which means that it is forty times as bright as the Sun and burns hydrogen at a rate forty times greater than the Sun does. How much hydrogen does Sirius burn each second?

9   Describe the Sun's interior, with attention to the main physical processes that occur at various depths within the Sun.

10   Describe the Sun's atmosphere. Be sure to mention some of the phenomena that occur at various altitudes above the solar surface.

11   What is a neutrino, and why are astronomers so interested in detecting neutrinos emanating from the Sun?

12   What is hydrogen burning? Why is hydrogen burning fundamentally unlike the burning of a log in a fireplace?

13   Why do astronomers say that the solar cycle is really 22 years long even though the number of sunspots varies with an 11-year period?

14   Suppose that you want to determine the Sun's rotation rate by observing its sunspots. Is it necessary to take the Earth's orbital motion into account? Why or why not?

15   When do you think the next sunspot maximum and minimum will occur? Explain.

16   It was once thought that sunspots were giant holes into the Sun's cooler interior. Why don't astronomers believe this anymore?

17   Why do you suppose the Sun generates energy only around its center?

18 Explain why studying the oscillations of the Sun's surface can give important, detailed information about physical conditions deep within the Sun.

Advanced questions

Tips and tools . . .
You may have to review Wien's law and the Stefan–Boltzmann law which were discussed in Chapter 5. By taking ratios, such as the ratio of the flux from a sunspot to that from the undisturbed photosphere, all the cumbersome constants cancel out, thereby simplifying your calculation.

---

19 What would happen if the Sun were not in a state of both hydrostatic and thermal equilibrium?

*20 (*Basic*) Assuming that the current rate of hydrogen burning in the Sun remains constant, what fraction of the Sun's mass will be converted into helium over the next 5 billion years? How will this affect the chemical composition of the Sun?

*21 (*Basic*) Calculate the wavelengths at which the photosphere, chromosphere, and corona emit the most radiation. Explain how the results of your calculations suggest the best way to observe these regions of the solar atmosphere. (*Hint:* Assume average temperatures of 50,000 K and 1.5×10^6 K for the chromosphere and corona, respectively.)

*22 Assuming that a sunspot is 1500 K cooler than the undisturbed photosphere, calculate how much less intense radiation from the sunspot is compared to the surrounding photosphere.

Discussion questions

23 Discuss the extent to which cultures around the world have worshiped the Sun as a deity throughout history. Why do you suppose such widespread veneration has occurred?

24 Discuss some of the difficulties of correlating solar activity with changes in the terrestrial climate.

25 Describe some of the advantages and disadvantages of observing the Sun (**a**) from space, and (**b**) from the South Pole. What kinds of phenomena and issues do solar astronomers want to explore from both the Earth-orbiting and the Antarctic observatories?

Observing projects

26 Use a telescope to view the Sun by projecting the Sun's image onto a screen or sheet of white paper. **DO NOT LOOK AT THE SUN! Looking directly at the Sun can cause blindness.** Do you see any sunspots? Sketch their appearance. Can you distinguish between the umbrae and penumbrae of the sunspots? Can you see limb darkening? Can you see any granulation?

27 If you have access to an H_α filter attached to a telescope specially designed for safe viewing of the Sun, use this instrument to examine the solar surface. How does the appearance of the Sun differ from that in white light? What do sunspots look like in H_α? Can you see any prominences? Can you see any filaments? Are the filaments in the H_α image near any sunspots seen in white light?

For further reading

Eddy, J. *A New Sun.* NASA SP-402, 1979 • This lavishly illustrated book gives a superb summary of our understanding of the Sun.

———. "The Case of the Missing Sunspots." *Scientific American*, May 1977 • This article by a noted solar astronomer discusses evidence for long periods when no sunspots were seen.

Foukal, P. "The Variable Sun." *Scientific American*, February 1990 • This up-to-date article discusses recent observations of the Sun's variability and its effects on Earth.

Frazier, K. *Our Turbulent Sun.* Prentice-Hall, 1983 • This well researched report by a noted science writer describes the irregularity and variability of the Sun's energy output.

Friedman, H. *Sun and Earth.* Scientific American Library, 1986 • This excellent, beautifully illustrated book conveys the excitement and adventure of solar research.

Leibacher, J. W., et al. "Helioseismology." *Scientific American*, September 1985 • This lucid article explains how astronomers can learn about the structure, composition, and dynamics of the Sun's interior from oscillations visible on its surface.

Harvey, J., et al. "GONG: To See Inside Our Sun." *Sky & Telescope*, November 1987 • This article describes the Global Oscillation Network Group (GONG), a worldwide program of scientists exploring the solar interior by means of the Sun's naturally occurring oscillations.

Levine, R. "The New Sun." In J. Cornell and P. Gorenstein, eds. *Astronomy from Space: Sputnik to Space Telescope.* MIT Press, 1983 • This article surveys the advantages of solar astronomy from space.

Mitton, S. *Daytime Star.* Scribner's, 1981 • This easily readable book presents a completely nontechnical introduction to the Sun.

Nichols, R. "Solar Max: 1980–89." *Sky & Telescope,* December 1989 • This brief article describes the Solar Maximum Mission (SMM) satellite, which made many important contributions to our understanding of the Sun during the 1980s.

Noyes, R. *The Sun, Our Star.* Harvard University Press, 1982 • This clear, well presented introduction to the Sun includes interesting chapters on the climate and solar energy.

Robinson, L. "The Disquieting Sun: How Big, How Steady?" *Sky & Telescope,* April 1982 • This article describes controversial observations that the Sun is shrinking.

Wallenhorst, S. "Sunspot Numbers and Solar Cycles." *Sky & Telescope,* September 1982 • This article discusses the counting of sunspots and the plotting of the solar cycle.

Wentzel, D. *The Restless Sun.* Smithsonian Institution Press, 1989 • This lucid, up-to-date, stimulating account of the Sun emphasizes the wealth of activity—from sunspots to flares—that our star so vividly displays.

Wolfson, R. "The Active Solar Corona." *Scientific American,* February 1983 • This article describes the dynamic processes that occur in the solar corona as rarified gases and magnetic field interact on a grand scale.

CHAPTER 19

The Nature of Stars

Although they appear only as brilliant pinpoints of light, stars are massive spheres of glowing gas, much like our Sun. Using straightforward techniques, astronomers have measured the distances to many of the nearer stars. From these data, they can calculate the true brightnesses of the stars. Basic information about stars also comes from their colors and spectra, which give clues to a wide range of such stellar properties as surface temperatures, chemical compositions, and luminosities. Stellar luminosities and surface temperatures come together in the all-important Hertzsprung–Russell diagram, which reveals the fundamental types of stars. Further essential facts come from binary stars, surprisingly common systems in which two stars orbit each other. By observing the motions of the two stars in a binary, astronomers can determine stellar masses. All this information gives us important insights into the essential nature of stars and forms the foundation of our understanding of the heavens.

A cluster of stars By analyzing starlight, we can determine a star's surface temperature, chemical composition, and luminosity. This photograph shows color differences in the star cluster NGC 3293. Reddish stars are comparatively cool, with surface temperatures around 3000 K. They are also quite luminous and have large diameters, typically 100 times larger than our Sun. Bluish and blue-white stars have much higher surface temperatures (15,000 to 30,000 K) but are roughly the same size as the Sun. (Anglo-Australian Observatory)

To the unaided eye, the night sky is spangled with thousands of stars, each appearing as a bright pinpoint of light. A telescope reveals many thousands of other stars too faint to be seen with the naked eye, but every star still appears only as a bright point of light in even the most powerful telescope. A star is a huge, massive ball of hot gas like our Sun, held together by its own gravity. We now know that some stars are larger than our Sun and some smaller; some stars are brighter than the Sun and some dimmer. Some stars are hotter than the Sun, but others are cooler.

The quest for information about the masses, luminosities, temperatures, and chemical compositions of the stars has been a major preoccupation of twentieth-century astronomers. In recent years, a remarkably complete picture of the stars has emerged. By understanding the stars, we gain insight into our relationship to the universe as a whole and our place in the cosmic scope of space and time.

19-1 Careful measurements of the positions and motions of stars reveal stellar distances and space velocities

Looking up into the nighttime sky, you can see hundreds upon hundreds of stars. Some appear quite bright, but most are rather dim. As you gaze up at this starry panorama, one of the questions you might ask is: How far away are the stars? Are they relatively nearby or inconceivably remote? The apparent brightnesses of the stars do not indicate their distances. A star that looks dim might actually be a brilliant star that happens to be extremely far away. To determine the distances to the stars, astronomers use geometric techniques that involve painstaking measurements of stellar positions and motions.

The most straightforward way of measuring stellar distances is based on **parallax**, which is the apparent displacement of an observed object because of a change in the observer's point of view. You experience parallax when nearby objects appear to shift their positions against a distant background as you move from one place to another (see Figure 19-1). Stars exhibit the same phenomenon. As the Earth orbits the Sun, nearby stars appear to move back and forth against the background of more distant stars.

The distance to a star can be determined by measuring the star's parallax. The parallax (*p*) of a star is half the angle through which the star's apparent position shifts as the Earth moves from one side of its orbit to the other (see Figure 19-2). If the angle *p* is measured in seconds of arc, then the distance *d* to the star in parsecs is given by

$$d = \frac{1}{p}$$

For example, a star whose parallax is 0.5 arc sec is 2 parsecs from Earth. This simple relationship between parallax and

Figure 19-1 Parallax *Imagine looking at some nearby object (a tree) as seen against a distant background (mountains). If you move from one location to another, the nearby object will appear to shift its location with respect to the distant background scenery. This familiar phenomenon is called parallax.*

distance in parsecs is one of the main reasons why astronomers usually measure cosmic distances in parsecs rather than light years. (Recall that 1 parsec is the distance at which the radius of the Earth's orbit about the Sun, 1 AU, subtends an angle of 1 second of arc. One parsec equals 3.26 light years, and 1 light year is nearly 10^{13} km. See Figure 1-13.)

The first parallax measurement was accomplished by the German astronomer-mathematician Friedrich Wilhelm Bes-

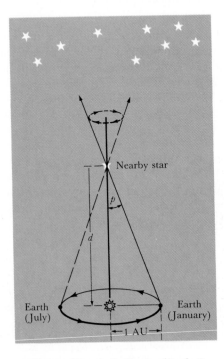

Figure 19-2 Stellar parallax *As the Earth orbits the Sun, a nearby star appears to shift its position against the background of distant stars. The parallax (p) of the star is equal to the angular radius of the Earth's orbit as seen from the star.*

sel. He found the parallax of 61 Cygni to be $\frac{1}{3}$ arc sec, and so its distance is about 3 pc from Earth. The nearest star, Proxima Centauri, has a parallax of 0.762 arc sec, and thus its distance is 1.31 pc. The parallax method therefore involves the measurement of extremely tiny angles. For instance, the parallax of Proxima Centauri is comparable to the angular diameter of a dime seen from a distance of 3 miles.

Parallaxes smaller than about $\frac{1}{20}$ arc sec are difficult to measure accurately from Earth-based observatories. Therefore the parallax method gives reliable distances only for stars nearer than about 20 pc. There are nearly 2000 stars within this range, half of which have had their parallaxes measured to a high degree of precision. Most of these nearby stars are far too dim to be seen with the naked eye. The majority of the familiar, bright stars in the nighttime sky are too far away to exhibit parallaxes measurable from the Earth's surface.

Parallax measurements made by an Earth-orbiting satellite would be unhampered by our atmosphere, permitting astronomers to determine the distances to stars well beyond the reach of ground-based observations. In 1989, the European Space Agency (ESA) launched a satellite called *Hipparcos* (an acronym for HIgh Precision PARallax COllecting Satellite), specifically designed to measure parallaxes as small as 0.002 arc second. Although the satellite failed to achieve its proper orbit, astronomers are still hopeful that *Hipparcos* will be able to complete much of its mission to measure the distances to stars as far away as 500 pc. During the 1990s, astronomers will increasingly turn to space-based observations to determine stellar distances.

In addition to annual cyclic motion due to parallax, many stars change their positions very slowly due to their own true motion through space. The component of this motion perpendicular to our line of sight is called **proper motion**. Because proper motion does not repeat itself yearly, it can be distinguished from the cyclic apparent motion due to paral-

lax. The proper motion of a star, usually designated by the Greek letter μ (mu), is measured in arc sec per year. For instance, the star with the largest proper motion, called Barnard's star after the American astronomer Edward E. Barnard, exhibits a proper motion of 10.3 seconds of arc per year (see Figure 19-3). It takes Barnard's star only 200 years to travel the apparent angular diameter of the full moon. This huge proper motion is far greater than that of any other star. Although the relative positions of stars are slowly changing, their proper motions are so small that the shapes and sizes of the constellations look the same as they did thousands of years ago.

In order to determine the proper motion of a star, an astronomer must carefully measure the star's position for many years. During a single night at a telescope, however, an astronomer can use the Doppler shift to determine a star's motion parallel to our line of sight. As discussed in Chapter 5 (recall Figure 5-20), the Doppler shift is a change in wavelength caused by the relative motion between a light source and an observer. If a star is approaching you, the wavelengths of all of its spectral lines are decreased (blueshifted); if the star is receding, the wavelengths are increased (redshifted). The size of the wavelength shift ($\Delta\lambda$) of a specific spectral line whose rest wavelength is λ_0 is given by

$$\frac{\Delta\lambda}{\lambda_0} = \frac{v_r}{c}$$

where v_r, called the star's **radial velocity**, is the component of the star's velocity parallel to our line of sight and c is the speed of light. A specific example showing how this equation is used to determine the radial velocity of the star Vega was given in Chapter 5 (recall Figure 5-21).

The proper motion and radial velocity of a star can be combined to give the size and direction of the total velocity of the star in space. As shown in Figure 19-4, a star's total velocity v consists of components parallel and perpendicular

August 24, 1894

May 30, 1916

Figure 19-3 Barnard's star These two photographs, which were taken 22 years apart, show the proper motion of Barnard's star in the constellation of Ophiuchus. In addition to having the largest known *proper motion (10.3" per year), Barnard's star is one of the nearest stars, being only 5.9 light years from Earth. (Yerkes Observatory)*

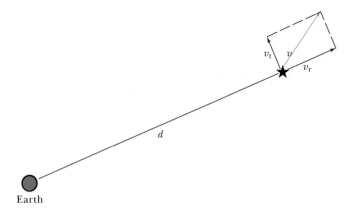

Figure 19-4 *Components of a star's space velocity It is useful to separate a star's velocity in space (v) into a component parallel to our line of sight plus a component perpendicular to our line of sight. The radial velocity (v_r) is determined from the Doppler shift of the star's spectral lines. The tangential velocity (v_t) can be determined from the star's proper motion if its distance from Earth is known. By combining these two components, the star's speed and direction of motion can be deduced.*

to our line of sight. The component parallel to our line of sight is just the radial velocity (v_r), determined from measurements of the Doppler shifts of the star's spectral lines. The component perpendicular to our line of sight, usually called the star's **tangential velocity** (v_t), can be directly related to the star's proper motion (μ) if the distance to the star is known. If the star's distance d is measured in parsecs and its proper motion is measured in arc seconds per year, then the tangential velocity in km/s is given by

$$v_t = 4.74\mu d$$

Finally we note that the tangential and radial components of a star's velocity form a right triangle. We can therefore use the Pythagorean theorem to write a concise formula for the star's total velocity in terms of its radial velocity and proper motion as follows:

$$v = \sqrt{v_r^2 + v_t^2} = \sqrt{v_r^2 + 22.5\mu^2 d^2}$$

Parallax is a straightforward method of distance determination that involves the simple geometry of surveying and the idea that the Earth goes around the Sun. A second technique of distance determination, called the **moving cluster method**, makes use of the fact that the stars in a cluster have the same average space velocity with respect to the Sun. As a result of this shared motion, all the stars in a cluster seem to be headed toward the same point on the celestial sphere. By knowing the convergent point and by measuring the proper motions and radial velocities of stars in a cluster, the cluster's distance from Earth can be determined as described in Box 19-1. The moving cluster method is an important technique because it gives reliable stellar distances out to about 100 pc. Unfortunately, there are only a few relatively nearby clusters where this method can be used with precision.

The measurement of the distances and motions of stars is a cornerstone of modern astronomy. Beginning with such measurements, astronomers go on to determine many properties of stars as well as the size, shape, and behavior of the entire Galaxy. Astronomers' attempts to determine the overall size, age, and structure of the entire universe are also based on measurements of stellar distances. Small inaccuracies in the measurements of the distances to nearby stars can translate into huge errors when the whole universe is concerned. For these reasons, astronomers are concerned about the accuracy of parallax measurements and are continually trying to perfect their distance-measuring techniques.

19-2 A star's luminosity can be determined from its apparent magnitude and distance

The system of magnitudes that astronomers use to denote the brightnesses of stars was invented in ancient Greece by the astronomer Hipparchus. The brightest stars Hipparchus could see in the sky he called first-magnitude stars. Those about one-half as bright he called second-magnitude stars, and so forth to sixth-magnitude stars, the dimmest ones he could see. After telescopes came into use, astronomers extended Hipparchus's magnitude scale to larger magnitudes to describe the dimmer stars now visible through their instruments.

In the nineteenth century, techniques were developed for measuring more exactly the amount of light energy arriving from a star. The branch of science dealing with such measurements is called **photometry**. Astronomers then set out to define the magnitude scale more precisely. Measurements showed that a first-magnitude star is about 100 times brighter than a sixth-magnitude star. In other words, it would take 100 stars of magnitude +6 to provide as much light energy as we receive from a single star of magnitude +1. As a result, the magnitude scale was redefined so that a magnitude difference of 5 corresponds exactly to a factor of 100 in the amount of light energy received. A magnitude difference of 1 therefore corresponds to a factor of 2.512 in light energy, because

$$2.512 \times 2.512 \times 2.512 \times 2.512 \times 2.512 = (2.512)^5 = 100$$

Thus, for example, it takes about $2\frac{1}{2}$ third-magnitude stars to provide as much light as we receive from a single second-magnitude star.

Figure 19-5 illustrates the modern apparent magnitude scale. Note, for example, that the dimmest stars visible through a pair of binoculars have a magnitude of +10 and the dimmest stars that can be photographed with a large telescope are magnitude +24. These magnitudes are properly called **apparent magnitudes** because they describe how bright an object appears to an Earth-based observer.

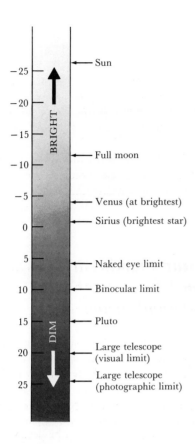

Apparent magnitude is a measure of the light energy arriving at the Earth. To understand the stars, astronomers need to know how bright the stars really are. Absolute magnitude is a measure of a star's energy output.

The **absolute magnitude** of a star is defined as the apparent magnitude it would have if it were located at a distance of exactly 10 parsecs from the Earth. For example, if the Sun were moved to a distance of 10 parsecs from the Earth, it would have an apparent magnitude of +4.8. The absolute magnitude of the Sun is thus +4.8. The absolute magnitudes of other stars range from roughly −10 for the brightest to +15 for the dimmest. The Sun's absolute magnitude is about in the middle of this range, providing us our first hint that the Sun is an average star.

Absolute magnitude is a very informative quantity, because it indicates the intrinsic brightness of a star. In contrast, apparent magnitude tells only how bright a star appears in the sky. To appreciate how astronomers determine the absolute magnitude of a star from its apparent magnitude, you must first realize that the farther away a source of light is, the dimmer it appears.

Imagine a source of light, such as a light bulb or a star. Suppose that L is the amount of energy emitted by the light source each second. This value of L is often called the *luminosity* of the source and is usually measured in joules per second or watts. For example, the Sun's luminosity is 3.90×10^{26} watts.

As the light energy moves away from its source, it spreads out over increasingly larger regions of space (see Figure 19-6). Imagine a sphere of radius r centered on the light source. The amount of energy passing through each square

Figure 19-5 The apparent magnitude scale Astronomers denote the brightnesses of objects in the sky by their apparent magnitudes. Most stars visible to the naked eye have magnitudes in the range +1 to +6. Photography through large telescopes can reveal stars as faint as magnitude +24.

Figure 19-6 The inverse-square law This drawing shows how the same amount of radiation from a light source must illuminate an ever-increasing area as distance from the light source increases. Because the light is spread out over an area that increases as the square of the distance, the apparent brightness decreases as the square of the distance.

Box 19-1 The moving cluster method

An **open cluster** is a loose collection of a few tens to a few hundreds of stars. The Pleiades in the constellation of Taurus (the bull) is a fine example of an open cluster easily seen with the naked eye. The nearest open cluster is the Hyades, whose stars comprise the V-shaped outline of the head of Taurus. The accompanying photograph shows the Pleiades and the Hyades. The brightest star in the V, Aldebaran, is not, however, a member of the cluster.

The distances to the nearest open clusters can be determined by the **moving cluster method**, which makes use of the fact that the stars in a cluster are all moving together in nearly the same direction in space. For instance, the star chart below shows the proper motions of the stars in the Hyades. The lengths of the arrows indicate how far each star will move over the next 10,000 years. Note that all the stars seem to be headed toward the same point. This convergence is the result of perspective, the same effect that explains why parallel train tracks stretching across a prairie seem to meet at the horizon.

Let θ (theta) be the angular distance between a particular star and the cluster's convergent point K, as shown in the diagram. Also shown are the star's space velocity v, radial velocity v_r, and tangential velocity v_t. The star's space velocity is parallel to our line of sight toward point K, and thus the angle between v and v_r is also θ. We can therefore write

$$v_r = v \cos\theta$$
$$v_t = v \sin\theta$$

As noted in the text, the radial velocity v_r can be measured from the Doppler shift of spectral lines in the star's spectrum. The tangential velocity v_t is related to the proper motion μ and the star's distance d from Earth by the equation

$$v_t = 4.74\mu d$$

where μ is measured in arc sec per year and d is measured in parsecs. A formula for the distance to the star in terms of the observable quantities v_r, μ, and θ can be derived as follows:

$$d = \frac{v_t}{4.74\mu} = \frac{v \sin\theta}{4.74\mu} = \frac{0.211 v_r}{\mu \tan\theta}$$

By this method, the distances of individual stars can be determined from the motion of the cluster as a whole. The distance to the Hyades obtained in this way is about 40 pc. The distance to the Hyades cluster is the most accurately determined of all stellar distances and provides the basis upon which all other astronomical distances are determined.

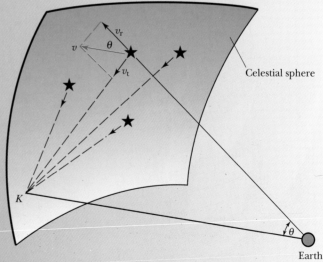

Celestial sphere

K

Earth

meter of the sphere each second is simply the total luminosity of the source (L) divided by the sphere's surface area ($4\pi r^2$). The resulting quantity, called the **apparent brightness** of the light (b), is measured in joules per square meter per second according to the following formula:

$$b = \frac{L}{4\pi r^2}$$

This relationship is called the **inverse-square law**, because the apparent brightness of light that an observer can see or measure is inversely proportional to the square of the observer's distance from the source. If you double your distance from a light source, its radiation is spread out over an area four times larger, so the apparent brightness you see is decreased by a factor of four. Similarly, at triple the distance the apparent brightness decreases by a factor of nine, as Figure 19-6 demonstrates.

Using the inverse-square law, astronomers have derived the following equation to relate a star's apparent magnitude (m), its absolute magnitude (M), and its distance (d, measured in parsecs) from the Earth:

$$m - M = 5 \log d - 5$$

It is important to realize that if you know any two of these quantities, such as apparent magnitude and distance, you can calculate the third one (in this case absolute magnitude). For instance, astronomers measure the apparent magnitude of a nearby star, find its distance by measuring its parallax, then calculate its absolute magnitude. (For the reader familiar with logarithms, Box 19-2 shows the derivation of the equation relating m, M, and d, with several examples.)

Many scientists prefer to speak of a star's luminosity rather than its absolute magnitude, because luminosity is a direct measure of the star's energy output. There is a simple relationship between absolute magnitude and luminosity (see Box 19-3), which astronomers can use to convert from one to the other as they see fit. For convenience, stellar luminosities are expressed as multiples of the Sun's luminosity ($L_\odot$), which equals 3.90×10^{26} watts. The most luminous stars (with an absolute magnitude of -10) have luminosities of 10^6 $L_\odot$, which means that each of these stars has the energy output of a million Suns. The least luminous stars (absolute magnitude = $+15$) have luminosities of 10^{-4} $L_\odot$.

19-3 A star's color reveals its surface temperature

One of the first things you notice when comparing stars in the nighttime sky is their differences in apparent magnitude. More careful examination, even with the naked eye, reveals that the stars also have different colors. For example, in the constellation of Orion you can easily note the difference between reddish Betelgeuse and bluish Rigel (examine Figure 2-1).

As discussed in Chapter 5, a star's color is directly associated with its surface temperature through such relationships as Wien's law. The intensity of light from a cool star peaks at long wavelengths, making the star look red (see Figure 19-7a). A hot star's intensity curve is skewed toward short wavelengths, so the star looks blue (see Figure 19-7c). The maximum intensity of a star with an intermediate temperature, such as the Sun, occurs near the middle of the visible spectrum, giving the star a yellowish color (see Figure 19-7b).

To measure accurately the colors of the stars, a technique called **photoelectric photometry** has been developed by as-

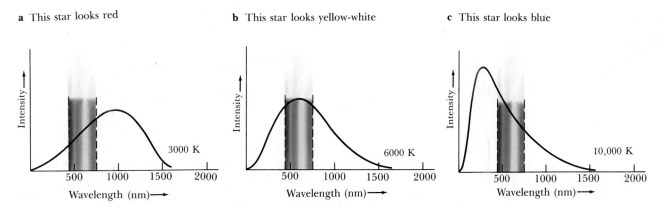

a This star looks red   **b** This star looks yellow-white   **c** This star looks blue

Figure 19-7 Temperature and color *This schematic diagram shows the relationship between the color of a star and its surface temperature. The intensity of light emitted by three hypothetical stars is plotted here against wavelengths (compare Figure 5-8). The range of visible wavelengths is indicated. The way in which a star's intensity curve is skewed determines the dominant color of its visible light.*

Box 19-2 Magnitude and brightness

The concept of a star's magnitude dates back to the days of ancient Greek astronomy. Modern astronomers often prefer to talk instead about a star's **apparent brightness**. The apparent brightness of a star is the number of joules of starlight energy arriving at each square meter on Earth each second; it is usually expressed in joules per square meter per second ($J\,m^{-2}\,s^{-1}$). As explained earlier in this chapter, each step in magnitude corresponds to a factor of 2.512 in brightness. In other words, your eyes receive 2.512 times more energy per square meter per second from a third-magnitude star than they do from a fourth-magnitude star.

Consider two stars with magnitudes m_1 and m_2 and brightnesses b_1 and b_2. The ratio of their apparent brightnesses (b_1/b_2) corresponds to a difference in their magnitudes ($m_2 - m_1$). Because each step in magnitude corresponds to a factor of 2.512 in brightness, we can construct the table on the right:

In general, the relationship between ($m_2 - m_1$) and (b_1/b_2) can be written as

$$\frac{b_1}{b_2} = 100^{(m_2 - m_1)/5}$$

Taking the logarithm of both sides of this equation and rearranging terms, we obtain

$$m_2 - m_1 = 2.5 \log \left(\frac{b_1}{b_2} \right)$$

The usefulness of these relationships and equations is best illustrated by examples.

Example: At greatest brilliancy, Venus has a magnitude of -4. Compare the apparent brightness of Venus with that of the dimmest stars visible to the naked eye.

The dimmest stars visible to the naked eye have a magnitude of $+6$. The magnitude difference is therefore $+6 - (-4) = 10$. This difference corresponds to a brightness ratio of $(2.512)^{10} = 10^4$. It would thus take 10,000 sixth-magnitude stars to shine as brilliantly as Venus.

Example: In August 1975, a nova appeared in the constellation of Cygnus (the swan). A nova is a violent outburst involving a certain kind of star. Its magnitude changed in just two days from $+15$ to $+2$. By what factor did its brightness increase?

The change in magnitude ($m_2 - m_1$) is 13. Thus the ratio of brightness is $b_1/b_2 = 100^{13/5} = 158,500$. In two days, the nova's brightness therefore increased by a factor of nearly 160,000.

Example: The variable star RR Lyrae periodically doubles its light output. How much does its magnitude change?

The brightness of the star varies by a factor of 2, so $b_1/b_2 = 2$. Thus $m_2 - m_1 = 2.5 \log(2) = 0.7$. The magnitude of RR Lyrae thus varies periodically by seven-tenths.

Magnitude difference ($m_2 - m_1$)	Ratio of apparent brightness (b_1/b_2)
1	2.512
2	$(2.512)^2 = 6.31$
3	$(2.512)^3 = 15.85$
4	$(2.512)^4 = 39.82$
5	$(2.512)^5 = 100$
10	$(2.512)^{10} = 10^4$
15	$(2.512)^{15} = 10^6$
20	$(2.512)^{20} = 10^8$

With these relationships, we can derive a useful equation between a star's apparent magnitude, its absolute magnitude, and its distance from Earth. Let m and b be the apparent magnitude and the apparent brightness, respectively, of a star at a distance d from Earth. Let M and B be the apparent magnitude and the apparent brightness, respectively, of the star if it were 10 parsecs from Earth. By definition, M is the star's absolute magnitude. From the general relationship we have established between apparent brightness and apparent magnitude,

$$m - M = 2.5 \log \left(\frac{B}{b} \right)$$

According to the inverse-square law, the apparent brightness of a light is inversely proportional to the square of the distance between the light and the observer. Thus,

$$\frac{B}{b} = \left(\frac{d}{10} \right)^2$$

Combining these two equations and rearranging terms, we obtain

$$M = m - 5 \log \left(\frac{d}{10} \right)$$

where d is measured in parsecs. This is often written as

$$m - M = 5 \log d - 5$$

where ($m - M$) is called the **distance modulus**.

Example: Consider Capella, a bright, nearby star. Its apparent magnitude is $+0.05$, and its distance is 14 parsecs. Thus, its absolute magnitude is $M = 0.05 - 5 \log (1.4) = -0.7$. Comparing this value to the Sun's absolute magnitude ($+4.8$), we see that Capella is actually 5.5 magnitudes brighter than the Sun. From $m_2 - m_1 = 5.5 = 2.5 \log (b_1/b_2)$, we find $b_1/b_2 = 158$. Thus, Capella emits about 160 times the light energy of our Sun.

Box 19-3　Bolometric magnitude and luminosity

In determining a star's absolute magnitude, astronomers must make allowances for nonvisible light and for the Earth's atmosphere. The apparent magnitude of a star seen in the sky could be misleading if the star happens to emit a significant fraction of its radiation at nonvisible wavelengths. For example, a luminous hot star with a surface temperature of 35,000 K appears deceptively dim to our eyes, simply because most of the star's light is emitted at ultraviolet wavelengths. Furthermore, the Earth's atmosphere is opaque to many nonvisible wavelengths, so a sizable fraction of the light from both the hottest and the coolest stars simply does not penetrate the air to reach our eyes or telescopes.

To cope with this difficulty, astronomers have defined the **bolometric magnitude** of a star, the star's apparent magnitude as measured above the Earth's atmosphere and over all wavelengths. In recent years, astronomical satellites have allowed us to determine the bolometric magnitudes of many stars.

The absolute magnitude of a star as deduced from its bolometric magnitude is called its **absolute bolometric magnitude** (M_{bol}). This quantity is always brighter than the star's absolute visual magnitude (M) as deduced from ground-based observations at visible wavelengths alone. By comparing satellite and ground-based data, astronomers have figured out how much they must subtract from a star's absolute visual magnitude to get its absolute bolometric magnitude. This factor, called the **bolometric correction** (BC), is given in the simple equation

$$M_{bol} = M - BC$$

The bolometric correction is quite large for both the hottest stars and the coolest stars. For example, for stars hotter than 20,000 K or cooler than 3000 K, the bolometric correction is 3 magnitudes or greater. For stars such as the Sun, which emit an overwhelming percentage of their radiation at visible wavelengths, the bolometric correction is almost zero. The Sun's absolute bolometric magnitude is +4.72, whereas its absolute visual magnitude is +4.83.

A star's absolute bolometric magnitude is directly related to the star's luminosity (L). From the equations in Box 19-1, it is possible to show that

$$M_{bol} = 4.72 - 2.5 \log (L/L_\odot)$$

By knowing a star's M_{bol} and using this equation, astronomers can calculate exactly how much energy is being released from the star's surface each second.

Example:　Consider Sirius, which has an apparent visual magnitude of −1.46, making it the brightest star in the night sky. The distance to Sirius is 2.7 parsecs, and Sirius's surface temperature is about 10,000 K, corresponding to a bolometric correction of 0.6 magnitude. Suppose that you want to know the luminosity of Sirius.

First, calculate the absolute visual magnitude of Sirius:

$$M = m - 5 \log \left(\frac{d}{10}\right)$$
$$= -1.46 - 5 \log 0.27$$
$$= -1.46 - 5(9.4314 - 10)$$
$$= 1.4$$

The absolute magnitude of Sirius is 1.4 and the bolometric correction is 0.6, so we find that the absolute bolometric magnitude of Sirius is $M_{bol} = M - BC = 1.4 - 0.6 = 0.8$. We know that $M_{bol} = 4.72 - 2.5 \log (L/L_\odot)$. This expression is the same as

$$\log (L/L_\odot) = \frac{4.72 - M_{bol}}{2.5}$$
$$= \frac{4.72 - 0.8}{2.5}$$
$$= 1.6$$

Finally, we obtain $L/L_\odot = 10^{1.6} = 40$. Thus, we have determined that Sirius is about 40 times more luminous than the Sun.

---

tronomers. This process uses a light-sensitive device (such as a CCD; recall Figure 6-16) at the focus of a telescope, with a standardized set of colored filters. The most commonly used filters are the UBV filters. Each of the three **UBV filters** is transparent in one of three broad wavelength bands: the ultraviolet (U), the blue (B), and the central yellow (V) region of the visible spectrum (see Figure 19-8). (The transparency of the V—for "visual"—filter mimics the sensitivity of the human eye.)

To do photometry, the astronomer aims a telescope at a star and measures the intensity of the starlight that passes through each of the filters. This procedure gives three apparent magnitudes, usually designated by the capital letters U,

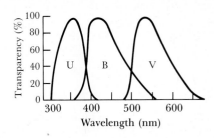

Figure 19-8　Light transmission through UBV filters　*This graph shows the wavelength ranges over which the standardized U, B, and V filters are transparent to light. The U filter is transparent to the near-ultraviolet. The B filter is transparent from about 380 to 550 nm, and the V filter is transparent from about 500 to 650 nm.*

Table 19-1 The UBV magnitudes and color indices of selected stars

Star	V	B	U	(B − V)	(U − B)	Apparent color
Bellatrix (γ Ori)	1.64	1.42	0.55	−0.22	−0.87	Blue
Regulus (α Leo)	1.35	1.24	0.88	−0.11	−0.36	Blue-white
Sirius (α CMa)	−1.46	−1.46	−1.52	0.00	−0.06	Blue-white
Megrez (δ UMa)	3.31	3.39	3.46	+0.08	+0.07	White
Altair (α Aql)	0.77	0.99	1.07	+0.22	+0.08	Yellow-white
Sun	−26.78	−26.16	−26.06	+0.62	+0.10	Yellow-white
Aldebaran (α Tau)	0.85	2.39	4.29	+1.54	+1.90	Orange
Betelgeuse (α Ori)	0.50	2.35	4.41	+1.85	+2.06	Red

B, and V, for the star. The astronomer then compares the intensity of starlight in neighboring wavelength bands by subtracting one magnitude from another to form the combinations (B − V) and (U − B), which are called the star's **color indices**. The UBV magnitudes and color indices for several representative stars are given in Table 19-1.

A color index tells you how much brighter or dimmer a star is in one wavelength band than in another. For example, the (B − V) color index tells you how much brighter or dimmer a star appears through the B filter than through the V filter.

The color index is important, because it tells you the star's surface temperature. If a star is very hot, its radiation is skewed toward the short-wavelength ultraviolet, which makes the star bright through the U filter, dimmer through the B filter, and dimmest through the V filter. The star

Regulus (see Table 19-1) is such an example. Alternatively, if the star is cool, its radiation peaks at long wavelengths, making the star brightest through the V filter, dimmer through the B filter, and dimmest through the U filter. The stars Aldebaran and Betelgeuse are examples.

The graph in Figure 19-9 gives the relationship between a star's (B − V) color index and its temperature. If you know a star's (B − V) color index, you can use this graph to find the star's surface temperature. For example, the Sun's (B − V) index is +0.62, which corresponds to a surface temperature of 5800 K.

A word of caution here is in order. The curve in Figure 19-9 was derived mathematically from the laws of blackbody radiation, but stars are not perfect blackbodies. The temperature deduced from a star's color index will thus be slightly different from temperatures deduced by other techniques. As a result, astronomers are careful to cite the particular method used to determine a given stellar temperature.

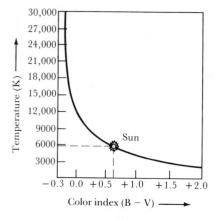

Figure 19-9 Blackbody temperature versus color index *The (B − V) color index is the difference between the B and V magnitudes of a star. If the star is hotter than about 10,000 K, it is a very bluish star with a (B − V) index of less than zero. If a star is cooler than about 10,000 K, its (B − V) index is greater than zero. The Sun's (B − V) index is about 0.62, which corresponds to a temperature of 5800 K. After measuring a star's B and V magnitudes, an astronomer can estimate the star's surface temperature from a graph like this one.*

19-4 Stars are classified according to the appearance of their spectra in a way that is directly related to their surface temperatures

The field of stellar spectroscopy was born in the 1860s when the Italian astronomer Angelo Secchi attached a spectroscope to his telescope and pointed it toward the stars. Secchi discovered that absorption lines are often found in the spectra of stars. He was also the first astronomer to classify stellar spectra into **spectral types**, according to their appearance.

At first glance, stellar spectra seem to come in a bewildering variety. Some stellar spectra prominently show the Balmer lines of hydrogen. Other spectra exhibit many absorption lines of calcium and iron. Still others are dominated by broad absorption features caused by molecules such as titanium oxide. To cope with this diversity, astronomers grouped similar-appearing stellar spectra in classes. According to one popular classification scheme that emerged in the late 1800s, a star was assigned a letter from A through P, according to the strength or weakness of the hydrogen Balmer lines in the star's spectrum. The A stars have the strongest Balmer lines, the P stars the weakest.

Astronomers who pioneered stellar spectroscopy were aware that stellar spectra must contain important information about stars, primarily because of the work of chemists Robert Bunsen and Gustav Kirchhoff. Bunsen and Kirchhoff had demonstrated that patterns of spectral lines are caused by chemicals (recall Figure 5-12). Specifically, Kirchhoff's third law explains that absorption lines are seen when the spectrum of a hot, glowing object is viewed through a cool gas (recall Figure 5-13). As we have seen, a star shines with nearly a perfect blackbody spectrum. This continuous spectrum of radiation, or **continuum**, is produced deep within a star's atmosphere where the gases are hot and dense. As this light moves outward through the cooler, less dense layers of the star's upper atmosphere, atoms absorb radiation at specific wavelengths, thereby producing spectral lines that astronomers observe.

Nineteenth-century science was not capable of explaining how the spectral lines of a particular chemical are affected by the temperature and density of the gas where the lines are formed. Nevertheless, a team of astronomers under the supervision of Edward C. Pickering at Harvard College Observatory forged ahead with a monumental project of examining the spectra of thousands of stars with the goal of developing a self-consistent system of spectral classification in which various spectral features change smoothly from one spectral class to the next.

Pickering's spectral classification project was financed by the estate of Henry Draper, a wealthy physician and amateur astronomer who, in 1872, first photographed stellar absorption lines. Principal researchers on the project were Williamina P. Fleming, Antonia C. Maury, and Annie Jump Cannon. As a result of their efforts, many of the original A-through-P classes were dropped and others were consolidated. The remaining spectral classes were reordered in the sequence: **OBAFGKM.** Students sometimes memorize this as "Oh, Be A Fine Girl, Kiss Me!"

Annie Cannon found it useful to further subdivide the original OBAFGKM sequence into finer steps. These additional steps are indicated by attaching an integer from 0 through 9 to the original letter . Thus, for example, we have . . . , F8, F9, G0, G1, G2, . . . , G9, K0, K1, K2, ... In going from one spectral type to the next, the strengths of spectral lines change in a smooth fashion. For instance, absorption lines of hydrogen become increasingly prominent as you go from spectral type B0 to A0. From A0 onward through F and G types, the hydrogen lines weaken and almost fade from view. Representative spectra of each spectral type are shown in Figure 19-10. The Sun, whose spectrum is dominated by calcium and iron, is a G2 star.

The Harvard project culminated in the *Henry Draper Catalogue* published between 1918 and 1924 by Cannon and Pickering. It listed 225,300 stars, each of which had been personally classified by Cannon. Meanwhile, physicists had been making important discoveries about the structure of atoms. Rutherford had demonstrated that atoms have nuclei (recall Figure 5-14) and Bohr made the brilliant hypothesis that electrons move along discrete orbits around the nuclei of atoms (recall Figure 5-17). These discoveries about

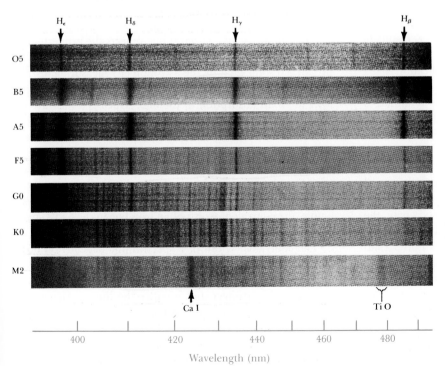

Figure 19-10 Principal types of stellar spectra A star's spectrum is shown in the middle of each of these seven strips. The hydrogen lines are strongest in A stars, which have surface temperatures of about 10,000 K. The spectra of G and K stars exhibit numerous lines caused by metals, indicating temperatures in the range of 4000 to 6000 K. The broad, dark bands in the spectrum of an M star are caused by titanium oxide, which can exist only when the temperature is below about 3500 K. (Courtesy of N. Houk, N. Irvine, and D. Rosenbush)

atoms gave scientists the conceptual and mathematical tools with which to interpret and understand stellar spectra and contributed to the later development of quantum mechanics.

In the late 1920s, Harvard astronomer Cecilia Payne and physicist Meghnad Saha in India succeeded in explaining how a star's spectrum is affected by the star's surface temperature. They demonstrated that the OBAFGKM spectral sequence is actually a sequence in temperature. The hottest stars are O stars. The absorption lines seen in the spectra of O stars could only occur if these stars have surface temperatures as high as about 35,000 K. M stars are the coolest stars. The spectral features seen in the spectrum of M stars are consistent with stellar surface temperatures of about 3000 K.

To see why the appearance of a star's spectrum is profoundly affected by the star's surface temperature, consider hydrogen. Hydrogen is by far the most abundant element in the universe, accounting for about three-quarters of the mass of a typical star. However, hydrogen lines do not necessarily show up in a star's spectrum. Recall that these Balmer lines are produced when an electron in the $n = 2$ orbit of hydrogen absorbs a photon having the energy to lift it to a higher orbit. If the star is much hotter than 10,000 K, high-energy photons pouring out of the star's interior will easily knock electrons out of the hydrogen atoms in the star's outer layers. This process ionizes the gas. When hydrogen's only electron is torn away, no spectral lines can be produced.

Conversely, if the star is much cooler than 10,000 K, the majority of the photons escaping from the star possess too little energy to boost many electrons up from the ground state to the $n = 2$ orbit of the hydrogen atoms. These unexcited atoms then also fail to produce Balmer lines. In summary, a star must be hot enough to excite the electrons out of their ground state, but not so hot that the atoms become ionized. A stellar surface temperature of 10,000 K results in the strongest Balmer lines.

A prominent set of Balmer lines is a clear indication that a star's surface temperature is about 10,000 K. At other temperatures, the spectral lines of other elements dominate a star's spectrum. For example, around 25,000 K the spectral lines of helium are strong; at this temperature, photons have enough energy to excite helium atoms without tearing away the electrons altogether.

Hydrogen atoms have only one electron to be removed during ionization, but an atom of a heavier element has two or more electrons. When one electron is knocked away, the remaining electrons can take over to produce a new, distinctive set of spectral lines. For example, in stars hotter than about 30,000 K, one of the two electrons in a helium atom is torn away. The remaining electron then produces a set of spectral lines that is recognizably different from the lines produced by un-ionized helium. When the spectral lines of singly ionized helium appear in a star's spectrum, we know that the star has a surface temperature greater than 30,000 K.

The strength of a particular spectral line depends on both the ionization and the excitation of the atom responsible for that line. Ionization and excitation of the atoms in a gas depend on the temperature of the gas. Astronomers designate an un-ionized atom with a Roman numeral I; thus, H I is neutral hydrogen. A Roman numeral II is used to identify an atom with one electron missing; thus, He II is singly ionized helium (He^+). Similarly, Si III is doubly ionized silicon (Si^{2+}), whose atoms are each missing two electrons. Using this notation, the main characteristics of the spectral classes are described in Figure 19-11.

A collection of stellar spectra is also displayed in Figure 19-11. Instead of recording a star's spectrum on a photographic plate at the focus of a spectrograph, astronomers today prefer to use a light-sensitive solid-state device, such as a CCD, because it is much more sensitive than photographic film. All the spectra in Figure 19-11 were obtained with such a device, called an Intensified Reticon Scanner, which generates a plot of intensity versus wavelength. Absorption lines are seen as dips on the curve of the continuum, which has a shape quite like the blackbody curve at the temperature of the star's surface.

It is enlightening to display a spectrum as a plot of intensity versus wavelength, because it is then possible to study the shapes of individual spectral lines. As Figure 19-12 shows, a typical spectral line consists of a "core" flanked by "wings." The detailed shape of a spectral line, called the **line profile**, contains important information about a star. For instance, if a star is rotating, light from the approaching side is slightly blueshifted while light from the receding side is comparably redshifted. As a result, the star's spectral lines are broadened in a characteristic fashion. By measuring the shape of the spectral lines, astronomers can deduce how fast the star is rotating.

The true shape of a spectral line reflects the properties of a star's atmosphere, but the observed line profile is also broadened somewhat by the astronomer's measuring instruments. Instrumental effects do not, however, significantly alter the *total* absorption of the line, which is a measure of the energy deleted from the continuum across the entire spectral line, and so does not depend on details of the line's shape. Astronomers express the total absorption of a spectral line in terms of its **equivalent width**, which is the width of a completely dark rectangular line with the same total absorption as the observed line (see Figure 19-12). For example, the equivalent width of an iron line in the Sun's spectrum is about 0.01 nm.

The equivalent width of a spectral line depends on how many atoms in the star's atmosphere are in a state in which they can absorb the wavelength in question. The more atoms there are, the stronger and broader the line is. Hence, by analyzing a star's spectrum, an astronomer can determine the star's chemical composition.

The main characteristics of the Henry Draper spectral classification scheme are summarized in Figure 19-13, which

Type O   Blue stars with surface temperatures of 20,000 to 35,000 K. Spectra show multiple ionized atoms, especially He II, C III, N III, O III, Si V. He I is visible, but H I is weak.

Type B   Blue–white stars with surface temperatures of about 15,000 K. He II lines have disappeared, He I is strongest at B2. H I lines getting stronger. O II, Si II, and Mg II lines are visible.

Type A   White stars with surface temperatures of about 9000K. The H I lines dominate the spectrum and are strongest at A0. He I no longer visible. Neutral metal lines begin to appear.

Type F   Yellow-white stars with surface temperatures of about 7000 K. The H I lines are getting weaker while Ca II are getting stronger. Many other metals such as Fe I, Fe II, Cr II, and Ti II are getting stronger.

Type G   Yellow stars like the Sun with surface temperatures of about 5500 K. H I lines still getting weaker while Ca II lines are strongest at G0. Metal lines are getting stronger.

Type K   Yellow-orange stars with surface temperatures of about 4000 K. Spectrum is dominated by metal lines. Ca I getting stronger. TiO bands become visible at K5.

Type M   Red stars with surface temperatures of about 3000 K. TiO bands are very prominent. Ca I at 423 nm is very strong. Many neutral metal lines are seen. For stars cooler than M4, absorption by TiO is so severe that it is very difficult to find the continuum.

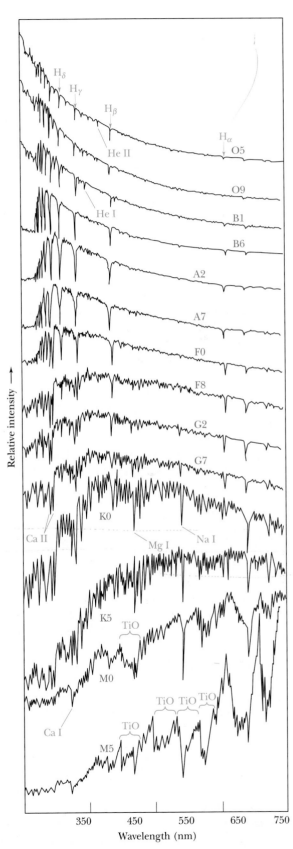

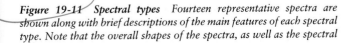

Figure 19-11 *Spectral types* *Fourteen representative spectra are shown along with brief descriptions of the main features of each spectral type. Note that the overall shapes of the spectra, as well as the spectral* lines, depend on the surface temperature of the stars. (Observations by G. Jacoby, D. Hunter, and C. Christian)

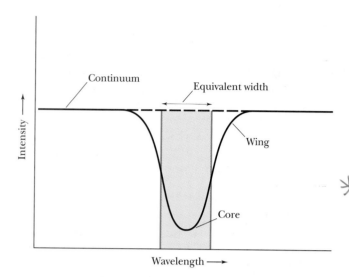

Figure 19-12 The profile of a spectral line *On a plot of intensity versus wavelength, an absorption line is a depression in a star's continuum spectrum. The equivalent width of a spectral line is the width of a completely dark rectangular line which has the same total absorption as the spectral line.*

plots the equivalent widths of spectral lines against spectral type. This graph consolidates the information that astronomers use to deduce the surface temperature or spectral type of a star from the intensity of the lines in its spectrum. For example, a star which exhibits strong Ca II and Fe I lines in its spectrum has a surface temperature around 4500 K.

19-5 Hertzsprung–Russell diagrams demonstrate that there are different kinds of stars

The first accurate measurements of stellar parallax were made in the mid-nineteenth century, at about the time astronomers had begun observing stellar spectra. During the

next half-century, observing techniques improved, and the spectral types and absolute magnitudes of many stars became known.

Around 1905, the Danish astronomer Ejnar Hertzsprung pointed out that a regular pattern appears when the absolute magnitudes of stars are plotted against their color indices on a graph. Almost a decade later, the American astronomer Henry Norris Russell discovered independently this same regularity, but in a graph using spectral types instead of color indices. Plots of this kind are today known as **Hertzsprung–Russell diagrams**, or **H–R diagrams**.

Figure 19-14 is a typical Hertzsprung–Russell diagram. Each dot represents a star whose absolute magnitude and spectral type have been determined. Bright stars are near the top of the diagram, dim stars near the bottom. Hot stars (O and B stars) are toward the left side of the graph; cool stars (M stars) are toward the right side.

The most striking feature of the H–R diagram is that the data points are not scattered randomly all over the graph but are grouped in several distinct regions. The band stretching diagonally across the H–R diagram includes many of the stars we see in the night sky. This band, called the **main sequence**, extends from the hot, bright, bluish stars in the upper-left corner of the diagram down to the cool, dim, reddish stars in the lower-right corner. A star whose properties place it in this region of an H–R diagram is called a **main-sequence star**. For example, the Sun (spectral type G2, absolute magnitude +4.8) is such a star.

Toward the upper-right side of the H–R diagram, there is a second major grouping of data points. Stars represented by these points are both bright and cool. From the Stefan–Boltzmann law, we know that a cool object radiates much less light per unit of surface area than does a hot object. In order for these stars to be as bright as they are, they must be huge, and so they are called **giants**. As explained in Box 19-4, these stars are around 10 to 100 times larger than the Sun. Most of these stars are around 100 to 1000 times more luminous than the Sun and have surface temperatures of about 3000 to 6000 K. Cooler members of this class of stars

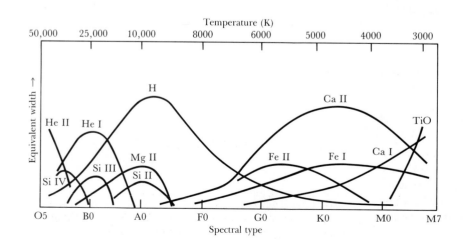

Figure 19-13 Spectral type and temperature *The strengths of the absorption lines of various elements are directly related to the temperature of the star's outer layers. For example, the Sun's spectrum has strong lines of singly ionized iron and calcium (Fe II and Ca II) that correspond to a spectral class of G2 and a surface temperature of about 5800 K. Note that hydrogen lines are strongest in A stars, whereas stars cooler than about 3500 K show absorption caused by titanium oxide.*

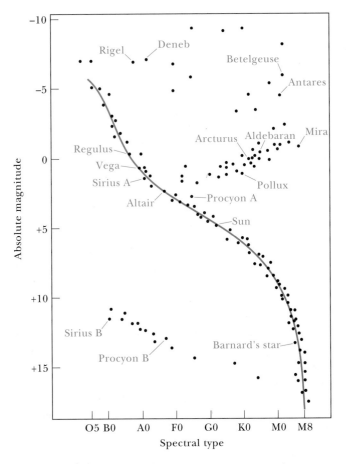

Figure 19-14 A Hertzsprung–Russell diagram An H–R diagram is a graph on which the absolute magnitudes of stars are plotted against their spectral types. Each dot on this diagram represents a star whose absolute magnitude and spectral class have been determined. Some well-known stars are identified. Note that the data points are grouped in specific regions on the graph. This pattern reveals the existence of different types of stars in the sky: main-sequence stars, giants, supergiants, and white dwarfs. The red curve indicates the location of the main sequence.

There is another useful way to exhibit an H–R diagram. Instead of using absolute magnitude, luminosity is plotted on the vertical axis of the graph; instead of spectral type, surface temperature is plotted on the horizontal axis. The resulting graph is still an H–R diagram, but the observational quantity (spectral type) has been replaced by the calculated quantity (temperature).

Figure 19-15 shows this type of H–R diagram. Note that on this graph the temperature scale on the horizontal axis increases toward the left. This practice came about when Hertzsprung and Russell drew their original diagrams with hot O stars on the left and cool M stars on the right, because the O stars have the simplest spectra. Having hot stars toward the left and cool ones to the right is a tradition no one has seriously tried to change.

The H–R diagram in Figure 19-15 shows data for nearly 200 of the nearest and brightest stars in the sky. Even though this diagram contains far fewer data points than does Figure 19-14, the main-sequence, white-dwarf, and giant regions are all still clearly delineated.

The stars are classified into spectral types on the basis of the most prominent lines in their spectra. However, there are subtle differences even among the spectral lines of stars having the same spectral type. For instance, for stars of spectral types B through F, the more luminous the star is, the deeper and narrower are its Balmer hydrogen lines. The Balmer lines are good indicators of luminosity because their line profiles are affected by the density and pressure of the gas in

(those with surface temperatures from about 3000 to 4000 K) are often called **red giants** because they appear reddish. Aldebaran in the constellation of Taurus and Arcturus in Boötes are good examples of red giants that can be easily seen with the naked eye.

A few rare stars are considerably bigger and brighter than typical red giants. These superluminous stars are, appropriately enough, called **supergiants**. Betelgeuse in Orion and Antares in Scorpius are examples of supergiants you can find in the nighttime sky.

Finally, there is a third distinct grouping of data points toward the lower-left corner of the Hertzsprung–Russell diagram. These stars are both hot and dim, so they must be small. They are appropriately called **white dwarfs**. These stars, which are roughly the same size as the Earth, can be seen only with the aid of a telescope.

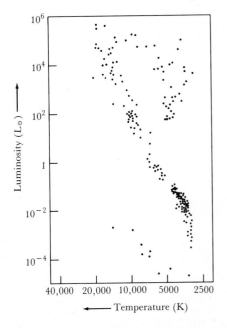

Figure 19-15 An H–R diagram of the nearest and brightest stars It is often informative to draw an H–R diagram by plotting the luminosities of stars against their surface temperatures. Data for nearly 200 of the nearest and brightest stars are plotted on this diagram.

Box 19-4 Stellar radii

Stars behave almost exactly like blackbodies, so to find out how much energy is being radiated from a star's surface each second, we can use the Stefan–Boltzmann law, which states that the energy flux from a blackbody is proportional to the fourth power of its temperature. Specifically, in Chapter 5 we saw that the energy flux E (usually measured in J m^{-2} s^{-1}) is related to the temperature T (in kelvins) by

$$E = \sigma T^4$$

where σ is a number called the Stefan–Boltzmann constant. From elementary geometry, we know that a sphere's surface area is $4\pi R^2$, where R is the sphere's radius. Because a star is spherical, we can use this expression for the surface area of a star. Multiplying the energy flux E by the star's surface area, we get the total energy output of the star per second, which is the star's luminosity L:

$$L = 4\pi R^2 \sigma T^4$$

By rearranging this equation, we can find the radius of a star in terms of its luminosity and surface temperature:

$$R = \frac{1}{T^2}\sqrt{\frac{L}{4\pi\sigma}}$$

Example: Consider the bright reddish star Betelgeuse in the constellation of Orion. Betelgeuse has a luminosity of 10,000 L$_\odot$ and a surface temperature of 3000 K. Substituting these values (and the value of the Sun's luminosity L$_\odot$) in the equation just derived, we find that the star's radius is 2.6×10^{11} m or nearly 2 AU. In other words, if Betelgeuse were located at the center of our solar system, this star would extend beyond the orbit of Mars.

In many calculations, it is convenient to relate everything to the Sun, which is a typical star. Specifically, for the Sun we have L$_\odot = 4\pi$R$_\odot^2\sigma$T$_\odot^4$, where L$_\odot$ is the Sun's luminosity, R$_\odot$ is the Sun's radius, and T$_\odot$ is the Sun's surface temperature (5800 K). Dividing the general equation for L by this specific equation for the Sun, we obtain

$$\frac{L}{L_\odot} = \left(\frac{R}{R_\odot}\right)^2\left(\frac{T}{T_\odot}\right)^4$$

This is a useful expression, because all the constants such as π and σ have canceled out, making the arithmetic much easier. We can also rearrange terms to arrive at an alternative useful equation:

$$\frac{R}{R_\odot} = \left(\frac{T_\odot}{T}\right)^2\sqrt{\frac{L}{L_\odot}}$$

Example: Again consider Betelgeuse, for which $L = 10^4$L$_\odot$ and $T = 3000$ K. Substituting these data into the previous equation, we get

$$\frac{R}{R_\odot} = \left(\frac{5800}{3000}\right)^2\sqrt{10^4} = 370$$

In other words, Betelgeuse's radius is 370 times larger than the Sun's radius.

The Hertzsprung–Russell diagram is a graph of luminosity versus temperature, so we can use the general equation relating L, T, and R to draw lines on the H–R diagram to represent various stellar radii, as shown.

Note that most of the stars on the main sequence are about the same size as the Sun. Giants have radii between 10 R$_\odot$ and 100 R$_\odot$, whereas the radii of supergiants range up to 1000 R$_\odot$. White dwarfs are roughly the same size as the Earth.

The radii of some stars have been measured with other techniques, some of which are discussed elsewhere in this text. These other techniques of measurement give radii consistent with those calculated by the method described in this box.

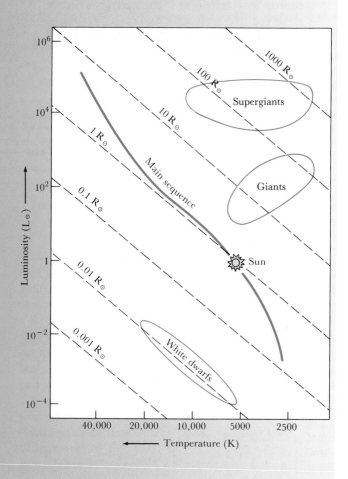

the star's atmosphere. The higher the density and pressure, the more frequently metal ions collide and interact with hydrogen atoms. These collisions shift the energy levels in the hydrogen atoms, which broadens the spectral lines.

The pressure and density in the atmosphere of a luminous giant star it quite low, because its mass is spread over a huge volume. Particles are relatively far apart and collisions between them are sufficiently infrequent that hydrogen atoms can produce Balmer lines that are deep and narrow. A main sequence star is, however, much more compact than a giant or supergiant. In the denser atmosphere of a main sequence star, frequent interatomic collisions perturb the energy levels in the hydrogen atoms, thereby giving rise to broader, shallower Balmer lines.

During World War II, W. W. Morgan, P. C. Keenan, and E. Kellman developed a system of **luminosity classes** based upon the subtle differences in spectral lines. When these luminosity classes are plotted on an H–R diagram (see Figure 19-16), they provide a useful subdivision of the star types in the upper-right half of the diagram. Luminosity class I includes all the supergiants, with luminosity class V being the main-sequence stars. The intermediate classes distinguish giant stars of various luminosities (see Table 19-2).

Astronomers commonly describe a star by combining its spectral type and its luminosity class into a sort of shorthand description; for example, the Sun is said to be a G2V star. This notation supplies a great deal of information. The spectral type is correlated with the star's surface temperature, and the luminosity class is correlated with its luminosity. Thus an astronomer knows immediately that any G2V star is a main-sequence star with a luminosity of about 1 $L_\odot$ and a surface temperature of nearly 6000 K. Similarly, a description of Aldebaran as a K5III star tells an astronomer that it is a red giant with a luminosity of around 500 $L_\odot$ and a surface temperature of about 4000 K.

The H–R diagram can be used to determine the distance to a star, no matter how far away it is. By examining a star's spectrum, you can determine its spectral type and luminosity class, which correspond to a particular point on the H–R diagram. From the location of that point on the H–R diagram, you can read off the star's absolute magnitude (M). Along with the star's apparent magnitude (m), which is easily measured at a telescope, you can write down the star's distance modulus, $m - M$. The difference between a star's apparent and intrinsic brightness is directly related to its distance from Earth (recall Box 19-2). This method of distance determination, which is based on information contained in a star's spectrum, is called **spectroscopic parallax**, even though no parallax is actually measured.

The fact that fundamentally different types of stars exist is the first important lesson to come from the H–R diagram. As we shall see in the following chapters, these different kinds of stars represent various stages of a star's life. We shall learn that a true appreciation of the H–R diagram involves an understanding of the life cycles of stars: how they are born and mature, and what happens when they die.

Table 19-2   Stellar luminosity classes

Luminosity class	Types of stars
I	Supergiants
II	Bright giants
III	Giants
IV	Subgiants
V	Main sequence

19-6 Binary stars provide crucial information about stellar masses

We now know something about the sizes, temperatures, and luminosities of stars. To complete our picture of the physical properties of stars, we need to know their masses. There is, however, no practical and direct way to measure the mass of an isolated star observed in the sky.

Fortunately for astronomers, about half of the visible stars in the night sky are not isolated individuals. They are instead revealed by the telescope to be multiple-star systems in which two or more stars orbit about each other. By ob-

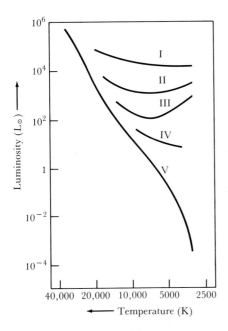

Figure 19-16   Luminosity classes   *It is convenient to divide the H–R diagram into regions corresponding to luminosity classes. This sundivision permits finer distinctions to be made between giants and supergiants. Luminosity class V encompasses the main-sequence stars, including the dim red stars called red dwarfs, toward the lower-right side of the H–R diagram.*

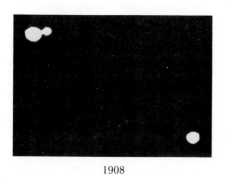
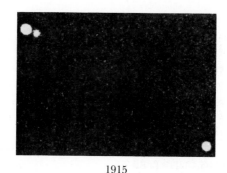

1908   1915   1920

Figure 19-17 *The binary star Kruger 60* *About half of the stars visible in the night sky have been revealed by the telescope to be double stars. This series of photographs shows the visual binary star Kruger 60 in the constellation Cepheus. The orbital motion of the two stars about* *each other is apparent. This binary system has a period of 44½ years. The maximum angular separation of the stars is about 2½ arc sec. Their apparent magnitudes are +9.8 and +11.4. (Yerkes Observatory)*

serving exactly how they orbit each other, astronomers can glean important information about their stellar masses.

A pair of stars located at nearly the same position in the night sky is called a **double star**. William Herschel made the first organized search for such pairs. Between 1782 and 1821 he published three catalogues, listing more than 800 double stars. Late in the nineteenth century, his son John Herschel discovered 10,000 more doubles. Many of these double stars are in fact true **binary stars**, or **binaries**, pairs in which the two stars are actually orbiting each other.

In cases where astronomers can actually see the two stars orbiting about each other, a binary is called a **visual binary** (see Figure 19-17). After many years of patient observation, astronomers can plot the orbits of the stars in a visual binary (see Figure 19-18).

A binary-star system is held together by gravity. Because the gravitational force between the two stars keeps them in orbit about each other, the details of their orbital motions can be described by Newtonian mechanics. Specifically, their orbits obey Kepler's third law (see Box 4-3). For the binary-star system, Kepler's third law can be written as

$$M_1 + M_2 = \frac{a^3}{P^2}$$

where M_1 and M_2 are the masses of the two stars expressed in solar masses, P is the orbital period in years, and a is the semimajor axis (measured in AU) of the elliptical orbit of one star about the other, plotted as in Figure 19-18.

In principle, it is easy to determine the orbital period of a visual binary. All that is necessary is to keep observing until the two stars return to their original positions relative to each other. As Figure 19-18 suggests, however, more than one lifetime of observations is usually needed to see a complete orbit.

Determining the semimajor axis of an orbit is somewhat more difficult. The angular separation between the stars can be determined by observation, but converting this angle into a linear distance (in AU) requires knowing the distance be-

tween the binary star and the Earth. This information may in some cases be obtained by making parallax measurements. Once the star-to-Earth distance is known, the small-angle formula (recall Box 1-1) can be used to convert the angular separation into a distance in AU. Some care may be needed, however, to correct for the angle of inclination at which the orbit is viewed from the Earth.

Once both P and a have been determined, Kepler's third law can be used to calculate $M_1 + M_2$, the sum of the masses of the two stars in the binary system. Note that this analysis

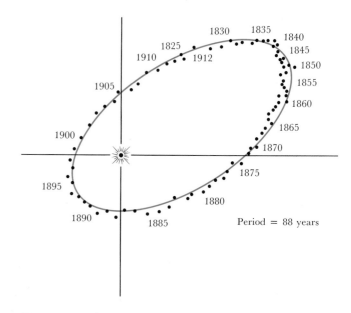

Figure 19-18 *The orbit of 70 Ophiuchi* *After plotting observations of a visual binary star over the years, astronomers can draw the orbit of one star with respect to the other. Once the orbit is known, Kepler's third law can be used to deduce information about the masses of the two stars. This illustration shows the orbit of a faint visual double star in the constellation Ophiuchus. In plotting the orbit, either star may be regarded as the stationary one—the shape and size of the orbit will be the same in either case.*

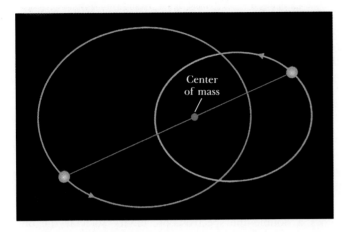

Figure 19-19 Star orbits in a binary system Both stars in a binary system follow elliptical orbits about their common center of mass. The center of mass is always nearer the more massive of the two stars.

provides no information about the individual masses of the two stars. To obtain the stars' individual masses, more data about the binary orbit are needed.

Each of the two stars in a binary system actually moves in an elliptical orbit about the **center of mass** of the system (see Figure 19-19). The concept of center of mass is analogous to that of two children on a seesaw. In order for the seesaw to balance properly, the center of mass of this two-child system must be located just above the support point, or fulcrum. As

you no doubt know from experience, this center of mass is offset from the midpoint between the two children toward the heavier child.

Similarly, the center of mass of a binary system could be imagined by placing the two stars at either end of a huge seesaw and determining where to place the fulcrum to balance the seesaw. The center of mass is always closer to the more massive star by an amount that is proportional to the ratio of the two masses.

In practice, the center of mass of a visual binary is determined by using the background stars as reference points. The separate orbits of the two stars can then be plotted as in Figure 19-19. The center of mass is located by finding the common focus of the two elliptical orbits. This information yields the ratio M_1/M_2. The sum $M_1 + M_2$ is already known, so the individual masses of the two stars can now be determined.

Years of careful, patient observations of binaries have slowly yielded the masses of many stars. As the data accumulated, an important trend began to emerge: For main-sequence stars, there is a direct correlation between mass and luminosity. The more massive the star, the more luminous it is. This **mass–luminosity relation** can be conveniently displayed as a graph (see Figure 19-20). Note that the range of stellar masses extends from about one-tenth of a solar mass to about 50 solar masses. The Sun's mass lies in the middle of this range, so we see again that our star is ordinary and typical.

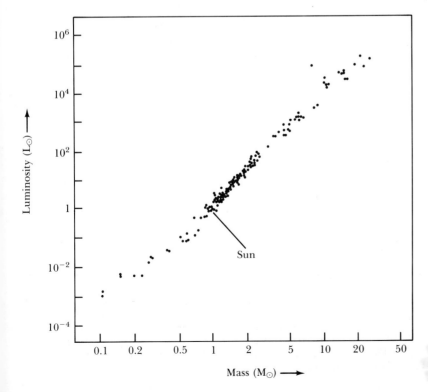

Figure 19-20 The mass–luminosity relation For main-sequence stars, there is a direct correlation between mass and luminosity. The more massive a star, the more luminous it is.

The mass–luminosity relation demonstrates that the main sequence on an H–R diagram is a progression in mass as well as in luminosity and surface temperature. The hot, bright, bluish stars in the upper-left corner of an H–R diagram are the most massive main-sequence stars in the sky. Likewise, the dim, cool, reddish stars in the lower-right corner of an H–R diagram are the least massive. Main-sequence stars of intermediate temperature and luminosity also have an intermediate mass. This relationship of mass to the main sequence will play an important role in our discussion of stellar evolution.

19-7 Binary systems that cannot be examined visually can still be detected and analyzed

Many binary stars are scattered throughout our galaxy, but only those that are nearby or that have enough separation between the two stars can be distinguished as visual binaries. The images of the two stars in a remote binary commonly blend together to produce the semblance of a single star. Spectroscopy provides evidence that many seemingly single stars are in fact double.

Occasionally, spectral analysis yields incongruous spectral lines for some stars. For example, the spectrum of what at first appears to be a single star may include both strong hydrogen lines (characteristic of a type-A star) and the strong absorption bands of titanium oxide typical of a type-M star. A single star could not have the differing physical properties of these two spectral types, so we must conclude that this star is actually a binary system that is too far away for resolution of its individual stars. A binary star detected in this way is called a spectrum binary.

If the orbital speeds of the two stars in a remote, unresolved binary are each more than a few kilometers per second, important additional information can be obtained from Doppler shift measurements. As one of the stars comes toward you, its spectral lines are displaced toward the short-wavelength (blue) end of the spectrum. Conversely, the spectral lines of a receding star are shifted toward the long-wavelength (red) end of the spectrum. If λ_0 is the unshifted wavelength of a spectral line and λ is the wavelength of that line in a star's spectrum, then

$$\frac{\lambda - \lambda_0}{\lambda_0} = \frac{v_\mathrm{r}}{c}$$

where v_r is the star's radial velocity toward or away from us along our line of sight and c is the speed of light. With this equation, we can convert a measurement of wavelength displacement into information about a star's motion. It is important to emphasize that the Doppler effect applies only to motion along the line of sight. Motion perpendicular to the line of sight does not affect the wavelengths of spectral lines.

Binaries discovered, and studied, through the periodic Doppler shifts in their spectra are called spectroscopic binaries. In many spectroscopic binaries, one of the stars is so dim that its spectral lines cannot be detected. Such a double-star system, which shows only a single set of spectral lines, is called a single-line spectroscopic binary. The fact that the star is a binary is obvious, however, because its spectral lines shift back and forth, thereby revealing the orbital motions of two stars about their center of mass. Less frequently, an apparently single star yields spectra in which two complete sets of spectral lines shift back and forth. Such stars are called double-line spectroscopic binaries. A single-line spectroscopic binary yields less information about its two stars than does a double-line spectroscopic binary such as the one shown in Figure 19-21.

Figure 19-21 shows two spectra of a spectroscopic binary taken a few days apart. In Figure 19-21a, two sets of spectral lines are visible, offset slightly in opposite directions from the normal positions of these lines. One star, which is moving toward the Earth, has its spectral lines blueshifted; the other star is moving away from the Earth, so its lines are redshifted. A few days later, the stars have progressed along their orbits so that one star is moving toward the left, the other toward the right. Neither star is moving toward or away from the Earth, so there is no Doppler shifting and both stars yield spectral lines at the same positions. That is why only one set of spectral lines appears in Figure 19-21b.

Significant information about the orbital velocities of the two stars in a spectroscopic binary can be deduced from measuring the shifts in spectral lines. This information is best displayed as a radial velocity curve, which graphs radial velocity versus time for the binary system (see Figure 19-22). Radial velocity is the portion of a star's motion that is parallel to the line of sight between the Earth and the star.

In Figure 19-22, note that the wavy pattern repeats with a period of about 15 days, which is the orbital period of the binary. Also note that the entire wavy pattern is displaced upward from the zero-velocity line by about 12 km/s, which is the overall motion of the binary system away from the Earth. Superimposed on this overall recessional motion are the periodic approaches and recessions of the two stars as they orbit about the center of mass.

The orbital speeds of the two stars in a binary are related to their masses by Kepler's laws and Newtonian mechanics. From a radial velocity curve, one can obtain a quantity involving the masses of the two stars. The individual masses of each star can be determined only if the tilt of the binary orbits is known. The tilt of these orbits determines how much of the true orbital speeds of the stars appears as radial velocity measured from Earth.

If the two stars are observed to eclipse each other periodically, then the orbit must be nearly edge-on, as viewed from the Earth. As we shall see next, individual stellar masses can be determined if a spectroscopic binary also happens to be an eclipsing binary.

Figure 19-21 A spectroscopic binary *A spectroscopic binary exhibits spectral lines that shift back and forth as the two stars revolve about each other. Shown are spectra of κ Arietis.* **(a)** *The stars are moving parallel to the line of sight, producing two sets of spectral lines.* **(b)** *Both stars are moving perpendicular to our line of sight. (Lick Observatory)*

19-8 Light curves of eclipsing binaries provide detailed information about the two stars

A small fraction of all binary systems are oriented so that the two stars periodically eclipse each other, as seen from the Earth. These eclipsing binaries can be detected even when the two stars cannot be resolved visually as two distinct images in the telescope. The apparent brightness of the image of the binary dims briefly each time one star blocks the light from the other.

Using a photoelectric detector at the focus of a telescope, an astronomer can measure the incoming light intensity quite accurately. The data for an eclipsing binary are most usefully displayed in the form of **light curves**, such as those shown in Figure 19-23. The overall shape of the light curve for an eclipsing binary reveals at a glance such information as whether the eclipse is total or partial (compare Figures 19-23a and 19-23b).

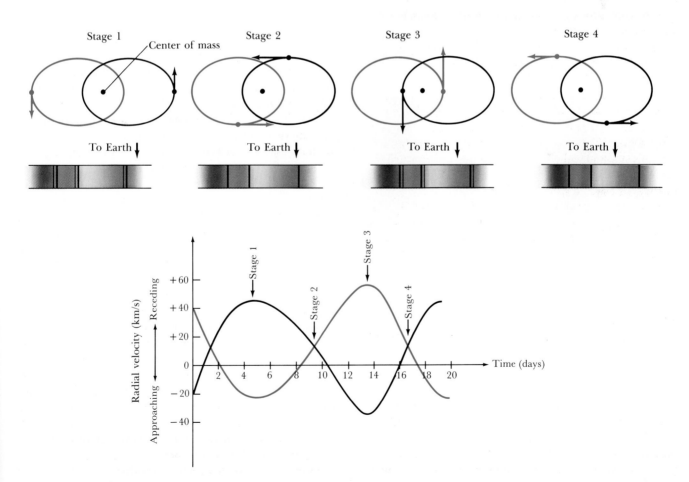

Figure 19-22 A radial velocity curve *The graph displays the radial velocity curves of the binary star HD 171978. The drawings above indicate the positions of the stars and their spectra at four selected moments during an orbital period.*

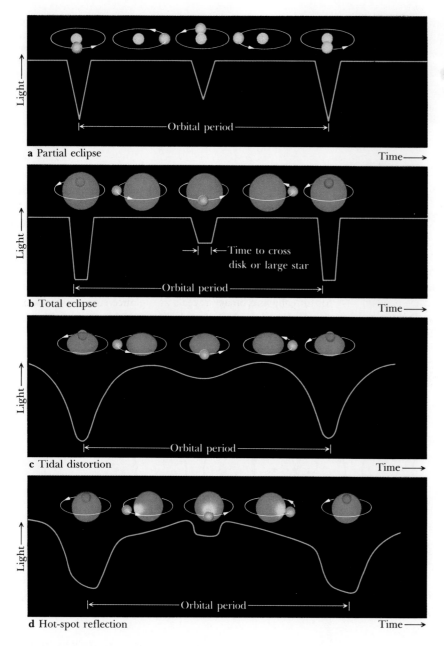

Figure 19-23 *Representative light curves of eclipsing binaries* The shape of its light curve usually reveals many details about an eclipsing binary. Illustrated here are (a) partial eclipse, (b) total eclipse, (c) tidal distortion, and (d) hot-spot reflection.

The light curve of a typical eclipsing binary can be analyzed to yield a surprising amount of information. For example, the ratio of the surface temperatures can be calculated directly from the ratio of the depths of the two eclipse minima. The duration of an eclipse depends on the inclination of the orbital plane of the binary to the line of sight to Earth, and on the size (radius) of the eclipsing star relative to the size (semimajor axis) of the orbit.

If the eclipsing binary is also a double-line spectroscopic binary, the combined analyses of the light curves and the velocity curves allow the astronomer to calculate the mass and radius of each star. The maximum amount of information about stellar masses and radii is obtained from studying double-line spectroscopic/eclipsing binaries. Unfortunately, very few binary stars are of this ideal type.

Additional details, such as the tidal distortion of one star by the other, are revealed by the shape of the light curve (see Figure 19-23c). If one of the stars in an eclipsing binary is very hot, its radiation may create a "hot spot" on its cooler companion star. Every time this hot spot is exposed to our view from Earth, we receive a little extra light energy, which produces a characteristic "bump" on the binary's light curve (see Figure 19-23d).

Information about stellar atmospheres can also be derived from light curves. Suppose that one star of a binary is a white dwarf and the other is a bloated red giant. By observing exactly how the light from the bright white dwarf is gradually cut off as it moves behind the edge of the red giant during the beginning of an eclipse, astronomers can infer the pressure and density in the upper atmosphere of the red giant.

A single star leads a straightforward, birth-to-death existence, but strange and exotic things can happen to stars in binary systems. One star in a binary might evolve rapidly, become a bloated red giant, and have its outer layers stripped away by the gravitational pull of its companion. The result would be an aging star with its interior exposed to view. As we shall see in later chapters, many curious variations are possible as the two stars in a binary affect each other's evolution.

Key words

- *absolute bolometric magnitude
- absolute magnitude
- *apparent brightness
- apparent magnitude
- binary star
- *bolometric correction
- *bolometric magnitude
- center of mass
- color index (plural indices)
- continuum
- *distance modulus

- double-line spectroscopic binary
- eclipsing binary
- equivalent width
- giant
- Hertzsprung–Russell diagram (H–R diagram)
- inverse-square law
- light curve
- line profile
- luminosity
- luminosity class

- main sequence
- mass–luminosity relation
- *moving cluster method
- OBAFGKM
- *open cluster
- parallax
- photometry
- proper motion
- radial velocity
- red giant
- single-line spectroscopic binary

- spectral type
- spectroscopic binary
- spectroscopic parallax
- spectrum binary
- supergiant
- tangential velocity
- visual binary
- white dwarf

Key ideas

- Distances to the nearer stars can be determined by parallax, the apparent shift of a star against the background stars observed as the Earth moves along its orbit.

  Distances to nearby clusters can be determined with the moving-cluster method.

- The absolute magnitude of a star is the apparent magnitude it would have if viewed from a distance of 10 parsecs; the absolute magnitude is calculated from the star's apparent magnitude and distance.

  Luminosity is the amount of energy escaping from the total surface area of a star each second; energy flux is the amount of energy emitted from each square meter of a star's surface each second; apparent brightness is the energy flux (joule m^{-2} s^{-1}) arriving at the Earth from a star.

- Photometry measures brightness through standard filters such as the UBV filters; the color indices of a star are the

differences between brightness values obtained through different filters.

  The color indices of a star are a measure of its surface temperature.

- Stars are classified into spectral types (O, B, A, F, G, K, or M) based on the major patterns of spectral lines in their spectra.

  The spectral type of a star is directly related to its surface temperature.

- The Hertzsprung–Russell (H–R) diagram is a graph plotting the absolute magnitudes of stars against their spectral types (or, equivalently, their luminosities against surface temperatures); it reveals the existence of such different types of stars as main-sequence stars, giants, supergiants, and white dwarfs.

- Binary stars are surprisingly common; those that can be resolved into two distinct star images by an Earth-based telescope are called visual binaries.

    The masses of the two stars in a binary system can be computed by measuring the orbital period and orbital dimensions of the system.

    Each of the two stars in a binary system moves in an elliptical orbit about the center of mass of the system.

- The mass–luminosity relation expresses a direct correlation between mass and luminosity for main-sequence stars.

- Some binaries can be detected and analyzed, even though the system may be so distant or the two stars so close together that the two star images are not resolvable by Earth-based telescopes.

    A spectrum binary is a system detected from the presence of spectral lines for two distinctly different spectral types in the spectrum of what appears to be a single star.

    A spectroscopic binary is a system detected from the periodic shift of spectral lines caused by the Doppler effect, as the orbits of the stars carry them first toward, then away from, the Earth.

    An eclipsing binary is a system whose orbits are viewed nearly edge-on from the Earth, so that one star periodically eclipses the other.

    Detailed information about the stars in an eclipsing binary can be obtained from a study of the binary's radial velocity curve and its light curve.

Review questions

1 What is the difference between apparent magnitude and absolute magnitude?

2 Briefly describe how you would determine the absolute magnitude of a nearby star. Of what value is knowing the absolute magnitude of various stars?

*3 How much dimmer does the Sun appear from Jupiter than from the Earth? (*Hint:* The average distance between Jupiter and the Sun is 5.2 AU.)

*4 Suppose that a dimly shinning star were located 1 million AU from the Sun. Find (a) the distance to the star in parsecs, and (b) the parallax angle of the star.

*5 If a star has a proper motion of 0.005"/year and a tangential velocity of 50 km/s, how far away is this star? What would its tangential velocity have to be in order for it to exhibit the same proper motion as Barnard's star?

*6 The space velocity of a certain star is 65 km/s and its radial velocity is 35 km/s. Find the star's tangential velocity.

7 Explain why the color index of a star is related to the star's surface temperature.

8 A red star and a blue star have the same size and the same distance from Earth. Which one looks brighter in the night sky? Explain why.

9 What are the most prominent absorption lines you would expect to find in the spectrum of (a) a hot star, (b) a cool star, and (c) a star like the Sun? Briefly describe why these stars have such different spectra even though they have essentially the same chemical composition?

10 Suppose that you want to determine the mass, temperature, diameter, and luminosity of a star. Which of these physical quantities require you to know the distance to the star? Explain.

11 Sketch a Hertzsprung–Russell diagram. Indicate the regions on your diagram occupied by (a) main-sequence stars, (b) red giants, (c) supergiants, (d) white dwarfs, and (e) the Sun.

12 What is the mass–luminosity relation?

13 Sketch the radial velocity curves of a binary consisting of two identical stars moving in circular orbits that are (a) perpendicular to and (b) parallel to our line of sight.

14 Sketch the light curve of an eclipsing binary consisting of two identical stars in highly elongated orbits oriented so that (a) their major axes are pointed toward the Earth and (b) their major axes are perpendicular to our line of sight.

*15 Estimate the mass of a main-sequence star that is 10,000 times as luminous as the Sun. What is the luminosity of a main-sequence star whose mass is one-tenth that of the Sun?

Advanced questions

Tools and tips . . .
Remember that a telescope's light gathering power is proportional to the area of its objective or primary mirror. Some of the problems concerning magnitudes may require facility with logarithms, because they make use of the equation that relates the apparent and absolute magnitudes of a star to its distance. Make use of the H–R diagrams given in this chapter to answer the question involving spectroscopic parallax.

*16 Suppose you can just barely see a twelfth-magnitude star through an amateur's 6-inch telescope. What is the magnitude of the dimmest star you could see through a 60-inch telescope?

*17 (Basic) Van Maanen's star, named after the Dutch astronomer who discovered it, is a nearby white dwarf whose parallax is 0.232 arc sec. How far away is the star?

*18 (Basic) Van Maanen's star has a proper motion of 2.95 arc sec yr^{-1} and a radial velocity of +54 km/s. What is the star's actual speed relative to the Sun?

*19 In the spectrum of a certain star, H_α has a wavelength of 656.50 nm. The laboratory value for the wavelength of H_α is 656.28 nm. Find (**a**) the star's radial velocity, and (**b**) the wavelength at which you would expect to find H_β in the star's spectrum, given that the laboratory wavelength of H_β is 486.13 nm.

*20 What would the distance between the Earth and the Sun have to be in order for the solar constant to be one watt per square meter (1 W/m^2)?

21 Explain why bolometric corrections are positive for both very hot and very cool stars.

*22 A certain type of variable star is known to have an average absolute magnitude of 0.0. Such stars are observed in a particular star cluster to have an average apparent magnitude of +16.0. What is the distance to that star cluster?

*23 The star Procyon in Canis Minor (the small dog) is a prominent star in the winter sky, because its apparent magnitude is +0.37. It is also one of the nearest stars, being only 3.51 parsecs from Earth. What is the absolute magnitude of Procyon? How many times brighter (or dimmer) than the Sun is it?

*24 Barnard's star, the star with the largest known proper motion, can be seen only with a telescope because its apparent magnitude is +9.54. Its distance from Earth is only 1.81 parsecs. How much closer to Earth would Barnard's star have to be in order for it to be visible to the unaided eye?

25 Derive the equation given in the text relating proper motion and tangential velocity. (*Hint:* To do this problem, you need to understand the concept of a radian and you will find it useful to know that 1 rad = 206,265 arc sec.)

*26 Suppose two stars have the same apparent magnitude, but one star is ten times farther away than the other. What is the difference in their absolute magnitudes?

*27 Suppose a star experiences an outburst in which its surface temperature doubles but its density decreases by a factor of eight. Find the new radius and luminosity of the star.

*28 (*Challenging*) The visual binary 70 Ophiuchi (see Figure 19-18) has a period of 87.7 years. The length of the semimajor axis is 4.5 arc seconds, and the parallax of the system is 0.2 arc sec. What is the sum of the masses of the two stars?

*29 An astronomer observes a binary star and finds one of the stars orbits the other once every 4 years at a distance of 2 AU. If the mass ratio of the system is 3.0, find the individual masses of the stars.

*30 The bright star Rigel in the constellation of Orion has a temperature about three times that of the Sun and its luminosity is 64,000 times that of the Sun. What is Rigel's radius compared to the radius of the Sun?

*31 The bright star Capella (α Aur) has a spectral type of G8 and an absolute magnitude of 0.7. What are the star's surface temperature, luminosity, and diameter?

*32 What is the distance to Regulus (α Leo), a B7V star whose apparent magnitude is 1.35?

Discussion questions

33 Why do you suppose that stars of the same spectral type but of a different luminosity class exhibit slight differences in their spectra?

34 Earth-based astronomers can measure with acceptable accuracy—10 percent or better—stellar parallaxes only if they are larger than about 0.05 arc sec. Discuss the advantages or disadvantages of making parallax measurements from a space telescope in a large solar orbit, say at the distance of Jupiter from the Sun. Assuming that this space telescope can also measure parallactic angles of 0.05 arc sec, what is the distance of the most remote stars that can be accurately determined? How much bigger a volume of space would be covered compared to Earth-based observations? How many more stars would you expect to be contained in that volume?

Observing projects

35 The following table lists well-known red stars and includes their coordinates (epoch 2000), apparent magnitudes, and (B − V) color indices. As indicated by the apparent magnitude column, all these stars are somewhat variable.

Star	R.A. (hr min)	Decl. (° ')	Apparent magnitude	(B − V)
Betelgeuse	5 55.2	+ 7 24	0.4–1.3	1.85
Y CVn	12 45.1	+45 26	5.5–6.0	2.54
Antares	16 29.4	−26 26	0.9–1.8	1.83
μ Cep	21 43.5	+58 47	3.6–5.1	2.35
TX Psc	23 46.4	+ 3 29	5.3–5.8	2.60

Star	R.A. (hr min)	Decl. (° ')	Apparent magnitudes	Angular separation (")	Spectral types
55 Psc	0 39.9	+21 26	5.4 and 8.7	6.5	K0 and F3
γ And	2 03.9	+42 20	2.3 and 4.8	9.8	K3 and A0
32 Eri	3 54.3	− 2 57	4.8 and 6.1	6.8	G5 and A2
14 Lyn	6 53.1	+59 27	5.7 and 6.9	0.4	F5 and A2
ι Cnc	8 46.7	+28 46	4.2 and 6.6	30.5	G5 and A5
γ Leo	10 20.0	+19 51	2.2 and 3.5	4.4	K0 and G7
24 Com	12 35.1	+18 23	5.2 and 6.7	20.3	K0 and A3
η Boo	14 45.0	+27 04	2.5 and 4.9	2.8	K0 and A0
α Her	17 14.6	+14 23	3.5 and 5.4	4.7	M5 and G5
59 Ser	18 27.2	+ 0 12	5.3 and 7.6	3.8	G0 and A6
β Cyg	19 30.7	+27 58	3.1 and 5.1	34.3	K3 and B8
δ Cep	22 29.2	+58 25	4* and 7.5	20.4	F5 and A0

*δ Cephei is a variable star of approximately the fourth magnitude.

Observe at least two of these stars both visually and through a small telescope. Is the reddish color of the stars readily apparent, especially in contrast to neighboring stars? Incidentally, Angelo Secchi named Y Canum Venaticorum "La Superba" and μ Cephei is often called William Herschel's "Garnet Star."

36 The table above of double stars includes vivid examples of contrasting star colors. The apparent magnitudes, spectral types, and the angular separation between the stars of each binary are given along with star names and coordinates (epoch 2000).

Observe at least four of the above double stars through a telescope. Use the spectral types listed above to estimate the difference in surface temperature of the stars in each pair you observe. Does the binary with the greatest difference in temperature seem to present the greatest color contrast? Based on what you see through the telescope and on what you know about the H–R diagram, explain why *all* the cool stars (spectral types K and M) listed above are probably giants or supergiants.

37 Observe the eclipsing binary Algol (β Persei), using nearby stars to judge its brightness during the course of an eclipse. Algol has an orbital period of 2.87 days and, with the onset of primary eclipse, its apparent magnitude drops from 2.1 to 3.4. It remains this faint for about 2 hours. The entire eclipse, from start to finish, takes about 10 hours. Consult the "Celestial Calendar" section of the current issue of *Sky & Telescope* for the predicted dates and times of the minima of Algol. Note that the schedule is given in Universal Time (the same as Greenwich Mean Time), so you will have to convert to your own time zone. Algol is normally the sec-

ond brightest star in the constellation of Perseus. Because of its northerly position (R.A. = $3^h08.2^m$, Decl. = +40°57'), Algol is readily visible from northern latitudes during the fall and winter months.

For further reading

Ashbrook, J. "Visual Double Stars for the Amateur." *Sky & Telescope*, November 1980 • This informative article discusses techniques for observing double stars.

Evans, D., et al. "Measuring Diameters of Stars." *Sky & Telescope*, August 1979 • This article explains how the technique of interferometry is used to determine the diameters of stars.

Gingerich, O. "A Search for Russell's Original Diagram." *Sky & Telescope*, July 1982 • The article in the "Astronomical Scrapbook" section gives entertaining insights about professional astronomy in the early 1900s.

Griffin, R. "The Radial-Velocity Revolution." *Sky & Telescope*, September 1989 • This article describes recent technological advances that astronomers use to perform extremely precise Doppler shift measurements.

Hodge, P. "How far away are the Hyades?" *Sky & Telescope*, February 1988 • This excellent article looks at the trials and tribulations of determining the distance to the Hyades cluster.

Kaler, J. "Origins of the Spectral Sequence." *Sky & Telescope*, February 1986 • This superb article, which traces the history of OBAFGKM, includes historical insights into the work of the astronomers who struggled to develop reliable and meaningful schemes of spectral classification.

_____. *Stars and Their Spectra.* Cambridge University Press, 1989 • This is an expanded version of Kaler's excellent series of articles on stellar spectroscopy that have appeared in *Sky & Telescope* over the past few years.

Labeyrie, A. "Stellar Interferometry: A Widening Frontier." *Sky & Telescope*, April 1982 • This article describes how a technique called interferometry can provide extraordinarily detailed images of stars.

Lovi, G. "A Stellar Teaching Tool." *Sky & Telescope*, March 1989 • In spite of a dreadfully dull title, this brief article gives a fascinating account of the motions of Sirius, including a star chart that shows Sirius's path among the constellations over the past 5 million years.

Mitton, J., and MacRobert, A. "Colored Stars." *Sky & Telescope*, February 1989 • This brief article in the "Celestial Calendar" section gives a superb overview of star colors and includes a listing of extremely red stars and colorful double stars.

Nielsen, A. "E. Hertzsprung—Measurer of Stars." *Sky & Telescope*, January 1968 • This excellent historical sketch describes the work of one of the inventors of the H–R diagram.

Phillip, A., and Green, L. "Henry N. Russell and the H–R Diagram." *Sky & Telescope*, April 1978; May 1978 • These two articles give many fascinating insights into the life and times of Henry Norris Russell.

Upgren, A. "New Parallaxes for Old: Coming Improvements in the Distance Scale of the Universe." *Mercury*, November/December 1980 • This well written article describes astronomers' ongoing quest for accurate stellar distances.

Reflection and emission nebulae The two main bluish objects, NGC 6589 (top) and NGC 6590 (below), are reflection nebulae surrounding main-sequence stars of spectral types B5 and B6. Interstellar dust around these two stars efficiently reflects their bluish light. Several smaller reflection nebulae are scattered around the large reddish patch of ionized hydrogen gas, IC 1283-4. Dust mixed with the gas dilutes the intense red emission with a soft blue haze. (Anglo-Australian Observatory)

C H A P T E R

The Birth of Stars

Observations of stars, along with calculations of stellar models, have given astronomers an understanding of stellar evolution. Stars are born in cold clouds of interstellar gas and dust that are scattered abundantly throughout our Galaxy. In response to some occurrence—perhaps an encounter with one of the Galaxy's spiral arms or the detonation of a nearby supernova—an interstellar cloud begins to contract under the pull of gravity and fragments into protostars. The protostars continue to contract gravitationally until temperatures in their cores are high enough to ignite hydrogen burning. The resulting outflow of energy from a protostar's core provides the pressure to resist further contraction, and a full-fledged main-sequence star is born. Resplendent stellar nurseries are created where clusters of young stars illuminate the surrounding interstellar gas and dust.

At a casual glance, the heavens seem eternal and unchanging; the sky that we see at night is virtually indistinguishable from the sky our ancestors saw. But this permanence is an illusion. Stars emit huge amounts of radiation: expenditures that must produce changes and cause the stars to evolve. With careful observation and calculation, astronomers have assembled a theory of **stellar evolution**. This theory explains how stars are born in great clouds of interstellar gas and dust. They mature, grow old, and some eventually blow themselves apart in death throes that enrich interstellar space with new material for future stellar generations. The stars seem unchanging to human beings only because of the colossal time scale over which these changes occur. Major stages in the life of a star can last for millions or even billions of years.

20-1 Significant amounts of interstellar gas and dust pervade the Galaxy

The German-born English astronomer William Herschel was one of the greatest astronomical observers of all time. During the late 1700s, he discovered numerous double stars, nebulae, star clusters, and the planet Uranus. Curiously, Herschel also reported sighting "holes in the heavens,"

places where there seemed to be far fewer stars than might be expected. In the late 1800s, Edward E. Barnard, an American astronomer who rose from extreme poverty to become one of the greatest photographers of the heavens, turned his talents to these "holes" (see Figure 20-1). His pictures suggested that these seemingly star-deficient regions are actually caused by vast clouds of interstellar matter that block the light from the more distant stars behind them.

Astronomers were not quick to accept the idea of matter between the stars. Indeed, until the 1930s, most astronomers believed that interstellar space was essentially a perfect vacuum. Evidence for the existence of both interstellar gas and dust, which together constitute the **interstellar medium**, came from various sources. (Actually, "smoke" would perhaps be a better name than "dust" because of the extremely small sizes of interstellar grains.)

Improvements in astrophotography helped convince many astronomers of the existence of interstellar matter. Photographs, like that of the Horsehead Nebula in Figure 20-2 or the Keyhole Nebula in Figure 20-3, clearly show clouds of dust obscuring background nebulosity (clouds of glowing gas). Indeed, the nebulosity itself, glowing with the characteristic red light of H_α emission, is direct evidence of interstellar gas.

Further evidence of interstellar atoms came from the discovery of stationary absorption lines in the spectra of certain spectroscopic binary stars. These lines, particularly those of calcium and sodium, remain at fixed wavelengths while the stars' spectral lines shift back and forth as the stars orbit their common center of mass. The stationary lines are therefore not associated with the double star, but rather are caused by sparse interstellar matter between us and the double star.

Some photographs show a bluish haze around certain stars (examine the photograph on the first page of this chapter). This haze, called a **reflection nebula**, is caused by fine grains of dust that scatter and reflect short-wavelength radiation more efficiently than long-wavelength radiation. When visible light from a star encounters interstellar dust, blue light is scattered by the dust grains, illuminating them and producing a reflection nebula. Red light manages to get through the dust without much scattering or deflection, provided the cloud is not so dense as to be opaque.

The most convincing evidence for interstellar matter came from the work of the American astronomer Robert Trumpler, who studied the brightnesses and distances of certain star clusters in the 1930s. Trumpler noticed that remote clusters seemed to be unusually dim, dimmer than would be expected from their distance alone. His observations demonstrated that light from remote stars is dimmed by its passage through sparse material in interstellar space, a process called **interstellar extinction**. In addition to being dimmed, light from remote stars is also reddened by its passage through the interstellar medium. Remote stars appear redder than usual because the blue component of their star-

Figure 20-1 A dark nebula *The dark nebula called Barnard 86 is visible in this photograph only because it blocks out light from the stars beyond it. This nebula is located in Sagittarius. The cluster of bluish stars to the left of the nebula is NGC 6520. (Anglo-Australian Observatory)*

Figure 20-2 The Horsehead Nebula   Dust grains block light from the background nebulosity whose glowing gases are excited by ultraviolet radiation from young, massive stars. This nebula is located in Orion, at a distance of roughly 1600 light years (490 parsecs) from Earth. The bright star near the top is Alnitak (ζ Ori), the easternmost star in the "belt" of Orion. (Royal Observatory, Edinburgh)

Figure 20-3 The Keyhole Nebula   A dark notch, called the "keyhole" because of its shape, is caused by interstellar dust silhouetted against this nebulosity in the southern constellation of Carina, the keel of the Argonauts' ship. A bright variable star, called η Carinae, is located at the center of the nebula. Both the star and the nebula are about 9000 light years from Earth. (Anglo-Australian Observatory)

light is scattered by interstellar dust. The effect is called **interstellar reddening**.

Measurements of both extinction and reddening in various directions in space indicate that interstellar gas and dust are largely confined to the plane of the Milky Way Galaxy. As we shall see in Chapter 25, our Galaxy is a rotating, disk-shaped collection of several hundred billion stars about 80,000 light years in diameter. The Sun is located about 25,000 light years from the Galaxy's center. If we could view our Galaxy from a great distance, it would look somewhat like the galaxy shown in Figure 20-4.

Interstellar gas and dust are the raw material from which stars are made. The disk of our Galaxy, where most of this matter is concentrated, is therefore an active site of ongoing star formation.

20-2 Protostars form in cold, dark nebulae and then quickly evolve to become young main-sequence stars

Stars are created by the action of gravity on cold, dark clouds of interstellar gas and dust. As we saw in Box 7-2, the temperature of a gas is directly related to the average speed of its atoms and molecules. If an interstellar cloud is warm, its atoms are moving about so rapidly that there is no chance for a star to condense out of these agitated gases. If the cloud's temperature is low, however, the atoms are moving slowly enough to allow denser portions of the cloud to contract under their own weight into clumps that form new stars.

Figure 20-4 A spiral galaxy Spiral galaxies, like our own Milky Way Galaxy, consist of stars, gas, and dust that are largely confined to a flattened, rotating disk. Spiral arms, consisting of bright stars and glowing gas, wind outward from the galaxy's center. This galaxy, called NGC 2997, is located in the southern constellation of Antlia (the air pump). (Anglo-Australian Observatory)

Cold interstellar clouds, called **dark nebulae**, appear as dark blobs obscuring background stars or nebulosities. Many dark nebulae discovered and catalogued around 1900 by E. E. Barnard are called **Barnard objects**. There are also some 200 relatively small, round, dark nebulae named **Bok globules**, after the Dutch-American astronomer Bart Bok, who first called attention to them in the 1940s.

A typical dark nebula contains a few thousand solar masses of gas and dust spread over a volume roughly 10 pc across. The chemical composition of this material is the standard "cosmic abundance" of about 74 percent (by mass) hydrogen, 25 percent helium, and 1 percent heavier elements (review Table 7-4). Emissions from molecules indicate that the cloud's internal temperature is about 10 K. At this temperature, the cloud is so cold and its atoms are moving so slowly that its gases do not provide enough internal pressure to support the cloud against its own weight. The cloud therefore contracts under the action of gravity and fragments into smaller lumps called **protostars**.

In the 1950s, astrophysicists such as L. Henyey in the United States and C. Hayashi in Japan performed calculations to describe the earliest stages of a protostar. At first, a protostar is merely a cool blob of gas several times larger than our solar system. Pressure inside the protostar is still incapable of supporting all this cool gas, and so the protostar contracts. As it does so, gravitational energy is converted into thermal energy (recall the discussion of Kelvin–Helmholtz contraction in Chapter 18), making the gases heat up and start glowing. Hayashi's calculations indicate that convection transports energy from the interior of the warming protostar to its surface. After only a few thousand years of gravitational contraction, the surface temperature has reached 2000 to 3000 K. At this point the protostar is

still quite large, so its glowing gases produce substantial luminosity. For instance, after only 1000 years of contraction, a protostar of 1 solar mass is 20 times larger in diameter and 100 times brighter than the Sun.

Astrophysicists use high-speed computers and equations similar to those for calculating stellar structure (described in Chapter 18) to determine the conditions inside a contracting protostar. The results tell how the protostar's luminosity and surface temperature change at various stages as it contracts. With this information, we can graph the **evolutionary track** of the protostar on a Hertzsprung–Russell diagram.

Protostars are relatively cool when they begin to shine at visible wavelengths. Thus, the evolutionary tracks of protostars begin near the right side of the H–R diagram (see Figure 20-5). However, continued gravitational contraction causes protostars to shift rapidly away from this region of the diagram. A protostar more massive than about five solar masses becomes hotter but shows little change in overall luminosity, because the effect of decreasing surface area

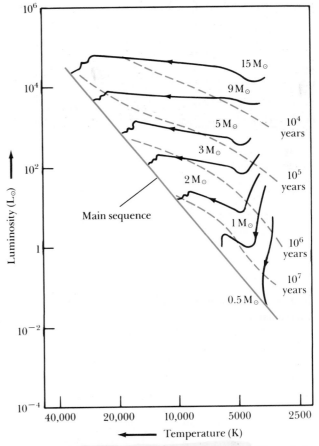

Figure 20-5 Pre–main-sequence evolutionary tracks The evolutionary tracks of seven stars having different masses are shown on this H–R diagram. The dashed lines indicate the stage reached after the indicated numbers of years of evolution. Note that all tracks terminate on the main sequence at locations that agree with the mass–luminosity relation. (Based on stellar model calculations by I. Iben)

(which by itself would decrease luminosity) is counterbalanced by an increase in surface temperature (which in itself would increase luminosity). Thus, the evolutionary tracks of massive protostars traverse the H–R diagram horizontally from right to left. In less massive contracting protostars, the increase in surface temperature is less rapid, so it does not fully compensate for the shrinking surface area and the luminosity therefore decreases.

A protostar continues to shrink until the temperature at its center reaches a few million degrees, when hydrogen burning begins. As we saw in Chapter 18, this thermonuclear process releases enormous amounts of energy. This outpouring of energy creates enough pressure inside the protostar to stop its gravitational contraction. Once hydrostatic and thermal equilibrium are established, a stable star is born. At this stage, the protostar's evolutionary track ends on the main sequence, as shown in Figure 20-5.

We now know that the main sequence represents stars in which hydrogen burning is occurring. This is quite a stable situation for stars. For example, our Sun will remain on or very near the main sequence, quietly burning hydrogen at its core, for a total of some 10 billion years.

Note that the evolutionary tracks in Figure 20-5 end at locations along the main sequence that agree with the stellar mass–luminosity relation (recall Figure 19-20). The most massive stars are the most luminous, while the least massive stars are the least luminous. Protostars less massive than about 0.08 solar masses can never manage to develop the necessary pressures and temperatures to start hydrogen burning at their cores. Instead, these small protostars contract and become planetlike objects. Protostars with masses greater than about 80 solar masses (80 $M_\odot$) rapidly develop such extremely high temperatures that radiation pressure tends to disrupt them. Main-sequence stars therefore have masses between about 0.08 and 80 $M_\odot$, with the high-mass stars being extremely rare.

The evolutionary tracks of protostars begin in the red-giant region of the H–R diagram, but protostars are not red giants. An H–R diagram like that in Figure 19-14 shows where stars spend *most* of their lives. Protostars spend only a tiny fraction of their existence in the red-giant region. A 15 $M_\odot$ protostar takes only 10,000 years to become a main-sequence star, and a 1 $M_\odot$ protostar takes a few million years to ignite hydrogen burning at its core. By astronomical standards, these intervals are so brief that pre–main-sequence stars are quite transitory.

We are unlikely to observe the birth of a star at visible wavelengths, because the surrounding interstellar cloud shields the protostar from view. The vast amount of visible light emitted by a protostar is absorbed by interstellar dust in its surrounding **cocoon nebula**, which becomes heated to a few hundred kelvins. The warmed dust then re-radiates its thermal energy at infrared wavelengths, to which the dust is relatively transparent. Consequently, infrared observations reveal what is going on inside a "stellar nursery."

Figure 20-6 shows views of a stellar nursery taken at both infrared and visible wavelengths. The dark cloud near the center of Figure 20-6b contains roughly 1000 solar masses of hydrogen and is a site of active star formation. This same region appears bright in the infrared view of Figure 20-6a. Although interstellar dust in the cloud absorbs nearly all the visible light emitted by the newborn stars, infrared radiation can still pass through the obscuring material. Infrared observations by U.S. astronomers Charles J. Lada and Bruce A. Wilking have revealed a cluster of 20 new stars embedded deep within the dust cloud. Astronomers are excited by the ease with which infrared observations can probe these dust clouds. Even better pictures and data will come from the Shuttle Infrared Telescope Facility (SIRTF), now scheduled to be launched in the late 1990s.

20-3 A vigorous ejection of matter often accompanies the birth of a star

Spectroscopic observations demonstrate that cool, young stars often eject substantial amounts of gas as they approach the main sequence. Such gas-ejecting stars are called **T Tauri stars**, after the first example discovered in the constellation of Taurus. Some astronomers suggest that the onset of hydrogen burning is preceded by vigorous chromospheric activity marked by enormous spicules and flares that propel the star's outermost layers back into space. In fact, an infant star going through its T Tauri stage can lose as much as 0.4 solar mass before it settles down on the main sequence.

T Tauri stars have relatively low masses of roughly 0.2 to 2.0 $M_\odot$. Young stars that are hotter and more massive than T Tauris propel their mass loss by vigorous stellar winds. Figure 20-7 is a dramatic photograph of a young star experiencing a significant mass loss of this sort.

Recent observations show that many young stars eject gas from two oppositely directed jets. This activity is precisely what one would expect from a star surrounded by a disk of material. Such disks have been observed (review Figure 7-15). Gas expelled by the star cannot easily push through this matter. Perpendicular to the disk, however, ejected gas meets very little resistance. Consequently, the circumstellar material channels the ejected gas into two beams or jets that emerge from one or the other face of the disk.

A photograph of a jet of gas from a young star is shown in Figure 20-8. The jet, which is hot and luminous as it emerges from the star, travels approximately 0.2 light year before cooling and becoming invisible. It travels on for another 0.5 light year and then strikes a clump of glowing nebulosity called HH 34. This nebulosity is a typical example of a **Herbig–Haro object**, named after the astronomers George Herbig and Guillermo Haro, who first discovered them in the 1940s. Herbig–Haro objects consist of several bright knots of material that change slightly in size, shape, and

a

b

Figure 20-6 *Formation of new stars near ρ Ophiuchi* (a) *Interstellar dust warmed by new stars dominates this wide-angle infrared view from IRAS that covers nearly 13° × 13° at a wavelength of 100 μm. The white rectangle indicates the smaller region covered by the photograph to the right.* (b) *This photograph was taken at visible wavelengths. Many stars in this region appear fuzzy because of reflection nebulae produced by interstellar dust. The bluish star near the top of this view is ρ Ophiuchi; Antares (α Scorpii) is at the lower left. The white circle indicates the location of at least 20 newborn stars that are probably less than 1 million years old. (National Aerospace Laboratory of the Netherlands; Royal Observatory, Edinburgh)*

Figure 20-7 *The mass-loss star HD 148937 This unusual star in the constellation of Norma is the brightest member of a triple system of stars in orbit about one another. This extremely hot, massive star is losing its outer layers continuously. Other vigorous outbursts in the past gave rise to the symmetric shells called NGC 6164 and NGC 6165 on either side of this star. (Anglo-Australian Observatory)*

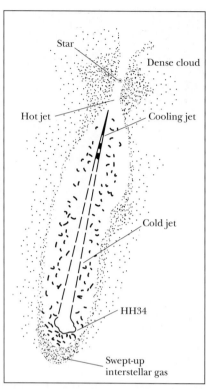

Figure 20-8 HH 34 and its surroundings *A young star is surrounded by circumstellar material that channels its ejected gases into a luminous jet. The jet travels about 0.7 light year and finally terminates in a Herbig–Haro object, a glowing gas cloud where the jet deposits its remaining energy. The photograph explains why Herbig–Haro objects are often found in the vicinity of T Tauri stars. (Courtesy of B. Reipurth, European Southern Observatory)*

brightness from year to year. They are often found in the vicinity of T Tauri stars. The photograph in Figure 20-8 proves that Herbig–Haro objects are produced where a high-velocity jet from a young star rams into interstellar gases, causing them to glow.

20-4 *Young star clusters are found in H II regions*

From the evolutionary tracks of protostars in Figure 20-5, we can see that high-mass stars evolve more rapidly than low-mass stars. Quite simply, the more massive the protostar, the sooner it develops the necessary central pressures and temperatures to begin hydrogen burning. These massive main-sequence stars, of spectral types O and B, are also the hottest, most luminous stars. Their surface temperatures are typically from 15,000 to 35,000 K, and thus—as indicated by Wien's law—they emit vast quantities of ultraviolet radiation.

The energetic ultraviolet photons from newborn massive stars easily ionize the surrounding hydrogen gas. For example, radiation from an O5 star can knock electrons loose from hydrogen atoms in a volume roughly 1000 light years across. This ionization has a dramatic effect on the dark nebula in which a cluster of stars is forming. While some hydrogen atoms are being knocked apart by ultraviolet photons,

other hydrogen atoms are being reassembled as free protons and electrons manage to get back together. During this **recombination** of hydrogen atoms, the captured electrons cascade downward through the atom's energy levels toward the ground state. These downward quantum jumps release numerous photons and many visible wavelengths, causing the nebula to glow. Particularly prominent is the transition from $n = 3$ to $n = 2$, which produces H_α photons at 656 nm in the red portion of the visible spectrum (review Figure 5-18). Thus, the nebulosity around a newborn star cluster shines with a distinctive reddish hue.

Figure 20-9 shows one of these **emission nebulae**. Because these nebulae are predominantly of ionized hydrogen, they are also called **H II regions** (several famous examples are listed in Box 20-1). The collection of a few hot, bright O and B stars near the core of the nebula that produces the ionizing ultraviolet radiation is called an **OB association**.

Observing individual stars in a young cluster can yield further information about stars in their infancy. Figure 20-10 shows a beautiful emission nebula surrounding the cluster NGC 2264. By measuring each star's magnitude and color index and knowing its distance, an astronomer can deduce its luminosity and surface temperature. The data for all the stars in the cluster can then be plotted on an H–R diagram, as in Figure 20-10. Note that the hottest stars, with surface temperatures around 20,000 K, are on the main sequence. These hot stars are the rapidly evolving, massive ones whose radiation is causing the surrounding gases to glow. The stars cooler than about 10,000 K, however, have

Figure 20-9 An H II region Because of its shape, this emission nebula is called the Eagle Nebula. It surrounds the star cluster called M16 or NGC 6611 in the constellation of Serpens, 6500 light years (2000 parsecs) from Earth. Several bright, hot O and B stars are responsible for the ionizing radiation that causes the nebula's gases to glow. (Anglo-Australian Observatory)

not yet quite arrived at the main sequence. These less-massive stars, which are in the final stages of pre–main-sequence contraction, are just now beginning to ignite thermonuclear reactions at their centers. The locations of the data points on Figure 20-10 suggest that this particular cluster is a few million years old.

Studies of H–R diagrams like that shown in Figure 20-9 reveal that a typical star cluster is created in several episodes of star birth rather than all at once. Furthermore, the most massive O and B stars are generally the last to form. Once created, these rapidly evolving stars produce brilliant ultraviolet radiation and powerful stellar winds that disrupt the surrounding nebulosity, thereby preventing any further star formation.

Figure 20-11 shows a young star cluster called the Pleiades that is easily visible to the unaided eye in the constella-

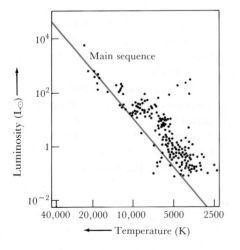

Figure 20-10 A young star cluster and its H–R diagram The photograph shows an H II region and the young star cluster NGC 2264 in the constellation of Monoceros. It is located about 2600 light years (800 parsecs) from Earth and contains numerous T Tauri stars. Each dot plotted on the H–R diagram represents a star in this cluster whose luminosity and surface temperature have been determined. Note that most of the cool, low-mass stars have not yet arrived at the main sequence. This star cluster probably started forming only 5 million years ago. (Based on observations by M. Walker, Anglo-Australian Observatory)

Box 20-1 Famous H II regions

Some of the most beautiful nebulae in the sky are H II regions. The table below lists some of the most famous H II regions.

Both the Lagoon and Trifid nebulae are shown in the photograph, which covers an area $2° \times 2\frac{1}{2}°$ in Sagittarius. The Lagoon Nebula is on the right, the smaller Trifid Nebula on the left.

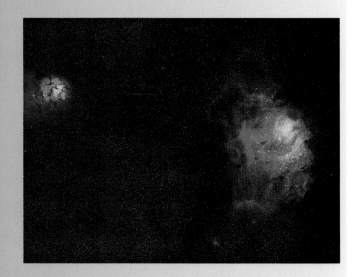

Nebula	Distance from Earth (light years)	Diameter (light years)	Mass ($M_\odot$)	Density (atoms/cm^3)
M8 (Lagoon)	4000	30	1000	80
M16 (Eagle)	6000	20	500	90
M17 (Omega)	5000	30	1500	120
M20 (Trifid)	3000	12	150	100
M42 (Orion)	1500	16	300	600

tion of Taurus. In contrast with the H–R diagram for NGC 2264, nearly all the stars in the Pleiades are on the main sequence. The cluster's age is about 100 million years, which is how long it takes for the least massive stars finally to begin hydrogen burning in their cores.

A loose collection of stars such as the Pleiades or NGC 2264 is referred to as an **open cluster** or **galactic clus-** ter. Such clusters possess barely enough mass to hold themselves together by gravitation. Occasionally, a star moving faster than average will escape, or "evaporate," from a cluster. Indeed, by the time the stars are a few billion years old, they may be so widely separated that a cluster no longer truly exists. If the group of stars is gravitationally unbound from the very beginning—that is, if the stars are moving

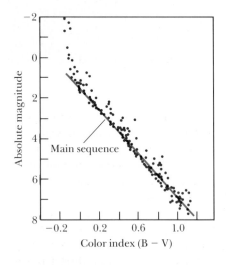

Figure 20-11 The Pleiades and its H–R diagram This open cluster called the Pleiades, which is about 400 light years from Earth, can easily be seen with the naked eye. Each dot plotted on the H–R diagram represents a star in the Pleiades whose absolute magnitude and color index have been measured. Note that all of the cool low-mass stars have arrived at the main sequence, indicating that hydrogen burning has begun in their cores. Substantial reflection nebulosity permeates this cluster, which has a diameter of about 5 light years and is about 100 million years old. (U.S. Naval Observatory)

Box 20-2   Interstellar molecules

A molecule is a combination of two or more atoms. A molecule vibrates and rotates at specific frequencies according to the laws of quantum mechanics. By either absorbing or emitting a photon, a molecule can speed up or slow down its rate of vibration or rotation. Astronomers can thus discover interstellar molecules by detecting the radiation from these vibrational or rotational transitions.

Although the first discovery of an interstellar molecule was in fact made at visual wavelengths, most molecules are strong emitters of radiation with wavelengths of around 1 to 10 mm. Consequently, observations with radio telescopes tuned to millimeter wavelengths have in recent years greatly increased the rate of discovery of interstellar molecules. Nearly 100 different kinds of interstellar molecules have been discovered so far, and the list is constantly growing. A representative partial listing is given below.

Note that the vast majority of these molecules contain carbon. Such substances are called organic molecules. Apparently,

organic chemistry is the chemistry of interstellar space as well as the chemistry of life on Earth. Astronomers have had such success in recent years in detecting interstellar molecules that they often boast of someday discovering every conceivable chemical somewhere in the universe. In fact, several organic molecules have been discovered in space that do not exist here on Earth, because they are too fragile to survive collisions with air molecules or with the walls of a container in a laboratory.

It is intriguing to speculate that interstellar molecules may have assembled somewhere to form living organisms. After all, it took no more than a billion years for the first primitive life forms to develop on Earth from the organic molecules in the primordial oceans. The interstellar clouds in our Galaxy have had 15 billion years to accomplish the same thing. Is it possible that the interstellar clouds drifting between the stars contain extraterrestrial life?

Two-atom molecules

CH	CO
H_2	SiO
CN	CS
OH	SO

Three-atom molecules

H_2O (water)	HCO
HCN	SO_2
H_2S	OCS

Four-atom molecules

NH_3 (ammonia)
H_2CO (formaldehyde)
HNCO
H_2CS
HC_2H (acetylene)

Five-atom molecules

H_2CHN
H_2NCN
HCOOH (formic acid)
HC_3N
H_2C_2O

Six-atom molecules

CH_3OH (methyl alcohol)
CH_3CN (methyl cyanide)
$HCONH_2$

Seven-atom molecules

CH_3NH_2
CH_3C_2H
$HCOCH_3$
H_2CCHCN (vinyl cyanide)

away from one another so rapidly that gravitational forces cannot keep them together—then the apparent cluster that initially exists is called simply a **stellar association**.

20-5 Star birth begins in giant molecular clouds

Hydrogen is by far the most abundant element in the universe. In the cold depths of interstellar space, hydrogen atoms combine to form hydrogen molecules (H_2). As described in Box 20-2, molecules vibrate and rotate at specific frequencies that are dictated by the laws of quantum mechanics. As a given molecule goes from one vibrational or

rotational state to another, it either emits or absorbs a photon. This process is analogous to what happens when an atom emits or absorbs a photon as an electron jumps from one energy level to another. Many interstellar molecules emit photons with wavelengths of several millimeters. Consequently, in recent years, radio telescopes tuned to wavelengths in this range have made observations that have greatly increased our knowledge of the interstellar medium.

Although hydrogen molecules are scattered abundantly across space, they are difficult to detect. The hydrogen molecule is symmetric, with two atoms of equal mass joined together, and such molecules do not emit photons efficiently at radio frequencies. Radio astronomers more easily detect asymmetric molecules such as carbon monoxide (CO), which consists of two atoms of unequal mass joined together. Carbon monoxide emits photons at a wavelength of

2.6 mm, which corresponds to a transition between two rates of rotation of the molecule.

The presence of carbon monoxide is especially useful when probing the interstellar medium. Astronomers have reason to believe that the ratio of carbon monoxide to hydrogen in space is reasonably constant: For every CO molecule, there are about 10,000 H_2 molecules. As a result, carbon monoxide is an excellent "tracer" for hydrogen gas. Wherever astronomers detect strong emission from CO, they know hydrogen gas must be abundant.

The first systematic surveys of our Galaxy looking for 2.6-mm CO radiation were undertaken in 1974 by Philip Solomon of the State University of New York and Nicholas Scoville, now at Caltech. In mapping the locations of CO emission, astronomers soon realized that vast amounts of hydrogen are concentrated in huge regions called **giant molecular clouds**. These clouds have masses in the range of 10^5 to 2×10^6 solar masses and diameters that range from about 50 to 300 light years. Inside one of these clouds, the density is about 200 hydrogen molecules per cubic centimeter. This is several thousand times larger than the average density of matter in the disk of our Galaxy, but 10^{17} times smaller than the density of molecules in the air we breathe. Astronomers now estimate that our Galaxy contains about 5000 of these enormous clouds.

The constellations of Orion and Monoceros include one of the most accessible regions of the sky for studying star formation and the interaction of young stars with the interstellar medium. Figure 20-12*a* shows a map of this region made with a radio telescope tuned to a wavelength of 2.6 mm. Note the extensive areas of the sky covered by giant molecular clouds. Comprehensive maps of CO emission, like that shown in Figure 20-12*a*, help astronomers understand how the large-scale structure of the interstellar medium is related to the formation of H II regions and OB associations.

Recently, U.S. astronomer Thomas M. Dame and his colleagues used CO emission to map the locations of giant molecular clouds in the inner regions of our Galaxy. The perspective drawing in Figure 20-13 shows the results of their work. They found that 17 molecular clouds clearly outline the spiral arm nearest the Sun, which is called the Sagittarius arm. These clouds lie roughly 3000 light years apart, strung along the spiral arms like beads on a string. This arrangement is much like the spacing of H II regions in some external galaxies (examine Figure 20-4).

When we discuss the details of our Galaxy in Chapter 25, we shall learn that spiral arms are caused by compressional waves that squeeze the interstellar gases passing through them. When one of these giant molecular clouds passes through a wave, it is compressed, causing vigorous star for-

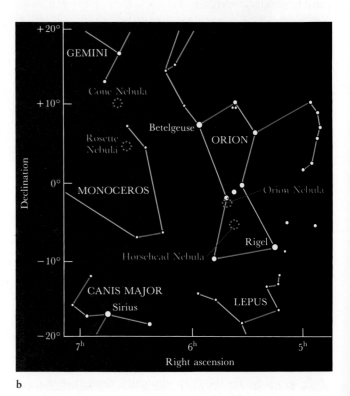

a

b

Figure 20-12 A map of carbon monoxide features in Orion (a) *This color-coded map of a large section of the sky shows the extent of giant molecular clouds in Orion and Monoceros. The intensity of CO emission is indicated by colors in the order of the rainbow, from violet for the weakest to red for the strongest. Black indicates no detectable emis-* sion. (b) *This star chart covers the same area as* (a). *The locations of four prominent star-forming nebulae are indicated. Note that the Orion and Horsehead nebulae are located at sites of intense CO emission.* (Courtesy of R. Maddalena, M. Morris, J. Moscowitz, and P. Thaddeus)

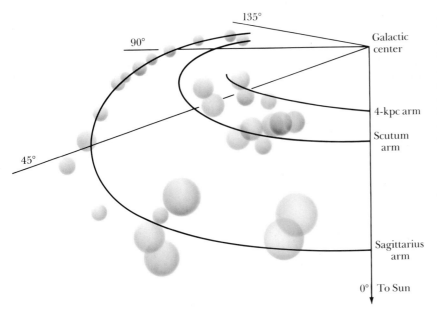

Figure 20-13 Giant molecular clouds in the Milky Way *This perspective drawing shows the locations of giant molecular clouds in an inner part of our Galaxy as seen from a vantage point above the Sun. Note how the Sagittarius spiral arm is outlined by giant molecular clouds that lie along it like beads on a string. The situation for the two inner spiral arms is not as clear. The distance from the Sun to the galactic center is about 25,000 light years. (Adapted from T. M. Dame and colleagues)*

mation to begin in the densest regions. The massive stars that are the first to form emit ultraviolet light, which soon ionizes the surrounding hydrogen, and an H II region is born. An H II region is thus a small, bright "hot spot" in a giant molecular cloud. The famous Orion Nebula (see Figure 20-14) is such an object. Four hot, massive O and B stars at the heart of the Orion Nebula are responsible for the ionizing radiation that causes the surrounding gases to glow. The Orion Nebula is embedded in a giant molecular cloud called Orion A, whose mass is estimated at 500,000 solar masses.

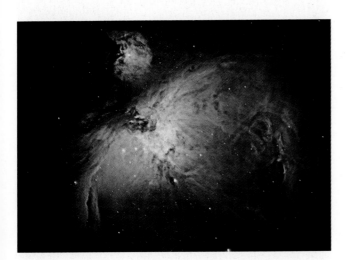

Figure 20-14 The Orion Nebula *This famous H II region can be seen with the naked eye in the middle of Orion's "sword." The nebula is located 1500 light years from Earth and has a diameter of roughly 16 light years. The mass of this nebulosity is estimated at 300 solar masses. Four bright, massive stars at the center of the nebula produce the ultraviolet light that causes the gases to glow. (Anglo-Australian Observatory)*

The OB association at the core of the H II region affects the rest of the giant molecular cloud. Vigorous stellar winds from the O and B stars carve out a cavity in the cloud, and the H II region, heated by the stars, expands into it. Much of this outflow is supersonic, so a shock wave forms where the outer edge of the expanding H II region impinges on the rest of the giant molecular cloud. This shock wave compresses the hydrogen gas through which it passes, thus stimulating a new round of star birth. The new O and B stars continue to power the expansion of the H II region still farther into the giant molecular cloud. Meanwhile, the older O and B stars, which were left behind, begin to disperse (see Figure 20-15). In this way, an OB association "eats into" a giant molecular cloud, "spitting out" stars in its wake.

Infrared observations reveal many features that resemble protostars in the swept-up layer immediately behind the shock wave from an OB association. For instance, Figure 20-16 shows both optical and infrared views of the core of the Orion Nebula. Four O and B stars, called the Trapezium, and glowing gas and dust dominate the view at visible wavelengths. Infrared observations penetrate this obscuring material to reveal dozens of infrared objects that may be cocoons of warm dust enveloping newly formed stars.

Radiation from newborn stars excites the OH and H_2O molecules in a giant molecular cloud in such a way that they become powerful sources of microwaves. The process starts when these molecules absorb photons, whose added energy then causes the molecules to spin faster. In the language of quantum mechanics, the molecules jump from their ground state to an excited rotational energy level, a process called *pumping* (see Figure 20-17). The excited molecules naturally drop back down to a lower energy level by emitting photons. The rules of quantum mechanics explain that some of these molecules will drop down into an energy level that is charac-

Figure 20-15 The evolution of an OB association
Ultraviolet radiation from young O and B stars produces a shock wave that compresses gas farther into the giant molecular cloud, stimulating new star formation deeper into the cloud. Meanwhile, older stars are left behind. (Adapted from C. Lada, L. Blitz, and B. Elmegreen)

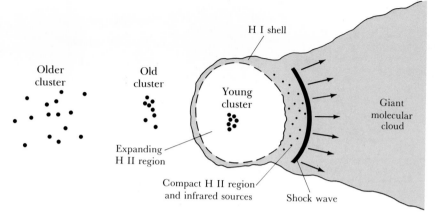

Older cluster

Old cluster

H I shell

Young cluster

Expanding H II region

Compact H II region and infrared sources

Shock wave

Giant molecular cloud

terized as being "meta-stable," meaning that the molecule will remain in that level for an exceptionally long time, because the probability of its dropping from there down to the ground state is very low.

The net effect of the pumping process in a giant molecular cloud is that regions develop where vast numbers of OH or H_2O molecules are stuck in low-lying excited states. Now imagine one of these regions being traversed by a microwave photon whose energy equals the energy difference between the meta-stable excited level and the ground level. The electromagnetic field of this photon acts to trigger many molecules along its path to lose energy, by emitting identical microwave photons traveling parallel to the original one. These new photons stimulate still more molecules to lose energy, resulting in a cascade that releases a vast amount of energy at microwave frequencies. The word **maser**, which is used to describe this phenomenon, is an acronym for **micro-wave amplification by stimulated emission of radiation**. A

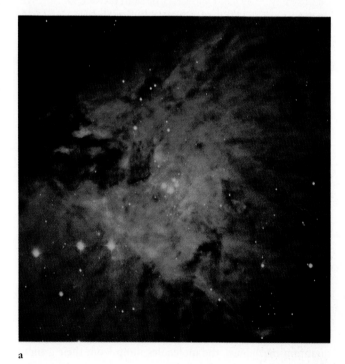

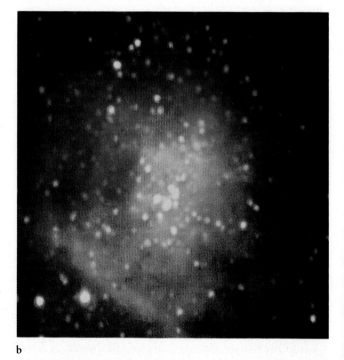

a

b

Figure 20-16 The core of the Orion Nebula (a) *This view at visible wavelengths shows the inner regions of the Orion Nebula (compare Figure 20-12). At the center are the four massive stars, called the Trapezium, which cause the nebula to glow. These stars are separated from each other by only 0.1 light year.* (b) *This infrared composite was taken at wavelengths of 1.2 and 2.2 μm, which can penetrate interstellar dust more easily than can visible photons. Numerous infrared objects, many of which are probably new stars in early stages of formation, are seen in this view. (Anglo-Australian Observatory)*

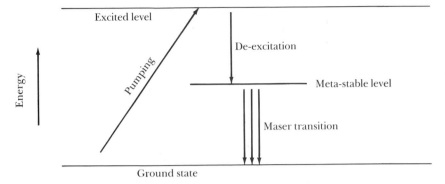

Figure 20-17 A maser *This energy level diagram shows how certain molecules like OH, H₂O, and SiO can be powerful sources of microwaves. After being excited by radiation from young stars, such molecules drop down into a meta-stable level, where they remain until a passing photon stimulates de-excitation down to the ground state.*

single maser lasts for only a few weeks or months. Near a site of active star formation, new masers are continually being turned on while old ones, having depleted their supplies of excited molecules, simply fade away.

20-6 Supernova explosions compress ③ the interstellar medium and thereby trigger star birth

Presumably, any mechanism that compresses interstellar clouds can trigger the birth of stars. As we shall see in Chapter 22, massive stars sometimes end their lives with a violent detonation called a **supernova explosion**. In a matter of seconds, the core of the doomed star collapses, releasing vast quantities of particles and energy that blow the star apart.

The star's outer layers are blasted into space at speeds of several thousand kilometers per second.

Astronomers find many nebulae across the sky that are the shredded funeral shrouds of these dead stars. Such nebulae, like the Cygnus Loop shown in Figure 20-18, are known as **supernova remnants**. Many supernova remnants have a distinctly arched appearance, as would be expected for an expanding shell of gas. This wall of gas is typically still moving away from the dead star at supersonic speeds. Its passage through the surrounding interstellar medium excites the atoms in it, causing its gases to glow.

Supersonic motion is always accompanied by a shock wave that abruptly compresses the medium through which it passes. If, for example, the expanding shell of a supernova remnant encounters an interstellar cloud, it can squeeze the cloud, stimulating star birth. This kind of star birth can be observed in the stellar association seen in Figure 20-19. This stellar nursery is located along a luminous arc of gas about

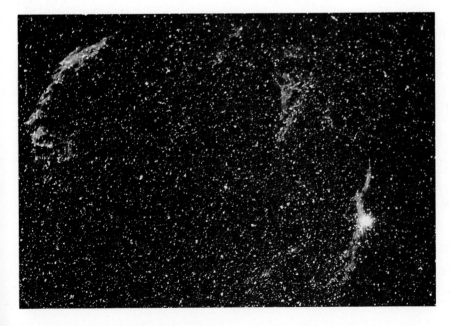

Figure 20-18 A supernova remnant *This remarkable nebula, called the Cygnus Loop, is the remnant of a supernova explosion that occurred about 20,000 years ago. This expanding spherical shell of gas now has a diameter of about 120 light years. (Courtesy of H. Vehrenberg)*

Figure 20-19 The Canis Major R1 association *This luminous arc of gas, about 100 ly long, is studded with numerous young stars. This stellar association illustrates the results of a shock wave from a supernova explosion: the triggering of star formation in the interstellar clouds through which the shock wave passes. (Courtesy of H. Vehrenberg)*

100 light years in length that is presumably the remnant of an ancient supernova explosion. In fact, this arc is part of an almost-complete ring of glowing gas with a diameter of about 200 light years. Spectroscopic observations of the stars along this arc reveal substantial T Tauri and chromospheric activity, which are indicative of newborn stars that are experiencing mass loss in their final stages of pre–main-sequence contraction.

As we learned in Chapter 17, there is strong evidence that the Sun was once a member of one of these stellar associations created by a supernova explosion. These loose associations do not remain intact for long, however. Individual stellar motions soon carry the stars in various directions away from their birthplaces. Nearly 5 billion years have passed since the birth of our star, so the Sun's brothers and sisters are now widely scattered across the Galaxy.

Our understanding of star birth has improved dramatically in recent years, primarily through infrared- and millimeter-wavelength observations. Nevertheless, many puzzles and mysteries remain. For example, astronomers have generally assumed that there must be a lot of interstellar dust shielding a stellar nursery, to protect it from the disruptive effects of external sources of ultraviolet light, which is what we seem to find in our own Galaxy. However, in the neighboring galaxy called the Large Magellanic Cloud (LMC), there are young OB associations with virtually no dust. Does the process of star birth differ slightly from one galaxy to another?

Another problem is that different modes of star birth tend to produce different percentages of different kinds of stars. For example, the passage of a spiral arm through a giant molecular cloud tends to produce an abundance of massive O and B stars. In contrast, the shock wave from a supernova

Figure 20-20 The core of the Rosette Nebula *The Rosette Nebula is a large, circular emission nebula located near one end of a sprawling giant molecular cloud in the constellation of Monoceros. Radiation from young, hot stars has blown gas away from the center of this nebula. Some of this gas has become clumped in dark globules that appear silhouetted against the glowing background gases. (Anglo-Australian Observatory)*

seems to produce fewer O and B stars, but many more of the less massive A, F, G, and K stars. We do not yet know why this is so.

There may be additional mechanisms of star birth that have yet to be discovered. For example, Robert Loren at the University of Texas has pointed out that a simple collision between two interstellar clouds should create new stars. When two such clouds collide, compression must occur at the interface, with vigorous star formation sure to follow. Another intriguing possibility, studied by American astronomer Bruce Elmegreen and others, is that stellar winds from a group of stars may exert strong enough pressure on interstellar clouds to cause compression, followed by star formation. The Rosette Nebula, shown in Figure 20-20, is probably an example of this process.

In spite of these and other unanswered questions, it is now clear that star birth involves mechanisms on a colossal scale, from the deaths of massive stars to the rotation of an entire galaxy. In many respects, we have just begun to appreciate these cosmic processes. The study of cold, dark stellar nurseries will certainly be an active and exciting area of astronomical research for many years to come.

Key words

Barnard object	galactic cluster	interstellar reddening	reflection nebula
Bok globule	giant molecular cloud	maser	stellar association
cocoon nebula	H II region	OB association	stellar evolution
dark nebula	Herbig–Haro object	open cluster	supernova explosion
emission nebula	interstellar extinction	protostar	supernova remnant
evolutionary track	interstellar medium	recombination	T Tauri star

Key ideas

- Interstellar gas and dust, which make up the interstellar medium, are concentrated in the disk of the Galaxy.

  Bluish reflection nebulae are produced when starlight is reflected from dust grains in the interstellar medium.

- Enormous cold clouds of gas called giant molecular clouds are scattered along the spiral arms of our Galaxy.

- Star formation begins when gravitational attraction causes a protostar to contract within a giant molecular cloud.

- A cloud that is visible as a dark blot against distant stars is called a dark nebula.

  Some dark nebula are named Barnard objects and Bok globules after the astronomers who discovered them.

- As a protostar grows by the gravitational accretion of gases, the process known as Kelvin–Helmholtz contraction causes it to heat and begin glowing. Its relatively low temperature and high luminosity place it in the upper right region on an H–R diagram.

  As the evolution of a protostar continues, it moves toward the main sequence on the H–R diagram. When its core temperatures become high enough to ignite hydrogen burning, it becomes a main-sequence star.

  In the final stages of pre–main-sequence contraction, when thermonuclear reactions are about to begin in the core of a protostar, the star may eject large amounts of gas into space; the low-mass stars that vigorously eject gas are called T Tauri stars.

  The most massive protostars rapidly become main-sequence O and B stars; they emit strong ultraviolet radiation that ionizes hydrogen in the surrounding cloud, thus creating the reddish emission nebulae called H II regions.

  Ultraviolet radiation and stellar winds from the OB association at the core of an H II region create shock waves that move outward through the gas cloud, compressing the gas to trigger the formation of more protostars.

Shock waves associated with the spiral arms of our Galaxy and with supernova explosions also compress gas clouds and trigger star formation.

- Circumstellar material may channel gas ejected from a young star into jets; clumps of glowing gas called Herbig–Haro objects are sometimes found at the ends of these jets.

- A collection of newborn stars may form an open or galactic cluster, in which stars are held together by gravity; occasionally, a star moving more rapidly than average will escape, or "evaporate," from such a cluster.

- A stellar association is a group of newborn stars that are moving apart so rapidly that their gravitational attraction for each other cannot pull them into orbit about each other.

Review questions

1 Why are low temperatures necessary in order for protostars to form inside dark nebulae?

2 What is a giant molecular cloud, and what role do these clouds play in the birth of stars?

3 What happens inside a protostar to slow and eventually halt its gravitational contraction?

4 Describe the energy source that causes a protostar to shine. How does this source differ from the energy source inside a true star?

5 Explain why thermonuclear reactions occur only at the center of a main-sequence star, never on its surface.

6 What is an H II region?

7 Why is the daytime sky blue? Why is the Sun red when seen near the horizon at sunrise or sunset? In what ways are your answers analogous to the explanations for the bluish color of reflection nebulae and the process of interstellar reddening?

8 What is an evolutionary track, and how can one help us interpret the H–R diagram?

9 Why are observations at infrared and millimeter wavelengths so much more useful in exploring interstellar clouds than are observations at visible wavelengths?

10 What is a maser?

11 Briefly describe four mechanisms that compress the interstellar medium and trigger star formation.

12 Speculate on why a shock wave from a supernova seems to produce relatively few high-mass O and B stars, compared to the lower-mass A, F, G, and K stars.

Advanced questions

> **Tips and tools . . .**
> You may find it helpful to review Box 19-2, which describes the relationship between magnitude and brightness. Remember that the Stefan–Boltzmann law relates the temperature of a blackbody to its energy flux, as described in Box 5-2.

13 If you looked at a spectrum of a reflection nebula, would you see absorption lines, emission lines, or no lines? Explain your answer, describing how the spectrum demonstrates that the light was reflected from nearby stars.

*14 *(Basic)* Find the density (in atoms per cubic meter) of a Bok globule having a radius of 1 light year and a mass of 100 solar masses. How does your result compare with the densities listed in Box 20-1? (Assume that the globule is made of pure hydrogen.)

*15 *(Challenging)* In the direction of a particular star cluster, interstellar extinction dims starlight by 2 magnitudes per kiloparsec. If the star cluster is 1.5 kiloparsecs away, what percentage of its photons survive the trip to Earth?

*16 At one stage during its birth, the protosun had a luminosity of 1000 $L_\odot$ and a surface temperature of about 1000 K. What was its radius?

17 How would you distinguish a newly formed protostar from a red giant, since they are both located in the same region on the H–R diagram?

*18 The concentration or abundance of ethyl alcohol (molecular weight = 46) in a typical molecular cloud is about 1 molecule per 10^8 m^3. What volume of such a cloud would contain enough alcohol to make a martini (about 10 g of alcohol)?

Discussion questions

19 What do you think would happen if our solar system were to pass through a giant molecular cloud? Do you think the Earth has ever passed through such clouds?

20 Speculate about the possibility of life forms and biological processes occurring in giant molecular clouds. In what ways might the conditions existing in giant molecular clouds favor or hinder biological evolution?

Observing projects

21 Use a telescope to observe at least two of the follow H II regions. Coordinates are for epoch 2000.

Nebula	R.A. (hr min)	Decl. (° ′)
M42 (Orion)	5 35.4	−5 27
M43	5 35.6	−5 16
M20 (Trifid)	18 02.6	−23 02
M8 (Lagoon)	18 03.8	−24 23
M17 (Omega)	18 20.8	−16 11

In each case, can you guess which stars are probably responsible for the ionizing radiation that causes the nebula to glow? Can you see any obscuration or silhouetted features that suggest the presence of interstellar dust? Draw a picture of what you see through the telescope and compare it with a photograph of the object. Take note of which portions of the nebula were not visible through your telescope.

22 On an exceptionally clear, moonless night, use a telescope to observe at least one of the following dark nebulae. These nebulae are very difficult to find because they are recognizable only by the *absence* of stars in an otherwise starry part of the sky. As usual, all coordinates are for epoch 2000.

Nebula	R.A. (hr min)	Decl. (° ′)
Barnard 72 (The Snake)	17 23.5	−23 38
Barnard 86	18 02.7	−27 50
Barnard 133	19 06.1	−6 50
Barnard 142 and 143	19 40.7	+10 57

Are you confident that you actually saw the dark nebula? Does the pattern of background stars suggest a particular shape to the nebula?

23 There are a few fine examples of objects covering such large regions of the sky that they are best seen with binoculars. If you have access to a high-quality pair of binoculars, observe the North America Nebula in Cygnus and the Pipe Nebula in Ophiuchus. Both nebulae are quite faint, so you should attempt to observe them only on an exceptionally dark, clear, moonless night.

The North America Nebula is a cloud of glowing hydrogen gas located about 3° east of Deneb, the brightest star in Cygnus. While searching for the North America Nebula, you may glimpse another diffuse H II region, the Pelican Nebula, located about 2° southeast of Deneb.

The Pipe Nebula is a 7°-long meandering dark nebula to the south and to the east of the star θ Ophiuchi, which is in a section of Ophiuchus that extends southward between the constellations of Scorpius and Sagittarius. Located about 12° east of the bright red star Antares, θ Ophiuchi is easily identified with the aid of star charts published during the summer months in such magazines as *Sky & Telescope* and *Astronomy*.

For further reading

Blitz, L. "Giant Molecular Cloud Complexes in the Galaxy." *Scientific American*, April 1982 • A fine discussion of interstellar molecules is the highlight of this article about the structure and properties of giant molecular clouds.

Bok, B. "Early Phases of Star Formation." *Sky & Telescope*, April 1981 • This informative article on star birth in the interstellar medium includes outstanding photographs of dark nebulae and Bok globules.

Boss, A. "Collapse and Formation of Stars." *Scientific American*, January 1985 • This article discusses how the process of star birth, although hidden from view, can nonetheless be modeled on supercomputers.

Cohen, M. *In Darkness Born: The Story of Star Formation.* Cambridge University Press, 1988 • This well written book presents an overview of our modern understanding of the early stages of stellar evolution.

Herbst, W. "Canis Major R1: A Stellar Nursery." *Mercury*, July/August 1979 • This brief article, which focuses on the Canis Major R1 association, explains how star formation can be triggered by a supernova explosion.

Lada, C. "Energetic Outflows from Young Stars." *Scientific American*, July 1982 • This article describes how radiation emitted by carbon monoxide molecules discloses bipolar outflow from newborn stars.

Loren, R., and Vrba, F. "Starmaking with Colliding Molecular Clouds." *Sky & Telescope*, June 1979 • This article explains how star formation can be triggered by collisions between giant molecular clouds.

Reipurth, B. "Bok Globules." *Mercury*, March/April 1984 • A superb collection of photographs complements this article on the relationship between Bok globules and the process of star formation.

Robinson, L. "Orion's Stellar Nursery." *Sky & Telescope*, November 1982 • This article explores the Orion Nebula and its environs with the aid of high-resolution infrared observations.

Rodriguez, L. "Searching for the Energy Source of the Herbig–Haro Objects." *Mercury*, March/April 1981 • This article, written when the idea of bipolar outflow from young stars was quite new, probes the mysteries of Herbig–Haro objects, which have baffled astronomers since their discovery in the 1950s.

Scoville, N., and Young, J. "Molecular Clouds, Star Formation, and Galactic Structure." *Scientific American*, April 1984 • This article describes how the process of star formation can affect the appearance of a galaxy.

Verschuur, G. "Interstellar Matters." Springer-Verlag, 1988 • This superb book gives an entertaining and informative introduction to the interstellar medium, replete with interesting historical anecdotes.

Wynn-Williams, G. "The Newest Stars in Orion." *Scientific American*, August 1981 • This fascinating article describes the star-forming processes that are occurring across a large portion of the constellation of Orion.

CHAPTER 21

Stellar Maturity and Old Age

Mature stars undergo a remarkable transformation after consuming all the hydrogen in their cores. With the cessation of hydrogen burning, a star leaves the main sequence and expands dramatically to become a red giant. Helium burning ignites in the hot, compressed core of a red giant. The more massive a star is, the more rapidly it consumes its thermonuclear fuels, and so the more rapidly it evolves. A cluster of stars, which naturally incorporates a range of stellar masses, therefore contains stars at various evolutionary stages, even though they are all roughly the same age. This distribution is best seen on an H–R diagram of the cluster, from which we can estimate the cluster's age. An aging red giant, which is so bloated that it constantly leaks gas into space, occasionally becomes unstable and pulsates. Fascinating scenarios occur when giant stars evolve in binary systems.

A mass-loss star Old stars become giants and supergiants whose bloated outer atmospheres shed matter into space. This star, HD 65750, is losing matter at a high rate and is surrounded by a reflection nebula (IC 2220) caused by starlight reflecting from dust grains. These dust grains may have condensed from material shed by the star. A typical red giant can lose 10^{-7} solar mass per year. Many red giants are surrounded by circumstellar shells of matter they have ejected. (Anglo-Australian Observatory)

Thermal equilibrium is a fundamental property of all stars: Energy liberated in their interiors is balanced by the energy radiated from their surfaces. A main-sequence star is one whose radiated energy comes from the thermonuclear process of hydrogen burning in its core. Eventually, however, all the hydrogen in the cores of main-sequence stars is used up. **Core hydrogen burning** then must cease, with dramatic effects upon the stars' equilibrium, structure, and evolution.

21-1 When core hydrogen burning ceases, a main-sequence star becomes a red giant

Hydrogen has been burning in the Sun's core for the past 4.6 billion years. Initially, the Sun's chemical composition was roughly 75 percent hydrogen and 25 percent helium, with a smattering of heavy elements. The ongoing fusion of hydrogen into helium in the Sun's core has dramatically altered the core's composition, however. As shown in Figure 21-1, there is now more helium than hydrogen at the Sun's center. Nevertheless, enough hydrogen remains in the Sun's core for another 5 billion years of core hydrogen burning. Therefore, the Sun's total lifetime on the main sequence will be about 10 billion years.

The length of time that a star remains on the main sequence depends critically on its mass (see Box 21-1 for mathematical details). The most massive main-sequence stars are the most luminous stars, and their rapid emission of energy corresponds to a rapid depletion of hydrogen in their cores. Even though a massive O or B star contains much more hydrogen fuel than a less massive main-sequence star, it also consumes its hydrogen far more rapidly. Table 21-1 shows how long stars take to exhaust the supplies of hydrogen in their cores. Note that high-mass stars gobble up their hydrogen fuel in only a few million years, but low-mass stars take hundreds of billions of years to use up their hydrogen.

Table 21-1 Main-sequence lifetimes

Mass ($M_\odot$)	Surface temperature (K)	Luminosity ($L_\odot$)	Time on main sequence (10^6 years)
25	35,000	80,000	3
15	30,000	10,000	15
3	11,000	60	500
1.5	7,000	5	3,000
1.0	6,000	1	10,000
0.75	5,000	0.5	15,000
0.50	4,000	0.03	200,000

As the supply of hydrogen at a star's center dwindles, the star begins to have difficulty supporting the weight of its outer layers. The burning of hydrogen decreases the number of particles in the star's core, since four hydrogen nuclei are consumed to make each helium nucleus. With fewer particles bouncing around to provide the pressure that supports the weight of the star's outer layers, the central core must contract slightly. This compression raises both the pressure and the temperature at the center of the star, keeping the star in hydrostatic and thermal equilibrium. The higher central temperature increases the rate of hydrogen burning, making the star more luminous, and also causes the region of hydrogen burning to expand outward from the core. Thus, during its final years on the main sequence a star makes a final attempt to maintain hydrostatic and thermal equilibrium by enlarging its hydrogen-burning region. There is still plenty of fresh hydrogen surrounding the star's center. By tapping this supply, the star manages to eke out a few million more years on the main sequence.

Finally, all the hydrogen in the core of an aging main-sequence star is used up, so hydrogen burning ceases in the star's core. However, hydrogen burning still continues within a thin spherical shell surrounding the core. This **shell hydrogen burning** initially occurs only in the hottest region

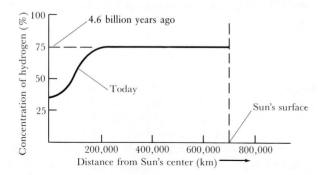

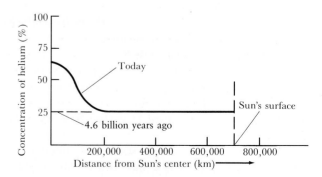

Figure 21-1 *The Sun's chemical composition* The Sun began with a composition of nearly 75 percent hydrogen and 25 percent helium. However, 4.6 billion years of thermonuclear reactions at the Sun's cen- ter have depleted the concentration of hydrogen and increased that of helium.

Box 21-1 Main-sequence lifetimes

During hydrogen burning, a portion of a star's mass is converted into energy. We can use Einstein's famous equation relating mass and energy to calculate how long a star will remain on the main sequence.

Suppose that M is the mass of a star and f is the fraction of the star's mass being converted into energy by hydrogen burning. The total energy E supplied by the hydrogen burning can be expressed as

$$E = fMc^2$$

where c is the speed of light.

This energy is released gradually over many years. Specifically, suppose that L is the star's luminosity and T is the total time over which the hydrogen burning occurs. Then

$$E = LT$$

From these two equations, we see that

$$T = \frac{fMc^2}{L}$$

In other words, a star's lifetime on the main sequence is proportional to its mass divided by its luminosity:

$$T \propto \frac{M}{L}$$

We can carry this analysis further by recalling that main-sequence stars obey the mass–luminosity relation (see Figure 19-20). This tells us that a star's luminosity is roughly proportional to the 3.5 power of its mass:

$$L \propto M^{3.5}$$

Substituting this relationship into the previous proportionality, we find that

$$T \propto \frac{1}{M^{2.5}} = \frac{1}{M^2 \sqrt{M}}$$

This approximate relationship can be used to obtain rough estimates of how long a star will remain on the main sequence. It is often convenient to relate these estimates to the Sun (a typical 1 $M_\odot$ star), which will spend 10^{10} years on the main sequence.

Example: Consider a star whose mass is 4 $M_\odot$. This star will be on the main sequence for

$$\frac{1}{4^{2.5}} = \frac{1}{4 \times 4 \times \sqrt{4}} = \frac{1}{32} \text{ solar lifetime}$$

Thus, a 4 $M_\odot$ star will burn hydrogen in its core for about $\frac{1}{3}$ billion years.

---

just outside the core, where the hydrogen fuel has not yet been exhausted.

No longer are there any thermonuclear reactions producing energy in the star's core. As a result, the heat flowing out of the hot core is not being replaced. The core thus gradually contracts, converting gravitational energy into thermal energy to maintain thermal equilibrium. The hydrogen-burning shell slowly works its way outward from the original core, thereby dumping more helium onto the core, which continues to contract and heat up. Over the course of a few million years, the core of a 1 $M_\odot$ star becomes compressed to about one-tenth of its original radius and its central temperature increases from around 15 million K to about 100 million K.

The high temperature of the star's core stimulates the hydrogen-burning shell, whose increased energy output increases the star's luminosity and causes the star's outer layers to expand, thereby increasing the overall size of the star. As the star's outer atmosphere expands farther and farther

into space, its gases cool. Soon the temperature of the star's bloated surface falls to about 3500 K, at which point the gases glow with a reddish hue, in accordance with Wien's law. The star is then appropriately called a **red giant**.

The Sun will take about 5 billion years more to finish converting hydrogen into helium at its core. As the Sun's core contracts, its atmosphere will expand to envelop first Mercury, then Venus, and finally our own planet. The red-giant Sun will swell to a diameter of about 1 AU, and its surface temperature will decline to about 3500 K. Although the Sun's surface temperature will be much lower than it is today, the Sun will be so huge that its luminosity will be much greater than today. As a full-fledged red giant (see Figure 21-2), our star will shine with the brightness of 2000 Suns. Some of the inner planets will be vaporized, and the thick atmospheres of the outer planets will boil away to reveal tiny, rocky cores. Thus, in its later years the aging Sun will destroy the planets that have accompanied it since its birth.

The Sun as a main-sequence star
(diameter = 1.4×10^6 km $\approx \frac{1}{100}$ AU)

The Sun as a red giant
(diameter $\approx$ 1 AU)

Figure 21-2 *The Sun today and as a red giant* Today, the Sun's energy is produced in a hydrogen-burning core whose diameter is about 300,000 km. When the Sun becomes a red giant, in some 5 billion years, it will draw its energy from a hydrogen-burning shell surrounding a compact helium-rich core. The helium core will have a diameter of only about 30,000 km.

21-2 Helium burning begins at the center of a red giant

Helium is the "ash" from hydrogen burning. When a star first becomes a red giant, its hydrogen-burning shell surrounds a small, compact core of almost pure helium. In a moderately low-mass red giant, which the Sun will be 5 billion years from now, the dense helium core is about twice the size of the Earth and the star's bloated surface has a diameter roughly half that of the Earth's orbit.

At first, thermonuclear reactions do not occur in the helium core of a red giant, because the temperature is too low to fuse helium nuclei. Each helium nucleus contains two protons and thus has twice the positive electric charge of a hydrogen nucleus. Compared to hydrogen, the helium nuclei must move faster to overcome their stronger electric repulsion and get close enough to fuse together, and this requires higher temperatures (recall Box 7-2). As the hydrogen-burning shell moves outward in the star, it adds mass to the helium core. The core slowly contracts, forcing the star's central temperature to climb.

When the central temperature finally reaches 100 million K, **helium burning** is ignited at the star's center. This new thermonuclear reaction occurs in two steps. First, two helium nuclei combine to form an isotope of beryllium:

$$^4\text{He} + {}^4\text{He} \rightarrow {}^8\text{Be}$$

This particular beryllium isotope is very unstable and quickly breaks down into two helium nuclei soon after it

forms. However, in the star's dense core a third helium nucleus may strike the ^{8}Be nucleus before it has a chance to fall apart. Such a collision creates a stable, common isotope of carbon:

$$^8\text{Be} + {}^4\text{He} \rightarrow {}^{12}\text{C} + \gamma$$

In this process, a gamma-ray photon (γ) is released.

During the pioneering days of nuclear-physics research, helium nuclei were called **alpha particles,** and the fusion of three helium nuclei to form a carbon nucleus is still called the **triple alpha process.** Some of the carbon created in this process can fuse with an additional helium nucleus to produce oxygen:

$$^{12}\text{C} + {}^4\text{He} \rightarrow {}^{16}\text{O} + \gamma$$

Thus, both carbon and oxygen make up the "ash" of helium burning.

The second step in the triple alpha process and the process of oxygen formation releases both energy and a gamma-ray photon. As a result, the aging star again has a central energy source for the first time since leaving the main sequence. This energy source, properly called **core helium burning** because of its central location, establishes thermal equilibrium, thereby preventing any further gravitational contraction of the star's core. A mature red giant burns helium in its core for about 20 percent as long as the time it spent burning hydrogen as a main-sequence star. For example, in the distant future the Sun will consume helium in its core for about 2 billion years.

The way in which helium burning begins at a red giant's center depends on the mass of the star. In high-mass stars (those with masses greater than about 3 $M_\odot$), helium burning begins gradually as temperatures in the star's core approach 100 million degrees. In low-mass stars (those with masses less than about 3 $M_\odot$), helium burning begins explosively and suddenly, in what is called the **helium flash.**

The helium flash occurs because of unusual conditions that develop in the core of a low-mass star on its way to becoming a red giant. To appreciate these conditions we must first understand how an ordinary gas behaves, then explore how the densely packed electrons at the star's center alter this behavior.

When a gas is compressed, it usually becomes denser and warmer. For convenience, scientists used the concept of a **perfect gas** that has a simple relationship between pressure, temperature, and density. Specifically, the pressure exerted by a perfect gas is directly proportional to both the density and the temperature of the gas. Many real gases in fact behave like a perfect gas over a wide range of temperatures and densities.

Under most circumstances, the gases inside a star act like a perfect gas: If the gas is compressed, it heats up, and if it expands, it cools down. This behavior serves as a safety valve to ensure that the star does not explode. For example, if energy production overheats the star's core, the core expands, cooling the gases and slowing the rate of thermonuclear reactions. Conversely, if too little energy is being created to support the star's overlying layers, the core becomes compressed and the increased temperatures that result speed up the thermonuclear reactions to increase the energy output.

In a low-mass red giant, the core must undergo considerable gravitational compression to drive the temperatures high enough to begin helium burning. At the extreme pressures and temperatures found inside the star, the atoms are completely ionized; thus, matter throughout most of the star consists of dissociated nuclei and electrons. In the highly compressed core, the free electrons are so closely crowded together that a law of quantum mechanics called the **Pauli exclusion principle** comes into play. This principle, formulated in 1925 by the Austrian physicist Wolfgang Pauli, explains that two identical particles cannot simultaneously occupy the same "quantum state." A quantum state is a particular set of circumstances concerning locations and speeds that are available to a particle. In the submicroscopic world of atoms and particles, the Pauli exclusion principle is analogous to saying you can't have two things in the same place at the same time.

Just before the onset of helium burning, the electrons in the core of a low-mass star have become so closely crowded together that any further compression would violate the Pauli exclusion principle. Because the electrons cannot be squeezed any closer together, they produce a powerful pressure that resists further core contraction.

This phenomenon, in which closely packed particles resist compression because of the Pauli exclusion principle, is called degeneracy. Astronomers say that the helium-rich core of a low-mass red giant is "degenerate" and is supported by **degenerate-electron pressure.** This degenerate pressure, unlike the pressure of a perfect gas, does not depend on temperature.

When the temperature in the core of a low-mass red giant reaches the high level required for the triple alpha reaction, energy begins to be released. The helium nuclei become heated, which causes the triple alpha process to proceed more rapidly. However, the pressure provided by the degenerate electrons is independent of the temperature, so the pressure does not change. Without the "safety valve" of increasing pressure, the star's core cannot expand and cool. The rising temperature causes the helium to burn at an ever-increasing rate, producing the helium flash. Eventually, however, the temperature becomes so high that electron degeneracy is no longer a factor. The electrons then behave like a perfect gas and the star's core expands, terminating the helium flash. These events occur extremely rapidly, so that the helium flash is over in only a few seconds.

21-3 Evolutionary tracks on the H–R diagram reveal the ages of star clusters

It is enlightening to follow the post–main-sequence evolution of mature stars by plotting their evolutionary tracks on a Hertzsprung–Russell diagram (see Figure 21-3). The **zero-age main sequence** (or **ZAMS**) is the location on an H–R diagram where stars first achieve hydrostatic equilibrium, a balance between the inward force of gravity and the outward pressure produced by hydrogen burning. In subsequent years, the evolutionary tracks slowly inch away from the ZAMS as the hydrogen-burning core grows in search of fresh fuel. The dashed line on Figure 21-3 shows the locations of the stellar models when all the core hydrogen has been consumed.

After the cessation of core hydrogen burning, the points representing high-mass stars move rapidly from left to right across the H–R diagram. During this transition, the star's core contracts and its outer layers expand in response to increased energy output from the star's hydrogen-burning shell. Although the star's surface temperature is decreasing, its surface area is increasing, so that its overall luminosity remains roughly constant.

Just before core helium burning begins, the evolutionary tracks of high-mass stars turn upward in the red-giant region of the H–R diagram. After the core helium burning begins, however, the evolutionary tracks back away from these temporary peak luminosities. The tracks then wander back and

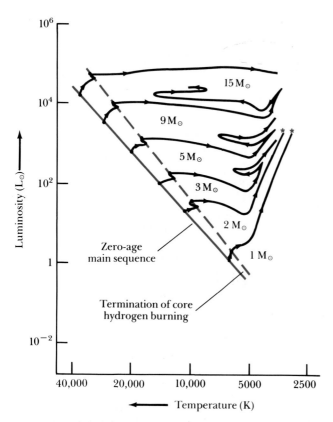

Figure 21-3 Post–main-sequence evolution *The evolutionary tracks of six stars are shown on this H–R diagram. In the high-mass stars, core helium burning ignites where the evolutionary tracks make a sharp downward turn in the red-giant region of the diagram. The evolutionary tracks for low-mass stars (1 M⊙ and 2 M⊙) are shown only up to the points, indicated by the red asterisks, where the helium flash occurs at their centers. (Adapted from I. Iben)*

forth in the red-giant region while the stars readjust to their new energy sources.

The evolutionary tracks of two low-mass stars also appear in Figure 21-3. However, these two tracks are shown only to the point where the helium flash occurs in these stars. The sudden flood of energy released by the helium flash does not disrupt the star's surface, but rather expends itself in altering conditions in the star's interior.

Immediately after the helium flash, a low-mass star's superheated core expands, because now degeneracy has been removed and it is again behaving like a perfect gas. Around the expanding core, temperatures fall, cooling the hydrogen-burning shell and reducing its energy output. As energy output declines, the star's outer layers contract and heat up. Consequently, a post–helium-flash star should be both smaller and hotter at the surface than a red giant.

Examples of these post–helium-flash stars are found in old star clusters, called **globular clusters** because of their spherical shape. A typical globular cluster, like that shown in Figure 21-4, contains up to 1 million stars in a volume less than 100 parsecs across. Astronomers know that such clus-

ters are old because they contain no high-mass main-sequence stars. If you measure the brightnesses and surface temperature of many stars in a globular cluster and plot the data on an H–R diagram, as shown in Figure 21-5, you discover that the upper half of the main sequence is missing. All the high-mass main-sequence stars evolved long ago into red giants, leaving behind only low-mass, slowly evolving stars that still have core hydrogen burning.

The H–R diagram of a globular cluster typically shows a horizontal grouping of stars in the left-of-center portion of the diagram. As seen in Figure 21-5, these stars form a horizontal row at luminosities of about 50 L⊙. These stars, called **horizontal-branch stars**, are post–helium-flash low-mass stars. In years to come, these stars will move back toward the red-giant region as their fuel is devoured by core helium burning and shell hydrogen burning.

An H–R diagram of a cluster can be used to determine the age of the cluster. In the diagram for a very young cluster (review Figure 20-11), the entire main sequence is intact. As a cluster gets older, though, stars begin to leave the main sequence. The high-mass, high-luminosity stars are the first to become red giants as the main sequence starts to burn down like a candle. Over the years, the main sequence gets shorter and shorter. The top of the surviving portion of the main sequence is known as the **turnoff point**. The stars at the point are just now exhausting the hydrogen in their cores, and their main-sequence lifetime is equal to the age of the cluster (recall Table 21-1). For example, in the case of the cluster M55 (see Figure 21-5), 0.8 M⊙ stars have just left the main sequence, indicating that the cluster's age is roughly 15 billion years.

Data for several star clusters are plotted on Figure 21-6, along with turnoff-point times from which the ages of the clusters can be estimated. The youngest clusters (those with most of their main sequences still intact) are said to be **metal**

Figure 21-4 A globular cluster *A globular cluster is a spherical cluster that typically contains a few hundred thousand stars. This cluster, referred to as M13, is located in the constellation of Hercules, roughly 25,000 light years from Earth. (U.S. Naval Observatory)*

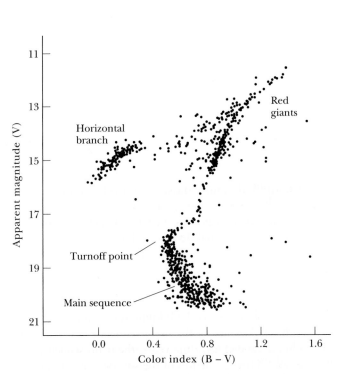

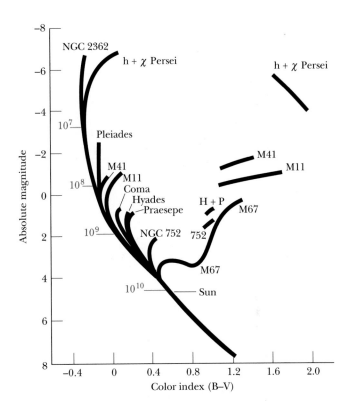

Figure 21-5 H–R diagram of the globular cluster M55 *Each dot represents a star whose V magnitude and B − V color index have been measured. The upper half of the main sequence is missing. The horizontal-branch stars, believed to be low-mass stars that recently experienced the helium flash, exhibit core helium burning and shell hydrogen burning. (Adapted from D. Schade, D. VandenBerg, and F. Hartwick)*

Figure 21-6 A composite H–R diagram *The shaded bands indicate where data from various open clusters fall on the H–R diagram. The ages of turnoff points (in years) are listed in red alongside the main sequence. The age of a cluster can be estimated from the location of the cluster's turnoff point, where the cluster's most massive stars are just now leaving the main sequence. (Adapted from A. Sandage)*

rich, because their spectra contain many prominent spectral lines of heavy elements. This material originally came from dead stars that exploded long ago, enriching the interstellar gases with the heavy elements formed in their cores. The Sun is a relatively young, metal-rich star.

Most of the oldest clusters are globular clusters. Such clusters are generally located outside the plane of our Galaxy, and their spectra show only weak lines of heavy elements. These ancient stars are thus said to be **metal poor,** because they typically contain only about 3 percent of the heavy elements relative to hydrogen as does the Sun. Such stars were created long ago from interstellar gases that had

not yet been substantially enriched with heavy elements. Spectra of a metal-poor star and of the Sun are compared in Figure 21-7.

The young, metal-rich stars, like our Sun, are commonly called **population I** stars. The old, metal-poor stars are **population II** stars. As we shall see in later chapters, hydrogen and helium were essentially the only two elements to emerge during the birth of the universe. Thus, the most ancient stars should have no spectral lines of any heavy elements. Astronomers have been searching for these very old, metal-free **population III** stars. So far only one candidate has been found.

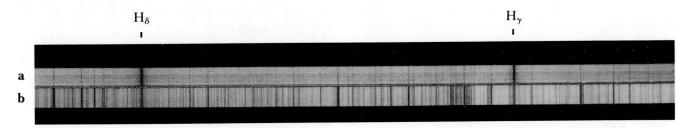

Figure 21-7 Spectra of a metal-poor and a metal-rich star *These spectra compare (a) a metal-poor star and (b) a metal-rich star (the Sun). Numerous spectral lines prominent in the solar spectrum are caused by* *elements heavier than hydrogen and helium. Note that corresponding lines in the metal-poor star's spectrum are weak or absent. Both spectra cover a wavelength range that includes H_γ and H_δ. (Lick Observatory)*

21-4 Supergiants and red giants typically show mass loss

Both supergiant and red-giant stars are so enormous that their bloated outer layers constantly leak gases into space. At times, this **mass loss** can be quite significant.

Mass loss can be detected spectroscopically. Escaping gases coming toward us exhibit narrow absorption lines that are slightly blueshifted. According to the Doppler effect (review Figure 5-20), this small shift toward shorter wavelengths corresponds to a speed of approximately 10 km/s. This value is characteristic of the expansion velocities with which gases leave the tenuous outer layers of red giants. A typical mass-loss rate for a red giant is roughly 10^{-7} solar masses per year. For comparison, the Sun's mass loss rate is only 10^{-14} $M_\odot$ per year.

Betelgeuse, in the constellation of Orion, is a good example of a red supergiant experiencing mass loss (see Figure 21-8). Betelgeuse is 310 light years away and has a diameter roughly equal to the diameter of Mars's orbit. Recent spectroscopic observations show that this star is losing mass at the rate of 1.7×10^{-7} solar masses per year and is surrounded by a huge **circumstellar shell** that is expanding at 10 km/s. These gases escaping from Betelgeuse have been detected out to distances of 10,000 AU from the star. Consequently, the expanding circumstellar shell has an overall diameter of $\frac{1}{3}$ light year.

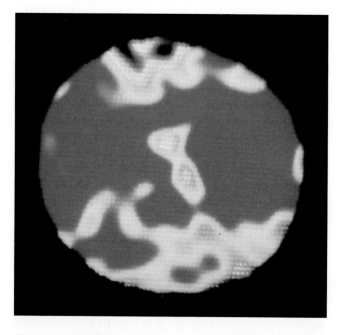

Figure 21-8 Betelgeuse *Betelgeuse (α Orionis) is one of the largest stars, with a diameter about 800 times the Sun's. This image, obtained by a technique called speckle interferometry, shows temperature differences on the surface of Betelgeuse. The bluish regions are cool, the red areas warmest. The two light blue spots near the center of the image are each about the same size as the Earth's orbit. (NOAO)*

Supergiant stars, which are brighter than 10^5 Suns, suffer mass loss throughout most of their existence, with mass-loss rates comparable to those of the red giants. Figure 21-9 shows a supergiant star losing mass. This particular star is a member of a class called **Wolf–Rayet stars,** named after two nineteenth-century astronomers who first drew attention to bright emission lines in their spectra. Wolf–Rayet stars have masses in the range of 30 to 50 $M_\odot$ and lie near the main sequence on the H–R diagram. They are thus fairly young stars. A large percentage of these rare and beautiful stars have been confirmed to be members of close binary systems. In fact, many Wolf–Rayet stars have nearby companions whose gravity plays an important role in detaching gases from the supergiants' outer atmospheres.

We shall see in the next chapter that dying stars eject vast quantities of material into space. Nevertheless, the mass loss from supergiants and red giants accounts for roughly one-fifth of all the matter returned by the stars to the interstellar medium.

21-5 Mass transfer in close binary systems can produce unusual double stars

When one member of a double star system becomes a red giant, it can dump gas onto its companion. This process, called **mass transfer,** occurs only if the red giant becomes sufficiently bloated and its companion star is near enough that the red giant's outer layers can be gravitationally captured by the companion star.

In the mid-1800s, the French mathematician Edouard Roche pointed out that you can draw a figure-eight curve around two stars in a binary to portray the gravitational domain of each star. This figure-eight curve (which is one of the equipotential contours described in Box 17-1) is often called the **critical surface.** Each half of the curve is known as a **Roche lobe.** The more massive star is always located inside the larger Roche lobe. If gas from a star leaks over its surrounding Roche lobe, it is no longer bound by gravity to that star and is free to escape into space or fall onto the companion star. When mass transfer occurs, gases flow through the point where the two Roche lobes touch, called the **inner Lagrangian point.**

In many binaries, the stars are so far apart that even during their red giant stage the stars' surfaces remain well inside their Roche lobes so that little mass transfer can occur. Each star thus lives out its life as if it were single and isolated.

A binary system in which each star is within its Roche lobe is referred to as a **detached binary** (see Figure 21-10). If the two stars are relatively close together, when one star expands to become a red giant, it may fill or overflow its Roche lobe, in which case the system is called a **semidetached**

Figure 21-9 *The Wolf–Rayet star HD 56925* *This beautiful nebulosity surrounds a supergiant star that is experiencing significant mass loss. The ejected material collides and interacts with the surrounding interstellar gas and dust, thereby producing the cosmic bubble seen here. This nebulosity, referred to as NGC 2359, is located in the constellation of Canis Major. (Anglo-Australian Observatory)*

binary. If both stars happen to fill their Roche lobes, the system is called a **contact binary,** because the two stars actually touch and may share a common envelope of gas. Semidetached and contact binaries are most easily detected if they happen also to be eclipsing binaries, because their

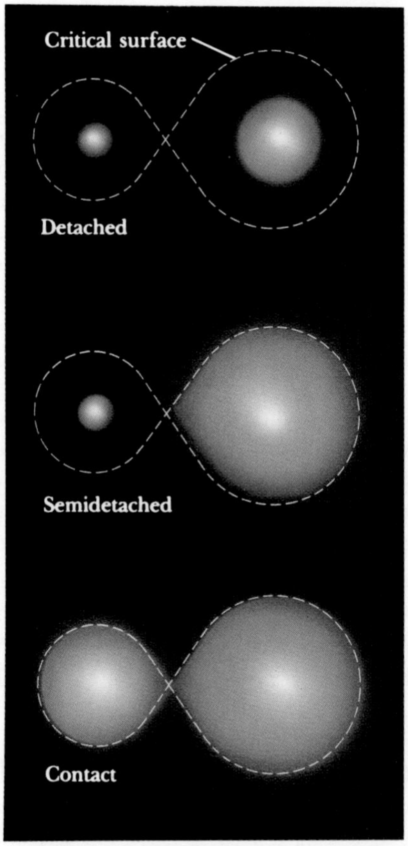

Figure 21-10 *Detached, semidetached, and contact binaries* *A double star is said to be a detached, semidetached, or contact binary, depending on whether neither, either, or both of the stars fills its Roche lobe. Mass transfer is often observed in semidetached binaries. The two stars in a contact binary share the same outer atmosphere.*

light curves then have a distinctly rounded appearance caused by the gravitationally distorted egg-shaped stars (recall Figure 19-23c).

Mass transfer in close binaries can produce unusual results. For instance, the eclipsing binary called Algol (from an Arabic term for demon) is a semidetached binary that can be easily seen with the naked eye in the constellation of Perseus. From Algol's light curve (see Figure 21-11a) and spectroscopic observations that give Doppler shifts, astronomers calculate that the detached star is on the main sequence and its less massive companion is filling its Roche lobe.

During the 1960s, astronomers labored to understand semidetached systems like Algol, which collectively came to be called Algol-type binaries. According to stellar evolution theory, the more massive a star is, the more rapidly it should evolve. But in Algol, the more massive primary is still on the main sequence, whereas the less massive secondary has evolved to become a red giant. The apparent contradiction of having the less massive star also be the more evolved star was thus called the Algol paradox.

The paradox was resolved when Zdenek Kopal at the University of Manchester and others proposed that the red giant in Algol-type binaries was originally the more massive star. As it left the main sequence to become a red giant, this star overflowed its Roche lobe, dumping gas onto the originally less massive companion. Because of the resulting mass transfer, that companion became the more massive star.

Another class of semidetached binaries, called β Lyrae variables after their prototype in the constellation of Lyra, may be related to the Algol-type binaries. As with Algol, the less massive star in β Lyrae fills its Roche lobe, but its light curve (see Figure 21-11b) and spectra demonstrate that the more massive detached star is severely under-luminous, contributing virtually no light at all to the visible radiation coming from the system. Furthermore, the spectra of β Lyrae are confounded by certain unusual features, some of which are caused by gas flowing between the stars and around the system as a whole.

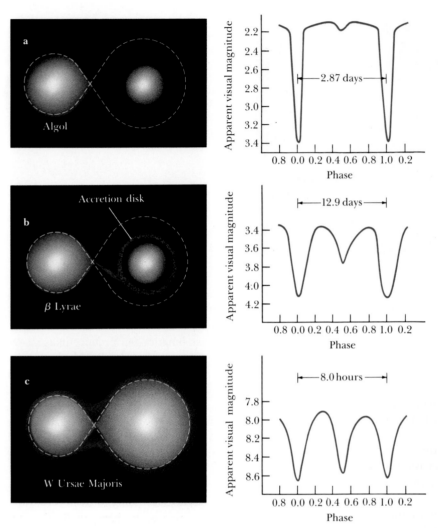

The mystery of β Lyrae was solved in 1963 when Su-Shu Huang of Northwestern University published his interpretation that the under-luminous star in β Lyrae is enveloped in a rotating disk of gas captured from its bloated companion. The disk is so thick that it completely shrouds the secondary star, making it impossible to observe at visible wavelengths.

Observations made since the 1960s have largely confirmed Huang's model, although the disk, which has come to be called an **accretion disk,** is now believed to be even thicker than Huang had supposed. The primary star is overflowing its Roche lobe, with gases streaming across the inner Lagrangian point onto the disk at the rate of 10^{-5} $M_\odot$ per year. Ultraviolet spectra taken from spacecraft in the 1970s revealed details about the gas flow between and around the two stars. There are apparently clouds of gas at the Lagrangian triangular points L_4 and L_5 (see Box 17-1). Some gas is constantly escaping altogether from the system.

The fate of an Algol or β Lyrae system depends on many factors, most notably on the stars' masses, which determine their rates of evolution. If the detached star is comparatively massive, it will evolve rapidly, expanding to fill its Roche lobe while the companion star is still filling its own Roche lobe. The result is a contact binary in which both stars share the same photosphere. Such binaries are often called W Ursae Majoris stars, after the prototype of this class (see Figure 21-11c). In Chapters 22 and 23 we shall see that mass transfer onto dead stars produces some of the most extraordinary objects in the sky.

21-6 Many mature stars pulsate

After core helium burning begins, mature stars move across the middle of the H–R diagram. Figure 21-3 showed the evolutionary tracks of high-mass stars crisscrossing the H–R diagram. Post–helium-flash low-mass stars on the horizontal branch also cross the middle of the H–R diagram as they return to the red-giant region.

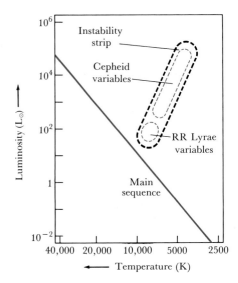

Figure 21-12 The instability strip *The instability strip occupies a region between the main sequence and the red-giant branch on the H–R diagram. A star passing through this region along its evolutionary track becomes unstable and pulsates.*

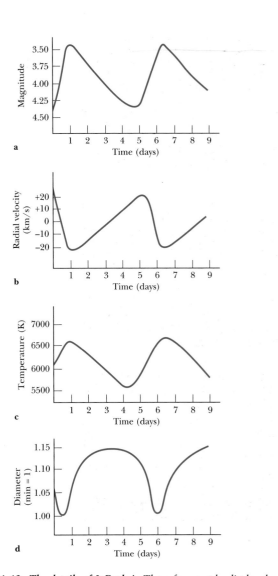

Figure 21-13 The details of δ Cephei *These four graphs display details about the pulsations of δ Cephei: (a) the star's light curve; (b) the velocity curve; (c) periodic variations in the star's surface temperature; and (d) periodic variations in the star's diameter.*

During these transitions across the H–R diagram, a star can become unstable and pulsate. In fact, there is a region on the H–R diagram between the main sequence and the red-giant branch that is called the **instability strip** (see Figure 21-12). When a star passes through this region as it evolves, the star pulsates. As it does so, its brightness varies periodically.

The prototype of an important class of pulsating stars called **Cepheid variables**, or simply "Cepheids," was discovered in 1784 by John Goodricke, a deaf, mute, 19-year-old English amateur astronomer. This star, δ Cephei, varies regularly in apparent magnitude from 4.3 to 3.4, with a period of 5.4 days.

A Cepheid variable is recognized by the characteristic way in which its light output varies: rapid brightening followed by gradual dimming. This behavior is best displayed in a light curve, like the one shown in Figure 21-13a.

A Cepheid variable brightens and fades because of cyclic expansion and contraction of the star's outer envelope. This behavior has been deduced from spectroscopic observations. In 1894, the Russian astronomer A. A. Belopolsky noticed that spectral lines in the spectrum of δ Cephei shift back and forth with the same 5.4-day period as that of the magnitude variations. From the Doppler effect, we can translate these wavelength shifts into speeds and draw a velocity curve (see Figure 21-13b). Negative speeds mean that the star's surface is expanding toward us; positive speeds mean that the star's surface is receding from us. Note that the light and velocity curves are mirror images of each other. The star is brighter than average as it expands and dimmer than average while it contracts.

When a Cepheid variable pulsates, the star's surface oscillates up and down like a spring. During these cyclical expansions and contractions, the star's gases alternately heat up and cool down. The temperature changes in the star's surface that result are displayed in Figure 21-13c. The periodic changes in the star's diameter are shown in Figure 21-13d.

Just as a bouncing ball eventually comes to rest, a pulsating star would soon stop pulsating without some sort of mechanism to keep its oscillations going. In 1941, the British astronomer Arthur Eddington suggested that a Cepheid pulsates because of a valvelike action involving the periodic ionization and deionization of gases in its outer layers. Eddington's valve mechanism requires that the star be more opaque or "light-tight" when compressed than when expanded. When the star is compressed, trapped heat can push the star's surface outward. When the star is expanded, the heat escapes and so the star's surface, which is no longer supported, can then fall inward.

Following up on Eddington's idea in the 1960s, American astronomer John Cox proved that helium is the gas that keeps Cepheids pulsating. Normally, when a star's helium is compressed, its temperature increases and the gas becomes more transparent. In certain layers near the star's surface, however, compression may ionize helium (remove one of its electrons) instead of raising its temperature. In this helium-ionization zone, the compressed helium is quite opaque and can trap the star's heat, which pushes the star's surface outward. Then, as the star expands, helium ions recombine with electrons, and so the gas becomes more transparent and releases the trapped energy. The star's surface then falls inward, recompressing the helium and the whole cycle begins all over again.

Cepheid variables are very important to astronomers, because there is a direct relationship between a Cepheid's period and its average luminosity. Dim Cepheid variables pulsate rapidly, with periods of one to two days and average brightness of a few hundred Suns. The most luminous Cepheids are the slowest variables, with periods of 100 days and average brightnesses equal to 10,000 Suns. This connection between period and brightness is known as the **period–luminosity relation.** This relationship plays an important role in measuring the overall size and structure of the universe, as we shall see in Chapter 26.

The details of a Cepheid's pulsation depend on the abundance of the heavy elements in its atmosphere. The average luminosity of metal-rich Cepheids is roughly four times greater than the average luminosity of metal-poor Cepheids having the same period. Thus, there are two classes: **Type I Cepheids,** which are the brighter, population I, metal-rich stars; and **Type II Cepheids,** which are the dimmer, population II, metal-poor stars. The period–luminosity relation for both types is shown in Figure 21-14.

The evolutionary tracks of mature, high-mass stars pass back and forth through the upper end of the instability strip on the H–R diagram. These stars become Cepheids when the ionization layers occur at just the right depth to drive the pulsations. For stars on the high-temperature side of the in-

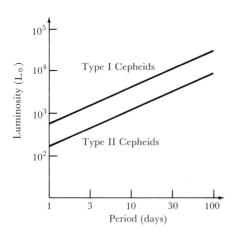

Figure 21-14 The period–luminosity relation *The period of a Cepheid variable is directly related to its average luminosity. The metal-rich (Type I) Cepheids are brighter than the metal-poor (Type II) Cepheids.*

stability strip, the helium ionization zone is too close to the surface and contains only an insignificant fraction of the star's mass. For stars on the cool side of the instability strip, convection becomes important in the star's outer layers and prevents the storage of the energy needed to drive the pulsations. Thus Cepheids exist only in a narrow range of effective temperatures on the H–R diagram.

Low-mass, post–helium-flash stars pass through the lower end of the instability strip as they move in the horizontal branch along their evolutionary tracks. These stars become **RR Lyrae variables,** named after their prototype in the constellation of Lyra. RR Lyrae variables all have periods shorter than one day, and have roughly the same average brightness as the stars on the horizontal branch. In fact, as shown in Figure 21-12, the RR Lyrae region of the instability strip is actually a segment of the horizontal branch.

Stellar pulsations can, in rare cases, be quite substantial. In some cases, the expansion velocity exceeds the star's escape velocity and the star's outer layers are ejected completely. As we shall see in the next chapter, significant mass ejection accompanies the death of stars in the sometimes violent process that renews and enriches the interstellar medium for future generations of stars.

Key words

accretion disk	Cepheid variable	contact binary	degeneracy
alpha particle	circumstellar shell	critical surface	degenerate-electron pressure

detached binary	mass loss	population I, II, III	Type I Cepheid
globular cluster	mass transfer	red giant	Type II Cepheid
helium burning	metal-poor star	Roche lobe	Wolf–Rayet star
helium flash	metal-rich star	RR Lyrae variable	zero-age main sequence (ZAMS)
horizontal-branch star	Pauli exclusion principle	semidetached binary	
inner Lagrangian point	perfect gas	triple alpha process	
instability strip	period–luminosity relation	turnoff point	

Key ideas

- The more massive a star is, the shorter is its main-sequence lifetime; the Sun has been a main-sequence star for about 4.6 billion years and should remain one for about another 5 billion years.

- Core hydrogen burning ceases when the hydrogen has been exhausted in the core of a main-sequence star, leaving a core of nearly pure helium surrounded by a shell from which hydrogen burning works its way outward in the star; the core shrinks and becomes hotter while the star expands to become a red giant.

- When the central temperature of a red giant reaches about 100 million K, the thermonuclear process of helium burning begins there; this process, also called the triple alpha process, converts helium to carbon and oxygen.

   In a more-massive red giant, helium burning begins gradually; in a less-massive red giant, it begins suddenly, in a process called the helium flash.

   As a star moves through the red-giant phase of its evolution, its evolutionary track moves rapidly to the right from the zero-age main sequence (ZAMS) on the H–R diagram at fairly constant luminosity.

- In a high-mass red giant, luminosity increases sharply just before core helium burning begins, then decreases sharply just after core helium burning begins; the star's evolutionary track then moves back and forth across the red-giant region of the H–R diagram. In a low-mass red giant, after the helium flash the star moves to the horizontal branch on the H–R diagram.

- The age of a stellar cluster can be estimated by plotting its stars on an H–R diagram; the age of the cluster is equal to the age of the main-sequence stars at the turnoff point (the upper end of the remaining main sequence); for an older cluster, the upper portion of the main sequence is missing, because more massive main-sequence stars will have become red giants.

- Relatively young population I stars are metal rich; ancient population II stars are metal poor.

- Supergiants and red giants undergo extensive mass loss, sometimes producing circumstellar shells of ejected material around the stars.

- Mass transfer in a close binary system can affect the appearance and evolution of both stars.

   Mass transfer occurs when one star in a close binary overflows its Roche lobe; gas flowing from one star to the other passes across the inner Lagrangian point.

- When a star's evolutionary track carries it through a region called the instability strip in the H–R diagram, the star becomes unstable and begins to pulsate.

   Cepheid variables are high-mass pulsating variables having a regular relationship between their periods of pulsation and their luminosities.

   RR Lyrae variables are low-mass pulsating variables with short periods.

Review questions

1 Why do you suppose that the majority of the stars we see in the sky are main-sequence stars?

2 On what grounds are astronomers able to say that the Sun has about 5 billion years remaining in its main-sequence stage?

3 What does it mean when an astronomer says that a star "moves" from one place to another on an H–R diagram?

*4 What are the main-sequence lifetimes of (a) a 25-$M_\odot$ star, and (b) a 5-$M_\odot$ star? Compare these lifetimes with that of the Sun.

5   What will happen inside the Sun 5 billion years from now when it begins to turn into a red giant?

6   What is the helium flash?

7   How is a degenerate gas different from ordinary gases?

8   Explain how and why the turnoff point on the H–R diagram of a cluster is related to the cluster's age.

9   Why do astronomers believe that globular clusters are made of old stars?

10   What is the difference between population I and population II stars?

11   What is the Algol paradox and how was it resolved?

12   Why do astronomers attribute the observed Doppler shifts of a Cepheid variable to pulsation, rather than to some other cause, such as orbital motion?

13   Suppose that an oxygen nucleus were to be fused with a helium nucleus. What element would be formed? Look up the relative abundance of this element and comment on whether such a process is likely.

Advanced questions

> **Tips and tools . . .**
> Recall that 6×10^{11} kg of hydrogen is converted into helium each second at the Sun's center, as explained in Chapter 18 (see Section 18.1). You may find it helpful to review the discussion of apparent magnitude, absolute magnitude, and luminosity found in Boxes 19-2 and 19-3. Assume that the bolometric correction for stars listed below is zero.

*14   (Basic)   Assuming that the Sun's luminosity remains roughly constant throughout its 10 billion years on the main sequence, what fraction of the Sun's hydrogen would be converted into helium?

*15   The earliest fossil records of life forms indicate that life appeared on the Earth about 2.8 billion years after the solar system was formed. If the Sun's lifetime on the main sequence is 10 billion years, what is the most mass that a star could have in order that its lifetime on the main sequence be long enough to permit life to form on one or more of its planets? Assume the evolutionary process would be similar to those which have occurred here on the Earth.

*16   When the Sun becomes a red giant, its luminosity will be 100 times greater than it is today. Assuming that this luminosity is caused only by the burning of the Sun's remaining hydrogen, calculate how long our star will be a red giant.

17   What observations would you make of a star to determine whether its primary source of energy was hydrogen or helium burning?

*18   (Challenging)   The star X Arietis is a RR Lyrae variable. Its apparent magnitude varies between 8.97 and 9.95 with a period of 0.65 day. Interstellar extinction dims the star by half a magnitude. Approximately how far away is the star?

*19   Polaris (α CMi) is a Type I Cepheid variable. Its apparent magnitude varies between 1.92 and 2.02 with a period of 3.97 days. Approximately how far away is the star?

Discussion questions

20   The half-life of the ^{8}Be nucleus is 2.6×10^{-16} second, which is the average time that elapses before this unstable nucleus decays into two alpha particles. How would the universe be different if the ^{8}Be half-life instead were zero? How would the universe be different if the ^{8}Be nucleus were stable?

21   What observational consequences would we find in H–R diagrams for star clusters if the universe had a finite age? Could we use these consequences to establish constraints on the possible age of the universe? Explain.

Observing projects

22   Observe several of the following red giants and supergiants with the naked eye and through a telescope. These stars are most easily found with the aid of star charts published every month in such magazines as *Sky & Telescope* and *Astronomy*. Epoch 2000 coordinates are given below.

Star	Spectral type	R.A. (hr min)	Decl. (° ′)
Almach (γ And)	K3II	2 03.9	+42 20
Aldebaran (α Tau)	K5III	4 35.9	+16 31
Betelgeuse (α Ori)	M2I	5 55.2	+07 24
Arcturus (α Boo)	K2III	14 15.7	+19 11
Antares (α Sco)	M1I	16 29.5	−26 26
Eltanin (γ Dra)	K5III	17 56.7	+51 29
Enif (ϵ Peg)	K2I	21 44.2	+09 52

Is the reddish color of these stars apparent in comparison with neighboring stars?

23 Several of the open clusters listed in Figure 21-6 can be seen quite well with a good pair of binoculars. Observe as many of these clusters as you can, using both a telescope and a pair of binoculars. Note the overall distribution of stars in each cluster. Which clusters are seen better through binoculars than through a telescope? Which clusters can you see with the naked eye?

Star cluster	R.A. (hr min)	Decl. (° ′)
h Persei	2 19.0	+57 09
χ Persei	2 22.4	+57 07
Pleiades	3 47.0	+24 07
Hyades	4 27	+16
Praesepe	8 40.1	+19 59
Coma	12 25	+26
M11	18 51.1	−06 16

24 There are many beautiful globular clusters scattered around the sky that can be easily seen with a small telescope. Several of the brightest and nearest globulars are listed below.

Globular cluster	R.A. (hr min)	Decl. (° ′)
M3 = NGC 5272	13 42.2	+28 23
M5 = NGC 5904	15 18.6	+2 05
M4 = NGC 6121	16 23.6	−26 32
M13 = NGC 6205	16 41.7	+36 28
M12 = NGC 6218	16 47.2	−1 57
M28 = NGC 6626	18 24.5	−24 52
M22 = NGC 6656	18 36.4	−23 54
M55 = NGC 6809	19 40.0	−30 58
M15 = NGC 7078	21 30.0	+12 10

Observe as many of these globular clusters as you can. How well can you distinguish individual stars toward the center of each cluster? Do you notice any differences in the overall distribution of stars between clusters?

For further reading

Hack, M. "Epsilon Aurigae." *Scientific American,* October 1984 • This article, which focuses on a particular eclipsing binary, demonstrates how astronomers use observations and an understanding of stellar evolution to deduce details of elaborate stellar systems.

Johnson, B. "Red Giant Stars." *Astronomy,* December 1976 • This well written article surveys the properties of red-giant stars.

Kafatos, M., and Michalitsianos, A. "Symbiotic Stars." *Scientific American,* July 1984 • This article discusses fascinating phenomena associated with double stars that consist of a red giant and a hot, compact companion.

Kippenhahn, R. *100 Billion Suns: The Birth, Life and Death of Stars.* Basic Books, 1983 • This classic book by a noted German astrophysicist describes the life cycles of stars.

MacRobert, A. "Epsilon Aurigae: Puzzle Solved?" *Sky & Telescope,* January 1988 • This brief article shows how astronomers struggle to understand the details of a close binary system in which complicated effects occur.

Percy, J. "Observing Variable Stars for Fun and Profit." *Mercury,* May/June 1979 • This entertaining article contains many important tips for a person interested in observing eclipsing binaries, Cepheids, and other variable stars.

———. "Pulsating Stars." *Scientific American,* June 1975 • Although it is a few years old, this article is still the best popular description of the physics of stellar pulsations.

Tomkin, J., and Lambert, D. L. "The Strange Case of Beta Lyrae." *Sky & Telescope,* October, 1987 • This fascinating article explores many of the mysteries associated with β Lyrae.

Weymann, R. "Stellar Winds." *Scientific American,* August 1978 • This article discusses ongoing mass loss experienced by many stars.

Wilson, O., et al. "The Activity Cycles of Stars." *Scientific American,* February 1981 • This article begins with a description of the solar cycle and shows how astronomers are discovering similar cycles in other stars.

Wyckoff, S. "Red Giants: The Inside Scoop." *Mercury,* January/February 1979 • This superb article, which includes many interesting details about red giants, emphasizes the effects of varying abundances of carbon and oxygen.

A planetary nebula Dying stars often eject their outer layers. A low-mass star can lose half its mass in a comparatively gentle process that produces a planetary nebula. The exposed stellar core typically has a surface temperature of about 100,000 K and is roughly one-tenth the Sun's size. UV radiation from the stellar core causes the surrounding gases to glow. The greenish color of this planetary nebula, NGC 6543, comes from doubly ionized oxygen. Its central star has exhausted all its fuel and is contracting into a white dwarf. (Lick Observatory)

C H A P T E R

22

The Deaths of Stars

After its old age as a red giant, a star approaches the end of its life. As stars die, they eject significant amounts of matter into space. Low-mass stars expel their outer layers relatively gently, producing planetary nebulae in the process. The burned-out cores of such stars contract to become dead stars called white dwarfs. In contrast, the core of a high-mass star can collapse suddenly, triggering a violent supernova explosion. A white dwarf ac-creting gas from a companion star in a close binary system can also become a supernova. Many different thermonuclear reactions occur during the final stages of stellar life, producing a wide variety of heavy elements. Astronomers are developing exotic instruments, including neutrino detectors and gravitational wave antennae, to learn more about the fascinating and sometimes violent death throes with which stars end their lives.

From infancy through adulthood, a star leads a fairly placid life, with hydrogen burning in its core. As old age approaches, however, the star takes on a seemingly schizophrenic character, with a compressed core but a bloated atmosphere, and becomes a red giant. Dramatic behavior becomes even more pronounced in the star as it devours its remaining nuclear fuels and begins to die.

22-1 Low-mass stars die by gently ejecting their outer layers, creating planetary nebulae

Carbon and oxygen make up the "ash" of helium burning. After the helium flash in a low-mass red giant, substantial amounts of these two elements begin to accumulate at the star's center as a result of core helium burning. Eventually, all the helium at the center of a low-mass star is used up, at which point core helium burning ceases. The star's core therefore contracts until stopped by the buildup of degenerate electron pressure in the core. This compression heats the helium-rich gases surrounding the degenerate carbon–oxygen core, and helium burning soon begins in a thin shell around the core. This process is called **shell helium burning**. The star's internal structure now consists of a helium-burning shell inside a hydrogen-burning shell, all within a volume roughly the size of the Earth. After a while, thermonuclear reactions in the hydrogen-burning shell temporarily cease, leaving the aging star's structure as shown in Figure 22-1.

When shell hydrogen burning first began, the outpouring of energy caused the star to expand and become a red giant. In describing the evolutionary track of the star on an H–R diagram, astronomers say that the star ascended the red-giant branch for the first time. Then came the helium flash, at which time the star shifted over to the horizontal branch. But now, the renewed outpouring of energy from shell helium burning causes the star to expand again. The star as-

cends the red-giant branch for a second and final time to become a **red supergiant.** Such stars are also commonly called **asymptotic giant branch stars,** or **AGB stars** for short, because of the region they occupy on an H–R diagram. These stars typically have diameters as big as the orbit of Mars and shine with the brightness of 10,000 Suns. A star experiencing this furious rate of energy loss cannot live much longer.

The star's impending death is signaled by instabilities that develop in its helium-burning shell. As with the helium flash discussed in Chapter 21, there is again a thermal runaway, but this time the details are quite different. The helium flash occurred because the star's core was degenerate. The helium-burning shell is not compressed to a high enough density to be degenerate, though. Instead, the thermal runaway occurs because the shell is thin. A slight increase in energy output from a thin shell increases the temperature of the helium there, but does little to relieve the pressure from the star's overlying layers. The higher temperature further increases the energy output, which then in turn drives up the temperature even more. This vicious circle produces a thermal runaway that ends only after the helium-burning shell becomes thick enough to relieve the pressure of the star's outer layers.

These **helium-shell flashes** were first investigated theoretically by M. Schwarzschild and R. Härm at Princeton University in the 1960s. The results of their calculations with stellar models are displayed in Figure 22-2. During the flashes, the energy output from the helium-burning shell jumps from that of 100 Suns to roughly 100,000 Suns in a rapid series of brief bursts referred to as **thermal pulses.** As shown in Figure 22-2, these bursts are separated by relatively quiet intervals lasting about 300,000 years. During the period of thermal pulses, thermonuclear reactions begin again in the star's hydrogen-burning shell.

During one of these periods of thermal oscillation, the dying star's outer layers can separate completely from its carbon–oxygen core. As the ejected material expands into space, dust grains condense out of the cooling gases. Radiation pressure from the star's hot, burned-out core acts on the

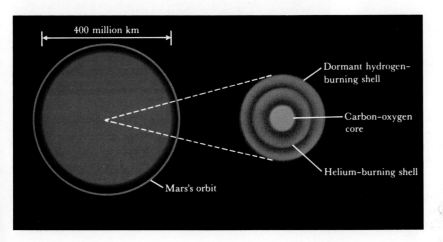

Figure 22-1 *The structure of an old low-mass star Near the end of its life, a low-mass star becomes a red supergiant whose diameter is almost as large as the diameter of the orbit of Mars. The star's dormant hydrogen-burning shell and active helium-burning shell are all contained within a volume roughly the size of the Earth.*

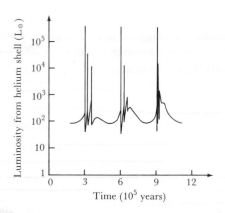

Figure 22-2 Helium-shell flashes This graph shows how the energy output of a helium-burning shell varies in a red supergiant. Brief periods of runaway helium burning produce thermal pulses that can eventually cause the star to eject its outer layers completely. (Adapted from M. Schwarzschild and R. Härm)

specks of dust to continue propelling them outward, and the star sheds its outer layers altogether. A star can lose more than half of its mass in this fashion.

As a dying star ejects its outer layers, the star's hot core becomes exposed and emits ultraviolet radiation intense enough to ionize and excite the expanding shell of ejected gases. These gases therefore glow, producing a so-called **planetary nebula,** or **planetary.** Planetary nebulae have nothing to do with planets; this misleading term was introduced in the nineteenth century because these glowing objects looked like distant planets when viewed through the small telescopes then available.

Many planetary nebulae, such as the one shown in Figure 22-3, have a distinctly spherical appearance, created by the symmetrical way in which the gases were ejected. In other cases, if the rate of expansion is not the same in all directions, the resulting nebula takes on an hourglass or dumbbell appearance, as shown in Figure 22-4.

Planetary nebulae are quite common. Astronomers estimate that there are 20,000 to 50,000 planetaries in our Galaxy alone. Several well-known examples are listed in Box 22-1. Spectroscopic observations of these nebulae show bright emission lines of ionized hydrogen, oxygen, and nitrogen. From the Doppler shifts of these lines, astronomers have concluded that the expanding shell of gas moves outward from a dying star at speeds from 10 to 30 km/s. A typical planetary nebula has a diameter of roughly 1 light year, indicating that it must have begun expanding about 10,000 years ago.

By astronomical standards, a planetary nebula is a very short-lived object. After about 50,000 years, the nebula has spread out so far from the cooling central star that its nebulosity simply fades from view. The gases then mix with the surrounding interstellar medium. Astronomers estimate that all the planetary nebulae in the Galaxy return a total of about 5 $M_\odot$ to the interstellar medium each year. This amount is about 15 percent of all the matter expelled by all the various sorts of stars each year. Because of this significant contribution, planetary nebulae play an important role in the chemical evolution of the Galaxy as a whole.

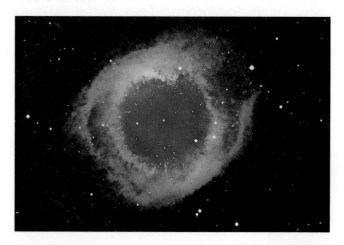

Figure 22-3 The planetary nebula NGC 7293 This beautiful object, often called the Helix Nebula, covers an area of the sky equal to half the full Moon. The star that ejected these gases can be seen at the center of the glowing shell. The bluish-green color comes from ionized oxygen, the pink and red from ionized nitrogen and hydrogen. The small radial blobs in the red shell (estimated to be about 150 AU across) give the object its alternate name, the Sunflower Nebula. It is in the constellation of Aquarius, 400 light years from Earth. (Anglo-Australian Observatory)

Figure 22-4 The planetary nebula NGC 6302 At the ends of their lives, low-mass stars shed much of their mass, leaving behind a tiny, hot, burned-out stellar core. Although most undergo a fairly symmetrical ejection of gases, there are many examples in which the gas has expanded unevenly. An irregular planetary nebula like NGC 6302 in Scorpius, shown here, is the result. Spectroscopic observations of NGC 6302 indicate that its gases are moving toward us at 400 kilometers per second, indicating that they had a particularly violent initial ejection. (Anglo-Australian Observatory)

Box 22-1   Famous planetary nebulae

Many planetary nebulae are scattered across the sky. Some of the most famous planetaries are listed in the table below with information about their distances and diameters.

Planetary nebulae are favorite objects for amateur astronomers to observe. In many cases (such as the Ring Nebula shown in the photograph), a small telescope can reveal the nebulosity but not the central star. It is this central star, destined to become a white dwarf, that supplies the ultraviolet radiation that causes the surrounding gases to glow.

Planetary nebula	Distance from Earth (ly)	Apparent magnitude of central star	Diameter (arc min)
Dumbbell (M27)	3,500	13	8
Ring (M57)	4,000	15	1
"Little Dumbbell" (M76)	15,000	17	1
Owl (M97)	12,000	13	3
Saturn (NGC 7009)	3,000	12	1
Helix (NGC 7293)	400	13	15
Clown (NGC 2392)	10,000	10	$\frac{1}{2}$

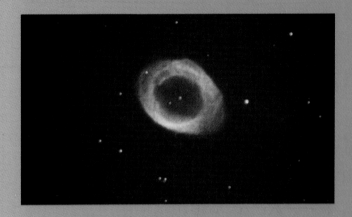

22-2 The burned-out core of a low-mass star cools and contracts until it becomes a white dwarf

Stars less massive than about 4 $M_\odot$ never develop the necessary central pressures or temperatures to ignite thermonuclear reactions that use carbon or oxygen as fuel. Instead, the process of mass ejection just strips most outer layers away from the carbon–oxygen core, which then simply cools off.

The evolutionary tracks of three burned-out stellar cores are shown in Figure 22-5. The initial red supergiants had masses between 0.8 and 3.0 $M_\odot$. During their final spasms, these dying stars eject up to 60 percent of their matter. In the mass ejection phase, the outward appearance of these stars changes rapidly. The resulting changes in luminosity and surface temperature cause the points that represent such stars on an H–R diagram to race along their evolutionary tracks, sometimes executing loops corresponding to thermal pulses. Finally, as the ejected nebulae fade and the stellar

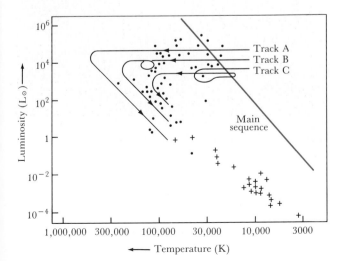

Figure 22-5 Evolution from red supergiants to white dwarfs *The evolutionary tracks of three low-mass red supergiants are shown as they eject planetary nebulae. The table gives the extent of mass loss in each case. The dots on the graph represent the central stars of planetary nebulae whose surface temperatures and luminosities have been determined. The crosses are white dwarfs for which similar data exist. (Adapted from B. Paczynski)*

Evolutionary track	Supergiant mass ($M_\odot$)	Mass of ejected nebula ($M_\odot$)	White dwarf mass ($M_\odot$)
A	3.0	1.8	1.2
B	1.5	0.7	0.8
C	0.8	0.2	0.6

cores cool, these dying stars' evolutionary tracks take a sharp turn downward toward the white-dwarf region of the H–R diagram.

There is no possibility of igniting any additional nuclear fuels inside one of these dead stars, so the crushing weight of its gases pressing inward from all sides severely compresses the stellar corpse. The density inside the dead star skyrockets until the electrons are so closely packed that they become degenerate throughout most of the star. Degenerate-electron pressure then produces the conditions necessary to support the star, and its gravitational contraction is halted. This state is reached when the star has contracted down to a sphere roughly the same size as the Earth. Such a star is known as a white dwarf.

The density of matter in one of these Earth-sized stellar corpses is typically 10^9 kg/m^3. In other words, a teaspoonful of white-dwarf matter brought to Earth would weigh nearly 5.5 tons, as much as an elephant. In addition, as we saw in Chapter 21, degenerate gases are governed by unusual relationships between pressure, density, and temperature. Consequently, these stars have some unique properties. For example, the radius of a white dwarf is inversely proportional to the cube root of its mass:

$$R \propto \frac{1}{M^{1/3}}$$

Thus, the more massive a white dwarf is, the smaller it is.

Figure 22-6 displays the **mass–radius relation** for white dwarfs. Note that the more degenerate matter you pile onto a white dwarf, the smaller it becomes. The electrons must be squeezed closer together to provide the greater pressure needed to support a more massive white dwarf. However, there is a limit to how much pressure degenerate electrons can produce. The rapidly moving electrons cannot travel faster than the speed of light, so the size of a white dwarf is smaller than predicted by the equation above. For this reason, there is an upper limit to the mass that a white dwarf can have. This maximum mass is called the **Chandrasekhar limit** after Subrahmanyan Chandrasekhar at the University of Chicago, who pioneered theoretical studies of white dwarfs. The Chandrasekhar limit is equal to 1.4 M$_\odot$. This amount is the maximum mass that can be supported by degenerate-electron pressure, meaning that all white dwarfs must have masses less than 1.4 M$_\odot$.

Many white dwarfs are found in the solar neighborhood, but all are too faint to be seen with the naked eye. One of the first white dwarfs to be discovered is a companion to the bright star Sirius. The binary nature of Sirius was first deduced in 1844 by the German astronomer Friedrich Bessel, who noticed that this star was moving back and forth slightly, as if it was being orbited by an unseen object. This companion, designated Sirius B, was first glimpsed in 1863 (see Figure 22-7). Recent satellite observations at ultraviolet wavelengths, where hot white dwarfs emit most of their light, demonstrate that the surface temperature of Sirius B is about 30,000 K.

The material inside a white dwarf consists of highly ionized atoms floating in a sea of degenerate electrons. As the dead star cools, the particles in this material slow down, and

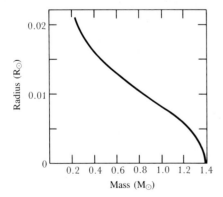

Figure 22-6 The mass–radius relationship for white dwarfs The more massive a white dwarf is, the smaller it is. This unusual relationship between mass and radius is a result of the degenerate-electron pressure that supports the star. The maximum mass of a white dwarf, called the Chandrasekhar limit, is 1.4 M$_\odot$. Incidentally, 0.01 R$_\odot$ = 6960 km = 1.09 R$_\oplus$, where R$_\oplus$ is the radius of the Earth.

Figure 22-7 Sirius and its white-dwarf companion Sirius, the brightest-appearing star in the sky, is actually a double star. The secondary star is a white dwarf, seen in this photograph at the five o'clock position, almost obscured by the glare of Sirius. The spikes and rays around Sirius are the result of optical effects within the telescope. (Courtesy of R. B. Minton)

electric forces between the ions begin to exert control over the previously random thermal motions. The ions no longer move freely. According to calculations by Hugh Van Horn and Malcolm Savedoff at the University of Rochester, the ions now arrange themselves in orderly rows to form a crystal lattice. This situation is roughly analogous to that in a cooling solution in which crystals form as the ions in the solution arrange themselves in orderly patterns dictated by the electric forces acting between ions. From this time on, you could say that the star is "solid." The degenerate electrons move around freely in this crystalline lattice, just as normal electrons move freely through an electrically conducting metal here on Earth. Thus, the material inside an old white dwarf has many properties similar to those of copper or silver. Furthermore, since a diamond is crystallized carbon, a cool carbon–oxygen white dwarf resembles an immense spherical diamond!

As a white dwarf cools off, its size remains constant because the degenerate-electron pressure supporting the dead star does not depend on its temperature. However, the white dwarf's luminosity and its surface temperature decline as it cools. Its energy now comes only from thermal energy (heat) as it cools while remaining at a constant size. Consequently, the evolutionary tracks of aging white dwarfs point toward the lower right corner of the H–R diagram (see Figure 22-8). As billions of years pass, white dwarfs grow dimmer and dimmer as their surface temperatures drop toward absolute zero. Such a fate will be what finally happens to our Sun as it evolves into a cold, dark, diamond sphere of carbon and oxygen, about the size of the Earth.

22-3 High-mass stars die violently by blowing themselves apart in supernova explosions

High-mass stars end their lives quite differently from low-mass stars. First, a high-mass star is capable of igniting a host of additional thermonuclear reactions in its core. After core helium burning ceases, gravitational compression drives the star's central temperature up to 600 million K, at which time **carbon burning** begins. This thermonuclear process produces neon, magnesium, oxygen, and helium according to the following reactions:

$$^{12}C + {}^{12}C \rightarrow {}^{20}Ne + {}^{4}He$$
$$^{12}C + {}^{12}C \rightarrow {}^{24}Mg + \gamma$$
$$^{12}C + {}^{12}C \rightarrow {}^{16}O + 2\,{}^{4}He$$

Only stars with masses greater than about 4 $M_{\odot}$ develop central temperatures high enough to ignite carbon burning. The more massive a star is, the greater the temperatures that can be reached in its core by gravitational compression.

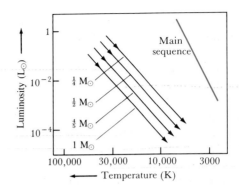

Figure 22-8 *White-dwarf "cooling curves"* The evolutionary tracks of four white dwarfs of different masses are shown here. As these dead stars lose heat at a constant size, they become dimmer and cooler, moving toward the lower right on the H–R diagram.

High temperatures are required for nuclear reactions involving heavy nuclei because the strong electric charges of these nuclei exert enough force to tend to keep the nuclei apart. Only at the great speeds associated with very high temperatures can the nuclei travel fast enough to penetrate each other's repulsive electric fields and fuse together.

If a star has a main-sequence mass of at least 9 $M_{\odot}$ (before mass ejection), its central temperature can rise to 1 billion K, at which temperature the process of **neon burning** would begin:

$$^{20}Ne + \gamma \rightarrow {}^{16}O + {}^{4}He$$
$$^{20}Ne + {}^{4}He \rightarrow {}^{24}Mg + \gamma$$

This process uses up the neon accumulated from carbon burning and further increases the concentrations of oxygen and magnesium in the star's core.

If the central temperature of the star reaches about 1.5 billion K, **oxygen burning** begins. The principal product of oxygen burning is sulfur:

$$^{16}O + {}^{16}O \rightarrow {}^{32}S + \gamma$$

However, as the star consumes increasingly heavier nuclei, the thermonuclear reactions produce a wider variety of products. Oxygen burning also produces two isotopes of silicon, as well as another isotope of sulfur, phosphorus, and more magnesium:

$$^{16}O + {}^{16}O \rightarrow \begin{cases} {}^{28}Si + {}^{4}He + \gamma \\ {}^{31}P + {}^{1}H + \gamma \\ {}^{31}S + n + \gamma \\ {}^{24}Mg + 2\,{}^{4}He \\ {}^{30}Si + 2\,{}^{1}H + \gamma \end{cases}$$

This example is typical of the way that thermonuclear reactions in the cores of massive stars create many different elements.

Note that one of the oxygen-burning reactions listed above produces neutrons (n). Many other reactions in the star's core also produce neutrons. You may recall that a neutron is very much like a proton except that it does not have an electric charge. Therefore, a neutron is not repelled by the positively charged nuclei. Consequently, neutrons can easily collide and combine with nuclei, creating new isotopes in the process. As described in Box 22-2, this absorption of neutrons can happen in a rapid **r process** producing **r-process isotopes** or in a slow **s process** producing **s-process isotopes**. In this way, **neutron capture** creates many elements and isotopes that are not produced directly in other nuclear-fusion reactions.

When a nuclear fuel is exhausted in the core of a massive star, the pressure declines, and gravitational contraction to ever-higher densities drives up the star's central temperature, thereby igniting the "ash" of the previous fusion stage, possibly along with the outlying shell of unburned fuel. Each successive thermonuclear reaction occurs with increasing rapidity. For example, calculations for a 25 $M_\odot$ star by Stanford E. Woosley and Thomas A. Weaver at Lawrence Livermore Laboratory demonstrate that carbon burning occurs for 600 years, neon burning for 1 year, but oxygen burning for only 6 months.

After half a year of core oxygen burning in a 25 $M_\odot$ star, a final gravitational compression forces the central temperature up toward 3 billion K, at which point **silicon burning** begins. This thermonuclear process happens so furiously that the entire core supply of silicon in a 25 $M_\odot$ star is used up in one day.

In order for an element to serve as a thermonuclear fuel, energy must be given off when its nuclei collide and fuse. This energy comes from packing together more tightly the neutrons and protons in the ash nuclei than in the fuel nuclei. Silicon burning involves many hundreds of nuclear reactions. The major final product, or "ash," of this process is a stable isotope of iron, ^{56}Fe. The 56 protons and neutrons inside the iron nuclei are so tightly bound together that no further energy can be extracted by fusing still more nuclei with iron. Because ^{56}Fe does not act as a fuel for any other thermonuclear reactions, the sequence of burning stages ends here.

The buildup of an inert, iron-rich core signals the impending violent death of a massive star. Successive layers of shell burning surrounding this iron core consume the star's remaining reserves of fuel (see Figure 22-9). The entire energy-producing region of the star is contained in a volume only as big as the Earth, whereas the star's enormously bloated atmosphere is nearly as big as the orbit of Jupiter.

Because iron does not "burn," the electrons in the star's core now must support its outer layers by the brute strength

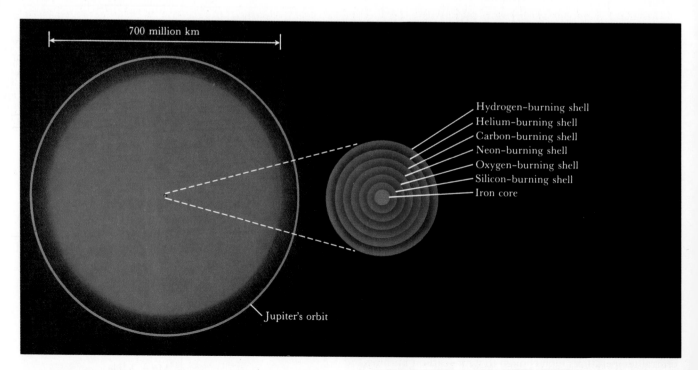

Figure 22-9 *The structure of an old high-mass star Near the end of its life, a high-mass star becomes a red supergiant almost as big as the orbit of Jupiter. The star's energy comes from six concentric burning shells, all contained within a volume roughly the same size as the Earth.*

Box 22-2 The s and r processes

A nucleus is composed of protons and neutrons. Generally, there are a few more neutrons than protons in a nucleus. This happens because neutrons have no electric charge, so they can be packed together more easily than protons. In fact, neutrons act as buffers between the protons, with their mutually repulsive electric fields. If a nucleus has either too many protons or too many neutrons, it is unstable. It will eject surplus particles until a stable balance of neutrons and protons has been restored.

There are many reactions in stellar cores that produce free neutrons. These neutrons can be absorbed by nuclei, thus transmuting one isotope into another and creating new elements as the unstable isotopes decay. For example, successive neutron captures by ^{110}Cd can produce four stable isotopes of cadmium:

$$^{110}Cd + {}^{1}n \rightarrow {}^{111}Cd$$
$$^{111}Cd + {}^{1}n \rightarrow {}^{112}Cd$$
$$^{112}Cd + {}^{1}n \rightarrow {}^{113}Cd$$
$$^{113}Cd + {}^{1}n \rightarrow {}^{114}Cd$$

The specific isotopes produced by a nucleus absorbing neutrons depends on how many neutrons are bombarding the nucleus. If the rate of neutron bombardment is low, then unstable nuclei have a chance to decay radioactively between neutron absorptions. For example, the absorption of a neutron by ^{114}Cd forms the unstable nucleus ^{115}Cd, which decays into an isotope of indium:

$$^{114}Cd + {}^{1}n \rightarrow {}^{115}Cd$$
$$^{115}Cd \rightarrow {}^{115}In + e^{-} + \overline{\nu}$$

Then the ^{115}In nucleus can absorb another neutron, creating an unstable isotope (^{116}In), which decays into a stable isotope of tin (^{116}Sn):

$$^{115}In + {}^{1}n \rightarrow {}^{116}In$$
$$^{116}In \rightarrow {}^{116}Sn + e^{-} + \overline{\nu}$$

The production of ^{115}In and ^{116}Sn occurs only if the rate of neutron absorption is low. This *slow* reaction is termed an **s process**, and these two isotopes are called **s-process isotopes**.

Occasionally, a flood of neutrons is released in a star's core so that a rapid absorption of neutrons can occur. In fact, a nucleus can soak up a large number of neutrons without having the chance to decay radioactively between absorptions. Such a *rapid* reaction, termed an **r process**, produces **r-process isotopes** that are often different from those created by the corresponding s process.

Example: Consider again ^{114}Cd, the stable isotope of cadmium. A deluge of neutrons could produce a highly unstable isotope:

$$^{114}Cd + 8 \; {}^{1}n \rightarrow {}^{122}Cd$$

This isotope decays into an unstable isotope of indium:

$$^{122}Cd \rightarrow {}^{122}In + e^{-} + \overline{\nu}$$

The indium then decays into a stable isotope of tin:

$$^{122}In \rightarrow {}^{122}Sn + e^{-} + \overline{\nu}$$

This particular ^{122}Sn isotope of tin can be produced only by the r process, just as ^{116}Sn can be produced only by the s process. Both isotopes are stable and are found on Earth.

By measuring the relative abundances of various isotopes in the world around us, astronomers can thus deduce conditions that once existed inside the stars that manufactured the elements of which our planet is composed.

---

of degeneracy pressure alone. Soon, however, the continued deposition of fresh iron from the silicon-burning shell causes the core's mass to exceed the Chandrasekhar limit. Electron degeneracy suddenly becomes unable to support the star's enormous weight, and the star's core begins to collapse.

Any star with an initial main-sequence mass greater than 10 $M_{\odot}$ is capable of developing an iron core that at some stage will exceed the Chandrasekhar limit, triggering a rapid series of cataclysms that will tear the star apart in a few seconds. For purposes of illustration, we shall examine the death of a 25 $M_{\odot}$ star.

In a 25 $M_{\odot}$ star, degenerate-electron pressure fails when the density inside the iron core reaches 1 trillion (10^{12}) kilograms per cubic meter. The core then immediately begins to collapse. Central core temperatures promptly soar to almost

inconceivable heights. In roughly a tenth of a second, the temperature will exceed 5 billion K. The gamma-ray photons associated with this intense heat have so much energy that they begin to break down the iron nuclei into alpha particles, a process called **photodisintegration**.

Within another tenth of a second, as the densities continue to climb, the electrons are forced to combine with protons to produce neutrons, in a process called **neutronization**, which also releases a flood of neutrinos (ν):

$$e^{-} + p \rightarrow n + \nu$$

A **neutrino** is a massless (or nearly massless) particle that is electrically neutral and easily passes through matter.

At about 0.25 second after the collapse begins, the density in the core reaches 4×10^{17} kg/m³, which is **nuclear density**, the density with which neutrons and protons are packed together inside nuclei.

At nuclear density matter is virtually incompressible. Thus, when the neutron-rich material of the core reaches nuclear density, it suddenly becomes very stiff and the core collapse comes to a sudden halt.

During this critical stage, the unsupported regions surrounding the core are plunging inward at speeds up to 15 percent of the speed of light. As this material crashes onto the now-rigid core, enormous temperatures and pressures develop, causing the falling material to bounce back. In just a fraction of a second, a wave of matter begins to move back out toward the star's surface, propelled in part by the flood of neutrinos trying to escape from the star's core. This wave rapidly accelerates as it encounters less and less resistance and soon becomes a shock wave. After a few hours, this shock wave reaches the star's surface, by which time the star's outer layers have begun to lift away from the core. In this way the star becomes a **supernova**.

This particular description of the death of a 25 M$_\odot$ star is based on detailed computer calculations of an evolving stellar model. The key stages in the star's evolution are summarized in Table 22-1.

The final stages in the evolution of other massive stars probably follow similar scenarios, though details may differ. This particular 25 M$_\odot$ star ejects 24 M$_\odot$ of material, leaving behind a 1 M$_\odot$ corpse called a **neutron star**. Under slightly different conditions, a massive star might blow itself completely apart, leaving no corpse at all. For example, the bounce and subsequent shock wave might occur at the center of the star rather than at the surface of its neutronized

core. Alternatively, with a different set of initial conditions inside the star, a massive burned-out core might gravitationally collapse to form a **black hole**. These exotic objects are discussed in the next two chapters.

As the outer layers of a massive dying star are blasted into space, the star's luminosity suddenly increases, by a factor of approximately 10^8, which is equivalent to a jump of 20 magnitudes in brightness. For a few days, a supernova can shine as brightly as an entire galaxy. Figure 22-10 shows a supernova that occurred in a nearby galaxy in 1987 and could easily be seen with the naked eye. Not since 1604 has a brighter-appearing supernova graced the heavens. Details about the 1987 supernova are discussed in Box 22-3.

Table 22-1 Evolutionary stages of a 25 M$_\odot$ star

Stage	Temperature (K)	Density (kg/m³)	Duration of stage
Hydrogen burning	4×10^7	5×10^3	7×10^6 years
Helium burning	2×10^8	7×10^5	7×10^5 years
Carbon burning	6×10^8	2×10^8	600 years
Neon burning	1.2×10^9	4×10^9	1 year
Oxygen burning	1.5×10^9	10^{10}	6 months
Silicon burning	2.7×10^9	3×10^{10}	1 day
Core collapse	5.4×10^9	3×10^{12}	$\frac{1}{4}$ second
Core bounce	2.3×10^{10}	4×10^{15}	milliseconds
Explosive	about 10^9	varies	10 seconds

Figure 22-10 Supernova 1987A In 1987, a supernova blazed forth in a nearby galaxy called the Large Magellanic Cloud, about 160,000 ly from Earth. The left photograph shows a small section of the Large Magellanic Cloud before the outburst, with the doomed star—a B3 supergiant—identified by an arrow. The right view shows the supernova a few days after the explosion. (Anglo-Australian Observatory)

Box 22-3   Supernova 1987A

On February 23, 1987, a supernova was discovered in the Large Magellanic Cloud (LMC), which is a companion galaxy to our own Milky Way. The supernova (designated SN 1987A because it was the first to be discovered that year) occurred near an enormous H II region in the LMC called 30 Doradus or the Tarantula Nebula because of its spiderlike appearance. For an overall view of the LMC without the supernova, see Figure 26-12.

This supernova was unusual because it did not promptly rise to its expected maximum brightness, but stopped at only a tenth of the luminosity typical for an exploding massive star (like the one described in Table 22-1). For 85 days following the detonation, the brightness of SN 1987A increased gradually; then it settled into a slow decline characteristic of an ordinary supernova (see accompanying graph). Pre-supernova observations of the doomed star provided important data that helped astronomers understand its unusual behavior.

SN 1987A was initially subluminous because the doomed star was a blue supergiant when it exploded, rather than a red supergiant. The doomed star was identified as Sk−69 202 in a catalogue of LMC stars produced by Nicholas Sanduleak at Case Western Reserve University. When this star was on the main sequence, its mass was about 20 $M_\odot$, although by the time it exploded it probably had shed a few solar masses.

The evolutionary track for such a high-mass star wanders back and forth across the top of the H–R diagram, so that late in its life the star shifts between being a red supergiant and a blue supergiant. When it happened to explode, Sk−69 202 was classified as a B3 I star, meaning that it had a surface temperature of approximately 15,000 K and a luminosity of about 10^5 Suns. Such a blue supergiant has a diameter only 10 times larger than the Sun's, whereas a red supergiant of the same luminosity would have a diameter 1000 times larger than the Sun's. Because Sk−69 202 was relatively small when it exploded, it reached only a tenth of the brightness that an exploding red supergiant would have had.

Calculations by Stanford Woosley, Thomas Weaver, and their colleagues at Lick Observatory suggest that the interior of Sk−69 202 contained 6 $M_\odot$ of helium and 2 $M_\odot$ of elements heavier than helium. At the time of detonation, the star's iron core had a mass of 1.5 $M_\odot$ at a temperature of 10 billion K. The declining brightness of the supernova that began 85 days after the outburst is accounted for by light from the decay of about 0.15 $M_\odot$ of radioactive isotopes that were created during the explosion. Some astronomers suspect that a neutron star may be uncovered in a few years.

Most of the energy of a supernova explosion is carried away by the neutrinos produced during the collapse of the star's core. Calculations show that at 0.1 second after core collapse, a burst of neutrinos with a luminosity of 10^{47} watts is emitted from the supernova. For comparison, the total optical luminosity of all the stars and galaxies in the observable universe is 10^{45} watts. Thus, for a fraction of a second, SN 1987A shone 100 times brighter than the entire rest of the universe!

Astronomers have been lucky with SN 1987A. Its progenitor, Sk−69 202, had been studied and its distance from Earth (160,000 ly) was known. Observations were facilitated by the LMC's location in an unobscured part of the heavens, as well as the supernova's timing in the middle of the southern-hemisphere summer with its dry, clear skies. Furthermore, neutrino detectors in Japan and the United States detected a burst of neutrinos three hours ahead of the light emitted from the exploding star's surface. And finally, the supernova was unusual, which advanced our understanding of supernovae in general. Because SN 1987A is such an important object, astronomers will be monitoring its progress for many years to come.

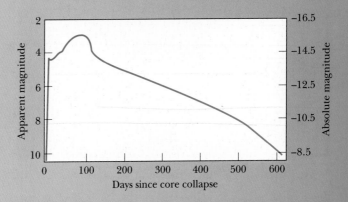

Days since core collapse

Type I supernova

22-4 *Accreting white dwarfs in close binary systems can also become supernovae*

Astronomers often find supernovae in distant galaxies (see Figure 22-11). Most of our understanding of supernovae comes from observing these remote outbursts. Such observations reveal that supernovae fall into two categories, designated Type I and Type II, which involve very different phenomena. A supernova's spectrum determines its type: Hydrogen lines are prominent in Type II, but are absent in Type I.

All supernovae begin with a sudden rise in brightness that occurs in less than a day (see Figure 22-12). A Type I super-

nova typically reaches an absolute magnitude of −19 at peak brightness, but a Type II supernova is usually about 2 magnitudes fainter. Type I supernovae then settle into a gradual decline that lasts for over a year, whereas the Type II light curve has a steplike appearance caused by alternating periods of steep and gradual declines in brightness.

As discussed above, a Type II supernova is caused by the death of a massive star. Gravitational energy powers a Type II supernova, because detonation is triggered by the collapse of the star's iron-rich core. Hydrogen lines dominate such a supernova's spectrum simply because that gas is abundant in the star's outer layers being blasted into space.

A Type I supernova begins with a carbon–oxygen-rich white dwarf in a close, semidetached binary system. To trigger the outburst, a swollen red giant companion overflows

a

b

Figure 22-11 A distant supernova Sometime during 1940, a supernova exploded in the galaxy NGC 4725 in the constellation of Coma Berenices. (**a**) *The galaxy before the outburst.* (**b**) *By the time this* photograph was taken in 1941, the supernova (at arrow) had faded from its maximum brightness. (Mount Wilson and Las Campanas Observatories)

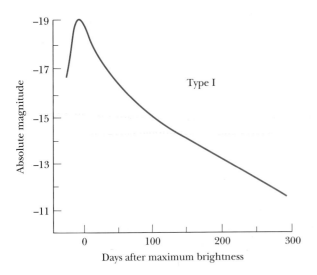

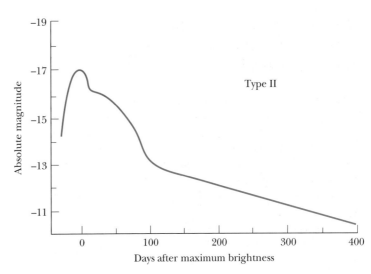

Figure 22-12 *Supernova light curves* *A Type I supernova, which exhibits a gradual decline in brightness, is caused by an exploding white dwarf in a close binary system. A Type II supernova usually has alter-* *nating intervals of steep and gradual decline. Type II supernovae are caused by the death of massive stars.*

its Roche lobe and dumps gas onto the white dwarf. When the white dwarf's mass gets close to the Chandrasekhar limit, carbon burning begins. Because the white dwarf is composed of degenerate matter, the usual "safety valve" between pressure and temperature does not operate. In a catastrophic runaway process reminiscent of the helium flash, the rate of carbon burning skyrockets, because increasing temperatures inside the white dwarf are not relieved by corresponding increases in pressure. Soon the temperature gets so high that the electrons are no longer degenerate and the star blows up.

If the carbon burning moves through the white dwarf faster than the speed of sound, a **detonation** rips the star apart. However, if the carbon burning propagates slower than the speed of sound, the white dwarf is consumed in a process called **deflagration**. Both processes supply about the same total amount of energy output.

A Type I supernova is powered by nuclear energy, and the resulting spectacle in the sky is simply the fallout from a thermonuclear explosion. A wide array of unstable isotopes are produced during the outburst, most notably ^{56}Ni, which has a half-life of 6.1 days. It decays into ^{56}Co, which has a half-life of 79 days and decays into a stable isotope of iron. The smooth decline of the light curve of a Type I supernova directly results from the gamma rays emitted by radioactive decay of ^{56}Co.

Various statistical studies of supernovae sightings indicate that a typical galaxy like our Milky Way has one supernova explosion every 20 to 50 years. Apparently, there are sightly more Type II supernovae than Type I. The fact that Type I and Type II supernovae reach nearly the same absolute magnitude at maximum brightness is a coincidence.

22-5 A supernova remnant is detectable at many wavelengths for many years after a supernova explosion

Astronomers find the debris of supernova explosions, called **supernova remnants**, scattered across the sky. A beautiful example is the Veil Nebula, seen in Figure 22-13. The

Figure 22-13 *The Veil Nebula* *This nebulosity is a portion of the Cygnus Loop (see Figure 20-16), which is the remnant of a supernova that exploded about 20,000 years ago. The distance to the nebula is about 500 pc, and the overall diameter of the loop is about 22 pc. (Palomar Observatory)*

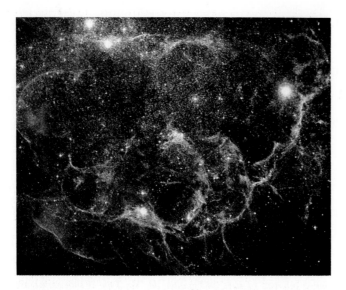

Figure 22-14 **The Gum Nebula** *The Gum Nebula, which spans 60° of the sky, has the largest angular size of any known supernova remnant. Only the central regions of the nebula are shown here. The nearest portions of this expanding nebula are only 100 pc (330 ly) from the Earth. The supernova explosion occurred about 11,000 years ago. The supernova remnant now has a diameter of about 700 pc (2300 ly). (Royal Observatory, Edinburgh)*

doomed star's outer layers were blasted into space with such violence that they are still traveling through the interstellar medium at supersonic speeds. As this expanding shell of gas plows through space, it collides with atoms in the interstellar medium, exciting this sparse gas and causing it to glow.

A few nearby supernova remnants cover sizable areas of the sky. The largest is the Gum Nebula, named after Colin

Gum, who first noticed its faint glowing wisps on photographs of the southern sky (see Figure 22-14). It is also called the Vela supernova remnant, because its 60° diameter is centered on the constellation of Vela.

The Gum Nebula looks big simply because it is quite nearby. Its near side is only about 330 ly (100 pc) from Earth, and the center of the nebulosity is just 1300 ly (460 pc) away. Studies of the nebula's expansion rate suggest that this supernova exploded around 9000 BC. It could have been witnessed by people then living in places such as Egypt and India. At maximum brilliance, the exploding star probably reached an apparent magnitude of −10, equal to the brightness of the Moon at first quarter.

Many supernova remnants are virtually invisible at optical wavelengths. However, when the expanding gases collide with the interstellar medium, they radiate energy at a wide range of wavelengths, from X rays through radio waves. For example, Figure 22-15 shows both X-ray and radio images of the supernova remnant Cassiopeia A. Optical photographs of this part of the sky reveal only a few small, faint wisps. In fact, radio searches for supernova remnants are more fruitful than optical searches. Only two dozen supernova remnants have been found on photographic plates, but more than 100 remnants have been discovered by radio astronomers.

From the expansion rate of the nebulosity in Cassiopeia A, astronomers conclude that this supernova explosion occurred about 300 years ago. Although telescopes were in wide use by the late 1600s, no one saw the outburst. In fact, the last supernova seen in our Galaxy, which occurred in 1604, was observed by Johannes Kepler. A few years earlier, in 1572, Tycho Brahe also recorded the sudden appearance of an exceptionally bright star in the sky. To find any other

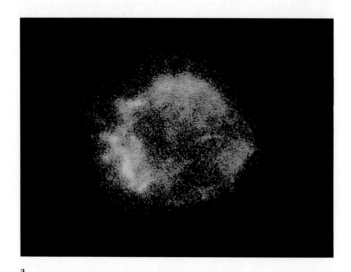

a

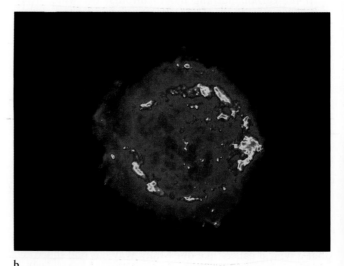

b

Figure 22-15 **Cassiopeia A** *Supernova remnants, such as Cassiopeia A, are typically strong sources of X rays and radio waves. (a) An X-ray picture of "Cas A" taken by the Einstein Observatory. (b) A corre-sponding radio image produced by the VLA. The supernova explosion that produced this nebula occurred 300 years ago, 10,000 light years from Earth. (Smithsonian Institution and the Very Large Array)*

accounts of nearby bright supernovae we must delve into astronomical records that are almost 1000 years old.

At first glance, this apparent lack of nearby supernovae may seem puzzling. Astronomers have seen more than 600 supernovae in distant galaxies. From this frequency, it is reasonable to suppose that a galaxy such as our own should have as many as five supernovae per century. Where have they been?

Vigorous stellar evolution in our Galaxy is occurring primarily in the disk, where the giant molecular clouds pass through a compressional density wave. Our Galaxy's disk is therefore the place where massive stars are born and where supernovae explode. This region is so rich in interstellar dust, however, that we simply cannot see very far into space in the directions occupied by the Milky Way. In other words, supernovae probably do in fact erupt every few decades in remote parts of our Galaxy, but their detonations are hidden from our view by intervening interstellar matter.

22-6 Neutrinos and gravity waves emanate from supernovae

In recent years, astronomers have devised an ingenious way of detecting supernovae. We have seen that a Type II supernova explosion is triggered by the collapse of a massive star's core. This collapse produces a flood of neutrinos that are created when electrons combine with protons to produce neutrons. Neutrinos are elusive particles that hardly interact with matter at all. In fact, under most conditions, matter is transparent to neutrinos. Consequently, when a supernova explodes, vast quantities of neutrinos pour out into space and stream across the Galaxy, passing easily through any gas, dust, or nebulae they happen to encounter. This deluge of neutrinos eventually rains down on—and through—the Earth. By detecting the arrival of these particles, astronomers could learn about the supernova.

Detecting neutrinos is a difficult, tricky business. The best available **neutrino "telescopes"** consist of large drums or tanks of water (see Figure 22-16) buried deep in the Earth. Water (H_2O) contains lots of protons, the nuclei of hydrogen atoms. Astronomers detect neutrinos by observing brief flashes of light when protons are struck by the neutrinos.

The neutrinos from a supernova explosion are traveling at or very near the speed of light and carrying a lot of energy, typically 20 MeV or more. (The energy units of electron volts, eV, were introduced in Box 5-3; 1 MeV = 10^6 electron volts.) Most of them pass right through the Earth as though it were not here. On rare occasions, however, a neutrino hits a proton in the water-filled drum. This collision produces a positron that then recoils at such a high speed that it emits a brief flash of light, known as **Cerenkov radiation**. This type of radiation, first observed in 1934 by the Russian physicist Pavel A. Cerenkov, occurs whenever a particle moves

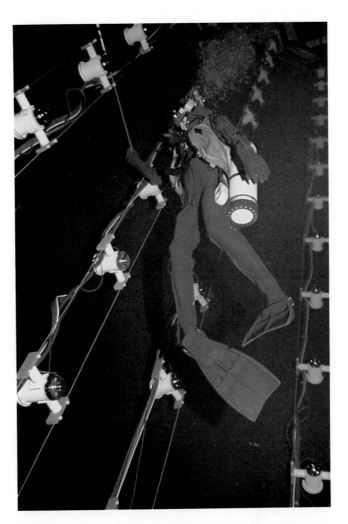

Figure 22-16 Inside a neutrino "telescope" *A diver is shown servicing one of the photomultiplier tubes in the Irvine–Michigan–Brookhaven neutrino detector. Neutrinos from Supernova 1987A were detected by this apparatus when they struck protons in the water, producing brief flashes of light that were picked up by the photomultipliers. (Courtesy of K. S. Luttrell)*

through a medium such as water at a speed that is faster than the speed of light in that medium. Such motion does not violate the tenet that the speed of light *in a vacuum* (3×10^8 m/s) is the ultimate speed limit in the universe. The speed of light through a substance like water or glass is much less than its speed in a vacuum. The recoiling electrons in the water-filled drums should easily exceed this reduced light speed, thus producing the flashes that reveal the existence of an erupting supernova somewhere in our Galaxy.

On February 23, 1987, a supernova exploded in the Large Magellanic Cloud (recall Figure 22-10 and Box 22-3). Teams of physicists working with neutrino detectors excitedly reported Cerenkov flashes from bursts of neutrinos 3 hours before astronomers saw the exploding star in the sky. At the Kamiokande II detector in Kamioka, Japan, scientists reported seeing flashes from 12 neutrinos at about the same time that 8 were detected by a team at the IMB (Irvine–Michigan–Brookhaven) detector in a salt mine under Lake Erie.

As we saw in Chapter 19, the mass of the neutrino is a controversial topic. The 1987 supernova gave scientists the opportunity to estimate that mass from the duration of the neutrino bursts. If neutrinos are massless, they would travel at the speed of light, and all the supernova's neutrinos should have arrived at Earth at essentially the same moment. The more massive the neutrino is, however, the greater the spread in arrival times. The interval over which the observed bursts were spread suggests that the mass of a neutrino is probably between 0 and 16 eV. For comparison, the mass of an electron is 511,000 eV.

The ability to detect neutrinos from a supernova offers new and exciting opportunities for astronomers. Ordinary telescopic observations of a supernova explosion can show us only the expanding outer layers of the dying star. Even X-ray or radio observations fail to see through the hot gases being blasted into space. Thus, ordinary techniques do not allow us to observe the extraordinary events occurring in and around the doomed star's core. However, neutrinos easily penetrate a star's outer layers. These particles carry information about the conditions under which they were created. By measuring the energy and momentum carried by the escaping neutrinos, we should be able to learn many facts about the star's collapsing core.

Another exotic technique being developed to probe supernova explosions involves gravitational radiation. During the process of core collapse and subsequent core bounce, vast amounts of matter crushed to nuclear density are hurled about at enormous speeds. This dense mass of material is surrounded by a strong gravitational field. Because this matter moves so rapidly during the explosion, the gravitational field around the star changes rapidly.

As we saw in Chapter 4, Einstein's general theory of relativity is the most accurate description of gravity we have. According to general relativity, gravity actually curves the fabric of space. During a supernova explosion, then, the sudden vigorous changes in gravity around the star can create ripples in the geometry of space. These ripples, which move outward from the supernova at the speed of light, are called **gravitational waves,** or **gravitational radiation.**

Like neutrinos, gravitational waves should bear the imprint of conditions in the star's collapsing core. Unfortunately, gravitational radiation carries very little energy, making the waves difficult to detect. Nevertheless, in the 1960s, Joseph Weber at the University of Maryland pioneered the construction of gravitational-wave antennas by gluing sensitive crystals onto a large aluminum cylinder. Because gravitational waves are ripples in the geometry of space, objects vibrate slightly when a gravitational wave passes through. If the cylinder oscillates, the crystals produce an electrical voltage that can be amplified and recorded. In this way, the passage of a gravitational wave may be detected.

Teams of physicists around the world are working hard to perfect new techniques for detecting gravitational waves. Many scientists, such as Kip Thorne at Caltech, are optimistic that highly sensitive antennas using lasers will be operational by the 1990s. These antennas should be able to detect bursts of gravitational waves from the collapsing cores of supernovae as far away as 30 million pc (about 100 million ly). This volume of space is so huge and contains so many galaxies that astronomers might be able to detect supernova explosions as frequently as once a month. By analyzing a burst of gravitational waves from a supernova, astronomers should be able to figure out many details concerning the creation of such objects as neutron stars and black holes.

Key words

asymptotic giant branch star (AGB star)	gravitational radiation	neutron star	red supergiant
black hole	gravitational waves	neutronization	s process
carbon burning	helium-shell flashes	nuclear density	s-process isotope
Cerenkov radiation	mass–radius relation	oxygen burning	shell helium burning
Chandrasekhar limit	neon burning	photodisintegration	silicon burning
deflagration	neutrino	planetary nebula	supernova
detonation	neutrino "telescope"	r process	supernova remnant
	neutron capture	r-process isotope	thermal pulses
			white dwarf

Key ideas

- A low-mass star becomes a red giant when shell hydrogen burning begins, a horizontal-branch star when core helium burning begins, and a red supergiant when the helium in the core is exhausted and shell helium burning begins.

  Thermal runaways in the helium-burning shell produce thermal pulses during which more than half a star's mass may be ejected into space.

  Ultraviolet radiation from a hot carbon–oxygen core ionizes and excites ejected gases, producing a planetary nebula.

- The burned-out core of a low-mass star becomes a degenerate, dense sphere about the size of the Earth called a white dwarf, which glows from thermal radiation; as the sphere cools, it becomes dimmer and eventually cold.

- A high-mass star undergoes a sequence of different thermonuclear reactions in its core and shells: carbon burning, neon burning, oxygen burning, and silicon burning.

  In the last stages of its life, a high-mass star has an iron-rich core surrounded by concentric shells hosting the various thermonuclear reactions.

  A high-mass star dies in a violent cataclysm in which its core collapses and most of its matter is ejected into space at high speeds; the luminosity of the star increases suddenly by a factor of around 10^8 during this explosion, producing a supernova.

  The ejected matter, moving at supersonic speeds through interstellar gases and dust, glows as a nebula called a supernova remnant.

  An accreting white dwarf in a close binary system can also become a supernova when carbon burning ignites explosively throughout such a degenerate star.

- Type I supernovae are produced by accreting white dwarfs in close binaries; Type II supernovae are created by the deaths of massive stars.

- Most supernovae occurring in our Galaxy are hidden from our view by interstellar dust and gases.

- Bursts of neutrinos and gravitational waves are emitted during a supernova explosion; the detection and study of these emissions will give important insight into phenomena occurring deep inside an exploding star.

Review questions

1 What is the difference between a red giant and a red supergiant?

2 How is a planetary nebula formed?

3 What is a white dwarf?

4 What is the significance of the Chandrasekhar limit?

*5 A 1-solar-mass white dwarf has a radius of about 15,000 km. What is the radius of a white dwarf whose mass is equal to the Chandrasekhar limit?

6 On an H–R diagram, sketch the evolutionary track that the Sun will follow as it leaves the main sequence to become a white dwarf. Approximately how much mass will the Sun have when it becomes a white dwarf? Where will the rest of the mass have gone?

7 Why do you suppose that all the white dwarfs known to astronomers are relatively close to the Sun?

8 What prevents thermonuclear reactions from occurring at the center of a white dwarf? If no thermonuclear reactions are occurring in its core, why doesn't the star collapse?

9 Why is the temperature in a star's core so important in determining which nuclear reactions can occur there?

10 Suppose that the brightness of a star becoming a supernova increases by 20 magnitudes. Show that this corresponds to an increase of 10^8 in luminosity.

11 What is the difference between Type I and Type II supernovae?

12 Why have radio searches for supernovae remnants been more fruitful than optical searches?

Advanced questions

Tips and tools for . . .
You may find it useful to review Box 19-4, which discusses stellar radii and their relationship to temperature and luminosity. Important data and a photograph of the Ring Nebula are found in Box 22-1. The formula for escape velocity was given in Box 7-2. The all-important relationship between absolute magnitude, apparent magnitude, and distance was discussed in Box 19-2.

*13 The central star in a newly formed planetary nebula has a luminosity of 1000 $L_\odot$ and a surface temperature of 100,000 K. How big is the star?

14 You want to determine the age of a planetary nebula. What observations should you make, and how would you use the resulting data?

*15 (Basic) The Ring Nebula is expanding at the rate of about 20 km/s. How long ago did the central star shed its outer layers?

16 What reasons can you think of to explain why the rate of expansion of the gas shells in some planetary nebulae is nonuniform?

*17 (Basic) Find the average density of a 1 M$_\odot$ white dwarf having the same size as the Earth. What speed is required to eject gas from the white dwarf's surface? Suppose some interstellar gas fell onto the white dwarf. With what speed would it strike the star's surface?

18 What kinds of stars would you monitor if you wished to observe a supernova explosion from its very beginning? Look up lists of the brightest and nearest stars. Which, if any, of these stars are possible supernova candidates? Explain.

*19 Suppose that the red supergiant star Betelgeuse became a supernova. At the height of the outburst, how bright would it appear in the sky? How would it compare with the brightness of the full moon?

*20 (Challenging) In mid-April, 1979, a supernova exploded in a galaxy called M100 (NGC 4321) in the constellation of Coma Berenices. It reached 12th magnitude at maximum brilliance and its spectrum showed hydrogen lines. How far away is the galaxy?

Discussion questions

21 Suppose that you discover a small, glowing disk of light while searching the sky with a telescope. How would you observationally decide if this object is a planetary nebula? Could your object be something else? Explain.

22 Why are astronomers interested in building neutrino "telescopes" and instruments to detect gravitational waves?

Observing projects

23 Although they represent a fleeting stage at the end of a star's life, planetary nebulae are found all across the sky. Some of the brightest are listed in the following table with epoch 2000 coordinates.

Observe as many of these planetary nebula as you can on a clear, moonless night using the largest telescope at your disposal. Note and compare the various shapes of the different nebulae. In how many cases can you see the central star? The so-called "Blinking planetary" (NGC 6826) affords an excellent demonstration of central versus peripheral vision:

Planetary nebula	R.A. (hr min)	Decl. (° ')
"Little Dumbbell" (M76)	1 42.4	+51 34
NGC 1535	4 14.2	−12 44
Eskimo	7 29.2	+20 55
Ghost of Jupiter	10 24.8	−18 38
Owl (M97)	11 14.8	+55 01
Ring (M57)	18 53.6	+33 02
Blinking Planetary	19 44.8	+50 31
Dumbbell (M27)	19 59.6	+22 43
Saturn	21 04.2	−11 22
NGC 7662	23 25.9	+42 33

The nebula seems to disappear when you look straight at it, but it reappears as soon as you look toward the side of the field of view. This effect occurs because the central region of the human retina, on which an image is focused when you look straight at it, is less sensitive than surrounding regions of the retina.

24 There are two supernova remnants that can be seen visually through modest-sized telescopes, one in the winter sky and the other in the summer sky. Both are quite faint, however, so you should schedule your observations for a moonless night.

The winter sky contains the Crab Nebula, which is discussed in detail in the next chapter. The coordinates are R.A. = 5^{h}34.5^m and Decl. = 22°00′, which places the object near the star marking the eastern horn of Taurus (the bull).

Whereas the entire Crab Nebula easily fits in the field of view of an eyepiece, the Veil or Cirrus Nebula in the summer sky is so vast that you can see only a small fraction of it at a time. The easiest way to find the Veil Nebula is to aim the telescope at the star 52 Cygni (R.A. = 20^{h}45.7^m and Decl. = +30°43′), which lies on one of the brightest portions of the nebula. Then move the telescope slightly north or south until 52 Cygni is just out of the field of view, and you should see faint wisps of glowing gas. (52 Cygni is the bright star toward the lower right corner of Figure 20-18, which shows the entire Cygnus Loop.)

25 Use a telescope to observe the remarkable triple star 40 Eridani whose coordinates are R.A. = 4^{h}15.3^m and Decl. = −7°39′. The primary, a 4.4 magnitude yellowish star like the Sun, has a 9.6 magnitude white dwarf companion, the most easily seen white dwarf in the sky. On a clear, dark night with a moderately large telescope you should also see that the white dwarf has an 11th-magnitude companion, which completes this most interesting trio.

For further reading

Bethe, H., and Brown, G. "How a Supernova Explodes." *Scientific American*, May 1985 • This enlightening article describes many fascinating details about the detonation of a Type II supernova.

Kaler, J. "Planetary Nebulae and Stellar Evolution." *Mercury*, July/August 1981 • This excellent article discusses the later stages of stellar evolution with particular emphasis on planetary nebulae.

Kahn, R. "Desperately Seeking Supernovae." *Sky & Telescope*, June 1987 • This article describes the trials and tribulations of developing an automated telescope to search for supernovae in galaxies.

Kwok, S. "Not with a Bang but a Whimper." *Sky & Telescope*, May 1982 • This brief article discusses intriguing aspects of planetary nebulae and the role they play in the deaths of low-mass stars.

Malin, D., and Allen, D. "Echoes of the Supernova." *Sky & Telescope*, January 1990 • Two noted Australian astronomers discuss their observations and photographs of "light echoes" caused by radiation from SN 1987A reflecting off dust clouds in the vicinity of the supernova.

Marschall, L. *The Supernova Story*. Plenum, 1988 • This superbly written book is an excellent choice for someone who wishes to learn more about supernovae, including SN 1987A.

Murdin, P., and Murdin, L. *Supernovae*. Cambridge University Press, 1985 • This book tells the story of our growing understanding of supernovae, pieced together from historical records, modern observations, and theoretical studies.

Reddy, F. "Supernovae: Still a Challenge." *Sky & Telescope*, December 1983 • This article explains how Type I and Type II supernovae explode.

Schorn, R. "Supernova Shines On." *Sky & Telescope*, May 1987 • This article contains many outstanding pictures of the early stages of SN1987A.

Seward, F., Gorenstein, P., and Tucker, W. "Young Supernova Remnants." *Scientific American*, August 1985 • This article focuses on X-ray observations of the remnants of supernovae whose detonations were seen within the past thousand years.

Tierney, J. "Quest for Order: Profile of S. Chandrasekhar." *Science 82*, September 1982 • This well written biographical sketch introduces the Indian astrophysicist who gave us our basic understanding of white dwarfs.

Wallerstein, G., and Wolff, S. "The Next Supernova?" *Mercury*, March/April 1981 • Written six years before SN 1987A, this article by two noted astronomers speculates that the double star HD 128220 will produce a supernova within the next million or so years.

Wheeler, C., and Harkness, R. "Helium-rich Supernovas." *Scientific American*, November 1987 • This article presents evidence of a third type of supernova, called Type Ib, whose progenitor is a massive star that has shed all its hydrogen prior to detonating, causing its spectrum to resemble that of a Type I.

Woosley, S., and Weaver, T. "The Great Supernova of 1987" *Scientific American*, August 1989 • Two preeminent authorities on supernovae teamed up to write this superb article, which describes many fascinating details of SN 1987A.

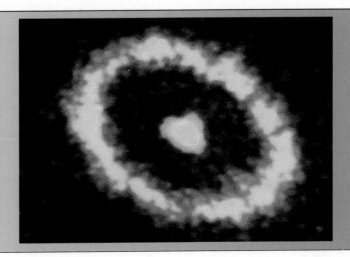

Box 22-4   HST observations of SN 1987A
This image from the Hubble Space Telescope clearly shows an elliptical, luminescent ring of gas about 1.3 light years across surrounding the still glowing center of the supernova explosion. This ring is a relic of a hydrogen-rich stellar envelope that was ejected by gentle stellar winds from the doomed star when it was a red giant, about 10,000 years ago. The diffuse gas was subsequently compressed into a narrow shell by high-speed stellar winds that flowed from the star when it became a blue supergiant. This picture, taken in August 1990 with the HST's Faint Object Camera, shows features as small as 0.1 arc second across.

Icko Iben, Jr., known world wide for his contributions to our understanding of stellar evolution, received his B.A. in physics from Harvard and his Ph.D. from the University of Illinois. After teaching at Williams College, Caltech, and MIT, Dr. Iben settled at the University of Illinois as head of the astronomy department and distinguished professor of astronomy and physics.

In his early research, Dr. Iben studied the evolution of low-mass stars in globular clusters and developed methods for dating them. A pioneer in the numerical solution of the equations of stellar structure, he created the theory of nucleosynthesis by slow neutron capture in asymptotic-branch giant stars. He then went on to show how the products of such nucleosynthesis are brought to the stellar surface. In 1989, the American Astronomical Society honored Dr. Iben with their highest prize, the Henry Norris Russell Lectureship.

Icko Iben, Jr. THE EVOLUTION OF CLOSE BINARY STARS

The evolution of close binary stars is quite different from that of single stars. The stars in a close binary system are so near each other that their mutual interactions dramatically affect their evolution. Some of these complex processes involved are rapid mass transfer between stars and a frictional egg-beaterlike action of the two stellar cores that churns up their common gaseous envelope formed during mass transfer. Tidal forces between the two stars can alter their rotation rates and produce significant tidal heating. A close binary can also lose angular momentum, which is carried away by a stellar wind or by gravitational radiation, eventually causing the two stars to merge.

This complexity prevents astrophysicists from constructing a simple set of equations, as they do with single stars, for calculating evolutionary tracks to compare with observation. Instead, researchers invent a set of qualitative ideas, which are disguised as mathematical formulas containing unknowns called "free parameters," and construct *scenarios* to describe the *possible* evolutionary behavior of a close binary system. Astronomers then use observations of real binaries to pin down some of the free parameters. Such restrictions on the range of free parameters place constraints on complex phenomena in close binaries, thereby permitting us to better understand the physics that is going on.

An example of such a scenario is shown in the accompanying illustration. Initially (see part *a*), the binary consists of two main-sequence stars of intermediate mass at a separation such that the more massive star reaches the asymptotic giant branch (AGB) stage before it begins to transfer mass to its companion. The scenario therefore involves two stars with masses in the range of 5 to 9 $M_\odot$, initially separated by 500 to 1500 $R_\odot$, revolving about each other with an orbital period of 1 to 6 years.

The more massive star swells as it becomes an AGB star and develops a compact electron-degenerate core (which is essentially a white dwarf) composed of carbon and oxygen surrounded by helium- and hydrogen-burning shells. When this star overflows its Roche lobe, mass transfer occurs so rapidly that the companion star cannot absorb the onslaught and a common envelope forms about the pair (part *b*). As mass from the primary continues to flow into the common envelope, friction generated between the gaseous envelope and the stars within it drives matter from the system. Mass loss continues until the primary is stripped of its hydrogen shell, exposing a cooling white dwarf, while the main-sequence companion remains relatively unchanged.

During the first common envelope stage, the separation between the two stars is typically reduced by a factor of ten.

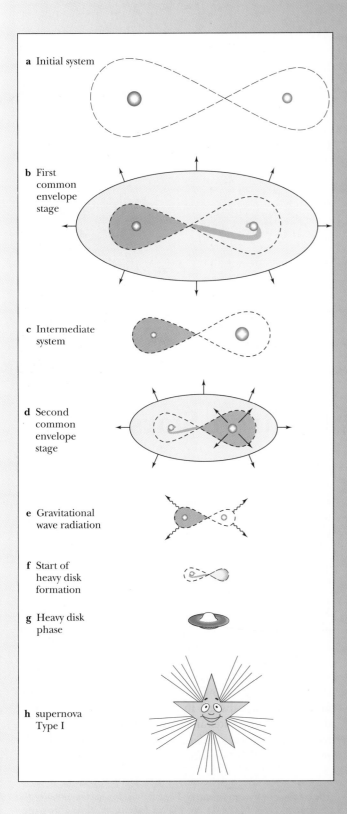

a Initial system

b First common envelope stage

c Intermediate system

d Second common envelope stage

e Gravitational wave radiation

f Start of heavy disk formation

g Heavy disk phase

h supernova Type I

Continuing hydrogen burning in the core of the secondary star leads to a contracting helium core and an expanding envelope. As the secondary expands and fills its Roche lobe, mass transfer again begins at a rate too high for the companion star to absorb, and a common envelope is again formed (part *d*). Friction again causes orbital shrinkage as matter is driven from the system, exposing the secondary's helium core. This core becomes degenerate after helium burning converts it to carbon and oxygen.

The system now consists of two white dwarfs (part *e*), the newly formed one being only slightly less massive than the first-formed white dwarf. Two phases of mass loss and orbital shrinkage have left the pair of white dwarfs in a very tight orbit. Over the next 10^8 to 10^{10} years, gravitational radiation extracts additional angular momentum from the system, forcing the stars to approach each other ever more closely. The critical surface about the stars shrinks as their orbits shrink, and eventually the less-massive white dwarf—which is the larger of the two stars—begins to overflow its Roche lobe (part *f*) and transfer mass to its companion. This process accelerates because, as the donor white dwarf becomes ever lighter, it also increases in size and so more readily overflows its Roche lobe. The lighter white dwarf dissolves into a thick envelope of nondegenerate material atop the heavier white dwarf, resulting in the heavy disk phase (part *g*).

What follows is a very complicated series of carbon-burning shell flashes interspersed by episodes of contraction. If the sum of the initial pair of white dwarfs exceeds the Chandrasekhar limit (~ 1.4 $M_\odot$), then one of the first contractions evolves into a core collapse that is halted and reversed by a star-disrupting thermonuclear explosion. During this cataclysm, at least half the matter is converted into iron, nickel, cobalt, and associated heavy elements, which are expelled into space at speeds near 10,000 km/s. The optical display is a Type I supernova (part *h*).

A particularly attractive feature of this scenario is that it accounts for the fact that Type I supernovae are the only kind occurring in elliptical galaxies, which are devoid of interstellar gas and dust, unlike spiral galaxies (Chapter 26). Type II supernovae, the end result of the evolution of stars more massive than ~ 10 $M_\odot$, are found only in spiral galaxies. Their absence in ellipticals demonstrates that these galaxies do not contain single massive stars. We know from studies of intermediate-mass stars in the Magellanic Clouds and our own Galaxy that, after shedding their hydrogen-rich envelope as a planetary nebula, single stars less massive than ~ 10 $M_\odot$ evolve into white dwarfs. Consequently, supernova explosions in elliptical galaxies can only result from intermediate-mass binaries that evolve into close pairs of CO white dwarfs having a combined mass larger than 1.4 $M_\odot$. Even long after star formation has ceased, Type I supernovae continue to occur as white-dwarf pairs formed at successively larger initial separations come into contact and merge.

This is due to the dissipation of energy by friction, which robs the stars of orbital energy and thereby causes orbital shrinkage. Now at an intermediate stage (part *c*), the two stars are separated by roughly 50 to 150 $R_\odot$ and are orbiting each other with a period of about 16 to 90 days.

Neutron Stars

A supernova explosion can produce a stellar corpse called a neutron star, which is an extremely dense sphere of degenerate neutrons. Depending on circumstances, neutron stars manifest a variety of phenomena. For instance, a pulsar is observed when a rapidly spinning neutron star with a powerful magnetic field sweeps beams of radiation around the sky. If a neutron star is in a binary system, pulsating X-ray sources, or bursters, can result when gas from the companion star falls onto the neutron star. Models of neutron stars devised by astronomers to explain these phenomena describe many properties of neutron stars, such as the superconductivity and superfluidity that dominate their interiors. These models also demonstrate that there is an upper limit to the mass of a neutron star, just as the Chandrasekhar limit sets the maximum mass of a white dwarf. In close binary systems, each of these two kinds of stellar corpses—white dwarfs and neutron stars—can undergo explosive thermonuclear reactions on their surfaces.

The creation of a neutron star?   The bright star near the center of this photograph is SN 1987A one week after detonation. Many astronomers suspect that the supernova's core may have collapsed to form a neutron star. In 1990, the supernova's light curve had leveled off, indicating the presence of an underlying, continual source of energy, possibly caused by gases falling back down onto a rapidly rotating neutron star. Astronomers are keeping a careful watch on the supernova's expanding layers in the hope of seeing pulsed radiation. This would confirm the presence of a spinning neutron star. (NOAO)

The brightening eastern sky was aglow with pink and red as Yang Wei-T'e waited patiently for the sunrise. As imperial astronomer to the Chinese court during the Sung dynasty, Yang was thoroughly familiar with the constellations. He knew at once that something quite extraordinary had happened that night. Preceding the Sun by only a few minutes, a dazzling object ascended above the eastern horizon. It was far brighter than Venus and more resplendent than any star he had ever seen. Yang dutifully recorded his observations and thoughts on this unusual celestial event:

> I make my kowtow. I observed the phenomenon of a guest star. Its color was slightly iridescent. Following an order of the Emperor, I respectfully make the prediction that the guest star does not disturb Aldebaran. This indicates that . . . the Empire will gain great power. I beg to store this prediction in the Department of Historiography.

The date would be officially recorded as "the day of Ch'ih Ch'iu in the fifth moon of the first year of the Shih-huo period"; we would call it July 4, 1054.

Yang's "guest star" was so brilliant that it could easily be seen during broad daylight for the rest of July. After a year, however, "it faded and became invisible." Actually, the supernova had occurred some 6500 years earlier, and its light had taken this long to reach the Earth. Today we realize that Yang Wei-T'e had witnessed the creation of a neutron star.

23-1 Pulsars are rapidly rotating neutron stars with intense magnetic fields

The neutron was discovered in 1932 during laboratory experiments. Within a year, two astronomers predicted the existence of neutron stars. Inspired by the realization that white dwarfs are supported by degenerate-electron pressure, Fritz Zwicky at the California Institute of Technology (Caltech) and his colleague Walter Baade at Mount Wilson Observatory proposed that a highly compact ball of neutrons could similarly produce a powerful pressure. This degenerate-neutron pressure also could support a stellar corpse, perhaps one even more massive than is permitted by the Chandrasekhar limit. "With all reserve," Zwicky and Baade theorized, "we advance the view that supernovae represent the transition from ordinary stars into **neutron stars**, which in their final stages consist of extremely closely

packed neutrons." In other words, there could be at least two types of stellar corpses: white dwarfs and neutron stars.

This prophetic proposal was politely ignored by most scientists for years. After all, a neutron star must be a rather weird object, so tiny that it would be virtually impossible to observe. As we saw in Chapter 22, in order to transform protons and electrons into neutrons, the density in the star must be equal to nuclear density, about 4×10^{17} kg/m³. Thus, a thimbleful of neutron-star matter brought back to Earth would weigh 100 million tons. Furthermore, an object compacted to nuclear density would be very small. A 1-$M_\odot$ neutron star would have a diameter of only 30 km, about the same size as San Francisco or Manhattan. The surface gravity on one of these neutron stars would be so strong that the escape velocity would equal one-half the speed of light. All of these conditions seemed so outrageous that few astronomers paid serious attention to the subject of neutron stars—until 1968.

As a young graduate student at Cambridge University, Jocelyn Bell spent many months helping construct an array of radio antennas covering $4\frac{1}{2}$ acres of the English countryside. The instrument was completed by fall of 1967, and Bell and her colleagues began detecting and scrutinizing radio emissions from various celestial sources. In November, Bell noticed that the antennas had detected regular pulses of radio noise from one particular location in the sky. Repeated careful observations demonstrated that the radio pulses were arriving with a regular period of 1.3373011 seconds (see Figure 23-1).

The regularity of this pulsating radio source was so striking that the Cambridge team suspected they might be detecting signals from an advanced alien civilization. This possibility was soon discarded, however, when several more of these pulsating radio sources, which came to be called pulsars, were discovered across the sky. In all cases, the periods were extremely regular, ranging from about $\frac{1}{4}$ s for the fastest to about $1\frac{1}{2}$ s for the slowest.

The discovery of pulsars was officially announced in early 1968, after which many astronomers around the world began proposing all sorts of explanations for these regular radio pulsations. Many of these theories were quite bizarre, leading to arguments that raged among astronomers for months. However, by late 1968, all this controversy was laid to rest with the discovery of a pulsar, now called the Crab pulsar, in the middle of the Crab Nebula.

Yang Wei-T'e and his colleagues had been quite precise about the location of the supernova of 1054. When we turn

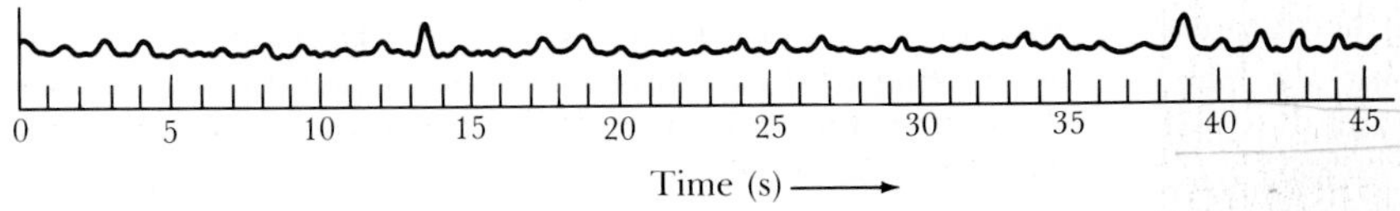

Figure 23-1 *A recording of the first pulsar* This chart recording shows the intensity of radio emission from the first known pulsar, CP1919 (this designation means "Cambridge pulsar at a right ascension of $19^h 19^m$"). Note that some pulses are weak, others strong. Nevertheless, the spacing between pulses is exactly 1.3373011 seconds. (Adapted from A. Hewish)

Figure 23-2 *The Crab Nebula* *This beautiful nebula, named for the armlike appearance of its filamentary structure, is the remnant of the supernova of AD 1054. The distance to the nebula is about 6500 ly (2000 pc), so its present angular size (4 by 6 arc min) corresponds to linear dimensions of about 7 by 10 ly. (Palomar Observatory)*

a telescope toward this location in the constellation of Taurus, we find the Crab Nebula, shown in Figure 23-2. This object, which looks like an exploded star, is a supernova remnant.

The fact that a pulsar is located in a supernova remnant tells us that pulsars probably are associated with dead stars. Before the discovery of pulsars, most astronomers believed that all stellar corpses were white dwarfs. There seemed to be a sufficient number of white dwarfs in the sky to account for all the stars that must have died since our Galaxy was formed. It was generally assumed that all dying stars—even the most massive ones, which produce supernovae—somehow manage to eject enough matter so that their corpses do not exceed the Chandrasekhar limit. The discovery of the Crab pulsar showed these conservative opinions to be in error.

The Crab pulsar is one of the fastest young pulsars ever discovered. Its period is 0.033 second, which means that it flashes 30 times each second. It was immediately apparent to astrophysicists that white dwarfs are too big and bulky to produce 30 signals per second. Calculations demonstrated that a white dwarf can neither rotate that fast nor vibrate that fast. The existence of the Crab pulsar clearly indicates that the stellar corpse at the center of the Crab Nebula is much smaller and more compact than a white dwarf. As Thomas Gold of Cornell University emphasized, astronomers would now have to face the possible existence of neutron stars seriously.

Exactly what would a neutron star be like? As we have seen, it would be small and dense. It should also be rotating rapidly. All stars rotate, but most of them do so in a leisurely fashion. For example, our Sun takes nearly a full month to rotate once about its axis. However, just as an ice skater doing a pirouette speeds up when she pulls in her arms, a collapsing star also speeds up as its size shrinks. This phenomenon is a direct consequence of a law of physics called the **conservation of angular momentum**, which was intro-

duced in Box 4-4 (see Box 23-1 for additional details). In fact, an ordinary star rotating once a month would be spinning faster than once a second by the time it was compressed to the size of a neutron star.

In addition to having a rapid rate of rotation, we would probably expect a neutron star to have an intense magnetic field. It seems safe to say that every star has a magnetic field. In some cases, such as the Sun, the strength of these stellar magnetic fields is typically quite low. A major reason for this weakness is that a star's magnetic field, which is firmly embedded in the ions and electrons in the star's outer layers, is spread out over millions upon millions of square kilometers of the star's surface. However, if a star of solar dimensions collapses down to a neutron star, its surface area (which is proportional to the square of its radius) shrinks by a factor of 10^9. Because the magnetic field is bonded to the star's gases, it becomes concentrated onto one-billionth of its original area, and so the strength of the magnetic field increases by a factor of 1 billion.

The strength of a magnetic field is usually measured in the unit called the gauss (G), named after the famous German mathematician and astronomer Karl Friedrich Gauss. For example, Zeeman splitting of lines in the solar spectrum reveals that the Sun's overall magnetic field has a strength in the range of 1 to 2 G, although fields thousands of times stronger have been measured in sunspots (recall Figure 18-17). The spectra of certain white dwarfs reveal Zeeman splitting that corresponds to field strengths of 1 million G or more, considerably stronger than anything ever produced in a laboratory. By our usual standards, the trillion-gauss (10^{12} G) fields surrounding neutron stars are extremely powerful.

Finally, we might expect the axis of rotation of a typical neutron star to be inclined at an angle to the magnetic axis connecting the north and south magnetic poles (see Figure 23-3). In 1969, Peter Goldreich at Caltech pointed out that the combination of a powerful magnetic field and rapid

Box 23-1   The conservation of angular momentum

The concept of angular momentum was introduced briefly in Box 4-4. **Angular momentum** is a measure of the momentum carried by an object because of its rotation or revolution about an axis. A mass m, moving with a speed v as it revolves about an axis, has an angular momentum L given by

$$L = mvr$$

where r is the distance from the mass to the axis of revolution, as shown in the diagram on the top right.

When speaking about angular motion, it is often convenient to define the **angular velocity** ω (omega) as

$$\omega = \frac{v}{r}$$

Thus, we can write

$$L = mr^2\omega$$

The conservation of angular momentum requires that both the magnitude and the direction of the angular momentum never change. For a mass revolving about an axis, this requires the value of L to remain constant and the direction of the axis of rotation to remain fixed in space.

For a rotating object of some size, the object's angular momentum is given by

$$L = I\omega$$

where I is the object's moment of inertia, an expression of the way matter is distributed throughout the object. For instance, the moment of inertia of a solid wheel of mass M and radius R is $\frac{1}{2}MR^2$. For a bicycle wheel, which has all its mass concentrated in a ring a distance R from the axis of rotation, the angular momentum is just MR^2.

These examples with wheels demonstrate a general property of moments of inertia: The more concentrated the mass is around the axis of rotation, the smaller the moment of inertia is. This trait helps us understand why an ice skater doing a pirouette speeds up as she pulls in her arms. When she pulls in her arms, her moment of inertia (I) decreases, and so her angular velocity (ω) must increase in order for her angular momentum ($I\omega$) to remain unchanged.

For a sphere of uniform density with mass M and radius R, the moment of inertia is given by

$$I_{\text{sphere}} = \tfrac{2}{5} MR^2$$

Thus, the angular momentum of a uniform sphere rotating with an angular velocity ω can be written as

$$L = \tfrac{2}{5} MR^2\omega$$

Example: Suppose that a star collapses from solar dimensions ($R = R_\odot = 7 \times 10^5$ km) down to neutron-star dimensions ($R =$

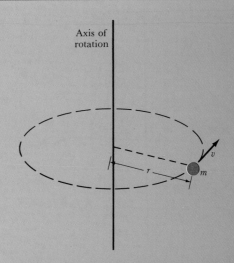

$R_* = 16$ km) and no mass is lost (so that M is constant). The conservation of angular momentum tells us that the original angular velocity (ω_0) is related to the final angular velocity (ω_f) by

$$L = \tfrac{2}{5} MR_\odot^2\omega_0 = \tfrac{2}{5} MR_*^2 \omega_f$$
$$R_\odot^2\omega_0 = R_*^2 \omega_f$$
$$\omega_0(7 \times 10^5)^2 = \omega_f(16)^2$$

or

$$\frac{\omega_f}{\omega_0} = \left(\frac{7 \times 10^5}{16}\right)^2 = 2 \times 10^9$$

In other words, the star is rotating 2 billion times faster after the collapse than it was before. For example, suppose the star rotates about once a month, as the Sun does. Because 1 month = 3×10^6 s, when the star becomes a neutron star it is rotating nearly 1000 times per second. In reality, however, material escaping from the collapsing star would carry away a significant amount of its angular momentum. The resulting neutron star would probably not be rotating quite as fast as this calculation suggests.

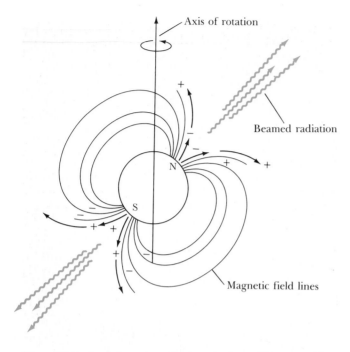

Figure 23-3 *A rotating, magnetized neutron star* It is reasonable to suppose that a neutron star is rotating rapidly and possesses a powerful magnetic field. Charged particles that have been accelerated near the star's magnetic poles produce two oppositely directed beams of radiation. As the star rotates, the beams sweep around the sky. If the Earth happens to lie in the path of the beams, we see a pulsar.

rotation should create extremely intense electric fields near the neutron star's surface, as with a giant electric generator. At its surface, there are plenty of protons and electrons, because the pressures there are too low to combine them into neutrons. The powerful electric fields acting on these charged particles cause them to flow out from the neutron star's polar regions along its curved magnetic field, as sketched in Figure 23-3. As the particles stream along the curved field, they become accelerated and emit energy. The final result is two thin beams of radiation pouring out of the neutron star's north and south magnetic polar regions.

A rotating, magnetized neutron star is somewhat like a lighthouse beacon. As the star rotates, the beams of radiation sweep around the sky. If the Earth happens to be located in the right direction, a brief flash is observed each time the beam sweeps through our line of sight.

When this scenario was first proposed in the late 1960s, a team of astronomers at the University of Arizona began wondering if pulsars might also emit flashes at wavelengths other than radio frequencies. After aiming a telescope at the center of the Crab Nebula, they used a spinning disk with a slit in it to "chop," or interrupt, incoming light at rapid intervals. To their surprise, they found that one of the stars at the center of the nebula is actually flashing on and off 30 times each second (see Figure 23-4).

Since those pioneering days, the Crab pulsar has been observed emitting flashes over a wide range of wavelengths.

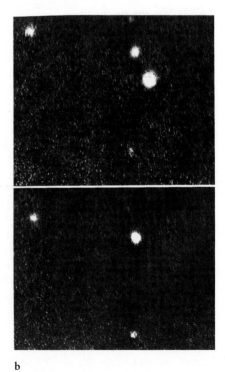

a   b

Figure 23-4 *The Crab pulsar* A pulsar is located at the center of the Crab Nebula. (a) The Crab pulsar is identified by the arrow. (b) These two photographs show the Crab pulsar in its "on" and "off" states. Like the radio pulses, these visual flashes have a period of 0.033 s. (Palomar Observatory; Lick Observatory)

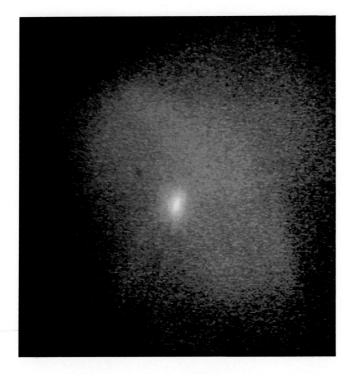

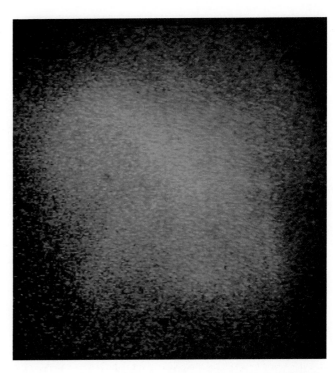

Figure 23-5 X-ray views of the Crab Nebula and pulsar These views
of the Crab Nebula were obtained with an X-ray telescope on the Earth-

orbiting Einstein Observatory. The pulsar is shown in its "on" (left) and
"off" (right) states. (Harvard–Smithsonian Center for Astrophysics)

For example, Figure 23-5 shows X-ray pictures from the
Einstein Observatory with the pulsar in its "on" and "off"
states. And the light curves of the Crab pulsar at optical,
X-ray, and radio wavelengths are displayed in Figure 23-6.
Note that, in addition to the main pulse, there is a secondary
pulse about halfway through the pulsar's period. Many pulsars in fact exhibit both main and secondary pulses, simply
because neutron stars have both north and south magnetic
poles. In general, one of the magnetic poles is more directly
aligned toward the Earth than the other. Hence we detect
one strong pulse and one weak one during each rotation of
the neutron star.

During the 1970s and 1980s, radio astronomers discovered about 300 pulsars scattered across the sky. Each one is
presumed to be the neutron-star corpse of a massive extinct
star. The Vela pulsar, located at the core of the Gum Nebula
(see Figure 23-7), is also visibly flashing like the Crab pulsar.
The Vela pulsar, whose period is 0.089 s, is the slowest pulsar ever detected at visible wavelengths.

The Crab pulsar is one of the youngest pulsars, its creation having been observed by Yang Wei-T'e some 900 years
ago. The Vela pulsar is also quite young, its creation having
occurred roughly 11,000 years ago. Pulsars slow down as
they get older. Only the very youngest pulsars are energetic
enough to emit both optical flashes and radio pulses.

Several supernovae have been seen during historical
times, but they did not necessarily produce pulsars. For instance, those observed by Tycho Brahe in 1572 and by
Johannes Kepler in 1604 were probably Type I supernovae,

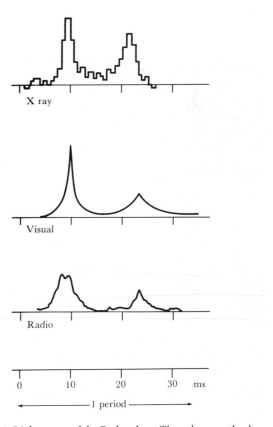

Figure 23-6 Light curves of the Crab pulsar These three graphs show
the intensity of radiation emitted by the Crab pulsar at X-ray, optical,
and radio wavelengths. One millisecond equals 10^{-3} s. Thus, the pulsar's period is 33 ms.

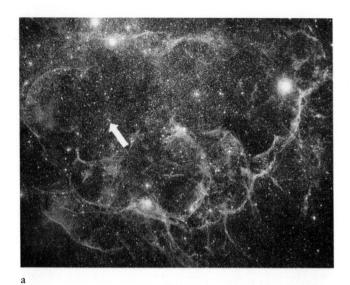

a

b

Figure 23-7 *The Vela pulsar in X rays* *The Vela pulsar, named for its location in the southern constellation of Vela, is at the middle of a supernova remnant called the Gum Nebula. Like the Crab pulsar, the Vela pulsar can be detected at X-ray, optical, and radio wavelengths.* **(a)** *The location of the pulsar is identified by the arrow.* **(b)** *This view was obtained from one of the X-ray telescopes on the Einstein Observatory. (Harvard–Smithsonian Center for Astrophysics)*

and it is difficult to imagine how an exploding white dwarf might become a neutron star. The primary source of pulsars is Type II supernovae, which involve the core-collapse of massive stars.

The great supernova of 1987 was a Type II, and so astronomers have been carefully observing its remains for signs of a pulsar. In 1990, SN 1987A's light curve had leveled off to a steady luminosity, just about what would be expected from gas falling onto a neutron star's surface. Meanwhile, NASA scientists are designing a satellite called the *X-ray Timing Explorer*, to be launched in 1995, which will carry X-ray sensors capable of seeing through much of the gas and dust surrounding the supernova to search for rapid X-ray variations indicative of a rapidly spinning neutron star. Obviously, SN 1987A will occupy astronomers' attention for many years to come.

23-2 Superfluidity and superconductivity are among the strange properties of neutron stars

Many of the details of how pulsars emit radiation are still poorly understood. Nevertheless, the concept of a rapidly rotating, magnetized neutron star has proved quite successful in explaining many phenomena that were once quite puzzling. For example, without a spinning neutron star it is virtually impossible to explain why the Crab Nebula shines the way it does.

The diffuse part of the Crab Nebula—not the reddish filaments—shines with an eerie light that has been identified by the Russian astronomer Iosif Shklovskii as synchrotron radiation. Synchrotron radiation was first observed in 1947 in a particle accelerator built at the General Electric Company in Schenectady, New York. This machine, called a synchrotron, was designed to accelerate electrons up to speeds close to the speed of light for experiments in high-energy nuclear physics. The electrons were whirled along a circular path, held in orbit by powerful magnets. Through a small window in the side of the doughnut-shaped machine, scientists noticed a strange light being emitted by the circulating beam of electrons. It is now known that this light, called **synchrotron radiation**, is emitted whenever high-speed electrons move along curved paths through a magnetic field. Because their velocities are near the speed of light, these electrons are said to be *relativistic*. Consequently, the Crab Nebula must contain quite a large number of relativistic electrons gyrating in an extensive magnetic field.

The total energy output of the Crab Nebula in synchrotron radiation is 3×10^{31} watts, while the Sun emits 4×10^{26} watts. Thus, the Crab Nebula is 75,000 times more luminous than the Sun. In 1966, two years before the discovery of pulsars, John A. Wheeler at Princeton University and Franco Pacini from Italy speculated that the ultimate source of this prodigious energy output might be a spinning neutron star. Their prophetic idea was confirmed by the discovery that the Crab pulsar is slowing down.

Although pulsars were first noted for their regular periods, careful timing measurements by radio telescopes soon revealed that all pulsars are gradually slowing down. The period of a typical pulsar increases by a few billionths of a second each day. The Crab pulsar, which has one of the shortest periods, is slowing more quickly than most other

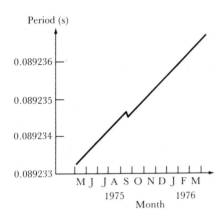

Figure 23-8 A glitch of the Vela pulsar This graph shows how the period of the Vela pulsar increased suddenly during September 1975. The speedup, called a glitch, is probably caused by a starquake, during which the star's crust cracks and settles.

pulsars: Its period increases by 3×10^{-8} seconds each day. Thirty-billionths of a second may sound like a trivial amount, but a rapidly spinning neutron star possesses so much rotational energy that even the slightest slowdown corresponds to a tremendous loss of energy. In fact, the observed rate of slowing for the Crab pulsar corresponds to an energy loss equal to the entire luminosity of the Crab Nebula. In other words, although the details are still poorly understood, the pulsar's rotational energy is constantly being converted into the radiant energy of synchrotron radiation, which is why the Crab Nebula shines.

While making timing observations, astronomers noticed that pulsars sometimes exhibit a sudden, unexpected speedup, which they called a **glitch**. For example, Figure 23-8 shows accurate period measurements of the Vela pulsar during 1975 and 1976. The pulsar's slowing down is obvious on this graph. In September 1975, however, there was an abrupt speedup, after which the pulsar continued to slow down at its usual rate. This glitch was the third one observed for the Vela pulsar; similar glitches have been observed for the Crab pulsar.

According to the law of the conservation of angular momentum, a rotating object speeds up when it contracts, as we have seen. A neutron star is so dense that a typical glitch (which involves a sudden speedup, causing the period to decrease by 10^{-7} s) corresponds to a settling of the star's crust by less than 1 mm. Because the speedup was sudden, the movement of the star's crust must have been equally abrupt. This phenomenon is called a **starquake**. The most powerful earthquakes have a Richter magnitude of $8\frac{1}{2}$. The glitches of the Vela and Crab pulsars, however, correspond to starquakes with magnitudes in the range of 23 to 25 on the Richter scale.

Starquakes tell us about the crust of a pulsar, but astrophysicists must use elaborate calculations to surmise conditions inside a neutron star. As we have seen, the interior of a

neutron star is dominated by degenerate-neutron pressure caused by closely packed neutrons obeying the Pauli exclusion principle. Detailed neutron-star models strongly suggest that this sea of densely packed neutrons can flow without any friction whatsoever. This phenomenon, which is called **superfluidity**, is observed in laboratory experiments with liquid helium at temperatures near absolute zero. Because it is frictionless, a superfluid exhibits such strange properties as being able to creep up the walls of a container, in apparent defiance of gravity. Rapid whirlpools, or vortices, of superfluid neutrons may develop and attach themselves to the crust above. A sudden "unpinning" of these vortices from the crust may be responsible for some of the glitches that arise from starquakes below the crust.

Although a neutron star is made up predominantly of neutrons, there are some protons and electrons scattered throughout the star's interior. Indeed, a pulsar's magnetic field is anchored to the neutron star by these charged particles. A neutron, of course, is electrically neutral, so neutrons cannot hold onto a star's magnetic field. Thus, without the protons and electrons in its interior, a neutron star would rapidly lose its magnetic field.

Different teams of astrophysicists have calculated slightly different details for the internal structure of a neutron star. Nevertheless, many of these models strongly suggest that the protons in a neutron star's core can move around without experiencing any electrical resistance whatsoever. This phenomenon, called **superconductivity**, is observed in the laboratory with certain metals at low temperatures.

As shown in Figure 23-9, the structure of a neutron star consists of a superfluid and superconducting core surrounded by a superfluid mantle, which in turn is surrounded by a brittle crust only 1 km thick.

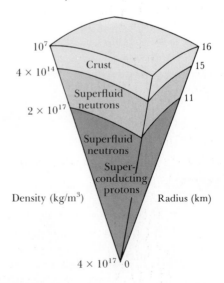

Figure 23-9 A model of a neutron star This possible model for a 1.3-$M_\odot$ neutron star has a superconducting, superfluid core 32 km in diameter. The core is surrounded by a superfluid mantle of neutrons 4 km thick. The star's crust is probably composed of heavy nuclei (such as iron) and free electrons.

23-3 *Pulsating X-ray sources are neutron stars in close binary systems*

During the 1960s, astronomers obtained tantalizing views of the X-ray sky during brief rocket and balloon flights that momentarily lifted X-ray detectors above the Earth's atmosphere. A number of strong X-ray sources were discovered, each being named after the constellation in which it is located. For example, Scorpius X-1 is the first X-ray source found in the constellation of Scorpius.

Astronomers were so intrigued by these preliminary discoveries that they began designing an Earth-orbiting, X-ray–detecting satellite that could make observations 24 hours a day. Their hopes and dreams were realized with the launch of *Explorer 42* on December 12, 1970 (see Figure 23-10). Because it was to be placed in an equatorial orbit, the satellite was launched from Kenya. In recognition of the hospitality of the Kenyan people, *Explorer 42* was christened *Uhuru*, which means "freedom" in Swahili.

Uhuru gave us our first comprehensive look at the X-ray sky. As the satellite slowly rotated, its X-ray detectors swept across the heavens. Each time an X-ray source came into view, signals were transmitted to receiving stations on the ground. Before its battery and transmitter failed in early 1973, *Uhuru* had succeeded in locating 339 X-ray sources.

The discovery of pulsars was still fresh in everyone's mind, so the *Uhuru* team, headed by Riccardo Giacconi, was very excited to detect X-ray pulses coming from Centaurus X-3 in early 1971. Figure 23-11 shows data from one sweep of *Uhuru*'s detectors across Centaurus X-3. The pulses have a regular period of 4.84 s. A few months later, similar pulses were discovered coming from a source designated Hercules X-1, which had a period of 1.24 s. Because the periods of these two X-ray sources are so short, astronomers began to suspect that they had found some rapidly rotating neutron stars.

It soon became clear, however, that systems such as Centaurus X-3 and Hercules X-1 are not ordinary pulsars like the Crab or Vela pulsars. Centaurus X-3 turns on and off periodically. Every 2.087 days, it turns off for almost 12 hours. This fact suggests that Centaurus X-3 is an eclipsing binary and that it takes nearly 12 hours for the X-ray source to pass behind its companion star.

The binary nature of Hercules X-1 is even more compelling. In addition to an "off" state corresponding to a 6-hour eclipse every 1.7 days, careful timing of the X-ray pulses from Hercules X-1 shows a periodic Doppler shifting every 1.7 days. This occurrence is direct evidence of orbital motion about a companion star: When the X-ray source is approaching us, its pulses are separated by slightly less than 1.24 seconds, and when the source is receding from us, slightly more than 1.24 seconds elapses between the pulses.

Careful optical searches of the location of Hercules X-1 soon revealed a dim star called HZ Herculis. The apparent magnitude of this star varies between 13 and 15, with a period of 1.7 days. Because this period is exactly the same as the orbital period of the X-ray source, astronomers have concluded that HZ Herculis is the companion star around which Hercules X-1 orbits.

Putting all the pieces together, astronomers now realize that systems such as Centaurus X-3 and Hercules X-1 are examples of double stars in which one of the stars is a neutron star. All these binaries have very short orbital periods. Consequently, the distance between the ordinary star and the neutron star in each of these systems must be small. This proximity puts the neutron star in position to capture gases escaping from the ordinary companion star.

Figure 23-10 **Uhuru** *Uhuru was a small satellite designed to detect astronomical sources of X rays. During three years of flawless operation, it located more than 300 different X-ray objects across the sky. (NASA)*

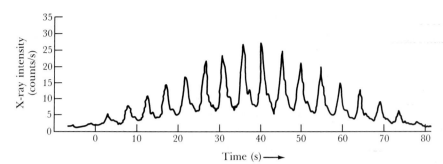

Figure 23-11 X-ray pulses from Centaurus X-3
This graph shows the intensity of X rays detected by Uhuru as Centaurus X-3 moved across the satellite's field of view. Successive pulses are separated by 4.84 s. The gradual variation in the height of the pulses from left to right is the result of the changing orientation of Uhuru's X-ray detectors toward the source as the satellite rotated. (Adapted from R. Giacconi and colleagues)

To explain pulsating X-ray sources such as Centaurus X-3 and Hercules X-1, astronomers assume that the ordinary giant star either fills or nearly fills its Roche lobe. In either case, matter escapes from the ordinary star. This mass transfer occurs by direct Roche-lobe overflow if the star fills its lobe, as does Hercules X-1, or it occurs by a stellar wind if the star's surface lies just inside its lobe, as is the case with Centaurus X-3 (see Figure 23-12). A typical rate of mass transfer from the ordinary star to the neutron star is roughly 10^{-9} solar masses per year.

The neutron star in a pulsating X-ray source, like an ordinary pulsar, is rotating rapidly and has a powerful magnetic field inclined to the axis of rotation (recall Figure 23-3). Because of its strong gravity, a neutron star easily captures much of the gas escaping from its companion star. As the gas falls toward the neutron star, its magnetic field funnels the incoming matter down onto the star's north and south mag-

netic polar regions. The neutron star's gravity is so strong that the gas is traveling at nearly half the speed of light by the time it crashes onto the star's surface. This violent impact creates hot spots at both poles having temperatures of about 10^8 K; these hot spots therefore emit abundant X rays. In fact, their X-ray luminosity is roughly 10^{31} watts, nearly 100,000 times brighter than the Sun. As the neutron star rotates, the beams of X rays from its polar caps sweep around the sky. If the Earth happens to be in the path of one of the two beams, we can observe a pulsating X-ray source. The neutron star's pulse period is thus equal to its rotation period. For example, the neutron star in Hercules X-1 is spinning at the rate of once every 1.24 seconds. The X rays from the neutron star can heat the exposed hemisphere of the companion star. One side of the companion star thus becomes hot and bright, whereas the other side is cooler and dimmer. We can now understand why the brightness of HZ

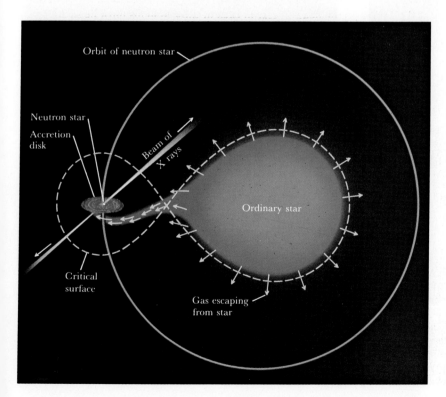

Figure 23-12 A model of a pulsating X-ray source
Gas escaping from the ordinary star is captured by the neutron star. The infalling gas is funneled down onto the neutron star's magnetic poles. It strikes the star with enough energy to create two X-ray–emitting hot spots. As the neutron star spins, beams of X rays from the hot spots sweep around the sky.

Herculis varies periodically: As the system rotates, the two hemispheres are alternately exposed to our view.

Matter spiraling down onto a neutron star can alter its rotation period. The impact of in-falling matter on the neutron star can actually cause the star to spin faster. This phenomenon may explain the new class of pulsars, called **millisecond pulsars,** discovered in the 1980s. One of them (called PSR 1937 + 214, which gives its position in the sky) is blinking on and off at visible wavelengths, like the Crab and Vela pulsars. These pulsars all have periods between 1 and 10 milliseconds, which means that their neutron stars are spinning at rates of 100 to 1000 rotations per second. Because they are often found in binary systems, astronomers believe these pulsars are neutron stars that have been spun up by mass falling in from the companion star. It would take a binary star system a very long time to reach this state, suggesting that millisecond pulsars are the oldest neutron stars in the universe. It may be that the companions of isolated millisecond pulsars have been worn away by the fierce emissions produced by the pulsar after it "turned on." The recently discovered "Black Widow Pulsar," caught in the act of destroying its companion, is a good candidate for such a process (see Figure 23-13).

23-4 High-speed jets of matter can be ejected from an accreting neutron star

In a binary system, gases captured by a neutron star's gravity may go into orbit about the neutron star, forming an accretion disk, as shown in Figure 23-12. Accretion disks have been detected in many close binaries where mass transfer is occurring between the two stars (recall Figure 21-10). Under certain circumstances, bizarre things can happen.

With ordinary pulsating X-ray sources, like Hercules X-1, the rate at which gas falls onto the neutron star from the inner edge of the accretion disk is low enough to allow the resulting X rays to escape. If the companion star is dumping vast amounts of material onto the neutron star, however, the resulting energy cannot escape easily. Instead, tremendous pressures build up in the gases crowding down onto the neutron star. These pressures meet with strong resistance in the plane of the accretion disk, where newly arrived gases are constantly spiraling in toward the neutron star. The path of least resistance lies along the rotation axis of the accretion disk. Consequently, pressure around the neutron star is re-

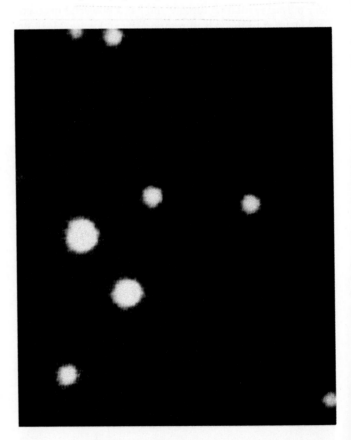

*Figure 23-13 **The Black Widow Pulsar*** *The optical counterpart of the eclipsing binary millisecond pulsar PSR 1957 +20 is one component of a close binary system. Most of the visible radiation comes from the side of the companion star facing the neutron star, which is spinning 622 times per second. Particles and radiation from the pulsar are evaporat-* *ing the companion star. The neutron star and its companion revolve about each other with an orbital period of 9.16 hours. An unrelated background star less than 1 arc sec to the northeast shines constantly. (Courtesy of Jan van Paradijs)*

lieved by squirting gases outward in a direction perpendicular to the accretion disk. The result can be two powerful beams of high-velocity hot gases. This is apparently the explanation for the weird star SS433. (SS433 is so named because it is the 433rd star on a list of similar objects published in 1977 by C. Bruce Stephenson and Nicholas Sanduleak of Case Western Reserve in Ohio.)

In the autumn of 1978, Bruce Margon and some of his colleagues at UCLA began taking a series of spectrograms of the star SS433, which has been noted for its strong emission lines. To everyone's surprise, the spectrum of SS433 contained several complete sets of spectral lines. One set was greatly redshifted away from its usual wavelengths and another set was comparably blueshifted. Somehow, SS433 seemed to be "coming and going" at the same time. To make matters even more puzzling, the wavelengths of these redshifted and blueshifted lines change dramatically from one night to the next.

Astronomers had never seen anything like this, and soon many were observing SS433. By mid-1979, it was clear that the system's redshifted and blueshifted lines actually move back and forth across the spectrum of SS433, with a period of 164 days (see Figure 23-14). The British astrophysicists Andrew Fabian and Martin Rees of Cambridge University were quick to point out that the two sets of spectral lines could be caused by two oppositely directed jets of gas, one pointing toward us, the other pointing away. Furthermore, the 164-day variation could be explained by a precession, or "wobble," of the accretion disk and its two jets. As the two jets circle about the sky every 164 days, we see a periodic variation in the Doppler shift.

All these features come together in the model sketched in Figure 23-15. To explain the large redshifts and blueshifts Margon discovered, the gas in the two oppositely directed jets must have a speed of 78,000 km/s, or roughly one-quarter the speed of light. In addition, the accretion disk must be tilted with respect to the orbital plane of the two stars in the binary system. Just as the tilt of the Earth's axis with respect to the plane of the ecliptic causes the Earth's rotation axis to precess, the changing tilt of the accretion disk results in the 164-day precession of the two jets.

Figure 23-16 is a high-resolution radio view of SS433. Note the two oppositely directed appendages emerging from the central source. As we shall see in Chapter 28, many quasars and peculiar galaxies have similar radio structures, though on a much larger scale. Quasars are incredibly far away, making them difficult to study. The real significance of SS433 may be that it gives us a miniature quasarlike object right in our own celestial backyard.

23-5 Explosive thermonuclear processes on white dwarfs and neutron stars produce novae and bursters

Occasionally, a star suddenly brightens by a factor of 10^6. This phenomenon is called a **nova** (not to be confused with a supernova, which involves a 10^8 increase in brightness). Novae are fairly common. Their abrupt rise in brightness is followed by a gradual decline that may stretch out over several months or more (see Figures 23-17 and 23-18).

Painstaking observations of numerous novae by Robert Kraft, Merle Walker, and their colleagues at Lick Observatory strongly suggest that all novae are members of close binary systems containing a white dwarf. Gradual mass transfer from the ordinary companion star (which presumably fills its Roche lobe) deposits fresh hydrogen onto the white dwarf. Because of the white dwarf's strong gravity, this hydrogen is compacted into a dense layer covering the hot surface of the white dwarf. As more gas is deposited and

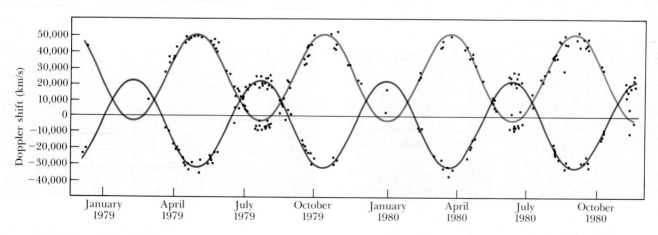

Figure 23-14 **The radial velocity curve of SS433** *This graph shows spectroscopic data on SS433 over a two-year interval. Two sets of spectral lines show Doppler shifts that vary with a period of 164 days. These spectral lines are probably caused by material ejected along jets that alternately point toward and away from the Earth. (Adapted from B. Margon)*

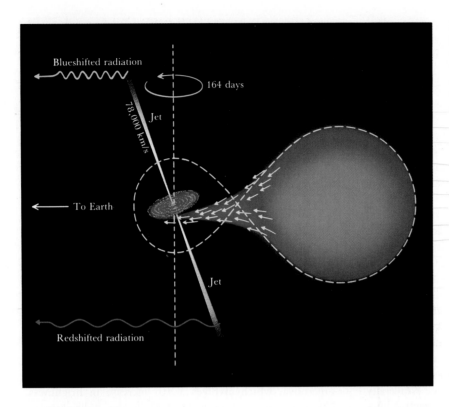

Figure 23-15 A model of SS433 *Gas from a normal star is captured into an accretion disk surrounding a neutron star. Two high-speed oppositely directed jets of gas are ejected from the faces of the disk. Because the disk is tilted, the gravitational pull of the normal star causes the jets to precess, with a period of 164 days.*

compressed, the temperature in the hydrogen layer increases. Finally, when the temperature reaches about 10^7 K, hydrogen burning ignites throughout the gas layer, embroiling the white dwarf's surface in a thermonuclear holocaust that we see as a nova.

A similar phenomenon occurs with neutron stars. Beginning in late 1975, astronomers analyzing data from X-ray satellites realized that their instruments had detected sudden, powerful bursts of X rays from objects in the sky. The record of a typical burst is shown in Figure 23-19. The source emits X rays at a constant low level until suddenly, without warning, there is an abrupt increase in X rays, followed by a more

gradual decline. Typically, an entire burst lasts for only 20 s. Sources that behave in this fashion are known as **bursters**. Several dozen bursters have been discovered, most being located toward the center of our Galaxy.

Bursters, like novae, are believed to involve close binaries experiencing mass transfer. With a burster, however, the stellar corpse is a neutron star rather than a white dwarf. Gases escaping from the ordinary companion star fall onto the neutron star. The burster's magnetic field probably is not strong enough to funnel the falling material toward the magnetic poles, so the gases are distributed more evenly over the surface of the neutron star. The energy released as these

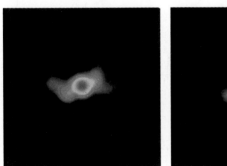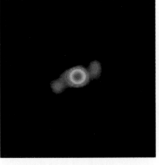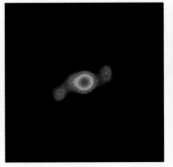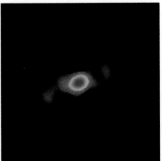

Figure 23-16 SS433 *These four views of SS433, made with the Very Large Array in early 1981, show jets extending out to one-sixth light year on either side of SS433. Three-quarters of the radio emission comes from SS433 itself (the red central blob). SS433 is located at the center of a supernova remnant roughly 5000 pc (16,000 ly) from Earth in the constellation of Aquila. At visual wavelengths, SS433 has an apparent magnitude of +14. (VLA)*

a

b

Figure 23-17 Nova Herculis 1934 *These two pictures show a nova* (a) *shortly after peak brightness as a magnitude +3 star, and* (b) *two months later, when it had faded to magnitude +12. Novae are named*

after the constellation and year in which they appeared. (Lick Observatory)

gases crash down onto the neutron star's surface produces the low-level X rays that are continuously emitted by the burster.

Most of the gas falling onto the neutron star is hydrogen, which becomes compressed against the hot surface of the star by its powerful surface gravity. In fact, temperatures and pressures in this accreting layer are so high that the arriving hydrogen is promptly converted into helium by the

hydrogen-burning process. This constant hydrogen burning soon produces a layer of helium that covers the entire neutron star.

Finally, when the helium layer is about 1 m thick, helium burning ignites explosively and we observe a sudden burst of X rays. In other words, whereas explosive hydrogen burning on a white dwarf produces a nova, explosive helium burning on a neutron star produces a burster. In both cases, the pro-

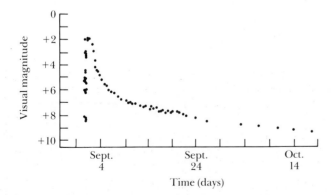

Figure 23-18 The light curve of Nova Cygni 1975 *This graph shows the history of a nova that blazed forth in the constellation of Cygnus in September 1975. Its rapid rise and gradual decline in magnitude are characteristic of all novae. This nova, officially called V1500 Cyg, was easily visible to the naked eye for nearly a week.*

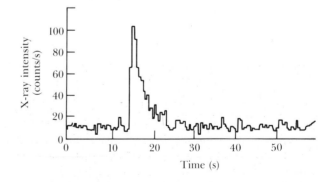

Figure 23-19 X rays from a burster *A burster emits a constant low intensity of X rays interspersed with occasional powerful bursts of X rays. This particular burst is typical. It was recorded on September 28, 1975, by an X-ray telescope on board the Astronomical Netherlands Satellite while its telescope was pointed toward the globular cluster NGC 6624. About one-third of all known bursters are located in globular clusters. (Adapted from W. Lewin)*

cess is explosive, because the fuel is compressed so strongly against the star's surface that it becomes degenerate, like the star itself. As we saw with the helium flash inside red giants, the ignition of a degenerate thermonuclear fuel always involves a sudden thermal runaway because the usual safety valve between temperature and pressure is not operating.

Just as there is an upper limit to the mass of a white dwarf, there is an upper limit to the mass of a neutron star. Above this limit, degenerate-neutron pressure cannot support the overpowering weight of the star's matter pressing inward from all sides. The Chandrasekhar limit for a white dwarf is 1.4 M$_\odot$. The corresponding upper limit for a neutron star is probably 2.5 to 3 M$_\odot$.

Before the discovery of pulsars, most astronomers believed all dead stars to be white dwarfs. Dying stars were thought to eject enough material somehow so that their corpses could be below the Chandrasekhar limit. Obviously,

this idea proved incorrect. Inspired by this lesson, astronomers soon began wondering what might happen if a dying massive star failed to eject enough matter to get below the upper limit for a neutron star. For example, what might a 5-M$_\odot$ stellar corpse be like?

The gravity associated with a neutron star is so strong that the escape velocity from it is roughly one-half the speed of light. With a stellar corpse greater than 3 M$_\odot$, there is so much matter crushed into such a small volume that the escape velocity actually exceeds the speed of light. Because nothing can travel faster than light, nothing—not even light—can leave this dead star. It therefore disappears from the universe, its powerful gravity leaving a hole in the fabric of space and time. In this way, the discovery of neutron stars inspired astrophysicists to examine seriously one of the most bizarre and fantastic objects ever predicted by modern science, the black hole.

Key words

*angular momentum

*angular velocity

burster

conservation of angular momentum

degenerate-neutron pressure

glitch

millisecond pulsar

neutron star

nova
(*plural* novae)

pulsar

starquake

superconductivity

superfluidity

synchrotron radiation

Key ideas

- A neutron star is a dense stellar corpse consisting of closely packed degenerate neutrons.

  A neutron star typically has a diameter of about 30 km, a mass less than 3 M$_\odot$, a magnetic field a trillion times stronger than that of the Sun, and a rotation period of roughly 1 second.

  A neutron star consists of a superfluid, superconducting core surrounded by a superfluid mantle and a thin, brittle crust.

  Energy pours out of the north and south polar regions of a neutron star in intense beams produced by streams of charged particles moving in the star's intense magnetic field.

- A pulsar is a source of radio radiation exhibiting periodic pulses of increased intensity. It is thought to be a neutron

star, with the pulses produced as a beam of radio waves from the star's polar regions sweeps past the Earth.

  The steady slowing of a pulsar's pulse rate presumably reflects the gradual slowing of the rotation rate of the neutron star; glitches (sudden speedups of the pulse rate) correspond to starquakes.

- Some X-ray sources exhibit regular pulses as well as the effects of orbital motion; these objects are neutron stars in close binary systems with ordinary stars.

  Gases from the ordinary star in a close binary system fall onto the dense neutron star at great speeds, creating hot spots near its poles; these hot spots then radiate intense beams of X rays.

- Gases falling onto the neutron star can form an accretion disk around the neutron star; beams of high-velocity gases may shoot out from opposite faces of the accretion disk.

- Material from the ordinary star in a close binary can fall onto the surface of the companion white dwarf or neutron star to produce a surface layer in which thermonuclear reactions can explosively ignite.

> Explosive hydrogen burning may occur in the surface layer of a companion white dwarf, producing the sudden increase in luminosity that we call a nova. The brightness increase in a nova is only a hundredth of that observed in a supernova.

> Explosive helium burning may occur in the surface layer of a companion neutron star, producing the sudden increase in X-ray radiation that we call a burster.

Review questions

*1 The distance to the Crab Nebula is about 2000 pc. When did the star actually explode?

2 Why do astronomers believe that pulsars are rapidly rotating neutron stars?

3 During the weeks immediately following the discovery of the first pulsar, one suggested explanation was that the pulses might be signals from an extraterrestrial civilization. Why do you suppose astronomers soon discarded this idea?

4 Compare a white dwarf and a neutron star. Which of these two types of stellar corpse is more common? Explain.

5 How does the law of the conservation of angular momentum help explain why neutron stars rotate so much more rapidly than ordinary stars?

6 Why do you suppose that most of the angular momentum contained in the solar system resides with the planets rather than with the far more massive Sun?

7 How do we know that the Crab pulsar is really embedded in the Crab Nebula and is not simply located at a different distance along the same line of sight?

8 How do you think astronomers have deduced that the Vela pulsar is about 11,000 years old?

9 If the model for Hercules X-1 discussed in the text is correct, at what orientation of the binary system do we see its maximum optical brightness? Explain your answer.

10 What is the difference between a nova and a Type I supernova?

11 What is SS433? In what ways does it resemble X-ray pulsars like Hercules X-1? In what ways is it unlike an X-ray pulsar?

Advanced questions

> **Tips and tools . . .**
> Recall that the Doppler effect was introduced in Chapter 5 (see Figure 5-20). The volume of a sphere of radius r is $\frac{4}{3}\pi r^3$. Recall that the small angle formula was given in Box 1-1. In Chapter 19 we saw that the tangential velocity (v_t) of an object is related to its proper motion (μ) by $v_t = 4.74\mu d$, where the tangential velocity is measured in km/s, the proper motion is in arc sec per year, and d is the object's distance in parsecs.

*12 Suppose the Sun were to expand into a red-giant star having a radius 400 times its present value, with no significant mass loss occurring in the process. If the Sun is currently rotating at a rate of once every 26 days, at what rate would a red-giant Sun rotate?

*13 To determine accurately the period of a pulsar, astronomers must take into account the Earth's orbital motion about the Sun. Explain why. Knowing that the Earth's orbital velocity is 30 km/s, calculate the maximum correction to a pulsar's period because of the Earth's motion. Explain why the size of the correction is greatest for pulsars located near the ecliptic.

*14 *(Basic)* The mass of a neutron is about 1.7×10^{-27} kg and its radius is about 10^{-15} m. Compare the density of matter in a neutron with the density of a neutron star.

15 Propose an explanation for the fact that X-ray pulsars are speeding up but ordinary (radio) pulsars are slowing down.

*16 From the data given in the caption to Figure 23-2, calculate the rate of expansion of the Crab Nebula. Assuming that your telescope can distinguish features as small as 1 arc sec, how long would you have to wait to see a change in the size of the Crab Nebula?

*17 *(Challenging)* The apparent expansion rate of the Crab Nebula is 0.23 arc sec per year, and emission lines in its spectra exhibit a Doppler shift indicating an expansion velocity of about 1200 km/s. The apparent size of the Crab Nebula is about 4 by 6 arc min. From these data, calculate the distance to the Crab Nebula and estimate the date on which it exploded. Do your answers agree with figures quoted in this chapter? If not, can you point to assumptions you made in your computations that lead to the discrepancies? Or do you think your calculations suggest additional physical effects are at work in the Crab Nebula, over and above a constant rate of expansion?

18 Consult recent issues of such magazines as *Sky & Telescope* and *Science News* for information about the latest observations of the stellar remnant at the center of SN 1987A. Has a pulsar been detected? Has the supernova's debris thinned out enough to give a clear view of the neutron star?

Discussion questions

19 Compare novae and bursters. What do they have in common? In what ways are they different?

20 How might astronomers be able to detect the presence of an accretion disk in a close binary system?

Observing projects

21 If you did not take the opportunity to observe the Crab Nebula as part of the previous chapter's exercises, do so now. The Crab Nebula is visible from late autumn through the beginning of spring. Its epoch 2000 coordinates are: R.A. = $5^h34.5^m$ and Decl. = 22°00′, which is near the star marking the eastern horn of Taurus (the bull). Be sure to schedule your observations for a moonless night. The larger the telescope you use, the better, because the Crab Nebula is quite dim.

22 Consult such publications as the current issue of *Sky & Telescope* or recent International Astronomical Union (IAU) *Circulars*, to see if any novae or supernovae have been sighted recently. If, by good fortune, one has been sighted, what is its magnitude? Is it within reach of a telescope at your disposal? If so, arrange to observe it. Draw what you see through the eyepiece, noting the object's brightness in comparison with other stars in the field of view. If possible, observe the same object a few weeks or months later to see how its brightness has changed.

For further reading

Anderson, L. "X-Rays from Degenerate Stars." *Mercury*, September/October 1976; November/December 1976 • These two articles describe some of the ways in which white dwarfs and neutron stars produce X rays.

Clark, D. *The Quest for SS433*. Dutton, 1985 • This excellent book describes the discovery and developing understanding of SS433 in the format of a detective story.

Greenstein, G. *Frozen Star*. Freundlich Books, 1984 • This eloquent book skillfully blends a discussion of pulsars, neutron stars, and black holes with an accurate portrait of astronomical research as a human endeavor.

Grindlay, J. "New Bursts in Astronomy." *Mercury*, September/October 1977 • This article describes the discovery of X-ray bursts from celestial sources.

Helfand, D. "Pulsars." *Mercury*, May/June 1977 • This article summarizes the most important observations and ideas to come from the first ten years of research on pulsars.

Lewin, W. "The Sources of Celestial X-ray Bursts." *Scientific American*, May 1981 • This informative article describes the ways in which matter falling on neutron stars can produce powerful bursts of X rays.

Margon, B. "The Bizarre Spectrum of SS433." *Scientific American*, October 1980 • The astronomer who discovered the unusual properties of SS433 discusses the star's spectrum and its interpretation.

Seward, F. "Neutron Stars in Supernova Remnants." *Sky & Telescope*, January 1986 • The author of this well written article proposed that differences in the structure of supernovae remnants may result from differences in the way neutron stars form.

Shaham, J. "The Oldest Pulsars in the Universe." *Scientific American*, February 1987 • This article describes the evolution of millisecond pulsars and concludes that they are extremely old neutron stars.

A black hole in a double star system This artist's rendition shows the close binary system that contains Cygnus X-1. Cygnus X-1 is a strong source of X rays and is widely believed by astronomers to be a black hole. Gas from the companion star is captured into orbit about the black hole, forming an accretion disk. As gases spiral in toward the black hole, they are heated by friction to high temperatures. At the inner edge of the accretion disk, the gases are so hot that they emit X rays. *(Courtesy of D. Norton, Science Graphics)*

Black Holes

A dying high-mass star can give rise to a stellar corpse too massive to be supported by degenerate pressure, and so it is doomed to collapse to a single point of infinite density. Such an object, called a black hole, is predicted by Einstein's general theory of relativity. A black hole is strange but simple. It is a place of inconceivably intense gravity from which nothing—not even light—can escape. Matter that falls into a black hole literally disappears forever from the universe. Nevertheless, a black hole is an uncomplicated object because its structure is completely specified by only three quantities: its mass, electric charge, and angular momentum. In recent years, astronomers have found evidence that certain binary systems may contain black holes. Each of these binaries is a powerful source of X rays presumably produced by hot gases in an accretion disk surrounding the black hole.

Suppose that the mass of a dying star's burned-out core exceeds 3 solar masses (3 $M_\odot$). This mass is well above the Chandrasekhar limit, so degenerate-electron pressure cannot support the resulting stellar corpse. Similarly, because 3 $M_\odot$ is also above the mass limit for neutron stars, degenerate-neutron pressure also is incapable of supporting the crushing weight of the burned-out matter pressing inexorably inward toward the dead star's center. If this massive stellar corpse can be neither a white dwarf nor a neutron star, what might it become?

As we saw in the previous chapter, a typical neutron star consists of roughly 1 $M_\odot$ of matter compressed by its own gravity to nuclear density in a sphere that is roughly 30 km in diameter. This neutron star's gravity is so strong that the escape velocity from its surface is equal to half the speed of light.

A massive stellar corpse, whose weight will overpower degenerate-neutron pressure, easily compresses its matter to densities greater even than nuclear density. It does not take very much further compression to cause the escape velocity from this stellar corpse to exceed the speed of light. For example, if 3 $M_\odot$ of matter is squeezed into a sphere 18 km in diameter, the escape velocity from the object will become greater than the speed of light. Because nothing can travel faster than the speed of light, nothing—not even light—can manage to escape from the dead star. The star has in a very real sense disappeared from the observable universe, although some of its effects can still be detected.

24-1 *The general theory of relativity describes gravity in terms of the geometry of space and time*

To appreciate fully the nature of a massive stellar corpse, we must use the best theory of gravity at our disposal. The gravitational field around one of these massive dead stars is so strong that Isaac Newton's theory of gravity gives the wrong answers. Instead, we must turn to Albert Einstein's general theory of relativity.

According to the classical physics of Newton, space is perfectly uniform and fills the universe like a rigid framework. Similarly, time passes at a monotonous, unchanging rate. It is always possible to know absolutely how fast you are moving through this rigid fabric of space and time, and the results of your observations depend on your state of motion. For example, according to Newton, a stationary person and a moving person will always measure different speeds for an object in motion.

Albert Einstein began a revolution in physics with his **special theory of relativity** in 1905. Einstein was guided in his thinking by one lofty idea: that neither our location in space and time nor our motion through space and time shall prejudice our description of physical reality. A surprising

experimental fact also played an important role in the development of relativity theory: Everyone who measures the speed of light gets the same answer (3×10^8 m/s), regardless of the person's state of motion. This conflicts with the Newtonian view that a stationary person and a moving person should measure different speeds. As explained in Chapter 4 (Section 4.7), as Einstein struggled to reconcile electromagnetic theory (which predicted the correct value for the speed of light) with these ideas, he arrived at a new understanding of the nature of space and time. The basic equations of the special theory of relativity logically follow from the fact that everyone agrees on the speed of light (see Box 24-1). These equations relate measurements by different observers and ensure, for instance, that both you on Earth and a friend in a rocketship traveling near the speed of light have the same complete description of electricity and magnetism, devoid of any pitfalls or paradoxes caused by your relative motion.

In developing the special theory of relativity, Einstein found that he had to abandon the old-fashioned, rigid notions of space and time. There is no absolute motion through space and time. Instead, there is only the relative motion between two observers. For example, imagine a friend whizzing across our solar system in a rocketship while you remain here on Earth. Einstein proved that in order for both of you to agree on the same coherent description of reality, you must say that your friend's clocks are ticking more slowly than your own and that rulers she might hold up parallel to the direction of motion have become shorter than yours. In brief, an observer always finds that moving clocks are slowed and moving rulers are shortened in the direction of motion. These details of the special theory of relativity, which are discussed in Box 24-1, are a direct consequence of the speed of light being an absolute constant.

After developing the special theory of relativity, Einstein turned his attention to gravity. He began by demonstrating that it is not necessary to think of gravity as a force. According to Newton's theory, an apple falls to the floor because the force of gravity pulls the apple down. Einstein pointed out that the apple would behave in exactly the same way in free space far from any gravity if the floor were to accelerate upward. In other words, in this case the floor actually comes up to meet the apple, as sketched in Figure 24-1.

This example of Einstein's **principle of equivalence** explains that, in a small volume of space, the downward pull of gravity can be accurately and completely duplicated by an upward acceleration of the observer. The two gentlemen watching an apple fall toward the floor of their closed compartments in Figure 24-1 have no way of telling who is at rest on the Earth and who is in the hypothetical elevator moving upward at a constantly increasing speed.

This approach allowed Einstein to focus entirely on motion, rather than force, in discussing gravity. From his special theory of relativity he knew exactly how rulers and clocks are affected by motion, and thus he could describe gravity entirely in terms of its effects on space and time. Far

The Earth

Interstellar space

Figure 24-1 The equivalence principle *The equivalence principle asserts that you cannot distinguish between being at rest in a gravitational field and being accelerated upward in a gravity-free environment. This idea was an important step in Einstein's quest to develop the general theory of relativity.*

from a source of gravity the acceleration is small, so its effect on clocks and rulers is small. Likewise, nearer a source of gravity the acceleration is larger, so the distortion of clocks and rulers is larger. In this way, Einstein "generalized" his special theory to arrive at his **general theory of relativity**.

The general theory of relativity describes gravity entirely in terms of the geometry of space and time. Far from a source of gravity, space is "flat" and clocks tick at their normal rate. As you move closer to a source of gravity (a mass), however, clocks slow down and space becomes increasingly curved, as seen by someone outside the region that is affected by the mass.

In a weak gravitational field, Einstein's general theory of relativity gives the same results as the classical theory of Newton. In stronger gravity, however, such as that near the Sun's surface, the two theories give different predictions. Examples include the precession of Mercury's perihelion and the deflection of a light ray grazing the Sun, discussed in Chapter 4 (review Figures 4-15 and 4-17). In these and other situations, general relativity has withstood numerous tests. It is by far the most elegant—and accurate—description of gravity yet devised.

24-2 A black hole is a simple object that has only a "center" and a "surface"

Imagine a dying star too massive to become either a white dwarf or a neutron star. The overpowering weight of the star's burned-out matter pressing inward from all sides causes the star to contract rapidly. The strength of gravity at the surface of this collapsing star increases dramatically as

the star's matter becomes compressed to enormous densities inside the rapidly shrinking sphere. According to the general theory of relativity, distortions of space and time become increasingly pronounced near the surface of such a dying star. Finally, the escape velocity from the star's surface equals the speed of light, and thus the star disappears from the universe. At this stage, space has become so severely curved that a hole is punched in the fabric of the universe. The dying star disappears into this hole in space, leaving behind only a **black hole**.

The geometry of space around a black hole is sketched in Figure 24-2. Note that space far from the hole is flat, because gravity is weak there. Near the hole, however, gravity is strong and the curvature of space is severe.

The location in space where the escape velocity from the black hole equals the speed of light is called the **event horizon**. This sphere is also sometimes thought of as the

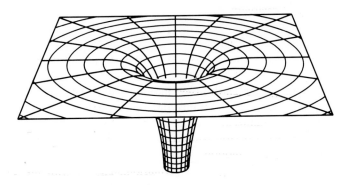

Figure 24-2 The geometry of a black hole *This diagram shows how the shape of space is distorted by the gravitational field of a black hole. Far from the hole, gravity is weak and space is thus "flat." Near the hole, gravity is strong and space is highly curved.*

Box 24-1   Some comments on special relativity

The special theory of relativity describes the way in which measurements of time, distance, and mass are affected by motion. Einstein proved that these measurements must depend on the speed of the observer in order for all people, whether moving or stationary, to agree on certain basic physical phenomena, especially those involving the behavior of light.

Imagine that you are standing on the Earth while a friend is traveling across our solar system at a high speed, as shown in the sketch. You set off a flash bulb that emits a sudden bright flash of light. The radiation moves away from you equally in all directions, and thus you "see" an expanding spherical shell of light.

What does your high-speed friend see? Einstein argued that this person must also see an expanding spherical shell of light. She does not see, for example, an expanding cube or an expanding football-shaped shell.

By requiring that both people observe a spherical shell, Einstein derived a series of equations to relate specific measurements of time and distance between two people. These equations are named the **Lorentz transformations**, after the famous Dutch physicist Hendrik Antoon Lorentz, a contemporary of Einstein's, who developed these equations indepen-

dently. These equations tell us exactly how a moving person's clocks slow down and how rulers shrink.

To appreciate the Lorentz transformations, again imagine that you are on the Earth while a friend is moving at a speed v with respect to you. Suppose that you both observe the same phenomenon on Earth, which appears to occur over an interval of time. According to your (stationary) clock, the phenomenon lasts for T_0 seconds; according to your friend's (moving) clock, the same phenomenon lasts for T seconds. The Lorentz transformation for time tells us that these two time intervals are related by

$$T = \frac{T_0}{\sqrt{1 - (v^2/c^2)}}$$

where c is the speed of light.

Example: Suppose that your friend is moving at 98 percent of the speed of light. Then

$$\frac{v}{c} = 0.98$$

so that

$$T = \frac{T_0}{\sqrt{1 - (0.98)^2}} = 5\,T_0$$

Thus, a phenomenon that lasts for 1 second on a stationary clock is stretched out to 5 seconds on a clock moving at 98 percent of the speed of light. This phenomenon is often called the **dilation of time.**

The Lorentz transformation for time is plotted in the graph opposite. This graph shows how 1 second on a stationary clock is stretched out, as measured by a moving clock. Note that significant differences between the recordings of the stationary and moving clocks occur only at speeds near the speed of light. For speeds less than about half the speed of light, the mathematical

Expanding spherical shell of light

You

Earth

Your friend

"surface" of the black hole. Once a massive dying star collapses to within its event horizon, it disappears permanently from the universe. The term *event horizon* is in fact quite appropriate, for this surface is literally a horizon in the geometry of space beyond which we cannot see any events.

In addition to making space curve, gravity causes time to slow down. If you stood at a safe distance and watched a friend fall toward a black hole, you would note that her clocks would begin ticking more and more slowly. In fact, when she reached the event horizon, you would conclude

that her clocks had stopped entirely. However, your friend would not notice this slowing down and stopping of time as she glances at her watch. From her point of view, she continues her fall through the event horizon into the black hole.

Once a dying star has contracted inside its event horizon, no forces in the universe can prevent the complete collapse of the star down to a single point at the center of the black hole. The star's entire mass is crushed to infinite density at this point, known as the **singularity**. The singularity corresponds to the center of a black hole.

factor of $\sqrt{1 - (v^2/c^2)}$ is almost exactly equal to 1, so stationary and slowly moving clocks tick at almost exactly the same rate.

In the language of relativity, we say that a clock at rest measures **proper time** (T_0) and a ruler at rest measures **proper distance** (L_0). According to the Lorentz transformations, distances perpendicular to the direction of motion are unaffected. However, a ruler of proper length L_0 held parallel to the direction of motion shrinks to a length L, given by

$$L = L_0 \sqrt{1 - \left(\frac{v^2}{c^2}\right)}$$

Example: If your friend is traveling at 98 percent of the speed of light relative to you, you thus conclude that her clocks are ticking only one-fifth as fast as yours and that her 1-ft ruler is only about $2\frac{1}{2}$ in. long when held parallel to the direction of motion:

$$L = 12 \text{ in. } \sqrt{1 - (0.98)^2} = 2\frac{1}{2} \text{ in.}$$

This shrinkage of length is often called **Fitzgerald–Lorentz contraction.**

Albert Einstein also demonstrated that measurements of mass are affected by the relative velocity of the observer. Specifi-

cally, suppose that an object has a **proper mass** M_0 when at rest. If this same object is moving with a velocity v, it appears to have a mass M, given by

$$M = \frac{M_0}{\sqrt{1 - (v^2/c^2)}}$$

Thus, a 1-kg brick traveling at 98 percent of the speed of light would appear to a stationary observer to behave as though it had a mass of 5 kg.

The mathematical expression $\sqrt{1 - (v^2/c^2)}$ becomes significantly different from 1 only when v is nearly as large as c. This explains why the effects of relativity on time, distance, and mass become noticeable only at extremely high speeds. Particle accelerators like cyclotrons and synchrotrons that propel electrons and protons to velocities near the speed of light have tested these predictions of special relativity to a high degree of accuracy. For instance, when a high-speed particle smashes into a stationary target, the resulting debris is scattered in a way that can only be understood from the viewpoint that the impacting particle had a mass higher than its proper mass by precisely the amount given by the above equation.

Finally, special relativity explains why it is impossible to travel faster than the speed of light. Suppose that a friend climbs aboard a rocketship containing an infinite supply of fuel. With rocket engines blazing, your friend attempts to "break the light barrier." You keep in touch by radio and, as she gets closer and closer to the speed of light, you begin to notice the effects of the dilation of time. For instance, her speech becomes drawn out to a leisurely drawl as her clock slows down relative to yours. In the same way, the rocket engines appear to you to be shutting down. The gallons-per-minute rate at which fuel pours into the rocket engines is directly associated with the passage of time and is therefore subject to time dilation. As the rocket's velocity approaches the speed of light, the factor $\sqrt{1 - (v^2/c^2)}$ approaches zero. Thus your friend never gets the chance to burn that last drop of fuel that would put her past the speed of light.

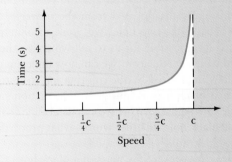

We now can see that the structure of a nonrotating black hole is quite simple. As sketched in Figure 24-3, it has only two parts: a singularity, or center, surrounded by an event horizon, or surface. The distance between the singularity

Figure 24-3 The structure of a black hole *A nonrotating black hole has only two parts: a singularity surrounded by an event horizon. The distance from the singularity to the event horizon is called the Schwarzschild radius, R_{Sch}. Inside the event horizon, the escape velocity exceeds the speed of light, so the event horizon is a one-way surface. Things can fall in, but nothing can get out.*

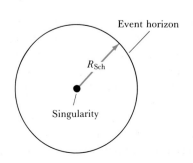

cosmic censorship

and the event horizon is called the **Schwarzschild radius** (R_{Sch}), after the German astronomer Karl Schwarzschild who in 1916 was the first to solve Einstein's equations of general relativity. The Schwarzschild radius is related to the mass M of the black hole by

$$R_{Sch} = \frac{2GM}{c^2}$$

where c is the speed of light and G the universal constant of gravitation. For example, for a 10-$M_\odot$ black hole, the Schwarzschild radius is 30 km.

To understand why the complete collapse of such a doomed star is inevitable, think about your own life here on Earth, far from any black holes. You have the freedom to move as you wish through the three dimensions of space: up and down, left and right, or forward and back. But you do not have the freedom to move at will through the dimension of time. Whether we like it or not, we are all carried inexorably from the cradle to the grave.

Inside a black hole, a powerful gravity distorts the structure of space and time so severely that the directions of space and time become interchanged. In a limited sense, inside a black hole you can have the freedom to move through time. This seeming gain does you no good, however, because you lose a corresponding amount of freedom to move through space. Whether you like it or not, you will be dragged inexorably from the event horizon toward the singularity. Just as no force in the universe can prevent the forward march of time from past to future outside a black hole, no force in the universe can prevent the inward march of space from event horizon to singularity inside a black hole.

At a black hole's singularity, the strength of gravity is infinite, so the curvature of space and time there is infinite. Space and time at the singularity are thus all jumbled up. They do not exist there as separate, identifiable entities.

This confusion of space and time has profound implications for what goes on inside a black hole. All the laws of physics require a clear, distinct background of space and time. Without this identifiable background, we could not speak rationally about the arrangement of objects in space or the ordering of events in time. Because space and time are all jumbled up at the center of a black hole, the singularity there does not obey the laws of physics. The singularity behaves in a random, capricious fashion, totally devoid of rhyme or reason.

Fortunately, we are shielded from the singularity by the event horizon. In other words, although random things do happen at the singularity, none of their effects manage to escape beyond the event horizon. Consequently, the outside universe remains understandable and predictable.

The chaotic, random behavior of the singularity has been so disturbing to physicists that in 1969 the British mathematician Roger Penrose and his colleagues proposed **the law of cosmic censorship**: "Thou shalt not have naked singulari-

ties." In other words, every singularity must be completely surrounded by an event horizon, because an exposed singularity could affect the universe in an unpredictable and random way.

24-3 The structure of a black hole can be completely described with only three numbers

In addition to shielding us from singularities, the event horizon prevents us from ever knowing much about anything that falls into a black hole. For example, there is no way we could ever discover the chemical composition of a massive star whose collapse has produced a particular black hole. Even if someone were to go into a black hole and make a measurement or chemical test, there is no way the observer could get any of this information back to the outside world. A black hole is in fact an "information sink," because infalling matter carries with it many properties, such as its chemical composition, texture, color, shape, and size, that are then forever removed from the universe.

Because a black hole removes information from the universe, there is thus no way that this information can affect the structure or properties of the hole. For example, consider two hypothetical black holes—one made from the gravitational collapse of 10 $M_\odot$ of iron, the other from the gravitational collapse of 10 $M_\odot$ of peanut butter. Obviously, quite different substances went into the creation of the two holes. Once the event horizons of these two black holes have formed, however, both the iron and the peanut butter will have permanently disappeared from the universe. As seen from the outside, the two holes will look absolutely identical, making it impossible for us to tell which ate the peanut butter and which ate the iron. We can now understand that a black hole is unaffected by the information it destroys.

Because a black hole is indeed an information sink, it is reasonable to wonder whether we can determine anything at all about a black hole. In other words, what properties characterize a black hole?

First, we can measure the mass of a black hole. One way to do this would be by placing a satellite in orbit about the hole. Kepler's third law (recall Box 4-3) relates a satellite's orbital period and semimajor axis to the mass around which it is moving. Thus, after measuring the size and period of the satellite's orbit, we could use Kepler's third law to determine the mass of the black hole. This mass is equal to the total mass of all the material that has gone into the hole.

Incidentally, science fiction abounds with nasty rumors that black holes are evil things that go around gobbling up everything in the universe. Not so! The bizarre effects created by highly warped space and time are limited to a region within only a few million kilometers around the hole—less than the distance from the Sun to Mercury. Farther from the hole, gravity is weak enough that Newtonian physics can adequately describe everything. For example, at a distance of only a few astronomical units from a 10-$M_\odot$ black hole, the behavior of gravity is identical to that of any ordinary 10-$M_\odot$ star.

We can also measure the total electric charge possessed by a black hole. Like the gravitational force, the electric force is a long-range interaction, making its effects felt in the space around the hole. Appropriate equipment on a space probe passing near the hole could measure the intensity of the electric field around the hole, and the electric charge can be determined from this information.

In reality, we would not expect a black hole to possess any appreciable electric charge. For instance, if a hole did happen to start off with a sizable positive charge, it would vigorously attract vast numbers of negatively charged electrons from the interstellar medium, which would soon neutralize the hole's charge. For this reason, astronomers neglect electric charge when discussing real black holes.

Although a black hole might theoretically have a tiny electric charge, it can have no magnetic field of its own whatsoever. Einstein's equations do not permit a north-pole–south-pole asymmetry in the geometry of space around a black hole. During the creation of an actual black hole, we would expect the collapsing star to possess an appreciable magnetic field. This magnetic field must be radiated away, in the form of electromagnetic and gravitational waves, before the dead star can settle down inside its event horizon. As we saw in Chapter 22, gravitational waves are ripples in the overall geometry of space. Some physicists are exploring the possibility of observing the creation of black holes by detecting bursts of gravitational radiation emitted by collapsing massive stars. A discussion of gravitational radiation appears in Box 24-2.

In addition to the mass and electric charge of a black hole, we can also measure a black hole's total angular momentum. Because of the conservation of angular momentum (recall Box 23-1), we expect a black hole to be spinning rapidly. Einstein's theory makes the startling prediction that this rotation causes space and time to be dragged around the hole. A spinning black hole is thus surrounded by space that rotates with the hole. In fact, around the event horizon of every rotating black hole, a region exists where this dragging of space and time is so severe that it is impossible to stay in the same place. No matter what you do, you get pulled around the hole, along with the rotating geometry of space and time. This region, where it is impossible to be at rest, is called the **ergosphere** (see Figure 24-4).

To measure a black hole's angular momentum, we could hypothetically place two satellites in orbit about the hole. Suppose that one satellite circles the hole in the same direction the hole rotates, the other in the opposite direction. One satellite is thus carried along by the geometry of space and time, but the other is constantly fighting its way "upstream." The two satellites will thus have different orbital periods.

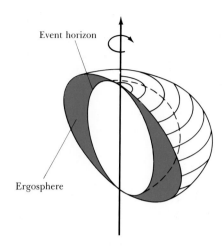

Figure 24-4 The ergosphere *A rotating black hole is surrounded by a region called the ergosphere, where the dragging of space and time around the hole is so severe that it is impossible for anything to remain at a fixed location. Because the ergosphere (the shaded area in this cross-sectional diagram) is outside the event horizon, this bizarre region is accessible to us and could be traversed by astronauts or asteroids without their disappearing into the black hole. Detailed calculations demonstrate that objects grazing the ergosphere could be catapulted back out into space at tremendous speeds. In other words, the ejected object leaves the ergosphere with more energy than it had initially, having extracted added energy from the hole's rotation. This is called the Penrose process after the British mathematician Roger Penrose, who proposed it.*

From a comparison of these two periods, the total angular momentum of the hole can be deduced.

And that is all. A black hole possesses no qualities other than mass, charge, and angular momentum. This simplicity is the essence of the famous **no-hair theorem** first formulated in the early 1970s: "Black holes have no hair." Any and all additional properties carried by the matter that has fallen into the hole have disappeared from the universe and thus can have no effect on the structure of the hole.

24-4 A black hole distorts the images of background stars and galaxies

Finding black holes in the sky is a difficult business. Obviously, since light cannot escape from inside the event horizon, you cannot observe a black hole directly in the same sense you can observe a star or a planet. The best you can hope for is to detect the effects of a black hole's powerful gravity.

One option to pursue is the distortion that a black hole creates in the appearance of background objects. For example, suppose that the Earth, a black hole, and a background star are in nearly perfect alignment, as sketched in Figure 24-5. Because of the warped space around the black hole,

Box 24-2 Gravitational radiation

A gravitational wave is a ripple in the overall geometry of space and time. As an example of how these ripples are produced, think of a man whose mass is 80 kg. All matter is a source of gravity and, according to general relativity, gravity curves space and slows down time. So the 80-kg man is surrounded by a slight warping of space and time commensurate with his mass.

Now suppose that this man begins waving his arms. Although his total mass does not change, the details of how his mass is distributed do change. The geometry of space and time must adapt to these changes, because the gravitational field of the man with his hands over his head is slightly different from that of the man when he has his hands at his sides. These minor readjustments appear as tiny ripples in the overall geometry of space and time surrounding the man. In the same way, a bouncing ball, the Moon going around the Earth, or two stars in a binary all produce gravitational waves. From the equations of general relativity, it is possible to prove that gravitational radiation moves outward from its source at the speed of light.

Gravitational waves are difficult to detect, because they carry very little energy. To appreciate how weak gravitational waves are, imagine two electrons separated by a short distance. Because they each possess mass and charge, these electrons exert both gravitational and electric forces on each other. The gravitational force is about 10^{42} times weaker than the electric force. If these two electrons are made to wiggle back and forth, they will radiate both gravitational and electromagnetic waves. Because gravity is so much weaker than electromagnetism, the resulting gravitational waves are subdued by a factor of 10^{-42} compared to the electromagnetic waves.

Processes involving dramatic changes in intense gravitational fields produce the strongest bursts of gravitational radiation.

For example, the collapse of a massive star's core during a supernova explosion emits substantial gravitational radiation. Of course, we cannot observe the actual core collapse with ordinary telescopes, because the outer layers of the supernova emit such an overpowering amount of light. However, gravitational waves from the collapsing core carry detailed information about how this dense matter is being rearranged. With a gravitational wave antenna, we should thus be able to observe directly the creation of a neutron star or black hole.

Although an actual burst of gravitational waves has not yet been conclusively detected, many astronomers believe that the effects of gravitational radiation have been observed. In 1974, Joseph Taylor and his colleagues at the University of Massachusetts discovered a pulsar in a binary system. The system apparently consists of two neutron stars separated by only 2.8 solar radii. One of the two stars emits radio pulses every 0.059 second, and the orbital period of the two stars about each other is only 7.75 hours. The average orbital velocity of these stars is thus about 0.1 percent of the speed of light. Because these two stars have strong gravitational fields and are moving so rapidly, this entire binary system should be a substantial source of gravitational waves. As gravitational radiation carries energy away from the system, the two stars should gradually spiral in closer and closer to each other, so the orbital period of the two stars should decrease. Because one of the stars is a pulsar, radio astronomers have been able to measure its orbital period with extreme accuracy. These observations demonstrate that the two stars are indeed spiraling in toward each other, because of their emission of gravitational waves.

---

there are two paths along which light rays can travel from the background star to us here on Earth. Thus, we should see two images of the star.

The distortion of background images by a powerful source of gravity is called a **gravitational lens**. It is virtually the only way we can hope to find an isolated black hole in our Galaxy. Unfortunately, in order for this effect to be no-

ticeable, the alignment between the Earth, the black hole, and a remote star must be almost perfectly straight. Without nearly perfect alignment, the secondary image of the background star is too faint to be noticed.

No one has ever found a gravitational lens within our Galaxy. However, several gravitational lenses have been discovered far from our Galaxy that involve the remote, lumi-

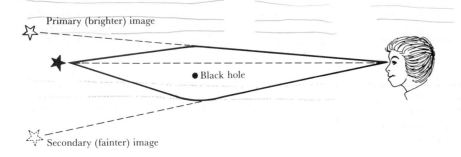

Primary (brighter) image

● Black hole

Secondary (fainter) image

Figure 24-5 A gravitational lens A black hole can deflect light rays from a distant star so that an observer sees two images of the star. Several so-called gravitational lenses have been discovered in which light from a remote quasar is deflected by an intervening galaxy. No gravitational lenses caused by black holes have yet been discovered, however.

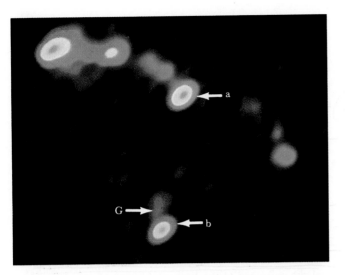

Figure 24-6 A "double" quasar Two images, (a) and (b), of the same quasar are seen in this radio view made by the Very Large Array. Light from the distant quasar is deflected to either side of a massive galaxy located between us and the quasar. A faint image of the deflecting galaxy (G) is seen directly above image (b). The jetlike feature protruding from the upper image (a) does not appear alongside the lower image because the jet is too far away from the required quasar–galaxy–Earth alignment. (VLA; NRAO)

nous objects called quasars. As we shall see in Chapter 27, a quasar is a very bright, starlike object. Typical quasars shine as brightly as a hundred galaxies and are located a few billion light years from Earth. Nearly 4000 quasars have been discovered across the sky, so that, on average, one quasar is found in roughly every 10 square degrees.

In 1979, astronomers Dennis Walsh, Robert Carswell, and Ray Weymann were surprised to find two quasars separated by only 6 arc sec. They took spectra of both quasars and discovered that they were nearly identical. They thus concluded that they were looking at two images of the same quasar. Further observations revealed that there was a galaxy between the quasar images (see Figure 24-6). The gravitational field of this galaxy bends the light from the remote quasar and thus acts like a gravitational lens.

By 1990, nearly two dozen candidates for gravitational lenses had been reported, of which six cases are quite convincing. All six involve a remote quasar whose starlike image is split by a galaxy located between us and the quasar. When such a galaxy deflects the light from a remote quasar, three images are often produced because the galaxy's mass is spread over a volume rather than being concentrated at a point, as in the case of a black hole. Searches for more gravitational lenses are under way.

Computers have been used to calculate the gravitational lensing of an extended object like a galaxy (see Figure 24-7). The primary and secondary images are elongated arcs that partly encircle the location of the deflecting black hole. These arcs connect to form a ringlike image when the galaxy, the black hole, and the observer are in nearly perfect alignment.

Several fine examples of arched gravitational images were discovered in the late 1980s. In 1987, Roger Lynds at Kitt Peak National Observatory and Vahe Petrosian of Stanford University found a huge luminous arc more than 300,000 light years long in a remote cluster of galaxies (see Figure 24-8). This arc is now known to be the light from an extremely remote galaxy or quasar that has been stretched out

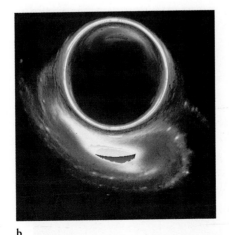

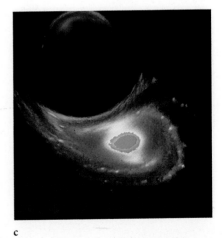

a

b

c

Figure 24-7 The gravitational lensing of a galaxy A computer was used to calculate these views of a galaxy seen through a gravitational lens consisting of a massive black hole (8×10^{12} M$_\odot$) located halfway between us and the galaxy. (a) An undistorted view of the galaxy. (b) A view of the galaxy seen with a nearly perfect alignment between the galaxy, the black hole, and the Earth. (c) A view of the galaxy seen with the black hole displaced slightly from perfect alignment. (Courtesy of E. Falco, M. Kurtz, and M. Schneps; Smithsonian Astrophysical Observatory)

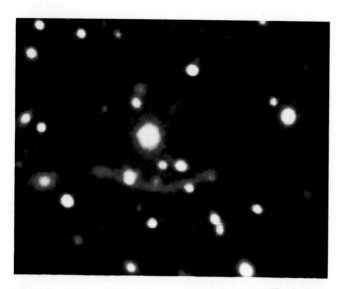

Figure 24-8 A giant luminous arc This luminous arc is about 300,000 light years long and is located in a cluster of galaxies, about 5 billion light years from Earth. Almost every fuzzy spot on this picture is a galaxy. Spectroscopic observations strongly suggest that this arc is the image of an extremely remote galaxy or quasar that has been stretched out into an arc by the gravitational field of an intervening galaxy. (NOAO)

into a curved image by the gravitational field of an intervening galaxy. In 1988, Jacqueline Hewitt of MIT reported her observations of a radio source called MG 1131+0456, which she believes is a ringlike image of a remote radio galaxy. Albert Einstein was first to point out that a ring-shaped image would be seen if a massive body were located directly between us and a remote source of light, and so the object

shown in Figure 24-9 is called an **Einstein ring**. Several additional examples of huge arcs and Einstein rings have recently been identified. Once its optical problems are solved, the Hubble Space Telescope may uncover many more cases of gravitational lensing as it examines remote galaxies and quasars.

24-5 Black holes have been discovered in binary-star systems

Binary stars offer the best chance of finding black holes in our Galaxy. For instance, if a black hole were to capture gas from its companion star, the fate of this material might reveal the existence of the hole.

Shortly after the launch of *Uhuru* in the early 1970s, astronomers became intrigued with an X-ray source designated Cygnus X-1. This source is highly variable and irregular. Its X-ray emission flickers on time scales that are as short as 10 milliseconds. One of the fundamental concepts in physics is that nothing can travel faster than the speed of light (recall Box 24-1). Because of this limitation, an object cannot vary its brightness or flicker faster than the time required for light to travel across the object. Because light travels 3000 kilometers in 10 milliseconds, Cygnus X-1 must be smaller than the Earth.

The X-ray detectors on the *Uhuru* satellite had poor angular resolution, so astronomers could not pinpoint the location of Cygnus X-1 from X-ray data alone. However, in the spring of 1971, Cygnus X-1 produced major X-ray fluctuations that were accompanied by the appearance of a weak

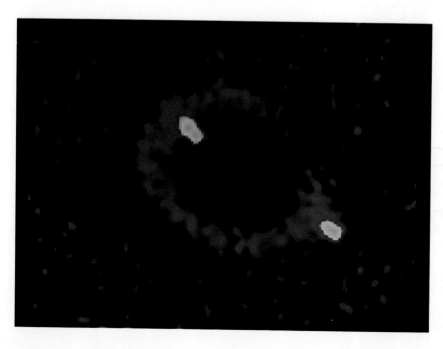

Figure 24-9 The Einstein ring A gravitational lens should produce a ringlike image if there is perfect alignment between the background light source, the observer, and the deflecting galaxy or black hole. This radio object in the constellation of Leo may be the first known example of a so-called Einstein ring. (VLA; NRAO)

Figure 24-10 HDE 226868 *This star is the optical companion of the X-ray source Cygnus X-1. The star is a B0 supergiant located 8000 ly from Earth. Many astronomers agree that Cygnus X-1 is probably a black hole. This photograph was taken with the 200-in. telescope at Palomar. (Courtesy of J. Kristian, Mount Wilson and Las Campanas observatories)*

From the mass–luminosity relation, HDE 226868 is estimated to have a mass of roughly 30 $M_\odot$. This information implies that Cygnus X-1 must also be fairly massive, or else it would not exert enough gravitational pull to make the B0 star wobble as much as it does, as deduced from the periodic Doppler shifting of its spectral lines. Specifically, Cygnus X-1 must have a mass greater than 6 $M_\odot$. This mass is too great for Cygnus X-1 to be either a white dwarf or a neutron star, so the only remaining possibility is that it is a black hole.

Of course, the X rays from Cygnus X-1 do not come from within the black hole itself. Gas captured from HDE 226868 goes into orbit about the hole, forming an accretion disk about 4 million km in diameter (see Figure 24-11). As material in the disk gradually spirals in toward the hole, friction heats the gas to temperatures approaching 2 million K. In the final 200 km above the hole, these extremely hot gases emit the X rays that we then detect with our satellites. The X-ray flickering is presumably caused by small "hot spots" on the rapidly rotating inner edge of the accretion disk. In this way, the black hole's existence is announced by doomed gases just before they plunge to oblivion.

Details about the structure of an accretion disk around a black hole are elucidated by supercomputer simulations. For instance, in the mid-1980s, Larry L. Smarr at the University of Illinois and John F. Hawley, now at the University of Virginia, used a supercomputer to solve the equations that describe how accreting gas is distributed around a black hole. The simulation begins with gas falling toward the black hole, as shown in Figure 24-12a. The in-falling gas is

radio source in the same part of the sky. Suspecting that the X-ray fluctuations and the radio source might be caused by the same object, astronomers turned to large radio telescopes capable of high resolution to locate Cygnus X-1. The 140-ft dish at the National Radio Astronomy Observatory was used to identify finally the star HDE 226868 (see Figure 24-10) at the site of the radio source.

Spectroscopic observations promptly revealed that HDE 226868 is a B0 supergiant. Such stars do not emit significant amounts of X rays, so HDE 226868 alone cannot be Cygnus X-1. Double-star systems are quite common, however, and as far as anyone could tell, HDE 226868 and Cygnus X-1 are at the same location. Thus, astronomers began to suspect that the visible star and the X-ray source might be in orbit about each other.

Further spectroscopic observations soon showed that the spectral lines in the spectrum of HDE 226868 shift back and forth with a period of 5.6 days. This behavior is characteristic of a single-line spectroscopic binary; the companion of HDE 226868 is just too dim to produce its own set of spectral lines. The clear implication is that HDE 226868 and Cygnus X-1 are the two components of a double-star system.

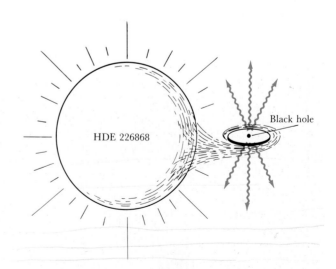

Figure 24-11 The Cygnus X-1 system *A stellar wind from HDE 226868 pours matter onto an accretion disk surrounding a black hole. The in-falling gases from the disk are heated to high temperatures as they spiral in toward the hole. At the inner edge of the disk, just above the black hole, the gases become so hot that they emit vast quantities of X rays. An artist's rendition of this system is seen at the beginning of this chapter.*

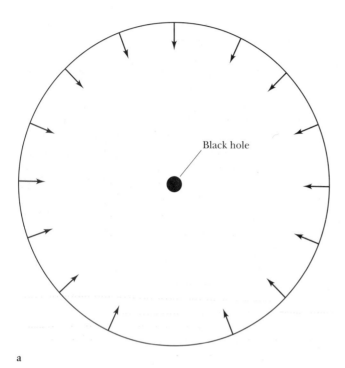

a

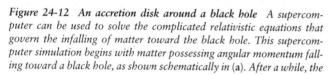

b

Figure 24-12   An accretion disk around a black hole   *A supercomputer can be used to solve the complicated relativistic equations that govern the infalling of matter toward the black hole. This supercomputer simulation begins with matter possessing angular momentum falling toward a black hole, as shown schematically in (a). After a while, the*

in-falling gases become captured into a doughnut-shaped accretion disk shown in cross section in (b). False color is used here to display density, from purple and blue for the least dense regions through the colors of the rainbow to red, which indicates the most dense regions. (Courtesy of L. L. Smarr and J. F. Hawley)

endowed with angular momentum, which causes the gas to orbit about the hole. Eventually, the orbiting gas settles in a thick disc centered about the black hole, as shown in cross-section in Figure 24-12b. These calculations demonstrate that a fat accretion disk (resembling a doughnut) is more

stable than a thin disk (like Saturn's rings). The inner edge of an accretion disk occurs where the outward-directed "centrifugal force" on the gas just balances the inward pull of gravity. Along this inner edge, the flow of gas is unstable and hot bubbles can develop (see Figure 24-13).

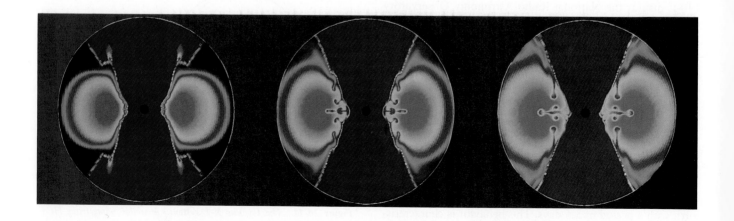

a   b   c

Figure 24-13   The inner edge of an accretion disk   *Three stages in the evolution of the inner edge of an accretion disk are shown here in cross section, using false color. The black hole is the dot in the center of each view. View (a) shows the undisturbed accretion disk. In (b), bubbles and*

fingers of hot gas develop because of instabilities along the inner edge of the disk. In (c), the bubbles have extended into the disk. (Courtesy of L. L. Smarr and J. F. Hawley)

In the early 1980s, a binary system similar to that of Cygnus X-1 was identified in a nearby galaxy called the Large Magellanic Cloud. This X-ray source, called LMC X-3, exhibits rapid fluctuations just like those of Cygnus X-1. Every 1.7 days, LMC X-3 circles a B3 main-sequence star. From the orbital data, astronomers conclude that the mass of the compact X-ray source is probably about 9 $M_\odot$. Once again, a black hole is the only reasonable candidate for such a massive, compact object.

A strong case can also be made for a spectroscopic binary in the constellation of Monoceros that contains the flickering X-ray source A0620−00 (the "A" refers to the British satellite *Ariel 5* that discovered the source; the numbers refer to its position in the sky). The visible companion of A0620−00 is an orange dwarf star of spectral type K called V616 Monocerotis, which orbits the X-ray source every 7.75 hours. From orbital data, astronomers estimate A0620−00 to be about 9 solar masses, whereas the visible companion's mass may be only about 0.5 solar masses.

There are several other rapidly flickering X-ray sources in binary systems; all are excellent black-hole candidates. They include Circinus X-1, and GX 339-4 in Scorpius. Some astronomers also argue that SS433 may contain a black hole rather than a neutron star. With these and similar objects, a great deal of observational effort is necessary to rule out all non−black-hole explanations of the data. Only then can we feel confident that additional black holes have been discovered.

Although finding black holes is a tedious and tricky business, it is becoming clear that they should be moderately common. Although a black hole is created by the gravitational collapse of a burned-out star whose mass exceeds 3 $M_\odot$, this is not the only way a black hole can form. A white dwarf or a neutron star in a binary system can be transformed into a black hole by accreting enough matter from its companion star. This transformation can occur when the companion star becomes a red giant and dumps a significant part of its mass over its Roche lobe. Another possibility is that two dead stars could coalesce to form a black hole. For example, imagine a binary system consisting of two neutron stars, such as the binary pulsar discussed in Box 24-2. Because of the emission of gravitational radiation, the two stars gradually spiral in toward each other and eventually merge. If their total mass exceeds 3 $M_\odot$, the entire system may become a black hole.

A black hole is one of the most bizarre and fantastic concepts ever to emerge from modern physical science. Although the idea of black holes initially met with skepticism, it is now clear that many of the stars we see in the sky are doomed to disappear from the universe someday, leaving only black holes behind. Even more astounding is the idea that enormous black holes, containing millions or even billions of solar masses, are located at the centers of many galaxies and quasars. As we shall see in the next chapter, one of these monstrosities may even be lurking at the center of our Milky Way, only 25,000 light years from the Earth.

Key words

black hole	*Fitzgerald contraction	no-hair theorem	Schwarzschild radius (R_{Sch})
*dilation of time	general theory of relativity	principle of equivalence	singularity
Einstein ring	gravitational lens	*proper distance	special theory of relativity
ergosphere	law of cosmic censorship	*proper mass	
event horizon	*Lorentz transformations	*proper time	

Key ideas

- The special theory of relativity asserts that an observer will note a slowing of clocks and a shortening of rulers that are moving with respect to the observer; this effect becomes significant only if the clock or ruler is moving with a speed near the speed of light.

- The general theory of relativity asserts that gravity causes space to become curved and time to slow down; these effects are significant only in the vicinity of large masses or compact objects.

- If a stellar corpse has a mass greater than about 3 $M_\odot$, gravitational compression will make the object so dense that the escape velocity from it exceeds the speed of light; the corpse then contracts rapidly to a single point called a singularity.

   The singularity is surrounded by a surface called the event horizon, where the escape velocity equals the speed of light; nothing—not even light—can escape from inside the event horizon.

   A black hole—a singularity surrounded by an event horizon—has only three physical properties: mass, electric charge, and angular momentum.

   In the region called the ergosphere, around the outside of the event horizon, space and time themselves are dragged along with the rotation of the black hole.

- The general theory of relativity predicts the existence of gravitational radiation; gravitational waves are ripples in the overall geometry of space and time that are produced by moving masses.

   It may be possible to study how stars collapse by measuring the gravitational waves emitted during these events; effective antennas for gravitational radiation are under development.

   An isolated black hole might be detected by how it works as a gravitational lens, distorting the image of a star or galaxy behind it.

- Some binary-star systems are thought to contain a black hole; in such a system, gases captured from the companion star by the black hole emit detectable X rays.

Review questions

1 Under what circumstances are degenerate-electron pressure and degenerate-neutron pressure incapable of preventing the complete gravitational collapse of a dead star?

∗2 Find the Schwarzschild radius for an object having a mass equal to that of the planet Jupiter.

3 In what way is a black hole blacker than black ink or a black piece of paper?

4 If the Sun suddenly became a black hole, how would the Earth's orbit be affected?

5 According to the general theory of relativity, why can't some sort of yet-undiscovered degenerate pressure prevent the matter inside a black hole from collapsing all the way down to a singularity?

6 What is the law of cosmic censorship?

7 What is a gravitational lens? Why do you suppose that no gravitational lenses have yet been discovered in our Galaxy?

8 What is the no-hair theorem?

9 What kind of black hole is surrounded by an ergosphere?

10 Why do you suppose that all the black-hole candidates mentioned in the text are members of very short-period binary systems?

11 As a binary system loses energy by emitting gravitational waves, why do its members speed up and why does the period become shorter?

Advanced questions

> **Tips and tools . . .**
> Remember that the density of an object is its mass divided by its volume. The volume of a sphere of radius r is $\frac{4}{3}\pi r^3$. Recall that Box 4-3 contains Newton's formulation of Kepler's third law, which explicitly includes masses.

∗12 *(Basic)* To what density must the matter of a dead 10-$M_\odot$ star be compressed in order for the star to disappear inside its event horizon?

∗13 Prove that the density of matter needed to produce a black hole is inversely proportional to the square of the mass of the hole. If you wanted to make a black hole from matter compressed to a density of water (1000 kg/m^3), how much mass would you need?

∗14 What is the Schwarzschild radius of a black hole whose mass is **(a)** the mass of the Earth, **(b)** the mass of the Sun, and **(c)** the mass of the Milky Way Galaxy, which is about 1.1×10^{11} $M_\odot$? In each case, also calculate what the density would be if the matter were spread uniformly throughout the volume of the event horizon.

∗15 *(Basic)* A clock on board a moving starship shows a total elapsed time of 2 minutes, while an identical clock on Earth shows a total elapsed time of 10 minutes. What is the speed of the starship?

∗16 *(Basic)* How fast should a meterstick be moving in order to appear to be a "centimeterstick"?

∗17 *(Basic)* To what speed should a particle be accelerated in order for its mass to appear to double to an at-rest observer?

*18 Find the total mass of the neutron-star binary system described in Box 24-2, for which the orbital period is 7.7 hours and the average distance between the neutron stars is 2.8 solar radii. Is your result reasonable for a pair of neutron stars? Explain.

19 (Challenging) The orbital period of the single-line spectroscopic binary containing A0620-00 is 0.32 day, and Doppler-shift measurements reveal that its radial velocity peaks at 457 km/s (about 1 million mi/hr). By scaling this system to the Sun–Earth system and using Kepler's third law, prove that the mass of the X-ray source must be at least 3.1 times the mass of the Sun. (Hint: Assume that the mass of the K5V visible star—about 0.5 $M_\odot$ from the mass–luminosity relationship—is negligible compared to that of the invisible companion.)

Discussion questions

20 Discuss why the fact that the speed of light is the same for all observers, regardless of their motion, requires us to abandon the Newtonian view of space and time.

21 Describe the kinds of observations you might make in order to locate and identify black holes.

22 Speculate on the effects you might encounter on a trip to the center of a black hole.

Observing project

23 As you well know, you cannot see a black hole with a telescope. Nevertheless, you might want to observe the visible companion of Cygnus X-1. The epoch 2000 coordinates of this ninth-magnitude star are R.A. = $19^h58.4^m$ and Decl. = $+35°12''$, which is quite near the bright star η Cygni. Compare what you see with the photograph in Figure 24-10.

For further reading

Chaffee, F. "The Discovery of a Gravitational Lens." *Scientific American*, November 1980 • This article, written soon after the first gravitational lens was discovered, de-scribes the lens phenomenon with the aid of excellent diagrams.

Jeffries, A., Saulson, P., Spero, R., and Zucker, M. "Gravitational Wave Observatories." *Scientific American*, June 1987 • This article, which includes an excellent description of gravitational radiation, explains how lasers can be used to detect gravitational waves.

Kaufmann, W. *Black Holes and Warped Spacetime*. W. H. Freeman and Company, 1979 • This slim book gives a nontechnical overview of black holes and general relativity.

———. *Cosmic Frontiers of General Relativity*. Little, Brown and Company, 1977 • This book describes many fascinating aspects of black holes, including views that might greet an astronaut who plunges through a worm hole connecting our universe with another.

McClintock, J. "Do Black Holes Exist?" *Sky & Telescope*, January 1988 • This well written article presents evidence for black holes in certain X-ray binaries, especially A0620−00.

Price, R., and Thorne, K. "The Membrane Paradigm for Black Holes." *Scientific American*, April 1988 • Two renowned physicists present a new way of picturing the interaction of a black hole with its environment.

Stokes, G., and Michalsky, J. "Cygnus X-1." *Mercury*, May/June 1979 • This brief article describes evidence that Cygnus X-1 is a black hole.

Thorne, K. "The Search for Black Holes." *Scientific American*, December 1974 • This superb article describes various aspects of black holes that may be useful in searches for these elusive objects.

Turner, E. "Gravitational Lenses." *Scientific American*, July 1988 • Beautiful illustrations make this article on gravitational lensing especially clear.

Weisberg, J., Taylor, J., and Fowler, L. "Gravitational Waves from an Orbiting Pulsar." *Scientific American*, October 1981 • This article explains how orbital changes in the binary pulsar PSR 1913 +16 result from gravitational radiation emitted by the system.

Our Galaxy This wide-angle photograph, taken from Australia, spans 180° of the Milky Way, from Cygnus at the far right to the Southern Cross at the left. The center of our Galaxy is in the constellation of Sagittarius, in the middle of this photograph. Figure 25-1 is a comparable photograph showing the northern Milky Way. (Courtesy of D. di Cicco)

C H A P T E R

Our Galaxy

The Milky Way, that hazy band of myriad faint stars stretching across the night sky, is our edge-on view of the disk of our own Galaxy. Throughout the twentieth century, astronomers have struggled to determine the size, shape, and rotation of our Galaxy. We now realize that we live in a vast disk-shaped assemblage of stars, gas, and dust roughly 80,000 light years in diameter. Huge spiral arms gracefully arching outward from the Galaxy's nucleus pose many puzzles for astronomers. Apparently, gravitational interactions with a nearby galaxy or random bursts of star formation within our own Galaxy can compress the interstellar medium and give rise to spiral structure. Most mysterious is the remarkable source of energy at the very center of the Galaxy. Vast amounts of radiation pour from this compact source, which may be a supermassive black hole. By exploring our Galaxy, we gain important insights about the properties of galaxies in general, thereby preparing ourselves to widen our perspective on the universe and ask fundamental questions on a cosmic scale.

On a clear, moonless night far from city lights, you can often see a hazy, luminous band stretching across the sky. Ancient peoples devised fanciful myths to account for this "milky way" among the constellations. Today we realize that this hazy band is actually our view from inside a vast disk-shaped assemblage of several hundred billion stars that includes the Sun.

In studying the Milky Way, we explore the universe on a grand scale. Instead of examining individual stars, we look at an entire system of stars. And instead of focusing on the location and life of an isolated star, we look at the overall arrangement and history of a huge stellar community of which the Sun is a member.

25-1 *The Sun is located in the disk of our Galaxy, about 25,000 light years from the galactic center*

Galileo was the first person to look at the Milky Way through a telescope. He immediately discovered it to be composed of countless dim stars. The Milky Way stretches all the way around the sky in a continuous band that is almost perpendicular to the plane of the ecliptic. Figure 25-1 is a wide-angle photograph showing roughly one-third of the Milky Way.

Because the Milky Way completely encircles us, astronomers began in the eighteenth century to suspect that the Sun and all the stars in the sky were part of an enormous disk-shaped assemblage called the **Milky Way Galaxy**. In the 1780s, William Herschel attempted to deduce the Sun's location in the Galaxy by counting the number of stars in each of 683 regions of the sky. Assuming that all stars have the same

luminosity, Herschel reasoned that he should see the greatest density of stars toward the Galaxy's center, whereas a lesser density of stars should be seen toward the edge of the Galaxy.

Herschel found roughly the same density of stars all along the Milky Way. Therefore, he concluded that we are at the center of our Galaxy (see Figure 25-2). In the 1920s, the Dutch astronomer J. C. Kapteyn analyzed the brightnesses and proper motions of a large number of stars and essentially confirmed Herschel's view. According to Kapteyn, the Milky Way was about 10 kpc in diameter and about 2 kpc thick, with the Sun near its center.

Both Herschel and Kapteyn were wrong about the Sun's being at the center of our Galaxy. The reason for their mistake was finally discovered in the 1930s by R. J. Trumpler. While studying star clusters, Trumpler discovered that the more remote clusters appear unusually dim—more so than would be expected from their distance alone. As a result, Trumpler concluded that interstellar space must not be a perfect vacuum: it must contain dust that absorbs light from distant stars. Like the stars themselves, this obscuring material is concentrated in the plane of the Galaxy. Great patches of this interstellar dust are clearly visible in such wide-angle photographs as Figure 25-1.

As we saw in Chapter 20, the dimming of light by the interstellar medium is called **interstellar extinction**. In the plane of our Galaxy, the average interstellar extinction is about 1 magnitude per kiloparsec. For example, a star in the Milky Way that is 5 kpc from Earth appears 5 magnitudes dimmer than it should from its distance alone. And in regions where there are dense interstellar clouds, such as toward the galactic center, the degree of extinction is even greater. In fact, at optical wavelengths the center of our Galaxy is totally obscured from view. It was this interstellar extinction that misled Herschel and Kapteyn. Because they

Figure 25-1 *The Milky Way* *This wide-angle photograph shows the northern Milky Way, from Cassiopeia on the left to Sagittarius on the right. Note the dark lanes and blotches. This mottling is caused by inter-* *stellar gas and dust obscuring the light from the background stars. (Steward Observatory)*

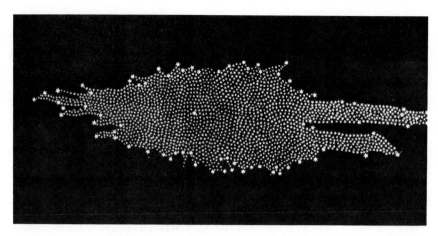

Figure 25-2 Herschel's map of our Galaxy William Herschel attempted to map the Milky Way Galaxy by counting the numbers of stars in various parts of the sky. Because of interstellar dust that blocked his view of distant stars, Herschel erroneously concluded that the Sun is at the center of the Galaxy. (Yerkes Observatory)

were actually seeing only the nearest stars in the Galaxy, they had no true idea of either the enormous size of the Galaxy or of the vast number of stars concentrated around the galactic center.

Because interstellar dust is concentrated in the plane of our Galaxy, interstellar extinction is strongest in those parts of the sky that are covered by the Milky Way. However, our view to either side of the Milky Way is relatively unobscured. Knowledge of our position in the Galaxy was to come from observations of globular clusters in these unobscured portions of the sky. Before we turn to those observations, we must review some background information on variable stars.

In 1912, the American astronomer Henrietta Leavitt reported her important discovery of the period–luminosity relation for classical (Type I) Cepheid variables. As we saw in Chapter 21 (recall Figure 21-13), Cepheid variables are pul-

sating stars that periodically vary in their brightness. Leavitt studied numerous Cepheids in the Small Magellanic Cloud (a small galaxy near the Milky Way) and found their periods to be directly related to their average luminosities (see Figure 25-3). Today, astronomers realize that there are two kinds of Cepheid variables: the metal-rich Type I Cepheids, which Leavitt studied, and metal-poor Type II Cepheids. As shown in Figure 21-14, the Type II Cepheids are slightly dimmer than those of Type I.

The period–luminosity law is an important tool in astronomy, because it can be used to determine distances. For instance, suppose you find a Cepheid variable in the sky. By measuring its period and using a graph such as that in Figure 25-3, you discover the star's average luminosity, which can be expressed as an absolute magnitude. Meanwhile, you can measure the star's apparent magnitude. Once you know the star's apparent and absolute magnitudes, you can calculate its distance using equations like those in Box 19-2.

Shortly after Leavitt's discovery of the period–luminosity law, Harlow Shapley, a young astronomer at the Mount Wilson Observatory in California, concluded that Cepheid variables are pulsating stars. Actually, Shapley was most interested in the closely related family of pulsating stars called the RR Lyrae variables, which we first discussed in Chapter 21. The light curve of an RR Lyrae variable (see Figure 25-4) is similar to that of a Cepheid, and Shapley believed RR Lyrae variables to be simply short-period classical Cepheids like those Leavitt had studied. RR Lyrae variables are commonly found in globular clusters (see Figure 25-5).

By 1915, Shapley had noticed a peculiar property of globular clusters. Ordinary stars and open star clusters are rather uniformly spread along the Milky Way. However, the majority of the 93 globular clusters that Shapley studied were preferentially located in one half of the sky, widely scattered around the portion of the Milky Way that is in the constellation of Sagittarius. Figure 25-6 shows two globular clusters in this part of the sky.

Shapley used the period–luminosity relation to determine the distances to the then-known 93 globular clusters in the

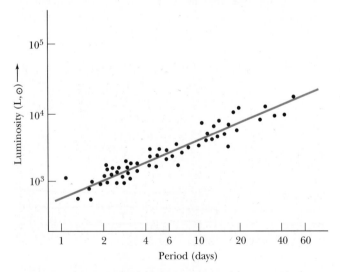

Figure 25-3 The period–luminosity relation for classical Cepheids Shown is the relationship between the periods and luminosities of Type I Cepheid variables. Each dot represents a Cepheid in the Small Magellanic Cloud whose brightness and period have been measured. The line is the "best fit" to the data. (Adapted from H. C. Arp)

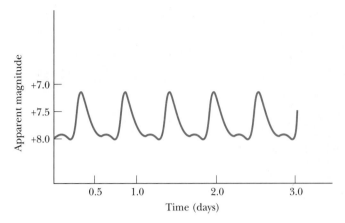

Figure 25-4 *The light curve of RR Lyrae* *An RR Lyrae variable is recognized by the characteristic way in which it varies its brightness. An RR Lyrae light curve is similar to that of a Cepheid variable (compare Figure 21-13). All RR Lyrae variables have periods of roughly ½ day and have brightnesses that vary by about 1 magnitude. The average luminosity of an RR Lyrae variable is approximately 100 Suns. The example shown here is the curve of RR Lyrae, the prototype of this class of stars.*

sky. From their directions and distances, he mapped out the three-dimensional distribution of these clusters in space. By 1917, Shapley had discovered that the globular clusters form a huge spherical system, not centered on the Earth, as Herschel and Kapteyn would have believed, but rather about a point in the Milky Way toward the constellation of Sagittarius. Shapley then made the bold conjecture, which was subsequently confirmed, that the globular clusters indicate the true size and extent of the Galaxy.

Since Shapley's pioneering observations, many astronomers have ventured to measure the distance from the Sun to the **galactic nucleus**, at the center of our Galaxy. Shapley's estimate was about three times larger than the now generally accepted distance of about 8.6 kpc (28,000 ly). Recent measurements based on radio observations of gas clouds containing powerful water masers (recall Figure 20-17) around the galactic center suggest a somewhat shorter distance of about 7 kpc (23,000 ly). In this book, we adopt a compromise distance of 7.7 kpc (25,000 ly) from the Sun to the galactic nucleus.

The distance to the center of the Galaxy establishes a scale from which the dimensions of other features can be determined. The **disk** of our Galaxy is about 25 kpc (80,000 ly) in diameter and about 0.6 kpc (2000 ly) thick (see Figure 25-7). The galactic nucleus is surrounded by a spherical distribution of stars, called the **central bulge**, that is about 4.6 kpc (15,000 ly) in diameter. The spherical distribution of globular clusters defines the **halo** of the Galaxy. Our Galaxy, seen edge on from a great distance, would probably look somewhat like NGC 4565, shown in Figure 25-8.

Different kinds of stars are found in the various components of our Galaxy. The globular clusters in the halo are composed of old, metal-poor, population II stars. The stars in the disk are mostly young, metal-rich, population I stars like the Sun. The disk of a galaxy appears bluish, as in Figure 25-8, because its light is dominated by radiation from young, hot O and B stars. The central bulge looks reddish because many of the older stars in this part of the Galaxy are red giants and red supergiants.

Figure 25-5 *The globular cluster M55* *Three RR Lyrae variables are identified with arrows in this globular cluster located in the constellation of Sagittarius. From their average apparent brightness (as seen in this photograph) and average true brightness (known to be roughly 100 Suns), astronomers have deduced that the distance to this cluster is 20,000 light years. (Harvard Observatory)*

Figure 25-6 *A view toward the galactic center* *More than 1 million stars are visible in this photograph. This view looks toward a relatively clear "window" just 4° south of the galactic nucleus in Sagittarius. There is surprisingly little matter to obscure the view in this tiny section of the sky. The two globular clusters are NGC 6522 and NGC 6528. (NOAO)*

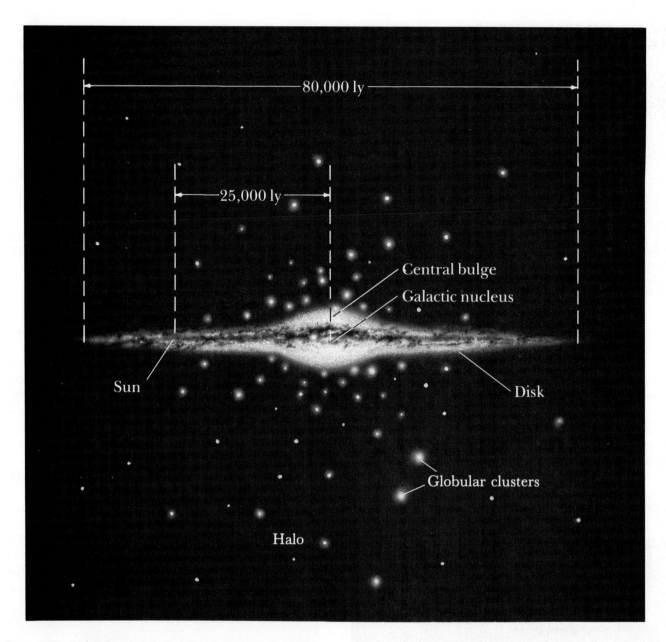

Figure 25-7 *Our Galaxy (schematic edge-on view)* *There are three major components of our Galaxy: a thin disk, a central bulge, and a halo. The disk contains gas and dust along with metal-rich (population I) stars. The halo is composed almost exclusively of old, metal-poor (population II) stars. The central bulge is a mixture of population I and population II stars.*

25-2 The spiral structure of our Galaxy has been plotted from radio and optical observations of star-forming regions

Because interstellar dust effectively obscures our visual view in the plane of our Galaxy, a detailed understanding of the structure of the galactic disk had to wait until the development of radio astronomy. Because of their long wavelengths, radio waves easily penetrate the interstellar medium without being scattered or absorbed. As we shall see in this section, radio and optical observations reveal that our Galaxy has spiral arms, spiral-shaped concentrations of gas and dust unwinding from the center in a shape reminiscent of a pinwheel.

We have seen that hydrogen is by far the most abundant element in the universe. Hence, by looking for concentrations of hydrogen gas we should be able to detect important clues about the structure of the disk of our Galaxy. Unfortunately, the major electron transitions in the hydrogen atom (review Figure 5-17) produce photons at ultraviolet and visible wavelengths, which do not penetrate the interstellar medium. How might we detect all this hydrogen?

Figure 25-8 Edge-on view of the galaxy NGC 4565 *If we could view our Galaxy edge-on from a great distance, it would probably look like this galaxy in the constellation of Coma Berenices. A layer of dust and gas is clearly visible in the plane of the galaxy. Also note the reddish color of the bulge that surrounds the galaxy's nucleus. (U.S. Naval Observatory)*

In addition to having mass and charge, particles such as protons and electrons possess a tiny amount of angular momentum commonly called **spin.** An electron or a proton can in fact be crudely visualized as a tiny spinning sphere. According to the laws of quantum mechanics, the electron and proton in a hydrogen atom can be spinning in either the same direction or opposite directions (see Figure 25-9), but they can have no other relative spin orientations. Physicists thus say that the particles' spin direction is "quantized." If,

for example, the electron flips over from one configuration to the other, the hydrogen atom must gain or lose a tiny amount of energy. For instance, in going from parallel to antiparallel spins, the atom emits a low-energy photon whose wavelength is 21 cm.

In 1944, the Dutch astronomer H. C. van de Hulst predicted that 21-cm radiation from this **spin–flip transition** in interstellar hydrogen would be detected as soon as appropriate radio telescopes were constructed. In 1951, a team of astronomers at Harvard did succeed in detecting faint 21-cm radio emission from interstellar hydrogen.

This detection of 21-cm radio radiation was a major breakthrough that permitted astronomers to probe the galactic disk. To see how they did so, suppose that you aim a radio telescope along a particular line of sight across the Galaxy, as sketched in Figure 25-10. Your radio receiver picks up 21-cm emission from hydrogen clouds at points 1,

Figure 25-9 The hyperfine structure of the hydrogen atom *In the lowest orbit of the hydrogen atom, the electron and the proton can be spinning in either the same or opposite directions. When the electron flips over, the atom either gains or loses a tiny amount of energy. This energy is either absorbed or emitted as a radio photon having a wavelength of 21 cm.*

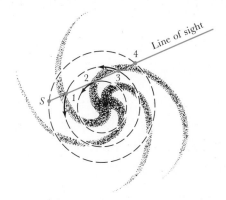

Figure 25-10 A technique for mapping our Galaxy *Hydrogen clouds at different locations along our line of sight are moving at slightly different speeds. As a result, radio waves from various gas clouds are subjected to slightly different Doppler shifts, permitting radio astronomers to sort out the gas clouds and map the Galaxy.*

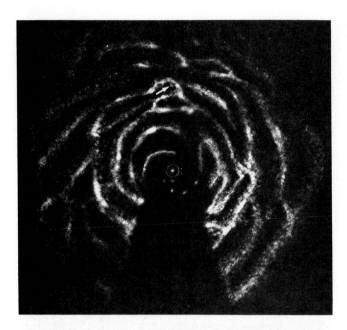

Figure 25-11 A map of neutral hydrogen in our Galaxy This map, obtained from radio-telescope surveys of 21-cm radiation, shows numerous arched lanes of gas that suggest a spiral structure for our Galaxy. The Sun's location is indicated by a white arrow. Details in the large, blank, wedge-shaped region toward the bottom of the map cannot be determined, because gas in this part of the sky is moving perpendicular to our line of sight and thus does not exhibit a detectable Doppler shift. (Courtesy of G. Westerhout)

2, 3, and 4 (point *S* is the location of the Sun). However, the radio waves from these various clouds are Doppler shifted by slightly different amounts, because they are moving at different speeds as they travel with the rotating Galaxy.

It is important to remember that the Doppler shift reveals only motion parallel to the line of sight (review Figure 5-20). Note that cloud 2 has the highest line-of-sight speed, because it is moving directly toward us. Consequently, the radio waves from cloud 2 exhibit a larger Doppler shift than those from any other clouds along our line of sight. Clouds 1 and 3 are at the same distance from the galactic center and thus have the same orbital speed. The fraction of their velocity parallel to our line of sight is also the same, so their radio waves exhibit the same Doppler shift, which is less than the Doppler shift of cloud 2. Cloud 4 is at the same distance from the galactic center as is the Sun. This cloud is thus orbiting the Galaxy at the same speed as the Sun, resulting in no net motion along the line of sight. Radio waves from cloud 4, as well as from hydrogen gas near the Sun, are not Doppler-shifted at all.

The various Doppler shifts that occur cause the 21-cm radiation to be smeared out over a range of wavelengths. Because radio waves from gases in different parts of the Galaxy arrive at our radio telescopes with slightly different wavelengths, it is possible to sort out the various gas clouds and thus produce a map of the Galaxy, such as that shown in Figure 25-11.

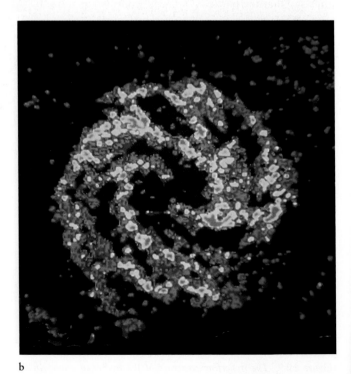

a

b

Figure 25-12 A spiral galaxy This galaxy, called M83, is in the southern constellation of Centaurus about 12 million light years from Earth. **(a)** *This photograph at visual wavelengths clearly shows the spiral arms illuminated by young stars and glowing H II regions.* **(b)** *This radio view at a wavelength of 21 cm shows the emission from neutral hydrogen gas (H I). Note that the spiral arms are better distinguished by emission from ionized hydrogen than neutral hydrogen. (Anglo-Australian Observatory; VLA, NRAO)*

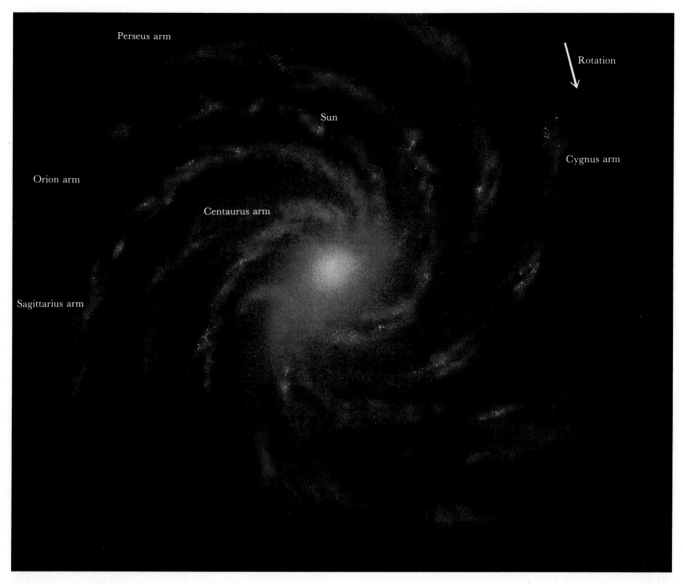

Figure 25-13 *Our Galaxy (face-on view)* *Our Galaxy has four major spiral arms and several shorter segments of arms. The Sun is located on the Orion arm, between two major spiral arms. The Galaxy's diameter is about 80,000 light years, and the Sun is about 25,000 light years from the galactic center.*

A 21-cm map of our Galaxy reveals numerous arched lanes of neutral hydrogen gas but gives only a vague hint of spiral structure. Photographs of other galaxies (see Figure 25-12) show spiral arms outlined by bright stars and emission nebulae indicative of active star formation. We can thus see that the best clues about the spiral structure of our Galaxy come from mapping the locations of star-forming complexes marked by OB associations, H II regions, and molecular clouds.

Interstellar absorption limits the range of visual observations in the plane of the Galaxy to less than 3 kpc (10,000 ly) from the Earth. Nevertheless, there are enough OB associations and H II regions visible in the sky to plot the spiral arms in the vicinity of the Sun. As we saw in Chapter 20, carbon monoxide (CO) is a good "tracer" of molecular clouds. Radio observations of this molecule have recently been used to plot remote regions of the Galaxy. Taken together, all these observations demonstrate that our Galaxy has four major spiral arms and several short arm segments (see Figure 25-13).

The Sun is located on a relatively short arm segment called the Orion arm, which includes the Orion Nebula (see Figure 20-14) and neighboring sites of vigorous star formation in that constellation (examine Figure 6-28). Two major spiral arms border either side of the Sun's position. The Sagittarius arm is on the side toward the galactic center. This is the arm you see during the summer months when you look at the portion of the Milky Way stretching across Scorpius and Sagittarius (see photograph on the first page of this chapter). During winter, when our nighttime view is directed away from the galactic center, we see the Perseus arm. The remaining two major spiral arms are usually referred to as the Centaurus arm and the Cygnus arm.

Radio astronomers are working to improve their distance measurements across the Milky Way, which will result in better maps of our Galaxy.

25-3 Moving at half a million miles per hour, the Sun takes about 200 million years to complete one orbit of our Galaxy

The presence of spiral arms suggests that our Galaxy rotates. Indeed, if the stars in our Galaxy were not orbiting the galactic center, they would all fall into the galactic center. However, measuring the rotation of our Galaxy is a difficult business.

Radio observations of 21-cm radiation from hydrogen gas give important clues about our Galaxy's rotation. By measuring Doppler shifts, astronomers can determine speeds parallel to our line of sight across the Galaxy. These observations clearly indicate that our Galaxy does not rotate like a rigid body but rather exhibits differential rotation: Stars at different distances from the galactic center travel at different orbital speeds about the Galaxy.

Further clues to this rotation come from examining the motions of stars in the sky. Because of differential rotation, the stars in the neighborhood of the Sun are like cars on a

circular freeway, with the fast lane on the inside of the curve and the slow lane on the outside. As sketched in Figure 25-14, stars in the fast lane are passing the Sun, making them appear to be moving in one direction, while stars in the slow lane are being overtaken by the Sun, making them seem to be moving backward in the opposite direction. This analysis was pioneered by the Dutch astronomer Jan H. Oort.

Unfortunately, like the 21-cm observations, this study of stellar motions reveals only how fast things are moving relative to the Sun. Of course, the Sun itself is also moving. To get a complete picture of our Galaxy's rotation, we must therefore find out how fast the Sun is traveling.

A method of doing this was proposed by the Swedish astronomer Bertil Lindblad. Not all the stars in the sky move in the orderly pattern sketched in Figure 25-14. Distant galaxies and some components of our own Galaxy, like globular clusters, do not participate in the general rotation of the Galaxy but have instead more or less random motions. Lindblad took the average of these random motions to establish a stationary background. He then used the Doppler shifts of the stars in this background to deduce that the Sun's speed along its orbit about the galactic center is 230 km/s, or about $\frac{1}{2}$ million miles per hour.

We know that we are 25,000 light years from the galactic center, so we can use the Sun's speed v to calculate the Sun's orbital period T, the time required for one trip about the Sun's orbit. For the sake of simplicity, we assume that the Sun travels along a circular orbit about the center of the Galaxy. The distance traveled by the Sun is the circumference of this circle, which is given by $2\pi r$ (with the r bring the radius of the circle). Consequently, the time required for one orbit is equal to the distance traveled divided by the Sun's speed:

$$T = \frac{2\pi r}{v} = \frac{2\pi \times 25{,}000 \text{ ly}}{230 \text{ km/s}} \times \frac{9.46 \times 10^{12} \text{ km}}{1 \text{ ly}}$$

$$= 6.5 \times 10^{15} \text{ s} = 2.0 \times 10^8 \text{ years}$$

Traveling at $\frac{1}{2}$ million miles per hour, it takes the Sun about 200 million years to complete one trip around the Galaxy. These results demonstrate how vast our Galaxy is.

Because we now know the basic features of our orbit around the Galaxy, we can use Kepler's third law, $P^2 = 4\pi^2 a^3/GM$ (recall Box 4-3), to estimate the mass of the Galaxy. Another version of this law, more useful for our purpose, is

$$M = \frac{rv^2}{G}$$

where M is the mass of the Galaxy within the Sun's orbit, r is the radius of the Sun's orbit, v is the Sun's orbital speed, and G is the gravitational constant. Putting in the numbers, we obtain a mass for the Galaxy of 9.4×10^{10} M$_\odot$.

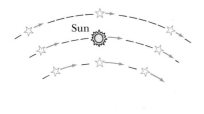

Sun

+
Galactic center

Figure 25-14 Differential rotation of our Galaxy *Different parts of the Galaxy are rotating at different speeds. Thus, stars traveling more quickly than the Sun appear to be moving in one direction, while stars traveling more slowly than it appear to be moving in the opposite direction.*

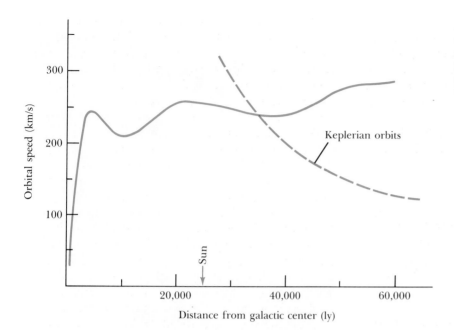

Figure 25-15 The Galaxy's rotation curve The orbital speeds of stars and gas are observed to increase out to a distance of at least 60,000 light years from the galactic center. Vast quantities of subluminous matter must therefore surround our Galaxy.

We know this estimate is too low, because Kepler's law gives us only the mass inside the Sun's orbit. The matter outside the Sun's orbit does not affect the Sun's motion and thus does not enter into Kepler's third law. Obviously, though, there is matter out there. Astronomers have been astonished in recent years to discover how much matter actually lies beyond the orbit of the Sun.

Because we know the true speed of the Sun, we can convert the Doppler shifts measured by radio astronomers into actual speeds for the spiral arms. This calculation gives us a **rotation curve**, a graph of the velocity of galactic rotation measured outward from the galactic center (see Figure 25-15). Note that the rotation curve keeps rising, even out to a distance of 60,000 light years from the galactic nucleus.

According to Kepler's third law, the orbital speeds of stars or gas clouds beyond the confines of most of the Galaxy's mass should decrease with increasing distance from the Galaxy's center, just as the orbital speeds of the planets decrease with increasing distance from the Sun. Instead, galactic orbital speeds continue to climb well beyond the visible edge of the galactic disk, which means that we still have not determined the actual edge of our Galaxy. A surprising amount of matter must surround our Galaxy in a presumably spherical distribution that extends far beyond the Galaxy's halo. Because of this matter, the total mass of our Galaxy could easily be at least 6×10^{11} M$_\odot$.

To make matters even more puzzling, this outlying matter is dark. No one has detected any radiation coming from it, and it does not show up on photographs. Various explanations of this dark matter have been proposed. It might, for instance, be composed of numerous Jupiter-like planets and dim, low-mass stars. Or it might be made of something much more exotic, such as black holes, neutrinos possessing mass, or a variety of hypothetical particles predicted by speculative theories at the forefront of physics. The issues raised by this dark matter constitute one of the most baffling mysteries in modern astronomy.

25-4 Self-sustaining star formation can produce spiral arms

That spiral arms should exist at all was another mystery that confounded astronomers for many years. Many galaxies exhibit the beautiful arching arms outlined by brilliant H II regions and OB associations (recall Figure 25-12). As we think through the effects of a galaxy's rotation, a dilemma arises. All spiral galaxies have rotation curves similar to our own. As Figure 25-15 demonstrated, the velocity of stars and gases is fairly constant over a large portion of a galaxy's disk. However, the farther that stars are from a galaxy's center, the farther they must, of course, travel to complete one orbit of the galaxy. Thus, stars and gases in the outskirts of a galaxy take much longer to complete an orbit than does material near the galaxy's center. Consequently, the spiral arms should eventually "wind up," wrapping themselves tightly around the galaxy's nucleus. After a few galactic rotations, the spiral structure should disappear altogether.

The appearance of spiral arms varies widely. In some galaxies, called **flocculent spirals** (see Figure 25-16a), the spiral arms are broad, fuzzy, chaotic, and poorly defined. In other galaxies, called **grand-design spirals** (see Figure 25-16b), the

b

Figure 25-16 *Variety in spiral arms The differences from one spiral galaxy to another suggest that more than one process can create spiral arms.* (a) *The galaxy NGC 7793 has fuzzy, poorly defined spiral arms.*

(b) *The galaxy NGC 628 has thin, well-defined spiral arms. (Courtesy of P. Seiden, D. Elmegreen, B. Elmegreen, and A. Mobarak; IBM)*

spiral arms are thin, delicate, graceful, and well defined. This range of shapes suggests that more than one mechanism can give rise to the spiral structure of a galaxy.

The theory of **self-propagating star formation** provides a straightforward explanation of spiral arms that takes into account the galaxy's differential rotation. Imagine that star formation begins in a dense interstellar cloud somewhere in the disk of a galaxy that does not yet have spiral arms. As soon as hot, massive stars form, their radiation compresses nearby nebulosity, triggering the formation of additional stars in that gas. The massive stars also become supernovae that produce shock waves, which further compress the surrounding interstellar medium, thus encouraging still more star formation. As the star-forming region grows, the galaxy's differential rotation drags the inner edges ahead of the outer edges. This conglomeration of bright O and B stars and glowing nebulae soon becomes stretched out in the form of a spiral arm.

Spiral arms produced by bursts of star formation come and go more or less at random across a galaxy. Bits and pieces of spiral arms appear where star formation has recently begun, but fade and disappear at other locations where all the massive stars have died off. Such galaxies thus have a chaotic appearance with poorly defined spiral arms, as was the case with NGC 7793 in Figure 25-16a. We must turn to an alternative explanation for spiral structure to account for the orderly appearance of other galaxies.

25-5 Spiral arms are caused by density waves that sweep around a galaxy

In the 1920s, Bertil Lindblad proposed that the spiral arms of a galaxy are a persistent pattern that moves among the stars. For instance, as waves on the ocean move across the surface of the water, the individual water molecules simply bob up and down in little circles. A cork in the water merely bobs up and down as the waves ripple by. The waves are simply a pattern that moves across the water; no water actually travels along with the wave pattern. Lindblad spoke of **density waves** in discussing a possible cause of spiral structure.

This density-wave theory was greatly elaborated upon and mathematically embellished in the mid-1960s by the American astronomers C. C. Lin and Frank Shu. Lin and Shu argued that density waves passing through the disk of a galaxy cause material to "pile up" temporarily. A spiral arm thus is simply a temporary enhancement or compression of the material in a galaxy.

The situation is somewhat analogous to a traffic jam. Imagine workers painting a line down a busy freeway. The cars normally cruise along the freeway at 55 mi/hr, but the crew of painters causes a temporary bottleneck. The cars must slow down temporarily to avoid hitting anyone. As

a Water wave

b Kinematic wave

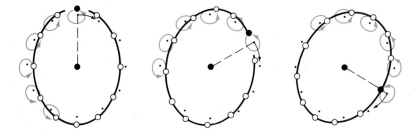

Figure 25-17 Water waves and kinematic waves (a) *In a water wave, each molecule rotates about a point on the undisturbed water level in a tiny ellipse.* (b) *Similarly, even a small disturbance in the orbit of a star can cause the star to oscillate in tiny ellipses about its original, nearly circular orbit (dots). The resulting path of the star is an ellipselike curve that precesses. (Adapted from A. Toomre)*

seen from the air, there is a noticeable congestion of cars around the painters. An individual car spends only a few moments in the traffic jam, though, before resuming its usual speed. The traffic jam itself lasts all day, however, inching its way along the road. The traffic jam, seen clearly from an airplane, is simply a temporary enhancement of the number of cars in a particular location.

To better understand how a density wave operates in a galaxy, think once again about the ocean. If the water molecules were left completely undisturbed, the surface of the ocean would be perfectly smooth. In reality, however, the molecules are constantly buffeted by small disturbances, called **perturbations**, such as the wind. A perturbation pushes one molecule which pushes the next one, which pushes the next, and so on. The result is a water wave. Individual molecules on the surface of the ocean move in tiny elliptical paths as the wave pattern moves across the water (see Figure 25-17a).

In a galaxy, stars are separated by such vast distances that they virtually never collide. Nevertheless, stars do interact, because they are affected by each other's gravity. In water waves and sound waves, molecular forces are responsible for orchestrating the motions of molecules. In a galaxy, the force of gravity controls the interactions between stars.

Seen from above, the undisturbed orbit of a star about the center of a galaxy would be a nearly perfect circle. However, other matter in the galaxy produces small gravitational perturbations that cause the star to deviate from its undisturbed orbit. Just as a water molecule bobs up and down on the surface of the ocean, the star oscillates back and forth about its undisturbed orbit. Lindblad demonstrated that these oscillations can be described by thinking of the star as being attached to a tiny epicycle. As seen in Figure 25-17b, the star rotates counterclockwise around the epicycle while the epicycle itself moves clockwise along the undisturbed path (shown as dots). The final path of the star lies along an ellipselike curve that slowly rotates, or precesses. Of course,

the gravity of this star affects the motions of its neighbors, creating a wave disturbance called a **kinematic wave** that propagates from one stellar orbit to the next.

In 1973, Agris J. Kalnajs in Australia bridged the gap between Lindblad's concept of orbits and the density-wave theory of Lin and Shu. Kalnajs argued that the precessing elliptical orbits of stars are not randomly oriented, as suggested in Figure 25-18a. He proposed that, instead, there is a strict correlation between orbits: Each precessing elliptical orbit is tilted with respect to its neighbor through a specific angle. The result, shown in Figure 25-18b, is a beautiful spiral pattern.

This spiral pattern arises in the locations where the ellipses are bunched closest together. Of course, stars in a galaxy are scattered randomly along their orbits. With correlated orbits, however, some of the stars happen to get close together along the huge, arching spiral arms. After all, the spiral arms are where the stars' orbits are closest together for the longest stretches.

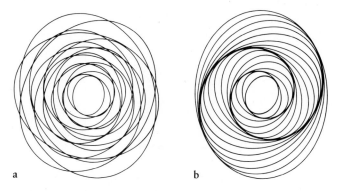

a   b

Figure 25-18 The origin of spiral density waves Both drawings have exactly the same number of ellipses, each one representing the orbit of a star. (a) *Randomly oriented ellipses.* (b) *Ellipses with a correlation between the orientations of adjacent ellipses.*

A temporary increase in the number of stars has a profound effect on the interstellar gas and dust. The presence of the extra stars creates an increased gravitational attraction all along the spiral. This force has almost no effect on the stars, which simply continue to lumber along their orbits. However, the atoms and molecules in the interstellar medium are readily sucked into the gravitational well along the spiral, to form the crest of the density wave.

As Kalnajs's spiral patterns precess, the density waves move through the material of a galaxy at a speed roughly 30 km/s slower than the speed of the stars and the interstellar medium. On its own, however, the interstellar gas can transport a disturbance, such as a slight compression, at a speed of only 10 km/s, which is the speed of sound in the interstellar medium. Thus, the density wave is supersonic, because its speed through the interstellar gas is greater than the speed of sound in that gas. As happens with a supersonic airplane traveling through the air, a shock wave builds up along the leading edge of the density wave. Shock waves are characterized by a sudden, abrupt compression of the medium through which they move, which is why you hear a sonic boom from a supersonic airplane. Similarly, the interstellar medium is violently compressed by the cosmic sonic boom of the density wave.

This density-wave theory explains many of the properties of orderly spiral structures. As the spiral density waves sweep through the plane of a galaxy, they recycle the interstellar medium. Old gases and dust left behind from ancient, dead stars are compressed into new nebulae in which new stars form. As first noted by M. Fujimoto and W. W. Roberts in the late 1960s, the sprawling dust lanes alongside the string of emission nebulae that outlines a spiral arm attest to the recent passage of a compressional shock wave (examine Figure 25-16b). Because the material left over from the death of ancient stars is enriched in heavy elements, new generations of stars are likely to be more metal-rich than were their ancestors.

Density waves expend an enormous amount of energy to compress the interstellar gas and dust. There must be a driving mechanism that keeps density waves going. Debra and Bruce Elmegreen at IBM have pointed out that grand-design spirals invariably have a central bar extending across their nuclei or a nearby companion galaxy. The asymmetric gravitational field of such a bar or companion pulls on the gas, stars, and dust of a galaxy in a way that generates density waves. In the next chapter we shall explore further the way in which gravitational interactions between galaxies induce spiral structure.

 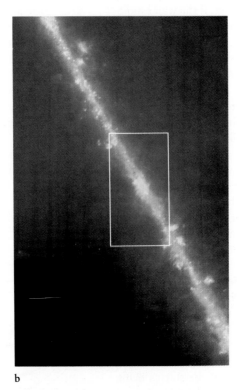 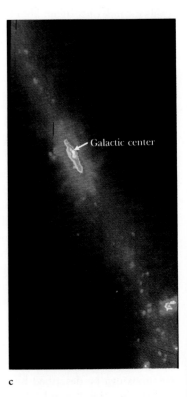

a   b   c

Figure 25-19   *The galactic center* (a) *This wide-angle photograph at visible wavelengths shows a 50° segment of the Milky Way centered on the nucleus of our Galaxy.* (b) *This wide-angle infrared view from IRAS covers the same area as (a). Black represents the dimmest regions of infrared emission, with blue the next dimmest, followed by yellow and* red, *and white for the strongest emission. The prominent band across this photograph is a layer of dust in the plane of the Galaxy.* (c) *This closeup infrared view of the galactic center covers the area outlined by the white rectangle in (b). Streamers of dust around the galactic center can be seen. (Courtesy of D. di Cicco; NASA)*

25-6 Infrared and radio observations are used to probe the galactic nucleus, whose nature is poorly understood

The nucleus of our Galaxy is an active, crowded place. The number of stars in Figure 25-6 gives some hint of the stellar congestion there. If you lived on a planet near the galactic center, you could see a million stars as bright as Sirius, the brightest single star in our own night sky. The total intensity of starlight from all those nearby stars would be equivalent to 200 of our full moons. In effect, night would never really fall on a planet near the center of our Galaxy.

Because of the severe interstellar absorption at visual wavelengths, some of our most important information about the galactic center comes from infrared and radio observations. Figure 25-19 shows three views of the center of our Galaxy. Figure 25-19a is a wide-angle photograph at visible wavelengths covering a 50° segment of the Milky Way through Sagittarius and Scorpius. The same field of view at infrared wavelengths, from the IRAS satellite, is seen in Figure 25-19b. The prominent band across this infrared image is a thin layer of dust in the plane of the Galaxy. The numerous knots and blobs along the dust layer are interstellar clouds heated by young O and B stars. Figure 25-19c is an IRAS view of the galactic center. Numerous streamers of dust (in blue) surround the galactic nucleus. The strongest infrared emission (in white) comes from Sagittarius A, which is a grouping of several powerful sources of radio waves. One of these sources, called Sagittarius A*, is believed to be the galactic nucleus.

Details of Sagittarius A can be seen in the high-resolution infrared photograph Figure 25-20, which covers a region at the galactic center that is only 1 light year across. The various colors represent the emission at different infrared wavelengths: blue is 1.6 μm, green is 2.2 μm, and red is 3.8 μm. The prominent white object near the top of the view is the red supergiant IRS 7 (for *infrared source* 7), which shines as brightly as 100,000 Suns. Most of the other bright objects are red giants. The bluish Y-shaped feature, called IRS 16, is located at the galactic nucleus, centered on Sagittarius A*.

Radio observations give a different picture of the center of our Galaxy. On a large scale, Doppler shift measurements of 21-cm radiation by Jan H. Oort and G. W. Rougoor in 1960 revealed two enormous expanding arms of hydrogen. One arm, which is located between us and the galactic center, is approaching us at a speed of 53 km/s. The other arm is on the other side of the galactic nucleus and is receding from us at a rate of 135 km/s. The total amount of hydrogen in these expanding arms is at least several million solar masses. Something quite extraordinary must have happened about 10 million years ago to expel such an enormous amount of gas from the center of our Galaxy.

In addition to analyzing 21-cm radiation from neutral hydrogen gas, astronomers also detect continuous-spectrum

Figure 25-20   The galactic center   *This composite infrared photograph shows details of the galactic center. The galactic nucleus is located at the center of the bluish Y-shaped feature just below the middle of the photograph. The area covered is only about 1 light year across. The colors represent the dominant infrared emission: Blue is 1.6 μm, green is 2.2 μm, and red is 3.8 μm. (Anglo-Australian Telescope).*

radio noise coming from the galactic center. This radio emission, which is produced by high-speed electrons spiraling around a magnetic field, is called synchrotron radiation, discussed in Chapter 23. In spite of its small size, Sagittarius A is one of the brightest sources of synchrotron radiation in the entire sky.

Some of the most detailed radio images of the galactic center come from the VLA. Figure 25-21a is a wide-angle view of Sagittarius A covering an area about 250 light years across. Huge filaments perpendicular to the plane of the Galaxy stretch 200 light years northward of the galactic center, then abruptly arch southward toward Sagittarius A. The orderly arrangement of these filaments, reminiscent of quiescent prominences on the Sun, suggests that a magnetic field may be controlling the distribution and flow of ionized gas.

The inner core of Sagittarius A, shown in Figure 25-21b, covers an area about 30 light years across. A pinwheel-like feature surrounds Sagittarius A* at the center of this view. One of the arms of this pinwheel is part of a ring of gas and dust orbiting the galactic center. Radio radiation from HCN molecules in the ring yields the view in Figure 25-22. Inclined by about 70° to our line of sight, this ring is quite

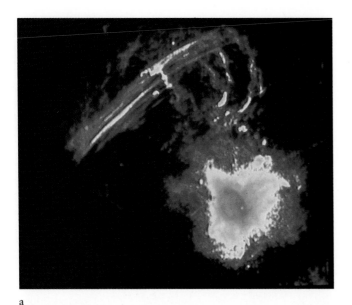

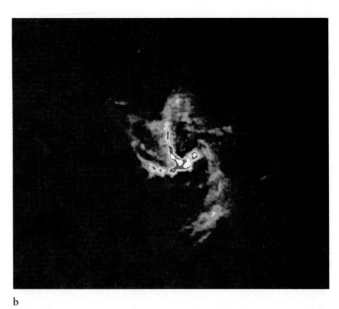

a

b

Figure 25-21 Two radio views of the galactic nucleus These two pictures show the appearance of the center of our Galaxy at radio wavelengths. The strongest radio emission is shown in red, weaker emission being colored green through blue. (a) This view covers an area of the sky about the same size as the full moon, corresponding to a distance of 250 ly across the picture. The parallel filaments may be associated with a magnetic field. The galactic nucleus is toward the lower right, at the center of strongest emission. (b) This high-resolution view shows details of the galactic center covering an area about 30 ly across. The pinwheel-like structure is centered on Sagittarius A*. (VLA; NRAO)

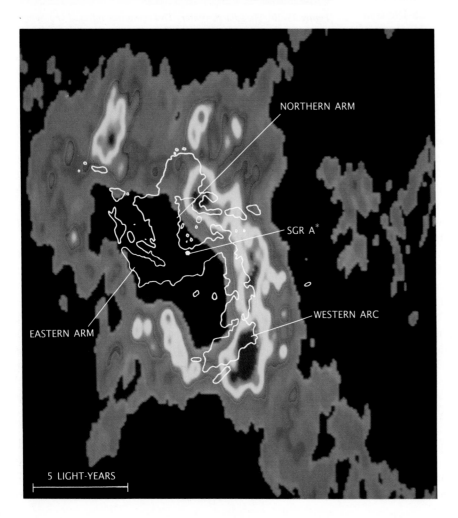

NORTHERN ARM

SGR A*

EASTERN ARM

WESTERN ARC

5 LIGHT-YEARS

Figure 25-22 The molecular ring around the galactic center This high-resolution map of HCN emission reveals a ring of neutral atomic and molecular gas surrounding the galactic nucleus. The ring is about 15 ly in diameter and consists of gas at a temperature of about 300 K, which is much warmer than the dust (80 K) in that region. The pinwheel feature in Figure 25-21b is outlined in white. Note that one of the pinwheel's arms (the western arc) coincides with the ring, but the other two arms do not. (Courtesy of R. Güsten and M.C.H. Wright)

lumpy and consists of turbulent clouds of gas. The total amount of gas, both atomic and molecular, orbiting the galactic nucleus along with this ringlike structure is estimated at $3 \times 10^4 M_\odot$. All the material less than about 6 light years from the galactic nucleus seems to have been ionized by a source that coincides with Sagittarius A*.

The motion of the gas clouds around the galactic center can be deduced from Doppler shift measurements of infrared spectral lines, such as emission from singly ionized neon at 12.80 μm. In the late 1970s, measurements by Charles H. Townes and his colleagues at the University of California revealed that the Ne II line is severely broadened, perhaps from the orbital speed of the gas around the galactic nucleus. The radiation from gas coming toward us is blueshifted, whereas the radiation from receding gas is redshifted. The final result is a smeared out spectral line covering a range of wavelengths corresponding to a range of line-of-sight velocities. On one side of the galactic nucleus, gas is coming toward us at speeds up to 200 km/s, but on the other side, it is rushing away from us at the same speeds.

Something must be holding this high-speed gas in orbit about the galactic center. Using Kepler's third law, we can estimate that a $10^6 M_\odot$ is needed to prevent this gas from flying off into interstellar space. Townes's observations thus suggest that an object with the mass of a million Suns is concentrated at Sagittarius A*. This object must be extremely compact—much smaller than a few light years across. Many astronomers, such as Martin Ryle of Cambridge University, argue that an object this massive and this compact could be only a black hole. Because of its enormous mass, it is called a **supermassive black hole**. As we shall see in Chapter 27, astronomers find extraordinary activity also occurring in the nuclei of many other galaxies, which indicates the possibility of supermassive black holes at their centers. The idea that a supermassive black hole is at the center of our own Galaxy would thus have many precedents.

Many astronomers disagree, however, with this idea of a supermassive black hole at the center of our Galaxy. For instance, the Australian astronomer David Allen and his colleagues, who produced the high-resolution infrared view of the galactic center in Figure 25-20, see nothing to suggest the existence of a supermassive black hole.

Astronomers are still groping for a better understanding of the galactic center. During the coming years, observations from Earth-orbiting satellites as well as from radio telescopes on the ground will certainly add to our knowledge and perhaps elucidate the nature of the core of the Milky Way.

Key words

central bulge (of a galaxy)

density wave

disk (of a galaxy)

flocculent spiral galaxy

galactic nucleus

grand-design spiral galaxy

halo (of a galaxy)

interstellar extinction

kinematic wave

Milky Way Galaxy

perturbation

rotation curve

RR Lyrae variable

self-propagating star formation

spin

spin–flip transition

spiral arm

supermassive black hole

Key ideas

- Our Galaxy has a disk about 80,000 ly in diameter and about 2000 ly thick, with a high concentration of interstellar dust and gas in the disk.

  The galactic nucleus is surrounded by a spherical distribution of stars called the central bulge; the entire Galaxy is surrounded by a spherical distribution of globular clusters and old stars called the halo of the Galaxy.

  OB associations, H II regions, and molecular clouds in the galactic disk outline huge spiral arms.

  From studies of the rotation of the Galaxy, astronomers estimate that the total mass of the Galaxy is about $6 \times 10^{11} M_\odot$, with much of this mass being in some nonvisible, unknown form spread beyond the edge of the luminous material in the Galaxy.

- The Sun is located about 25,000 ly from the galactic nucleus, between two major spiral arms.

  The Sun moves in its orbit at a speed of about $\frac{1}{2}$ million miles per hour and takes about 200 million years to complete one orbit about the Galaxy.

- Interstellar dust obscures our view at visual wavelengths along lines of sight that lie in the plane of the galactic disk.

    Hydrogen clouds can be detected despite the intervening interstellar dust by the 21-cm radio waves emitted in the spin–flip transition.

    The galactic nucleus has been studied through its radio and infrared emissions, which also pass readily through the intervening interstellar dust.

- According to the theory of self-propagating star formation, spiral arms are caused by the birth of stars over an extended region in a galaxy. Differential rotation of the galaxy stretches the star-forming region into an elongated arch of stars and nebulae.

- According to the density-wave theory, spiral arms are created by density waves that sweep around the Galaxy.

    Each star moves in a precessing ellipse about the galactic nucleus; because the orbits are correlated, a spiral pattern is created.

    The gravitational field of this spiral pattern compresses the interstellar clouds through which it passes, thereby triggering the formation of the OB associations and H II regions that illuminate the spiral arms.

- Infrared and radio observations have revealed many details of the galactic nucleus, but astronomers are still puzzled by the processes occurring there.

    A supermassive black hole with a mass of about 10^6 $M_\odot$ may exist at the galactic center.

Review questions

1   Why do you suppose the Milky Way is far more prominent in July than in December?

2   How did interstellar extinction mislead astronomers into believing that we are at the center of our Galaxy?

3   How would the Milky Way appear if the Sun were relocated to the edge of the Galaxy?

4   Why don't astronomers detect 21-cm radiation from the hydrogen in giant molecular clouds?

5   Why do astronomers believe that vast quantities of dark matter surround our Galaxy?

6   Describe the rotation curve you would obtain if the Galaxy were rotating like a rigid body.

7   Explain why globular clusters spend most of their time in the galactic halo, even though their eccentric orbits take them close to the galactic center.

*8   A gas cloud located in the spiral arm of a distant galaxy is observed to have an orbital velocity of 400 km/s. If the cloud is 20,000 pc from the center of the galaxy and is moving in a circular orbit, find (a) the orbital period of the cloud, and (b) the mass of the galaxy which is contained within the cloud's orbit.

*9   Approximately how many times has our solar system orbited the center of our Galaxy since the Sun and planets were formed approximately five billion years ago?

10   How would you estimate the total number of stars in our Galaxy?

11   Compare the kinds of spiral arms produced by density waves with those produced by stochastic, self-propagating star formation. Examine a map of our Galaxy (see Figure 25-13) and cite evidence that both processes probably occur in our Galaxy.

Advanced questions

> **Tips and tools . . .**
> Several of the following questions make extensive use of Kepler's third law, and so you might find it helpful to review Box 4-3. According to the Pythagorean theorem, an isosceles right triangle has a hypotenuse that is longer than its sides by a factor of $\sqrt{2}$.

12   (*Basic*)   What can you surmise about galactic evolution from the fact that the galactic halo is dominated by population II stars, whereas population I stars are predominantly found in the galactic disk?

*13   The mass of our Galaxy interior to the Sun's orbit is calculated from the radius of the Sun's orbit and its orbital speed. By how much would this mass be in error if the calculated distance to the galactic center were off by 10 percent? By how much would this mass be in error if the calculated orbital velocity were off by 10 percent?

14   Speculate on the reasons for the rapid rise in the Galaxy's rotation curve (see Figure 25-15) at distances close to the galactic center.

*15   According to the Galaxy's rotation curve in Figure 25-15, a star 60,000 ly from the galactic center has an orbital velocity of about 300 km/s. What is the mass within that star's orbit?

*16   (*Challenging*)   Suppose you were to use a radio telescope to measure the Doppler shift of 21-cm radiation in the plane of the Galaxy. At an angle of 45° from the galactic center, you detect 21-cm radiation from a cloud of atomic hydrogen that has the highest Doppler shift along your line of sight. How far is that cloud from you and from the galactic center?

*17 The Galaxy is about 25,000 parsecs in diameter and 600 pc thick. If supernovae occur randomly in the Galaxy at the rate of about three each century, how often, on average, should we expect to see a supernova within 300 pc (1000 ly) of the Sun?

18 (Challenging) Show that the form of Kepler's third law stated in Box 4-3 ($P^2 = 4\pi^2 a^3/GM$) is equivalent to $M = rv^2/G$, which is the form of the law used in this chapter, provided the orbital eccentricity is zero (that is, the orbit is a circle).

Discussion questions

19 From what you know about stellar evolution, the interstellar medium, and the density-wave theory, explain the appearance and structure of the spiral arms of grand-design spiral galaxies.

20 What observations would you propose making to determine the nature of the hidden mass in our Galaxy's halo?

Observing project

21 Examine the Milky Way on a clear, moonless night with a high-quality pair of binoculars or a telescope with a wide field of view. During the summer months, look toward the galactic center in Sagittarius and follow the Milky Way arching across the sky through the constellations of Aquila and Cygnus. In the winter, follow the Milky Way as it sweeps across such recognizable constellations as Canis Major, Auriga, and Perseus. Take your time to note details in the structure of the Milky Way. For instance, toward the galactic center you should be able to see a few H II regions along with extensive star clouds. Other stretches of the Milky Way clearly display dark patches and mottling where clouds of interstellar gas and dust have blocked the light from background stars. Enjoy the spectacle.

For further reading

Bok, B. "A Bigger and Better Milky Way." *Astronomy*, January 1984 • This entertaining and informative article was written by a renowned Dutch-American astronomer who made significant contributions to our modern understanding of the Milky Way.

———, and Bok, P. *The Milky Way.* 5th ed. Harvard University Press, 1981 • This classic book presents an up-to-date, in-depth survey of the Milky Way.

——————. "The Milky Way Galaxy." *Scientific American*, March 1981 • This article paints a clear picture of how astronomers have struggled to understand the structure and dynamics of our Galaxy.

Chaisson, E. "Journey to the Center of the Galaxy." *Astronomy*, August 1980 • This well written article describes many of the mysteries associated with the galactic center.

Geballe, T. "The Central Parsec of the Galaxy." *Scientific American*, July 1979 • This article explains how astronomers explore the swirling mass of stars, gas, and dust that envelops the galactic center.

Kraus, J. "The Center of our Galaxy." *Sky & Telescope*, January 1983 • This brief article explains how radio astronomers investigate the galactic center.

Palmer, E. S. "Unveiling the Hidden Milky Way." *Astronomy*, November 1989 • Beautiful illustrations grace this article, which explains how radio astronomers trace the spiral structure of our Galaxy.

Scoville, N., and Young, J. "Molecular Clouds, Star Formation and Galactic Structure." *Scientific American*, April 1984 • This article shows how star formation and giant molecular clouds can affect the structure of the Galaxy.

Smith, R. "The Great Debate Revisited." *Sky & Telescope*, January 1983 • This historical article describes the circumstances surrounding the Shapley–Curtis Debate.

Weaver, H. "Steps Toward Understanding the Large-Scale Structure of the Milky Way." *Mercury*, September/October 1975; November/December 1975; January/February 1976 • This wonderful series by a noted astronomer traces the development of our understanding of the Galaxy.

Wynn-Williams, G. "The Center of our Galaxy." *Mercury*, September/October 1979 • This article examines infrared observations of the galactic center.

26

C H A P T E R

A spiral galaxy This galaxy, called NGC 1365, is the largest and most impressive member of a cluster of galaxies about 60 million light years from Earth. NGC 1365 is classified as a barred spiral because of the "bar" crossing through its nucleus. From the ends of the bar two distinct and quite open spiral arms branch off. The yellow color of the nucleus and the bar shows that these parts of the galaxy are dominated by old, relatively cool stars. The blue color of the arms is caused by light from young, hot stars. (European Southern Observatory)

Galaxies

The universe is populated with galaxies, which come in many shapes and sizes. Some are disk-shaped, like our own Milky Way, with arching spiral arms that are active sites of star formation. Others are featureless, ellipse-shaped agglomerations of stars, virtually devoid of interstellar gas and dust. Some galaxies are comparatively small, having only a hundredth of the size and one ten-thousandth of the mass of the Milky Way. Others are veritable giants, typically five times the size and fifty times the mass of the Milky Way. Most galaxies are grouped in clusters, which stretch across the universe in huge, lacy patterns. Galaxies occasionally collide, producing such spectacular phenomena as a violent burst of star formation and enormous trails of stars and gas cast outward into intergalactic space. Most importantly, remote galaxies are receding from us with speeds that are proportional to their distances from our Galaxy. This relationship between distance and recessional velocity, called the Hubble law, reveals that we live in an expanding universe.

William Parsons was the third Earl of Rosse in Ireland. He was rich, he liked machines, and he was fascinated with astronomy. Accordingly, he set about building gigantic telescopes. In February 1845, his pièce de résistance was finished. The telescope's massive mirror measured 6 ft in diameter and was mounted at one end of a 60-ft tube controlled by cables, straps, pulleys, and cranes. For many years, this triumph of nineteenth-century engineering enjoyed the reputation of being the largest telescope in the world.

With this new telescope, Lord Rosse examined many of the nebulae that had been discovered and catalogued by William Herschel. Lord Rosse observed that some of these nebulae have a distinct spiral structure. One of the best examples is M51 (also called NGC 5194, the 5194th object in the **New General Catalogue**, which is a list of all the nebulae and star clusters observed by William Herschel, his son John Herschel, and others).

Lacking photographic equipment, Lord Rosse had to make drawings of what he saw. Figure 26-1 shows a drawing he made of M51, and Figure 26-2 shows a modern photograph of it. Views such as this inspired Lord Rosse to echo the famous German philosopher Immanuel Kant, who in 1755 had suggested that these objects might be "island universes"—vast collections of stars far beyond the confines of the Milky Way.

Many astronomers did not subscribe to this notion of island universes. A considerable number of the objects listed in the NGC are in fact nebulae and star clusters scattered throughout the Milky Way, and it seemed likely that these intriguing spiral nebulae could also be components of our Galaxy.

The astronomical community became increasingly divided over the nature of the spiral nebulae. Finally, in April 1920, a debate was held at the National Academy of Sciences in Washington, D.C. On one side was Harlow Shapley, a young, brilliant astronomer renowned for his recent determination of the size of the Milky Way Galaxy. Shapley believed the spiral nebulae to be relatively small, nearby objects scattered around our Galaxy like the globular clusters he had studied. Opposing Shapley was Heber D. Curtis of the Lick Observatory near San Jose, California. Curtis championed the island-universe theory, arguing that each of these spiral nebulae is a rotating system of stars much like our own Galaxy.

The Shapley–Curtis debate generated much heat, but little light. Nothing was decided, because no one could present conclusive evidence to demonstrate exactly how far away the spiral nebulae are. Astronomy desperately needed a definitive determination of the distance to a spiral nebula. Such a measurement became the first great achievement of a young man who abandoned his law practice in Kentucky to move to Chicago and study astronomy. His name was Edwin Hubble.

26-1 Hubble measured the distances to galaxies and devised a system for classifying them

Edwin Hubble joined the staff of the Mount Wilson Observatory in Pasadena, California, and on October 6, 1923, took an historic photograph of the supposed Andromeda

Figure 26-1 Lord Rosse's sketch of M51 Using a large telescope of his own design, Lord Rosse of England was able to distinguish spiral structure in this "spiral nebula." (Courtesy of Lund Humphries)

Figure 26-2 The spiral galaxy M51 (also called NGC 5194) This spiral galaxy in the constellation of Canes Venatici is often called the Whirlpool Galaxy, because of its distinctive appearance. Its distance from Earth is about 15 million light years. The "blob" at the end of one of the spiral arms is the companion galaxy NGC 5195. (Courtesy of D. Elmegreen, B. Elmegreen, and P. Seiden; IBM)

Figure 26-3 *The Andromeda Galaxy (called M31 or NGC 224) This nearby galaxy covers an area of the sky roughly five times as large as the full moon. Under good observing conditions, the galaxy's bright central bulge can be glimpsed with the naked eye in the constellation of Andromeda. The distance to this galaxy is 2¼ million light years. The white rectangle outlines the area shown in Figure 26-4. (Palomar Observatory)*

Figure 26-4 *Cepheid variables in the Andromeda Galaxy Two Cepheid variables are identified (see arrows) in this view showing the outskirts of the Andromeda Galaxy. Because these stars appear so faint, although they are intrinsically luminous, Edwin Hubble successfully demonstrated that the Andromeda "Nebula" is extremely far away. (Mount Wilson and Las Campanas Observatories)*

"Nebula," one of the spiral nebulae around which controversy raged. A modern photograph of this object appears in Figure 26-3. Hubble carefully examined his photographic plate and discovered what he at first thought to be a nova. Referring to previous plates of that region, he soon realized that the object was actually a Cepheid variable star. Further scrutiny of additional plates over the next several months revealed many other Cepheids, two of which are identified in Figure 26-4.

As we have seen, Cepheid variables help astronomers determine distances. From the period–luminosity relation (recall Figure 21-13), astronomers can determine the average absolute magnitude of a Cepheid variable. From both its absolute magnitude M and its apparent magnitude m, a star's distance d in parsecs can be deduced from

$$d = 10^{(m-M+5)/5}$$

(recall Box 19-2).

Cepheid variables are intrinsically bright. They typically have luminosities of a few thousand Suns. Hubble realized that, for these luminous stars to appear as dim as they were on his photographs of the Andromeda "Nebula," they must be extremely far away. Straightforward calculations using modern data about Cepheids demonstrate that M31 is 2¼ million light years from Earth, thus proving it is not a traditional nebula but an enormous stellar system far beyond the confines of the Milky Way. Today, the Andromeda "Nebula" is properly called the Andromeda Galaxy.

Hubble's results, which were presented at a meeting of the American Astronomical Society on December 30, 1924, settled the Shapley–Curtis debate once and for all. The universe was recognized to be far larger and populated with far bigger objects than anyone had thus far seriously imagined. Hubble had discovered the realm of the galaxies.

There are millions of galaxies spread across the sky. They can be seen in every unobscured direction. A typical spiral galaxy contains 100 billion stars and measures about

Sa   **Sb**   **Sc**

Figure 26-5   Various spiral galaxies (face-on)   Edwin Hubble classified spiral galaxies according to the winding of their spiral arms and the *relative size of their central bulges. Three examples are shown here. (Sa courtesy of R. Schild; Sb and Sc courtesy of P. Seiden)*

100,000 light years in diameter. Galaxies are the biggest individual objects in the universe.

Hubble found that galaxies can be classified into four broad categories. These categories form the basis for what is now referred to as the **Hubble classification scheme.** The four types of galaxies are the spirals, barred spirals, ellipticals, and irregulars.

Both M51 and M31, seen in Figures 26-2 and 26-3 respectively, are examples of **spiral galaxies,** characterized by arched lanes of stars and glowing nebulae. While studying spiral galaxies, Hubble noted that this class can be further subdivided, according to the relative size of the central bulge and the winding of the spiral arms (see Figure 26-5). Spirals with tightly wound spiral arms and a prominent, fat central bulge are called **Sa galaxies.** Those with moderately wound

spiral arms and a moderate-sized central bulge, such as M51 and M31, are **Sb galaxies.** And galaxies with loosely wound spirals and a tiny central bulge are **Sc galaxies.**

Fortunately, the size of the central bulge and the degree of winding of the spiral arms go hand in hand, so we can classify the spiral galaxies that are viewed nearly edge-on. For instance, M104 (see Figure 26-6a) must be an Sa galaxy, because of its huge central bulge. An Sb galaxy, such as NGC 4565 in Figure 25-8, has a smaller central bulge. The tiny central bulge of an Sc (see Figure 26-6c) is barely noticeable in an edge-on view.

In **barred spiral galaxies,** the spiral arms originate at the ends of a bar running through the galaxy's nucleus rather than from the nucleus itself (see Figure 26-7). As with ordinary spirals, Hubble subdivided barred spirals according to

a   b   c

Figure 26-6   Various spiral galaxies (edge-on)   (a) Because of the large size of its central bulge, this galaxy is classified as an Sa. If we could see it face on, we would find that its spiral arms are tightly wound around *its voluminous bulge. (b) Note the relatively smaller central bulge in this Sb galaxy. (c) This galaxy is an Sc because of its insignificant central bulge. (Sa from NOAO; Sb and Sc courtesy of R. Schild)*

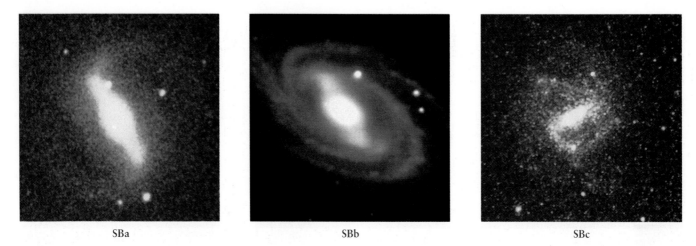

SBa   SBb   SBc

Figure 26-7 **Various barred spiral galaxies** *As with spiral galaxies, Edwin Hubble classified barred spirals according to the winding of their spiral arms and the relative size of their central bulges. Three examples are shown here. (SBa and SBc courtesy of P. Seiden; SBb courtesy of R. Schild)*

the relative size of their central bulge and the winding of their spiral arms. An SBa galaxy has a large central bulge and tightly wound spiral arms. Likewise, a barred spiral with a moderate central bulge and moderately wound spiral arms is an SBb galaxy, and an SBc galaxy has loosely wound spiral arms and a tiny central bulge.

The development of a bar across a galaxy's nucleus is apparently related to how stars in the galaxy's disk are moving. Computer simulations suggest that gravitational interactions between stars in a huge rotating disk can cause these stars occasionally to "pile up" along a barlike structure. A bar is thus a fleeting occurrence, which may explain why ordinary spiral galaxies outnumber barred spirals. Recent computations further suggest that the presence of a massive

extended halo may also inhibit the formation of a bar, so it may be that only those spiral galaxies without significant extended halos can develop bars.

Elliptical galaxies, so named because of their distinctly elliptical shapes, have no spiral arms. Hubble subdivided elliptical galaxies according to how round or flattened they look. The roundest elliptical galaxies he called E0 galaxies; the flattest ones are E7 galaxies. Elliptical galaxies with intermediate amounts of flattening receive intermediate designations (see Figure 26-8).

Of course, an E1 or E2 galaxy might actually be a very flattened disk of stars that just happens to be viewed face-on, and a cigar-shaped E7 galaxy might look spherical if seen end-on. The Hubble scheme classifies galaxies entirely by

E2   E5   E7

Figure 26-8 **Various elliptical galaxies** *Edwin Hubble classified elliptical galaxies according to how round or flattened they look. An E0 galaxy is round. The flattest elliptical galaxies are E7s. Three examples are shown here. (Yerkes Observatory)*

Figure 26-9 A giant elliptical galaxy This huge galaxy sits near the center of a rich cluster of galaxies in the constellation of Coma Berenices. Many normal-sized galaxies surround this giant elliptical one, which is about 2 million light years in diameter. (Courtesy of R. Schild)

Figure 26-10 A dwarf elliptical galaxy This nearby dwarf elliptical, called Leo I, is about 1 million light years from Earth. It is a satellite of the Milky Way and is thus a member of the cluster to which we belong. (Anglo-Australian Observatory)

how they appear from Earth. Statistical studies have shown, however, that elliptical galaxies of various actual degrees of ellipticity do exist.

Elliptical galaxies look far less dramatic than their spiral and barred spiral cousins because elliptical galaxies are virtually devoid of interstellar gas and dust. In addition, there is no evidence of young stars in most elliptical galaxies, and there is no material from which stars could have formed recently. Star formation in elliptical galaxies ended long ago.

Elliptical galaxies exist in a wide range of sizes and masses. Both the biggest and the smallest galaxies in the universe are elliptical. Figure 26-9 shows an example of a **giant elliptical galaxy** that is about 20 times larger than a normal galaxy. This giant elliptical is located near the middle of a large cluster of galaxies in the constellation of Coma Berenices (Berenice's hair).

Giant ellipticals are rather rare, but **dwarf elliptical galaxies** are quite common. Dwarf ellipticals are only a fraction the size of their normal counterparts and contain so few stars—only a few million—that these galaxies are completely transparent. You can actually see straight through the center of a dwarf galaxy and out the other side, as shown in Figure 26-10.

Several years ago, many astronomers suspected that the flattened appearance of some elliptical galaxies (for instance, E5s, E6s, and E7s) might be the result of rotation. Recent spectroscopic observations demonstrate that this idea is wrong. Because of the Doppler effect, the average motions of

stars in a galaxy can be deduced from detailed examination of the galaxy's spectral lines. These studies demonstrate that the motions of stars in elliptical galaxies are quite random. In the roundest elliptical galaxies, this randomness can be **isotropic**, meaning *equal in all directions*. Because the stars are whizzing around equally in all directions, the result is a galaxy that is genuinely spherical. In flattened elliptical galaxies, however, the randomness of the stellar motions is **anisotropic**, which means that the range of star speeds is different in different directions. This anisotropy explains why E5s, E6s, and E7s are fatter in one direction than in another.

Edwin Hubble connected the three types of galaxies—spirals, barred spirals, and ellipticals—in his now-famous tuning-fork diagram (see Figure 26-11). According to this scheme, the S0 and SB0 galaxies, also called **lenticular galaxies**, are midway in appearance between ellipticals and the two kinds of spirals. Although they look somewhat like ellipticals, lenticular galaxies have both a central bulge and a disk like spiral galaxies.

The tightness of a galaxy's spiral windings is related to the dust and gas content of the galaxy. Sa and SBa galaxies tend to have lower abundances of gas and dust than do Sc and SBc galaxies. As stars form, galaxies gradually exhaust their gas and dust content. Although supernovae and planetary nebulae return some of this material to the interstellar medium, as a galaxy ages, more and more of its gas and dust become locked inside stars. To some astronomers, these

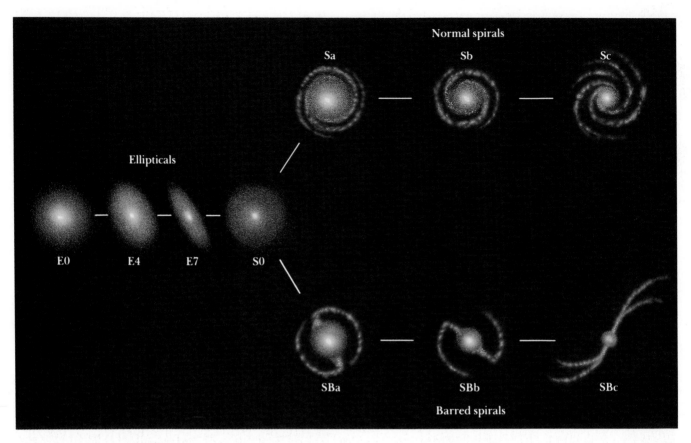

Elstandsals

Normal spirals

Sa　Sb　Sc

E0　E4　E7　S0

SBa　SBb　SBc

Barred spirals

Figure 26-11 Hubble's tuning-fork diagram Hubble summarized his classification scheme for regular galaxies with this tuning-fork diagram. An S0 galaxy is midway in appearance type between ellipticals and spi- *rals. Astronomers now realize that this diagram has no profound physical or evolutionary significance.*

circumstances suggest an evolutionary trend. S0 and SB0 galaxies may be former spirals and barred spirals that have lost most of their interstellar material, either by having converted it into stars or by having had it swept away in a collision with another galaxy.

Galaxies that do not fit into Hubble's scheme of spirals, barred spirals, and ellipticals are usually referred to as **irregular galaxies**. Astronomers today recognize two types of irregulars. **Irr I galaxies**, which usually resemble underdeveloped spiral galaxies, contain many OB associations and H II regions. **Irr II galaxies** have asymmetrical, distorted shapes that seem to have been caused by collisions with other galaxies or by violent activity in their nuclei.

The best-known examples of Irr I galaxies are the Large Magellanic Cloud (LMC) and the Small Magellanic Cloud (SMC), which are nearby companions of our Milky Way that can be seen with the naked eye from southern latitudes. Telescopic views of the Magellanic clouds can easily resolve individual stars (see Figures 26-12 and 26-13). The SMC does not exhibit any of the geometric symmetry characteristic of spirals or ellipticals, but the LMC does have a faint barlike structure. The Magellanic clouds are the nearest members of a cluster of galaxies to which the Milky Way and Andromeda galaxies belong.

26-2 Galaxies are grouped in clusters that are members of superclusters

Galaxies are not scattered randomly throughout the universe but are grouped in **clusters**. A typical cluster, called the Fornax cluster because it is located in the constellation of Fornax (the furnace), is seen in Figure 26-14.

A cluster is said to be either **poor** or **rich**, depending on how many galaxies it contains. For example, the Milky Way Galaxy, the Andromeda Galaxy, and the Large and Small Magellanic clouds belong to a poor cluster familiarly known as the **Local Group**. The Local Group contains nearly two dozen galaxies, as listed in Box 26-1, most of which are dwarf ellipticals.

The nearest fairly rich cluster is the Virgo cluster. It is a sprawling collection of over 1000 galaxies covering a 10° × 12° area of the sky. The distance to the Virgo cluster is about 50 million light years, too far away for Cepheid variables to be seen from our Galaxy. Instead, the distance to the Virgo cluster has been determined by the apparent faintness of O and B supergiant stars and by the brightnesses of globular clusters surrounding some of the galaxies. The overall diameter of the Virgo cluster is about 7 million light years.

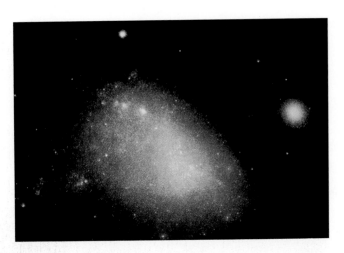

Figure 26-12 The Large Magellanic Cloud (LMC) *At a distance of only 160,000 light years, this galaxy is the nearest companion of our Milky Way Galaxy. Note the huge H II region (called the Tarantula Nebula or 30 Doradus) toward the left side of the photograph. Its diameter of 800 light years and mass of 5 million Suns make it the largest known H II region. (Anglo-Australian Observatory)*

Figure 26-13 The Small Magellanic Cloud (SMC) *The SMC is only slightly farther away from us than the LMC is. Because of its sprawling, asymmetrical shape, the SMC is classified as an irregular galaxy. Note that the SMC is rich in young blue stars. (Anglo-Australian Observatory)*

Figure 26-14 A cluster of galaxies *This cluster of galaxies, called the Fornax cluster, is about 60 million light years from Earth. Both elliptical and spiral galaxies are easily identified. The barred spiral galaxy at the lower right is NGC 1365, the largest and most impressive member of the cluster. (Anglo-Australian Observatory)*

Box 26-1   The Local Group

The Andromeda Galaxy, M31, is the biggest and most massive galaxy in the Local Group. Our Milky Way Galaxy takes second place. Scattered around these two primary galaxies are 20 smaller ones, which are listed in the table opposite (the letter *d* preceding the Hubble type indicates a dwarf galaxy). The nine galaxies closest to us are satellites of the Milky Way Galaxy. Similarly, the Andromeda Galaxy has eight satellites. The diagram shows the galaxies of the Local Group. Additional dwarf galaxies are occasionally discovered.

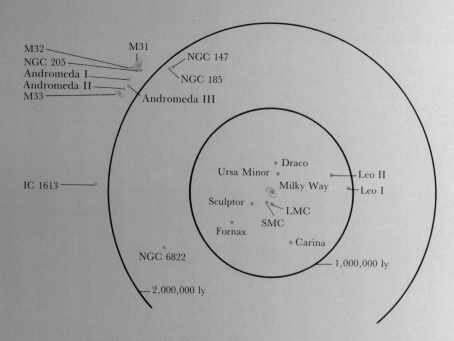

The center of the Virgo cluster is dominated by three giant elliptical galaxies. Two of them appear in Figure 26-15. These enormous galaxies are 2 million light years in diameter, 20 times as large as an ordinary elliptical or spiral. In other words, one giant elliptical is roughly the same size as the entire Local Group!

In addition to using the rich-versus-poor classification, astronomers often categorize clusters of galaxies as **regular** or **irregular**, depending on the overall shape of the cluster. The Virgo cluster, for example, is called irregular, because of

Figure 26-15   The center of the Virgo cluster The Virgo cluster is a rich, sprawling collection of galaxies about 50 million light years from Earth. Only the center of this huge cluster appears in this photograph. Note the two giant elliptical galaxies (M84 and M86). (Royal Observatory, Edinburgh)

Galaxy	Type	Distance (10^3 ly)	Diameter (10^3 ly)	Luminosity (10^6 L$_\odot$)	Mass (10^6 M$_\odot$)
Milky Way	Sb	—	80	20,000	200,000
LMC	Irr	170	30	3,000	25,000
SMC	Irr	190	25	600	6,000
Draco	dE2	220	3	1	0.1
Ursa Minor	dE4	220	3	0.4	0.1
Sculptor	dE3	270	8	2	3
Carina	dE3	500	4	?	?
Fornax	dE3	650	20	23	20
Leo I	dE4	700	6	3	4
Leo II	dE0	700	4	1	1
NGC 6822	Irr	1630	10	100	1,000
NGC 147	E6	2000	10	70	350
NGC 185	E2	2000	8	90	450
IC 1613	Irr	2100	15	65	250
NGC 205	E5	2250	16	330	3,000
NGC 221 (M32)	E3	2250	8	250	2,000
Andromeda (NGC 224 = M31)	Sb	2250	130	25,000	300,000
Andromeda I	dE0	2250	2	2	2
Andromeda II	dE0	2250	2	2	2
Andromeda III	dE0	2250	2	2	2
NGC 598 (M33)	Sc	2350	60	4,000	40,000
Maffei I	S0	3300	100	10,000	200,000

the asymmetrical way its galaxies are scattered throughout a sprawling region of the sky. Our own Local Group is also an irregular cluster. In contrast, a regular cluster has a distinctly spherical appearance, with a marked concentration of galaxies at its center. This distribution is what would be expected for a system of objects whose mutual gravitational interactions over the ages have equitably distributed the system's energy among its members.

The nearest example of a rich, regular cluster is the Coma cluster, located about 300 million light years from us toward the constellation of Coma Berenices. Despite the great distance to this cluster, more than 1000 bright galaxies within it are easily visible on photographic plates. Certainly, there must be many thousands of dwarf ellipticals that are too faint to be detected from our distance. The total membership of the Coma cluster may then be as many as 10,000 galaxies. The core of the Coma cluster is dominated by two giant ellipticals surrounded by many normal-sized galaxies (see Figure 26-16).

A correlation exists between the regular-versus-irregular classification scheme and the dominant types of galaxies in a cluster. Rich, regular clusters, like the Coma cluster, contain mostly elliptical and S0 galaxies. Only 15 percent of the Coma cluster's galaxies are spirals and irregulars. Irregular clusters, such as the Virgo cluster and the Hercules cluster (see Figure 26-17), have a more even mixture of galaxy types. For instance, of the 200 brightest galaxies in the Hercules cluster, 68 percent are spirals, 19 percent are ellipticals, and the rest are irregulars. We shall see later that the differences between regular and irregular clusters may reflect different rates of collisions and close encounters between their member galaxies.

Clusters of galaxies are themselves grouped together, in huge associations called **superclusters**. A typical supercluster

Figure 26-16 The center of the Coma cluster This rich, regular cluster is about 300 million light years from Earth. Only the cluster's center, dominated by two giant galaxies, appears in this view. Regular clusters are composed mostly of elliptical and S0 galaxies and are common sources of X rays. (Courtesy of R. Schild)

Figure 26-17 The Hercules cluster This irregular cluster, which is about 700 million light years from Earth, is much less dense than the Coma cluster. The Hercules cluster contains a high proportion of spiral galaxies, often associated in pairs and small groups. (Courtesy of R. Schild)

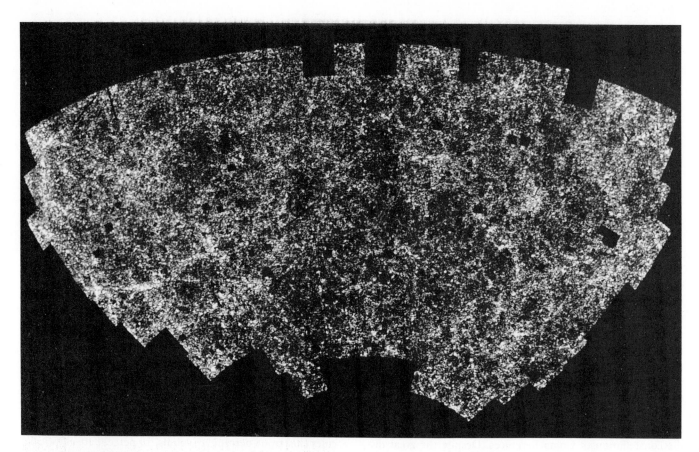

Figure 26-18 Two million galaxies This map shows the distribution of roughly two million galaxies over about 10 percent of the sky. Each dot on the map is shaded according to the number of galaxies it contains. Dots are black where there are no galaxies, white where there are more than 20, and grey for a number between 1 and 19. Note that the clusters are not smoothly distributed across the sky but rather form a lacy, filamentary structure. The small, bright patches are individual gal-axy clusters. The larger, elongated bright areas are superclusters and filaments, which generally surround darker voids containing few galax-ies. Statistical analysis of this map shows that galaxies clump together on distance scales up to about 150 million light years. Over distances greater than this, the distribution of galaxies in the universe appears to be roughly uniform. (S. J. Maddox, W. J. Sutherland, G. P. Efstathiou, and J. Loveday; Oxford Astrophysics)

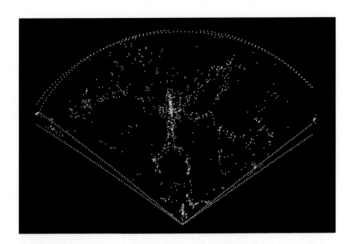

Figure 26-19 *A slice of the universe* *This map shows the locations of 1099 galaxies in a thin slice of the universe. To produce this map, a team of astronomers painstakingly measured the positions of galaxies out to a distance of nearly a billion light years from Earth in a strip of the sky 6° wide and 153° long. Note that most galaxies surround voids that are roughly circular. This distribution of galaxies is like a slice cut through suds in a bubble bath. (Courtesy of V. de Lapparent, M. Geller, and J. Huchra)*

contains dozens of individual clusters spread over a region of space up to 100 million light years across. Patterns in the distribution of galaxies can be seen in maps such as Figure 26-18, which covers many hundreds of square degrees. Note the delicate, filamentary structure spread across the sky.

The arrangement of superclusters in space was clarified in the early 1980s when astronomers began discovering enormous **voids** where exceptionally few galaxies are found. These voids are roughly spherical and measure 100 million to 400 million light years in diameter. Recent surveys reveal that galaxies are concentrated on the surfaces of these voids (see Figure 26-19). The distribution of clusters of galaxies is therefore said to be "sudsy" because it resembles a collection of giant soap bubbles. Galaxies surround voids in the same way a soap film is concentrated on the surface of bubbles. As we shall see in Chapter 29, many astronomers suspect that this sudsy pattern contains important clues about conditions shortly after the Big Bang that led to the formation of clusters of galaxies.

26-3 Most of the matter in the universe has yet to be discovered

A cluster of galaxies must be a gravitationally bound system. In other words, there must be matter enough in the cluster to produce gravity sufficient to prevent the galaxies from wandering away. Nevertheless, careful examination of a rich cluster, like the Coma cluster, typically reveals that the mass of the visually luminous matter is not in fact sufficient to bind the cluster gravitationally. The observed line-of-sight speeds of the cluster galaxies, measured by Doppler shifts, are so large that more mass than has been observed is needed to keep them bound in orbit about the center of the cluster. This dilemma is called the **dark-matter problem.** A lot of nonluminous matter must be contained within each of the clusters, or else the galaxies would long ago have dispersed in random directions and the cluster would no longer exist today. Analyses demonstrate that the total mass needed to bind a typical rich cluster is 10 times greater than the mass of material that shows up on visual photographs.

Some of this mystery has been solved recently by X-ray astronomers. Satellite observations of rich clusters have revealed that X rays pour from the space between galaxies in rich clusters (see Figure 26-20). This flow is evidence of substantial amounts of hot intergalactic gas at temperatures between 10 and 100 million kelvins. Analyses show that the mass of this hot intergalactic gas is typically as great as the combined mass of all the visible galaxies in the cluster.

Unfortunately, this discovery of hot intergalactic gas in rich clusters solves only a small part of the dark-matter problem. Most astronomers agree that a great deal of matter still remains to be discovered in rich clusters. One popular speculation is that these clusters may contain a lot of undetected dim stars. These faint stars could be located in extended halos surrounding individual galaxies and be scattered throughout the spaces between the galaxies of a cluster.

Evidence to support the idea of massive extended halos comes from the rotation curves of galaxies. As mentioned in Chapter 25, many galaxies have rotation curves similar to

Figure 26-20 *An X-ray view of a cluster of galaxies* *This view from the Einstein Observatory shows the rich cluster A1367 (that is, the 1367th cluster in a list of rich clusters catalogued by UCLA astronomer George O. Abell). The X-ray emission shown here comes from hot gas between the galaxies. (Harvard–Smithsonian Center for Astrophysics)*

that of our Milky Way Galaxy (recall Figure 25-15). These rotation curves remain remarkably flat out to surprisingly great distances from the galaxy's center. For example, Figure 26-21 shows the rotation curves of four spiral galaxies. In all cases, the orbital speed remains roughly constant out to a distance beyond which the galaxies' stars and H II regions are so dim and widely scattered that reliable measurements are not possible. Nevertheless, we have still not detected the true edges of these and many similar galaxies. In the outer portions of a galaxy we should see a decline in orbital speed, in accordance with Kepler's third law. Because this decline has not been observed, astronomers conclude that there must be a considerable amount of dark material extending well beyond the visible portion of a galaxy's disk. Many proposals have been made to account for this hidden matter. Some of the exotic suggestions made so far include black holes, WIMPs, and different types of massive particles predicted by various speculative theories. The more mundane suggestions include dim stars and Jupiter-like planets. The nature of this unseen matter is one of the greatest mysteries in modern astronomy.

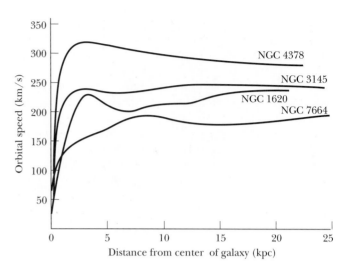

Figure 26-21 *The rotation curves of four spiral galaxies* This graph shows the orbital speed of material in the disks of four spiral galaxies. Many galaxies have flat rotation curves, indicating the presence there of extended halos of low-luminosity material. (Adapted from V. Rubin and K. Ford)

26-4 Colliding galaxies produce starbursts, spiral arms, and other spectacular phenomena

The galaxies in a cluster are all in orbit about their common center of mass, and occasionally two galaxies pass close enough to each other to collide. When they do collide, their stars themselves pass by each other. There is so much space between the stars that the probability of two stars crashing into each other is extremely small. However, the galaxies' huge clouds of interstellar gas and dust cannot interpenetrate, and so slam into each other, producing strong shock waves. The stars keep right on going, but the colliding interstellar clouds are stopped in their tracks. In this way, two colliding galaxies may be stripped of their interstellar gas and dust. The violence of the collision between the interstel-

a

b

Figure 26-22 *A starburst galaxy* Prolific star formation is occurring at the center of this galaxy, called M82 or NGC 3034, located about 10 million light years from Earth. This activity was probably triggered by the collision of M82 with a huge gas cloud produced by tidal interaction with a neighboring galaxy. (a) This wide-field view shows the extent of M82. (b) This CCD image shows details of the turbulent interstellar gas and dust heated by the numerous young stars around the galaxy's center. (Lick Observatory; CCD image courtesy of R. Schild)

lar clouds heats the gas stripped from these galaxies to extremely high temperatures. This process may be a major source of the hot intergalactic gas often observed in rich, regular clusters.

In a less-violent collision or a near miss between two galaxies, the compressed interstellar gas may have enough time to cool sufficiently to allow many protostars to form. Such collisions can thus stimulate prolific star formation, which may account for the **starburst galaxies** that blaze with the light of numerous newborn stars. These galaxies are characterized by bright centers surrounded by clouds of warm interstellar dust, indicating a recent, vigorous episode of star birth (see Figure 26-22). The warm dust is so abundant that starburst galaxies are among the most luminous objects in the universe at infrared wavelengths.

Gravitational interaction between colliding galaxies can hurl thousands of stars out into intergalactic space, along huge, arcing streams. These shapes are dramatically illustrated in supercomputer simulations such as those by Joshua Barnes at the Canadian Institute for Theoretical Astrophysics. One of his simulations is shown in Figure 26-23, which displays the collision and eventual merging of two galaxies. The simulation begins with the two galaxies approaching each other along parabolic orbits. The first passage (at a simulated time of 250 million years) of the galaxies through each other generates strong tidal forces that severely distort the galaxies, throwing out a pair of extended tails. Both galaxies develop bars across their nuclei as a result of this interaction. Another result is orbital decay; instead of traveling away from each other on their original parabolic trajecto-

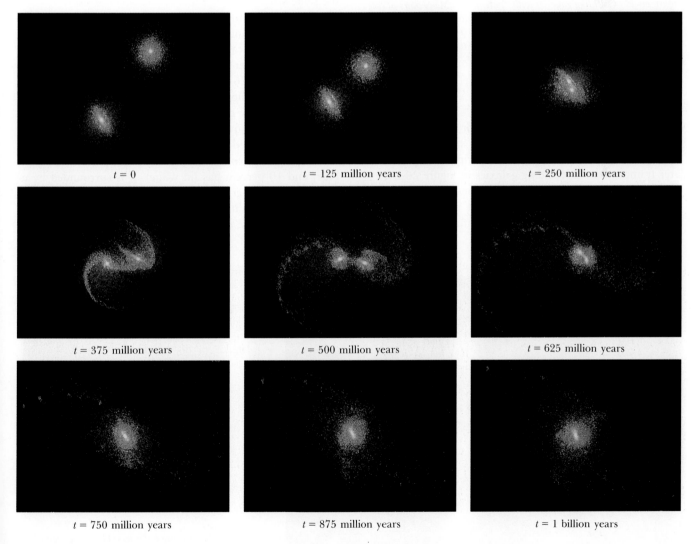

$t = 0$ $t = 125$ million years $t = 250$ million years

$t = 375$ million years $t = 500$ million years $t = 625$ million years

$t = 750$ million years $t = 875$ million years $t = 1$ billion years

Figure 26-23 A simulated collision between two galaxies *These frames from a supercomputer simulation show the collision and merging of two disk-shaped galaxies. Stars in the disk of each galaxy are colored blue, while stars in their central bulges are yellow. Red indicates dark matter that surrounds each galaxy. The pictures display progress at 125-million-year intervals. Compare the final frames with the photograph of NGC 2623 in Figure 26-24. (Courtesy of J. Barnes; CITA)*

Figure 26-24 Colliding galaxies with "antennae" *Many pairs of colliding galaxies exhibit long "antennae" of stars ejected by the collision. This particular system, called NGC 2623, is also a significant source of radio emission. Supercomputer simulations, like the one shown in Figure 26-23, give important insights into possible histories of such systems. (Palomar Observatory)*

ries, the two galaxies fall back together for a second encounter (at 625 million years). They merge shortly thereafter, leaving a single object. Note the similarity of some of the later frames with the colliding galaxies in Figure 26-24.

As this simulation illustrates, while some of the stars are flung far and wide, scattering material into intergalactic space, other stars lose energy and momentum. As these stars

slow down, the galaxies may merge. Several dramatic examples of such **galaxy mergers** have been discovered recently (see Figure 26-25).

In a rich cluster there must be many near-misses between galaxies. If galaxies are surrounded by extended halos of dim stars, these near-misses could strip the galaxies of their outlying stars. In this way, a loosely dispersed sea of dim stars might come to populate the space between galaxies in a cluster. Searching for these dim stars in extended halos and intergalactic space is one of the projects for the Hubble Space Telescope in the 1990s.

When two galaxies merge, the result is a bigger galaxy. If this new galaxy is located in a rich cluster, it may capture and devour additional galaxies and grow to enormous dimensions. This phenomenon is called **galactic cannibalism**. Cannibalism differs from mergers in that the dining galaxy is bigger than its dinner, whereas merging galaxies are about the same size.

Many astronomers suspect that galactic cannibalism is the reason that giant ellipticals are so huge. As we have seen, giant galaxies typically occupy the centers of rich clusters. In many cases, smaller galaxies are located around these giants (examine Figure 26-9). As they pass through the extended halo of a giant elliptical, these smaller galaxies slow down and are eventually devoured by the larger galaxy.

Figure 26-26 shows a supercomputer simulation in which a large, disk-shaped galaxy devours a small satellite galaxy. The large galaxy consists, by mass, of 90 percent stars (in blue) and 10 percent gas (in white). It is surrounded by a halo of dark matter having a mass about 3.3 times that of the disk. The satellite galaxy, which has a tenth of the mass of the large galaxy, contains only stars (in orange). Initially,

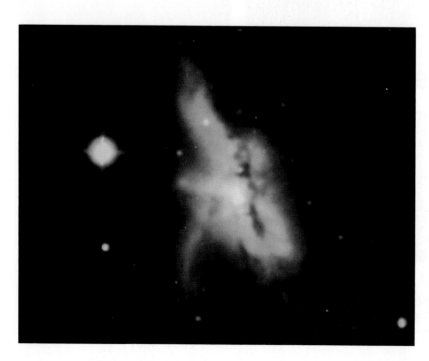

Figure 26-25 Merging galaxies *This contorted object in the constellation of Ophiuchus consists of two spiral galaxies in the process of merging. The collision between the two galaxies has triggered an immense burst of star formation. (Courtesy of W. C. Keel)*

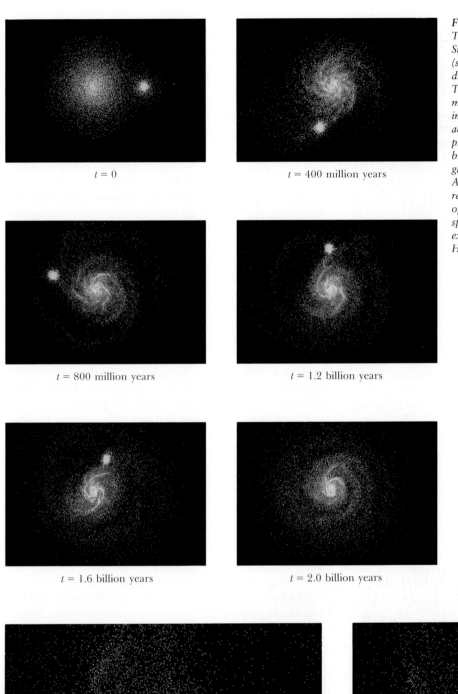

t = 0

t = 400 million years

t = 800 million years

t = 1.2 billion years

t = 1.6 billion years

t = 2.0 billion years

Figure 26-26 A simulated galactic cannibalism *This simulation, performed at the Pittsburgh Supercomputing Center, shows a small galaxy (stars in orange) being devoured by a larger, disk-shaped galaxy (stars in blue, gas in white). The upper six pictures display progress at 400-million year intervals. Note how spiral arms are induced in the disk galaxy as a result of its interaction with the satellite galaxy. The lower two pictures, which cover the interval from 1.8 to 1.9 billion years, are closeup views of the satellite galaxy plunging into the core of the disk galaxy. As the satellite galaxy sweeps through the inner regions of the disk galaxy, a significant amount of gas becomes concentrated along one of the spiral arms. Vigorous star formation would be expected in these gas clouds. (Courtesy of L. Hernquist)*

t = 1.8 billion years

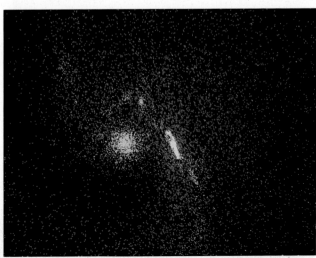

t = 1.9 billion years

the satellite is in circular orbit about the large galaxy. Spiral arms appear in the large galaxy as the collision proceeds. Two billion years elapse as the satellite spirals in toward the core of the large galaxy. Although much material is striped from the satellite, most of its stars plunge into the nucleus of the large galaxy.

Close encounters between galaxies provide a third way of forming spiral arms, in addition to density waves and self-propagating star formation discussed in the previous chapter. Supercomputer simulations clearly demonstrate that spiral arms can be created during a collision, either by drawing out long streamers of stars or by compressing clouds of interstellar gas. For instance, the spiral arms of M51 (examine Figure 26-2) may have been produced by a close encounter with a second galaxy. The disruptive galaxy is now located at the end of one of the spiral arms created by the collision. Some astronomers argue that the spiral arms of our Milky Way Galaxy were similarly produced by a close encounter with the Large Magellanic Cloud.

As we shall see in the next chapter, some astronomers suspect very massive black holes may be located at the centers of certain peculiar galaxies that are unusually powerful sources of energy. A cannibalistic collision of the type shown in Figure 26-26 would "feed" the black hole at the core of such a galaxy. Astrophysicists calculate that the in-falling matter would release enormous amounts of energy, thereby explaining the galaxy's power output. Supercomputer simulations are becoming an increasingly important technique for exploring these exotic processes.

26-5 Galaxies were formed billions of years ago from the gravitational contraction of huge clouds of primordial gas

Astronomers can probe the past to gain important clues about galactic evolution simply by looking deep into space. The more distant a galaxy is, the longer its light takes to reach us. Consequently, as we examine galaxies that are at increasing distances from Earth, we are actually looking farther and farther back in time, seeing galaxies at increasingly earlier stages of their lives.

By observing remote galaxies, astronomers have discovered that galaxies were bluer and brighter in the past than they are today. These changes in color and brightness suggest that a newly formed galaxy has an abundance of young, bright, hot, massive stars. As the galaxy ages, these short-lived O and B stars become red giants and eventually die off. The galaxy therefore gradually becomes somewhat redder and dimmer.

To make use of the color and luminosity changes exhibited by galaxies, astronomers construct theoretical models of galaxies, which they then try to match to their observations. For example, Figure 26-27a shows a Hertzsprung–Russell diagram of a hypothetical young galaxy whose stars are all on the main sequence. The H–R diagram is divided into cells, each covering a small range in luminosity and surface temperature. Each cell contains a number of stars, based on the percentage of stars that would be expected within the range covered by the cell. Although there are few O and B stars, light from these hot, bright stars dominates the galaxy's colors (see the simulated spectrum below the H–R diagram), making the galaxy appear bluish. As the galaxy ages, the O and B stars are the first to evolve away from the main sequence, causing the galaxy's color and luminosity to change. Figure 26-27b shows the same hypothetical galaxy 10 billion years later. Note that all the O and B stars are now gone, and so the galaxy has become dimmer and redder. Matching the colors of a theoretical model of an evolving galaxy to those of a real galaxy allows the age of the real one to be estimated.

From age-dating studies for various types of galaxies, astronomers now know that ellipticals formed nearly all their stars in one vigorous burst of activity that lasted for only about a billion years. In contrast, spiral galaxies have been forming stars continuously over the past 15 billion years, although at a gradually decreasing rate. Figure 26-28 compares the rates at which spiral and elliptical galaxies form stars.

The difference in star-formation rates between ellipticals and spirals may be the result of the close encounters and collisions between galaxies in clusters. Recall that ellipticals are especially abundant in rich, regular clusters whose spherical shapes are evidence of numerous close encounters that made random the velocities of galaxies in all directions. Billions of years ago, these encounters stimulated an initial burst of star formation in the gas-rich infant galaxies then populating the cluster. This era of prolific star birth was soon terminated, however, as continued encounters and collisions swept away the remaining interstellar gas, leaving gas-free, dust-free ellipticals. In irregular clusters, where spiral galaxies predominate, there were fewer collisions in the past, which allowed time for the gas in young galaxies to settle into a disk where star formation continues even today at a leisurely pace.

The most obvious feature of a spiral galaxy is a rotating disk of gas and dust in which new stars are constantly being formed. Of course, there are many old stars in the disk that are left over from earlier generations of stellar birth, but the disk's light is dominated by radiation from the young, hot, massive stars that outline the spiral arms. Surrounding the galaxy and concentric with its nucleus are a central bulge and a halo consisting of old stars, such as red giants and red supergiants, whose light dominates these outer regions. The differences between the disk and the surrounding spherical portion of a spiral galaxy are important clues that guide astronomers in formulating a theory of galactic evolution.

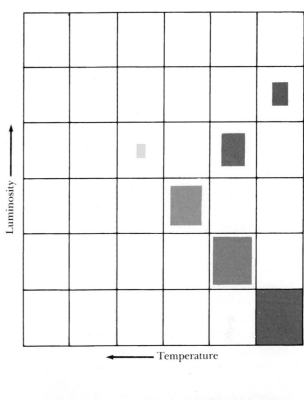

a   Spectrum initially

b   Spectrum after 10 billion years

Figure 26-27 An evolutionary model of a galaxy *These schematic H–R diagrams and the associated spectra show changes that occur in the colors of a galaxy as a result of the evolution of its stellar content. (a) At its birth, the galaxy is simulated by a collection of stars on the main sequence. The number of stars in each spectral range (indicated by the size of the colored rectangles) reflects the fact that low-mass stars are much more abundant than high-mass stars. (b) After 10 billion years, all the stars more massive than the Sun have evolved away from the main sequence. In the absence of these hot, blue stars, the galaxy has become dimmer and redder.*

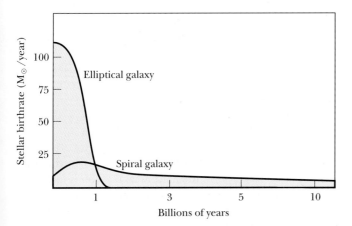

Figure 26-28 The stellar birthrate in galaxies *Most of the stars in an elliptical galaxy were created in a brief burst of star formation when the galaxy was very young. In spiral galaxies, star formation occurs at a more leisurely pace that may extend over billions of years. (Adapted from J. Silk)*

It seems reasonable to suppose that galaxies formed from huge clouds of primordial hydrogen and helium roughly 15 billion years ago. This birth date for both spirals and ellipticals is deduced from the ages of the most ancient, metal-poor stars that both types contain. Under the action of gravity, these pregalactic clouds of gas started to contract and form protogalaxies studded with the first generation of stars.

The rate of star formation in a protogalaxy determines whether it becomes a spiral or an elliptical. If the star-formation rate is low, then the gas has plenty of time to settle in a flattened disk. A flattened disk is the natural consequence of the overall angular momentum possessed by the original gas cloud. Star formation continues in the disk because it contains an abundance of hydrogen, and thus a spiral galaxy is created. But if the stellar birthrate is high, then virtually all of the gas is used up in the formation of stars before a disk has time to form. In this case, an elliptical galaxy is created. These contrasting scenarios are depicted in Figure 26-29.

Figure 26-29 *The formation of spiral and elliptical galaxies* A galaxy begins as a huge cloud of primordial gas that contracts under the influence of gravity. (a) If the rate of star birth is low, then much of the gas settles into a disk and a spiral galaxy is created. (b) If the rate of star birth is high, then the gas is converted into stars before a disk can form, resulting in an elliptical galaxy.

Galactic evolution is a difficult and controversial subject in which many questions remain. For instance, our discussion of the birth of a galaxy began with a pregalactic cloud of the right mass and size to guarantee that it would contract to become a galaxy. But where did this cloud come from? And what happened in the early universe to cause the primordial hydrogen and helium to clump up in clouds destined to evolve into galaxies, instead of becoming objects a million times bigger or smaller? Even more troublesome is the issue of the dark matter. We know that the observable stars, gas, and dust in a galaxy constitute only 10 percent of the mass associated with the galaxy. We have no idea what the remaining 90 percent looks like, how it's distributed in space, or what it's made of. With this level of ignorance, some astronomers argue that any discussion of the evolution of just the visible 10 percent that comprises a galaxy is woefully inadequate.

26-6 The Hubble law is a simple relationship between the redshifts of galaxies and their distances from Earth

Whenever an astronomer finds an object in the sky that can be seen or photographed, the natural inclination is to attach a spectrograph to a telescope and record the spectrum. As long ago as 1914, V. M. Slipher, working at the Lowell Observatory in Arizona, began taking spectra of "spiral nebulae." He was surprised to discover that, of the 15 spiral nebulae he studied, 11 had their spectral lines shifted toward the red end of the spectrum, indicating substantial recessional velocities for those objects. This marked dominance

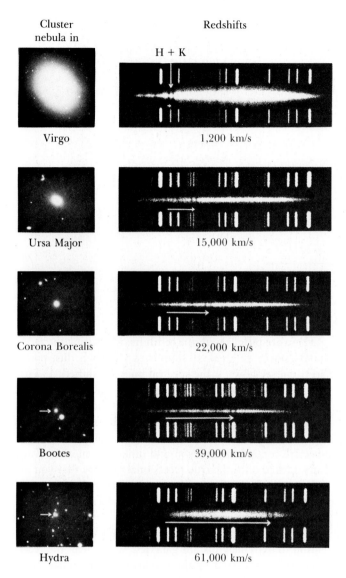

Cluster nebula in	Redshifts
Virgo	1,200 km/s
Ursa Major	15,000 km/s
Corona Borealis	22,000 km/s
Bootes	39,000 km/s
Hydra	61,000 km/s

Figure 26-30 *Five galaxies and their spectra The photographs of these five elliptical galaxies all have the same magnification. They are arranged, from top to bottom, in order of increasing distance from Earth. The spectrum of each galaxy is the hazy band between the comparison spectra. In all five cases, the so-called H and K lines of singly ionized calcium can be seen. The recessional velocity, calculated from the Doppler shifts of the H and K lines, is given below each spectrum. Note that the more distant a galaxy is, the greater is its redshift. (Mount Wilson and Las Campanas observatories)*

from us slowly, and more distant galaxies are rushing away from us much more rapidly. This universal recessional movement is sometimes referred to as **Hubble flow**.

Using various techniques, Hubble estimated the distances to a number of galaxies. With the Doppler formula (review Figure 5-20), Hubble calculated the speed with which each galaxy is receding from us. When he plotted the data on a graph of distance versus speed, he found that the points lie near a straight line. Figure 26-31 is a modern version of Hubble's graph based on recent data.

This relationship between the distances to galaxies and their redshifts is one of the most important astronomical discoveries of the twentieth century. As we shall see in Chapter 28, this relationship tells us that we are living in an expanding universe. In 1929 Hubble published this discovery, which is now known as the **Hubble law**.

The Hubble law is most easily stated as a formula:

$$v = H_0 r$$

where v is the recessional velocity, r is the distance, and H_0 is a constant commonly called the **Hubble constant**. This formula is equivalent to the straight line displayed in Figure 26-31. The Hubble constant H_0 is the slope of the line in Figure 26-31. From the data plotted on this graph we find that

$$H_0 = 15 \text{ km s}^{-1} \text{ Mly}^{-1}$$

where Mly stands for 10^6 light years (say "fifteen kilometers per second per million light years"). In other words, for each million light years to a galaxy, the galaxy's speed away from us increases by 15 km/s. For example, a galaxy located 100 million light years from Earth should be rushing away from us with a speed of 1500 km/s.

Incidentally, most astronomers prefer to speak of millions of parsecs, termed megaparsecs (Mpc), rather than millions of light years (Mly). Using that unit,

$$H_0 = 50 \text{ km s}^{-1} \text{ Mpc}^{-1}$$

(say "fifty kilometers per second per megaparsec"). In addition, some people prefer to write the units of the Hubble constant with slashes rather than exponents:

$$H_0 = 15 \text{ km/s/Mly} = 50 \text{ km/s/Mpc}$$

The exact value of the Hubble constant is a topic of heated debate among astronomers today. The data plotted in Figure 26-31, as well as the values of H_0 given above, are all from recent work by Allan Sandage and Gustav Tammann, who did most of their observations with the 200-in. Palomar telescope. However, other prominent astronomers, such as Sidney van den Bergh in Canada, Gerard de Vaucouleurs of the University of Texas, and John Huchra at Harvard, argue strongly that the Sandage–Tammann value for H_0 is too low. Independent, meticulous observations by these three astronomers and their colleagues have led them

of redshifts was presented by Curtis in the Shapley–Curtis debate as evidence that these spiral nebulae could not be ordinary nebulae in our Milky Way Galaxy.

During the 1920s, Edwin Hubble and Milton Humason photographed the spectra of many galaxies with the 100-in. telescope on Mount Wilson. Five representative elliptical galaxies and their spectra are shown in Figure 26-30. As indicated by this illustration, there seems to be a direct correlation between the distance to a galaxy and the size of its redshift. In other words, nearby galaxies are moving away

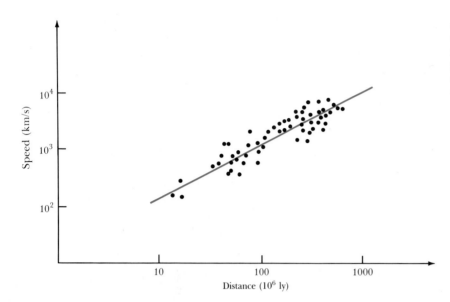

Figure 26-31 The Hubble law The distances and recessional velocities of 60 Sc spiral galaxies are plotted on this graph. The straight line is the "best fit" for the data. This linear relationship between distance and speed is called the Hubble law. (Adapted from Sandage and Tammann)

to infer that the true value for H_0 is about 100 km/s/Mpc (= 30 km/s/Mly). Many astronomers simply use a number between the two extremes, such as 75 km/s/Mpc.

The Hubble constant is one of the most important numbers in all astronomy. It expresses the rate at which the universe is expanding and thus tells us the age of the universe. Furthermore, the Hubble law can be used to determine the distances to extremely remote galaxies and quasars. If the redshift of a galaxy is known, the Hubble law can be used to convert that redshift into a distance from Earth, as described in Box 26-2.

Since the value of the Hubble constant is controversial, many astronomers prefer not to take the final step of converting redshifts to distances. The Hubble law tells us that the redshift of a remote galaxy is proportional to its distance from Earth, and so you can think of redshift as distance. This is the approach of several teams of astronomers who, in recent years, have undertaken monumental projects to map the three-dimensional distribution of galaxies in space. On the resulting maps, distances from Earth are expressed in km/s (the velocity inferred from the redshifts of the galaxies). A fine example of such work, which reveals large-scale structure in the universe, is described in Box 26-3.

Teams of astronomers are making concerted efforts to determine an accurate value for the Hubble constant, which is a difficult task. To appreciate the difficulties, think about what you would have to do to measure H_0. First, you would have to observe many galaxies. For each galaxy, you would have to determine its recessional velocity from the redshift of its spectral lines. By some means, you would also have to determine the distance to each galaxy. Finally, you would need to plot all the data on a graph like the one in Figure 26-31. The slope of the line that best passes through the data points gives you H_0. Unfortunately, you would encounter numerous pitfalls along the way.

There are pitfalls with the redshifts. Measuring the recessional velocity of a galaxy sounds easy enough: Just take a spectrum, see how far the spectral lines are shifted, and use the Doppler formula to get the speed. An example is worked out in Box 26-2. Galaxies come in clusters, however, with all the galaxies in a cluster being in orbit about their common center of mass. Because of these orbital motions, some galaxies are coming toward you and others are moving away. Thus, tacked onto the true recessional velocity that you are trying to determine is an additional velocity (either toward you or away from you) produced by the galaxy's own orbital motion within its cluster.

One way of coping with this problem is to turn your attention to extremely distant galaxies. With them, the recessional velocity caused by the expansion of the universe far exceeds the galaxies' orbital motions within their clusters. Your redshift measurements of remote galaxies will thus more accurately reflect the Hubble flow. Astronomers trying to determine H_0 do commonly restrict their observations to galaxies farther than 100 Mly from Earth. Unfortunately, at this distance a new pitfall arises: You can no longer be exactly sure how far it is to these remote galaxies.

The distances to nearby galaxies can be determined by fairly reliable methods. For example, Cepheid variables can be seen out to 20 Mly from Earth. The distances to galaxies in this nearby volume of space can thus be determined from the period–luminosity law.

Beyond 20 Mly, even the brightest Cepheid variables, which have absolute magnitudes of about −6, fade from view. Astronomers then turn to more luminous stars. The brightest red supergiants have absolute magnitudes of −8, the brightest blue supergiants have absolute magnitudes of −9. These two types of stars can be seen out to distances of 50 million and 80 million light years, respectively. Out to these limits, you can determine the distances to

Box 26-2 The Hubble law as a distance indicator

Suppose that you aim a telescope at an extremely distant galaxy. You take a spectrum of the galaxy and find that the spectral lines are shifted toward the red end of the spectrum.

Example: A spectral line that normally has a wavelength λ_0 is observed at a longer wavelength λ. Thus, the spectral line has been shifted by an amount $\Delta\lambda = \lambda - \lambda_0$. The **redshift** of the galaxy (usually denoted by z) is given by

$$z = \frac{\Delta\lambda}{\lambda_0} = \frac{\lambda - \lambda_0}{\lambda_0}$$

According to the Doppler effect (review Figure 5-20), this wavelength shift corresponds to a speed v, where

$$z = \frac{v}{c}$$

and c is the speed of light.

According to the Hubble law, the recessional velocity v of a galaxy is related to its distance r from Earth by

$$v = H_0 r$$

where H_0 is the Hubble constant.

Since $v = zc$ and $v = H_0 r$, then $H_0 r = zc$. Thus, the distance to a galaxy is related to its redshift by

$$r = \frac{zc}{H_0}$$

Example: Suppose that you observe the giant elliptical galaxy NGC 4889 in the Coma cluster (recall Figure 26-16). The so-called K line of singly ionized calcium normally has a wavelength of 393.3 nm. In the spectrum of NGC 4889, you find this spectral line at 401.8 nm. Thus, the redshift of the galaxy is

$$z = \frac{401.8 - 393.3}{393.3} = 0.0216$$

The galaxy is therefore moving away from us with a speed of

$$v = zc = (0.0216)(3 \times 10^5 \text{ km/s}) = 6500 \text{ km/s}$$

Using $H_0 = 50$ km/s/Mpc, the Hubble law gives you the distance to the galaxy:

$$r = \frac{zc}{H_0} = \frac{6500}{50} = 130 \text{ Mpc} = 420 \text{ Mly}$$

Note that if you had used $H_0 = 100$ km/s/Mpc, you would have concluded that the distance is

$$r = \frac{zc}{H_0} = \frac{6500}{100} = 65 \text{ Mpc} = 210 \text{ Mly}$$

Unfortunately, this uncertainty is the kind that astronomers must deal with. We look forward to an accurate determination of the Hubble constant. Most astronomers suspect that the true value of H_0 lies between the extremes of Sandage (50 km/s/Mly) and van den Bergh (100 km/s/Mly). Thus, the true distance to NGC 4889 is probably somewhere between 210 and 420 million light years. A compromise distance of about 300 million light years has been adopted for this text.

galaxies from the apparent magnitudes of these luminous supergiants.

Beyond 80 Mly, individual stars are no longer discernible. Astronomers therefore turn to entire star clusters and nebulae. The brightest globular clusters, which have a total absolute magnitude of about -10, can be seen out to 130 Mly from Earth. The brightest H II regions have absolute magnitudes of -12 and can be detected out to 300 Mly. From the measured apparent magnitudes of these clusters and nebulae, distances to remote galaxies can be estimated.

Finally, to get beyond 300 Mly, astronomers must wait for supernova explosions. The brightest supernovae reach an absolute magnitude of -19 at the peak of their outbursts (see Figure 26-32). These brilliant outbursts can, in theory, be seen out to distances of 8 billion light years from Earth. However, no two supernovae are identical, a fact that introduces uncertainty into the distances calculated from them.

These varied objects—Cepheid variables and the most luminous supergiants, as well as globular clusters, H II regions, and supernovae—are commonly called **standard candles**. A good standard candle should have three characteristics. First of all, it must be bright so that you can see it out to great distances. Secondly, it must have a well defined luminosity so that you know its intrinsic brightness. And finally, it must be relatively common so that you can use it to determine the distances to many different galaxies. As you might suspect, astronomers go to great lengths to check the reliability of their standard candles. After all, a tiny mistake in the absolute magnitude of a supergiant star or globular cluster can lead to an error of many millions of light years in calculating the distance to a remote galaxy.

The major obstacle in determining the Hubble constant is that the farther we look into space, the fewer standard candles we have. For example, the distance to a nearby galaxy

Box 26-3 Exploring the large-scale structure of the universe

Mapping the large-scale structure of the universe is an exciting field of research that investigates the distribution of clusters and superclusters of galaxies, often revealing huge structures across vast reaches of space. Most astrophysicists believe that these structures contain important clues about the earliest stages of the universe, when matter first began to coalesce into coherent structures that eventually evolved into stars and galaxies.

During the 1980s, several teams of astronomers undertook monumental projects of measuring the redshifts of many galax-

ies across large segments of the sky. For instance, Martha Haynes at Cornell and Riccardo Giovanelli at Arecibo Observatory in Puerto Rico examined a region of the sky that includes the Pisces–Perseus supercluster, named for the constellations over which it extends. A map of the area covered in their survey is shown below. Each dot represents a galaxy whose redshift is in the range of 4000 to 6500 km/s. Note the prominent ridge extending across the upper half of the map. This is a very large structure, extending all the way from the constellation of Per-

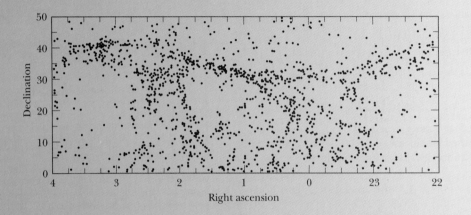

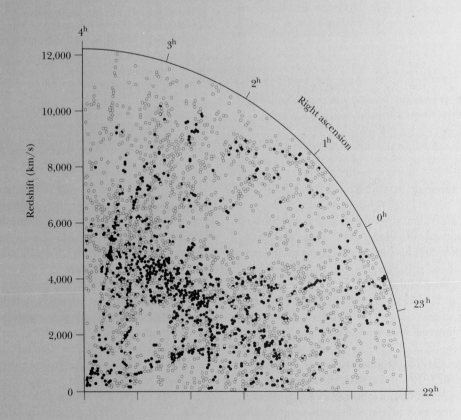

seus at the upper left, across Pisces, and through Pegasus on the right.

To explore this supercluster, Haynes and Giovanelli used radio telescopes such as the 1000-ft dish at Arecibo (recall Figure 10-6) to measure the redshift of 21-cm emission from neutral hydrogen in the spiral galaxies. Elliptical galaxies do not possess detectable amounts of neutral hydrogen, and so their redshifts were obtained using a spectrograph at an optical telescope. Altogether, Haynes and Giovanelli measured the redshifts of nearly 5000 galaxies.

The Haynes–Giovanelli survey covered a wedge-shaped piece of the sky extending 90° along the celestial equator. To get an indication of where the galaxies lie in space, they plotted the redshift and the right ascension of each of the nearly 5000 galaxies on a wedge-shaped graph (opposite page, below). The Earth is located at zero redshift, at the apex of the wedge. Redshift increases radially outward, away from the Earth. Right ascension, from 22^h to 4^h, is indicated along the outside of the wedge. All galaxies, regardless of their declinations, are plotted on this graph, so this illustration "squashes" all the galaxies in the survey onto a single plane.

Several significant features stand out in this wedge-shaped map. First is the main ridge of the supercluster at a redshift of about 5000 km/s. If the Hubble constant is 75 km/s/Mpc, then this ridge is at least 60 Mpc long and about 10 Mpc wide. Second is the prominent void between us and the supercluster. This void must be tubelike rather than spherical, because it shows up even though 50° of declination are plotted onto a single plane. A less prominent void is located on the far side of the supercluster. The nearer void and the main ridge are also seen in the accompanying three-dimensional plot of the data, which shows a pyramid-shaped volume of space extending outward from the Earth.

Large, coherent structures extending vast distance across space also appear in an impressive survey by Margaret Geller, John Huchra, and their colleagues at the Harvard–Smithsonian Center for Astrophysics. These astronomers focused their efforts on a region of the sky that includes the Coma cluster, which is approximately in the opposite direction to the Pisces–Perseus supercluster. Using optical telescopes, the redshifts of more than 10,000 glaxies were measured. One 6°-wide slice of their survey showing 1099 galaxies appears in Figure 26-19. By combining data from several adjacent slices, Geller and Huchra discovered that many galaxies at roughly the distance of the Coma cluster lie in an enormous thin sheet, now called the "Great Wall," shown in the final illustration (below). If the Hubble constant equals 75 km/s/Mpc, then this "wall" covers an area at least 80 by 230 Mpc (about $\frac{1}{4}$ by $\frac{3}{4}$ billion light years).

In recent years, astronomers have time and again been startled to find vast structures whose sizes seem to be limited only by the area of the sky covered in their surveys. Indeed, when larger areas are explored, larger features are found. Furthermore, some of these features exhibit significant movement, indicating large-scale motions in the universe (see the essay by Alan Dressler that follows this chapter).

At present, astrophysicists are at a loss to explain how these enormous structures came into existence. Ongoing redshift surveys to explore even larger volumes of the universe may well uncover more surprises during the next few years.

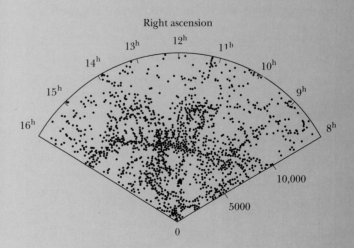

Figure 26-32 A supernova in NGC 4303 *In 1961, a supernova erupted in the spiral galaxy NGC 4303 (also designated M61), which is a member of the Virgo cluster. Supernovae, which can be seen even in extremely remote galaxies, are important "standard candles" used to determine the distances to these faraway galaxies. (Lick Observatory)*

As we turn to more distant galaxies, however, we can see fewer and fewer standard candles. For example, the nearest rich, regular cluster is the Coma cluster (recall Figure 26-16). These galaxies are so far away that it is impossible to see individual stars in them. With fewer standard candles, fewer cross-checks can be made. The distance to these remote galaxies thus becomes less certain. Unfortunately, remote galaxies are precisely the objects whose distances we must determine to find the value of the Hubble constant. Uncertainty in determining distances is the cause of our uncertainty in the value of H_0.

Recently, two astronomers discovered an important correlation, called the **Tully–Fisher relation**, between the width of the hydrogen 21-cm line and the absolute magnitude of spiral galaxies. This correlation may be the best distance indicator currently available. That such a relationship should exist can be seen as follows. Radiation from the approaching side of a galaxy is blueshifted while that from the galaxy's receding side is redshifted. Thus the 21-cm line is Doppler broadened by an amount directly related to how fast a galaxy is rotating, which is related to the galaxy's mass by Kepler's third law. Furthermore, the more massive a galaxy is, the more stars it contains, and so the brighter it is. Consequently, the width of a galaxy's 21-cm line is directly related to the galaxy's luminosity. Since line widths can be measured quite accurately, astronomers can use the Tully–Fisher relation to determine the luminosities of spiral galaxies.

Astronomers are hopeful that new techniques, such as the Tully–Fisher relation, along with instruments like the Hubble Space Telescope and the Keck Telescope, will solve many of the problems of determining H_0. For instance, after its optical defects are corrected, the HST is expected to produce exceptionally sharp views of galaxies, which will allow astronomers to identify standard candles out to distances well beyond the reach of earlier generations of telescopes. Advances such as these will contribute significantly to our understanding of the structure and evolution of galaxies and the universe.

can be cross-checked in many ways. The distance computed from the period–luminosity relation can be compared to the distance determined from the magnitudes of the most luminous supergiants. Then these results can be compared with the magnitudes of the galaxy's globular clusters and the angular sizes of its H II regions. After all these steps, the results from all these methods can be averaged to obtain a distance to the galaxy in which astronomers can have confidence.

Key words

anisotropic	giant elliptical galaxy	irregular galaxy	standard candle
barred spiral galaxy	Hubble classification scheme	isotropic	starburst galaxy
cluster (of galaxies)	Hubble constant	lenticular galaxy	supercluster
dark-matter problem	Hubble flow	Local Group	Tully–Fisher relation
dwarf elliptical galaxy		poor cluster (of galaxies)	void
elliptical galaxy	Hubble law	regular cluster (of galaxies)	
galactic cannibalism	irregular cluster (of galaxies)	rich cluster (of galaxies)	
galaxy merger		spiral galaxy	

Key ideas

• Galaxies can be grouped into the Hubble classification of four major categories: spirals, barred spirals, ellipticals, and irregulars.

Spiral galaxies like our own are sites of active star formation; barred spiral galaxies appear to be temporary stages in the normal development of spiral galaxies.

Elliptical galaxies are virtually devoid of interstellar gas and dust; no star formation is occurring in these galaxies.

• Galaxies are grouped into clusters rather than being scattered randomly through the universe.

A rich cluster contains hundreds or even thousands of galaxies; a poor cluster may contain only a few dozen galaxies.

A regular cluster has a nearly spherical shape with a central concentration of galaxies; an irregular cluster has an asymmetrical distribution of its galaxies.

Our Galaxy is a member of a poor, irregular cluster called the Local Group.

Rich, regular clusters contain mostly elliptical and S0 galaxies; irregular clusters contain more spiral and irregular galaxies.

Giant elliptical galaxies are often found near the centers of rich clusters.

• The observable mass of a cluster of galaxies is not large enough to account for the observed motions of the galaxies; a large amount of unobserved mass must be present between the galaxies; this situation is referred to as the dark-matter problem.

Hot intergalactic gases emit X rays in rich clusters; massive extended halos probably surround all the galaxies.

• When two galaxies collide, their stars pass each other, but their interstellar media collide violently, either stripping the gas and dust from the galaxies or triggering prolific star formation.

The gravitational effects during a galactic collision can throw stars out of their galaxies into intergalactic space.

Galactic mergers may occur; a large galaxy in a rich cluster may tend to grow steadily through galactic cannibalism, perhaps producing in the process a giant elliptical galaxy.

• Galaxies probably formed roughly 15 billion years ago by the gravitational contraction of huge clouds of hydrogen and helium.

Collisions between galaxies in rich, regular clusters could have stimulated prolific star birth in protogalaxies that then evolved to become ellipticals.

Less frequent collisions would permit the gas in a young galaxy to settle into a disk, thus forming a spiral galaxy.

• There is a simple linear relationship between the distance from the Earth to a galaxy and the redshift of that galaxy (which is a measure of the speed with which it is receding from us); this relationship is the Hubble law, $v = H_0 r$.

Because of difficulties in measuring the distances to galaxies, the value of the Hubble constant, H_0, is not known with certainty; this situation leads to uncertainties in our knowledge about the rate at which the universe is expanding and about the age of the universe.

• Standard candles, such as Cepheid variables and the most luminous supergiants, globular clusters, H II regions, and supernovae in a galaxy, are used in estimating intergalactic distances.

• The Tully–Fisher relation, which correlates the width of the 21-cm line of hydrogen in a spiral galaxy with its absolute magnitude, is one of the best available methods of distance determination.

Review questions

1 What types of galaxies are most likely to have new stars forming? Describe the observational evidence that supports your answer.

2 Which type of galaxy includes the biggest? Which type includes the smallest? Which type is most common?

3 How is it possible that galaxies in our Local Group still remain to be discovered? In what part of the sky would these galaxies be located? What sorts of observations might reveal these galaxies?

4 Are there any galaxies besides our own that can be seen with the naked eye? If so, which one(s)?

5 How would you distinguish star images from unresolved images of remote galaxies on a photographic plate?

6 Explain why the dark matter in galaxy clusters could not be neutral hydrogen.

7 Outline a process that could have caused a protogalaxy to evolve into an elliptical galaxy. What would have had to

occur differently for the protogalaxy to have become a spiral galaxy?

8 Why do some galaxies in the Local Group exhibit blueshifted spectral lines? Is this phenomenon a violation of the Hubble law? Explain.

*9 A certain galaxy is observed to be receding from the sun at a rate of 75,000 km/s. The distance to this galaxy is measured independently and found to be 1.4×10^9 pc. What is the value of the Hubble constant for these data?

10 Why do you suppose there are many discordant determinations of H_0?

11 What kinds of stars would you expect to find populating space between galaxies in a cluster?

Advanced questions

Tips and Tools . . .

The relationship between apparent magnitude, absolute magnitude, and distance is discussed in Box 19-2. Type I and Type II Cepheid variables are compared in Figure 21-14. As explained in Chapter 25, a useful form of Kepler's third law is $M = rv^2/G$, where M is the mass within an orbit of radius r where the orbital speed is v; G is the gravitational constant.

---

*12 *(Basic)* In 1937, a Type I supernova was observed in the galaxy IC 4182. At maximum brilliance, this supernova reached an apparent magnitude of +8. Assuming that the absolute magnitude at maximum light was −19, calculate the distance to the galaxy.

*13 *(Challenging)* Suppose that you mistook a Type I Cepheid for a Type II Cepheid in a distant galaxy. How great an error would you make in determining the distance to that galaxy?

*14 The average radial velocity of galaxies in the Hercules cluster pictured in Figure 26-17 is 10,800 km/s. How far away is the cluster? How does your answer depend on the value of the Hubble constant?

*15 *(Challenging)* The rotation curve of the Sa galaxy NGC 4378 is shown in Figure 26-21. Using data from that graph, calculate the orbital period of stars 20 kpc from the galaxy's center. What is the mass of the galaxy out to 20 kpc from its center?

16 How might you determine what part of a galaxy's redshift is caused by the galaxy's orbital motion about the center of mass of its cluster?

Discussion questions

17 Discuss the advantages and disadvantages of using the various "standard candle" distance indicators to obtain extragalactic distances.

18 Discuss whether the various Hubble types of galaxies actually represent some sort of evolutionary sequence.

Observational projects

19 Using a telescope with an aperture of at least 30 cm (12 in.), observe as many of the following spiral galaxies as you can. Many of these galaxies are members of the Virgo cluster, which can best be seen during the Spring months. Since all galaxies are quite faint, be sure to schedule your observations for a moonless night. The best view is obtained when a galaxy is near the meridian.

Spiral galaxy	R.A. (h m)	Decl. (° ')	Hubble type
M31 (NGC 224)	0 42.7	+41 16	Sb
M58 (NGC 4579)	12 37.7	+11 49	Sb
M61 (NGC 4303)	12 21.9	+4 28	Sc
M63 (NGC 5055)	13 15.8	+42 02	Sb
M64 (NGC 4826)	12 56.7	+21 41	Sb
M74 (NGC 628)	1 36.7	+15 47	Sc
M83 (NGC 5236)	13 37.0	−29 52	Sc
M88 (NGC 4501)	12 32.0	+14 25	Sb
M90 (NGC 4569)	12 36.8	+13 10	Sb
M91 (NGC 4548)	12 35.4	+14 30	SBb
M94 (NGC 4736)	12 50.9	+41 07	Sb
M98 (NGC 4192)	12 13.8	+14 54	Sb
M99 (NGC 4254)	12 18.8	+14 25	Sc
M100 (NGC 4321)	12 22.9	+15 49	Sc
M101 (NGC 5457)	14 03.2	+54 21	Sc
M104 (NGC 4594)	12 40.0	−11 37	Sab
M108 (NGC 3556)	11 11.5	+55 40	Sc

While at the eyepiece, make a sketch of what you see. Can you distinguish any spiral structure? After completing your observations, compare your sketches with photographs found in such popular books as *Galaxies* by Timothy Ferris (Sierra Club Books, 1980).

20 Using a telescope with an aperture of at least 30 cm (12 in.), observe as many of the following elliptical galaxies as you can. As in the previous exercise, be sure to schedule your observations for a moonless night, when the galaxies you wish to observe will be near the meridian. Do these elliptical galaxies differ in appearance from spiral galaxies?

Elliptical galaxy	R.A. (h m)	Decl. (° ')	Hubble type
M58 (NGC 4472)	12 29.8	+8 00	E4
M59 (NGC 4621)	12 42.0	+11 39	E3
M60 (NGC 4649)	12 43.7	+11 33	E1
M84 (NGC 4374)	12 25.1	+12 53	E1
M86 (NGC 4406)	12 26.2	+12 57	E3
M89 (NGC 4552)	12 35.7	+12 33	E0
M110 (NGC 205)	00 40.4	+41 41	E6

21 Using a telescope with an aperture of at least 30 cm (12 in.), observe as many of the following interacting galaxies as you can. As in the previous exercises, be sure to schedule your observations for a moonless night, when the galaxies you wish to observe will be near the meridian.

M51 and NGC 5195
M51 (NGC 5194)	$13^h29.9^m$	$+47°12'$
NGC 5195	$13^h30.0^m$	$+47°16'$

M65, M66, and NGG 3628
M65 (NGC 3623)	$11^h18.9^m$	$+13°05'$
M66 (NGC 3627)	$11^h20.2^m$	$+12°59'$
NGC 3628	$11^h20.3^m$	$+13°36'$

M81 and M82
M81 (NGC 3031)	$9^h55.6^m$	$+69°04'$
M82 (NGC 3034)	$9^h55.8^m$	$+69°41'$

M95, M96, and M105
M95 (NGC 3351)	$10^h44.0^m$	$+11°42'$
M96 (NGC 3368)	$10^h46.8^m$	$+11°49'$
M105 (NGC 3379)	$10^h47.8^m$	$+12°35'$

While at the eyepiece, make a sketch of what you see. Can you distinguish hints of interplay between the galaxies? After completing your observations, compare your sketches with photographs found in such popular books as *Galaxies* by Timothy Ferris (Sierra Club Books, 1980).

For further reading

de Vaucouleurs, G. "The Distance Scale of the Universe." *Sky & Telescope*, December 1983 • This article gives fascinating historical insights into the trials and tribulations of determining distances to galaxies.

Field, G. "The Hidden Mass in Galaxies." *Mercury*, May/June 1982 • This well written article describes evidence for dark matter in the universe.

Gorenstein, P., and Tucker, W. "Rich Clusters of Galaxies." *Scientific American*, November 1978 • This article summarizes the main properties of rich clusters of galaxies.

Hartley, K. "Elliptical Galaxies Forged by Collision." *Astronomy*, May 1989 • This brief article summarizes arguments for the idea that collision and merging of spiral galaxies may give birth to elliptical galaxies.

Hodge, P. "The Andromeda Galaxy." *Scientific American*, January 1981 • This excellent article gives an in-depth look at one of our nearest galactic neighbors.

_____. *Galaxies*. Harvard University Press, 1986 • Written in a friendly, informal style, this superb book covers galaxies, galactic evolution, and closely related topics.

Keel, W. "Crashing Galaxies, Cosmic Fireworks." *Sky & Telescope*, January 1989 • This article discusses the many galactic forms and phenomena that result from collisions between galaxies.

Osterbrock, D., Brashear, R., and Gwinn, J. "Young Edwin Hubble." *Mercury*, January/February 1990 • This fascinating article chronicles the adolescence and early adulthood of one of the greatest astronomers of all time.

Rubin, V. "Dark Matter in Spiral Galaxies." *Scientific American*, June 1983 • This fine article on dark matter gives many insights into how astronomers practice their profession.

Schroeder, M., and Comins, N. "Galactic Collisions on Your Computer." *Astronomy*, December 1988 • This article includes a simple, short BASIC program that simulates collisions between galaxies. If you have access to a personal computer, you can watch the interaction that produced the Whirlpool Galaxy.

Silk, J. "Formation of the Galaxies." *Sky & Telescope*, December 1986 • Taking clues from the many properties of galaxies, the author of this article presents an overview of how galaxies probably form.

Smith, R. *The Expanding Universe: Astronomy's Great Debate*. Cambridge University Press, 1982 • This book gives a detailed, meticulously documented history of the birth and development of extragalactic astronomy.

Tully, R. "Unscrambling the Local Supercluster." *Sky & Telescope*, June 1982 • This article explains how astronomers discover and try to understand large-scale structures like the Local Supercluster.

Wray, J. *The Color Atlas of Galaxies*. Cambridge University Press, 1988 • This gorgeous book contains a magnificent collection of color photographs of more than six hundred galaxies.

Alan Dressler is an astronomer at the Observatories of the Carnegie Institution of Washington, in Pasadena, California. As a teenager he polished mirrors for 4- and 8-inch telescopes; he later studied physics at the University of California at Berkeley and received his Ph.D. in astronomy from the University of California at Santa Cruz.

In addition to his research in large-scale structure of the universe, Dressler has focussed his studies on the nature of galaxies—their present-day structure and stellar populations and how these have evolved. Dressler and colleague James E. Gunn of Princeton University have pursued the difficult task of obtaining electronic pictures and spectra of very faint, high-redshift galaxies. Looking far into space is a view to earlier cosmic times, so Dressler and Gunn have been able to study the evolution of galaxies directly, finding that many galaxies formed stars at a much higher rate 5 to 10 billion years ago.

Dressler was one of the first to find observational evidence for massive black holes in the centers of galaxies. He found the speeds of stars increase rapidly as they cross the very center of the Andromeda galaxy (M31), indicating a central mass concentration of tens of millions of solar masses in a region only a few light years across.

Alan Dressler THE GREAT ATTRACTOR

Our gaze into a starry sky scarcely reveals the magnificent forms that nature has sculpted into the distribution of matter and galaxies over vast intergalactic distances. To see these features we must look beyond our Milky Way to the billions of other galaxies that, interwoven with mysterious invisible matter, make up our universe.

A true three-dimensional view of where the galaxies are in space requires some knowledge of their distances, which can be estimated by measuring galaxy redshifts and applying Hubble's discovery that redshifts are approximately proportional to distance. Advances in telescopes have greatly increased the number of galaxy redshifts that have been measured, and a new picture is emerging of the distribution of galaxies in space. More like a sculpture than spatter, galaxies are found linked in long chains and giant walls surrounding nearly empty regions called voids.

Such regularity probably descends from the Big Bang itself, when the physical properties of unimaginably hot matter carved these patterns into the primeval matter distribution. As cosmic time passed, these once-small fluctuations grew. Gravity drew more and more matter into denser regions and emptied out the less dense areas. Thus, today's distribution of galaxies should be a contrast-enhanced picture of the universe's structure at birth. These patterns engraved by the Big Bang may be our best opportunity to test theories of particle physics that describe the nature of matter in its ultimate, high-density, high-temperature form.

However, the theories predict the distribution of *all* matter, and most of the matter in the universe might not be in galaxies but *in between* them in a form that gives off little or no light. Evidence for this view has built steadily since the 1930s when physicist Fritz Zwicky noted that galaxies in great clusters move very rapidly, with typical speeds of 1000 km/s. If gravity is to prevent their escape, there must be much more matter in clusters of galaxies than can be seen in the galaxies themselves. This "dark matter" is probably in a very different form than the "ordinary matter" of protons, neutrons, and electrons from which the stars, our world, and we ourselves are fashioned. A popular model suggests that dark matter is in the form of elementary particles like the neutrino, which interact only through the relatively weak force of gravity. Such particles, if they exist, are immune to electromagnetism, nature's strongest force, and therefore would not clump together to form such things as stars, planets, or puppy dogs. In fact, these mysterious particles would pass right through them!

To test these theories, we must map the distribution of mass in at least a representative chunk of the universe. Galaxies may be markers of where the dark matter has collected due to gravity, but we will not know this for certain until we make a map of the *mass* distribution in some region of space and compare it with the *galaxy* distribution. Fortunately, the pull of gravity from dark matter allows us to map its distribution even though we cannot see it directly. This is because galaxies, which we can see, are attracted to regions of great mass density and evacuate the low-density zones. Thus, by measuring the *motions* of galaxies, in addition to their positions in space, we can infer the distribution of dark matter simply by finding out which way galaxies are moving.

Such observations are not simple because the dominant motion of a galaxy is its flight in the Hubble expansion of the universe. However, if the distance to a galaxy can be estimated, the expansion velocity can be removed and any remaining "peculiar velocity" due to the pull of unseen matter will be uncovered.

In 1986 I and six of my colleagues extended the work done by others before us and made the largest such map of peculiar motions of galaxies. We discovered that our Milky Way Galaxy and its neighbors are streaming toward a distant point, roughly in the direction marked by the Centaurus constellation. Galaxies over nearly one-hundred million light years are cruising at speeds of 400 to 1000 km/s, and are converging toward a region roughly 150 million light years from our Galaxy. We identified the cause of this galaxy streaming as the gravitational pull of a *Great Attractor*, an extensive continent of dark matter and galaxies whose higher-than-average density is tugging on galaxies over a huge volume of space. Though the Great Attractor was discovered from the *motions* of galaxies, subsequent maps showed that the number of galaxies in this region is also far enhanced. Our observations verified that dark matter accounts for at least 10 times as much mass as is contained in the visible galaxies, which supports the idea that most of the mass in the Great Attractor is in a form other than "ordinary matter." The total mass in the Great Attractor is approximately 10^{17} times the mass of our Sun and equivalent to the most massive *superclusters* of galaxies, with an extent of roughly 500 million light years.

Our discovery of the Great Attractor confirms the general picture that the universe is very lumpy, as full of large coherent piles of dark matter as of highly clustered galaxies. This lumpiness probably originated during the very earliest moments of the universe. If so, there must be many structures like the Great Attractor that we have yet to find. Further exploration of the large-scale structure of the universe promises to teach us more about how the universe was born and how our destinies were forged in the first moments of creation.

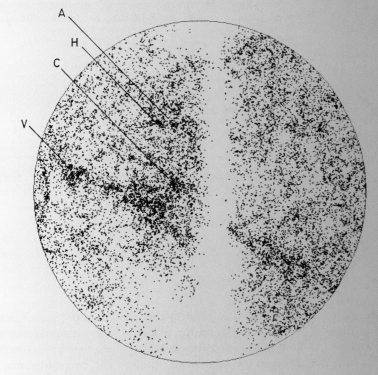

This map, centered approximately on the Great Attractor, shows the distribution of galaxies over half of the sky. The dark vertical band is obscured by the Milky Way. The Virgo (V), Centaurus (C), Hydra (H), and Antlia (A) clusters are indicated. The great concentration of galaxies just below the Centaurus cluster is obvious, which suggests that some of the matter associated with the Great Attractor might be visible.

The core of a radio galaxy *This artist's rendition shows a scenario that many astronomers believe is responsible for double radio sources. A supermassive black hole at the center of the galaxy is surrounded by an accretion disk. In the inner regions of the accretion disk, matter crowding toward the hole is diverted outward along two oppositely directed beams, which deposit energy into two huge radio-emitting lobes on either side of the galaxy. A similar scenario may cause the enormous power output of quasars, active galaxies, and blazars. (Astronomy)*

C H A P T E R

27

Quasars and Active Galaxies

Since the 1960s, astronomers have discovered hundreds of objects in the sky whose large redshifts indicate that they are extremely far away, typically more than a billion light years from Earth. To be observable at all at such huge distances, these quasars and active galaxies must be exceptionally luminous. Indeed, observations indicate that a typical quasar emits the energy output of 100 galaxies from a volume roughly the size of our solar system. Some quasars and the nuclei of active galaxies also exhibit puzzling features, such as jets that propel matter outward at relativistic speeds. The preponderance of evidence leads many astronomers to believe that supermassive black holes are responsible for the energy production of these luminous objects.

The development of radio astronomy in the late 1940s ranks among the most important scientific accomplishments of the twentieth century. Until then, everything known about the distant universe had to be gleaned from visual observations. Radio telescopes could provide a view of the universe in a wavelength range far removed from that of visible light. Many unexpected and surprising discoveries emerged with this new ability to examine the previously invisible universe.

The first radio telescope was built in 1936 by an amateur astronomer, Grote Reber, in his backyard in Illinois. By 1944, Reber had detected strong radio emission from Sagittarius, Cassiopeia, and Cygnus. Two of these sources, nicknamed Sgr A and Cas A, happen to be in our own Galaxy—they are the galactic nucleus and a supernova remnant. The exact location of the third source, called Cygnus A (Cyg A), was finally established in 1951, using a newly constructed radio interferometer. Armed with precise coordinates, Walter Baade and Rudolph Minkowski used the 200-in. optical telescope on Palomar Mountain to discover a strange-looking galaxy at that position. Figure 27-1 is a photograph of the optical counterpart of Cyg A.

The peculiar galaxy associated with Cygnus A is very dim. Baade and Minkowski nevertheless managed to photograph its spectrum, which shows a number of bright (i.e.,

emission) spectral lines, all shifted by 5.7 percent toward the red end of the spectrum. A redshift of $z = 0.057$ corresponds to a speed of 17,000 km/s. According to the Hubble law, this speed corresponds to a distance of 1 billion light years.

The enormous distance to Cygnus A astounded astronomers, because Cyg A is one of the brightest radio sources in the sky. Although Cyg A is barely visible through the 200-in. telescope at Palomar, its radio waves can be picked up by amateur astronomers with backyard equipment. The energy output in radio waves from Cyg A must therefore be enormous. In fact, Cyg A shines with a radio luminosity that is 10^7 times as great as that of an ordinary galaxy like M31 in Andromeda. Obviously, the object that comprises Cyg A must be something quite extraordinary.

27-1 Quasars look like stars but have huge redshifts

During the late 1950s and early 1960s, radio astronomers were busy making long lists of all the radio sources they kept finding across the sky. One of the most famous of these lists, titled the *Third Cambridge Catalogue* (the first two catalogues produced by the British team were incomplete and had some inaccuracies), was published in 1959. It lists 471 radio sources. Even today, astronomers often refer to sources by their "3C numbers." With the discovery of the extraordinary luminosity of Cyg A (designated 3C 405, because it is the 405th source on the Cambridge list), astronomers were eager to learn whether any other sources in the 3C catalogue have similarly extraordinary properties.

One interesting case is 3C 48. In 1960, Allan Sandage used the 200-in. telescope to discover a "star" at the location of this radio source (see Figure 27-2). Because ordinary stars are not strong sources of radio emission, astronomers knew 3C 48 must be something unusual. Indeed, its spectrum showed a series of bright spectral lines that no one could identify. Although 3C 48 was clearly an oddball, many astronomers thought at first that it was just another strange star in our own Galaxy.

However, another such "star," 3C 273, was discovered in 1962. In a continuing effort to identify the radio sources in the *Third Cambridge Catalogue*, astronomers at the Australian National Radio Observatory observed several lunar occultations of 3C 273. (Recall that a lunar occultation occurs when the Moon eclipses a background object.) Australian astronomers managed to determine the exact location of 3C 273 by noting when its radio waves were blocked by the Moon.

The radio source 3C 273 was found to have several peculiar characteristics. For one, a luminous jet protrudes from one side of 3C 273, as shown in Figure 27-3. And, as was the case with 3C 48, this object contains a series of bright spectral lines that no one could identify.

Figure 27-1 Cygnus A (3C 405) *This strange-looking galaxy was discovered at the location of the radio source Cygnus A. This galaxy has a substantial redshift (z = 0.057), which means that it must be extremely far away (nearly 1 billion light years from Earth). Because Cyg A is one of the brightest radio sources in the sky, the energy output of this remote galaxy must be enormous. (Palomar Observatory)*

Figure 27-2 The quasar 3C 48 *For several years, astronomers believed erroneously that this object was simply a peculiar star nearby that happened to emit radio waves. Actually, the redshift of this starlike object is so great (z = 0.367) that, according to the Hubble law, it must be roughly 5 billion light years away. (Palomar Observatory)*

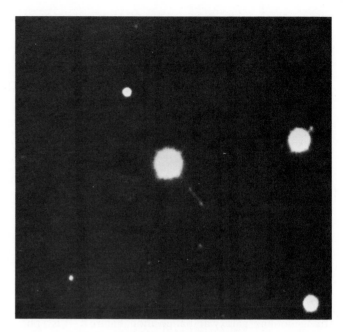

Figure 27-3 The quasar 3C 273 *This greatly enlarged view shows the starlike object associated with the radio source 3C 273. Note the luminous jet projecting from the object's core and pointing toward the lower right. By 1963, astronomers discovered that the redshift of this object is so great (z = 0.158) that its distance from Earth, according to the Hubble law, is about 2 billion light years. (NOAO)*

The reason astronomers had difficulty identifying the emission lines in the spectra of 3C 48 and 3C 273 was that they assumed these starlike objects to be peculiar stars nearby in our Galaxy. They certainly did look like stars.

In 1963, a breakthrough occurred when Maarten Schmidt at the California Institute of Technology was examining the spectrum of 3C 273 and realized that four of its brightest spectral lines are positioned relative to one another with exactly the same spacing as four familiar spectral lines of hydrogen. However, the emission lines of 3C 273 are at much longer wavelengths than the usual positions of the Balmer lines of hydrogen. In other words, the light from 3C 273 has a substantial redshift.

Stellar spectra exhibit comparatively small Doppler shifts, because a star in our Galaxy cannot move extremely fast in relation to the Sun without soon escaping from the Galaxy. Schmidt thus conjectured that 3C 273 might not be a nearby star after all. Pursuing this hunch, he promptly identified all four spectral lines as being hydrogen lines that have suffered an enormous redshift of almost 16 percent, corresponding to a speed of 45,000 km/s (i.e., 15 percent of the speed of light). According to the Hubble law, this huge redshift implies that the distance to 3C 273 is roughly 3 billion light years.

Because of their strong radio emission and starlike appearance, 3C 48 and 3C 273 were dubbed **quasi-stellar radio sources**. This term soon became shortened to **quasars**.

Figure 27-4 shows the spectrum of 3C 273. Instead of using photography to record a spectrum, many observatories now use a charge-coupled device (CCD) at the focus of a

spectrograph. As discussed in Chapter 6 (review Figure 6-21), the output of this device is a graph of intensity versus wavelength, on which emission lines appear as peaks. The graph in Figure 27-4 was obtained in this way. The four hydrogen lines are identified.

The spectral lines of 3C 273 are brighter than the intensity of the background radiation at other wavelengths. The background is called the **continuum**. The bright lines are emission lines caused by excited atoms that are emitting radiation at specific wavelengths. Spectra of ordinary galaxies (recall Figure 26-30) are dominated by dark absorption lines. Most quasars and many peculiar galaxies exhibit strong emission lines in their spectra, a sign that something unusual is going on.

Once the four hydrogen lines in the spectrum of 3C 273 were identified, Schmidt recognized that the remaining spectral lines were those of carbon and oxygen. Two other Caltech astronomers, Jesse Greenstein and T. A. Matthews, then proceeded to identify the spectral lines of 3C 48 as having suffered a redshift of $z = 0.367$. That shift corresponds to a velocity of nearly one-third the speed of light, which places 3C 48 twice as far away as 3C 273, or approximately 5 billion light years from Earth, according to the Hubble law.

Incidentally, these distances to quasars assume that the Hubble constant (H_0) equals 50 km/s/Mpc (15 km/s/Mly). As discussed in the previous chapter, some astronomers believe that this value is too low and would prefer to use 100 km/s/Mpc (30 km/s/Mly) for H_0. Doubling the Hubble constant would then halve these quasar distances.

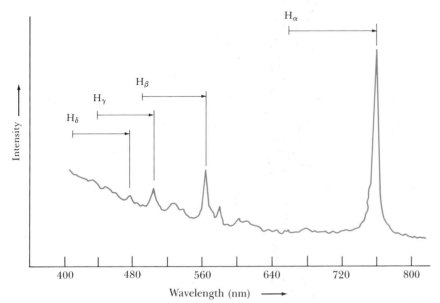

Figure 27-4 The spectrum of 3C 273 *Four bright emission lines of hydrogen dominate the spectrum of 3C 273. The arrows indicate how far these spectral lines are redshifted from their usual wavelengths.*

Hundreds of quasars have been discovered since the pioneering days of the early 1960s. All quasars look like stars, and all have enormous redshifts. For example, the quasar OH 471, shown in Figure 27-5, has a redshift of $z = 3.4$, which corresponds to a speed slightly greater than 90 percent of the speed of light. Figure 27-6 shows the spectrum of the quasar PKS 2000-330, whose huge redshift ($z = 3.78$)

means that its Lyman-alpha line is shifted all the way from its usual wavelength of 121.6 nm in the far-ultraviolet into the middle of the visible spectrum. In recent years, astronomers have discovered a few quasars with redshifts greater than 4. When this book went to press, the record holder was PC 1158+4635 with a redshift of $z = 4.73$.

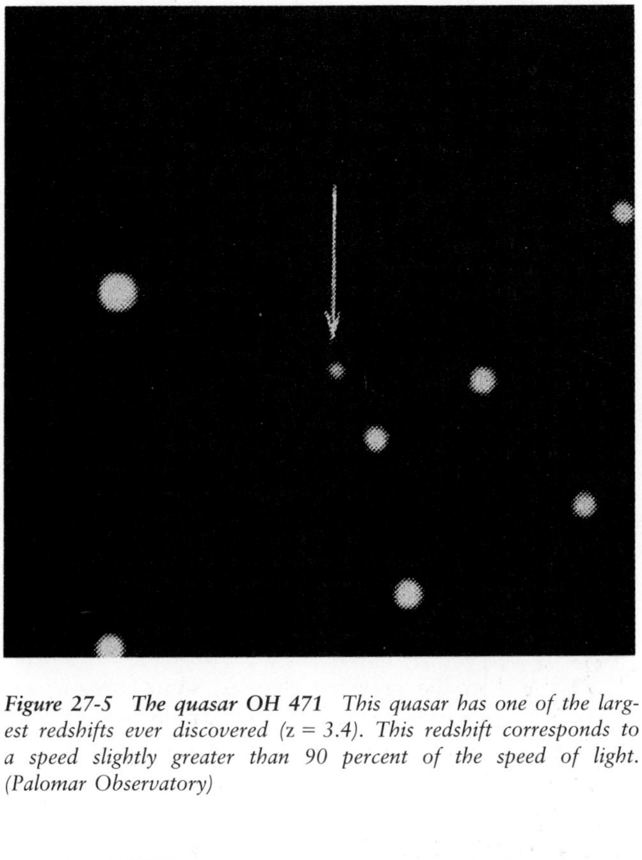

A value of z greater than 1 does not mean that a quasar is receding from us faster than the speed of light. At high speeds, the formula expressing the redshift must be modified by the special theory of relativity, as explained in Box 27-1. According to the relativistic formula, as the speed of a receding source approaches the speed of light, its redshift (z) increases without bound. Indeed, $z = \infty$ corresponds to a speed equal to the speed of light.

It is also important to know that, for high redshifts, the relationship between redshift and distance from Earth depends on details of the evolution of the universe. To compute the distance to a high-redshift quasar, you cannot simply use the formulae given in Box 26-2, which are accurate only for low redshifts ($z < 1$). Distances corresponding to high redshifts are so large that they are affected by the expansion of the universe.

As we shall see in Chapter 28, the Hubble law reveals that the universe is expanding. In other words, if you could watch the motions of two widely separated clusters of galaxies over millions of years, you would see them gradually migrating away from each other. The rate of recession is given by the Hubble constant. For instance, if H_0 is 50 km/s/Mpc, then a galaxy 100 Mpc away should exhibit a recessional speed of 5000 km/s due to the expansion of the universe. This expansion is noticeable only over distances larger than about 100 Mpc; over smaller distances it is masked by the orbital motions of galaxies and clusters of galaxies.

The gravity of all the matter in the universe is causing the expansion of the universe to slow down. The rate of this

Figure 27-5 The quasar OH 471 *This quasar has one of the largest redshifts ever discovered ($z = 3.4$). This redshift corresponds to a speed slightly greater than 90 percent of the speed of light. (Palomar Observatory)*

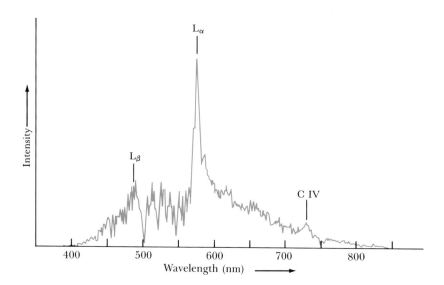

Figure 27-6 The spectrum of PKS 2000-330 The light from this quasar is so highly redshifted ($z = 3.78$) that spectral lines normally in the far-ultraviolet (L_α and L_β) can be seen at visible wavelengths. Note the large number of deep absorption lines on the short-wavelength side of L_α. These lines, collectively called the "Lyman-alpha forest," are probably caused by remote clouds of gas along our line of sight to the quasar. Hydrogen in these clouds absorbs photons from the quasar's continuum at wavelengths corresponding to redshifted L_α.

slowing is described by the so-called **deceleration parameter,** q_0. The distance to a high-redshift object depends on both H_0 and q_0, neither of which is accurately known. We have seen that H_0 probably lies in the range of 50–100 km/s/Mpc, and most astronomers suspect that q_0 is probably between 0 and 1. If q_0 is larger than $\frac{1}{2}$, the expansion of the universe will eventually halt and reverse itself, while if q_0 is smaller than $\frac{1}{2}$, the expansion will continue forever.

Figure 27-7 shows the relationship between redshift and distance for two representative cases of H_0 and q_0. The largest possible redshift ($z = \infty$) corresponds to looking back through space and time all the way back to the Big Bang, the creation event during which the universe came into existence. Further discussion of the deceleration parameter and the Big Bang is found in Chapter 28.

It is not particularly enlightening to speak about the distance to a remote quasar, because of the uncertainty in the value of the Hubble constant. It is more meaningful to talk about the factor by which the universe has expanded since some ancient time. As shown in Chapter 28, an object of redshift z emitted its light when the distance scale of the universe was $1/(1 + z)$ of what it is today. For example, a quasar with a redshift of $z = 4$ emitted its light when the distances between clusters of galaxies were $1/(1 + 4)$, or one-fifth of what they are today. At that time, galaxies were more densely crowded together than now. Since density is inversely proportional to the cube of a volume's dimensions, the density of galaxies in space back then was 5^3, or 125 times the value today.

Figure 27-7 [right] The relationship between distance and redshift The distance to a high-redshift object depends on both the Hubble constant (H_0) and the deceleration parameter (q_0). This illustration graphically displays the distance–redshift relationship for two cases having the same value of H_0 but different values of q_0. Redshifts are tabulated on the left side of each scale and the corresponding distances from Earth, in billions of light years, are given on the right side.

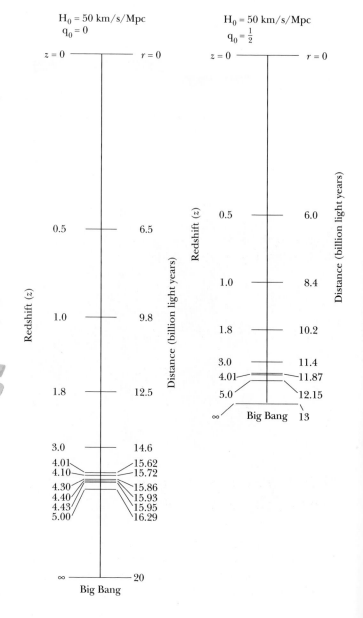

Box 27-1   The relativistic redshift

By definition, the redshift (z) is

$$z = \frac{\lambda - \lambda_0}{\lambda_0}$$

where λ is the observed wavelength of a spectral line in the spectrum of a star, galaxy, or quasar, and λ_0 is the unshifted wavelength of that same spectral line, as deduced from laboratory experiments.

Example: Consider the quasar called PKS 2000-330, whose spectrum is shown in Figure 27-6. The strongest spectral line is the Lyman-alpha line of hydrogen, observed at a wavelength of 582.5 nm. From laboratory measurements, we know that this spectral line is normally seen in the ultraviolet at a wavelength of 121.6 nm. Thus, this quasar has a redshift of

$$z = \frac{582.5 - 121.6}{121.6} = 3.79$$

The average redshift for this quasar, determined from measurements of a number of spectral lines, is $z = 3.78$.

For low velocities, we used the following nonrelativistic equation for the Doppler shift:

$$z = \frac{v}{c}$$

where c is the speed of light. Thus, for example, a 5 percent shift in wavelength ($z = 0.05$) corresponds to a velocity of 5 percent of the speed of light ($v = 0.05\,c$). For high velocities, however, we must use the relativistic equation

$$z = \sqrt{\frac{c + v}{c - v}} - 1$$

An equivalent and useful form of this relationship is

$$\frac{v}{c} = \frac{(z + 1)^2 - 1}{(z + 1)^2 + 1}$$

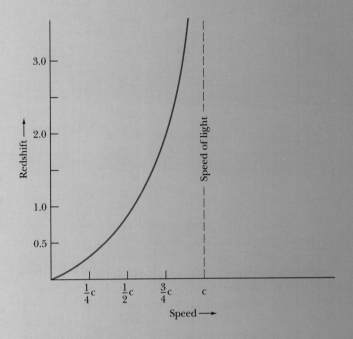

This relativistic relationship between z and v is displayed in the graph. Note that z approaches infinity as v approaches the speed of light. For very low velocities, these complicated relativistic equations both reduce to the simpler, nonrelativistic equation

$$v = cz$$

Example: As stated earlier, quasar PKS 2000-330 has a redshift of $z = 3.78$. Using this value and applying the full, relativistic equation to find the radial velocity for the quasar, we obtain

$$\frac{v}{c} = \frac{(4.78)^2 - 1}{(4.78)^2 + 1} = \frac{21.85}{23.85} = 0.92$$

In other words, this quasar appears to be receding from us with a velocity of 92 percent of the speed of light.

27-2 A quasar emits a huge amount of energy from a small volume

Galaxies are big and bright. A typical large galaxy, like our own Milky Way, contains several hundred billion stars and shines with a luminosity of 10 billion Suns. The most gigantic and most luminous galaxies, such as the giant ellipticals, are only 10 times brighter and shine with the brilliance of 100 billion Suns. A redshift of $\frac{1}{2}$ corresponds to a distance of roughly 6 billion light years, beyond which even bright galaxies may be too faint to be detected. Many ordinary galaxies are too dim to be detected at half this distance.

Although it is difficult to find high-redshift galaxies, high-redshift quasars are quite common. Quasars must therefore be incredibly luminous, far more so than galaxies. A typical

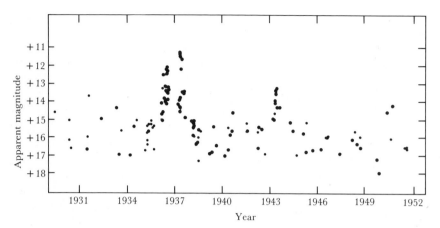

Figure 27-8 The brightness of 3C 279 This graph shows magnitude variations of the quasar 3C 279. The data were obtained by carefully examining old photographic plates in the files of the Harvard College Observatory. Note the large outburst in 1937, and the somewhat smaller one in 1943. (Adapted from L. Eachus and W. Liller)

quasar is in fact 100 times brighter than a typical bright galaxy like our Milky Way. The luminosity of some quasars is as much as 10^{39} watts, or about a trillion times the output of the Sun. For instance, the infrared luminosity of 3C 48 is 5×10^{12} $L_\odot$. Recent X-ray observations from the Einstein Observatory show that the X-ray luminosity of quasars is even more impressive, perhaps as great as 10^4 watts.

In the mid-1960s, several astronomers discovered that some of the newly identified quasars had been photographed inadvertently in the past. For example, 3C 273 was found in numerous photographs, including one taken in 1887. By carefully examining the images of quasars on these old photographs, astronomers found that some quasars fluctuate in brightness, occasionally flaring up. See, for example, the data from old photographs of another quasar, 3C 279, plotted in Figure 27-8. Note the prominent outbursts that occurred around 1937 and 1943. During these outbursts, the luminosity of 3C 279 increased by a factor of at least 25. Because of the enormous distance to 3C 279, at the peak of each of these outbursts, this quasar must have been shining with a brilliance at least 10,000 times as great as that of the entire Andromeda Galaxy.

Fluctuations in the brightness of quasars allow astronomers to place strict limits on the maximum sizes of quasars, because an object cannot vary in brightness faster than light's travel time across that object. For example, an object that is 1 ly in diameter cannot vary significantly in brightness within a period of less than 1 year.

To understand this limitation, imagine an object that measures 1 ly across, as shown in Figure 27-9. Suppose the entire object emits a brief flash of light. Photons from that part of the object nearest the Earth arrive at our telescopes first. Photons from the middle of the object arrive at Earth six months later. And finally, light from the far side of the object arrives a year after the first photons. Although the object emitted a sudden flash of light, we observe a gradual brightness variation that lasts a full year. In other words, the flash is stretched out over an interval equal to the difference in the light travel time between the nearest and most remote observable regions of the object.

The brightness of many quasars varies over periods of only a few weeks or months. Some quasars actually fluctuate from night to night. Recent X-ray data from the Einstein Observatory even reveal large variations in as little as three hours for a few quasars. This rapid flickering means that quasars must be small. The energy-emitting region of a typical quasar—the "powerhouse" that blazes with the luminosity of 100 galaxies—is

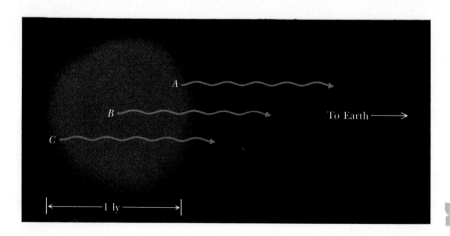

Figure 27-9 A limit in the speed of brightness variation The rapidity with which the brightness of an object can vary significantly is limited by the light travel time across the object. Even if the object, shown here to be 1 ly in size, emits a sudden flash of light, photons from point A arrive at Earth one year before photons from point C. Thus, as seen from Earth, the sudden flash is observed over a full year.

less than 1 light-day in diameter. If quasars are in fact at the huge distances indicated by their redshifts, they must be producing the luminosity of 100 galaxies in a volume whose dimensions are not much larger than the diameter of our solar system.

27-3 Does the Hubble law apply to quasars?

That something so small as a quasar should produce so much energy is an enigma so troublesome that some astronomers speculate that quasars may not be located at the vast distances inferred from their redshifts. If the Hubble law does not, after all, apply to quasars, they could be much nearer to us than their redshifts have led us to believe. If quasars are actually nearby, they need not be extremely luminous, which would relieve astronomers of the difficult task of accounting for their enormous energy output.

A new problem then arises, however: If quasars are in fact nearby, what causes their large redshifts? The known laws of physics cannot explain how a nearby quasar could have such a large redshift. From the traditional viewpoint, the only thing that makes sense is the Hubble-law interpretation: A large redshift means a large distance. Consequently, maverick astronomers who profess to hold the non–Hubble-law interpretation of quasar redshifts argue that we must be at the brink of discovering new laws of physics. This "new physics" would then explain how quasars can be

nearby yet still have large redshifts. This alternative interpretation of the quasar enigma is not widely accepted.

Nonetheless, these astronomers have made some interesting observations indicating that something might indeed be wrong with the traditional interpretation. For example, in 1972 Halton C. Arp of the Mount Wilson and Las Campanas observatories found a quasarlike object that seems to be attached to a nearby galaxy by a luminous bridge of gas (see Figure 27-10). The galaxy (NGC 4319) has a small redshift ($z = 0.006$), corresponding to a speed of only 1800 km/s. According to the Hubble law, NGC 4319 is thus only 120 Mly away. The quasarlike object (called Markarian 205 or Mk 205), however, has a much larger redshift ($z = 0.07$), equal to a speed of 21,000 km/s, which would indicate a distance of 1400 Mly. If Markarian 205 is really connected to NGC 4319, this quasarlike object must be much closer to us than the distance implied by its redshift and the Hubble law.

Most astronomers argue that this is a case of chance alignment, and that the bridge may be a photographic effect. They say the galaxy NGC 4319 is where it belongs, at the distance indicated by its redshift. And the quasar Markarian 205, which has a redshift 11 times larger than that of the galaxy, is where it belongs, 11 times farther away. They only look connected, because they happen to lie in nearly the same direction in the sky.

Undaunted, Arp and his colleagues continue to find intriguing cases across the sky. For example, Figure 27-11 shows the barred spiral galaxy NGC 1073. Three quasars are apparently nestled in its spiral arms. Arp argues that the probability of finding three quasars this bright in this small a

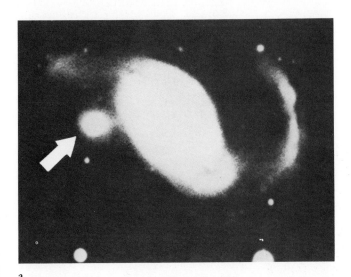

a

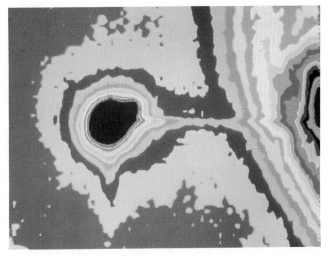

b

Figure 27-10 NGC 4319 and Markarian 205 (a) *The galaxy NGC 4319 has a small redshift that corresponds to a relatively nearby distance of 120 Mly. The quasarlike object Mk 205, indicated by the arrow, has a redshift 11 times larger than the redshift of the galaxy, but* *appears to be attached to the galaxy.* (b) *This computer-enhanced image shows details of the bridge between NGC 4319 and Mk 205. If these two objects are really connected, they would be evidence of a major violation of the Hubble law. (Courtesy of H. C. Arp; NOAO)*

Hawaii published his observations of the regions surrounding 27 low-redshift quasars. Because of their low redshifts, these quasars should be comparatively nearby, perhaps close enough to detect ordinary galaxies in their vicinity. Indeed, in eight cases Stockton did find galaxies huddled around the quasars (see Figure 27-13). In each of these situations, the redshifts of the galaxies are virtually the same as the redshift of the quasar around which they are clustered. Apparently quasars sometimes exist in poor clusters of galaxies, just as giant elliptical galaxies sometimes exist at the centers of rich clusters. Because the Hubble law works for ordinary galaxies, and because each of Stockton's eight quasars has nearly the same redshift as the galaxies surrounding it, the redshifts of the quasars themselves almost certainly also obey the Hubble law.

These kinds of observations may be the deciding factor in the long-standing redshift debate between the two schools of thought. Hubble discovered Cepheid variables scattered around several enigmatic spiral nebulae and thereby settled the Shapley–Curtis debate. Likewise, many astronomers expect the Hubble Space Telescope to find more examples of ordinary galaxies with high redshifts grouped around distant quasars, thus settling the redshift debate.

Figure 27-11 NGC 1073 and three quasars *This photograph, taken with the 200-in. telescope at Palomar, shows the beautiful barred spiral galaxy NGC 1073 in the constellation of Cetus. Three quasars, identified by arrows, appear to be nestled in the galaxy's spiral arms. Most astronomers believe that this is just a "projection effect" created by three very remote quasars that happen to be in the same part of the sky as the galaxy, as seen from Earth. (Courtesy of H. C. Arp)*

region of the sky is exceedingly low and therefore the quasars must somehow be associated with the galaxy.

Arp and his colleagues have also discovered several cases in which normal galaxies with widely differing redshifts appear to be located together in space. Figure 27-12 shows Seyfert's sextet, a grouping of six galaxies first studied by the American astronomer Carl Seyfert in 1954. Although these galaxies seem clustered together, one has a redshift more than four times as large as the redshifts of the others. According to the Hubble law, then, the high-redshift galaxy must be a background object, four times as far away as the other five galaxies. However, the apparent tight grouping seen in Figure 27-12 makes some people wonder whether something may be wrong with the Hubble law.

Most astronomers nevertheless feel that the Hubble law is still a valid indicator of distance and, consequently, that quasars are indeed located at the enormous distances indicated by their huge redshifts. These researchers accept the challenge of trying to explain how the luminosity of 100 galaxies can be generated within a volume with dimensions not much greater than the diameter of our solar system.

Powerful evidence supporting the prevailing interpretation came in 1978 when Alan Stockton of the University of

Figure 27-12 Seyfert's sextet *Five of the galaxies in this cluster have nearly the same redshift (z = 0.015), but the remaining galaxy, indicated by an arrow, has a much higher redshift (z = 0.067). (Palomar Observatory)*

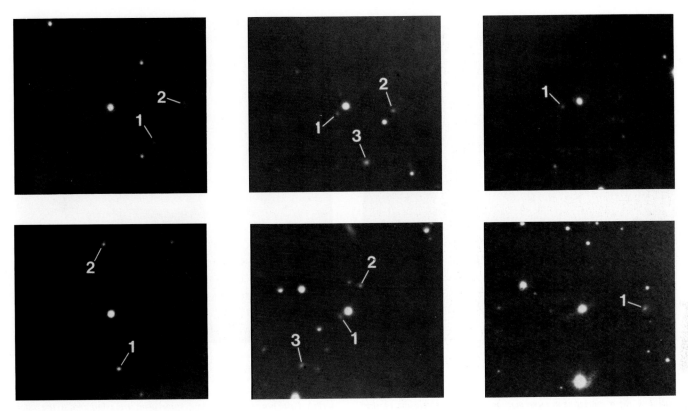

Figure 27-13 *Quasars in remote clusters of galaxies A quasar is located at the center of each of these photographs. Very distant galaxies, identified by the numbers, are faintly visible near each of the quasars. In* *each case, the redshift of the quasar is virtually the same as the redshifts of the galaxies surrounding it. (Courtesy of A. Stockton)*

27-4 *Active galaxies bridge the gap between ordinary galaxies and quasars*

In the 1960s, the gap in energy output between ordinary galaxies and quasars seemed so huge that some astronomers preferred to challenge the Hubble law rather than accept the existence of such highly luminous objects. In recent years, however, astronomers have discovered various kinds of peculiar galaxies whose luminosities fall between those of ordinary galaxies and quasars. Some of these strange galaxies have unusually bright, starlike nuclei. Others have strong emission lines in their spectra. Still more are highly variable. Some have jets or beams of radiation emanating from their cores. Most of these objects are more luminous than ordinary galaxies. All are called **active galaxies**.

The first active galaxies were discovered in 1943, during a survey of spiral galaxies, by Carl Seyfert at the Mount Wilson Observatory. Called **Seyfert galaxies**, these luminous objects have bright, starlike nuclei and strong emission lines in their spectra. For instance, NGC 4151 (see Figure 27-14) has an extremely rich emission spectrum; 28 percent of this galaxy's light is concentrated in its emission lines. These emission lines include Fe X and Fe XIV (iron atoms with either 9 or 13 electrons stripped away), indicating that NGC 4151 contains a quantity of extremely hot gas. Seyfert galaxies also commonly exhibit variability in brightness; for instance, the magnitude of NGC 4151 changes every few months.

Many more Seyfert galaxies have been discovered in recent years, largely through the efforts of Wallace Sargent at Caltech. Approximately 10 percent of the most luminous galaxies in the sky have been identified as Seyfert galaxies. Some of the brightest Seyfert galaxies shine as brightly as faint quasars, which has led many astronomers to suspect that remote Seyfert galaxies could easily be misidentified as quasars.

Seyfert galaxies are divided into two categories, based on their spectra. In the spectrum of a **Type 1 Seyfert galaxy**, the hydrogen lines are much broader than the other emission lines, probably because of turbulence at the galaxy's center, where the hydrogen emission lines are produced. In the spectrum of a **Type 2 Seyfert galaxy**, the hydrogen lines have roughly the same widths as the other emission lines. The Seyfert galaxy NGC 4151 in Figure 27-14 is an example of a Type 1 Seyfert; NGC 1068, shown in Figure 27-15, is a Type 2 Seyfert.

Figure 27-14 *The Type I Seyfert galaxy NGC 4151 This particular galaxy is one of the best studied Seyfert galaxies. Because of its bright, starlike nucleus and strong emission-line spectrum, NGC 4151 might be mistaken for a quasar if it were very far away. The redshift of this galaxy is 0.0033, suggesting it is at a distance of 66 million light years from Earth according to the Hubble law. (Palomar Observatory)*

Figure 27-15 *The Type 2 Seyfert galaxy NGC 1068 This Type 2 Seyfert galaxy, also called M77 or 3C 71, is renowned for its extraordinary infrared luminosity, which varies over intervals as short as a few weeks. Note how bright the inner spiral arms are, compared to the outer spiral arms. NGC 1068 is an example of how the nucleus of an active galaxy can stimulate star formation. The redshift of this galaxy is 0.0036, which means that it is probably about the same distance from Earth as is NGC 4151 in Figure 27-12. (Courtesy of R. Schild)*

An interesting feature of NGC 1068 is its extraordinary brightness at infrared wavelengths. The total infrared luminosity of this galaxy equals 10^{11} Suns. Frank Low of the University of Arizona has detected variations in infrared brightness in NGC 1068 of 7×10^9 Suns over only a few weeks. In other words, the infrared power output of the nucleus of NGC 1068 rises and falls by an amount equal to the total luminosity of our entire galaxy.

Studies of NGC 1068 by Rudolph Schild and his colleagues at Harvard–Smithsonian Center for Astrophysics show that a prolific episode of star formation has recently begun in the inner regions of that galaxy. Note how bright the inner spiral arms are in Figure 27-15. The outer spiral arms are heavily obscured by dust, which makes them appear darker at visible wavelengths than the inner arms. Several bluish "knots" that line up along an arc across the galaxy are very young sites (age $\simeq 10^7$ yr) of vigorous star formation, probably resulting from activity in the nucleus.

Some Seyfert galaxies seem to exhibit the vestiges of explosive phenomena in their nuclei. For instance, NGC 1275 in Figure 27-16 has filaments of gas tens of thousands of light years long protruding from it in all directions. Spectroscopic studies indicate that the gas is being blasted away from the galaxy's nucleus at approximately 3000 km/s. The

nucleus of this galaxy is a strong source of X rays (see Figure 27-17) and radio waves. In 1977, Vera Rubin and her colleagues at the Carnegie Institution of Washington reported observations demonstrating that NGC 1275 actually consists of two galaxies. As we saw in the previous chapter, a collision or close encounter between two galaxies can result in the ejection of matter into intergalactic space.

The high-speed ejection of matter from active galaxies can often be seen best at non-visible wavelengths. For instance, see Figure 27-18a, a photograph of the peculiar galaxy NGC 5128 in the southern constellation of Centaurus. An unusual broad dust lane is the only obvious hint of the extraordinary activity occurring at this galaxy's center.

The galaxy NGC 5128, one of the brightest sources of radio waves in the sky, was one of the first sources discovered when radio telescopes were first built in Australia. As a radio source, it is called Centaurus A, because of its location in the sky. In part, the brightness of Centaurus A at radio wavelengths comes from its proximity to Earth, only 13 Mly away. As shown in Figure 27-18b, radio waves pour from two regions, called **radio lobes**, on either side of the galaxy's dust lane. Farther from the dust lane is a second set of radio lobes that spans a volume 2 Mly across. Recent X-ray pictures of NGC 5128 reveal an X-ray jet (see Figure 27-18c)

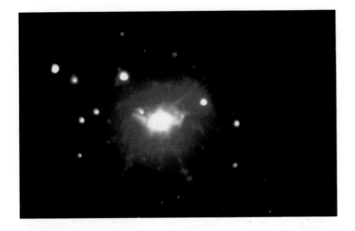

Figure 27-16 *The exploding galaxy NGC 1275* *This Seyfert galaxy, also called 3C 84, located in the Perseus cluster, is a strong source of X rays and radio radiation. The galaxy's redshift is 0.018, suggesting a distance of about 360 Mly from Earth. Note the streamers of gas. Spectroscopic observations confirm significant mass ejection from the galaxy's center. (NOAO)*

Figure 27-17 *An X-ray image of NGC 1275* *This picture from the Einstein Observatory shows the X-ray appearance of the Seyfert galaxy NGC 1275. Most of its X-ray emission comes from a point source at the galaxy's nucleus. (Harvard–Smithsonian Center for Astrophysics)*

sticking out of the galaxy's nucleus. This jet, which is perpendicular to the galaxy's dust lane, is aimed toward one of the radio lobes.

By 1970, radio astronomers had discovered dozens of objects similar to Centaurus A that are now called double radio sources. An active galaxy, usually resembling a giant elliptical, is often found between the two radio lobes. For instance, the visible galaxy associated with Cygnus A (recall Figure 27-1) is located between the two radio lobes shown in Figure 27-19. Cygnus A is a fine example of a small double radio source that is comparatively far away. Generally, the nearer to us a double radio source is, the larger it is, sometimes spanning a volume as big as an entire cluster of galaxies. This relationship of size to distance—and thus to a retrospective look-back time into the past—suggests that double radio sources get bigger as they grow older.

a

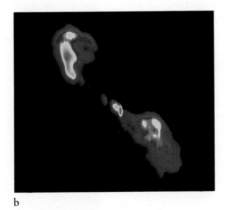

b

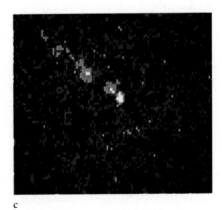

c

Figure 27-18 *The peculiar galaxy NGC 5128 (also called Cen A)* *This extraordinary galaxy, located in the southern constellation of Centaurus, is roughly 13 million light years from Earth. These three views show visible, radio, and X-ray images of this galaxy. (a) This photograph at visible wavelengths shows a broad dust lane across the face of the galaxy. Vast quantities of radio radiation pour from extended regions of the sky on either side of the dust lane. (b) Oppositely directed radio jets emanating from the core of NGC 5128 are nearly perpendicular to the galaxy's dust lane. The structure shown here is usually referred* *to as the inner lobes, to differentiate it from the much larger lobes that extend much farther out from the galaxy. (c) This X-ray image from the Einstein Observatory shows that NGC 5128 has a bright X-ray nucleus. An X-ray jet protrudes from this nucleus along a direction perpendicular to the galaxy's dust lane. Diffuse X-ray emission comes from the regions surrounding the galaxy's center. (Cerro Tololo Inter-American Observatory, NOAO; VLA, NRAO; Harvard-Smithsonian Center for Astrophysics)*

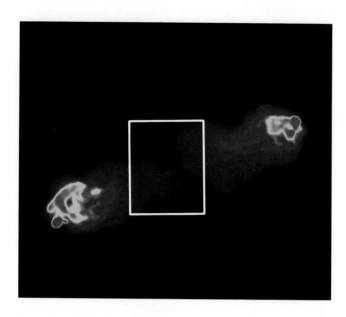

Figure 27-19 A radio image of Cygnus A This radio image shows that most of the radio emission from Cygnus A comes from the radio lobes located on either side of the visible peculiar galaxy. These two radio lobes are each about 160,000 light years from the optical galaxy. Each lobe contains a brilliant, condensed region of radio emission. The white rectangle indicates the area shown in Figure 27-1. (VLA; NRAO)

At radio wavelengths, double radio sources are among the brightest objects in the universe. Using some basic physics, astronomer Geoffrey Burbidge at the University of California in San Diego demonstrated that the energy in the radio lobes of a typical source must exceed 10^{54} joules, which is equal to the energy released by 10 billion supernova explosions.

A final class of active galaxies is the **N galaxies,** so named because of their bright nuclei. Certain extreme examples of these galaxies are the **BL Lacertae objects,** named after their prototype, BL Lac, in the constellation of Lacerta (the lizard). BL Lacertae objects are also sometimes called **blazars.**

BL Lacertae (see Figure 27-21) was discovered in 1929, when it was at first mistaken for a variable star, largely because its brightness varies by a factor of 15 within only a few months. BL Lac's most intriguing characteristic is its totally featureless spectrum that exhibits neither absorption nor emission lines.

Careful examination of BL Lac revealed some "fuzz" (visible in Figure 27-21) around its bright, starlike core. In the early 1970s, Joseph Miller at the Lick Observatory blocked out the light from the bright center of BL Lac and managed to obtain a spectrum of the fuzz. This spectrum, which contains many spectral lines, strongly resembles the spectrum of

All double radio sources seem to have some sort of central "engine" that squirts electrons and a magnetic field outward along two oppositely directed jets at speeds very near the speed of light. After traveling for many thousands or even millions of light years, this ejected material slows down, allowing the electrons and the magnetic field to produce the radio radiation that we detect. As we saw in Chapter 23 in the discussion of the Crab Nebula, a specific type of radio emission called synchrotron radiation occurs whenever electrons spiral around a magnetic field. The velocities of these electrons are near the speed of light. The radio waves that come from the lobes of a double radio source have all the characteristics of synchrotron radiation.

The idea that a double radio source may involve powerful jets of relativistic particles (recall that relativistic means *traveling near the speed of light*) is supported by the existence of **head–tail sources,** so named because each such source appears to have a "head" of concentrated radio emission, with a weaker "tail" trailing behind it. A good example is the active elliptical galaxy NGC 1265 in the Perseus cluster of galaxies. This galaxy is known to be moving at a high speed (2500 km/s) relative to the cluster as a whole. Figure 27-20 is a radio map of NGC 1265. Note that the radio emission from it has a distinctly windswept appearance. Just as smoke pouring from an old-fashioned steam locomotive trails behind a rapidly moving train, particles ejected along two jets from this galaxy are deflected by the galaxy's passage through the sparse intergalactic medium.

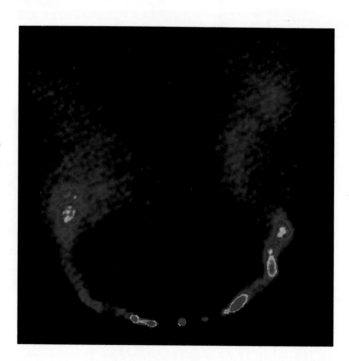

Figure 27-20 The head–tail source NGC 1265 The active elliptical galaxy NGC 1265 would probably be an ordinary double radio source except that this galaxy is moving at a high speed through the intergalactic medium. Because of this motion, its two jets trail behind the galaxy, giving this radio source its distinctly windswept appearance. (NRAO)

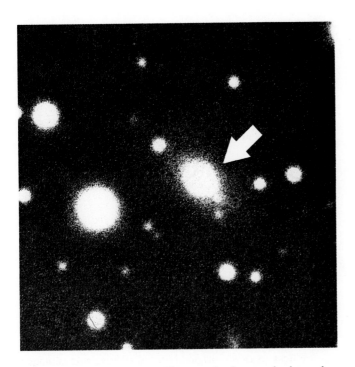

Figure 27-21 BL Lacertae *This superb photograph shows fuzz around BL Lacertae itself. BL Lacertae objects appear to be giant elliptical galaxies with bright, starlike nuclei, much as Seyfert galaxies are spiral galaxies with quasarlike nuclei. BL Lacertae objects contain much less gas and dust than do Seyfert galaxies. (Courtesy of T. D. Kinman; Kitt Peak National Observatory)*

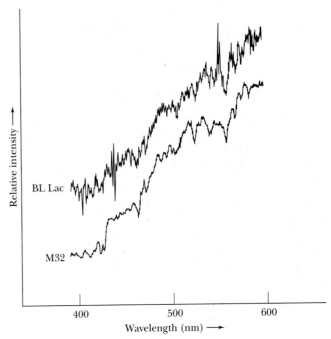

Figure 27-22 The spectrum of BL Lacertae *The spectrum of the fuzz surrounding BL Lac is shown here, along with the spectrum of M32, a small elliptical galaxy in the Local Group. The slight differences between these two spectra can be explained by assuming that BL Lac is a giant elliptical galaxy. The spectrum of BL Lac is redshifted by an amount corresponding to a distance of 1 billion light years from Earth. (Adapted from J. S. Miller, H. B. French, and S. A. Hawley)*

an elliptical galaxy (see Figure 27-22). In other words, a BL Lacertae object is an elliptical galaxy with a bright, starlike center, much as a Seyfert galaxy is a spiral galaxy with a quasarlike center.

These discoveries have prompted many astronomers to suspect that quasars may be the superluminous centers of very distant, very active galaxies. Painstaking observations by Susan Wyckoff, Peter Wehinger, and others have in fact revealed faint galaxylike fuzz around several quasars. Of course, quasars are extremely far away and thus difficult to observe. Therefore, clues about the "engine" that powers a quasar might come from studies of such relatively nearby active galaxies.

27-5 Supermassive black holes may power quasars and active galaxies

How do quasars and active galaxies produce such enormous amounts of energy from such small volumes? As long ago as 1968, the British astronomer Donald Lynden-Bell, working at Caltech, pointed out that an extremely massive black hole might be the "engine" powering a quasar or active galaxy. After all, nothing could possibly be more compact than a black hole and, as explained in Box 27-2, extremely massive black holes might in fact be quite common in the universe. At the center of a quasar or an active galaxy, nature may be drawing upon the tremendous amounts of energy tied up in a black hole's powerful gravitational field.

As we saw in Chapter 24, finding black holes is a difficult business. At best, we can see only the effects of the hole's gravity and try to rule out all non–black-hole explanations of the data. This general approach has been applied to several nearby galaxies, including M31 and M32.

The Andromeda Galaxy (M31) is the largest, most massive galaxy in the Local Group. At a distance of only 2.2 million light years from Earth, M31 is so close to us that details in its core as small as 1 parsec across can be resolved under the best seeing conditions. A wide-field view of M31 is seen in Figure 26-3. A photograph of the core of M31 is shown in Figure 27-23.

In the 1980s, several astronomers used high-resolution spectroscopy and photometry of M31's core to discover the motions of stars around the galaxy's center. For instance, John Kormendy used the 3.6-m Canada–France–Hawaii

Figure 27-23 The central bulge of M31 The central bulge of the Andromeda Galaxy can be seen with the naked eye on a clear, moonless night. A photograph of the entire galaxy is shown in Figure 26-3. Spectroscopic observations suggest that a supermassive black hole is located at the center of this large, nearby galaxy. (NOAO)

telescope at the summit of Mauna Kea, where the excellent seeing permitted him to measure the Doppler shift of spectral lines at closely spaced positions across M31's nucleus. Results of these observations are plotted in Figure 27-24, which is a rotation curve of the innermost 80 arc seconds of

the galaxy. (At M31's distance, 80 arc sec equals 850 ly.) Note that the rotation curve does not follow the trend set in the outer core and gradually drops to zero at the galaxy's center. Rather, there are sharp peaks—one on the approaching side of the galaxy and the other on the receding side—within 5 arc seconds of the galaxy's center.

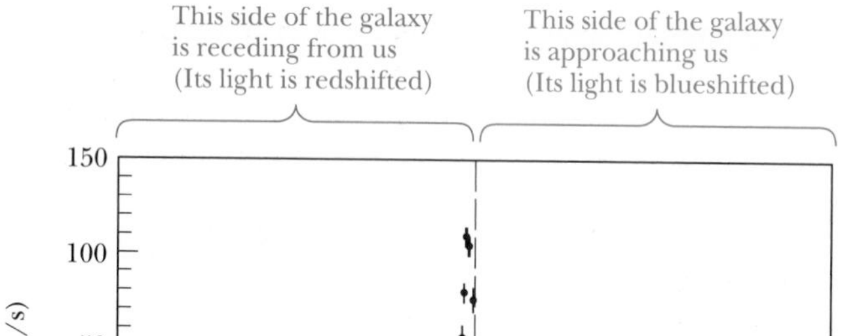

The most straightforward interpretation of these remarkably symmetrical peaks in M31's velocity curve is that they are caused by radiation coming from a disk of material orbiting M31's center. The highest observed radial velocity (at 1.1 arc sec from the galaxy's center; see Figure 27-24) is 110 km/s, which corresponds to an orbital velocity of about 150 km/s when the tilt of M31 to our line of sight is taken into account. This is surely an underestimate, because seeing prohibited the detection of features smaller than about 0.5 arc sec across.

The high speed of material orbiting M31's center indicates the presence of a massive central object. For example, a straightforward application of Kepler's third law using the observed orbital velocity (150 km/s) demonstrates that there must be about 20 million solar masses within 3.6 pc of the galaxy's center. That much matter confined to such a small volume strongly suggests the presence of a supermassive black hole.

Located near M31 is a small companion galaxy called M32, which has a bright starlike nucleus (see Figure 27-25). High-resolution spectroscopy of this galaxy also indicates that stars quite close to the galaxy's center are orbiting the galaxy's nucleus at unusually high speeds. In this case, the orbital motions suggest the presence of an 8-million solar mass black hole at the center of M32.

Astronomers have searched for evidence of supermassive black holes in other, more distant galaxies. For instance,

This side of the galaxy is receding from us (Its light is redshifted)

This side of the galaxy is approaching us (Its light is blueshifted)

Figure 27-24 Rotation curve of the core of M31 The radial velocity of matter in the core of M31 is plotted against the angular distance from the galaxy's center. Note the sharp peaks, one blueshifted and one redshifted, within 5 arc sec of the galaxy's center. At the distance of M31, 1 arc sec corresponds to 10.7 ly. (Adapted from J. Kormendy)

Figure 27-25 The elliptical galaxy M32 This small galaxy is a satellite of M31, a portion of which is seen toward the left. Both galaxies are about 2.2 million light years from Earth. Spectroscopic observations suggest that an 8-million solar mass black hole may be located at the center of M32. (Palomar Observatory)

high-resolution spectroscopy of M104 (see Figure 27-26) reveals high-speed orbital motions inside a bright, starlike nucleus. These motions suggest that a billion solar masses lie within 3.5 arc sec of the galaxy's center. Assuming that M104 is 60 million light years from Earth, all this material must be jammed into a region only 2000 ly across. Once again the observations imply the presence of a supermassive black hole.

Observations pointing to the existence of supermassive black holes have so far been successful only with rather ordinary galaxies. Although no nearby active galaxies exhibit clear-cut evidence of supermassive black holes, the giant elliptical galaxy M87 in the Virgo cluster has suggestive attributes (see Figure 27-27). In 1918, H. D. Curtis made a brief photographic exposure of M87 that revealed a bright, starlike nucleus from which a jet protrudes. A modern, computer-enhanced photograph of the jet appears in Figure 27-28. Figure 27-29 shows radio emission from M87 in blue, with starlight in red.

Inspired by observations of quasars and active galactic nuclei, astrophysicists have devised various realistic scenarios about how to tap the gravitational energy of a supermassive black hole. For instance, schemes worked out by Richard Lovelace at Cornell University and Roger Blandford at Caltech are essentially scaled-up versions of the explanation given for Cygnus X-1 in Chapter 24. In brief, a large black hole is needed to ensure a large energy output.

Imagine a black hole with a mass of 10^6 to 10^9 $M_\odot$ sitting at the center of a galaxy. Because the centers of galaxies are congested places, we expect this **supermassive black hole** to be surrounded by a huge accretion disk of matter captured by the hole's gravity. According to Kepler's third law, the inner regions of this accretion disk would orbit the hole more rapidly than would the outer parts. Thus, the rapidly spinning inner regions would be constantly rubbing against

Figure 27-26 The Sombrero Galaxy This spiral galaxy in Virgo is nearly edge on to our Earth-based view. Spectroscopic observations suggest that a billion-solar mass black hole is located at the galaxy's center. (ESO)

Box 27-2   The plausibility of extremely massive black holes

Despite their exotic properties, massive black holes do not necessarily require exotic circumstances for their creation. To see why this is so, we shall examine the density of material needed to produce a massive black hole. However, first a word of caution: Remember that space around a black hole is highly curved. Simple geometric equations (such as a volume $V = \frac{4}{3}\pi R^3$ for a sphere of radius R) are thus not exactly correct. Nevertheless, they are accurate enough to ensure that the following arguments remain valid.

We want to know how tightly we must compress a mass M inside a sphere of radius R in order to create a black hole. Just before the creation of the hole, the average density (ρ) of the compressed matter is the mass M divided by the volume $\frac{4}{3}\pi R^3$:

$$\rho = \frac{3M}{4\pi R^3}$$

As explained in Chapter 24, however, the radius of a black hole is related to its mass by Schwarzschild's equation:

$$R = \frac{2GM}{c^2}$$

where c is the speed of light and G is the gravitational constant. Substituting this expression for R into the previous equation, we obtain

$$\rho = \frac{3c^6}{32\pi G^3 M^2}$$

The important point is that the density required to create a black hole is inversely proportional to the square of the mass of the hole. In other words, as the mass increases, the density needed to make a black hole drops dramatically.

Example: To make a 1 $M_\odot$ black hole, the equation tells us that we must compress this one solar mass to a density of roughly 10^{19} kg/m^3, which is about twenty times the typical density inside a neutron star.

To make a 10^9 $M_\odot$ black hole, however, the required average density is lowered by a factor of 10^{18}. Thus, we need to squeeze the matter to a density of only 10 kg/m^3, only one-hundredth the density of water.

It is interesting to think about the biggest possible black hole. As we shall see in Chapter 28, the age of the universe can be estimated as $1/H_0$. Thus, the size of the universe is

$$\frac{c}{H_0} = 20 \text{ billion light years}$$

Using this distance as the radius R of a black hole, we can obtain the density in terms of the Hubble constant as follows. First we solve the Schwarzschild equation for M:

$$M = \frac{Rc^2}{2G}$$

We then substitute this value for M into the original equation for density to get

$$\rho = \frac{3c^2}{8\pi GR^2}$$

Finally, by substituting c/H_0 for R, we obtain

$$\rho = \frac{3H_0^2}{8\pi G}$$

Using a Hubble constant of 50 km/s/Mpc, we find

$$\rho = \frac{3H_0^2}{8\pi G} = 4.5 \times 10^{-27} \text{ kg/m}^3$$

This density is equivalent to about three hydrogen atoms per cubic meter of space. As we shall see in Chapter 28, that is roughly the average density of matter across the universe. In other words, we just might be living inside the biggest possible black hole! This universe-sized black hole would, however, be fundamentally different from the black holes we have considered in previous chapters. Many of the properties of an ordinary black hole are related to the fact that the hole is embedded in an exterior spacetime that is flat far from the hole. The universe itself contains all spacetime, and so there is neither space nor time outside of the Schwarzschild radius of the universe-sized black hole. Physicists therefore argue that the concept of the universe as a black hole has no real merit.

---

the more slowly moving gases in the outer regions. The resulting friction would heat up the gases as they spiral toward the hole.

Because of the constant inward crowding of hot gases, the gas pressure and radiation pressure surrounding the hole

would be tremendous. Of course, some of the inflowing material might be swallowed by the hole, but the pressures around the center of the accretion disk would be so great that most of the hot gases would never really get near the hole. Furthermore, particles orbiting the inner edge of the

The details of an accretion disk orbiting a supermassive black hole were first elucidated in 1984 by astrophysicists Larry L. Smarr and John F. Hawley, using a supercomputer. The setup for their supercomputer simulation is shown in Figure 27-31*a*. A black hole, at the center of the field of view, is surrounded by the inner regions of a rotating accretion disk (the axis of rotation is vertical). Matter flows inward toward the hole, as indicated by the arrows around the edge of the field of view. To account for the luminosity of a quasar (10^{40} watts), a 10^9 $M_\odot$ black hole would have to accrete matter at the rate of 10 $M_\odot$ per year.

The program that runs the supercomputer divides space around the black hole into small boxes, some of which can be seen near the upper and lower edges of Figure 27-31*b*. Using the laws of physics, the supercomputer goes from one box to the next, calculating such physical quantities as the pressure, temperature, and density of the in-falling matter. The laws of physics also delineate how these quantities change with time, so the computer can determine how conditions around the black hole evolve.

Supercomputer simulations have demonstrated that in-falling matter in an accretion disk is accelerated to super-sonic speeds. Near the hole, however, this inward rush of gas is abruptly stopped by a "centrifugal barrier" caused by gas rapidly orbiting the hole. This abrupt halt produces a **standing shock wave,** which marks the inner edge of the accretion disk, as shown in Figure 27-31*b*. The funnel-shaped cavity surrounding the axis of rotation is kept empty by the centrifugal effect.

The continuous deluge of matter crashing onto the standing shock wave propels gas along the walls of the evacuated funnel, as shown in Figure 27-32. How is this ejected matter confined in narrow jets? Recent studies in the physics of

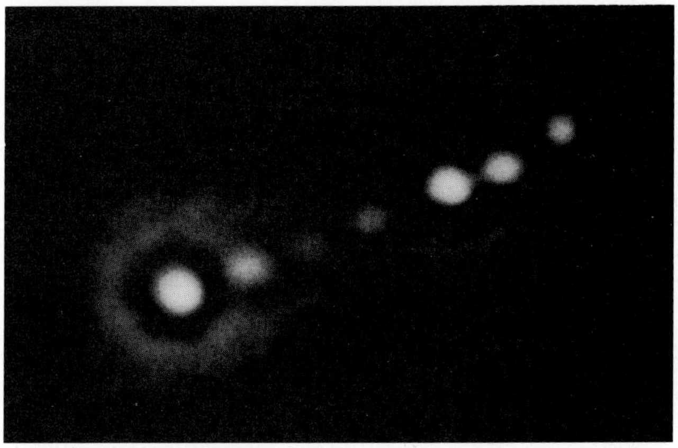

Figure 27-27 The giant elliptical galaxy M87 This giant elliptical is located near the center of the Virgo cluster about 50 million light years from Earth. Note the numerous globular clusters that surround this galaxy. M87 is one of the most luminous galaxies known to astronomers and is about 40 times more massive than our Milky Way Galaxy. (Anglo-Australian Observatory)

accretion disk would experience an outward, "centrifugal force," which alone might be strong enough to prevent further any inflow into the black hole, thus adding to the congestion surrounding the hole. In an attempt to relieve this congestion, matter would be violently ejected along the perpendicular to the accretion disk, that being the direction along which the hot gases experience the least resistance. The result would be two oppositely directed beams of relativistic particles. The overall scenario is sketched in Figure 27-30 and an artist's rendition is seen on the opening page of this chapter.

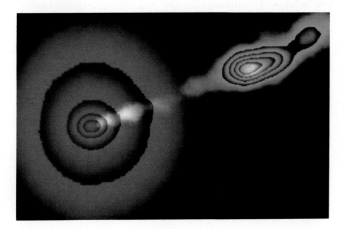

Figure 27-28 The visible jet of M87 This short-time exposure covers only the central regions of M87. A luminous jet 6500 light years long can be seen surging out of the galaxy's bright, starlike nucleus. Several photographic plates taken with the Palomar 200-in. telescope were combined with the aid of a computer to produce this image. Note the even spacing between the "blobs" in the jet. (Courtesy of H. C. Arp and J. J. Lorre)

Figure 27-29 The radio jet of M87 This computer-processed photograph is colored to show the galaxy's stars in red and the light emitted by relativistic electrons in blue. The field of view is the same as that in Figure 27-30. (Smithsonian Institution)

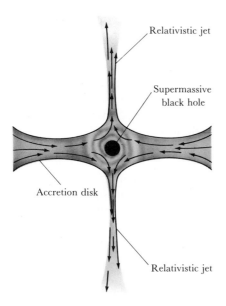

Figure 27-30 A supermassive black hole *The energy output of active galaxies and quasars may involve extremely massive black holes that accrete matter from their surroundings. In the scenario depicted here, the inflow of material through an accretion disk produces two oppositely directed, powerful, relativistic jets.*

fluids demonstrate that there is a natural "self-focusing" effect whenever any high-speed matter penetrates a medium instead of squirting out into empty space. For instance, consider the action of an ordinary garden hose to which a nozzle is attached. As water squirts out the nozzle, the unconfined stream broadens and the spray fans out through a wide angle. If the nozzle is placed in a swimming pool, however, the stream of water does not fan out as much in the water as it did in the air. Similarly, as two jets of hot gas leave the vicinity of a black hole, they must blast their way through some gas that is still crowding inward. Passage through this material causes the jets to become extremely narrow, concentrated beams. This self-focusing effect helps explain not only the double radio sources but also the jets and beams we see protruding from active galaxies and some quasars.

The model of a supermassive black hole surrounded by an accretion disk with oppositely directed high-speed jets has received considerable attention in recent years because it apparently offers a single explanation for a wide variety of quasars and active galaxies. Specifically, Dutch astronomer Peter Barthel has proposed that the main difference between double radio sources, quasars, and blazars is the angle at which the "central engine" is viewed. As Figure 27-33

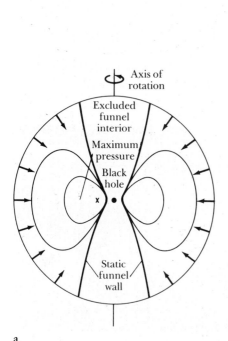

a

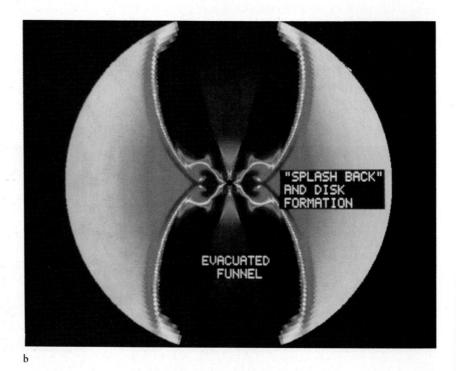

b

Figure 27-31 Standing shock waves in an accretion disk *These cross-sectional diagrams show (a) a schematic and (b) the color-coded results of a supercomputer simulation of an accretion disk around a black hole. Gases plunge at supersonic speed toward the black hole at the center. "Centrifugal forces" exclude matter from cone-shaped funnels sur-* *rounding the disk's axis of rotation. The black arrow-shaped features in the densest (red) part of the disk are standing shock waves created where in-falling gas slows abruptly as it encounters the "centrifugal barrier" at the walls of the evacuated funnel. (Courtesy of J. F. Hawley and L. L. Smarr)*

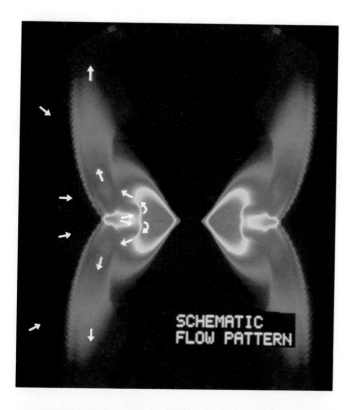

Figure 27-32 [left] Flow patterns along the inner edge of an accretion disk This computer-generated picture shows the flow pattern of gas in an accretion disk surrounding a black hole. The black hole itself is located at the center of this cross-sectional diagram, where the broad-angled inner edges of the accretion disk point. The colors display gas pressure, from red for high pressure across the spectrum to blue for low pressure. The flow pattern, indicated by white arrows, shows how the in-falling gas is channeled into two oppositely directed jets. (Courtesy of J. F. Hawley and L. L. Smarr)

shows, an observer sees a double radio source when the accretion disk is viewed nearly edge on, so that the jets are nearly in the plane of the sky. At a steeper angle, the observer sees a quasar. If one of the jets is aimed almost directly at Earth, a blazar or N galaxy is seen.

The possible existence of supermassive black holes in quasars and active galaxies is one of the most exciting topics in modern astronomy. Once its optical problems are solved, the Hubble Space Telescope will be used to probe the nuclei of galaxies with much higher resolution than is possible from the ground. These observations will undoubtedly reveal many details about the powerful "engines" at the cores of active galaxies and quasars.

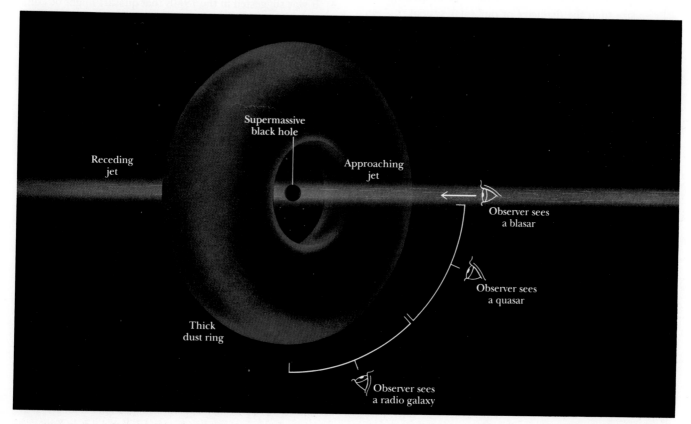

Figure 27-33 The "central engine" Double radio sources, quasars, and blazars may be the same type of object viewed from different directions. The "central engine" that powers these sources is believed to be a supermassive black hole surrounded by an accretion disk from which two oppositely directed relativistic jets flow.

Key words

active galaxy

BL Lacertae object

blazar

continuum

deceleration parameter

double radio source

head–tail source

N galaxy

quasar

quasi-stellar radio source

radio lobes

Seyfert galaxy

standing shock wave

supermassive black hole

Type 1 Seyfert galaxy

Type 2 Seyfert galaxy

Key ideas

- A quasar, or quasi-stellar object, is an object in the sky that looks like a star but has a huge redshift. This corresponds to an extreme distance from the Earth, according to the Hubble law.

  To be seen from Earth, a quasar must be quite luminous, typically about 100 times brighter than an ordinary bright galaxy.

  Relatively rapid fluctuations in the brightness of quasars indicate that they cannot be much larger in size than the diameter of our solar system.

- An active galaxy is an extremely luminous galaxy that has one or more unusual features: an unusually bright, starlike nucleus; strong emission lines in its spectrum; rapid variations in its luminosity; or jets or beams of radiation emanating from its core.

  An active galaxy with a bright, starlike nucleus and strong emission lines in its spectrum is called a Seyfert galaxy.

  Most double radio sources seem to have an active galaxy located between the two radio lobes that distinguish this type of radio source.

  A head–tail radio source seems to show evidence of relativistic particle jets emerging from an active galaxy.

  Two types of active galaxies, N galaxies and BL Lacertae objects, have bright nuclei whose cores show relatively rapid variations in luminosity.

  A BL Lacertae object is probably an elliptical galaxy with a quasarlike center; a Seyfert galaxy is probably a spiral galaxy with a quasarlike center. Quasars may be very distant objects much like active galaxies.

- The strong energy emission from quasars, active galaxies, and double radio sources may be produced as matter falls toward a supermassive black hole at the center of the object.

Review questions

1 Suppose you saw an object in the sky that you suspected might be a quasar. What sort of observations might you perform to find out if it was indeed a quasar?

2 Explain why astronomers do not use any of the standard candles described in Chapter 26 to determine the distances to quasars.

3 How would you distinguish between thermal and nonthermal radiation?

4 It was suggested in the 1960s that quasars might be compact objects ejected at high speeds from the centers of nearby ordinary galaxies. Why does the absence of blueshifted quasars disprove this hypothesis?

5 Why do you suppose there are no quasars relatively near our Galaxy?

6 Compare and contrast SS433 (review Figures 23-14 and 23-15) with a typical double radio source.

7 How can it be that a black hole, from which nothing—not even light—can escape, could be responsible for the extraordinary luminosity of a quasar?

Advanced questions

Tips and tools...

Relativistic redshift is discussed in Box 27-1. Relationships between the density of matter needed to form a black hole and the Schwarzschild radius are given in Box 27-2. Relationships between apparent magnitude, absolute magnitude, and distance are discussed in Boxes 17-2 and 17-3. You may find it useful to know that 1 ly = 63,240 AU and 1 km/s = 0.211 AU/yr.

*8 *(Basic)* What redshift would be observed for an object receding from the Sun at a rate of 99 percent of the speed of light?

*9 *(Basic)* The quasar PC 1158+4635 discovered in 1989 in the constellation of Ursa Major has a redshift of $z = 4.73$. At what speed does this quasar seem to be receding from us?

*10 *(Basic)* Suppose that someday an astronomer discovers a quasar with a redshift of 6.0. With what velocity does this quasar seem to be receding from us?

*11 How much water at a density of 1 g/cm^3 would it take to make a black hole?

*12 What density is required to make a black hole whose mass is equal to the Earth's?

*13 Calculate the Schwarzschild radius of a 10^9 M$_\odot$ black hole. How does your answer compare with the size of our solar system?

14 Verify by direct calculation the statement in the text that, for distances beyond about 10 billion ly, many of the brightest galaxies are too faint to be detected.

15 Explain how the existence of gravitational lenses involving quasars (review Figure 24-6) constitutes important evidence that quasars are located at the great distances from Earth inferred from the Hubble law.

16 *(Challenging)* Verify by direct calculation the statement in the text that Kepler's third law applied to the observed orbital velocity (150 km/s) at 1.1 arc sec from the center of M31 demonstrates the presence of about 20 million solar masses within 3.6 pc of the galaxy's center.

Discussion questions

17 Some quasars show several sets of absorption lines whose redshifts are less than the redshift of the quasars' emission lines. For example, the quasar PKS 0237–23 has five sets of absorption lines with redshifts in the range of 1.364 to 2.202, whereas the quasar's emission lines have a redshift of 2.223. Propose an explanation for these sets of absorption lines.

18 Speculate on the possibility that quasars, double radio sources, giant elliptical galaxies, and so on form some sort of evolutionary sequence.

Observing projects

19 Use a telescope with an aperture of at least 20 cm (8 inches) to observe the Seyfert galaxy NGC 1068. Located in the constellation of Cetus (the whale), this galaxy is most

conveniently seen during autumn and winter months (September through January). The epoch 2000 coordinates are: R.A. = 2^{h}42.7^m and Decl. = $-0°01'$. Sketch what you see. Is the galaxy's nucleus diffuse or starlike? How does this compare with other galaxies you have observed?

20 Use a telescope with an aperture of at least 20 cm (8 in.) to observe the two companions of the Andromeda Galaxy, M32 and NGC 205. Both are small elliptical galaxies, but only one is suspected of harboring a supermassive black hole. They are located on opposite sides of the Andromeda Galaxy and are most conveniently seen during autumn and winter months (September through January). The epoch 2000 coordinates are:

Galaxy	R.A. (h m)	Decl. (° ')
M32 (NGC 221)	0 42.7	+40 52
NGC 205	0 40.4	+41 41

Make a sketch of each galaxy. Can you see any obvious difference in the appearance of these galaxies? How does this difference correlate with what you know about these galaxies?

21 If you have access to a telescope with an aperture of at least 30 cm (12 in.), you should definitely observe the Sombrero Galaxy, M104. It is located in Virgo and can be most conveniently seen during the spring and early summer (March through July). The epoch 2000 coordinates are: R.A. = 12^{h}40.0^m and Decl. = $-11°37'$. Make a sketch of the galaxy. Can you see the dust lane? How does the nucleus of M104 compare with the centers of other galaxies you have observed?

22 If you have access to a telescope with an aperture of at least 30 cm (12 in.), observe M87 and compare it with the other two giant elliptical galaxies (M84 and M86) that dominate the central regions of the Virgo cluster. The epoch 2000 coordinates are:

Galaxy	R.A. (h m)	Decl. (° ')
M84 (NGC 4374)	12 25.1	+12 53
M86 (NGC 4406)	12 26.2	+12 57
M87 (NGC 4486)	12 30.8	+12 24

The negative print from the Palomar Sky Survey shown on the next page may be helpful in identifying the galaxies. This photograph covers an area 3° × 3°; north is at the top.

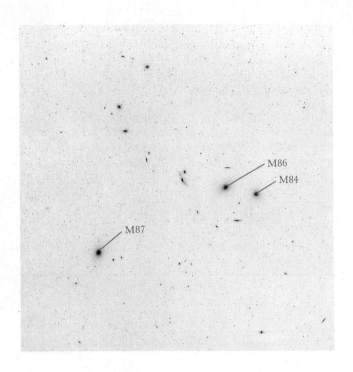

M86

M84

M87

23 If you have access to a telescope with an aperture of at least 40 cm (16 in.), you might try to observe the brightest appearing quasar, 3C273, which has an apparent magnitude of nearly +13. It is located in Virgo at coordinates R.A. = $12^h29^m07^s$ and Decl. = $+2°03'07''$. You may find it helpful to consult Mood's article in *Astronomy*, listed below.

For further reading

Arp, H. "Related Galaxies with Different Redshifts." *Sky & Telescope*, April 1983 • This brief article includes puzzling photographs that show high-redshift galaxylike objects apparently connected to low-redshift galaxies.

————. *Quasars, Redshifts and Controversies*. Interstellar Media, 1987 • In this book, which has both technical and nontechnical sections, Arp sets out his controversial theories with considerable conviction.

Blandford, R., et al. "Cosmic Jets." *Scientific American*, May 1982 • This article, which describes jets emanating from the centers of active galactic nuclei, includes an explanation of superluminal (i.e., faster-than-light) motions sometimes observed in the structure of such jets.

Burns, J., and Price, R. "Centaurus A: The Nearest Active Galaxy." *Scientific American*, November 1983 • By examining the nearby active galaxy NGC 5128, the authors

piece together a rather comprehensive model of active galactic nuclei and double radio sources.

Downes, A. "Radio Galaxies." *Mercury*, March/April 1986 • This superb article on double radio sources includes an interesting overview of the techniques used by radio astronomers.

Ferris, T. "The Spectral Messenger: The Redshift Controversy." *Science 81*, October 1981 • This well written article describes the major issues involved in the redshift controversy.

McCarthy, P. "Measuring Distances to Remote Galaxies and Quasars." *Mercury*, January-February 1988 • This brief article presents an exceptionally clear summary of how the distances to remote objects depend on such factors as the Hubble constant.

Mood, J. "Star Hopping to a Quasar." *Astronomy*, May 1987 • If you have access to a fairly large telescope and want to hunt for the brightest appearing quasar in the sky, you might consult this article on 3C273.

Preston, R. *First Light*. The Atlantic Monthly Press, 1987 • This book takes the reader behind the scenes at the Palomar Observatory. It includes an exciting and well written section on Maarten Schmidt's pioneering research on quasars.

Shipman, H. *Black Holes, Quasars, and the Universe*, 2nd ed. Houghton Mifflin, 1980 • This clear, cogent text has an excellent section on quasars.

Smith, D. "Mysteries of Cosmic Jets." *Sky & Telescope*, March 1985 • This one-page article describes recent observations of the jet in M87, contrasting it to the jet in 3C273.

Sulentic, J. "Are Quasars Far Away?" *Astronomy*, October 1984 • This article by one of Arp's co-workers raises some interesting questions about the distances to quasars.

Tananbaum, H., and Lightman, A. "Cosmic Powerhouses: Quasars and Black Holes." In Cornell, J., and Lightman, A., eds. *Revealing the Universe*. MIT Press, 1982 • This chapter describes observations and theoretical work that lead astronomers to believe that quasars are powered by supermassive black holes.

Verschuur, G. *The Invisible Universe Revealed*. Springer Verlag, 1987 • This fascinating and well written book has four excellent chapters summarizing our knowledge of radio galaxies and quasars.

Weedman, D. *Quasar Astronomy*. Cambridge University Press, 1986 • This somewhat technical monograph summarizes our current understanding of quasars.

Near-infrared image of the Milky Way This portrait of our Galaxy was obtained by infrared detectors on board the COBE satellite. It combines images at wavelengths of 1.2, 2.2, and 3.4 μm. The image is redder where the absorption of starlight by interstellar dust is strongest. The scattered white dots are individual stars near the Sun. The primary purpose of the COBE satellite is to examine details of the cosmic microwave background, a profusion of photons left over from the Big Bang. (Goddard Space Flight Center; NASA)

C H A P T E R 28

Cosmology: The Creation and Fate of the Universe

We live in an expanding universe. This universal expansion—which involves the actual expansion of space—began with an explosive event at the beginning of time called the Big Bang. The universe now is full of microwave photons, ghostly relics of the primordial fireball that filled all space shortly after the Big Bang. Today barely detectable, this radiation field dominated the universe will expand forever, or collapse in a Big Crunch, depends on the density of matter throughout space. Data pointing to the ultimate fate of the universe can be gleaned from the motions of extremely remote galaxies. If the universe expands forever, phenomena such as black hole evaporations may dominate the extremely distant future.

As foolish as it may seem, one of the most profound questions you can ask is, "Why is the sky dark at night?" This question apparently haunted Johannes Kepler as long ago as 1610. It was brought to public attention in the early 1800s by the German amateur astronomer Heinrich Olbers.

To appreciate the problem, assume that the universe is infinite and stars are scattered more or less randomly across this infinite expanse of space. Isaac Newton argued that no other assumption could make sense. If the universe were not infinite, or if stars were grouped in only one part of the universe, the gravitational forces acting between the stars would soon cause all this matter to fall together into a compact blob. Obviously, this has not happened. Thus, as classical Newtonian mechanics would have it, we must be living in a universe that is both infinite and static. Only then does each star feel a uniform gravitational pull from every part of the sky, from all the other stars in the universe. According to this model, the universe is infinitely old and will continue to exist forever without major changes occurring in its structure.

Imagine looking out into space in such a static, infinite universe. Because space goes on forever, with stars scattered throughout it, your line of sight must eventually hit a star. No matter where you look in the sky, you should ultimately see a star. The entire sky should thus be as bright as an average star. Even at night, the entire sky should be blazing like the surface of the Sun. That this is not so is the dilemma called **Olbers's paradox.**

28-1 We live in an expanding universe

Olbers's paradox tells us that there is something wrong with Newton's idea of an infinite, static universe. According to the classical, Newtonian picture of reality, space is laid out in all directions like a great, flat sheet of inflexible, rectangular graph paper. This rigid, flat, three-dimensional space stretches on and on, totally independent of stars or galaxies or anything else. Similarly, a Newtonian clock ticks steadily and monotonously forever, never slowing down or speeding up. Furthermore, Newtonian space and time are unrelated to each other; measurements made with rulers are independent of measurements made with clocks.

Albert Einstein demonstrated that this view of space and time is wrong. In his special theory of relativity (recall Box 24-1), he proved that measurements with clocks and rulers depend on the motion of the observer. As explained in Chapter 24, Einstein's general theory of relativity tells us that the shape of space is profoundly influenced by the masses occupying that space. As John A. Wheeler at the University of Texas puts it, "matter tells space how to curve, and curved space tells matter how to move."

Think about viewing the stars at night. If you observe a star that is 100 light years away, you are actually seeing how that star looked 100 years ago, because its light took that long to reach your eyes. This observation reveals the inti-

mate connection between time and space: as you gaze out into space, you are also looking back in time. A relativistic view is the only one that correctly describes the intimate relationships between space, time, and matter.

Shortly after formulating his general theory of relativity in 1915, Einstein applied his ideas to the structure of the universe. The prevailing Newtonian view was that the universe is infinite and static. To Einstein's dismay, his calculations could not produce a truly static universe; they indicated instead that the universe must be either expanding or contracting. In desperation, he added a term called the **cosmological constant** (Λ) to his equations to ensure that his calculations would predict a static universe. This cosmological constant represents a pressure that balances gravitational attraction in the static universe and prevents it from collapsing. "It was the greatest blunder of my life," Einstein reportedly later lamented. Because he doubted his original equations, he missed the opportunity of postulating that we live in an expanding universe. He could have beat Hubble to the punch by at least ten years.

Edwin Hubble is usually credited with discovering that we live in an expanding universe. It was Hubble who discovered the simple linear relationship between the distances to remote galaxies and the redshifts of those galaxies' spectral lines (review Figure 26-31). This relationship, now called the Hubble law, states that the greater the distance to a galaxy, the greater is the galaxy's redshift. Thus, remote galaxies are moving away from us with speeds proportional to their distances. Specifically, the recessional velocity v of a galaxy is related to its distance r from Earth by the equation

$$v = H_0 r$$

where H_0 is the Hubble constant. Because the clusters of galaxies are getting farther and farther apart as time goes on, astronomers say that the universe is expanding.

Cosmology is unique among the sciences in being strongly dependent upon purely philosophical assumptions. Since there is only one observable universe, we cannot carry out controlled experiments or even make comparisons: We must accept some assumptions or abandon hope of making progress in understanding the nature of the universe. The Hubble law provides a classic example; it could be interpreted as implying that we are at the center of the universe. We reject this interpretation, however, because it violates a cosmological extension of Copernicus's belief that we do not occupy a special location in space.

When Einstein began applying his general theory of relativity to cosmology, he made a daring speculation that gives precise meaning to the idea that we do not occupy a special location in space. Einstein assumed that over very large distances the universe is *homogeneous* (that is, every region is the same as every other region) and *isotropic* (that is, the universe looks the same in every direction). These assumptions constitute the **cosmological principle.** Models of the universe based on the cosmological principle have proved to be surprisingly successful in describing the structure and evolution of the universe and in interpreting observational data. All of our discussion about the universe in this chapter and the next assumes that the universe is homogeneous and isotropic.

What does it actually mean to say that the universe is expanding? According to general relativity, space is not rigid. The amount of space between widely separated locations can in fact change over time. A good analogy is that of a person blowing up a balloon, as sketched in Figure 28-1. Small coins, each representing a galaxy, are glued onto the surface of the balloon. As the balloon expands, the amount of space between the coins gets larger and larger. In the same way, as the universe expands, the amount of space between widely separated galaxies increases. The expansion of the universe *is* the expansion of space.

The expanding-balloon analogy reveals several important characteristics about the expanding universe. For example, imagine that you are stationed on one of the coins in Figure 28-1. As the balloon expands, you see all the other coins moving away from you. Specifically, you observe that the nearby coins are moving away from you slowly, whereas the more distant coins are moving away more rapidly. You find this relationship, which is just like the Hubble law, no matter which coin you decide to call home. Your coin is not at the center of the balloon, of course. Indeed, the surface of the balloon does not actually have a center: You could move around the surface indefinitely without ever finding a center. Likewise, the surface of the balloon has no edge; you could explore every inch of it and never find an edge.

Just as the surface of the balloon has neither center nor edge, our universe has no center or edge. No matter which galaxy you call home, all the other galaxies are receding from you. No one is ever at the "center" of the universe. Questions such as "What is beyond the edge of the universe?" or "What is the universe expanding into?" are as meaningless as asking "What is north of the Earth's north pole?"

The fact that space is continually expanding explains how photons from remote galaxies are redshifted. Imagine a photon coming toward us from a distant galaxy. As the photon travels through space, the space is expanding, so the pho-

ton's wavelength becomes stretched. When the photon reaches our eyes, we see a longer wavelength than usual, which is the redshift. The longer the photon's journey, the more its wavelength will have been stretched. Thus, photons from distant galaxies have larger redshifts than those of photons from nearby galaxies, as expressed by the Hubble law.

A redshift caused by the expansion of the universe is properly called a **cosmological redshift**, to distinguish it from a Doppler shift. Doppler shifts are caused by an object's *motion through* space, whereas a cosmological redshift is caused by the *expansion of* space.

We can calculate the factor by which the universe has expanded since some ancient time from the redshift of light emitted by objects at that time. Redshift, z, is defined as

$$z = \frac{\lambda - \lambda_0}{\lambda_0}$$

where λ_0 is the unshifted wavelength of a photon, while λ is the wavelength we observe. For instance, λ_0 could be the wavelength of a particular spectral line in the spectrum of light leaving the surface of a remote quasar. As the quasar's light travels through space, its wavelength is stretched by the expansion of the universe so that at our telescopes we observe the spectral line to have a wavelength λ. The ratio λ/λ_0 is a measure of the amount of stretching. By rearranging terms in the preceding equation, we can solve for this ratio to obtain

$$\frac{\lambda}{\lambda_0} = 1 + z$$

For example, consider a quasar with a redshift of $z = 3$. Since the time that light left that quasar, the universe has expanded by a factor of $1 + 3 = 4$. In other words, when the light left that quasar, representative distances between widely separated galaxies were only one-quarter as large as they are today. A representative volume of space, which is proportional to the cube of its dimensions, was only $1/4^3$ or $\frac{1}{64}$th as large as it is today. Thus the density of matter in such a volume was 64 times greater than it is today.

Finally, it is important to realize that the expansion of space occurs primarily in the intergalactic space that separates the various clusters of galaxies. Just as the coins in

 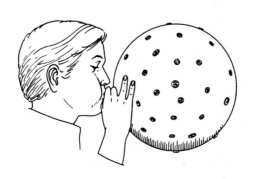

Figure 28-1 The expanding-balloon analogy The expanding universe can be compared to the expanding surface of an inflating balloon. All the coins on the balloon recede from one another as the balloon expands, just as all the galaxies recede from one another as the universe expands. (Adapted from C. Misner, K. Thorne, and J. Wheeler)

Figure 28-1 do not expand as the balloon inflates, galaxies themselves do not expand. Einstein and others have established that an object that is held together by its own gravity, such as a galaxy, is always contained within a patch of non-expanding space. A galaxy's gravitational field produces this nonexpanding region, which is indistinguishable from the flat, rigid space described by Newton. Thus, the Earth and your body, for instance, are not getting any bigger. Only the distance between widely separated galaxies increases with time.

28-2 The expanding universe probably emerged from an explosive event called the Big Bang

The universe has been expanding for billions of years, so there must have been a time in the ancient past when all the matter in the universe was concentrated in a state of infinite density. Presumably, some sort of colossal explosion initiated the expansion of the universe. This explosion, commonly called the **Big Bang,** marks the creation of the universe.

To calculate the time elapsed since the Big Bang, imagine watching a movie of any two galaxies separated by a distance r and receding from each other with a velocity v. Now run the film backward, and observe the two galaxies approaching each other as time runs in reverse. We can calculate the time T_0 it will take for the galaxies to collide by using the simple equation

$$T_0 = \frac{r}{v}$$

Employing the Hubble law, $v = H_0 r$, to replace the velocity v in this equation, we get

$$T_0 = \frac{1}{H_0} = \frac{1}{50 \text{ km/s/Mpc}} = 20 \text{ billion years}$$

Because the separation r has cancelled, this time T_0 is the same for all galaxies. This is the time in the past when all galaxies were crushed together, the time back to the Big Bang.

Actually, the true age of the universe is probably between 13 and 18 billion years. Because of their mutual gravitational attraction, the various galaxies have not been flying away from each other at a constant velocity. Gravity has caused the speed at which galaxies have been separating to decrease gradually since the Big Bang. In other words, the expansion rate of the universe has been decreasing. An age of 20 billion years for the universe assumes no deceleration, but this value would be valid only in an empty universe with no matter and thus no gravity to slow its expansion.

The finite age of the universe offers a resolution of Olbers's paradox. The reason the entire sky is not as bright as the surface of the Sun is that we simply cannot see any objects that are more than 20 billion light years away. The universe may indeed be infinite, with galaxies scattered throughout its limitless expanse. However, the universe is less than 20 billion years old, so the light from stars more than 20 billion light years away has just not had enough time to get here.

You can think of the Earth as being at the center of an enormous sphere having a radius of roughly 20 billion light years (see Figure 28-2). The surface of this sphere is called the **cosmic particle horizon.** Our entire **observable universe** is located inside this sphere. We cannot see anything beyond the cosmic particle horizon, because the travel time for light coming from these incredibly remote distances is greater than the age of the universe. Throughout the observable universe, galaxies are distributed sparsely enough that most of our lines of sight hit no stars, which explains why the night sky is dark.

The finite age of the universe gives us one way out of Olbers's paradox. However, there is a second effect that also contributes significantly to the darkness of the night sky: the redshift. The Hubble law tells us that the redshift of a galaxy is directly related to its distance from Earth—the greater the distance, the greater the redshift. Recall from Chapter 5,

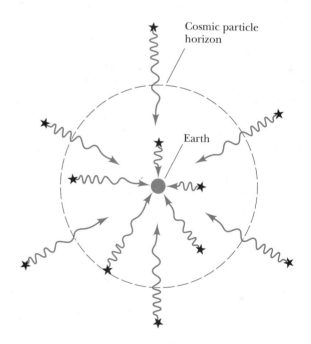

Figure 28-2 The observable universe *The radius of the cosmic particle horizon is equal to the distance that light has traveled since the Big Bang. This event occurred about 20 billion years ago, so the cosmic particle horizon is about 20 billion light years away. We cannot see objects beyond the cosmic particle horizon, because their light has not had enough time to reach us.*

however, that the energy E carried by a photon is related to its wavelength λ by the expression

$$E = \frac{hc}{\lambda}$$

where h is Planck's constant and c is the speed of light.

When a photon is redshifted, its wavelength is lengthened, thus decreasing its energy. The greater the redshift, the less the energy. Consequently, although there are many galaxies far from the Earth, they have large redshifts and their light does not carry much energy. A galaxy nearly at the cosmic particle horizon has in fact a nearly infinite redshift, meaning that its light carries practically no energy at all. This decrease in photon energy because of the expansion of the universe decreases the brilliance of remote galaxies, thus contributing to the darkness of the night sky.

The concept of a Big Bang origin for the universe is a straightforward, logical consequence of having an expanding universe. If you can just imagine far enough back into the past, you can arrive at a time nearly 20 billion years ago when the density of matter throughout the universe was infinite. The entire universe was in effect like the center of a black hole. For this reason, a better name for the Big Bang is the cosmic singularity.

There are many misconceptions about the Big Bang. For one thing, it was not at all like an exploding bomb. When a bomb explodes, pieces of debris fly off *into space* from a central location. If you could trace all the pieces back to their origin, you could find out exactly where the bomb had been. This process is not possible with the universe, though, because the universe itself always has and always will consist of all space. There is genuinely nothing—not even space—beyond the "edge" of the universe. As we have seen, the universe logically cannot have an edge.

Comparing the Big Bang to the center of a black hole can help us appreciate certain aspects of the creation of the universe. As we saw in Chapter 24, matter at the center of a black hole is crushed to infinite density. This location, called the singularity, is characterized by infinite curvature in which space and time are all tangled up. Without a clear background of space and time, such concepts as *past, future, here,* and *now* cease to have meaning.

At the moment of the Big Bang, a state of infinite density filled the universe. Throughout the universe, space and time were completely jumbled up in a condition of infinite curvature like that at the center of a black hole. Thus, we cannot use the laws of physics to tell us exactly what happened at the moment of the Big Bang. And we certainly cannot use science to tell us what existed before the Big Bang. These things are fundamentally unknowable. The phrases "*before the Big Bang*" or "at the *moment* of the Big Bang" are meaningless, because time did not really exist until *after* that moment.

A very short time after the Big Bang, space and time did begin to behave in the way we think of them today. This

interval, called the **Planck time** (t_P), is given by the expression

$$t_P = \sqrt{\frac{Gh}{c^5}} = 1.35 \times 10^{-43} \text{ s}$$

where G is the constant of gravitation, h is Planck's constant, and c is the speed of light (see Box 28-1).

From the Big Bang, at time $t = 0$, to the Planck time 10^{-43} second later, all known science fails us. We do not know how space, time, and matter behaved in that brief span of time. Nevertheless, some physicists have speculated that space and time as we know them today burst forth from a seething, foamlike, space–time mishmash during the Planck time. We can think of the Big Bang as an explosion of space at the beginning of time.

28-3 The microwave radiation that fills all space is evidence for a hot Big Bang

One of the major successes in modern astronomy involves discoveries about the origin of the heavy elements. We know today that all the heavy elements are created in the infernos at the centers of stars (review Box 22-2 for typical details) and in supernovae. As astronomers began to understand the details of thermonuclear synthesis in the 1960s, a new problem arose: There is too much helium around. For example, the Sun consists of about 74 percent hydrogen and 25 percent helium, leaving only 1 percent for all the remaining heavier elements combined. This 1 percent can be understood as material produced inside earlier generations of massive stars that long ago cast these heavy elements out into space during supernovae. Some freshly made helium certainly accompanied these heavy elements, but this amount was not nearly enough to account for one-quarter of the Sun's mass.

Shortly after World War II, Ralph Alpher and Robert Hermann proposed that the universe immediately following the Big Bang must have been so incredibly hot that thermonuclear reactions could have occurred everywhere throughout space. Following up this idea in 1960, the Princeton physicists Robert Dicke and P. J. E. Peebles discovered that they could indeed account for today's high abundance of helium by assuming that the early universe had been at least as hot as the Sun's center, where helium is currently being produced. The early universe must therefore have been filled with many high-energy, short-wavelength photons, which formed a radiation field whose temperature can be given by Planck's blackbody law (review Figure 5-8).

The universe has expanded so much since those ancient times that all those short-wavelength photons have now been so stretched that they have become low-energy, long-

Box 28-1 The Planck time: A limit of knowledge

In the general theory of relativity, space and time are treated together as a continuum. Because of quantum effects, however, this smooth picture of space and time breaks down at extremely small distances or short time intervals. To see why this breakdown occurs, we must first examine some basic ideas in quantum mechanics.

Quantum concepts were introduced around 1900 to explain fundamental properties of light, particularly details of blackbody radiation and the photoelectric effect. As we saw in Chapter 5, this new understanding revealed that light has *both* wavelike and particlelike properties. If you perform an experiment to study the wavelike properties of light (e.g., Young's double-slit experiment, shown in Figure 5-4), you find that light behaves like waves. If you perform an experiment to study the particlelike properties of light (e.g., the photoelectric effect shown in Figure 5-10), you find that light behaves like particles. This chameleonlike behavior is called **wave–particle duality**.

As experiment and theory continued to progress through the 1920s, physicists discovered that particles, such as electrons, can behave like waves. In other words, wave–particle duality applies to particles as well as photons. Experiments performed by American physicist Arthur H. Compton with X-ray photons striking matter helped confirm the wave–particle duality of nature. Because massive particles have wavelike properties, they typically cannot be precisely located over a distance λ_C given by

$$\lambda_C = \frac{h}{mc}$$

where h is Planck's constant, m is the particle's mass, c is the speed of light, and λ_C is today called the **Compton wavelength**. For example, an electron has a Compton wavelength of 0.243 nm, which is a very tiny distance.

The Compton wavelength is a measure of how fuzzy the physical world is at the quantum level. Quantum effects limit our ability to locate a particle on scales smaller than its Compton wavelength, which is a characteristic distance over which the particle exists.

General relativity, the second great revolution in physics to emerge from the early twentieth century, tells us that, near a mass m, events within distances shorter than the Schwarzschild radius are hidden from the outside world (recall Figure 24-3).

This limiting distance (D) is on the order of

$$D = \frac{Gm}{c^2}$$

where G is the constant of gravitation.

To see why general relativity breaks down, imagine a black hole whose mass is so tiny that its Schwarzschild radius is smaller than its Compton wavelength. Because of quantum-mechanical uncertainty, we cannot be sure that the singularity is inside the event horizon. Indeed there is a probability that the singularity spends some time outside the event horizon.

The limiting case of the smallest black hole that does not suffer from this breakdown is obtained by equating D and λ_C. The resulting equation can be solved for the **Planck mass** (m_P):

$$m_P = \sqrt{\frac{hc}{G}} = 5.46 \times 10^{-8} \text{ kg}$$

The Planck mass must not be confused with the mass of a proton, which is 10^{19} times smaller. A black hole whose mass is less than the Planck mass is not correctly described by general relativity.

The lengths D and λ_C, corresponding to $m = m_P$, are called the **Planck length** (λ_P):

$$\lambda_P = \sqrt{\frac{Gh}{c^3}} = 4.05 \times 10^{-35} \text{ m}$$

Note that the size of a proton (about 10^{-15} m) is enormous compared to the Planck length.

The light-travel time across this length is called the **Planck time** (t_P):

$$t_P = \frac{\lambda_C}{c} = \sqrt{\frac{Gh}{c^5}} = 1.35 \times 10^{-43} \text{ s}$$

Because general relativity breaks down for times earlier than the Planck time, no one knows how to describe the universe between $t = 0$ and $t = t_P$. At the moment $t = t_P$, the mass contained within the cosmic particle horizon at each point in space was of the order of the Planck mass.

wavelength photons. The temperature of this cosmic radiation field is now quite low, only a few degrees above absolute zero. By Wien's law, at such a low temperature radiation should have its peak intensity at microwave wavelengths of a few centimeters. In the early 1960s, Dicke, Peebles, and their colleagues began designing an antenna to detect this microwave radiation.

Meanwhile, just a few miles from Princeton University, Arno Penzias and Robert Wilson of Bell Telephone Laboratories were working on a new microwave horn antenna designed to relay telephone calls to Earth-orbiting communication satellites (see Figure 28-3). Penzias and Wilson were deeply puzzled when, no matter where in the sky they pointed their antenna, they detected faint back-

ground noise. Thanks to a colleague, they happened to learn about the work of Dicke and Peebles, and came to realize that they had discovered the cooled-down cosmic background radiation left over from the hot Big Bang.

Since those pioneering days, scientists have made many measurements of the intensity of this background radiation, at a variety of wavelengths. The most accurate measurements come from the Cosmic Background Explorer satellite (COBE), which was placed in orbit about the Earth in 1989 (see Figure 28-4). Data from COBE's spectrometer shown in Figure 28-5 demonstrate that this ancient radiation has the spectrum of a blackbody with a temperature of just a little

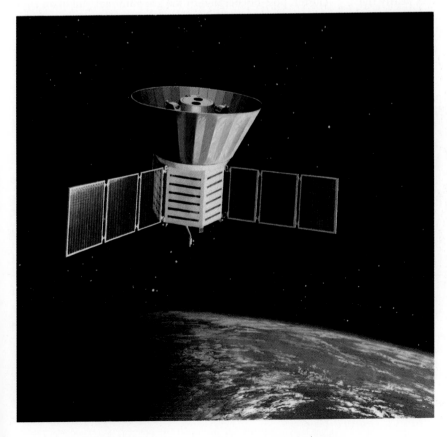

Figure 28-4 The Cosmic Background Explorer (COBE) This satellite, launched in 1989, is measuring the spectrum and angular distribution of the cosmic microwave background over a wavelength range of 1 μm to 1 cm. COBE (pronounced CO-bee) is designed to detect deviations from a perfect blackbody spectrum and from perfect isotropy. (Courtesy of J. Mather; NASA)

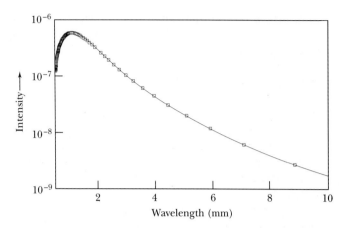

Figure 28-5 *The spectrum of the cosmic microwave background* *The little squares on this graph are COBE's measurements of the brightness of the cosmic microwave background plotted against wavelength. The data fall along a blackbody curve for 2.735 K to a remarkably high degree of accuracy. Note that the peak of the curve occurs at a wavelength of 1.1 mm, in accordance with Wien's law. (Courtesy of E. Cheng; NASA COBE Science Team)*

less than 3 K. Because of its various properties, this radiation field, which fills all of space, is commonly called the **3-degree cosmic microwave background.**

An important feature of the microwave background is its intensity, which is almost perfectly isotropic (meaning *the same in all directions*). In other words, we detect nearly the same background intensity from all parts of the sky. However, extremely accurate measurements first made from high-flying airplanes and more recently from COBE reveal a slight anisotropy. The microwave background is slightly warmer than average toward the constellation of Leo, and in the opposite direction (toward Aquarius) the microwave background is slightly cooler than average. Between the warm spot in Leo and the cool spot in Aquarius, the background temperature declines smoothly across the sky. A map of the microwave sky showing this anisotropy is seen in Figure 28-6.

This temperature variation can be explained as a result of the Earth's overall motion through the cosmos. If we were at rest with respect to the microwave background, the radiation would be truly isotropic. Because we are moving through this radiation field, however, we see a Doppler shift. Specifically, we see shorter-than-average wavelengths in the direction toward which we are moving, as sketched in Figure 28-7. A decrease in wavelength corresponds to an increase in photon energy and thus to an increase in temperature. The temperature excess observed corresponds to a speed of 390 km/s.

Conversely, we see longer-than-average wavelengths in that part of the sky from which we are receding. An increase in wavelength corresponds to a decline in photon energy and hence a decline in temperature. We are thus traveling away from Aquarius toward Leo at a speed of 390 km/s. Taking into account the known velocity of the Sun around the center of our Galaxy, we find that the entire Milky Way Galaxy is moving at 600 km/s toward the Hydra–Centaurus supercluster, possibly because of the gravitational pull of an enormous mass, dubbed the **Great Attractor,** lying in that direction.

Figure 28-6 *The microwave sky* *This map of the microwave sky was produced from data taken by instruments on board COBE. The galactic center is in the middle of the map, and the plane of the Milky Way runs horizontally across the map. Color indicates temperature: magenta is warm and blue is cool. The temperature variation across the sky is caused by the Earth's motion through the microwave background. (NASA)*

28-4 Cosmic background radiation dominated the universe during the first million years

It is enlightening to examine what exists in the universe today, then look back in time toward the Big Bang. Everything in the universe falls into one of two categories: matter or energy. The matter is contained in such objects as stars, planets, and galaxies, composed of particles such as electrons, protons, and neutrons. If we take a large volume V of space and add up the total mass M of all the stars and galaxies we can see in it, we can calculate the **average density of matter** (ρ_m) in the universe at the present time simply by dividing the volume into the mass: $\rho_m = M/V$. Such measurements yield the value

$$\rho_m = 5 \times 10^{-28} \text{ kg/m}^3$$

In other words, if we took all the luminous matter we could find and smeared it out uniformly throughout space, we

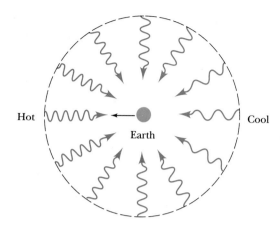

Figure 28-7 *Our motion through the microwave background* Because of the Doppler effect, the microwave background is slightly warmer in that part of the sky toward which we are moving. Recent measurements indicate that our Galaxy, along with the rest of the Local Group, is moving in the general direction of the Virgo and Hydra clusters.

would have 5×10^{-28} kilogram in every cubic meter. The mass of a hydrogen atom is 1.7×10^{-27} kg, so ρ_m is equivalent to about one hydrogen atom in every 3 cubic meters of space.

This density refers only to the luminous matter that we can actually observe. Many astronomers strongly suspect that there is a lot of nonluminous matter in space. For example, the famous missing-mass problem discussed in Chapter 26 applies to clusters of galaxies: The matter we can actually detect is not sufficient to hold clusters together gravitationally. The true density of the matter in the universe, including both the luminous and the nonluminous forms, may be as high as 10^{-26} kg/m³. This value is equivalent to about six hydrogen atoms per cubic meter of space. As we shall see in the final chapter, a density this great has profound implications for the ultimate fate of the universe.

The radiation energy in the universe consists of photons. There are, of course, many starlight photons traveling across space, but the vast majority of photons in the universe belong to the 3-K microwave background.

To compare the matter in the universe with the radiation in the universe, recall Einstein's famous equation $E = mc^2$, which tells us that we can think of the energy in the universe as being equal to a mass multiplied by the square of the speed of light. This connection between matter and energy allows us to speak of the **mass density of radiation** (ρ_{rad}). Einstein's equation ($E = mc^2$) can be combined with the Stefan–Boltzmann law (review pages 87–88) to give

$$\rho_{rad} = \frac{4\sigma T^4}{c^3}$$

where σ is the Stefan–Boltzmann constant. For $T = 3$ K, this equation yields

$$\rho_{rad} = 7 \times 10^{-31} \text{ kg/m}^3$$

Note that ρ_m is about a thousand times larger than ρ_{rad}. In other words, the density of matter (which we have probably underestimated) is clearly much greater than the mass density of radiation. For this reason we say we are living in a **matter-dominated universe**.

Matter prevails over radiation today only because the energy carried by microwave photons is so small. Nevertheless, the number of photons in the microwave background is astounding. From the physics of blackbody radiation it can be demonstrated that there are today 550 million photons in every cubic meter of space. In other words, the photons in space outnumber atoms by roughly a billion to one. In terms of total number of particles, the universe thus consists almost entirely of microwave photons. This radiation field no longer has much "clout," though, because its photons have been redshifted to long wavelengths and low energies, after nearly 20 billion years of being stretched by the expansion of the universe.

Although matter dominates the universe today, this was not always the case. To see why, think back toward the Big Bang. As we do so, we find the universe increasingly compressed, so that the density of matter was greater. The photons in the background radiation also were crowded more closely together then, but an additional effect must be considered. In those times, the photons were less redshifted and thus had shorter wavelengths and higher energy than they do today. Because of this added energy, the mass density of radiation (ρ_{rad}) increases more quickly than the density of matter (ρ_m) as we go back toward the Big Bang. In fact, as shown in Figure 28-8, there was a time in the ancient past when ρ_{rad} equaled ρ_m. Before this time, $\rho_{rad} > \rho_m$, and there was a **radiation-dominated universe**.

This transition from a radiation-dominated universe to a matter-dominated universe occurred nearly a million years after the Big Bang, at a time that corresponds to a redshift of about $z = 1000$. In other words, since that time the wavelengths of photons have been stretched by a factor of 1000. Today these microwave photons typically have wavelengths of about 1 mm, but when the universe was about a million years old, they had wavelengths of about 0.001 mm.

We can use Wien's law to calculate the temperature of the cosmic background radiation at the time of this transition from a radiation-dominated universe to a matter-dominated one. Just as a peak wavelength λ_{max} of 1 mm corresponds to a blackbody temperature of 3 K, a peak wavelength of 0.001 mm corresponds to 3000 K. In other words, the temperature of the radiation background is proportional to the redshift and has been declining over the ages, as shown in Figure 28-9. It just so happens that $T = 3000$ K when $z = 1000$ and $\rho_m = \rho_{rad}$.

The nature of the universe changed in a fundamental way when z was equal to 1000. To understand this change, recall that hydrogen is by far the most abundant element in the universe—hydrogen atoms outnumber helium atoms about 12 to 1. Of course, a hydrogen atom consists of a single

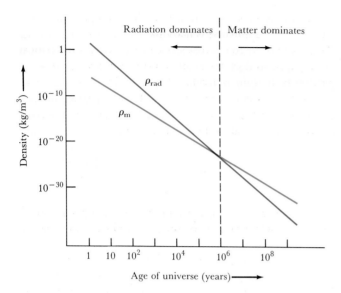

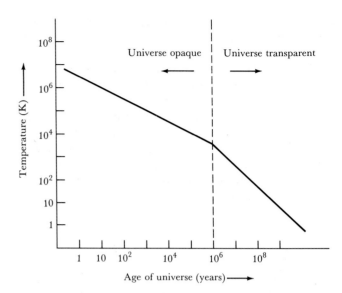

Figure 28-8 *The evolution of density* *During roughly the first million years, the mass density of radiation (ρ_{rad}) exceeded the matter density (ρ_m), and the universe was radiation-dominated. Later, however, continued expansion of the universe caused ρ_{rad} to become less than ρ_m, at which point the universe became matter-dominated.*

Figure 28-9 *The evolution of temperature* *As the universe expands, the photons in the radiation background become increasingly redshifted and the temperature of the radiation falls. Roughly 1 million years after the Big Bang, when the temperature fell below 3000 K, hydrogen atoms formed and the radiation field "decoupled" from the matter in the universe.*

proton orbited by a single electron. It takes only a little energy to knock the proton and electron apart. In fact, a radiation field warmer than about 3000 K easily ionizes hydrogen. Thus, hydrogen atoms could not have existed before $z = 1000$ (that is, at times earlier than about 1 million years after the Big Bang). As sketched in Figure 28-10a, the background photons prior to $t = 1$ million years had energies great enough to prevent electrons and protons from binding to form hydrogen atoms. Only since $t = 1$ million years have these photons been redshifted enough to permit hydrogen atoms to exist (see Figure 28-10b).

Prior to $t = 1$ million years, the universe was completely filled with a shimmering expanse of high-energy photons

colliding vigorously with protons and electrons. This state of matter, called a **plasma**, is opaque, just as the glowing gases inside a discharge tube (like a neon advertising sign) or a fluorescent light bulb are opaque. P. J. E. Peebles coined the term **primordial fireball** to describe the universe during this time.

After $t = 1$ million years, the photons no longer had enough energy to keep the protons and electrons apart. As soon as the temperature of the radiation field fell below 3000 K, protons and electrons began combining to form hydrogen atoms. Because hydrogen is transparent, the universe suddenly became transparent! All the photons that had just a few seconds earlier been vigorously colliding with

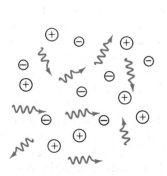

a Before recombination

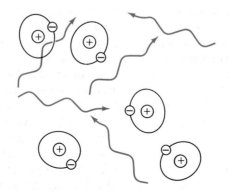

b After recombination

Figure 28-10 *The era of recombination* (a) *Before recombination, photons in the cosmic background prevented protons and electrons from forming hydrogen atoms.* (b) *As soon as hydrogen atoms could survive, the universe became transparent. This transition from opaque to transparent occurred roughly 1 million years after the Big Bang.*

charged particles could now stream unimpeded across space. Today we see these same photons in the microwave background.

This dramatic moment, when the universe went from being opaque to being transparent, is referred to as the **era of recombination**. Because the universe was opaque prior to $t = 1$ million years, we cannot see any farther into the past than the era of recombination. The microwave background, whose photons have suffered a redshift of $z = 1000$, contains the most ancient photons we shall ever be able to observe. This microwave background is today only a ghostly relic of its former dazzling splendor.

28-5 *The future of the universe will be determined by the average density of matter in it*

During the 1920s, general relativity was applied to cosmology by Alexandre Friedmann in Russia, Georges Lemaître in France, Willem de Sitter in the Netherlands, and, of course, Einstein himself. The resulting picture of the structure and evolution of the universe, called **relativistic cosmology**, is in surprisingly good agreement with our intuitive notion that gravity should be slowing the cosmological expansion. In the discussion that follows, we shall assume that Einstein's infamous cosmological constant is precisely zero ($\Lambda = 0$).

To envision the effect of gravity on the expansion of the universe, consider a cannonball shot upward from the surface of the Earth. If the cannonball's speed is less than the escape velocity from the Earth of 11.2 km/s (review Box 7-2), the ball will fall back to Earth. If the cannonball's speed equals the escape velocity, it will just barely escape falling back to Earth. And if the ball's speed exceeds the escape velocity, it can easily leave the Earth and never fall back or stop, despite the relentless pull of gravity.

An analogous set of circumstances surrounds the ultimate fate of the universe. Instead of talking about some sort of "mutual escape velocity" between galaxies, however, astronomers speak of the average density of matter in the universe.

If the average density of matter throughout space is low, the gravity associated with this matter is weak and the expansion of the universe will continue forever. Even infinitely far into the future, galaxies will continue to rush away from each other. In such a case, we say that the universe is **unbounded**, or open.

Conversely, if the density of matter throughout space is high, the resulting gravity will be strong enough eventually to halt the expansion of the universe. At some point, the universe will reach a maximum size, then begin contracting as gravity starts to pull the galaxies back toward each other. In such a case, we say that the universe is **bounded**, or closed.

Separating these two scenarios is the situation in which we say that the universe is **marginally bounded**. In this case, the density of matter across space exactly equals the **critical density** (ρ_c), so that the galaxies just barely manage to keep moving away from each other. This situation is analogous to that of a cannonball leaving the Earth with a speed exactly equal to the escape velocity. As discussed in Box 27-2, the critical density is given by the expression

$$\rho_c = \frac{3\,H_0^2}{8\,\pi G}$$

where H_0 is the Hubble constant and G is the gravitational constant. Using a Hubble constant of 50 km/s/Mpc, we get

$$\rho_c = 4.8 \times 10^{-27} \text{ kg/m}^3$$

This density is equivalent to about three hydrogen atoms per cubic meter of space.

As we saw earlier in this chapter, the average density of the luminous matter we see in the sky seems to be about 5×10^{-28} kg/m^3, which is only about one-tenth of the critical density. However, the missing-mass problem suggests that the true density of matter across space may be equal to or greater than ρ_c. In short, present-day observations and estimates of density are not accurate enough to tell us yet whether the universe is bounded or unbounded.

Cosmological models of the universe become somewhat more complicated if the cosmological constant, Λ, is not equal to zero. Like Einstein, most theoreticians suspect that Λ is zero, or at least extremely small. Box 28-2 is supplied for the reader who wishes to get a taste of some of the complexities introduced by a non-zero cosmological constant.

28-6 *The rate of deceleration of the universe can be determined from observations of extremely distant galaxies*

Another way of determining the future of the universe is to measure the rate at which the cosmological expansion is gradually slowing down. The deceleration of the universe shows up as a deviation from the straight-line relationship predicted by the Hubble law. To see why, reexamine the Hubble-law curve displayed in Figure 26-31. This graph shows the usual linear relationship between the distance to a galaxy and its recessional velocity from the Earth. Note, however, that the graph is based on data extending to only a billion light years from Earth. Over a distance of less than a billion light years, the deceleration is virtually undetectable. All we can see in Figure 26-31 is a straight line, which only tells us that we live in an expanding universe.

Box 28-2 Cosmology with a non-zero cosmological constant

Some interesting effects come from adding the cosmological constant Λ to the equations of general relativity that describe cosmological models. The cosmological constant behaves like a constant energy density, present even if the universe is totally devoid of matter and radiation. Λ can therefore be thought of as the energy density of the vacuum.

Einstein introduced Λ to provide a cosmological repulsion that can balance the self-gravity of the universe, thereby making the universe static. The critical value of the cosmological constant for which this balance occurs is called Λ_c.

If Λ is only slightly larger than Λ_c, the universe starts with a Big Bang but then has a long "quasi-stationary" phase with gravity and cosmological repulsion almost in balance, before cosmological repulsion wins and the expansion continues. This scenario is sometimes called a Lemaître model, after the French cosmologist Georges Lemaître, who discovered it. If Λ is much larger than Λ_c, then the universe expands continually forever.

If Λ is less than Λ_c, there are two possibilities: a bounce model or an oscillating model. A bounce universe contracts from an infinitely extended state (no Big Bang) until cosmological repulsion halts the contraction and the universe rebounds out to infinity again. An oscillating universe expands from a Big Bang, reaches a maximum size, and then collapses, only to erupt again in another Big Bang. Whether the universe bounces or oscillates depends on factors such as the overall curvature of space. If the cosmological constant is negative ($\Lambda < 0$), then only oscillating models are possible. These various scenarios are sketched in the accompanying graph.

Astronomers who have attempted to estimate Λ typically suggest that its value is somewhere between 2×10^{-51} m^2 and -2×10^{-51} m^2, which is a range extremely close to zero. Most researchers find it convenient to assume that $\Lambda = 0$.

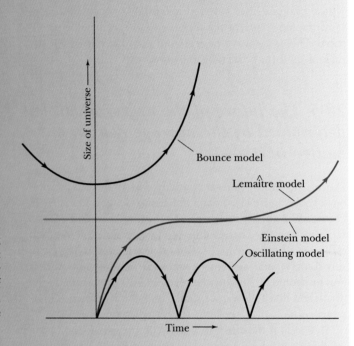

Suppose, however, that you were to measure the redshifts of galaxies several billion light years from Earth. The light from these galaxies has taken billions of years to arrive at your telescope, so what your measurements will reveal is how fast the universe was expanding billions of years ago. Because the universe was then expanding faster than it is today, your data will deviate slightly from the straight-line Hubble law.

The graphs in Figure 28-11 display the relationship between the deceleration of the universe and the Hubble law, extended to great distances. Astronomers denote the amount of deceleration with the **deceleration parameter** (q_0). Appropriately, $q_0 = 0$ corresponds to no deceleration at all. This is possible if the universe is completely empty and thus has no gravity to slow down the expansion. As sketched in Figure 28-11b, a $q_0 = 0$ universe expands forever at a constant rate. As explained earlier in this chapter, the present age of this empty universe, with no deceleration, is exactly $1/H_0$, or about 20 billion years.

The case $q_0 = \frac{1}{2}$ corresponds to a marginally bounded universe. Such a universe just barely manages to expand for-

ever. If the cosmological constant, Λ, is zero (so there is no pressure propelling the expansion of the universe), then the universe contains matter at the critical density ρ_c. In this case, the present age of the universe equals $\frac{2}{3}(1/H_0)$, or about 13 billion years. For simplicity, we shall continue to assume that $\Lambda = 0$.

If q_0 is in the range between 0 and $\frac{1}{2}$, the universe is unbounded and will continue to expand forever. Such a universe contains matter at less than the critical density, and has a present age of 13 to 20 billion years.

If q_0 is greater than $\frac{1}{2}$, the universe is bounded and is filled with matter having a density greater than ρ_c. The present age of such a universe is less than $\frac{2}{3}(H_0)$. Such a universe is doomed to collapse in upon itself, in the extremely distant future.

Some astronomers prefer to characterize the behavior of the universe in terms of the **density parameter** (Ω_0), which is the density of matter in the universe divided by the critical density. If $\Lambda = 0$, then $\Omega_0 = 2q_0$. For instance, the marginally bounded case, $q_0 = \frac{1}{2}$, corresponds to $\Omega_0 = 1$, so that the universe is filled with matter at the critical density. Table

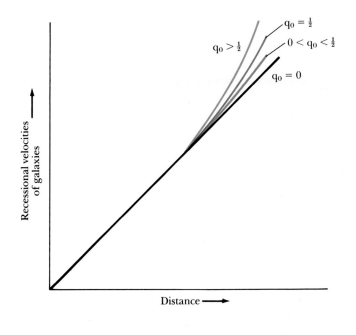

a

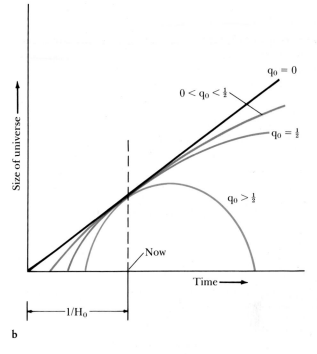

b

Figure 28-11 Deceleration and the Hubble diagram These two graphs compare the appearance of the Hubble diagram with the evolution of the universe. (a) A schematic Hubble diagram shows the relationship between recessional velocity and distance for various values of q_0. (b) The size of the universe is plotted for various values of q_0. The case $q_0 = 0$ is an empty universe $(\rho = 0)$. The case $q_0 = \frac{1}{2}$ is a marginally bounded universe $(\rho = \rho_c)$. If $0 < q_0 < \frac{1}{2}$, then the universe is unbounded and will expand forever. If $q_0 > \frac{1}{2}$, the universe is bounded and will someday collapse.

28-1 shows how the deceleration parameter, the density parameter, and the age of the universe are related.

In principle, it should be possible to determine the nature of the universe by measuring the redshifts and distances of many remote galaxies, then plotting the data on a Hubble diagram like the one in Figure 28-11a. If the data points fall above the $q_0 = \frac{1}{2}$ line, the universe is bounded. If the data points fall between the $q_0 = 0$ and $q_0 = \frac{1}{2}$ lines, the universe is unbounded.

Unfortunately, such observations are extremely difficult to make. Galaxies nearer than a billion light years are of no help in determining q_0. And beyond a billion light years, uncertainties cloud distance determinations. The main problem is that the objects used as standard candles cease to be reliable distance indicators, primarily because of galaxy evolution, which is poorly understood. Nearby galaxies and their components, which are used as standard candles, have been subjected to a longer duration of evolutionary effects than remote galaxies, which we observe at an earlier stage simply because of the travel time of light to Earth. Such evolutionary effects, as well as the ravages of galaxy collisions, alter the chemical compositions, luminosities, and sizes of these standard candles, reducing their usefulness. For example, Figure 28-12 shows data obtained by Allan Sandage of the Mount Wilson and Las Campanas Observatories. Note how the data points are scattered about the $q_0 = 1$ line. This scatter prevents us from knowing conclusively whether the universe is bounded or unbounded. Nevertheless, because

Table 28-1 Deceleration, density, and the age of the universe ($\Lambda = 0$)

Deceleration parameter (q_0)	Density parameter (Ω_0)	Average density	Age of universe (T_0)
Greater than $\frac{1}{2}$	Greater than 1	Greater than ρ_c	Less than $\frac{2}{3}$ $(1/H_0)$
$\frac{1}{2}$	1	ρ_c	$\frac{2}{3}$ $(1/H_0)$
Between 0 and $\frac{1}{2}$	Between 0 and $\frac{1}{2}$	Less than ρ_c	Between $\frac{2}{3}$ $(1/H_0)$ and $(1/H_0)$
0	0	0	$(1/H_0)$

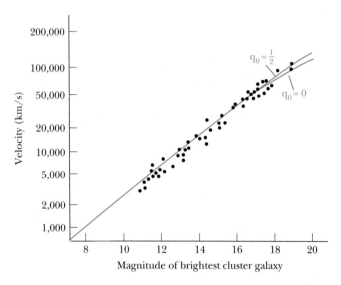

Figure 28-12 *The Hubble diagram This graph shows the Hubble diagram extended to include extremely remote galaxies. The magnitude of the brightest galaxy in a cluster is directly correlated with the distance to the cluster. If the data points fall between the curves marked $q_0 = 0$ and $q_0 = \frac{1}{2}$, the universe is unbounded. If the data points fall above the curve marked $q_0 = \frac{1}{2}$, the universe is bounded. (Adapted from A. Sandage)*

the data lie close to the $q_0 = 1$ line, we suspect that the universe might be close to being marginally bounded. Many astronomers look forward to using the Hubble Space Telescope, after its optical problems are solved, to do a much better job of determining the value of q_0.

28-7 The shape of the universe is related to the rate of deceleration and to the average matter density

There is another way of tackling the issue of the fate of the universe. Our present understanding of the universe is based on Einstein's general theory of relativity, which explains that gravity curves the fabric of space. The gravity of all matter scattered across space should thus give the universe some overall shape (or geometry). Gravity also determines the fate of the universe. Thus, by measuring the shape of space, we should be able to discover whether the universe is bounded or unbounded.

To see what astronomers mean by the geometry of the universe, imagine shining two powerful laser beams out into space. Suppose that we can align these two beams so that they are perfectly parallel as they leave the Earth. Finally, suppose that nothing gets in the way of these two beams, so that we can follow them for billions of light years across the universe, across the space whose curvature we wish to detect.

With this arrangement there are only three possibilities. First, we might find that our two beams of light remain perfectly parallel, even after traversing billions of light years. In this case, space would not be curved: The universe would have **zero curvature, and space would be flat**.

Alternatively, we might find that our two beams of light gradually converge. Gradually they would get closer and closer together as they moved across the universe and might even intersect at some enormous distance from Earth. In such a case, space would not be flat. Recall that the lines of longitude on the Earth's surface are parallel at the equator but intersect at the poles. Thus, in this case the three-dimensional geometry of the universe would be analogous to the two-dimensional geometry of a spherical surface. We would then say that space is **spherical and that the universe has positive curvature**.

The third and final possibility is that the two initially parallel beams of light would gradually diverge, becoming farther and farther apart as they moved across the universe. In this case also the universe would have to be curved, but in the opposite sense from that in the spherical example. We would then say that the universe had **negative curvature**. In the same way that a sphere is a positively curved two-dimensional surface, a saddle is a good example of a negatively curved two-dimensional surface. Parallel lines drawn on a sphere always converge, but parallel lines drawn on a saddle always diverge. Mathematicians say that saddle-shaped surfaces are hyperbolic. Thus, in a negatively curved universe we would describe space as **hyperbolic**.

These three situations are sketched in Figure 28-13. To illustrate these cases, three surfaces are drawn: a sphere, a plane, and a saddle. Of course, real space is three dimensional, but it is much easier to visualize an analogous two-dimensional surface. Thus, as you examine the drawings in Figure 28-13, remember that the real universe has one more dimension. For example, if the universe is in fact hyperbolic, then the geometry of space must be the three-dimensional analogue of the two-dimensional surface of a saddle. Three-dimensional space curves through a fourth dimension that is perpendicular to each of the other three, something that few people can visualize except by analogy.

Each of the three possible geometries corresponds to a different behavior and fate of the universe. Flat space corresponds to the marginally bounded case ($q_0 = \frac{1}{2}$) in which the galaxies just barely manage to keep receding from each other. This flat-space scenario divides the positive-curvature cases from the negative-curvature cases. If the density across space is greater than the critical density, then $q_0 > \frac{1}{2}$, and space is positively curved. Conversely, if the density across space is less than the critical density, then $0 < q_0 < \frac{1}{2}$, and space is negatively curved. These relationships are summarized in Table 28-2.

Note that both the flat and the hyperbolic universes are infinite. They extend forever in all directions, meaning that they have neither an "edge" nor a "center." In contrast, a

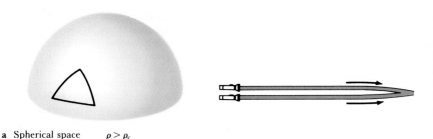

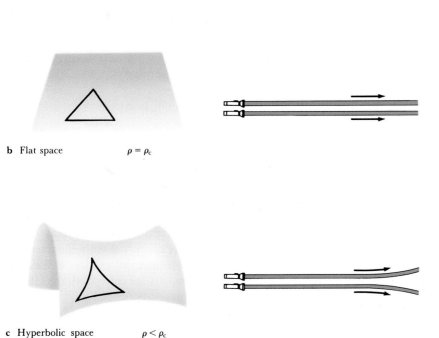

a Spherical space $\rho > \rho_c$

b Flat space $\rho = \rho_c$

c Hyperbolic space $\rho < \rho_c$

Figure 28-13 The geometry of the universe The "shape" of space is determined by the matter contained in the universe. The curvature is either positive (a), zero (b), or negative (c), depending on whether the density is greater than, equal to, or less than the critical density. In theory, the curvature of the universe could be determined by measuring the angles of a huge triangle in space.

spherical universe is finite. However, it also lacks both a center and an edge. A good analogy is the Earth. Even though the Earth has a finite surface area (511 million km²), you could walk forever around it without ever finding a center or an edge. Likewise, relativistic cosmology strictly rules out any possibility of a center or edge to the universe.

There is no easy way of measuring the curvature of space across the universe. In theory, you might imagine drawing an enormous triangle whose sides are each a billion light years long (see Figure 28-13). You could then measure the three angles of the triangle. If their sum equaled 180°, space would be flat. If the sum is greater than 180°, space would be

spherical. And if the sum of the three angles is less than 180°, space would be hyperbolic. Unfortunately, this direct method for measuring the curvature of space is not practical.

The curvature of space should affect the density of galaxies we see at great distances from Earth. To understand why this is so, imagine three surfaces—a sphere, a plane, and a saddle, as shown in Figure 28-14—on which grains of sand have been randomly sprinkled. For each surface, suppose you make a map of the locations of the sand grains simply by plotting the distance of each grain from some reference point (the cross) onto flat paper. As shown at the bottom of Figure 28-14, the resulting maps exhibit noticeable differences.

Table 28-2 The geometry and fate of the universe ($\Lambda = 0$)

Geometry of space	Curvature of space	Average density throughout space	Deceleration parameter	Type of universe	Ultimate future of the universe
Spherical	Positive	Greater than the critical density	Greater than $\frac{1}{2}$	Closed	Eventual collapse
Flat	Zero	Exactly equal to the critical density	Exactly equal to $\frac{1}{2}$	Flat	Perpetual expansion (just barely)
Hyperbolic	Negative	Less than the critical density	Between 0 and $\frac{1}{2}$	Open	Perpetual expansion

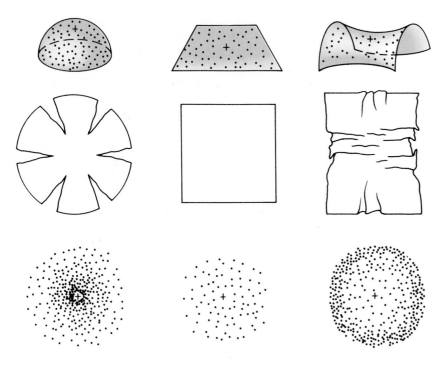

Figure 28-14 Galaxy counts and the curvature of space This illustration shows, by analogy, why the number density of galaxies is affected by the geometry of space. Three surfaces representing the three possible geometries are sprinkled with sand grains. In each case, we measure the distances d from a reference point (the cross) to the grains of sand. On the spherical surface, the number of grains within distance d increases more slowly than d². On the flat surface, the number of grains increases precisely as the square of the distance. On the saddle-shaped surface, the number of grains increases more rapidly than d².

Only the map of the flat plane still shows a uniform distribution of dots just like the random distribution of sand grains. The dots on the map of the spherical surface appear to thin out, because making a flat map of a sphere amounts to stretching the spherical surface until it flattens out. Conversely, the density of dots on the map of the saddle surface increases near the edges, because the surface in effect piles up there when mapped onto a flat piece of paper.

Instead of making maps, astronomers count the number of galaxies per cubic megaparsec at various distances from Earth. The rate at which this number density declines with increasing distance depends on the shape of space and hence on Ω_0, as shown in Figure 28-15. Also shown are the recent data of Princeton physicists Edwin Loh and Earl Spillar, who measured the redshifts of 1000 galaxies in five small patches of the sky. Their results favor Ω_0 nearly equal to 1; and thus we again see that our universe is nearly flat and very close to being marginally bounded.

There is yet another method of determining the geometry of the universe. Recall that light rays from distant galaxies are bent slightly by the curvature of space. This deflection of light rays thus distorts the appearance of remote galaxies. Normally, the more distant an object is, the smaller it appears. However, for the extremely remote galaxies the bending of light by the curvature of the universe can actually magnify images. An extremely distant galaxy can thus actually appear bigger than would normally be expected.

The graph in Figure 28-16 shows how the diameters of the images of galaxies depend on Ω_0. Unfortunately, we must examine galaxies with redshifts greater than $z = 1$ to decide between various possible values of Ω_0. These galaxies

are so far away that they appear only as faint, hazy blobs through Earth-based telescopes. Furthermore, because galaxies evolve, we are forced to compare galaxies at different stages of evolution. After repairs are completed, the Hubble

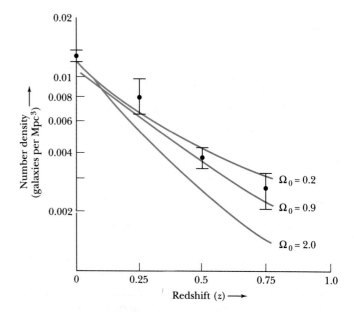

Figure 28-15 The number density of galaxies The number density of galaxies, expressed in galaxies per cubic megaparsec, is plotted here against redshift. Curves for three different values of Ω_0 are shown. The data imply that the density of matter in the universe is nearly equal to the critical density and thus the universe is nearly flat. (Adapted from E. Loh and E. Spillar)

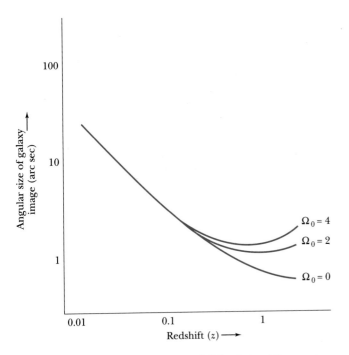

Figure 28-16 The angular-diameter-redshift relation The curvature of the universe can act to magnify the images of remote galaxies. The amount of this magnification depends on the distance to a given galaxy and the value of Ω_0. This effect is noticeable only for galaxies with redshifts greater than about z = 1.

The contraction of the universe will also cause the temperature of the radiation background to increase. Just as the expansion of the universe is ultimately the expansion of space, so the contraction of the universe is actually the contraction of space. Photons traveling across this contracting space will be subjected to a shortening of their wavelengths. Their energies will then increase, and thus the temperature of the radiation field will increase.

Just as the expanding universe began with a Big Bang, the contracting universe will end with a Big Crunch. About 70 million years before the Big Crunch, the temperature of the background radiation field will be up to 300 K, at which point galaxies will be so closely crowded together that the night sky will be as bright as day.

At one million years before the Big Crunch, the radiation temperature will be up to 3000 K. The photons will then possess enough energy to begin dissociating atoms and molecules. With three weeks to go before the Big Crunch, the temperature will be 10 million K, and the stars and planets will dissolve. With only three minutes to go, the temperature will climb above 10 billion K, and the resulting extremely energetic photons will break up nuclei. Finally, all matter and radiation will be crushed out of existence, as a cosmic singularity comes to envelop all space and time.

Just as we cannot use physics to tell us what existed before the Big Bang, so science cannot reveal what might happen after the Big Crunch. The infinite warping of space and time in a cosmic singularity prevents us from extrapolating any further into the future.

Space Telescope will help us understand galactic evolution and, through its sharp images of galaxies, allow us to use tests such as the diameter–redshift relation to determine the ultimate fate of the cosmos.

28-8 *If the universe is bounded, we are living inside a universe that will ultimately collapse in a Big Crunch*

If the average density of matter across space is greater than the critical density, the universe will someday stop expanding. Eventually, the mutual gravitational attraction between galaxies will start them moving back toward each other. As the universe begins to contract, the redshifts we see today in galactic spectra will be replaced by blueshifts.

The specific events leading up to the death of a contracting universe mimic those of the birth of the universe, but in reverse. For many billions of years, this contraction will be noticeable only to astronomers. At first, only the nearest clusters of galaxies will exhibit blueshifts in their spectra, because these are the galaxies we see in the most recent past. Further into the future, however, increasing numbers of more remote galaxy clusters will have their redshifts replaced by blueshifts.

28-9 *If the universe expands forever, black-hole evaporations will occur in the extremely distant future*

If the average density of matter across space is less than or equal to the critical density, the universe will expand forever. Extrapolating into the extremely distant future is a risky business. However, we can still reasonably predict a few significant events that will occur in a universe that exists forever.

Although the universe today consists mostly of hydrogen and helium, these gases will eventually be used up in the thermonuclear fires of generation after generation of stars. Astronomers estimate that this depletion will occur in about 10^{12} years. Star birth will then cease because there will be neither the hydrogen nor the helium required for it in the interstellar medium. Galaxies will grow dim as the final generation of stars dies off. Eventually, roughly a trillion years after the Big Bang, all the matter in the universe will consist of either dead stars (white dwarfs, neutron stars, and black holes) or cold lumps (meteorites, used spaceships, rocks, and so forth).

Box 28-3   The evaporation of black holes

During the early 1970s, Stephen W. Hawking of Cambridge University proposed that numerous small black holes, called **primordial black holes**, could have been created during the Big Bang. Theoretically, these black holes could have had masses as small as the Planck mass: 5×10^{-8} kg.

Most astronomers feel that primordial black holes probably do not exist. Only very special kinds of density fluctuations during the earliest moments of the universe could have led to the copious creation of small black holes. We have no reason to suspect that the early universe obliged us with the necessary special conditions. Nevertheless, theoretical investigations of the properties of small black holes have produced some surprising and significant discoveries.

First, it is instructive to realize how tiny low-mass black holes indeed are. As we saw in Chapter 24, the Schwarzschild radius of a black hole is given by

$$R_{Sch} = \frac{2\,GM}{c^2}$$

where G is the gravitational constant, M is the mass of the hole, and c is the speed of light. For example, if the Earth, with its mass of 6×10^{24} kg, were crushed to become a black hole, it would be about the size of a Ping-Pong ball. A typical large asteroid with a mass of 10^{17} kg crushed to become a black hole would be about the size of an atom. A black hole made from a billion metric tons of matter (roughly 10^{12} kg) would be about the size of a proton.

Because low-mass black holes are submicroscopic, we must use quantum mechanics to understand their properties. The **Heisenberg uncertainty principle** is a basic tenet of submicroscopic physics. This principle states that you cannot determine precisely both the position and the speed of a subatomic particle. Over extremely short distances or times, a certain amount of "fuzziness" is built into the nature of reality.

The Heisenberg uncertainty principle leads logically to the concept of **virtual pairs**, which explains that at every point in space, pairs of particles and antiparticles are constantly being created and destroyed. As we shall see in greater detail in the next chapter, an antiparticle is quite like an ordinary particle,

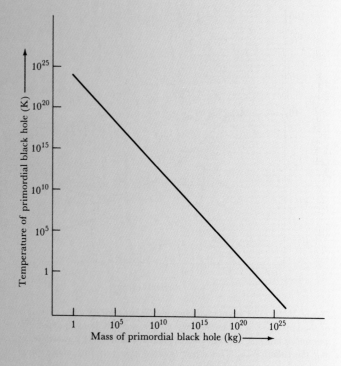

except that it has an opposite electric charge and can annihilate an ordinary particle so that both disappear. In the case of virtual pairs, the process of creation and annihilation occurs over such incredibly brief time intervals that these virtual particles and antiparticles are not directly observable.

Think about a tiny black hole. Furthermore, think about the momentary creation of a virtual proton–antiproton pair just outside the hole's event horizon. It may happen that one of the particles falls into the black hole. Its partner is then deprived of a counterpart with which to annihilate and must therefore become a real particle. To accomplish this conversion, some of the energy of the black hole's gravity must be converted into matter, according to $E = mc^2$. This decreases the mass of the black hole

Still further into the future, the effects of highly improbable events will begin to become important. For example, as noted in an earlier chapter, the probability of two stars colliding somewhere in our Galaxy (or any other galaxy) is extremely small. Indeed, the probability of two stars even passing near each other is almost infinitesimal. Over trillions upon trillions of years, however, these events will sooner or later occur.

A close encounter between two stars in a galaxy is likely to be significant because one of the stars can easily transfer enough energy and momentum to the other to kick it out of its galaxy. The remaining star would then fall into a lower-energy orbit closer to the galaxy's nucleus. Eventually, because of the emission of gravitational radiation (recall Box 24-2), these remaining stars would gradually spiral into the galaxy's center, coalescing into an enormous black hole. The

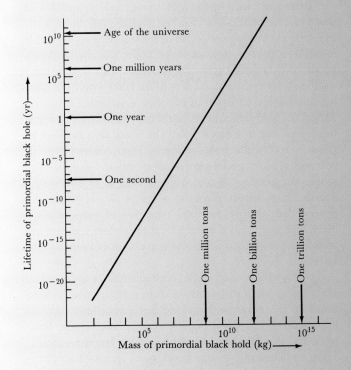

where h is Planck's constant, k is the Boltzmann constant, and M is the mass of the hole. For example, a 1-billion-ton black hole emits particles and energy as if its temperature were nearly 10^{12} K. This relationship between the mass of a black hole and its temperature is graphed in the diagram on the left.

For ordinary black holes, such as Cygnus X-1, this effect is negligible over time spans of billions of years. For example, the temperature of a 10 $M_\odot$ black hole is about 10^{-7} K, barely above absolute zero. In other words, particles hardly ever manage to escape from ordinary black holes.

As particles escape from a small black hole, the mass of the black hole decreases, making its temperature go up. As its temperature rises, still more particles escape, further decreasing the hole's mass and forcing the temperature still higher. This runaway process finally causes the black hole to evaporate completely. During its final seconds of evaporation, the hole gives up the last of its mass in a violent burst of energy equal to the detonation of a billion 1-megaton hydrogen bombs.

The smaller a black hole is, the hotter it is and the faster it evaporates. Hawking proved mathematically that the lifetime t of a black hole is

$$t = 10{,}240\ \pi^2\left(\frac{G^2 M^3}{hc^4}\right)$$

This value tells us that the very low mass primordial black holes evaporated fairly soon after the Big Bang. For example, a 1-million-ton black hole lasts for only 3000 years. The relationship between the mass of a black hole and its lifetime is graphed in the diagram on the right.

Because the universe is about 20 billion years old, it is possible to calculate the minimum mass needed for a primordial black hole to have survived to the present. The answer is roughly 2×10^{11} kg (about 200 million tons). In other words, only those primordial black holes with masses greater than 200 million tons are cool enough and thus evaporating slowly enough to have survived to the present time.

Although few people think that such small black holes actually exist, the ideas and formulas developed by Stephen Hawking reveal that there are fundamental interrelations among thermodynamics, quantum mechanics, and general relativity.

by a corresponding amount, and the particle is free to escape from the hole. In this way, particles can quantum-mechanically "leak" out of a black hole, carrying some of the hole's mass with them.

The smaller a black hole is, the more easily particles can leak out through its event horizon. Jacob Bekenstein and Stephen Hawking proved mathematically that you can speak of the *temperature* of a black hole as a way of describing the amounts of energy carried away by particles "leaking" out of the black hole. This temperature is given by

$$T = \frac{hc^3}{16\ \pi^2 kGM}$$

mass of the resulting **galactic black hole** would be roughly 10^{11} $M_\odot$. The time needed for all this to occur is about 10^{27} years. Thus, when the universe is a billion billion billion years old, galaxies will consist of enormous black holes surrounded by dead stars (see Figure 28-17).

The orbital motions of galaxies in clusters also emit gravitational radiation. As a result of this energy loss, entire galaxies spiral toward each other, eventually to coalesce into

supergalactic black holes. It will take roughly 10^{31} years for these colossal objects to form, each of which might contain 10^{15} $M_\odot$.

Even further into the future, an extraordinary process called **black-hole evaporation** will become important. As discussed in Box 28-3, the laws of quantum mechanics allow a black hole to emit particles. Although this quantum-mechanical phenomenon is significant only for low-mass

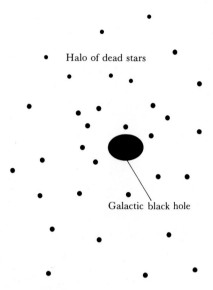

Halo of dead stars

Galactic black hole

Figure 28-17 A dead galaxy *Close encounters between stars can cause their ejection from a galaxy. Over a span of 10^{27} years, a galaxy might lose up to 99 percent of its stars in this manner. Meanwhile, the remaining stars would coalesce into an enormous black hole. This galactic black hole might typically have a mass of about 10^{11} M$_\odot$.*

black holes at normal time scales, even the most massive black holes will eventually evaporate completely if the universe exists forever. From the equations given in Box 28-3, we find that ordinary black holes (such as Cygnus X-1) evaporate after roughly 10^{67} years. Galactic black holes disappear after 10^{97} years. Finally, supergalactic black holes evaporate after 10^{106} years. All of these evaporations end the same way—in a burst of particles and radiation equivalent

to the detonation of a billion 1-megaton hydrogen bombs, releasing a total of 10^{24} joules.

Astronomers usually think of a black hole as an **information sink**: material falls in, but nothing gets out. Specifically, as explained in Chapter 24, all of the information carried by in-falling matter (such as its shape, color, and chemical composition) is removed from the universe. This conception is valid only for ordinary, massive black holes where the quantum mechanical emission of particles is negligible. However, during the final stages of evaporation, vast quantities of particles pour out of a black hole. The black hole in effect behaves like a **white hole**, dumping new material into the universe. This material carries information (such as color, shape, size, and so on), so a white hole is an **information source**. The British physicist Stephen W. Hawking has proved that, during the final moments of evaporation, a black hole is actually a white hole that randomly pumps new matter and information into the universe in an unpredictable fashion.

To emphasize the totally unpredictable way in which evaporating black holes can randomly dump new information into the universe, Hawking has humorously—though improbably—proposed that one of these holes "could emit a television set or the works of Proust in ten leather volumes." Thus, as we stand at the frontiers of human knowledge, groping dimly into the remote future for an understanding that might not even exist, it is perhaps amusing to recall the words of Mark Twain:

There is something fascinating about science. One gets such a wholesale return of conjecture out of such a trifling investment of fact.

—*Life on the Mississippi* (1883)

Key words

Big Bang

Big Crunch

black-hole evaporation

bounded universe

*Compton wavelength

cosmic microwave
  background

cosmic particle horizon

cosmic singularity

cosmological constant

cosmological principle

cosmological redshift

critical density

deceleration parameter

density parameter

era of recombination

Great Attractor

*Heisenberg uncertainty
  principle

marginally bounded
  universe

mass density of radiation

matter-dominated universe

negative curvature

observable universe

Olbers's paradox

*Planck length

*Planck mass

Planck time

plasma

positive curvature

*primordial black hole

primordial fireball

radiation-dominated
  universe

relativistic cosmology

supergalactic black hole

unbounded universe

*virtual pair

*wave–particle duality

white hole

zero curvature

Key ideas

- The universe began as an infinitely dense cosmic singularity that expanded explosively in the event called the Big Bang, which can be described as an explosion of space at the beginning of time.

  The Hubble law describes the continuing expansion of space; it is meaningless to speak of an edge or center to the universe.

  The observable universe extends about 20 billion light years in every direction from the Earth; we cannot see objects beyond the cosmic particle horizon at the distance of 20 billion years, because light from these objects has not had enough time to reach us.

  Before the Planck time (about 10^{-43} second after the calculated time of the Big Bang), the universe was so dense that the known laws of physics do not properly describe the behavior of space, time, and matter.

- The 3-K cosmic microwave background radiation, corresponding to radiation from a blackbody at a temperature of nearly 3 K, is the greatly redshifted remnant of the hot universe that existed about 1 million years after the Big Bang.

  During the first million years of the universe, matter and energy formed an opaque plasma called the primordial fireball.

  By 1 million years after the Big Bang, the expansion of the universe had caused the temperature of the primordial fireball to fall below 3000 K so that protons and electrons could combine to form hydrogen atoms; this event is called the era of recombination.

  The universe became transparent during the era of recombination, meaning that the 3-K microwave background radiation contains the oldest photons in the universe.

  The universe today is matter-dominated, but during the first million years, before the background radiation became greatly redshifted, the universe was radiation-dominated.

- If the average density of matter in the universe is greater than the critical density ρ_c, space is spherical (with positive curvature), the density parameter Ω_0 has a value greater than 1, and the universe is closed (bounded) and will ultimately collapse in a Big Crunch.

- If the average density of matter in the universe is less than ρ_c, space is hyperbolic (with negative curvature), Ω_0 has a value less than 1, and the universe is open (unbounded) and will continue to expand forever.

- If the average density of matter in the universe is exactly equal to ρ_c, space is flat (with zero curvature), Ω_0 is exactly equal to 1, the universe is marginally bounded, and its expansion will just barely continue forever (decreasing toward a zero rate of expansion at an infinite time in the future).

- Present techniques do not permit the measurement of ρ or Ω_0 precisely enough to determine which of the preceding cases exists; the best evidence now available suggests that the average density is very near ρ_c and thus Ω_0 is near 1.

  If the universe is closed (bounded), it will collapse in a Big Crunch during which the events following the Big Bang will recur, but in a reverse sequence.

  If the universe is open (unbounded) or marginally bounded (flat), black holes will eventually evaporate, pouring matter and energy randomly back into the universe.

Review questions

1 What does it mean when astronomers say that we live in an expanding universe? What is actually expanding?

2 What is Olbers's paradox, and how can it be resolved?

3 Explain the difference between a Doppler shift and a cosmological redshift.

4 How does modern cosmology preclude the possibility of either a "center" or "edge" to the universe?

5 Suppose that the universe will expand forever. What will eventually become of the microwave background radiation?

6 What does it mean to say that the universe is matter-dominated? When was the universe radiation-dominated?

7 What was the era of recombination? What significant events occurred in the universe during this era?

8 What is meant by the critical density of the universe? Why is this quantity important to cosmologists?

9 Describe how the critical density, the deceleration parameter, and the geometry of space are related.

10 Under what circumstances does a black hole behave like a white hole?

Advanced questions

> **Tips and tools . . .**
> You may find it useful to know that 1 pc equals
> 3.26 light years and a year contains 3.16×10^7
> seconds. The lifetimes of black holes are discussed
> in Box 28-3.

*11 Calculate the maximum age of the universe for a Hubble constant of 50 km/s/Mpc, 75 km/s/Mpc, and 100 km/s/Mpc. In view of your answers, explain how the ages of globular clusters could be used to place a limit on the maximum value of the Hubble constant.

*12 (Basic) The so-called creationists claim that the universe came into being about 6000 years ago. Find the Hubble constant for such a universe. Is this a reasonable value for H_0? Explain.

*13 (Basic) If the matter density of radiation in the universe were 300 times larger than it is now, what would the background temperature be?

*14 (Basic) What would the critical density of matter in the universe be if the value of the Hubble constant were 100 km/s/Mpc?

15 (Challenging) Prior to the discovery of the 3-K cosmic microwave background, it seemed possible that we might be living in a "steady-state universe" whose overall properties do not change with time. The steady-state model, like the Big Bang model, assumes an expanding universe, but not a "creation event." Instead, in the steady-state theory, matter is assumed to be created continuously everywhere in space to ensure that the average density of the universe remains constant. Explain why the existence of the cosmic microwave background is a fatal blow to the steady-state theory.

16 (Challenging) How can astronomers be certain that the 3-K cosmic microwave background is not some sort of localized phenomenon?

*17 Find the Compton wavelength of a one-kilogram mass. Do you think that this is a detectable wavelength? Explain.

18 Using the values for h, c, and G given in the text, verify that the Planck mass is about 5×10^{-8} kg.

19 With a diagram, show how you would expect the observed number of galaxies to depend on the distance from Earth for various values of q_0. Discuss some of the problems of using such a diagram to determine q_0.

*20 What is the lifetime of a black hole with a mass of 10 $M_\odot$?

*21 Find the mass of a black hole having a temperature of 3 K.

*22 Using the values of h, c, and G given in the text, calculate the value of the Planck time.

*23 What is the mass of a black hole that has a lifetime equal to the age of the universe?

Discussion questions

24 Suppose we were living in a radiation-dominated universe. Discuss how such a universe would be different from what we now observe.

25 Discuss the theological implications of the idea that we cannot use science to tell us what existed before the Big Bang.

26 Suppose that numerous primordial black holes had been produced during the Big Bang. What kinds of observations might you perform to search for these black holes? Is it absurd to suggest that one of these black holes might be relatively nearby (perhaps in our own solar system!) and have escaped being discovered by astronomers? How might you use data from a gamma-ray survey of the sky to place an upper limit on the number of primordial black holes created during the Big Bang?

27 Do you think there can be "other universes," regions of space and time that are not connected to our universe? Should astronomers be concerned with such possibilities? Why or why not?

For further reading

Cohen, N. *Gravity's Lens: Views of the New Cosmology.* John Wiley, 1988 • This appealing introduction to cosmology includes glimpses of recent research on the Big Bang, the cosmic microwave background, gravitational lenses, quasars, active galaxies, and dark matter.

Dicus, D., et al. "The Future of the Universe." *Scientific American*, March 1983 • This fascinating article, which speculates about the future of the universe through the year 10^{100}, includes descriptions of proton decay and the evaporation of black holes.

Disney, M. *The Hidden Universe.* Macmillan, 1984 • This book on dark matter in the universe illustrates how astronomers develop and investigate new ideas.

Dressler, A. "The Large-Scale Streaming of Galaxies." *Scientific American*, September 1987 • This article describes the strategy that astronomers used to piece together observations that led to the discovery of the Great Attractor.

Gott, J., Gunn, J., Schramm, D., and Tinsley, B. "Will the Universe Expand Forever?" *Scientific American*, March 1976 • This article presents a nice summary of the theoretical and observational consequences of the expansion of the universe.

Gribbin, J. *In Search of the Big Bang.* Bantam, 1986 • This very readable introduction to modern cosmology describes the history, observations, and ideas leading to our modern understanding of the creation and structure of the universe.

Harrison, E. *Darkness at Night.* Harvard University Press, 1987 • This definitive book on Olbers's Paradox uses eloquent, nontechnical language to explain why the night sky is dark.

Hawking, S. "The Quantum Mechanics of Black Holes." *Scientific American*, January 1977 • This lucid article explains how matter can quantum mechanically escape from black holes, causing them to evaporate.

Islam, J. "The Ultimate Fate of the Universe." *Sky & Telescope*, January 1979 • This fascinating article uses clear explanations of quantum-mechanical phenomena to speculate about the extremely distant future.

Kippenhahn, R. *Light from the Depths of Time.* Springer Verlag, 1987 • This delightful introduction to cosmology by a noted German astronomer features the space-travel daydreams of an "Everyman" named Herr Meyer and good analogies to explain such phenomena as curved space and the Big Bang.

Shu, F. "The Expanding Universe and the Large-Scale Geometry of Space–Time." *Mercury*, November/December 1983 • This lucid article describes the relationships between gravitation, relativity, and the expansion of the universe.

Trefil, J. *The Dark Side of the Universe.* Scribner, 1988 • This book presents an excellent summary of the current status of research on such cosmological issues as dark matter, the ultimate fate of the universe, and cosmic strings.

Wagoner, R., and Goldsmith, D. *Cosmic Horizons: Understanding the Universe.* W. H. Freeman and Company, 1983 • This superb introduction to cosmology is a clear, concise guide to modern ideas about the origin and evolution of the universe.

Webster, A. "The Cosmic Background Radiation." *Scientific American*, August 1974 • This article describes the radiation, ranging from radio waves to gamma rays, that fills the universe.

Weinberg, S. *The First Three Minutes*, 2nd ed. Basic Books, 1988 • This updating of a classic 1977 introduction to cosmology includes a lucid description of the events immediately following the Big Bang.

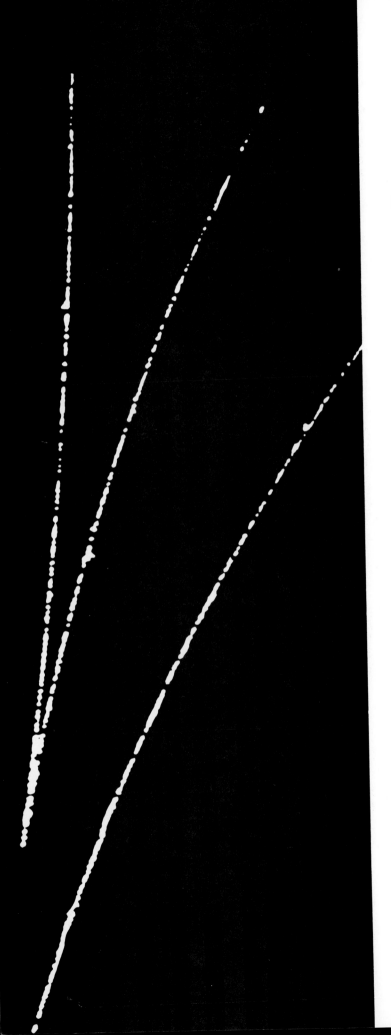

29

The Physics of the Early Universe

The laws of physics are the tools with which astrophysicists explore the earliest moments of the universe. Quantum mechanics, which reveals that empty space is actually seething with activity, leads one to speculate that all the matter and energy in the universe may have burst forth from a vacuum during the Big Bang. The smoothness of the cosmic microwave background suggests that the newborn universe may have suddenly "inflated" to many billions of times its original size. As it expanded, the four physical forces became separate and distinct. To understand the origins of these forces, researchers have invented speculative theories that predict such oddities as "cosmic strings" and ten dimensions of space. Also, they are using information about large-scale structure of the universe along with supercomputer simulations to try to understand how clusters of galaxies formed.

The creation of matter from energy   A gamma ray (from below) enters a bubble chamber, a device filled with liquid hydrogen and is designed to make the path of a charged particle visible. Near the bottom of the photograph, the energy carried by the gamma ray is converted into an electron and an antielectron. Because of a magnetic field surrounding the bubble chamber, the electron is deflected to the right and the antielectron veers toward the left. The path of a stray electron can also be seen. The path of the gamma ray is not visible, because photons are electrically neutral. (Courtesy of Lawrence Berkeley Laboratory)

In recent years there has been a productive marriage of two apparently diverse fields of science. Cosmology, the study of the universe on the largest scale, has become intimately tied to particle physics, the study of matter on the smallest scale. This union has already yielded extraordinary insights into the creation of the universe and the fundamental nature of matter and physical forces.

The reason for this union of cosmology and particle physics stems from the extreme conditions that existed throughout space immediately after the Big Bang. The universe then was so dense and hot that particles were everywhere constantly colliding at speeds nearly equal to the speed of light. These collisions were violent enough to shatter particles, breaking them into their most fundamental constituents, in much the same way that physicists explore the structure of matter by using particle accelerators. Thus, particle physics is an important resource for the astronomer wishing to understand the early universe.

Similarly, advances in cosmology benefit particle physics. Accelerators are limited to energies that are quite low compared to those during the Big Bang. Intriguing phenomena have been predicted at energies that are much higher than any we can achieve on Earth. Because of the extreme energies with which particles interacted during the Big Bang, the earliest moments of the universe are a physicist's laboratory, in which the most deeply hidden secrets of matter and forces are explored.

A few words of caution are in order. First, many of the topics in this chapter are quite speculative and are presented primarily to give the reader a flavor of some of the ideas at the frontiers of science. Second, much of this material is conceptually difficult and intellectually challenging.

29-1 During the Big Bang, all the mass and energy in the universe may have burst forth from a vacuum

Where did all the matter and radiation in the universe come from in the first place? Recent intriguing theoretical research by such physicists as Steven Weinberg of Harvard and Ya. B. Zeldovich in Moscow suggests that the universe began as a perfect vacuum and all the particles of the material world were created from the expansion of space. To see how these events might have happened, we must first understand what quantum mechanics tells us about empty space.

Quantum mechanics is the branch of physics that deals with submicroscopic phenomena such as the behavior of individual particles and photons. This field of physics tells us, for example, how to calculate the structure of atoms and the interactions between nuclei.

There are significant differences between the ordinary world around us and the submicroscopic world of quantum mechanics. In the ordinary world we have no trouble knowing where things are. You know where your house is; you know where your car is; you know where this book is. In the subatomic world of electrons and nuclei, however, you can no longer speak with this same confidence. A certain amount of fuzziness and uncertainty enters into the description of reality. At the incredibly small dimensions of the quantum world, things are no longer sharp and distinct.

To appreciate the reasons for this uncertainty, imagine trying to measure the position of a single electron. To find out where it is located, you must observe it. And to observe it, you shine a light on it. However, the electron is so tiny and has such a small mass that the photons in your beam of light possess enough energy to give the electron a mighty kick. As soon as a photon strikes the electron, to let you observe the reflected photon and measure the electron's position, this photon imparts momentum to the electron, causing it to recoil in some unpredictable direction. Consequently, no matter how carefully you try to measure the precise location of an electron, you necessarily introduce some uncertainty into the speed and momentum of that electron.

These ideas are at the heart of the famous Heisenberg uncertainty principle, first formulated by Werner Heisenberg in 1927. This principle states that there is a reciprocal uncertainty between position and momentum. The more precisely you try to measure the position of a particle, the more unsure you become of how the particle is moving. Conversely, the more accurately you determine the momentum (mass × velocity) of a particle, the less sure you are of its location.

There is an analogous uncertainty involving energy and time. You cannot know with infinite precision the exact energy of a quantum-mechanical system at every moment in time. Over short time intervals, there can be great uncertainty about the amounts of energy in the subatomic world. Specifically, let ΔE be the uncertainty in energy measured over a short interval of time Δt. Then

$$\Delta E \times \Delta t \geq \frac{h}{2\pi}$$

where h is Planck's constant.

We might look upon the Heisenberg uncertainty principle as being merely an unfortunate limitation on our ability to know everything with infinite precision. Alternatively, we can explore how this principle may be helpful in our quest to understand the universe.

We have seen that one of the important conclusions of Einstein's special theory of relativity is the equivalence of mass and energy: $E = mc^2$. There is nothing uncertain about the speed of light (c). Therefore any uncertainty in the energy of a physical system can, according to Einstein's equation, be attributed to an uncertainty Δm in the mass. Thus

$$\Delta E = c^2 \Delta m$$

Combining this expression with the previous equation, we obtain

$$\Delta m \times \Delta t \geq \frac{h}{2\pi c^2}$$

This result is astonishing. It means that, in a very brief interval Δt of time, we cannot be sure how much matter there is in a particular location, even in "empty space." During this brief moment, matter can spontaneously appear, then disappear. Whenever this happens, each particle that is created is accompanied by an antiparticle.

There is nothing mysterious about antimatter. A particle and an antiparticle are identical in almost every respect; their main distinction is that they carry opposite electric charges. For example, an antielectron (or positron, e^+) has a positive charge, whereas an ordinary electron (e^-) carries a negative charge. Particles and antiparticles are always created or destroyed in equal numbers, thus ensuring that the total electric charge in the universe remains constant, regardless of how many pairs of particles and antiparticles come and go. Physicists often refer to this kind of balance between matter and antimatter as a **symmetry**.

The time interval over which a particle–antiparticle pair can exist is incredibly brief. For example, consider an electron and an antielectron, each with a mass 9.1×10^{-31} kg. To see how long an electron–antielectron pair might exist without violating the principle of conservation of mass, we substitute their combined mass in the previous equation for Δm and we find that Δt equals 6.5×10^{-22} s. In other words, during a time interval shorter than 6.5×10^{-22} s, an electron and an antielectron can spontaneously appear, then disappear without violating any laws of physics.

This process can happen absolutely anywhere and at any time. It can also occur with more massive particles, but because of the "reciprocal" nature of the uncertainty principle, such massive particles can exist only for correspondingly shorter time intervals. For example, the proton is about 2000 times heavier than the electron. Pairs of protons and antiprotons can appear and disappear spontaneously, but they exist for only $\frac{1}{2000}$ as long as the pairs of electrons and antielectrons exist.

One of the fundamental doctrines of subatomic physics, sometimes called the **Principle of Totalitarianism**, is that if a quantum mechanical process is not strictly forbidden, then it must occur. Pairs of every conceivable particle and antiparticle are constantly being created and destroyed at every location across the universe. We have no way of observing these pairs of particles and antiparticles directly, which is forbidden by the uncertainty principle. The particle–antiparticle pairs exist for such short intervals that direct observation of them is impossible in theory as well as in practice. For this reason, these particle–antiparticle pairs are called **virtual pairs**. They don't "really" exist; they "virtually" exist.

Although these virtual pairs of particles and antiparticles cannot be observed directly, their effects have been detected.

Imagine, for instance, an electron in orbit about the nucleus of an atom, such as a hydrogen atom. Ideally, the electron should follow its orbit in a smooth, unhampered fashion. However, because of the constant brief appearance and disappearance of pairs of particles and antiparticles, minuscule electric fields exist for extremely short intervals of time. These tiny, fleeting electric fields cause the electron to jiggle slightly in its orbit. The effect of this jiggling was first detected in 1947 by W. E. Lamb and R. C. Retherford, who noticed a tiny shift in the spectrum of light from the hydrogen atom. This phenomenon, known as the **Lamb shift**, provides powerful support for the idea that every point in space, all across the universe, is seething with virtual pairs of particles and antiparticles. This constant appearance and disappearance of particles and antiparticles is sketched schematically in Figure 29-1.

The concept of virtual pairs also provides a ready explanation for the phenomenon of pair production. It has been known for years that highly a energetic gamma ray (photon) can convert its energy into pairs of particles and antiparticles according to the equation $E = mc^2$. Quite simply, the gamma ray disappears upon colliding with a second photon, and a particle and an antiparticle appear in its place. This process is routinely observed in high-energy nuclear experiments. Indeed, it is one of the ways in which physicists can manufacture many of the more exotic species of particles and antiparticles. This process of **pair production** and its inverse process of **annihilation** are sketched in Figure 29-2.

These particles and antiparticles come from nature's ample supply of virtual pairs. The gamma rays provide a virtual pair with so much energy that the virtual particles can materialize and appear as real particles in the real world. The only requirement is that nature's balance sheet be satisfied. To create a particle and an antiparticle having a total mass M, the incoming gamma-ray photons must possess an amount of energy E that satisfies the condition $E \geq Mc^2$. If

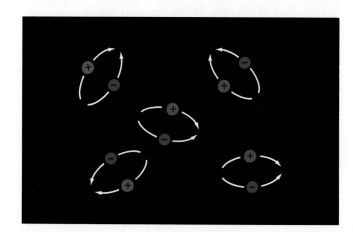

Figure 29-1 *Virtual pairs* *Pairs of particles and antiparticles can appear then disappear anywhere in space, provided that each pair exists only for a very short time interval, as dictated by the uncertainty principle.*

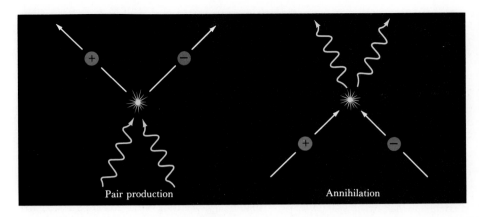

Figure 29-2 *Pair production and annihilation* *Pairs of virtual particles can be converted into real particles by high-energy gamma-ray photons. Conversely, a particle and an antiparticle can annihilate each other and be transformed into energy in the form of gamma rays.*

Pair production

Annihilation

the photons carry too little energy, pair production will not occur. Likewise, the more energetic the photon, the more massive are the particles and antiparticles that can be manufactured.

Around 1980, physicists began applying these ideas to their thinking about the creation of the universe. Immediately after the Big Bang, space was expanding with explosive vigor. Yet, as we have seen, all space is seething with virtual pairs of particles and antiparticles. Normally, a particle and an antiparticle have no trouble getting back together in a time interval (Δt) short enough as required by the uncertainty principle. During the Big Bang, however, the universe was expanding so fast that particles were rapidly separated from their corresponding antiparticles. Deprived of the opportunity to recombine, these virtual particles had to become real particles in the real world.

Where did the energy come from to achieve this materialization? Recall that the Big Bang was like the center of a black hole. A vast supply of gravitational energy was associated with the intense gravity of this cosmic singularity. This resource provided ample energy to fill the universe completely with all the conceivable kinds of particles and antiparticles. Thus, immediately after the Planck time, the universe became flooded with particles and antiparticles created by the violent expansion of space.

29-2 As the primordial fireball cooled off, most of the matter and antimatter in the early universe annihilated each other

As soon as the flood of matter and antimatter appeared in the universe, collisions between particles and antiparticles began to produce numerous high-energy gamma rays. An equilibrium was then quickly arrived at. For example, for every proton (p) and antiproton ($\bar{p}$) that annihilated each other to create gamma rays, two gamma rays collided else-

where to produce a proton and an antiproton. In other words, reactions such as

$$p + \bar{p} \rightarrow \gamma + \gamma$$

and

$$\gamma + \gamma \rightarrow p + \bar{p}$$

proceeded with equal vigor. This condition, wherein the number of particles and antiparticles is just high enough so that precisely as many are being created each second as are being destroyed, is called thermal equilibrium.

Now let us think forward in time from the Big Bang. As the universe continued to expand, all the gamma-ray photons became increasingly redshifted, so the temperature of the radiation field fell. The temperature eventually became so low that the gamma rays no longer had enough energy to create particular kinds of particles and antiparticles. Collisions between particles and antiparticles did add photons to the cosmic-radiation background, but collisions between photons could no longer replenish the supply of particles and antiparticles.

For example, consider again protons and antiprotons, each having a mass of 1.7×10^{-27} kg. Substituting this mass into the equation $E = mc^2$, we calculate the **rest energy** associated with the mass of each of these particles to be 938 MeV, where MeV stands for "million electron volts," a unit of energy commonly used by physicists (review Box 5-3). This value tells us how much energy the colliding gamma-ray photons must have in order to create a proton and an antiproton. If the combined energy of the two photons is less than 2×938 MeV, the reaction $\gamma + \gamma \rightarrow p + \bar{p}$ will not occur.

A photon energy of 938 MeV corresponds to a particular temperature, according to the equation

$$E = kT$$

where k is the Boltzmann constant (k = 1.38×10^{-23} J/K = 8.6×10^{-5} eV/K). For gamma-ray photons having energy

Box 29-1 Properties of some elementary particles

Physicists classify all particles into two categories, **leptons** (which do not feel the strong nuclear force) and **hadrons** (which do feel the strong force). Leptons typically are low-mass particles related to electrons. Hadrons, a category that includes protons and neutrons, typically are more massive. Most physicists believe that all hadrons are composed of still more basic particles called **quarks**.

The rest energy of a particle equals its mass times the square of the speed of light. It is the energy that would be released if all the mass of the particle were to be converted into energy. The threshold temperature of a particle is its rest energy divided by Boltzmann's constant. It is the temperature above which a particle can be freely created out of thermal radiation.

The table that follows lists several types of particles, along with their rest energies and threshold temperatures.

Particle	Symbol	Rest energy (MeV)	Threshold temperature (10^9 K)
Neutrino	$\nu, \bar{\nu}$	0.00001 (?)	0.0001
Electron	e^-, e^+	0.5110	5.930
Muon	μ^-, μ^+	105.66	1,226.2
Pi meson	π°	134.96	1,556.2
	π^+, π^-	139.57	1,619.7
Proton	p, $\bar{p}$	938.26	10,888
Neutron	n, $\bar{n}$	939.55	10,903

of 938 MeV, the temperature is 10.9 trillion K. Thus, as soon as the temperature of the radiation field falls below 10.9×10^{12} K, proton–antiproton pairs can no longer be created. This temperature is therefore known as the **threshold temperature** for the creation of protons and antiprotons.

There are many different kinds of particles, each having its own threshold temperature. For example, electrons and antielectrons have masses about 1800 times smaller than protons and antiprotons. Thus, the threshold temperature for the creation of electron–antielectron pairs is $\frac{1}{1800}$ that for proton–antiproton pairs. Consequently, when the radiation temperature falls below 5.9 billion K, electron–antielectron pairs can no longer be created by the reaction $\gamma + \gamma \rightarrow e^+ + e^-$. We thus say that the threshold temperature for electrons and antielectrons is 5.9×10^9 K. Several common types of particles, with their masses and threshold temperatures, are listed in Box 29-1.

When the universe was about 0.0001 second old (that is, at time $t = 10^{-4}$ s after the Big Bang), the temperature of the radiation fell below 10^{13} K. At that moment, the universe became cooler than the threshold temperatures of both protons and neutrons, so the annihilation of protons by antiprotons and of neutrons by antineutrons occurred vigorously everywhere throughout space. This wholesale annihilation dramatically lowered the matter content (particles and antiparticles) of the universe, while simultaneously increasing the radiation (photon) content.

Similarly, when the universe was about 1 second old, its temperature fell below 6×10^9 K, the threshold temperature for electrons and antielectrons. A similar annihilation of pairs of electrons and antielectrons began occurring everywhere, further decreasing the matter content of the universe while raising its radiation content.

Now we have a dilemma. If the symmetry between particles and antiparticles were truly valid, then for every proton there should have been an antiproton. For every electron, there should likewise have been an antielectron. Consequently, by the time the universe was 1 second old, every particle would have been annihilated by an antiparticle, leaving no matter at all in the universe.

Obviously, this did not happen. The planets, stars, and galaxies we see in the sky are made of matter, not antimatter. Thus there must have been an excess of matter over antimatter immediately after the Big Bang. Physicists therefore say that a **symmetry breaking** occurred during the earliest moments of the universe, in which the number of particles emerging from the Big Bang was not exactly equaled by the number of antiparticles. As noted in Chapter 28, there are roughly a billion photons today in the microwave background for each proton and neutron in the universe. Consequently, for every billion antiprotons, a billion plus one ordinary protons were created. And for every billion antielectrons, a billion plus one ordinary electrons were created.

29-3 A background of neutrinos and most of the helium in the universe are relics of the primordial fireball

The early universe must have been populated with vast numbers of neutrinos (ν) and antineutrinos ($\bar{\nu}$). These particles are the ones involved in the nuclear reaction that transforms neutrons into protons and vice versa. For example, under laboratory conditions, the neutron decays into a proton by emitting an electron and an antineutrino:

$$n \rightarrow p + e^- + \bar{\nu}$$

The half-life for this radioactive decay is about 10.5 minutes, which is why we do not find free neutrons floating around in the universe today.

In the early universe, however, collisions between particles kept the number of protons approximately equal to the number of neutrons, according to the two-way reactions

$$p + e^- \rightleftharpoons n + \nu$$

and

$$p + \bar{\nu} \rightleftharpoons n + e^+$$

This balance could be maintained only as long as the density of matter in the universe remained high. By the time the universe was about 2 seconds old, however, matter was thinned out enough that the neutrinos and antineutrinos no longer collided with the protons and neutrons in significant numbers. The natural tendency for neutrons to decay into protons then took over, and the number of neutrons began to decline.

Before all the neutrons could decay into protons, the primordial fireball had cooled to the extent that helium nuclei could survive without being broken apart. The first step in creating helium involves combining a proton and a neutron to produce deuterium (^2H), sometimes called "heavy hydrogen," according to the reaction

$$p + n \rightarrow {}^2H$$

The proton and the neutron do not stick together very well, however. In the early universe, high-energy gamma rays easily broke down deuterons into independent protons and neutrons. Because deuterons could not survive, helium could thus not be created. This block to helium creation is called the **deuterium bottleneck**.

Finally, when the universe was about 3 minutes old, the photons in the background radiation became so redshifted that they could no longer break up the deuterium. By this time, the decaying of neutrons into protons had shifted the neutron–proton balance to about 14 percent neutrons and 86 percent protons. Because deuterons could now survive, all these remaining neutrons combined with protons and, through the usual series of reactions (review Box 18-2), rapidly produced helium. The final result is the proportion of 1 helium atom for every 10 hydrogen atoms that we find in the universe today.

What happened to all those primordial neutrinos and antineutrinos that had interacted so vigorously with the protons and neutrons before the universe was 2 seconds old? As mentioned in the discussion of the Sun, neutrinos and antineutrinos are very difficult to detect because they do not interact strongly with matter. The Earth itself is virtually transparent to the neutrinos from the Sun.

By the time the universe was about 2 seconds old, matter was sufficiently spread out so that the universe had become transparent to neutrinos and antineutrinos. From that time on, neutrinos and antineutrinos could travel across the universe unimpeded.

This "decoupling" of the neutrino–antineutrino background from the matter in the universe was very much like the decoupling of the photon background during the era of recombination (recall Figure 28-9). In fact, physicists estimate that these neutrinos and antineutrinos may be about as populous today as the photons in the microwave background ($5.5 \times 10^8 \text{ m}^{-3}$). The neutrino–antineutrino background should be slightly cooler than the photon background, which received extra energy from the electron–antielectron annihilations. Physicists estimate that the current temperature of the neutrino–antineutrino background is about 2 K, as opposed to 3 K for the microwave background.

Until recently, no one had paid much attention to this elusive neutrino–antineutrino background. Physicists generally believed these particles, like photons, to be massless. Because they interact so weakly with ordinary matter, they just did not seem very important. However, around 1980, experiments by E. T. Tretyakov and colleagues in the Soviet Union, as well as Frederick Reines and his colleagues in the United States, suggested that neutrinos and antineutrinos have mass after all. It is an extremely tiny mass, probably less than 10^{-4} of the mass of the electron. More recent experiments have failed to confirm the existence of the neutrino mass, however. As we saw in Chapter 22, analyses of the burst of neutrinos detected from the 1987 supernova constrain the neutrino mass to be less than about 10 eV, or about $\frac{1}{50,000}$ the mass of an electron.

There are so many neutrinos and antineutrinos spread throughout space that these particles could account for most of the matter in the universe if this tiny mass does exist. Indeed, the matter contained in neutrinos and antineutrinos may be ten times greater than all the matter in the stars, planets, and galaxies combined. Consequently, the neutrino–antineutrino background is a candidate for the "missing mass" that seems to pervade the universe.

29-4 The flatness of the universe and the isotropy of the microwave background suggest that a period of vigorous inflation followed the Big Bang

Ever since Edwin Hubble discovered that the universe is expanding, astronomers have struggled to determine its rate of deceleration. During the 1960s and 1970s, various teams of astronomers determined different values for Ω_0, some slightly larger than $\Omega_0 = 1$ and some smaller. Because $\Omega_0 = 1$ is the special case of a flat universe that separates a bounded cosmological model from an unbounded one, the predicted fate of the universe swung back and forth. According to some data, the universe seems to be just barely open and infinite, whereas other data indicate that it is just barely closed and is doomed ultimately to collapse.

Motivated by the fact that Ω_0 is nearly equal to 1, physicists began looking at the special conditions associated with a flat universe in which the average density of matter is exactly equal to the critical density, ρ_c. Specifically, suppose that the average density of matter during the Big Bang were slightly larger, or smaller, than the critical density. How would this deviation grow, or decrease, as the universe evolved?

In the previous chapter (recall Box 28-1) we saw that the earliest understandable moment in the universe was the Planck time, about 10^{-43} second after the Big Bang. Between $t = 0$ and $t = 10^{-43}$ second, the universe was so dense and particles were interacting so violently that no known theory can properly describe what was happening then. However, immediately after the Planck time, the fate of the universe became sensitively dependent on the density of matter. Calculations demonstrate that the slightest deviation from the precise critical density would mushroom very rapidly, multiplying itself every 10^{-35} second. If the density had been slightly less than ρ_c, the universe would soon have become wide open and virtually empty. If, on the other hand, the density had been slightly greater than ρ_c, the universe would soon have become tightly closed and so packed with matter that the entire cosmos would rapidly collapse in a Big Crunch. In other words, immediately after the Big Bang the fate of the universe hung in the balance, like a pencil teetering on its point, so that the tiniest deviation from the precise equality $\rho = \rho_c$ would have rapidly propelled the universe away from the special case of $\Omega_0 = 1$.

Observations reveal that Ω_0 is today approximately 1. Consequently, the density of the universe immediately after the Big Bang must have been equal to the critical density to an incredibly high order of precision. Calculations demon-

strate that, in order for Ω_0 to be roughly 1 today, the value of ρ right after the Big Bang must have been equal to ρ_c to more than 50 decimal places!

What could have happened immediately after the Planck time to ensure that $\rho = \rho_c$ to such an astounding degree of accuracy? Because $\rho = \rho_c$ means that space is flat, this enigma is called the **flatness problem**.

A second enigma, closely related to the flatness problem, is the isotropy of the 3-K cosmic microwave background. As we saw in Chapter 28, the microwave background is so incredibly uniform across the sky that sensitive temperature measurements can reveal our own motion through this radiation field (recall Figure 28-5). Subtracting the effects of our own motion, we find that the temperature of the microwave background is the same in all parts of the sky to an accuracy of 1 part in 10,000.

To appreciate this so-called **isotropy problem**, think about microwave radiation coming toward us from two opposite parts of the sky. This radiation left over from the primordial fireball has been traveling for nearly 20 billion years. The total distance between the opposite sides of the observable universe is roughly 40 billion light years, so these widely separated regions have absolutely no connection with each other. Why, then, do these unrelated parts of the universe have the same temperature?

In the early 1980s, Alan Guth at Stanford University offered a remarkable solution to the problems of the flatness of the universe and the isotropy of the microwave background. Guth analyzed the suggestion that the universe might have experienced a brief period of extremely rapid expansion shortly after the Planck time (see Figure 29-3). During this **inflationary epoch**, as it is called, the universe ballooned outward in all directions to become many billions of times its original size. This inflation has moved much of the material that was originally near our location far beyond the edge of the observable universe today. Thus, the observable universe is now expanding *into* space containing matter and radiation that was once in close contact with our location.

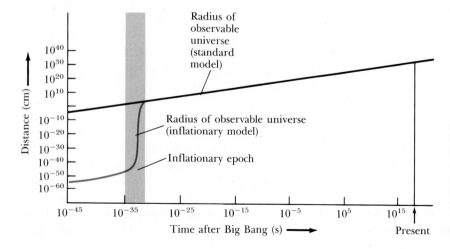

Figure 29-3 The observable universe with and without inflation According to the inflationary model, the universe expanded by a factor of about 10^{50} shortly after the Big Bang. This sudden growth in the size of the observable universe probably occurred during a very brief interval, as indicated by the vertical shaded area on the graph. For comparison, the projected size of the universe without inflation is also shown. (Adapted from A. Guth)

In Guth's mathematical analysis, the violent expansion of the universe during the inflationary epoch occurred because, for a brief interval, the cosmological constant Λ was huge. Recall that the cosmological constant was introduced by Einstein to hold up a static universe. Today Λ is very tiny and may, in fact, be zero (see Box 28-2 for some details). During the inflationary epoch, however, Λ was so large that it provided repulsive forces 10^{120} times larger than Einstein needed in his static cosmology to prop up the universe. This cosmic repulsion drove a violent expansion of space in all directions, literally producing the event we today call the Big Bang.

Inflation accounts for the flatness of the universe. To see why, think about a small portion of the Earth's surface, such as your backyard. For all practical purposes, it is impossible to detect the Earth's curvature over such a small area, so your backyard looks flat. Similarly, the observable universe is such a tiny fraction of the entire inflated universe that any overall curvature in it is virtually undetectable. Like your backyard, the segment of space we can observe looks flat.

The existence of an inflationary epoch would also account for the isotropy of the microwave background. In examining microwaves that are from opposite parts of the sky, what we are seeing is radiation from parts of the universe that were originally in intimate contact with each other. This common origin is why they have the same temperature.

Finally, it is important to note that the concept of inflation does not violate Einstein's dictum that nothing can travel faster than the speed of light. Remember that the expansion of the universe is the expansion of space and does not involve the motion of objects through space. The inflationary epoch was a brief moment during which the distances between particles suddenly increased by an enormous amount. The important point is that this expansion was accomplished entirely by a sudden vigorous *expansion of space*. Particles did not move through space; space inflated.

29-5 Grand unified theories (GUTs) explain that all the forces had the same strength immediately after the Big Bang

All of the behavior and interactions of everything in the universe can be understood as the result of just *four* physical forces: gravity, electromagnetism, and the strong and weak nuclear forces. We are all intimately familiar with the force of gravity, the long-range force that dominates the universe over astronomical distances. The electromagnetic force is also a long-range force (in principle, its influence extends to infinity, as gravity's does), but it is much stronger than the gravitational force. Just as the force of gravity holds the Moon in orbit about the Earth, the electromagnetic force

holds electrons in orbit about the nuclei in atoms. We do not generally observe the long-range effects of the electromagnetic force, however, because usually there is a negative electric charge for every positive charge and a south magnetic pole for every north magnetic pole. Thus, over great volumes of space the effects of electromagnetism effectively cancel out. No similar canceling occurs with gravity because there is no equivalent "negative mass."

Both the strong and the weak nuclear forces are said to be short-range because their influence extends only over distances that are less than about 10^{-15} m. The **strong nuclear force** holds protons and neutrons together inside the nuclei of atoms. Without the strong nuclear force, nuclei would disintegrate because of the electromagnetic repulsion of the positively charged protons. The strong nuclear force in fact overpowers the electromagnetic forces inside nuclei.

The weak nuclear force is so weak that it cannot hold anything together. Instead, the **weak nuclear force** is at work in certain kinds of radioactive decay, such as the transformation of a neutron (n) into a proton (p), with the release of an electron (e^-) and an antineutrino ($\bar{\nu}$): $n \rightarrow p + e^- + \bar{\nu}$.

Numerous experiments in nuclear physics have strongly suggested that protons and neutrons are composed of more basic particles called **quarks**, the most common varieties being "up" (u) quarks and "down" (d) quarks. A proton is composed of two up quarks and one down quark, a neutron of two down quarks and one up quark.

In the 1970s, the concept of quarks gave rise to a more fundamental description of the strong and weak nuclear forces. In its most basic form, the strong nuclear force is what holds quarks together. Similarly, the weak nuclear force is at work whenever a quark changes from one variety to another. For example, when a neutron decays into a proton, one of the neutron's down quarks changes into an up quark. Thus, the weak nuclear force is responsible for transformations such as

$$d \rightarrow u + e^- + \bar{\nu}$$

In the 1940s, the physicists Richard P. Feynman, Julian S. Schwinger, and Sinitiro Tomonaga succeeded in developing a basic description of what we mean by "force." Focusing their attention on the electromagnetic force, they tried to describe exactly what happens when two charged particles interact. According to their theory, now called **quantum electrodynamics**, charged particles interact by exchanging virtual photons. Virtual photons, like virtual particles, cannot be observed directly, because they exist for extremely short time intervals. The process of exchanging virtual photons is often described with a **Feynman diagram** (named after its inventor), as in Figure 29-4, which schematically shows two electrons interacting by exchanging a photon. The directions in which the electrons are traveling is altered by the exchange.

Quantum electrodynamics has proven to be one of the most successful theories in modern physics. It accurately

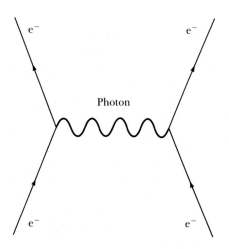

Figure 29-4 A Feynman diagram of the electromagnetic force According to quantum electrodynamics, electric and magnetic forces occur as the result of virtual photons jumping back and forth between charged particles. This exchange, shown schematically here, occurs during time intervals so short that the photons cannot be directly detected or observed while in flight between the charged particles.

describes many details of the electromagnetic interaction between charged particles. Inspired by these successes, physicists have tried to develop similar theories for the other three forces. Specifically, the weak nuclear force occurs when particles exchange **intermediate-vector bosons**, the

gravitational force occurs when particles exchange **gravitons**, and quarks stick together by exchanging **gluons**. The four physical forces can thus be summarized as indicated in Table 29-1.

In recent years, important progress has been made in understanding the weak nuclear force, primarily through the efforts of physicists Steven Weinberg, Sheldon Glashow, and Abdus Salaam. They proposed a theory that predicted the existence of three types of intermediate-vector bosons (W^+, W^-, and Z^0), which are exchanged in various manifestations of the weak force (see Figure 29-5). These three particles were discovered in high-energy nuclear-physics experiments in the early 1980s, thus providing strong support for this theory.

One of the startling predictions of the Weinberg–Glashow–Salaam theory is that the weak force and the electromagnetic force should be identical to each other at energies greater than 100 GeV. In other words, if particles are slammed together with a total energy greater than 100 billion electron volts, then electromagnetic interactions become indistinguishable from weak interactions. We can thus say that "above 100 GeV" the electromagnetic force and the weak force are "unified."

This unification occurs because the three types of intermediate-vector bosons behave just like photons above 100 GeV. Physicists describe this similarity by saying that "symmetry is restored" above 100 GeV. In the world around

Table 29-1   The four forces

Force	Relative strength	Particles exchanged	Particles acted upon	Range	Example
Strong	1	Gluons	Quarks	10^{-15} m	Holds nuclei together
Electromagnetic	1/137	Photons	Charged particles	Infinite	Holds atoms together
Weak	1/10,000	Intermediate-vector bosons	Quarks, electrons, neutrinos	10^{-16} m	Radioactive decay
Gravity	6×10^{-39}	Gravitons	Everything	Infinite	Holds the solar system together

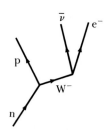

a Decay of neutron
($n \rightarrow p + c^- + \bar{\nu}$)

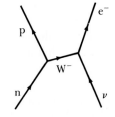

b Neutron collides with neutrino
($n + \nu \rightarrow p + c^-$)

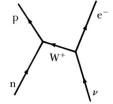

c Neutron collides with neutrino
($n + \nu \rightarrow p + c^-$)

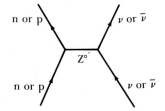

d Neutron or proton collides
with neutrino or antineutrino

Figure 29-5 Feynman diagrams of the weak nuclear force The weak nuclear force exists when particles emit or exchange intermediate-vector bosons (W^+, W^-, Z^0). (a) A neutron (n) decays into a proton (p) by emitting a W^-, which rapidly turns into an electron (e^-) and an antineutrino ($\bar{\nu}$). (b) When a neutron collides with a neutrino (ν), a W^- is exchanged between the particles, transforming the neutron to a proton and the neutrino to an electron. However, in a Feynman diagram an outgoing antineutrino is exactly equivalent to an incoming ordinary neutrino. Therefore, this reaction is exactly equivalent to the neutron decay shown in (a). (c) This diagram is also equivalent to those in (a) and (b) because a W^+ going from right to left is equivalent to a W^- going from left to right. (d) A neutron or proton interacts with a neutrino or antineutrino through exchanging a Z^0.

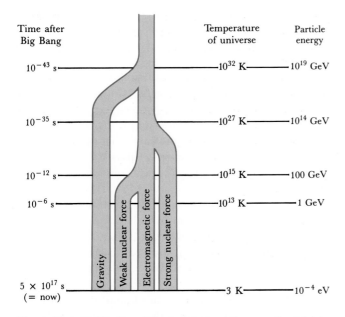

Figure 29-6 *Unification of the four forces* The strength of the four physical forces depends on the speed or energy with which particles interact. As shown in this schematic diagram, the higher the energy, the more the forces resemble each other. Also included here are the temperature of the universe and the time after the Big Bang when the strengths of the forces are thought to have been equal.

In the 1970s, Sheldon Glashow and Howard Georgi proposed a **grand unified theory** (or **GUT**), which predicts that the strong, weak, and electromagnetic forces are unified at energies above 10^{14} GeV. In other words, if particles were to collide at energies greater than 100 trillion GeV, the strong, weak, and electromagnetic interactions would be indistinguishable from each other.

Many physicists suspect that all four forces may be unified at energies greater than 10^{19} GeV (see Figure 29-6). If particles were to collide at these colossal energies, there would be no difference between the gravitational, electromagnetic, and nuclear forces. However, no one has yet succeeded in working out the details of such a **supergrand unified theory**, which is sometimes also called a **theory of everything** (or **TOE**).

Particle accelerators do exist that have energy enough to test the unification of the weak and electromagnetic forces around 100 GeV. Physicists see no hope, though, of ever constructing machines capable of slamming particles together with energies in the trillions of GeV. It may thus be impossible to test directly the grand and supergrand unified theories in a laboratory. However, the universe immediately after the Big Bang was so hot and its particles were moving so fast that they did indeed collide with energies in the trillions of GeV. The earliest moments of the universe have thus become a laboratory in which scientists can try to test some of the most elegant and most sophisticated theories in physics.

Many of the ideas connecting particle physics and cosmology are new and still quite speculative. Nevertheless, it is possible to summarize our understanding with the aid of Figure 29-7. During the Planck time (from $t = 0$ to $t = 10^{-43}$ s), particles collided with energies greater than 10^{19} GeV, and all four forces were unified. Because we do not yet have a TOE that properly describes the behavior of gravitons and

us, however, particles interact with much less energy than this. Below 100 GeV, intermediate-vector bosons behave like massive particles, but photons are always massless. Because the similarity between intermediate-vector bosons and photons does not exist at low energies, we say that "symmetry is broken" below 100 GeV, which is why the electromagnetic and the weak forces behave so differently in the world around us.

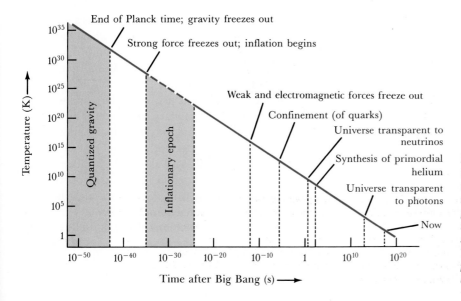

Figure 29-7 *The early history of the universe* As the universe cooled, the four forces "froze out" of their unified state as a result of spontaneous symmetry breaking. The inflationary epoch lasted from 10^{-35} to 10^{-24} s after the Big Bang. Neutrons and protons "froze out" of the hot "quark soup" one millionth of a second after the Big Bang. The universe became transparent to light (that is, photons "froze out") when the universe was a million years old.

quantized gravity, we remain ignorant of what was going on during the first 10^{-43} second of the universe's existence. We know, however, that by the end of the Planck time the energy of particles in the universe had fallen to 10^{19} GeV, below which gravity is no longer unified with the other three forces. We can thus say that, at $t = 10^{-43}$ s, there was a **spontaneous symmetry breaking** in which gravity was "frozen out" of the otherwise unified hot soup that filled all space. As we saw earlier, energy (E) is related to temperature (T) by $E = kT$, where k is the Boltzmann constant (roughly 10^{-4} eV/K). Thus, the temperature of the universe was 10^{32} K when gravity emerged as a separate force.

At $t = 10^{-35}$ s, the energy of particles in the universe had fallen to 10^{14} GeV, equivalent to a temperature of 10^{27} K, below which the strong nuclear force is no longer unified with the electromagnetic and weak nuclear forces. Thus, at $t = 10^{-35}$ s, there was a second spontaneous symmetry breaking, at which time the strong nuclear force made its appearance, freezing out of the otherwise unified hot soup that had already given us gravity. Calculations suggest that the inflationary epoch lasted from $t = 10^{-35}$ s to about $t = 10^{-24}$ s, during which the universe increased its size by a factor of between 10^{20} and 10^{30} over what it would otherwise have been.

At $t = 10^{-12}$ s, the temperature of the universe had dropped to 10^{15} K, the energy of the particles had fallen to 100 GeV, and there was a final spontaneous symmetry breaking and "freeze-out" that separated the electromagnetic force from the weak nuclear force. From that moment on, all four forces have interacted with particles essentially as they do today.

The next significant event occurred at $t = 10^{-6}$ s, when the temperature was 10^{13} K and particles were colliding with energies of roughly 1 GeV. Prior to this moment, particles collided so violently that individual protons and neutrons could not exist, being constantly fragmented into quarks. After this time, appropriately called the period of **confinement**, quarks were finally able to stick together to form individual protons and neutrons.

The remaining significant events proceeded as previously outlined. The universe became transparent to neutrinos at $t = 2$ s, and all the primordial helium was produced by $t = 3$ min. Finally, the universe became transparent to photons at $t = 1$ million years. At this point, we face the difficult problem of explaining why the matter in the universe came to be concentrated in galaxies.

29-6 Cosmic strings and other oddities may be relics of the early universe

We have seen that, according to quantum mechanics, the vacuum of empty space is in fact seething with activity. Just as an atom can be excited into a higher energy level, GUTs explain that the vacuum of space can exist in an excited state. This excited or **false vacuum** would look identical to what we normally regard as empty space (a true vacuum) except that it would contain a significant amount of energy. Alan Guth has proposed that inflation occurred when the universe went from the high-energy symmetric state of a false vacuum to the low-energy symmetry-broken state of a true vacuum. The energy released during this transition gave rise to a powerful "negative pressure" that caused space to balloon outward in all directions. This is the source of the large value of Einstein's cosmological constant Λ during the inflationary epoch.

An analogy using pencils may help to illustrate the nature of the transition from a symmetric false vacuum to an asymmetric true vacuum. A pencil balanced on its point is said to be in a symmetric state, because no particular horizontal direction is singled out over any other. When the pencil falls over, however, this symmetry is broken, because the pencil is then pointed toward a specific direction of the compass. In this asymmetric state one direction has become singled out over all others.

Now imagine thousands of pencils precariously balanced on their points, as shown in Figure 29-8a. This arrangement is a very unstable one, and soon the pencils fall over, with an end result such as that shown in Figure 29-8b. Note that symmetry has been broken everywhere, because every pencil is pointed in a particular horizontal direction.

As soon as the universe cooled to the temperature of 10^{27} K at which the strong nuclear force froze out, the false vacuum that then filled space became unstable, like thousands of pencils balanced on their points. The rapid transition to an asymmetric true vacuum is analogous to the pencils falling over. But note that something odd might occur. As thousands upon thousands of pencils knock each other over, it could happen that waves of falling pencils might fall together in such a way that a single pencil remains upright, as shown in Figure 29-8c. For that pencil, symmetry would not have been broken. An analogous relic left over from the early universe, called a **cosmic string**, is a long, thin, massive wire of unbroken symmetry along which the strong, weak, and electromagnetic forces remain unified.

Many physicists suspect that cosmic strings might actually exist in the universe today. They would be incredibly massive (about a trillion tons per millimeter), and would be detectable only by their gravitational effects. Thus, cosmic strings are an ideal candidate for the "missing mass" in the universe. Perhaps more importantly, though, they may have served as the gravitational focus along which clusters of primordial galaxies could have condensed billions of years ago. This suspicion arises partly from the way in which clusters of galaxies seem to be scattered across the sky.

Figure 29-9 shows the locations of 400,000 galaxies, covering approximately one-quarter of the sky. On this map, the sky is divided into tiny squares each measuring about 10 arc min across; colors are used to indicate the number of

  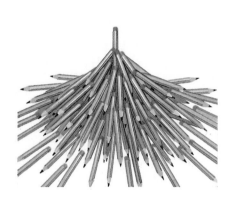

Figure 29-8 *Symmetry breaking and cosmic strings* The symmetry that existed in the early universe can be likened to orderly rows of pencils balanced on their points, as sketched in (**a**). The state of broken symmetry that exists today is analogous to all the pencils having fallen over, as sketched in (**b**). A situation could arise, however, in which the falling pencils might topple toward each other so that one pencil is supported upright, as in (**c**). Some physicists theorize that comparable structures, called cosmic strings, might be left over from the symmetry breaking that occurred during the earliest moments of the universe.

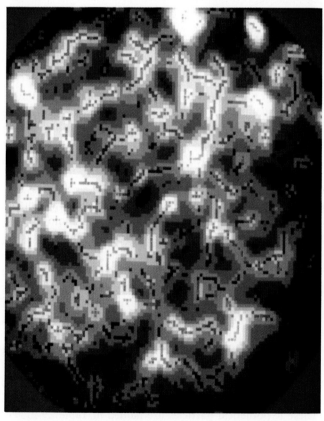

Figure 29-9 *The distribution of 400,000 galaxies* This map shows the locations of nearly half a million galaxies spanning 100° of the sky. The north pole of our Galaxy (in the constellation of Coma Berenices) is at the center of the map. Each small square covers 10 × 10 arc min. The color indicates the number of galaxies found in each square. Light colors indicate a high density of galaxies, and darker regions indicate few galaxies. Green squares emphasize filaments along which galaxies are concentrated, and the single red squares (near the centers of the white regions) indicate the points where the galaxy counts reach a maximum. (Courtesy of E. L. Turner, J. E. Moody, and J. R. Gott)

galaxies within each square. The light-colored regions indicate high densities of galaxies, whereas the darker sections denote areas where very few galaxies are found. The brown and black areas are **voids**, that measure 100 million to 400 million light years across and are virtually free of galaxies. Note that many of the white regions seem to be connected to each other by bridges, called **filaments**, as emphasized by the green squares. Filaments are typically 100 Mly long and contain up to 1 million galaxies with a mass of roughly 10^{16} M$_\odot$. Densely populated clumps seem to occur where filaments intersect. Some physicists suspect that cosmic strings may be responsible for these filaments.

29-7 Kaluza–Klein theories and supergravity predict that the universe may have eleven dimensions

In relativity theory, it is customary to combine time with the three dimensions of ordinary space, resulting in a four-dimensional combination called **space–time**. In 1919, a then virtually unknown mathematician, Theodor Franz Éduard Kaluza, proposed that the four dimensions of space–time should be supplemented with a fifth dimension. Kaluza's aim in introducing a fifth dimension was to give a unified geometric description of the only known forces of nature then: gravity and electromagnetism. Kaluza had discovered that many of the effects of both these forces could be fully described as the curvature of five-dimensional space–time, just as Einstein had proved that the effects of gravity alone are manifested in the curvature of four-dimensional space–time.

Kaluza's hypothetical fifth dimension exists at every point in ordinary space but, like a very tiny loop, is curled up so

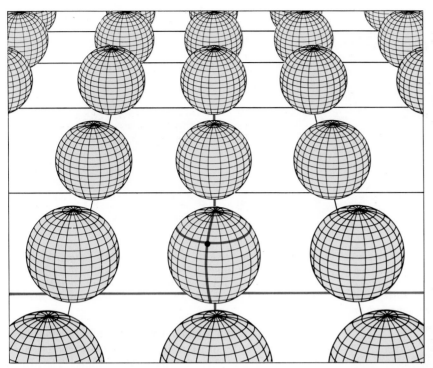

Figure 29-10 Hidden dimensions of space *Hidden dimensions of space might exist provided they are curled up so tightly that we cannot observe them. This drawing shows how an ordinary two-dimensional plane might contain two additional dimensions. At every point on the plane, there is a very tiny sphere so small that it cannot be seen. To pinpoint a particular location, you need to give not only a position on the plane but also a position on the sphere that is tangent to the plane at that point, as indicated by the red lines. (Adapted from D. Freedman and P. Van Nieuwenhuizen)*

tightly that it is not directly observable. A particle always follows the straightest possible path in five dimensions, but when viewed in the three dimensions of ordinary space, the path appears curved, exactly as if the particle had been deflected by gravitational and electromagnetic forces.

In 1926, the Swedish physicist Oskar Klein expanded upon Kaluza's five-dimensional theory to make it compatible with quantum mechanics. In doing so, Klein discovered that particles could be identified with particular vibrations of the compact loop of the fifth dimension. Today, any quantum-mechanical theory that uses more than four dimensions in an attempt to produce a unified description of the forces of nature is called a **Kaluza–Klein theory.**

Today we know of four physical forces, so modern Kaluza–Klein theories must have even more than five dimensions. The best agreement between theory and experiment is in fact obtained if space–time has eleven dimensions. In other words, at every point in ordinary space and at every

moment of time there must be a very compact seven-dimensional structure far too tiny for us to detect (see Figure 29-10). Particles travel along the straightest possible paths in this eleven-dimensional space, but their paths in ordinary three-dimensional space appear curved, as if the particles were being deflected by the four forces.

The choice of eleven dimensions is supported by a super-grand unified theory called **supergravity,** which treats different kinds of particles on an equal footing. Edward Witten at Princeton has proven that at least seven hidden dimensions must be added to ordinary space–time to account for the four forces. But if more than eleven dimensions are used, the partnership between the various types of particles breaks down. Although eleven dimensions would seem to be best choice, some theorists have recently argued in favor of as many as 26 dimensions.

A surprising prediction of supergravity theories is that every type of ordinary particle has a "supersymmetric partner," meaning that there may be many particles scattered throughout the universe that have never been detected. These hypothetical particles are named in Table 29-2.

Some of these supersymmetric particles might be quite massive. They may be the WIMPs that were proposed to account for the observed under-abundance of neutrinos from the Sun. Or they could be the material of the "missing mass" that pervades the universe and holds clusters of galaxies together. Nevertheless, most scientists are cautiously skeptical about these recent theoretical developments. These speculative multidimensional theories could well have absolutely nothing to do with reality.

Table 29-2 Some particles and their supersymmetric partners

	Ordinary particle	Supersymmetric partner
Basic particles of matter	Electron	Selectron
	Quark	Squark
	Neutrino	Sneutrino
Particles that mediate forces	Graviton	Gravitino
	Photon	Photino
	Gluon	Gluino

29-8 Galaxies were formed from density fluctuations that existed in the early universe

The distribution of matter in the universe today is quite "lumpy." Stars are grouped together in galaxies, galaxies into clusters, and clusters into **superclusters** that stretch across 100 million light years.

Although there is a lot of lumpiness in the universe today, the early universe must have been exceedingly smooth. To see why, think back to the era of recombination that occurred roughly 1 million years after the Big Bang. Before recombination, high-energy photons were constantly colliding vigorously with charged particles throughout all of space. After recombination, the universe became transparent, and these photons stopped interacting with the matter in the universe. Astronomers say that matter "decoupled" from radiation during the era of recombination. Because, as we have seen, the 3-K microwave background is extremely isotropic, we can conclude that the matter with which these photons once collided so frequently must also have been spread smoothly across space.

Although the distribution of matter across space during the early universe must have been quite smooth, it could not have been perfectly uniform. If it had been absolutely uniform billions of years ago, it would still have to be absolutely uniform today; there would now be neither stars nor galaxies, only a few atoms per cubic meter throughout space. Consequently, a slight lumpiness, or **density fluctuations**, in the distribution of matter, must have existed in the early universe. Through the action of gravity, these fluctuations eventually grew to become the galaxies and clusters of galaxies that we see today throughout the universe.

Our understanding of how gravity can amplify density fluctuations dates back to 1902, when the British physicist James Jeans solved a problem first proposed by Isaac Newton. Suppose that you have a uniform, static distribution of matter having density ρ and temperature T. Also, suppose that you perturb this medium by introducing slight fluctuations in density, as shown in Figure 29-11. These regions of enhanced density will then have an enhanced gravity that will attract nearby material. As this happens, however, gas pressure inside these regions will increase, thus making it resist further compression and growth. The question then becomes: Under what conditions does gravity overwhelm gas pressure so that a permanent object can condense out of the medium?

James Jeans proved that an object will grow gravitationally from a density perturbation, provided that the overall size of the original disturbance is greater than the **Jeans length** L_J given by

$$L_J = \sqrt{\frac{\pi k T}{m G \rho}}$$

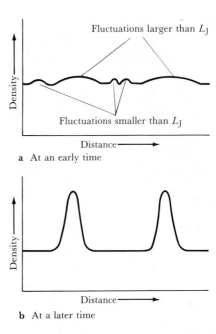

a At an early time

b At a later time

Figure 29-11 *The growth of density fluctuations* (a) *Small density fluctuations in the distribution of matter shortly after the era of recombination.* (b) *If the size of a fluctuation is greater than the Jeans length* (L_J), *it becomes gravitationally unstable and can grow in amplitude.*

where k is the Boltzmann constant, G the gravitational constant, and m the mass of a single particle in the medium.

Using the conditions prevalent during the era of recombination ($T = 3000$ K, $\rho = 10^{-18}$ kg/m^3) and m as the mass of the hydrogen atom ($m = 2 \times 10^{-27}$ kg), we find that $L_J = 100$ ly, the diameter of a typical globular cluster. Furthermore, the mass contained in this volume is $\rho L_J^3 = 5 \times 10^5 M_\odot$, the mass of a typical globular cluster. For these reasons, Robert Dicke and P. J. E. Peebles at Princeton have proposed that globular clusters (see Figure 29-12) were among the first objects to form after matter decoupled from radiation. Indeed, as we saw in Chapter 21, globular clusters are composed of the most ancient stars we can find in the sky. But, interesting as these calculations may be, they do not shed light on how galaxies and clusters of galaxies came to be formed.

An important clue about the formation of clusters and superclusters came in the early 1980s when astronomers began discovering enormous voids in space, where exceptionally few galaxies are found. The first of these voids was noticed by Robert Kirshner and his colleagues at the Kitt Peak National Observatory, who were involved in measuring the redshifts of galaxies in several regions of the sky. In one section of Boötes they found only one galaxy in the redshift range of 12,000 to 18,000 km/s. This redshift range corresponds to a volume roughly $\frac{1}{3}$ billion light years across, using a Hubble constant of 50 km/s/Mpc.

Numerous additional voids soon were discovered by teams of astronomers carefully analyzing the locations of

Figure 29-12 A globular cluster *A typical globular cluster contains 10^5 to 10^6 stars. Cluster diameters range from about 20 to 400 ly. Because these parameters are comparable to the Jeans length (L_J) during the era of recombination, astronomers believe that globular clusters must have been among the first objects to form in the universe. (NASA)*

thousands of galaxies in the sky. The results from recent surveys reveal that galaxies are concentrated along enormous filaments that seem to surround voids that are roughly spherical and measure between 100 million and 400 million light years across (see Figure 29-13 and and Box 26-3). These surveys demonstrate that the distribution of clusters of galaxies looks "sudsy," because it resembles a collection of giant soap bubbles. Galaxies surround spherical voids in the same way a soap film is concentrated on the surface of bubbles.

Astronomers are now working on a variety of ideas to explain this large-scale bubbly structure of the universe. Jeremiah Ostriker and Lennox Cowie at Princeton have proposed that giant explosions dominated the universe shortly after the first galaxies and stars formed. These explosions, which could have involved colossal supernovae or some sort of quasarlike activity, would have been strong enough to clear out bubblelike voids, while sweeping aside and compressing material for the next generation of stars and galaxies. Unfortunately, it seems that these explosions would have produced only small voids, one-tenth as large as those that have been observed.

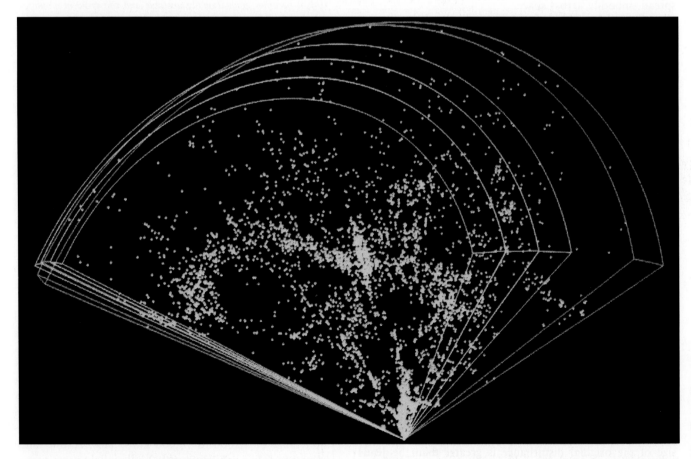

Figure 29-13 The large-scale distribution of galaxies *This map shows the distribution of nearly 4000 galaxies in four slices extending out to a redshift of z = 15,000 km/s. Yellow dots indicate galaxies in the right ascension range 8^h to 17^h and declination range 26.5° to 44.5°; orange dots cover the declination range 8.5° to 14.5°. The "Great Wall" crosses the survey nearly parallel to the outer boundary. (Courtesy of M. J. Geller; Smithsonian Astrophysical Observatory)*

Several proposals to explain the observed voids employ unseen matter and thus also solve the mystery of the missing mass. We have already seen cosmic strings proposed in connection with the filamentary structure of clusters of galaxies. Another possibility involves neutrinos. If a neutrino has no mass, it will always travel at the speed of light, just as a photon does. If, however, it has a little mass, it must slow down as the universe cools. Eventually, neutrinos begin to accumulate at the sites of density excesses present in the density fluctuations left over from the Big Bang. The gravitational pull of these neutrinos on surrounding matter eventually leads to the formation of clusters of galaxies. In this way, neutrinos could have played an important role in helping fluctuations grow into superclusters.

Today there should be roughly 100 million neutrinos, on the average, per cubic meter. If each neutrino has a mass of about 10 eV, then the cosmic neutrino background alone has an average density very nearly equal to the critical density. This assumption would explain why Ω_0 is nearly equal to 1, even though the density of matter that we can see in the universe is much less than the critical density. Also, because neutrinos possessing mass would eventually be captured gravitationally in clusters of galaxies, these particles could be the solution to the missing-mass problem.

In recent years, astronomers and astrophysicists have performed a variety of supercomputer simulations to show how primordial fluctuations could have grown into features identified with clusters and superclusters. For instance, Adrian Melott at the University of Kansas has been particularly successful with simulations involving neutrinos that produce maps of the universe containing filaments and voids quite like those actually observed.

Instead of using neutrinos, some researchers assume that the universe is filled with **cold dark matter** consisting of slowly moving elementary particles that interact only through gravity. The primary candidates for such particles include WIMPs, photinos, gluinos, and gravitinos, predicted by supergravity. These hypothetical particles are assumed to be moving slowly, because supercomputer simulations involving rapidly moving particles (i.e., hot dark matter) fail to produce galaxies.

The computational setup for a representative supercomputer simulation involving cold dark matter is shown in Figure 29-14. This simulation follows the motions of more than 2 million particles as a volume of space expands along with the expansion of the universe. Thus the computational box was smaller in the past and will be larger in the future. At the present time, the box measures 51.2 Mpc on a side, provided the Hubble constant is 50 km/s/Mpc.

The simulation begins with particles perturbed slightly from an absolutely uniform distribution, thereby mimicking slight departures from uniformity that probably characterized the universe immediately after inflation magnified quantum-level fluctuations to macroscopic sizes. A supercomputer programmed with Newton's laws in an expanding

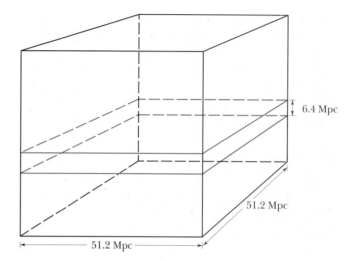

Figure 29-14 A computational box *This diagram shows the computational setup used the supercomputer simulation presented in Figure 29-15. The volume (a cube) used in the supercomputer simulation expands with the expansion of the universe. The dimensions shown on this drawing apply to the present time. The box contains nearly 2.1 million "particles" (128^3, to be precise), each representing a dark-matter cloud 4 billion times as massive as the Sun. An IBM supercomputer at the Cornell National Supercomputing Facility was used to calculate the paths of the particles as the universe expands.*

universe is then used to calculate the motions of the nearly 2.1 million particles. At selected intervals, a slice is taken through the box and the distribution of particles is examined. The slice through today's universe is 6.4 Mpc thick.

Nine representative slices through the computational box are shown in Figure 29-15. Each slice is designated by an "expansion factor" a, which is scaled so that the final slice at the end of the simulation is set at $a = 1$. Thus, for instance, the first frame ($a = 0.04$) shows a slice of the universe when the computational box was only 4 percent of its size in the last frame. The first four slices ($a = 0.04$ through $a = 0.10$) show the appearance of the universe at very early times, while the last five views ($a = 0.20$ through $a = 1.00$) are from much later times. The slice labeled $a = 0.60$ seems to resemble the universe today, so the simulation covers both the past and the future.

Note how the tiny fluctuations barely visible in the first slice gradually become more pronounced as time passes. These features merge to form larger structures as gravity continues to draw matter together, overcoming the expansion of the universe. The development of both voids and filaments is clearly seen. Ordinary matter, which follows the course set by the cold dark matter, accumulates along density enhancements forming galaxies and clusters of galaxies. The slice $a = 0.20$ seems to represent the era during which significant galaxy formation occurred.

Supercomputer simulations, along with redshift surveys that map the large-scale structure of the universe, are providing important insights into fundamental issues about the

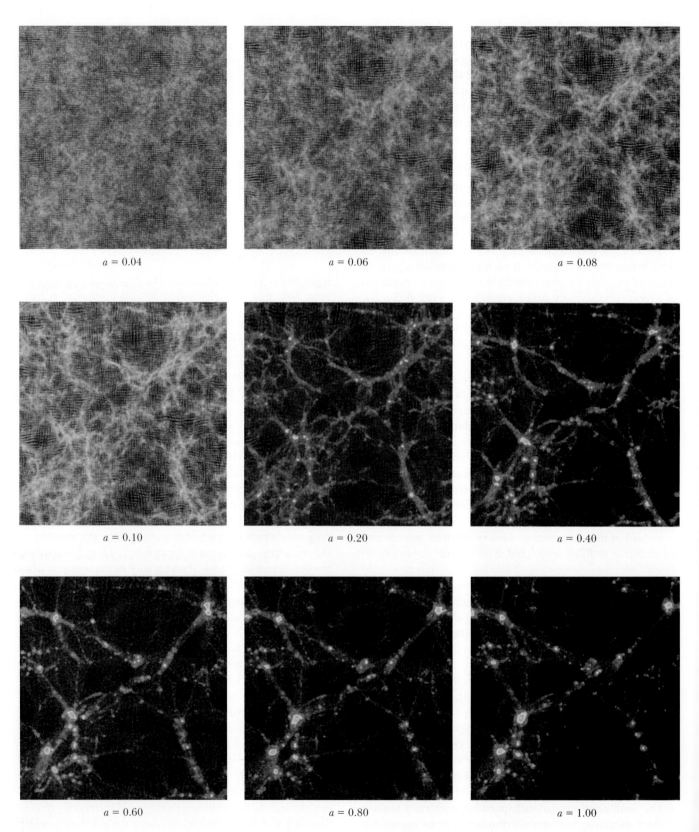

a = 0.04

a = 0.06

a = 0.08

a = 0.10

a = 0.20

a = 0.40

a = 0.60

a = 0.80

a = 1.00

Figure 29-15 A cold-dark-matter simulation *These nine views show slices through a large volume of space as the universe expands. The "expansion factor" a indicates the relative size of the computational box at the time when a slice was taken. The first four views represent early epochs within the first few billion years after the Big Bang. The last five views are from much later times, with the* a = 0.60 *view possibly representing the present time. Galaxy formation probably occurs around the time of the* a = 0.20 *slice. (Courtesy of J. M. Gelb and E. Bertschinger; MIT)*

evolution of the universe. It now seems that the universe may be composed mostly of bizarre, undetected particles whose existence suggests hidden dimensions of space. If this turns out to be the case, the subject of astronomy may soon become more fantastic than the most fanciful science fiction.

We will first understand
How simple the universe is
When we realize
How strange it is.
Anonymous

Key words

cold dark matter	graviton	pair production	supergrand unified theory
cosmic string	*hadron	Principle of Totalitarianism	supergravity
density fluctuation	inflationary epoch	quantum electrodynamics	symmetry breaking
deuterium bottleneck	isotropy problem	quark	theory of everything (TOE)
false vacuum	intermediate-vector boson	rest energy	thermal equilibrium
Feynman diagram	Jeans length	spontaneous symmetry breaking	threshold temperature
flatness problem	Kaluza–Klein theory	strong nuclear force	weak nuclear force
gluon	Lamb shift	supercluster	
grand unified theory (GUT)	*lepton		

Key ideas

- Heisenberg's uncertainty principle states that the amount of uncertainty in the momentum of a particle increases as its position is known more precisely, and vice versa; a similar reciprocal relationship exists between mass and time.

- Because of the uncertainty principle, particle–antiparticle pairs can form spontaneously and disappear again within a fraction of a second; these pairs, which can never be detected directly, are called virtual pairs.

  A virtual pair can become a real particle–antiparticle pair when photons collide in the process of pair production; the photons disappear, then their energy is replaced by the mass of the particle–antiparticle pair.

  A corresponding process of annihilation involves the disappearance of a colliding particle–antiparticle pair and the appearance of photons.

- Just after the Planck time, the universe was filled with particles and antiparticles formed by pair production and with numerous high-energy photons formed by annihilation; a state of thermal equilibrium existed in this hot plasma.

  As the universe expanded, its temperature decreased; as the temperature fell below the threshold temperature required for producing each kind of particle, annihilation dominated production for that kind of particle.

The present domination of matter over antimatter results because particles and antiparticles were not created in exactly equal numbers just after the Planck time; this imbalance is an example of the process called symmetry breaking.

Production of helium could not begin until the cosmological redshift eliminated most of the high-energy photons that created a deuterium bottleneck by breaking down deuterons before they could combine to form helium.

That the universe is nearly flat and the 3-K microwave background is almost perfectly isotropic may be explained as results of the brief period of rapid expansion called the inflationary epoch, immediately after the Big Bang; during this tiny fraction of a second, the universe expanded to a size many times larger than it would have reached through its normal expansion rate.

During the inflationary period, much of the material originally near our location moved far beyond the limits of our observable universe; this observable universe today is therefore expanding into space containing matter and radiation that was originally in close contact with our matter and radiation during the first instant after the Big Bang.

- Four basic forces—gravity, electromagnetism, the strong nuclear force, and the weak nuclear force—explain all of the interactions observed in the universe.

  Grand unified theories (GUTs) are attempts to explain two or more forces in terms of a single consistent set of physical laws; a superGUT would explain all four forces.

  GUTs suggest that all four physical forces were equivalent just after the Big Bang; however, because we have no satisfactory superGUT, we can say nothing about the nature of the universe during this period before the Planck time.

- At the Planck time ($t = 10^{-43}$ s after the Big Bang), gravity "froze out," to become a distinctive force in a spontaneous symmetry breaking; at $t = 10^{-35}$ s, there was a second spontaneous symmetry breaking, in which the strong nuclear force became a distinct force by itself.

  At $t = 10^{-12}$ s, a final spontaneous symmetry breaking separated the electromagnetic force from the weak nuclear force; from that moment on, the universe behaved as it does today.

  At $t = 10^{-6}$ s, the confinement of quarks occurred, permitting the quarks to combine into protons and neutrons.

  At $t = 2$ s, the universe became transparent to neutrinos; all of the original helium had been produced by $t = 3$ s; the universe became transparent to photons at $t = 10^6$ year.

- Cosmic strings, long, massive remnants of the early universe, may yet exist.

- Kaluza–Klein theories predict that there may be hidden dimensions of space.

- Supergravity predicts that many types of undiscovered particles (the supersymmetric partners of ordinary particles) may be abundantly scattered throughout the universe; these particles are candidates for the "missing mass."

- Galaxies are generally located on the surfaces of roughly spherical voids; thus, the arrangement of clusters and superclusters of galaxies has a "sudsy" or bubbly appearance.

- Computer simulations that best reproduce the observed arrangements of clusters and superclusters assume that most of the mass of the universe is cold dark matter, possibly consisting of the supersymmetric particles predicted by supergravity.

Review questions

1 What is the Heisenberg uncertainty principle, and how does it lead to the idea that all space is filled with virtual particle–antiparticle pairs?

2 What is the difference between an electron and an antielectron?

3 In what way would it be possible to say that the universe was matter dominated immediately after the Big Bang? How did this matter-dominated period differ from the matter-dominated universe in which we live today?

4 Where did most of the photons in the 3-K microwave background come from?

5 What is meant by the threshold temperature of a particle?

6 Why is it reasonable to suppose that all space is filled with a "neutrino background" analogous to the 3-K microwave background?

7 What is the "deuterium bottleneck," and why was it important during the formation of heavy nuclei in the early universe?

8 Describe an example of each of the four basic types of interactions in the physical universe. Do you think it possible that a fifth force might someday be discovered? Explain your answer.

9 What is the observational evidence for (a) the Big Bang, (b) the inflationary epoch, (c) the era of recombination, and (d) the confinement of quarks?

10 Describe the large-scale structure of the universe as revealed by the distribution of clusters and superclusters of galaxies.

11 Why is it reasonable to say that the inflationary epoch was in fact the phenomenon called the Big Bang?

Advanced questions

Tips and tools...
The mass of a proton is 1.67×10^{-27} kg. The mass of an electron is 9.11×10^{-31} kg. As explained in Chapter 28, the critical density is proportional to the square of the Hubble constant and is equal to about 5×10^{-27} kg/m^3 if H_0 is 50 km/s/Mpc. It is also useful to know that 1 m$^3 = 10^6$ cm^3.

*12   An electron has a lifetime of 1.0×10^{-8} s in a given energy state before it makes a downward transition to a lower state. What is the uncertainty in the energy of the photon emitted in this process?

*13   (Basic)   How long can a proton–antiproton pair exist without violating the principle of the conservation of mass?

*14   (Basic)   Calculate the threshold temperature for electrons and antielectrons.

15   Using the physical conditions present in the universe during the era of recombination ($T = 3000$ K and $\rho = 10^{-18}$ kg/m^3), verify that the Jeans length for the universe at that time was about 100 ly and that the total mass contained in a sphere with this diameter is about 4×10^5 solar masses.

*16   (Challenging)   Suppose that the average density of neutrinos throughout space is 100 neutrinos per cubic centimeter and that these neutrinos are responsible for giving the universe a density equal to the critical density. What range of neutrino masses corresponds to the often-quoted range of the Hubble constant, namely 50–100 km/s/Mpc?

Discussion questions

17   Some versions of GUTs predict that the proton is unstable. What would it be like to live at a time when protons were decaying in large numbers?

18   Suppose we were now living in a radiation-dominated universe. How would such a universe be different from what we currently observe?

For further reading

Barrow, J., and Silk, J. *The Left Hand of Creation: Origin and Evolution of the Universe.* Basic Books, 1983   •   This authoritative book tells the exciting story of how space, time, and matter may have been created.

Burns, J. "Very Large Structures in the Universe." *Scientific American*, July 1986   •   This article describes observations that led to the discovery of enormous superclusters and immense voids in space.

Carringan, R., and Trower, W. *Particle Physics in the Cosmos* and *Particles and Forces: At the Heart of Matter.* W. H. Freeman and Company, 1989   •   These two superb collections of readings from *Scientific American* consist of articles that cover many exciting discoveries and developments in particle physics and cosmology during the 1980s.

Close, F. *The Cosmic Onion.* American Institute of Physics, 1983   •   This well written book uses clever drawings to explain some details of particle physics.

Davies, P. *Superforce.* Simon & Schuster, 1984   •   This lucid and entertaining treatise examines the quest for a grand unified theory of physical reality.

_____. *The Forces of Nature*, 2nd ed. Cambridge University Press, 1986   •   This is perhaps one of the best introductions to particle physics and grand unified field theory that exists.

Freedman, D., and Van Nieuwenhuizen, P. "The Hidden Dimensions of Spacetime." *Scientific American*, March 1985   •   This challenging article describes speculative work by theoretical physicists who think there might be more than three dimensions of space.

Gregory, S., and Thompson, L. "Superclusters and Voids in the Distribution of Galaxies." *Scientific American*, March 1982   •   This article shows how redshift surveys establish the existence of superclusters and voids.

Guth, A., and Steinhardt, P. "The Inflationary Universe." *Scientific American*, May 1984   •   This interesting article explains why the universe probably experienced a sudden but brief period of vigorous expansion very early in its history.

Haber, H., and Kane, G. "Is Nature Supersymmetric?" *Scientific American*, June 1986   •   This article explains how theoretical physicists use geometrylike concepts in their quest for a deeper understanding of physical reality.

Kaufmann, W., ed. *Particles and Fields.* W. H. Freeman and Company, 1980   •   This collection of readings on particle physics from *Scientific American* explores the fundamental nature of matter.

Krauss, L. "Dark Matter in the Universe." *Scientific American*, December 1986   •   This article surveys evidence for dark matter and speculates about its nature.

Peat, F. *Superstrings and the Search for the Theory of Everything.* Contemporary Books, 1988   •   This interesting book explores the idea that all elementary particles— including protons, neutrons, and electrons—are actually tiny strings vibrating in ten dimensions.

Quigg, C. "Elementary Particles and Forces." *Scientific American*, April 1985   •   This article surveys the quest for a grand-unified-field theory from an experimental viewpoint.

Silk, J. *The Big Bang.* W. H. Freeman and Company, 1989   •   This updated edition of a fascinating book describes many of the processes that made the universe the way it is.

Silk, J., Szalay, A., and Zeldovich, Ya. B. "The Large-Scale Structure of the Universe." *Scientific American*, October 1983   •   The article shows how the large-scale structure of the universe may have resulted from perturbations in the matter density that emerged from the Big Bang.

Stephen Hawking is a theoretical physicist at the University of Cambridge. He was born in Oxford in 1942 and took his B.A. at Oxford in 1962 and went on to take his doctorate working on gravity and cosmology under Dr. Dennis W. Sciama at Cambridge. He continued his work on these subjects at Cambridge and, during 1974 and 1975, at California Institute of Technology. A Fellow of the Royal Society since 1974, he has received many honors and awards, including the Eddington Medal of the Royal Astronomical Society, The Dannie Heinemann Prize of the American Institute of Physics and the American Physical Society, the Maxwell Medal and Prize of the Institute of Physics, and the Einstein Medal given in Berne, Switzerland in 1979. Also in 1979 he was elected to Newton's old chair, Lucasian Professor at Cambridge.

His contributions have been in the areas of general relativity, gravitation, and quantum theory as it relates to black holes and their thermodynamics. Though afflicted by a progressive nervous disease since 1961, which has confined him to a wheelchair for a decade, he continues to be phenomenally productive as both a writer and a researcher.

Stephen W. Hawking THE EDGE OF SPACETIME

From the dawn of civilization people have asked questions like: "Did the universe have a beginning in time? Will it have an end? Is the universe bounded or infinite in spatial extent?" In this essay I shall outline some of the answers to these questions which are suggested by modern developments in science. Most of what I describe is now fairly generally accepted though some of it was controversial. My final conclusion, however, is based on some very recent work on which there has not yet been time to reach a consensus.

In most of the early mythological or religious accounts the universe, or at least its human inhabitants, was created by a Divine Being at some date in the fairly recent past like 4004 BC. Indeed the necessity of a "First Cause" to account for the creation of the universe was used as an argument to prove the existence of God. The Greek philosophers like Plato and Aristotle, on the other hand, did not like the thought of such direct Divine intervention in the affairs of the world and so mostly preferred to believe that the universe had existed and would exist forever. Most people in the ancient world believed that the universe was spatially bounded. In the earliest cosmologies the world was a flat plate with the sky as a pudding basin overhead. The Greeks, however, realized that the world was round. They con-

structed an elaborate model in which the Earth was a sphere surrounded by a number of spheres that carried the Sun, Moon, and the planets. The outermost sphere carried the so-called fixed stars which maintain the same relative positions but which appear to rotate across the sky.

This model with the Earth at the center was adopted by the Christian Church. It had the great attraction that it left plenty of room outside the sphere of the stars for Heaven and Hell, though quite how these were situated was never clear. It remained in favor until the seventeenth century when the observations of Galileo showed that this model of the universe had to be replaced by the Copernican model in which the Earth and the other planets orbited around the Sun. Not only did this get rid of the spheres but it also showed that the "fixed stars" must be at a very great distance because they did not show any apparent movement as the Earth went round the Sun, apart from that caused by the rotation of the Earth about its own axis. Having realized this and having abandoned the belief that the Earth was at the center of the universe, it was fairly natural to postulate that the stars were other Suns like our own and that they were distributed roughly uniformly throughout an infinite universe. This, however, raised a problem: according to

Newton's theory of gravity, published in 1687, each star would be attracted toward every other star in the universe. Why, then, did not the stars all fall together to a single point? Newton himself tried to argue that this would indeed happen for a bounded collection of stars but that if one had an infinite universe, the gravitational force on a star caused by the attraction of the stars on one side of it would be balanced by the force arising from the stars on the other side. The net force on any star would therefore be zero and so the stars could remain motionless. This argument is in fact an example of the fallacies one can fall into when one adds up an infinite number of quantities: by adding them up in different orders one can get different results. We now know that an infinite distribution of stars cannot remain motionless if they are all attracting each other; they will start to fall toward each other. The only way that one can have a static infinite universe is if the force of gravity becomes repulsive at large distances. Even then, the universe is unstable because if the stars get slightly nearer each other, the attraction wins out over the repulsion and the stars fall together. On the other hand, if they get slightly farther away from each other, the repulsion wins and they move away from each other.

Despite these and other difficulties, nearly everyone in the eighteenth and nineteenth centuries believed that the universe was essentially unchanging in time. For such a universe the question of whether it had a beginning was metaphysical: one could equally well believe that it had existed forever or that it had been created in its present form a finite time ago. The belief in a static universe still persisted in 1915 when Einstein formulated his general theory of relativity which modified Newton's theory of gravity to make it compatible with discoveries about the propagation of light. He therefore added a so-called cosmological constant, which produced a repulsive force between particles at a great distance. This repulsive force could balance the normal gravitational attraction and allow a static uniform solution for the universe. This solution was unstable but it had the interesting property that in it space was finite but unbounded, just as the surface of the Earth is finite in area but does not have any boundary or edge. Time in this solution, however, could be infinite.

Einstein's static model of the universe was one of the great missed opportunities of theoretical physics: if he had stuck to his original version of general relativity without the cosmological constant, he could have predicted that the universe ought to be either expanding or collapsing. As it happened, however, it was not realized that the universe was changing with time until astronomers like Slipher and Hubble began to observe the light from other galaxies. Visible light is made up of waves, like radio waves only with a much shorter wavelength or distance between wave crests. If one passes the light through a triangular shaped piece of glass called a prism, it is decomposed into its constituent wavelengths—colors like a rainbow. Slipher and Hubble found the same characteristic patterns of wavelengths or colors as for the light from stars in our own Galaxy but the patterns were all shifted toward the red, or longer, wavelength end of the rainbow or spectrum. The only reasonable explanation of this was that the galaxies were moving away from us. In this case the distance between the wave crests would be crowded up and the wavelength would be reduced. This effect, known as the Doppler shift, is used by the police to measure the speed of cars.

During the 1920s Hubble observed the remarkable fact that the red shift was greater the further the galaxy was from our own. This meant that other galaxies were moving away from us at rates that were roughly proportional to their distance from us. The universe was not static as had been previously thought but was expanding. The rate of expansion is very low: it will take something like twenty thousand million years for the separation of two galaxies to double but it completely changes the nature of the discussion about whether the universe has a beginning or an end. This is not just a metaphysical question as in the case of a static universe: as I shall describe, there may be a very real physical beginning or end of the universe.

The first model of an expanding universe that was consistent with Einstein's general theory of relativity and Hubble's observations of red shifts was proposed by the Russian physicist and mathematician Alexander Friedmann in 1922. However, it received very little attention until similar models were discovered by other people toward the end of the 1920s. The Friedmann model and its later generalizations assumed that the universe was the same at every point in space and in every direction. This is obviously not a good approximation in our immediate neighborhood; there are local irregularities like the Earth and the Sun and there are many more visible stars in the direction of the center of our Galaxy than in other directions. However, if we look at distant galaxies, we find that they are distributed roughly uniformly throughout the universe, the same in every direction. Thus it does seem to be a good approximation on a large scale. Even better evidence comes from observations of the background of microwave radiation that was discovered in 1965 by two scientists at the Bell Telephone Laboratories. The universe is very transparent to radio waves of a few inches wavelength so this radiation must have traveled to us from very great distances. Any large-scale irregularities in the universe would cause the radiation reaching us from different directions to have different intensities. Yet the observed intensity is the same in every direction to a very high degree of accuracy.

There are three kinds of generalized Friedmann models. In one of them, the galaxies are moving apart sufficiently slowly that the gravitational attraction between them will eventually stop them moving apart and start them approaching each other. The universe will expand to a maximum size and then re-collapse. In the second model, the galaxies are moving apart so fast that gravity can never stop them and the universe expands forever. Finally, there is a third model in which the galaxies are moving apart at just the critical rate to avoid re-collapse. In principle we could determine which model corresponds to our universe by comparing the present rate of expansion with the present average mass density. The mass of the matter in the universe that we can observe directly is not enough to stop the expansion. However, we have indirect evidence that there is more mass that we cannot see. Whether this "invisible" mass could be enough to stop the expansion eventually remains an open question.

In the Friedmann model which re-collapses eventually, space is finite but unbounded, like in the Einstein static model. In the other two Friedmann models, which expand forever, space is infinite. Time, on the other hand, has a boundary or edge. In all the models the expansion starts from a state of infinite density called the Big Bang singularity. In the model that re-collapses there is another singularity called the Big Crunch at the end of the re-collapse. Singularities are places where the curvature of spacetime is infinite and the concepts of space and time cease to have any meaning. Scientific theories are formulated on a spacetime background so they will all break down at a singularity. If there were events before the Big Bang, they would not enable one to predict the present state of the universe because predictability would break down at the Big Bang. Similarly, there is no way that one can determine what happened before the Big Bang from a knowledge of events after the Big Bang. This means that the existence or non-existence of events before the Big Bang is purely metaphysical; they have no consequences for the present state of the universe. One might as well apply the principle of economy, known as Occam's razor, to cut them out of the theory and say that time began at the Big Bang. Similarly, there is no way that we can predict or influence any events after the Big Crunch, so one might as well regard it as the end of time. This beginning and possible end of time that are predicted by the Friedmann solutions are very different from earlier ideas. Prior to the Friedmann solutions, the beginning or end of time was something that had to be imposed from outside the universe; there was no necessity for a beginning or an end. In the Friedmann models, on the other hand, the beginning and end of time occur for dynamical reasons. One could still imagine the universe being created by an external agent in a state corresponding to some time after the Big Bang but it would not have any meaning to say that it was created *before* the Big Bang. From the present rate of expansion of the universe we can estimate that the Big Bang should have occurred between ten and twenty thousand million years ago.

Many people disliked the idea that time had a beginning or an end because it smacked of Divine Intervention. There were therefore a number of attempts to avoid this conclusion. One of these was the "steady state" model of the universe proposed in 1948 by Herman Bondi, Thomas Gold, and Fred Hoyle. In this model it was proposed that, as the galaxies moved further away from each other, new galaxies were formed in-between out of matter that was being "constantly created." The universe would therefore look more or less the same at all times and the density would be roughly constant. This model had the great virtue that it made definite predictions that could be tested by observations. Unfortunately, observations of radio sources by Martin Ryle and his collaborators at Cambridge in the 1950s and early 1960s showed that the number of radio sources must have been greater in the past, contradicting the steady state model. The final nail in the coffin of the steady state theory was the discovery of microwave background radiation in 1965. There was no way this radiation could be accounted for in the model.

Another attempt to avoid a beginning of time was the suggestion that maybe the singularity was simply a consequence of the high degree of symmetry of the Friedmann solutions. This restricted the relative motion of any two galaxies to be along the line joining them. It would therefore not be surprising if they all collided with each other at some time. However, in the real universe, the galaxies would also have some random velocities perpendicular to the line joining them. These transverse velocities might be expected to cause the galaxies to miss each other and to allow the universe to pass from a contracting phase to an expanding one without the density ever becoming infinite. Indeed, in 1963 two Russian scientists claimed that this would happen in nearly every solution of the equations of general relativity. They based this claim on the fact that all the solutions with a singularity that they constructed had to satisfy some constraint or symmetry. They later realized, however, that there was a more general class of solutions with singularities that did not have to obey any constraint or symmetry.

This showed that singularities *could* occur in general solutions of general relativity but it did not answer the question of whether they necessarily *would* occur. However, between 1965 and 1970 a number of theorems were proved which showed that any model of the universe which obeyed general relativity, satisfied one or two other reasonable assumptions, and contained as much matter as we observe in the universe, must have a Big Bang singularity. The same theorems predict that there will be a singularity that will be an end of time if the whole universe re-collapses. Even if the universe is expanding too fast to collapse in its entirety, we nevertheless expect some localized regions, such as massive burnt-out stars, to collapse and form black holes. The theorems predict that the black holes will contain singularities which will be an end of time for anyone unfortunate or foolhardy enough to fall in.

Einstein's general theory of relativity is probably one of the two greatest intellectual achievements of the twentieth century. It is, however, incomplete because it is what is called a classical theory; that is, it does not incorporate the uncertainty principle of the other great discovery of this century, quantum mechanics. The uncertainty principle states that certain pairs of quantities, such as the position and velocity of a particle, cannot be predicted simultaneously with an arbitrary high degree of accuracy. The more accurately one predicts the position of the particle, the less accurately one will be able to predict its velocity and vice versa. Quantum mechanics was developed in the early years of this century to describe the behavior of very small systems such as atoms or individual elementary particles. In particular, there was a problem with the structure of the atom, which was supposed to consist of a number of electrically charged particles called electrons orbiting around a central nucleus, like the planets around the Sun. The previous classical theory predicted that the electron would radiate light waves because of their motion. The waves would carry away energy and so would cause the electrons to spiral inwards until they collided with the nucleus. However, such behavior is not allowed by quantum mechanics because it would violate the uncertainty principle: if an electron were to sit on the nucleus, it would have both a definite position and a definite velocity. Instead, quantum mechanics predicts that the electron does not have a definite position but that the probability of finding it is spread out over some region around the nucleus with the probability density remaining finite even at the nucleus.

The prediction of classical theory of an infinite probability density of finding the electron at the nucleus is rather similar to the prediction of classical general relativity that there should be a Big Bang singularity of infinite density. Thus one might hope that if one were able to combine general relativity and quantum mechanics into a theory of quantum gravity, one would find that the singularities of gravitational collapse or expansion were smeared out like in the case of the collapse of the atom. The first indication that this might be the case came with the discovery that black holes, formed by the collapse of localized regions such as stars, were not completely black if one took into account the uncertainty principle of quantum mechanics. Instead, a black hole would radiate particles and radiation like a hot body with a temperature higher than the smaller mass of the black hole. The radiation would carry away energy and so would reduce the mass of the black hole. This in turn would increase the rate of emissions. Eventually, it seems that the black hole will disappear completely in a tremendous burst of emission. All the matter that collapsed to form the black hole and any astronaut who was unlucky enough to fall into the black hole would disappear, at least from our region of the universe. However, the energy that corresponded to his mass by Einstein's famous equation $E = mc^2$ would be emitted by the black hole in the form of radiation. Thus the astronaut's mass-energy would be recycled to the universe. However, this would be rather a poor sort of immortality as the astronaut's subjective concept of time would almost certainly come to an end and the particles out of which he was composed would not in general be the same as the particles that were re-emitted by the black hole. Still, black hole evaporation did indicate that gravitational collapse might not lead to a complete end of time.

The real problem with spacetime having an edge or boundary at a singularity is that the laws of science do not determine the initial state of the universe at the singularity but only how it evolves thereafter. This problem would remain even if there were no singularity and time continued back indefinitely: the laws of science would not fix what the state of the universe was in the infinite past. In order to pick out one particular state for the universe from among the set of all possible states that are allowed by the laws, one has to supplement the laws by boundary conditions that say what the state of the universe was at an initial singularity or in the infinite past. Many scientists are embarrassed at talking about the boundary conditions of the universe because they feel that it verges on metaphysics or religion. After all, they might say, the universe could have started off in a completely arbitrary state. That may be so, but in that case it could also have evolved in a completely arbitrary manner. Yet all the evidence that we have suggests it evolves in a well-determined way according to certain laws. It is therefore not unreasonable to suppose that there may also be simple laws that govern the boundary conditions and determine the state of the universe.

In the classical general theory of relativity, which does not incorporate the uncertainty principle, the initial state of the universe is a point of infinite density. It is very difficult to define what the boundary conditions of the universe should be at such a singularity. However, when quantum mechanics is taken in account, there is the possibility that the singularity may be smeared out and that space and time together may form a closed four-dimensional surface without boundary or edge, like the surface of the Earth but with two extra dimensions. This would mean that the universe was completely self-contained and did not require boundary conditions. One would not have to specify the state in the infinite past and there would not be any singularities at which the laws of physics would break down. One could say that the boundary conditions of the universe are that it has no boundary.

It should be emphasized that this is simply a *proposal* for the boundary conditions of the universe. One cannot deduce them from some other principle but one can merely pick a reasonable set of boundary conditions, calculate what they predict for the present state of the universe, and see if they agree with observations. The calculations are very difficult and have been carried out so far only in simple models with a high degree of symmetry. However, the results are very encouraging. They predict that the universe must have started out in a fairly smooth and uniform state. It would have undergone a period of what is called exponential or "inflationary" expansion during which its size would have increased by a very large factor but the density would have remained the same. The universe would then have become very hot and would have expanded to the state that we see it today, cooling as it expanded. It would be uniform and the same in every direction on very large scales but would contain local irregularities that would develop into stars and galaxies.

What happened at the beginning of the expansion of the universe? Did spacetime have an edge at the Big Bang? The answer is that if the boundary conditions of the universe are that it has no boundary, time ceases to be well-defined in the very early universe just as the direction "north" ceases to be well-defined at the North Pole of the Earth. Asking what happens before the Big Bang is like asking for a point one mile north of the North Pole. The quantity that we measure as time had a beginning but that does not mean spacetime has an edge, just as the surface of the Earth does not have an edge at the North Pole, or at least, so I am told: I have not been there myself.

If spacetime is indeed finite but without boundary or edge, this would have important philosophical implications. It would mean that we could describe the universe by a mathematical model which was determined completely by the laws of science alone; they would not have to be supplemented by boundary conditions. We do not yet know the precise form of the laws: at the moment we have a number of partial laws which govern the behavior of the universe under all but the most extreme conditions. However, it seems likely that these laws are all part of some unified theory that we have yet to discover. We are making progress and there is a reasonable chance that we will discover it by the end of the century. At first sight it might appear that this would enable us to predict everything in the universe. However our powers of prediction would be severely limited, first by the uncertainty principle that states that certain quantities cannot be exactly predicted but only their probability distribution, and secondly, and even more importantly, by the complexity of the equations which makes them impossible to solve in any but very simple situations. Thus we would still be a long way from Omniscience.

The prevalence of life This painting, entitled DNA Embraces the Planets, *artistically expresses the suspicion of many scientists that carbon-based life may be a common phenomenon in the universe. Other scientists argue, however, that we may be unique and no intelligent alien civilizations exist. In either case, humanity has the clear mandate to preserve and protect the abundance of life-forms with which we share our planet. (Courtesy of J. Lomberg)*

A F T E R W O R D

The Search for Extraterrestrial Life

The heavens inspire us to contemplate a variety of profound topics, including the creation of the universe, the nature of the stars, and the formation of the Earth. Of all the fascinating subjects we might explore, perhaps none is as compelling as the question of extraterrestrial life. Are we alone? Does life exist elsewhere in the universe? What are the chances that we might someday make contact with an alien civilization?

There are no firm answers to such questions. Earth is the only planet on which life is known to exist, and our probes to other worlds have so far failed to detect any life-forms on them. This lack of data does not, however, undermine the possibility of extraterrestrial biology.

As you have seen throughout this book, one of the great lessons of modern astronomy is that our circumstances are quite ordinary. Contrary to the beliefs of our ancestors, we do not occupy a special location, like the "center of the universe." Over the past four centuries it has become clear that we inhabit one of nine planets orbiting an unremarkable star—just one of billions in an undistinguished galaxy. Is it possible that we are also biologically commonplace? The answer to this question is sought by scientists involved in SETI, the *s*earch for *e*xtra*t*errestrial *i*ntelligence.

Although the possibility cannot be ruled out that an alternative form of biochemistry exists, scientists for now confine their search to life as we now know it. All terrestrial life is based on the unique properties of the carbon atom. Carbon is an extremely versatile element whose atoms are capable of forming chemical bonds that result in especially long and complex molecules. Among these carbon-based compounds, called **organic molecules**, are the molecules of which living organisms are made.

Organic molecules can be linked together to form elaborate structures, such as chains, lattices, and fibers. Some of these structures are capable of complex, self-regulating chemical reactions. Furthermore, the primary constituents of organic molecules—carbon, hydrogen, nitrogen, oxygen, sulfur, and phosphorus—are among the most abundant elements in the universe. Indeed, the versatility and abundance of carbon suggest that extraterrestrial biology may also be based on organic chemistry.

Organic molecules are scattered abundantly throughout the Galaxy. In interstellar clouds, carbon atoms have combined with other elements to produce an impressive variety of organic compounds. Beginning in the 1960s, radio astronomers have detected telltale microwave emission lines from interstellar clouds that help identify dozens of these carbon-based chemicals. Examples include ethyl alcohol (CH_3CH_2OH), formaldehyde (H_2CO), methyl cyanoacetylene (CH_3C_3N), and acetaldehyde (CH_3CHO), to name just a few.

Further evidence of extraterrestrial organic molecules comes from newly fallen meteorites called carbonaceous chondrites (see Figure A-1), which are often found to contain a variety of organic substances. As noted in Chapter 17, carbonaceous chondrites are ancient meteorites dating from the formation of the solar system. So it seems reasonable to conclude that, even from their earliest days, the planets have been continually bombarded with organic compounds.

Interstellar space is not the only source of organic material. In a classic experiment performed in 1952, American chemists Stanley Miller and Harold Urey demonstrated that simple chemicals can combine to form prebiological compounds under supposedly primitive Earthlike conditions. In a closed container, they subjected a mixture of hydrogen, ammonia, methane, and water vapor to an electric arc (to simulate lightning bolts) for a week. At the end of this period, the inside of the container had become coated with a reddish-brown substance rich in compounds essential to life.

Scientists today tend to believe that Earth's primordial atmosphere probably was composed of carbon dioxide, nitrogen, and water vapor outgassed from volcanoes, along with some hydrogen. Modern versions of the Miller–Urey experiment (see Figure A-2) using these common gases also succeed in producing a wide variety of organic compounds.

It is important to emphasize that scientists have not created life in a test tube. Biologists have yet to figure out, among other things, how these organic molecules gathered themselves into cell-like arrangements and managed to develop systems for self-replication. Nevertheless, since so many chemical components of life are so easily synthesized under conditions that simulate the primordial Earth, it seems reasonable to suppose that life could have originated as the result of chemical processes involving materials from the Earth or from space.

A widespread abundance of organic precursors does not guarantee that life is commonplace throughout the universe. If a planet's environment is hostile, life may never get started or quickly becomes extinct. It seems quite likely, however, that there are planets orbiting other stars. Perhaps conditions on some of these worlds are sufficiently suitable for life as we know it.

Figure A-1 A carbonaceous chondrite Carbonaceous chondrites are ancient meteorites that date back to the formation of the solar system. Chemical analyses of newly fallen specimens disclose that they are rich in organic molecules, many of which are the chemical building blocks of life. This sample is a piece of the Allende meteorite, a large carbonaceous chondrite that fell in Mexico in 1969. (From the collection of Ronald A. Oriti)

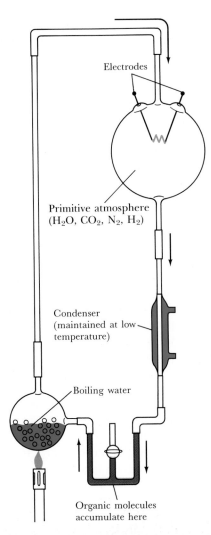

Electrodes

Primitive atmosphere
(H_2O, CO_2, N_2, H_2)

Condenser
(maintained at low
temperature)

Boiling water

Organic molecules
accumulate here

Figure A-2 The Miller–Urey experiment (updated) Modern versions of this classic experiment prove that numerous organic compounds important to life can be synthesized from gases that were present in Earth's primordial atmosphere. This experiment supports the hypothesis that life on Earth arose as a result of ordinary chemical reactions.

The development of life on Earth seems to suggest that extraterrestrial life, including intelligent species, might evolve on habitable planets, given sufficient time and appropriately hospitable conditions. How might we ascertain whether such worlds exist, given the tremendous distances that indubitably separate us from them? Many astronomers hope to learn about extraterrestrial civilizations by detecting radio transmissions from them. As we have seen, radio waves can travel immense distances without being significantly degraded by the gas and dust through which they pass. Because of this ability to penetrate the interstellar medium, radio waves are a logical choice for interstellar communication.

Over the past several decades, astronomers have proposed various ways to search for alien radio transmission

and several limited searches have been undertaken. One of the first searches occurred in 1960, when Frank Drake used a radio telescope at the National Radio Astronomy Observatory in West Virginia to "listen" to two Sunlike stars, τ Ceti and ϵ Eridani, without success. About forty similar unsuccessful searches have taken place since then using radio telescopes in both the United States and the Soviet Union. In 1973, for instance, astronomers "listened" to 600 nearby solar-type stars for $\frac{1}{2}$ hour each, but no unusual signals were detected. Since 1983, a radio telescope belonging to Harvard University has been used along with a sophisticated computer program to scan a wide range of frequencies over a large portion of the sky. So far, nothing has turned up.

Should we be discouraged by this lack of success? What are the chances that a radio astronomer might someday detect radio signals from an extraterrestrial civilization? The first person to tackle such issues was Frank Drake, now at the University of California at Santa Cruz. Drake proposed that the number of technologically advanced civilizations in the Galaxy (designated by the letter N) could be estimated with the equation:

$$N = R_* f_p n_e f_l f_i f_c L$$

where

R_* = the rate at which solar-type stars form in the Galaxy

f_p = the fraction of stars that have planets

n_e = the number of planets per solar system suitable for life

f_l = the fraction of those habitable planets on which life actually arises

f_i = the fraction of those life-forms that evolve into intelligent species

f_c = the fraction of those species that develop adequate technology and then choose to send messages out into space

L = the lifetime of that technologically advanced civilization

The Drake equation is enlightening because it expresses the number of extraterrestrial civilizations in a series of terms, some of which can be estimated from what we know about stars and stellar evolution. For instance, the first two factors, R_* and f_p, could be determined by observation. In estimating R_*, we should probably exclude massive stars (those larger than about 1.5 $M_\odot$), because they have main-sequence lifetimes shorter than the time it took to develop intelligent life here on Earth. Life on Earth originated some 3.5 to 4.0 billion years ago. If that is typical of the time needed to evolve higher life-forms, then a massive star probably becomes a red giant or a supernova before intelligent creatures appear on any of its planets. While low-mass stars have much longer lifetimes, they, too, seem unsuited for life because they are so cool. Only planets very near a low-mass star would be sufficiently warm for life as we know it, and a planet that close can become tidally coupled to the star, with one side continually facing the star, while the other is in perpetual frigid darkness. This leaves us with main-sequence stars like the Sun, those with spectral types between F5 and M0. Based on statistical studies of star formation in the Milky Way, some astronomers estimate that roughly one of these Sun-like stars forms in the Galaxy each year, thus setting R_* at 1 per year.

We learned in Chapter 7 that the planets in our solar system formed as a natural consequence of the birth of the Sun, and have seen evidence suggesting that similar processes of planetary formation may be commonplace around single stars. Yet, so far, no planet outside our solar system has been discovered. Nevertheless, many astronomers give f_p a value of 1, meaning they believe it likely that most Sunlike stars have planets.

Unfortunately, the rest of the terms in the Drake equation are very uncertain. Let's play with some hypothetical values. The chances that a planetary system has an Earthlike world are not known. Were we to consider our own solar system as representative, we could put n_e at 1. Let's be more conservative, however, and suppose that one in ten solar-type stars is orbited by a habitable planet, making $n_e = 0.1$. From what we know about the evolution of life on Earth we might assume that, given appropriate conditions, the development of life is a certainty, which would make $f_l = 1$. This is, of course, an area of intense interest to biologists. For the sake of argument, we might also assume that evolution might naturally lead to the development of intelligence (a conjecture that is hotly debated) and also make $f_i = 1$. It's anyone's guess as to whether these intelligent extraterrestrial beings would attempt communication with other civilizations in the Galaxy, but were we to assume they would, f_c would be put at 1 also. The last variable, L, involving the longevity of

civilization, is the most uncertain of all, and certainly cannot be subjected to testing! Looking at our own example, we see a planet whose atmosphere and oceans are increasingly polluted by creatures that possess nuclear weapons. If we are typical, perhaps L is as short as 100 years. Putting all these numbers together, we arrive at

$$N = 1/\text{year} \times 1 \times 0.1 \times 1 \times 1 \times 1 \times 100 \text{ years} = 10$$

In other words, out of the hundreds of billions of stars in the Galaxy, we would estimate that there are only ten technologically advanced civilizations from which we might receive communications.

A wide range of values has been proposed for the terms in the Drake equation, and these various guesses produce vastly different estimates of N. Some scientists argue that there is exactly one advanced civilization in the Galaxy and that we are it. Others speculate that there may be hundreds or thousands of planets inhabited by intelligent creatures.

If extraterrestrial beings were purposefully sending messages into space, it seems reasonable that they might choose a frequency that is fairly free of interference from extraneous sources. SETI pioneer Bernard Oliver has pointed out that a range of relatively noise-free frequencies exists in the neighborhood of the microwave emission lines of hydrogen and OH (see Figure A-3). This region of the microwave spectrum is called "the water hole," a humorous reference to the H and OH lines being so close together. Or, perhaps, they would choose to transmit at a wavelength of 21 cm, because astronomers studying the distribution of hydrogen around the Galaxy would already have their radio telescopes tuned to that wavelength (recall Figure 25-9).

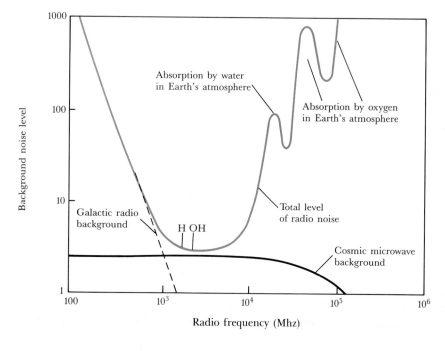

Figure A-3 The "water hole" *The so-called water hole is a range of radio frequencies (from about 1000 to 10,000 Mhz) that happens to have relatively little cosmic noise. Some scientists suggest that this noise-free region would be well suited for interstellar communication. (Adapted from C. Sagan and F. Drake)*

Figure A-4 Project Cyclops *Project Cyclops was the grandest plan ever seriously considered for the purpose of searching for extraterres-* *trial civilizations. It would have consisted of 1000 to 2500 radio antennae spread over an area 16 km in diameter. (NASA)*

Even if there are only a few alien civilizations scattered across the Galaxy, we have the technology to detect radio transmission from them. The most ambitious plan ever proposed is Project Cyclops, which would consist of 1000 to 2500 radio antennae, each 100 feet in diameter (see Figure A-4). This colossal array would be so sensitive that it could detect signals from virtually anywhere in the Galaxy. Unfortunately, it seems unlikely that such a project will ever be funded.

On a more modest scale, NASA hopes to fund the SETI project. From 1992 through 1998, existing radio telescopes will be used along with sophisticated computers to examine all of nearly 800 known stars like the Sun (spectral types F, G and K, luminosity class V) within a distance of 25 parsecs (82 light years) from Earth. In addition, the plan calls for a less sensitive survey of the entire sky.

The detection of a message from an alien civilization would be one of the greatest events in human history. Such a message could dramatically change the course of civilization, whether it were to share scientific information or bring about some sort of social or humanistic enlightenment. In a few short years, our technology, industry, and social structure might advance the equivalent of centuries into the future. Such changes would touch every person on Earth. Mindful of these profound implications, scientists push ahead with the search for extraterrestrial communication.

For further reading

Baugher, J. *On Civilized Stars: The Search for Intelligent Life in Outer Space.* Prentice-Hall, 1985 • This non- technical book by a physicist speculates on extraterrestrial intelligence and the possibility of our communicating with it.

Beatty, J. K. "The New, Improved SETI." *Sky & Telescope*, May 1983 • This article discusses new directions in SETI research that emerged in the early 1980s.

Davies, P. *The Cosmic Blueprint*. Simon & Schuster, 1988 • This extraordinary book examines connections between physics and biology in order to understand the process of life more fully.

Dawkins, R. *The Blind Watchmaker*. W. W. Norton, 1987 • This well written book surveys our modern understanding of the process of evolution.

Feinberg, G., and Shapiro, R. *Life Beyond Earth: The Intelligent Earthling's Guide to Life in the Universe*. Morrow, 1980 • This excellent introduction to exobiology includes fascinating speculation about possible forms of life.

Finney, B., and Jones, E. *Interstellar Migration and the Human Experience*. University of California Press, 1985 • This proceedings of a conference of astronomers, anthropologists, and philosophers draws upon large-scale human migrations on Earth to speculate about galactic colonization.

Goldsmith, D., ed. *The Quest for Extraterrestrial Life: A Book of Readings*. University Science Books, 1980 • This compilation includes articles by Carl Sagan, Frank Drake, Fred Hoyle, Bernard Oliver, to name just a few.

————, and Owen, T. *The Search for Life in the Universe*. Benjamin/Cummings, 1980 • This book sets the stage with sections on astronomy, biology, and planetary evolution before dealing with the search for extraterrestrial life.

Gould, S. *Wonderful Life*. W. W. Norton, 1989 • This fascinating book explains how a recent reinterpretation of fossils from the "Cambrian explosion"—a sudden proliferation of life-forms that occurred about 600 million years ago—is changing the way we think about evolution.

Hart, M., and Zuckerman, B., eds. *Extraterrestrials—Where Are They?* Pergamon Press, 1982 • This thought-provoking volume examines the implications of our failure to observe extraterrestrials.

Horowitz, N. *To Utopia and Back: The Search for Life in the Solar System*. W. H. Freeman and Company, 1986 • This concise, nontechnical introduction to the processes and origin of life pays particular attention to the results from the Viking landers on Mars.

Kutter, G. S. *The Universe and Life*. Jones and Bartlett, 1987 • The first half of this well written textbook covers astronomy, while the second half is devoted to biology and evolution.

Marx, G., ed. *Bioastronomy—The Next Steps*. Kluwer, 1988 • Although this book consists of the proceedings of a technical conference on SETI, many of its papers are easily understandable by someone who has read this textbook.

Papagiannis, M. D. "Bioastronomy: The search for extraterrestrial life." *Sky & Telescope*, June 1984 • This article eloquently but briefly examines various issues involved in SETI research.

Regis, E. *Extraterrestrials: Science and Alien Intelligence*. Cambridge University Press, 1985 • This engrossing collection of essays explores philosophical and moral issues surrounding the search for alien intelligence.

Rood, R., and Trefil, J. *Are We Alone?* Scribner, 1981 • This book takes an informal, friendly, occasionally skeptical look at modern ideas about extraterrestrial life.

Sagan, C., and Drake, F. "The Search for Extraterrestrial Intelligence." *Scientific American*, May 1975 • This classic article explores the possibility and methodology of communicating with extraterrestrials.

Schorn, R. A. "Extraterrestrial Beings Don't Exist." *Sky & Telescope*, September 1981 • This one-page article looks at the possibility that intelligent life might be extremely rare.

Shklovskii, I. S., and Sagan, C. *Intelligent Life in the Universe*, Holden-Day, 1966 • This somewhat dated, but nevertheless classic, book was one of the first to examine the question of extraterrestrial life in the light of our modern understanding of astronomy and biology.

Wilson, A. C. "The Molecular Basis of Evolution," *Scientific American*, October 1985 • This interesting article surveys recent developments in our understanding of evolution at the molecular level.

Appendixes

1 The planets: Orbital data

Planet	Semimajor axis (AU)	Semimajor axis (10⁶ km)	Sidereal period (tropical years)	Sidereal period (days)	Synodic period (days)	Mean orbital speed (km/s)	Orbital eccentricity	Inclination of orbit to ecliptic (°)
Mercury	0.3871	57.9	0.2408	87.97	115.88	47.9	0.206	7.00
Venus	0.7233	108.2	0.6152	224.70	583.96	35.0	0.007	3.39
Earth	1.0000	149.6	1.0000	365.26	—	29.8	0.017	0.0
Mars	1.5237	227.9	1.8809	686.98	779.94	24.1	0.093	1.85
(Ceres)	2.7671	414	4.603		466.6	17.9	0.077	10.6
Jupiter	5.2028	778	11.86		399	13.1	0.048	1.30
Saturn	9.529	1426	29.41		378	9.6	0.056	2.49
Uranus	19.192	2871	84.04		370	6.8	0.046	0.77
Neptune	30.061	4497	164.79		367	5.4	0.010	1.77
Pluto	39.529	5914	248.6		367	4.7	0.248	17.15

The terrestrial worlds This montage of photographs taken by various spacecraft shows the terrestrial planets and six large moons at the same scale. (Prepared for NASA by S. P. Meszaros)

2 The planets: Physical data

Planet	Equatorial diameter (km)	Equatorial diameter (Earth = 1)	Mass (Earth = 1)	Mean density (g/cm³)	Rotation period (days)	Inclination of equator to orbit	Surface gravity (Earth = 1)	Albedo	Brightest visual magnitude	Escape velocity (km/s)
Mercury	4,878	0.38	0.055	5.42	58.6	0.0	0.38	0.106	−1.9	4.3
Venus	12,104	0.95	0.82	5.24	−243.0	177.4	0.91	0.65	−4.4	10.4
Earth	12,756	1.00	1.00	5.50	0.997	23.4	1.00	0.39	—	11.2
Mars	6,794	0.53	0.107	3.94	1.026	25.2	0.38	0.15	−2.0	5.0
Jupiter	142,800	11.19	317.8	1.3	0.41	3.1	2.53	0.52	−2.7	60
Saturn	120,000	9.26	95.3	0.7	0.43	26.7	1.07	0.76	−0.3	36
Uranus	51,120	4.01	14.6	1.3	−0.65	97.9	0.92	0.51	+5.6	21
Neptune	49,528	3.88	17.2	1.7	0.67	29.6	1.12	0.35	+7.7	24
Pluto	2,290	0.18	0.002	2.0	6.387	122.5	0.06	0.4	+15.1	1

3 Satellites of the planets

Planet	Satellite	Discovered by	Mean distance from planet (km)	Sidereal period (days)	Orbital eccentricity	Diameter or size of satellite* (km)	Approximate magnitude at opposition
Earth	Moon	—	384,400	27.322	0.055	3476	−12.7
Mars	Phobos	A. Hall (1877)	9,380	0.319	0.015	28 × 23 × 20	+11
	Diemos	A. Hall (1877)	23,460	1.263	0.001	16 × 12 × 10	12
Jupiter	Almalthea	Barnard (1892)	181,300	0.498	0.003	270 × 200 × 155	13
	Io	Galileo (1610)	421,600	1.769	0.000	3630	5
	Europa	Galileo (1610)	670,900	3.551	0.000	3138	6
	Ganymede	Galileo (1610)	1,070,000	7.155	0.002	5262	5
	Callisto	Galileo (1610)	1,880,000	16.689	0.008	4800	6
	Leda	Kowal (1974)	11,094,000	238.72	0.15	(16)	20
	Himalia	Perrine (1904)	11,480,000	250.57	0.16	(180)	15
	Lysithea	Nicholson (1938)	11,720,000	259.22	0.11	(40)	18
	Elara	Perrine (1905)	11,737,000	259.65	0.21	(80)	17
	Aranke	Nicholson (1951)	21,200,000	631	0.17	(30)	19
	Carme	Nicholson (1938)	22,600,000	692	0.21	(44)	18
	Pasiphae	Melotte (1908)	23,500,000	735	0.38	(70)	17
	Sinope	Nicholson (1914)	23,700,000	758	0.28	(40)	18
Saturn	Mimas	Herschel (1789)	185,500	0.942	0.020	390	13
	Enceladus	Herschel (1789)	238,000	1.370	0.004	500	12
	Tethys	Cassini (1684)	294,700	1.888	0.000	1060	10
	Dione	Cassini (1684)	377,400	2.737	0.002	1120	10
	Rhea	Cassini (1672)	527,000	4.518	0.001	1530	10
	Titan	Huygens (1655)	1,222,000	15.945	0.029	5150	8
	Hyperion	Bond (1848)	1,481,000	21.277	0.104	410 × 260 × 220	14
	Iapetus	Cassini (1671)	3,561,000	79.331	0.028	1460	11
	Phoebe	Pickering (1898)	12,954,000	550.45	0.163	220	16
Uranus	Miranda	Kuiper (1948)	129,900	1.414	0	480	17
	Ariel	Lassell (1851)	190,900	2.520	0.003	1160	14
	Umbriel	Lassell (1851)	266,000	4.144	0.004	1190	15
	Titania	Herschel (1787)	436,300	8.706	0.002	1610	14
	Oberon	Herschel (1787)	583,400	13.463	0.001	1550	14
Neptune	Triton	Lassell (1846)	354,800	5.877	0.000	2760	14
	Nereid	Kuiper (1949)	5,513,400	360.16	0.749	170	19
Pluto	Charon	Christy (1978)	19,640	6.387	0	1280	17

Note: This table does not include several very small satellites about Jupiter, Saturn, Uranus, and Neptune that were discovered by Voyagers 1 and 2.
*A diameter of a satellite given in parentheses is estimated from the amount of sunlight it reflects.

4 The nearest stars

Name	Parallax (arc sec)	Distance (ly)	Spectral type	Radial velocity (km/s)	Proper motion (arc sec/yr)	Apparent visual magnitude	Luminosity (Sun = 1.0)
Sun			G2 V			−26.72	1.0
Proxima Cen	0.772	4.2	M5e	−16	3.85	11.05	0.00006
α Cen A	0.750	4.3	G2 V	−22	3.68	−0.01	1.6
B			K0 V			1.33	0.45
Barnard's star	0.545	5.9	M5 V	−108	10.31	9.54	0.00045
Wolf 359	0.421	7.6	M8e	+13	4.70	13.53	0.00002
BD+36° 2147	0.397	8.1	M2 V	−84	4.78	7.50	0.0055
Luyten 726-8A	0.387	8.4	M6e	+29	3.36	12.52	0.00006
B(UV Ceti)			M6e	+32		13.02	0.00004
Sirius A	0.377	8.6	A1 V	−8	1.33	−1.46	23.5
B			wd			8.3	0.003
Ross 154	0.345	9.4	M5e	−4	0.72	10.45	0.00048
Ross 248	0.314	10.3	M6e	−81	1.60	12.29	0.00011
ε Eri	0.303	10.7	K2 V	+16	0.98	3.73	0.30
Ross 128	0.298	10.8	M5	−13	1.38	11.10	0.00036
61 Cyg A	0.294	11.2	K5 V	−64	5.22	5.22	0.082
B			K7 V			6.03	0.039
ε Ind	0.291	11.2	K5 V	−40	4.70	4.68	0.14
BD+43° 44A	0.290	11.6	M1 V	+13	2.90	8.08	0.0061
BD+43° 44B			M6 V	+20		11.06	0.00039
Luyten 789-6	0.290	10.8	M7e	−60	3.26	12.18	0.00014
Procyon A	0.285	11.4	F5 IV–V	−3	1.25	0.37	7.65
B			wd			10.7	0.00055
BD+59° 1915A	0.282	11.5	M4	0	2.29	8.90	0.0030
BD+59° 1915B			M5	+10	2.27	9.69	0.0015
CD−36° 15693	0.279	11.7	M2 V	+10	6.90	7.35	0.0013
G 51−15	0.278	11.7			1.27	14.81	0.00001
τ Ceti	0.277	11.9	G8 V	−16	1.92	3.50	0.45
BD+5° 1668	0.266	12.2	M5	+26	3.77	9.82	0.0015
L725-32 (YZ Ceti)	0.261	12.4	M5e	+28	1.32	12.04	0.0002
CD−39° 14192	0.260	12.5	M0 V	+21	3.46	6.66	0.028
Kapteyn's star	0.256	12.7	M0 V	+245	8.72	8.84	0.0039
Kruger 60 A	0.253	12.8	M3	−26	0.86	9.85	0.0016
B			M5e			11.3	0.0004

5 The brightest stars

Star	Name	Apparent visual magnitude	Spectral type	Absolute magnitude	Distance (ly)	Radial velocity (km/s)	Proper motion (arc sec/yr)
α CMa	Sirius	−1.46	A1 V	+1.4	8.8	−8	1.324
α Car	Canopus	−0.72	F0 I	−8.5	1200	+21	0.025
α Boo	Arcturus	−0.04	K2 III	−0.2	36	−5	2.284
α Cen	Rigil Kentaurus	0.00	G2 V	+4.4	4.2	−25	3.676
α Lyr	Vega	0.03	A0 V	+0.3	26	−14	0.345
α Aur	Capella	0.08	G8 III	−0.6	42	+30	0.435
β Ori	Rigel	0.12	B8 Ia	−7.1	910	+21	0.001
α CMi	Procyon	0.38	F5 IV	+2.6	11	−3	1.250
α Eri	Achernar	0.46	B5 IV	−1.6	85	+19	0.098
α Ori	Betelgeuse	0.50	M2 Iab	−5.6	310	+21	0.028
β Cen	Hadar	0.61	B1 II	−5.1	460	−12	0.035
α Aql	Altair	0.77	A7 IV–V	+2.2	17	−26	0.658
α Tau	Aldebaran	0.85	K5 III	−0.3	68	+54	0.202
α Sco	Antares	0.96	M1 Ib	−4.7	330	−3	0.029
α Vir	Spica	0.98	B1 V	−3.5	260	+1	0.054
β Gem	Pollux	1.14	K0 III	+0.2	36	+3	0.625
α PsA	Fomalhaut	1.16	A3 V	+2.0	23	+7	0.367
α Cyg	Deneb	1.25	A2 Ia	−7.5	1800	−5	0.003
β Cru	Mimosa	1.25	B0.5 III	−5.0	420	+20	0.049
α Leo	Regulus	1.35	B7 V	−0.6	85	+4	0.248

6 Some important astronomical quantities

Astronomical unit	1 AU = 1.496×10^{11} m
Parsec	1 pc = 3.086×10^{16} m = 3.262 ly
Light year	1 ly = 9.460×10^{15} m = 63,240 AU
Solar mass	1 $M_\odot$ = 1.989×10^{30} kg
Solar radius	1 $R_\odot$ = 6.960×10^{8} m
Solar luminosity	1 $L_\odot$ = 3.90×10^{26} W

7 Some important physical constants

Speed of light	$c = 2.998 \times 10^{8}$ m/s
Gravitation constant	$G = 6.668 \times 10^{-11}$ N m^2 kg^{-2}
Planck constant	$h = 6.625 \times 10^{-34}$ J s = 4.136×10^{-15} eV s
Boltzmann constant	$k = 1.380 \times 10^{-23}$ J K^{-1} = 8.617×10^{-5} eV K^{-1}
Stefan–Boltzmann constant	$\sigma = 5.669 \times 10^{-8}$ W m^{-2} K^{-4}
Mass of electron	$m_e = 9.108 \times 10^{-31}$ kg
Mass of ^{1}H atom	$m_H = 1.673 \times 10^{-27}$ kg

Glossary

absolute bolometric magnitude The absolute magnitude of a star measured above the Earth's atmosphere over all wavelengths. (page 377)

absolute magnitude The apparent magnitude a star would have at a distance of 10 pc. (pages 373, 376–377)

absolute zero A temperature of $-273°C$ (or 0 K) where all molecular motion stops; the lowest possible temperature. (page 86)

absorption line spectrum Dark lines superimposed on a continuous spectrum. (page 93)

acceleration A change in velocity. (page 68)

accretion The gradual accumulation of matter in one location, typically due to the action of gravity. (page 142)

accretion disk A disk of gas orbiting a star or black hole. (page 424)

accretion hypothesis The theory that planetary atmospheres came from ices and gas trapped in planetesimals. (page 216)

active galactic nucleus (AGN) The center of an active galaxy.

active galaxy A galaxy that is emitting exceptionally large amounts of energy; a Seyfert galaxy or a quasar. (page 539)

active Sun The Sun during times of frequent solar activity such as sunspots, flares, and associated phenomena.

AGB star An asymptotic giant branch star. (page 431)

albedo The fraction of sunlight that a planet, asteroid, or satellite reflects. (page 193)

alpha decay The decay of a radioactive isotope by the emission of alpha particles. (page 184)

alpha particle The nucleus of a helium atom, consisting of two protons and two neutrons. (pages 93, 418)

amino acid The chemical building blocks of proteins. (page 336)

angstrom (Å) A unit of length equal to 10^{-10} m. (page 11)

angular diameter The angle subtended by the diameter of an object. (page 7)

angular measure The size of an angle, usually expressed in degrees, minutes of arc, and seconds of arc. (page 7)

angular momentum A measure of the momentum associated with rotation. (pages 72, 453)

angular size The angle subtended by an object. (page 7)

angular velocity The speed with which an object revolves about an axis. (page 453)

anisotropic Unequal properties in different directions; not isotropic. (page 505)

annihilation The process by which the masses of a particle and antiparticle are converted into energy. (page 579)

annular eclipse An eclipse of the Sun in which the Moon is too distant to completely cover the Sun, so that a ring of sunlight is seen around the Moon at mid-eclipse. (page 47)

anorthosite Rock commonly found in ancient, cratered highlands on the Moon. (page 181)

anticyclone The circulation of winds around a high-pressure system in a planet's atmosphere. (page 256)

antielectron A positron. (page 578)

antimatter Matter consisting of antiparticles such as antiprotons, antielectrons (positrons), and antineutrons. (page 578)

aperture The diameter of an opening; the diameter of the primary lens or mirror of a telescope.

aphelion The point in its orbit where a planet is farthest from the Sun. (page 64)

apogee The point in its orbit where a satellite or the Moon is farthest from the Earth. (page 47)

Apollo asteroid An asteroid whose orbit brings it closer to the Sun than the distance of the Earth's orbit. (pages 329–330)

apparent brightness The flux of star's light arriving at the Earth. (page 376)

apparent magnitude A measure of the brightness of light from a star or other object as measured from Earth. (pages 49, 372–373)

apparent solar day The interval between two successive transits of the Sun's center across the local meridian. (page 29)

apparent solar time Time reckoned by the position of the Sun in the sky. (page 29)

ascending node The point along an orbit where an object crosses a reference plane (usually the ecliptic or the celestial equator) from south to north. (page 42)

association A loose cluster of stars whose spectra, motions, or positions suggest that they had a common origin. (page 405)

asteroid One of tens of thousands of small, rocky, planet-like objects in orbit about the Sun. (pages 132, 327)

asteroid belt A region between the orbits of Mars and Jupiter which encompasses the orbits of many asteroids. (pages 327–328)

asthenosphere A warm, plastic layer of the mantle beneath the lithosphere of the Earth. (page 156)

astigmatism An optical defect of a lens or mirror whereby light rays in different planes do not focus at the same point.

astrometry The branch of astronomy dealing with the precise determination of the positions and motions of celestial objects.

astronomical unit (AU) The semimajor axis of the Earth's orbit; the average distance between the Earth and the Sun. (page 9)

astronomy The branch of science dealing with objects and phenomena that lie beyond the Earth's atmosphere.

astrophysics That part of astronomy dealing with the physics of astronomical objects and phenomena.

asymptotic giant branch star A red supergiant star that has ascended the giant branch for the second and final time. (page 431)

atmosphere (atm) A unit of atmospheric pressure. (page 159)

atmospheric pressure The force per unit area exerted by a planet's atmosphere. (page 159)

atom The smallest particle of an element that has the properties characterizing that element. (pages 93, 95, 136)

atomic mass unit (amu) A unit of mass (1.67×10^{-24} g) nearly equal to the mass of a hydrogen atom.

atomic number The number of protons in the nucleus of the atom of a particular element. (page 135)

atomic weight The mass of an atom in atomic mass units.

aurora Light radiated by atoms and ions in the Earth's upper atmosphere, mostly in the polar regions. (page 163)

autumnal equinox The intersection of the ecliptic and the celestial equator where the Sun crosses the equator from north to south. (page 26)

average density The mass of an object divided by its volume. (page 132)

Balmer lines Emission or absorption lines in the hydrogen spectrum involving electron transitions between the second and higher energy levels. (page 95)

bar A unit of pressure. (page 159)

Barnard object A dark nebula discovered by E. E. Barnard. (page 399)

barred spiral galaxy A spiral galaxy in which the spiral arms begin from the ends of a "bar" running through the nucleus rather than from the nucleus itself. (pages 503–504)

belt A dark band in Jupiter's atmosphere. (page 247)

Big Bang An explosion of all space, roughly 20 billion years ago, from which the universe emerged. (pages 6, 556)

Big Crunch The fate of the universe if it is bounded and ultimately collapses in upon itself. (page 569)

binary asteroid Two asteroids that revolve about each other as they orbit the Sun. (page 330)

binary star Two stars revolving about each other. (pages 385–387)

biosphere The layer of soil, water, and air surrounding the Earth in which living organisms thrive. (pages 162–163)

bipolar sunspot group A sunspot group that has roughly comparable areas covered by north and south magnetic polarity. (page 349)

BL Lacertae object A type of active galaxy whose nucleus does not exhibit spectral lines. (pages 542–543)

black hole An object whose gravity is so strong that the escape velocity exceeds the speed of light. (pages 5, 467–479)

black hole evaporation The process be which black holes emit particles. (pages 570–571)

blackbody A hypothetical perfect radiator that absorbs and re-emits all radiation falling upon it. (page 88)

blackbody radiation The radiation emitted by a blackbody; thermal radiation. (page 89)

blazar A BL Lacertae object. (page 542)

blueshift A decrease in the wavelength of photons emitted by an approaching source of light. (page 97)

Bode's law A numerical sequence that gives the approximate average distances of the planets from the Sun in astronomical units. (page 326)

Bohr atom A model of the atom, described by Niels Bohr, in which electrons revolve about the nucleus in certain allowed circular orbits. (page 96)

Bok globule A small, roundish dark nebula. (page 399)

bolometric correction The difference between the visual and bolometric magnitudes of a star. (page 377)

bolometric magnitude A measure of the brightness of a star or other object as detected by a device above the Earth's atmosphere and sensitive to all wavelengths of radiation. (page 377)

bounded universe A universe throughout which the average density exceeds the critical density. (page 563)

breccia A rock formed by the sudden amalgamation of various rock fragments under pressure. (pages 181–182)

burster A nonperiodic X-ray source that emits powerful bursts of X rays. (pages 5, 462–463)

butterfly diagram A plot of sunspot latitude versus time. (page 359)

caldera The crater at the summit of a volcano. (page 231)

capture theory The hypothesis that the Moon was gravitationally captured by the Earth. (page 183)

carbon burning The thermonuclear fusion of carbon to produce heavier nuclei. (page 435)

carbon cycle A series of nuclear reactions, involving carbon as a catalyst, by which hydrogen is transformed into helium; the CNO cycle. (page 349)

carbonaceous chondrite A type of meteorite that has a high abundance of carbon and volatile compounds. (page 335)

Cassegrain focus An optical arrangement in a reflecting telescope in which light rays are reflected by a secondary mirror to a focus behind the primary mirror. (page 107)

Cassini division An apparent gap between Saturn's A and B rings. (pages 284, 295)

CCD Charge-coupled device; a type of solid-state silicon wafer designed for the detection of photons. (page 112)

celestial equator A great circle on the celestial sphere 90° from the celestial poles. (page 23)

celestial mechanics The branch of astronomy dealing with the motions and gravitational interactions of objects in the solar system.

celestial poles Points about which the celestial sphere appears to rotate. (page 24)

celestial sphere A sphere of very large radius centered on the observer; the apparent sphere of the sky. (pages 23–24)

center of mass That point in an isolated system that moves at a constant velocity in accordance with Newton's first law. (pages 70, 387)

central bulge A spherical distribution of stars around the nucleus of a spiral galaxy. (pages 485–486)

Cepheid variable One of two types (Type I and Type II) of yellow, supergiant, pulsating stars. (page 425)

Cerenkov radiation The radiation emitted by a particle traveling through a medium at a speed greater than the speed of light in that medium. (page 443)

Ceres The largest asteroid and the first to be discovered. (pages 326, 328)

Chandrasekhar limit The maximum mass of a white dwarf. (page 434)

chemical differentiation The process by which the heavier elements in a planet sink toward its center while lighter elements rise toward its surface. (page 143)

chemical element A substance that cannot be decomposed by chemical means into simpler substances. (page 91)

chromatic aberration An optical defect whereby different colors of light passing through a lens are focused at different locations. (pages 104–105)

chromosphere A layer in the solar atmosphere between the photosphere and the corona. (page 354)

circumstellar shell Gases immediately surrounding a star from which they have escaped. (page 442)

cluster of galaxies A collection of galaxies containing a few to several thousand member galaxies. (pages 506–510)

CNO cycle A series of nuclear reactions in which carbon is used as a catalyst to transform hydrogen into helium. (page 349)

cocoon nebula The nebulosity surrounding a protostar. (page 400)

cocreation theory The hypothesis that the Earth and the Moon formed at the same time from the same material. (page 183)

cold dark matter Slowly moving, weakly interacting particles presumed to constitute the bulk of matter in the universe. (page 591)

collisional ejection theory The hypothesis that the Moon formed from material ejected from the Earth by the impact of a large asteroid. (page 183)

color index The difference between the magnitudes of a star measured in two different spectral regions. (page 378)

color–magnitude diagram A plot of the magnitudes of stars in a cluster against their color indices. (page 404)

coma (of a comet) The diffuse gaseous component of the head of a comet. (pages 336–337)

coma (optical) The distortion of off-axis images formed by a parabolic mirror. (page 110)

comet A small body of ice and dust in orbit about the Sun. While passing near the Sun, a comet's vaporized ices give rise to a coma and tail. (pages 132–133, 336–340)

comet–asteroid hypothesis The theory that planetary atmospheres came from the impacts of comets and asteroids. (page 216)

Compton wavelength A characteristic length associated with the size of an elementary particle. (page 558)

condensation temperature The temperature at which a particular substance in a low-pressure gas condenses into a solid. (page 140)

conduction The transfer of heat by directly passing energy from atom to atom. (page 350)

conic section The curve of intersection between a circular cone and a plane; this curve can be a circle, ellipse, parabola, or hyperbola. (pages 69, 71)

conjunction The geometric arrangement of a planet in the same part of the sky as the Sun; a planet with an elongation of 0°. (page 60)

conservation of angular momentum A law of physics stating that the total amount of angular momentum in an isolated system remains constant. (pages 72, 453)

conservation of energy A law of physics stating that the total energy of an isolated system remains constant. (page 72)

conservation of momentum A law of physics stating that the total momentum of an isolated system remains constant. (page 72)

constellation A configuration of stars, often named after an object, person, god, or animal. (page 19)

contact binary A double star in which both members fill their Roche lobes. (page 423)

continental drift The gradual movement of the continents over the surface of the Earth due to plate tectonics. (page 155)

continuous spectrum A spectrum of light over a range of wavelengths without any spectral lines. (page 93)

continuum The intensity of a continuous spectrum. (pages 338, 382, 532)

convection The transfer of energy by moving currents of fluid or gas containing that energy. (pages 156, 215, 350)

convection cell A circulating loop of gas or liquid that transports heat from a warm region to a cool region. (page 216)

convection zone The region in a star where convection is the dominant means of energy transport. (page 351)

coorbital satellites Satellites that share the same orbit. (page 298)

Coriolis effect The deflection of moving objects on a rotating surface. (page 256)

corona The Sun's outer atmosphere, which has a high temperature and a low density. (pages 355–356)

coronagraph An instrument for photographing the solar corona in which an occulting disk inside the telescope produces an artificial eclipse. (page 355)

coronal hole A region in the Sun's corona that is deficient in hot gases. (page 357)

cosmic microwave background An isotropic radiation field with a blackbody temperature of about 2.7 K that permeates the entire universe. (pages 557–560)

cosmic particle horizon An imaginary sphere, centered on the Earth, whose radius equals the distance light has traveled since the Big Bang. (page 556)

cosmic rays Atomic nuclei (mostly protons) that strike the Earth with extremely high speeds.

cosmic singularity The Big Bang. (page 557)

cosmic string A hypothetical long, thin, wirelike concentration of matter left over from the Big Bang. (pages 586–587)

cosmological constant (Λ) A quantity in the equations describing the universe that proved a pressure that helps the universe expand or inhibits its contraction. (pages 554, 564)

cosmological model A specific theory about the structure and evolution of the universe.

cosmological principle The assumption that the universe is homogeneous and isotropic on the largest scale. (page 554)

cosmological redshift A redshift that is caused by the expansion of the universe. (page 555)

cosmology The study of the structure and evolution of the universe.

coudé focus A reflecting telescope in which a series of mirrors direct light to a remote focus away from the moving parts of the telescope. (page 107)

crater A circular depression on a planet or satellite, caused by the impact of a meteoroid. (page 171)

crescent moon One of the phases of the Moon in which less than one-half of its illuminated surface is visible from the Earth. (pages 38–39)

critical density The average density throughout the universe at which space is flat and galaxies just barely continue receding from each other infinitely far into the future. (page 563)

critical surface The surface of the Roche lobes in a double star system. (page 422)

crust (of a planet) The surface layer of a planet. (page 158)

cyclone The circulation of winds around a low-pressure system in a planet's atmosphere. (page 256)

cyclonic motion Circular wind motion in a planet's atmosphere due to the Coriolis effect. (page 256)

dark-matter problem The enigma that most of the matter in the universe is severely underluminous. (pages 511–512)

dark nebula A cloud of interstellar gas and dust that obscures the light of more distant stars. (pages 397, 399)

daughter isotope An isotope that results from the radioactive decay of another isotope. (page 184)

decametric radiation Radiation whose wavelength is about ten meters. (page 250)

decay series A sequence of isotopes that radioactively decay one into another. (page 184)

deceleration parameter (q_0) A number whose value indicates how fast the expansion of the universe is slowing down. (pages 534, 564–565)

decimetric radiation Radiation whose wavelength is about a tenth of a meter. (page 250)

declination Angular distance of a celestial object north or south of the celestial equator. (page 24)

deferent A stationary circle in the Ptolemaic system along which another circle (an epicycle) moves, carrying a planet, the Sun, or the Moon. (page 57)

deflagration A sudden, violent burning. (page 441)

degeneracy The phenomenon, due to quantum mechanical effects, whereby the pressure exerted by a gas does not depend on its temperature. (page 419)

degenerate-electron/neutron pressure The pressure exerted by degenerate electrons or neutrons. (pages 419, 434, 464)

degenerate gas A gas in which all the allowed states for particles (electrons or neutrons) have been filled, thereby causing the gas to behave differently than ordinary gases do. (page 419)

degenerate pressure The pressure exerted by a degenerate gas. (page 419)

degree A basic unit of angular measure, designated by the symbol °. (page 7)

degree (Fahrenheit/Celsius) A basic unit of temperature, designated by the symbol °. (page 86)

density The ratio of the mass of an object to its volume. (page 11)

density parameter (Ω_0) The ratio of the average density of the universe to the critical density. (page 564)

density-wave theory An explanation of spiral arms in galaxies proposed by C. C. Lin and colleagues. (pages 492–494)

descending node A point along an orbit where an object crosses a reference plane (usually the ecliptic or celestial equator) from north to south. (page 42)

detached binary A binary in which neither star fills its Roche lobe. (pages 422–423)

deuterium An isotope of hydrogen whose nuclei each contain one proton and one neutron; heavy hydrogen. (page 349)

deuterium bottleneck The situation about 3 minutes after the Big Bang when deuterium inhibited the formation of heavier elements. (page 581)

differential rotation The rotation of a nonrigid object in which parts adjacent to each other do not always stay close together. (pages 248, 357)

differentiation (geological) The separation of different kinds of material in different layers inside a planet. (page 143)

diffraction The spreading out of light passing the edge of an opaque object.

diffraction grating A piece of glass containing thousands of closely spaced lines that is used to disperse light into a spectrum. (pages 113–114)

diffuse nebula A reflection or emission nebula consisting of interstellar gas and dust.

dilation of time The slowing of time due to relativistic motion. (page 470)

direct motion The apparent eastward movement of a planet seen against the background stars. (page 57)

disk (of a galaxy) The disk-shaped distribution of population I stars that dominates the appearance of a spiral galaxy. (page 485)

dissociative recombination The capture of an electron by a positive molecular ion, which results in breaking the molecule into two neutral atoms. (page 235)

distance modulus The difference between the apparent and absolute magnitudes of an object. (page 376)

diurnal Daily. (page 23)

diurnal motion Motion in one day. (page 23)

Doppler effect The apparent change in wavelength of radiation due to relative motion between the source and the observer along the line of sight. (page 98)

double-line spectroscopic binary A spectroscopic binary in which spectral lines of both stars can be seen. (pages 388–389)

double radio source An extragalactic radio source characterized by two large regions of radio emission, typically located on either side of an active galaxy. (pages 541–542)

dust tail The tail of a comet that is composed primarily of dust particles. (page 339)

dwarf elliptical galaxy A low-mass elliptical galaxy that contains only a few million stars. (page 505)

dynamo effect The creation of a magnetic field by a moving electric current. (page 161)

dyne A unit of force; the force needed to accelerate 1 g by 1 cm/s². (page 69)

eccentricity (of an ellipse) The value $\sqrt{1 - b^2/a^2}$, where a is the semimajor axis of the ellipse and b is its semiminor axis. (pages 70, 129)

eclipse The cutting off of part or all the light from one celestial object by another. (page 41)

eclipse path The track of the tip of the Moon's shadow along the Earth's surface during a total or annular solar eclipse. (page 45)

eclipse season A period during the year when a solar or lunar eclipse is possible.

eclipse year The interval between successive passages of the Sun through the same node of the Moon's orbit.(page 48)

eclipsing binary A binary system in which, as seen from Earth, the stars periodically pass in front of each other. (pages 389–390)

ecliptic The apparent annual path of the Sun on the celestial sphere. (page 25)

Einstein ring The circular image of a remote light source produced by a gravitational lens having nearly a perfect alignment between the source, the observer, and the deflecting mass. (page 476)

electromagnetic radiation Radiation consisting of oscillating electric and magnetic fields including gamma rays, X rays, visible light, ultraviolet and infrared radiation, radio waves, and microwaves. (page 84)

electromagnetic spectrum The entire array or family of electromagnetic radiation. (page 85)

electromagnetic theory The branch of physics that deals with electricity, magnetism, and electromagnetic radiation. (page 74)

electron A negatively charged subatomic particle usually found in orbits about the nuclei of atoms. (pages 90, 136)

electron volt The energy acquired by an electron accelerated through an electric potential of one volt. (page 92)

ellipse A conic section obtained by cutting completely through a circular cone with a plane. (page 64)

elliptical galaxy A galaxy with an elliptical shape and no conspicuous interstellar material. (pages 504–505)

elongation The angular distance between a planet and the Sun as viewed from Earth. (page 59)

emission line A bright spectral line. (page 93)

emission line spectrum A spectrum that contains emission lines. (page 93)

emission nebula A glowing gaseous nebula whose light comes from fluorescence caused by a nearby star. (pages 402–403)

energy The ability to do work. (page 72)

energy flux The rate of energy flow, usually measured in joules per square meter per second. (page 87)

energy level (in an atom) A particular amount of energy possessed by an atom above the atom's least energetic state. (page 96)

energy-level diagram A diagram showing the arrangement of an atom's energy levels. (page 96)

Encke division A narrow gap in Saturn's A ring. (pages 292–293)

epicenter The location on the Earth's surface directly over the focus of an earthquake. (page 153)

epicycle A moving circle in the Ptolemaic system about which a planet revolves. (page 57)

epoch A date and time selected as a fixed reference. (page 29)

equation of state An equation relating the temperature, pressure, and density of a gas.

equation of time The difference between apparent and mean solar time. (page 30)

equations of stellar structure The equations used by astrophysicists to calculate the structure, properties, and evolution of a star. (page 350)

equinox One of the intersections of the ecliptic and the celestial equator. (page 25)

equipotential contour A surface along which gravitational potential energy is constant. (page 331)

equivalent width A measure of the amount of light subtracted from the continuous spectrum by an absorption line. (pages 380, 382)

era of recombination The moment, approximately 1 million years after the Big Bang, when the universe had cooled sufficiently to permit the formation of hydrogen atoms. (page 563)

erg A unit of energy; the work done by a force of 1 dyne moving through a distance of 1 cm.

ergosphere The region of space immediately outside the event horizon of a rotating black hole where it is impossible to remain at rest. (page 473)

escape velocity The speed needed by one object to achieve a parabolic orbit away from a second object and thereby permanently move away from the second object. (page 139)

event horizon The location around a black hole where the escape velocity equals the speed of light; the surface of a black hole. (pages 469–471)

evolutionary track The path on an H–R diagram followed by a star as it evolves. (page 399)

excitation The process of imparting energy to an atom or ion.

exponent A number placed above and after another number to denote the power to which the latter is to be raised, as n in 10^n. (page 8)

extinction The attenuation of light due to absorption by material between the source and the observer. (page 397)

extragalactic Beyond our Galaxy.

eyepiece A magnifying lens used to view the image produced at the focus of a telescope. (page 104)

false vacuum Empty space with a high-energy density that possibly filled the universe immediately after the Big Bang. (page 586)

favorable opposition A Martian opposition that affords good Earth-based views of the planet. (pages 226–227)

Feynman diagram A schematic diagram showing interactions between particles. (pages 583–584)

filtergram A photograph of the Sun taken through special filters that are transparent only to a narrow range of wavelengths.

first quarter moon The phase of the Moon that occurs when the Moon is 90° east of the Sun. (page 38)

fission theory The hypothesis that the Moon was pulled out of a rapidly rotating proto-Earth. (page 183)

Fitzgerald contraction The shrinking of distances due to relativistic motion; Fitzgerald–Lorentz contraction. (page 471)

flare A sudden, temporary outburst of light from an extended region of the solar surface. (page 361)

flatness problem The dilemma posed by the fact that the average density throughout the universe is very nearly equal to the critical density. (page 582)

flocculent spiral galaxy A spiral galaxy with fuzzy, poorly defined spiral arms. (page 491)

flux The number of particles or the amount of energy flowing across a given area per unit time. (page 352)

focal length The distance from a lens or mirror to the point where converging light rays meet. (pages 103–104, 107)

focus (of an earthquake) The location below the Earth's surface where an earthquake occurs. (page 153)

focus (of an ellipse) One of two points inside an ellipse such that the combined distance from the two foci to any point on the ellipse is a constant. (page 64)

focus (optical) The point where light rays converged by a lens or mirror meet. (pages 103–104, 107)

force That which can change the momentum of an object. (page 68)

frequency The number of wave crests or troughs that cross a given point per unit time; the number of vibrations per unit time. (page 85)

full moon A phase of the Moon during which its full daylight hemisphere can be seen from Earth. (page 38)

fusion crust The coating on a stony meteorite caused by the heating of the meteorite as it descended through the Earth's atmosphere. (page 333)

galactic cannibalism A collision between two galaxies of unequal mass and size in which the smaller galaxy seems to be absorbed into the larger galaxy. (pages 514–516)

galactic cluster A loose association of young stars in the disk of our Galaxy. (page 404)

galactic equator   The intersection of the principal plane of the Milky Way with the celestial sphere.

galactic nucleus   The center of a galaxy. (pages 485, 495–496)

galaxy   A large assemblage of stars, nebulae, and interstellar gas and dust. (page 6)

galaxy merger   A collision between two galaxies that causes them to coalesce. (page 514)

Galilean satellite   Any one of the four large moons of Jupiter. (page 265)

gamma rays   The most energetic form of electromagnetic radiation. (page 85)

gauss   A unit of magnetic field strength. (page 360)

general theory of relativity   A description of gravity formulated by Albert Einstein, which explains that gravity affects the geometry of space and the flow of time. (pages 76, 468–469)

geocentric cosmology   An Earth-centered theory of the universe. (page 57)

geomagnetic   Referring to the Earth's magnetic field.

giant   A star whose diameter is typically 10 to 100 times that of the Sun and whose luminosity is roughly 100 Suns. (pages 382, 384)

giant elliptical galaxy   A massive elliptical galaxy containing many billions of stars. (page 505)

giant molecular cloud   A large cloud of interstellar gas and dust. (page 406)

gibbous moon   A phase of the Moon in which more than one-half, but not all, of the Moon's daylight hemisphere is visible from Earth. (pages 38–39)

glitch   A sudden speedup in the period of a pulsar. (page 457)

globular cluster   A large spherical cluster of stars, typically found in the outlying regions of a galaxy. (page 420)

globule   A small, dense, dark nebula. (page 399)

gluon   A particle that is exchanged between quarks. (page 584)

grand design spiral galaxy   A galaxy with well-defined spiral arms. (pages 491–492)

grand unified field theory (GUT)   A theory that describes and explains the four physical forces. (page 585)

granulation   The "rice-grain"–like structure of the solar photosphere. (page 353)

granule   A convective cell in the solar photosphere. (page 354)

gravitation   The tendency of matter to attract matter.

gravitational lens   The deflection of light from a remote source caused by the presence of an intervening mass. (page 474)

gravitational radiation/waves   Oscillations of space produced by changes in the distribution of matter. (page 444)

graviton   The particle that is responsible for the gravitational force. (page 584)

gravity   The force with which matter attracts matter. (page 68)

Great Attractor   A huge concentration of matter toward which many galaxies, including the Milky Way, appear to be moving. (pages 528–529, 560)

Great Dark Spot   A prominent high-pressure system in Neptune's southern hemisphere. (page 314)

Great Red Spot   A prominent high-pressure system in Jupiter's southern hemisphere. (pages 247, 254)

greatest elongation   The largest possible angle between the Sun and an inferior planet. (pages 59–60)

greenhouse effect   The trapping of infrared radiation near a planet's surface by the planet's atmosphere. (pages 164, 212)

Greenwich meridian   The meridian of longitude that passes through the old Royal Greenwich Observatory near London; the longitude of 0°. (page 281)

H I region   A region of neutral hydrogen in interstellar space.

H II region   A region of ionized hydrogen in interstellar space. (pages 402–403)

hadron   A particle composed of quarks. (page 580)

half-life   The time required for one-half of the radioactive nuclei of an isotope to disintegrate. (page 184)

halo (of a galaxy)   A spherical distribution of globular clusters and population II stars that surround a spiral galaxy. (pages 485–486)

harmonic law   Kepler's third law. (page 65)

head–tail source   A radio source consisting of a bright "head" and a long low-brightness "tail." (page 542)

Heisenberg uncertainty principle   A principle of quantum mechanics that places limits on the precision of simultaneous measurements. (pages 570, 577)

helio-   A prefix referring to the Sun.

heliocentric cosmology   A Sun-centered theory of the universe. (page 59)

helium burning   The thermonuclear fusion of helium to form carbon and oxygen. (page 418)

helium flash The nearly explosive beginning of helium burning in the dense core of a red-giant star. (page 419)

helium-shell flash A brief thermal runaway that occurs in the helium-burning shell of a red supergiant. (page 431)

Helmholtz contraction The contraction of a gaseous body, such as a star or nebula, during which gravitational energy is transformed into thermal energy; Kelvin–Helmholtz contraction. (page 140)

Herbig–Haro object A small, luminous nebula associated with the end point of a jet emanating from a young star. (pages 400, 402)

Hertzsprung–Russell (H–R) diagram A plot of the absolute magnitude (or luminosity) of stars against their spectral type (of surface temperature). (pages 382–384)

heterogeneous accretion theory A theory of planetary formation which argues that the composition of planetesimals changed as the planets formed. (page 143)

high-pressure system A region of high atmospheric pressure. (page 256)

highlands Ancient, heavily cratered terrain on the Moon. (page 172)

Hirayama family A group of asteroids that have nearly identical orbits about the Sun. (page 330)

homogeneous accretion theory A theory of planetary formation which argues that the planets formed from planetesimals of generally the same composition. (page 143)

horizontal branch A group of stars on the Hertzsprung–Russell diagram of a typical globular cluster, near the main sequence and having roughly constant absolute magnitude. (pages 420–421)

hot-spot volcanism Volcanic activity that occurs over a hot region buried deep within a planet. (pages 157, 218)

Hubble classification scheme A method of classifying galaxies as spirals, barred spirals, ellipticals, or irregulars according to their appearance. (pages 503–506)

Hubble constant (H_0) The constant of proportionality in the relation between the recessional velocities of remote galaxies and their distances. (page 520)

Hubble flow The recessional motions of remote galaxies caused by the expansion of the universe. (page 519)

Hubble law The empirical relationship stating that the redshifts of remote galaxies are directly proportional to their distances from Earth. (pages 518–520)

hydrocarbon Any one of a variety of chemical compounds composed of hydrogen and carbon. (page 296)

hydrogen burning The thermonuclear conversion of hydrogen into helium. (page 348)

hydrogen envelope A huge, tenuous sphere of gas surrounding the head of a comet. (page 337)

hydrostatic equilibrium A balance between the weight of a layer in a star and the pressure that supports it. (page 249)

hyperbola A conic section formed by cutting a circular cone with a plane at an angle steeper than the sides of the cone. (page 69)

hypothesis An idea or collection of ideas that seems to explain a specified phenomenon; a conjecture. (page 2)

igneous rock A rock that formed from the solidification of molten lava or magma. (page 152)

impact breccia A type of lunar rock consisting of rock fragments cemented together by the impact of a meteoroid. (pages 181–182)

impact crater A crater formed by the impact of a meteoroid. (page 330)

inertia The property of matter that requires a force to act on it to change its state of motion. (page 68)

inferior conjunction The configuration when an inferior planet is between the Sun and Earth. (pages 59–60)

inferior planet A planet that is closer to the Sun than the Earth is. (page 59)

inflationary epoch A brief period shortly after the Big Bang during which the scale of the universe increased very rapidly. (page 582)

infrared radiation Electromagnetic radiation of wavelength longer than visible light, yet shorter than radio waves. (page 85)

inner Lagrangian point The point between two stars in binary where their Roche lobes touch; the point across which mass transfer can occur. (page 422)

instability strip A region of the H–R diagram occupied by pulsating stars. (page 425)

interferometry A technique of combining the observations two or more telescopes to produce images better than one telescope alone could do. (page 115)

intermediate-vector boson The particle that is responsible for the weak nuclear force. (page 584)

interplanetary medium The sparse distribution of gas and dust particles in interplanetary space.

interstellar dust Microscopic solid grains of various compounds in interstellar space. (pages 387–398)

interstellar extinction The dimming of starlight as it passes through the interstellar medium. (pages 397, 483)

interstellar gas Sparse gas in interstellar space. (pages 397–398)

interstellar medium Interstellar gas and dust. (pages 397–398)

interstellar reddening The reddening of starlight passing through the interstellar medium, caused by the fact that blue light is scattered more than red. (page 398)

inverse-square law The statement that the apparent brightness of a light source varies inversely with the distance from the source. (pages 373–375)

Io torus A doughnut-shaped ring of gas circling Jupiter at the distance of Io's orbit. (page 275)

ion An atom that has become electrically charged due to the addition or loss of one or more electrons. (page 213)

ion tail (of a comet) The relatively straight tail of a comet produced by the solar wind acting on ions. (page 339)

ionization The process by which an atom loses electrons. (page 96)

ionization potential The energy required to remove an electron from an atom.

ionopause The boundary around a planet like Venus where the pressure exerted by ions counterbalances the pressure of the solar wind. (page 213)

ionosphere A layer in the Earth's upper atmosphere in which many of the atoms are ionized.

iron meteorite A meteorite composed primarily of iron. (pages 333–334)

irregular cluster (of galaxies) A sprawling collection of galaxies whose overall distribution in space does not exhibit any noticeable spherical symmetry. (page 508)

irregular galaxy An unsymmetrical galaxy having neither spiral arms nor an elliptical shape. (page 506)

isochron A line on a graph indicating the abundance ratios of various isotopes for a rock of a certain age. (page 185)

isotope Any of several forms for the same chemical element whose nuclei all have the same number of protons but different numbers of neutrons. (page 136)

isotropic The same in all directions. (page 505)

isotropy problem The dilemma posed by the fact that the cosmic microwave background is isotropic. (page 582)

Jeans length The smallest scale over which a density fluctuation in a medium will contract to form a gravitationally bound object. (page 589)

joule (J) A unit of energy. (page 86)

Jovian planet Any of the four largest planets: Jupiter, Saturn, Uranus, or Neptune. (page 132)

Kaluza–Klein theory A theory that describes the four physical forces in terms of the geometry of an eleven-dimensional space–time. (pages 587–588)

kelvin (K) A unit of temperature on the Kelvin temperature scale. (page 86)

Kelvin–Helmholtz contraction The contraction of a gaseous body, such as a star or nebula, during which gravitational energy is transformed into thermal energy. (page 348)

Kepler's laws Three statements, discovered by Johannes Kepler, that describe the motions of the planets. (pages 64–65)

kiloparsec (kpc) One thousand parsecs; about 3260 light years. (page 10)

kinematic wave A wavelike disturbance in the orbits of stars about the center of a galaxy. (page 493)

kinetic energy The energy possessed by an object because of its motion. (pages 72, 138)

Kirchhoff's laws Three statements about circumstances that produce absorption line, emission line, and continuous spectra. (page 93)

Kirkwood's gaps Gaps in the spacing of asteroid orbits discovered by Daniel Kirkwood. (pages 328–329)

Lamb shift A tiny shift in the spectral lines of hydrogen caused by the presence of virtual particles. (page 578)

Lagrangian points Five points in the orbital plane of two bodies revolving about each other in circular orbits where a third object of negligible mass can remain in equilibrium. (pages 329, 331)

law of cosmic censorship The hypothesis that all singularities must be surrounded by an event horizon. (page 472)

law of equal areas Kepler's second law. (page 64)

law of inertia Newton's first law. (page 68)

laws of physics A set of physical principles with which we can understand natural phenomena and the nature of the universe. (page 3)

leap year A calendar year with 366 days. (page 30)

lenticular galaxy A galaxy with a central bulge and a disk, but no spiral structure; an S0 galaxy (page 505)

lepton Any member of a class of particles that includes the electron and neutrino. (page 580)

libration A slight rocking of the Moon in its orbit whereby an Earth-based observer can, over time, see slightly more than one-half the Moon's surface. (page 170)

light Electromagnetic radiation. (page 83)

light curve A graph that displays variations in the brightness of a star or other astronomical object. (pages 389–390)

light-gathering power A measure of the amount of radiation brought to a focus by a telescope. (page 108)

light year (ly) The distance light travels in a vacuum in one year. (page 9)

limb (of Sun or Moon) The apparent edge of the Sun or Moon as seen in the sky.

limb darkening The phenomenon whereby the Sun is darker near its limb than near the center of its disk. (page 354)

limiting magnitude The faintest magnitude that can be observed with a certain telescope under certain conditions.

line of apsides The major axis of an elliptical orbit. (page 43)

line of nodes A line connecting the nodes of an orbit. (page 42)

line profile The shape of a spectral line. (pages 380–382)

lithophile element Low-density elements usually found in crustal rocks. (page 335)

lithosphere The solid, upper layer of the Earth; essentially the Earth's crust. (pages 156–157)

LMC The Large Magellanic Cloud. (page 507)

Local Group The cluster of galaxies of which our Galaxy is a member. (pages 508–509)

long-period comet A comet that takes hundreds of thousands of years to complete one orbit of the Sun. (page 339)

Lorentz transformations Equations that relate the measurements of different observers who are moving relative to each other at high speeds. (page 470)

low-pressure system A region of low atmospheric pressure. (page 256)

luminosity The rate at which electromagnetic radiation is emitted from a star or other object. (pages 88, 377)

luminosity class A classification of a star of a given spectral type according to its luminosity. (page 385)

lunar Referring to the Moon.

lunar eclipse An eclipse of the Moon by the Earth; a passage of the Moon through the Earth's shadow. (page 44)

lunar month The time it takes to complete one cycle of lunar phases; the synodic month. (page 47)

Lyman series A series of spectral lines of hydrogen produced by electron transitions to and from the lowest energy state of the hydrogen atom. (page 96)

Magellanic clouds Two nearby galaxies visible to the naked eye from southern latitudes. (page 507)

magma Molten rock beneath a planet's surface. (page 152)

magnetic-dynamo model A theory that explains the solar cycle as a result of the Sun's differential rotation acting on the Sun's magnetic field. (page 362)

magnetic field A region of space near a magnetized body within which significant magnetic forces can be detected.

magnetogram An artificial picture of the Sun that shows regions of magnetic polarity in various colors. (page 361)

magnetometer A device for measuring magnetic fields. (page 199)

magnetopause That region of a planet's magnetosphere where the magnetic field counterbalances the pressure from the solar wind. (pages 161, 204)

magnetosheath The region of turbulent flow between a planet's magnetopause and bow shock. (pages 161, 204)

magnetosphere The region around a planet occupied by its magnetic field. (pages 161, 204)

magnifying power The number of times larger in angular diameter an object appears through a telescope than when viewed with the naked eye. (pages 104, 107)

magnitude A measure of the amount of light received from a star or other luminous object. (pages 49, 372–373)

magnitude scale A system for denoting the brightnesses of astronomical objects. (page 49)

main sequence A grouping of stars on the Hertzsprung–Russell diagram extending diagonally across the graph from the hottest, brightest stars to the dimmest, coolest stars. (pages 382–384)

major axis (of an ellipse) The longest diameter of an ellipse. (pages 64, 70)

mantle (of a planet) That portion of a terrestrial planet located between its crust and core. (pages 153–154)

mare Latin "sea;" a large, relatively crater-free plain on the Moon; plural, maria. (page 170)

mare basalt A type of lunar rock commonly found in the mare basins. (page 181)

marginally bounded universe A universe throughout which the average density equals the critical density. (page 563)

mascon A localized concentration of dense material beneath the lunar surface.

maser A device or naturally occurring situation that amplifies microwaves. (pages 408–409)

mass   A measure of the total amount of material in an object. (page 68)

mass density of radiation   The energy possessed by a radiation field per unit volume divided by the square of the speed of light. (page 561)

mass function   A numerical relationship involving the masses of the stars in a binary system and the angle of inclination of their orbit in the sky.

mass loss   A process by which a star gently loses matter. (page 422)

mass–luminosity relation   A relationship between the masses and luminosities of main-sequence stars. (page 387)

mass–radius relation   A relationship between the masses and radii of white-dwarf stars. (page 434)

mass transfer   The flow of gases from one star in a binary to the other. (page 422)

matter-dominated universe   A universe in which the average density of matter exceeds the mass density of radiation. (page 561)

Maunder butterfly diagram   A plot of sunspot latitude versus time. (page 359)

maximum eastern elongation   The configuration of an inferior planet at its greatest angular distance eastward of the Sun. (page 59)

maximum western elongation   The configuration of an inferior planet at its greatest angular distance westward of the Sun. (page 60)

mean solar day   The interval between successive meridian passages of the mean Sun; the average length of a solar day. (page 30)

mean solar time   Time reckoned by the location of the mean Sun. (page 30)

mean Sun   A fictitious object that moves eastward at a constant speed along the celestial equator, completing one circuit of the sky with respect to the vernal equinox in one tropical year. (pages 29–30)

mechanics   The branch of physics dealing with the behavior and motions of objects acted upon by forces. (page 71)

megaparsec (Mpc)   One million parsecs. (page 10)

meridian (local)   The great circle on the celestial sphere that passes through an observer's zenith and the north and south celestial poles. (page 29)

meridian transit   The crossing of the meridian by any astronomical object. (page 29)

mesosphere   A layer in a planet's atmosphere above the stratosphere. (page 160)

metal-poor star   A star which, compared to the Sun, is underabundant in elements heavier than helium. (page 421)

metal-rich star   A star whose abundance of heavy elements is roughly comparable to that of the Sun. (page 421)

metamorphic rock   A rock whose properties and appearance have been transformed by the action of pressure and heat beneath the Earth's surface. (page 152)

meteor   The luminous phenomenon seen when a meteoroid enters the Earth's atmosphere; a "shooting star." (pages 332–333)

meteor shower   Many meteors that seem to radiate from a common point in the sky. (pages 340–341)

meteorite   A fragment of a meteoroid that has survived passage through the Earth's atmosphere. (page 332–335)

meteoroid   A small rock in interplanetary space. (page 332)

micrometeorite   A very small meteoroid; a grain of interplanetary dust.

microwaves   Short-wavelength radio waves. (page 85)

Milky Way   Our Galaxy; the band of faint stars seen from the Earth in the plane of our Galaxy's disk. (pages 482–483)

millibar   A unit of atmospheric pressure; one thousandth of a bar. (page 159)

millisecond pulsar   A pulsar with a period of roughly 1 to 10 milliseconds. (page 460)

mineral   A naturally occurring solid composed of a single element or chemical combination of elements, often in the form of crystals. (page 158)

minor axis (of an ellipse)   The smallest diameter of an ellipse. (pages 64, 70)

minor planet   An asteroid. (page 327)

minute of arc   One sixtieth of a degree, designated by the symbol ′. (page 7)

model   A hypothesis that has withstood experimental or observational tests. (page 2)

model (of the Sun or a star)   The results of a theoretical calculation that gives the values of temperature, pressure, density, and so forth throughout the Sun or a star. (page 350)

model atmosphere   The results of a theoretical calculation that gives the values of temperature, pressure, density, and so forth throughout the outer layers of a star.

molecule   A combination of two or more atoms. (page 135)

momentum   A measure of the inertia of an object; an object's mass multiplied by its velocity. (page 72)

monochromatic   Of one wavelength or color.

moving cluster method   A technique for determining the distance to a cluster of stars from the motions of the cluster's members. (page 374)

muon   A subatomic particle that behaves like a heavy electron. (page 580)

N galaxy   A galaxy with a bright, starlike nucleus. (page 542)

nadir   The point on the local meridian 180° from the zenith.

nanosecond   One-billionth (10^{-9}) second.

neap tide   An ocean tide that occurs when the Moon is near first-quarter or third-quarter phase. (page 180)

nebula   A cloud of interstellar gas and dust. (page 4)

negative curvature   A surface or space in which parallel lines diverge and the sum of the angles of a triangle is less than 180°. (pages 566–567)

neon burning   The thermonuclear fusion of neon to produce heavier nuclei. (page 435)

neutrino   A subatomic particle with no electric charge and little or no mass, yet one that is important in many nuclear reactions. (pages 349, 351, 443)

neutrino telescope   A device capable of detecting neutrinos from supernovae. (page 443)

neutron   A subatomic particle with no electric charge and with a mass nearly equal to that of the proton. (pages 94, 136)

neutron capture   The buildup of neutrons inside nuclei. (page 436)

neutron star   A very compact, dense star composed almost entirely of neutrons. (pages 450–464)

neutronization   The combining of protons and electron to produce neutrons in a star's core. (page 437)

new moon   The phase of the Moon when the dark hemisphere of the Moon faces the Earth. (pages 37–38)

Newtonian focus   An optical arrangement in a reflecting telescope in which a small mirror reflects converging light rays to a focus on one side of the telescope tube. (page 107)

Newtonian mechanics   The branch of physics based on Newton's law that deals with gravitation. (pages 3, 71)

Newtonian reflector   A reflecting telescope that uses a small mirror to deflect the image to one side of the telescope tube. (page 107)

Newton's laws of motion   Three statements about the nature of physical realty on which Newtonian mechanics is based. (pages 67–69)

no-hair theorem   A statement of the simplicity of black holes. (page 473)

node   The intersection of an orbit with a reference plane such as the plane of the celestial equator or the ecliptic. (page 42)

nonradiogenic   Refers to an isotope that is not created by the radioactive decay of another isotope. (page 185)

nonthermal radiation   Radiation emitted by charged particles moving through a magnetic field; synchrotron radiation. (page 250)

north celestial pole   The point directly above the Earth's north pole where the Earth's axis of rotation, if extended, would intersect the celestial sphere. (page 24)

northern lights   Aurorae; aurorae borealis. (page 162)

nova   A star that experiences a sudden outburst of radiant energy, temporarily increasing its luminosity roughly a thousandfold. (pages 461–463)

nuclear bulge   The central region of our Galaxy; the central bulge. (page 486)

nuclear density   The density of matter in an atomic nucleus; about 4×10^{17} kg/m³. (page 438)

nucleus (of an atom)   The massive part of an atom, composed of protons and neutrons, about which electrons revolve. (pages 94, 136)

nucleus (of a comet)   A collection of ices and dust that constitute the solid part of a comet. (pages 336–337)

nucleus (of a galaxy)   The concentration of stars and dust at the center of a galaxy. (page 486)

nutation   A small periodic wobbling of the Earth's axis superimposed on precession.

OB association   A grouping of hot, young, massive stars, predominantly of spectral types O and B. (pages 402, 408)

OBAFGKM   The temperature sequence of spectral types. (page 379)

objective lens   The principle lens of a refracting telescope. (page 104)

oblate   Flattened at the poles. (page 248)

oblateness   A measure of how much a flattened sphere (or spheroid) differs from a perfect sphere. (page 151)

obliquity (of the ecliptic)   The angle between the planes of the celestial equator and the ecliptic (about $23\frac{1}{2}°$).

obscuration (interstellar)   The absorption of starlight by interstellar dust.

observable universe   That portion of the universe inside the cosmic event horizon. (page 556)

Occam's razor   The notion that a straightforward explanation of a phenomenon is more likely to be correct than a convoluted one. (page 58)

occultation   The eclipsing of an astronomical object by the Moon or a planet. (page 265)

oceanic rift   A crack in the ocean floor that exudes lava. (page 157)

Olber's paradox   The dilemma associated with the fact that the night sky is dark. (page 554)

Oort cloud   A presumed accumulation of comets and cometary material surrounding the Sun at a distance of roughly 50,000 to 100,000 AU. (page 339)

opacity   The ability of a material to impede the passage of light.

open cluster   A loose association of young stars in the disk of our Galaxy; a galactic cluster. (pages 374, 404)

opposition   The configuration of a planet when it is at an elongation of 180° and thus appears opposite the Sun in the sky. (page 60)

optical window   The range of visible wavelengths to which the Earth's atmosphere is transparent. (page 117)

optics   The branch of physics dealing with the behavior and properties of light. (pages 103–111)

orbit   The path of an object that is moving about a second object or point.

outgassing   Volcanic processes by which gases escape from a planet's crust into its atmosphere.

oxygen burning   The thermonuclear fusion of oxygen to produce heavier nuclei. (page 435)

pair production   The creation of a particle and its antiparticle from energy. (pages 578–579)

Pallas   The second asteroid to be discovered. (page 327)

Pangaea   The name of a hypothetical continent that fragmented into several of the continents on Earth today. (page 155)

parabola   A conic section formed by cutting a circular cone at an angle parallel to one of the sides of the cone. (page 69)

parallax   The apparent displacement of an object due to the motion of the observer. (pages 62, 370)

parent isotope   An unstable isotope that radioactively decays into another isotope. (page 184)

parsec (pc)   A unit of distance; 3.26 light years. (pages 9–10)

partial eclipse   A lunar or solar eclipse in which the eclipsed object does not appear completely covered. (pages 44–45)

partial lunar eclipse   A lunar eclipse in which the Moon does not appear completely covered. (page 44)

partial solar eclipse   A solar eclipse in which the Sun does not appear completely covered. (page 45)

pascal (Pa)   A unit of pressure. (page 159)

Paschen series   A series of spectral lines of hydrogen produced by electron transitions to and from the third (n = 3) energy level of the hydrogen atom. (page 96)

Pauli exclusion principle   A principle of quantum mechanics stating that two identical particles cannot have the same position and momentum. (page 419)

penumbra   The portion of a shadow in which only part of the light source is covered by an opaque body. (page 44)

penumbra (of a sunspot)   The grayish region that usually surrounds the umbra of a sunspot. (page 357)

penumbral eclipse   A lunar eclipse in which the Moon passes only through the Earth's penumbra. (page 44)

perfect gas   An idealized gas that obeys a very simple equation of state. (page 419)

perigee   The point in its orbit where a satellite or the Moon is nearest the Earth. (page 47)

perihelion   The point in its orbit where a planet is nearest the Sun. (page 64)

period (of a planet)   The interval of time between successive geometric arrangements of a planet and an astronomical object, such as the Sun. (page 60)

period–luminosity relation   A relationship between the period and average density of a pulsating star. (page 426)

periodic table   A listing of the chemical elements according to their properties, invented by Dmitri Mendeleev. (page 135)

perturbation   A small disturbing effect. (page 493)

phases of the Moon   The appearances of the Moon at different times as it orbits the Earth. (pages 38–39)

photodisintegration   The breakup of nuclei by high-energy gamma rays. (page 437)

photoelectric effect   The phenomenon whereby certain metals fire off electrons when exposed to short-wavelength light. (page 90)

photometry   The measurement of light intensities. (pages 375, 377–378)

photon   A discrete unit of electromagnetic energy. (page 90)

photosphere   The region in the solar atmosphere from which most of the visible light escapes into space. (pages 353–354)

pixel A picture element. (page 112)

plage A bright region in the solar atmosphere as observed in the monochromatic light of a spectral line.

Planck length A fundamental length defined by basic physical constants. (page 558)

Planck mass A fundamental mass defined by basic physical constants. (page 558)

Planck time A fundamental interval of time defined by basic physical constants. (pages 557–558)

Planck's constant (h) The constant of proportionality between a photon's energy and its frequency. (page 91)

Planck's law The relationship between the energy of a photon and its wavelength or frequency; $E = h\nu$. (page 91)

planetary nebula A luminous shell of gas ejected from an old, low-mass star. (pages 430, 432)

planetesimal A small body of primordial dust and ice from which the planets formed. (page 142)

plasma A hot ionized gas. (pages 252, 360, 562)

plate tectonics The motions of large segments (plates) of the Earth's surface over the underlying mantle. (page 156)

polymer A long molecule consisting of many smaller molecules joined together. (page 297)

poor cluster (of galaxies) A cluster of galaxies with very few members. (page 506)

population I star A star whose spectrum exhibits spectral lines of many elements heavier than helium; a metal-rich star. (page 421)

population II star A star whose spectrum exhibits comparatively few spectral lines of elements heavier than helium; a metal-poor star. (page 421)

population III star A star virtually devoid of elements heavier than helium. (page 421)

positive curvature A surface or space in which parallel lines converge and the sum of the angles of a triangle is greater than 180°. (pages 566–567)

positron An electron with a positive rather than negative electric charge; an antielectron. (page 349)

potential energy The energy possessed by an object because of its elevated position in a gravitational field. (page 72)

powers of ten A shorthand method of writing numbers, involving 10 followed by an exponent. (page 8)

precession (of the Earth) A slow, conical motion of the Earth's axis of rotation caused by the gravitational pull of the Moon and Sun on the Earth's equatorial bulge. (page 28)

precession (of the equinoxes) The slow westward motion of the equinoxes along the ecliptic due to precession of the Earth. (page 29)

prime focus The point in a telescope where the objective focuses light. (page 107)

primordial black hole A hypothetical black hole that may have been created during the Big Bang. (page 570)

primordial fireball The extremely hot gas that filled the universe immediately following the Big Bang. (page 562)

principle of equivalence A principle of general relativity stating that, in a small volume, it is impossible to distinguish between the effects of gravitation and acceleration. (pages 468–469)

principle of totalitarianism The notion in quantum mechanics that if a process is not strictly forbidden, it must occur. (page 578)

prism A wedge-shaped piece of glass that is used to disperse white light into a spectrum. (page 83)

prominence Flamelike protrusions seen near the limb of the Sun and extending into the solar corona. (pages 360–361)

proper distance A length measured by rulers at rest with respect to an observer. (page 471)

proper mass The mass of an object measured at rest. (page 471)

proper motion The angular rate of change in the location of a star on the celestial sphere, usually expressed in seconds of arc per year. (page 371)

proper time A time interval measured with clocks at rest with respect to an observer. (page 471)

proto- A prefix referring to the embryonic stage of a young astronomical object (planet, star, etc.) that is still in the process of formation. (pages 140, 142)

proton A heavy, positively charged subatomic particle that is one of two principle constituents of atomic nuclei. (pages 94, 136)

proton–proton chain A sequence of thermonuclear reactions by which hydrogen nuclei are built up into helium nuclei. (page 349)

pulsar A pulsating radio source believed to be associated with a rapidly rotating neutron star. (pages 5, 451–456)

pulsating variable A star that pulsates in size and luminosity.

quadrature The configuration in which a planet has an elongation of 90°. (page 60)

quantum electrodynamics A theory that describes details of how charged particles interact by exchanging photons. (pages 583–584)

quantum mechanics The branch of physics dealing with the structure and behavior of atoms and their interaction with light.

quark One of several hypothetical particles presumed to be the internal constituents of certain heavy subatomic particles such as protons and neutrons. (page 583)

quarter moon A phase of the Moon when it is located 90° from the Sun in the sky, so that one-half of its daylit hemisphere is visible from the Earth. (pages 38–39)

quasar A starlike object with a very large redshift; a quasi-stellar object or quasi-stellar source. (pages 6, 532–533)

quasi-stellar radio source A quasar that emits detectable radio radiation. (page 532)

r process A process of nuclear transformation initiated when certain isotopes rapidly capture large numbers of neutrons. (page 437)

r-process isotope An isotope that is created only through the r process. (page 437)

radar A technique of reflecting radio waves from a distant object. (page 221)

radial velocity That portion of an object's velocity parallel to the line of sight. (pages 98, 371)

radial-velocity curve A plot showing the variation of radial velocity with time for a binary star or variable star. (page 389)

radiant (of a meteor shower) The point in the sky from which meteors of a particular shower seem to originate. (pages 341–342)

radiation Electromagnetic energy; photons.

radiation-dominated universe A universe in which the mass density of radiation exceeds the average density of matter. (page 561)

radiation pressure The transfer of momentum carried by radiation to an object on which the radiation falls. (page 339)

radiative diffusion The random-walk migration of photons from a star's center toward its surface. (page 350)

radiative zone A region within a star where radiative diffusion is the dominant mode of energy transport. (page 351)

radio astronomy That branch of astronomy dealing with observations at radio wavelengths. (pages 115–116, 531)

radio galaxy A galaxy that emits an unusually large amount of radio waves. (page 531)

radio lobe A region near an active galaxy from which significant radio radiation emanates. (page 540)

radio telescope A telescope designed to detect radio waves. (pages 114–115)

radio waves The longest wavelength electromagnetic radiation. (page 85)

radio window The range of radio wavelengths to which the Earth's atmosphere is transparent. (page 117)

radioactivity The process whereby certain atomic nuclei naturally decompose by spontaneously emitting particles.

ray (lunar) Any one of a system of bright, elongated streaks on the lunar surface. (page 171)

recombination The process in which an electron combines with a positively charged ion. (page 402)

recurrent nova A nova that has erupted more than once.

red giant A large, cool star of high luminosity. (pages 383, 417)

red supergiant An extremely large, cool star of luminosity class I. (page 431)

reddening (interstellar) The reddening of starlight as it passes through the interstellar medium. (page 398)

redshift The shifting to longer wavelengths of the light from remote galaxies and quasars; the Doppler shift of light from a receding source. (page 98)

reflecting telescope A telescope in which the principal optical component is a concave mirror. (page 107)

reflection The return of light rays by a surface. (page 106)

reflection grating A diffraction grating that produces a spectrum when light is reflected off it. (page 114)

reflection nebula A comparatively dense cloud of dust in interstellar space that is illuminated by a star. (page 397)

reflector A reflecting telescope. (page 107)

refracting telescope A telescope in which the principal optical component is a lens. (page 104)

refraction The bending of light rays passing from one transparent medium to another. (page 103)

refractor A refracting telescope. (page 103)

refractory element An element with high melting and boiling points. (page 183)

regolith The layer rock fragments covering the lunar surface. (page 181)

regression of the line of nodes The slow motion of the line of nodes of the Moon due to the gravitational pull of the Sun. (page 43)

regular cluster (of galaxies) A spherical cluster of galaxies. (pages 508–509)

regular orbit An orbit in the plane of a planet's equator along which a satellite travels in the same direction that the planet rotates. (page 297)

relativistic cosmology A cosmology based on the general theory of relativity. (page 563)

relativistic particle A particle moving at nearly the speed of light.

resolution The degree to which fine details in an optical image can be distinguished. (pages 108–109)

resolving power A measure of the ability of an optical system to distinguish, or resolve, fine details in the image it produces. (page 108)

retrograde motion The apparent westward motion of a planet with respect to background stars. (page 57)

retrograde rotation The rotation of an object in the direction opposite to which it is revolving about another object. (page 210)

rest energy The mass of a particle that is not moving multiplied by the square of the speed of light. (page 579)

revolution The motion of one body about another.

rich cluster (of galaxies) A cluster of galaxies containing many members. (page 506)

right ascension A coordinate for measuring the east–west positions of objects on the celestial sphere. (page 24)

rille A trenchlike depression on the lunar surface. (page 171)

ringlet One of many narrow bands of particles of which Saturn's ring system is composed. (page 292)

Roche limit The smallest distance from a planet or other object at which a second object can be held together by purely gravitational forces. (page 287)

Roche lobe A teardrop-shaped volume surrounding a star in a binary inside which gases are gravitationally bound to that star. (pages 422–423)

rock A mineral or combination of minerals. (page 158)

rotation The turning of a body about an axis passing through the body.

rotation curve A plot of the orbital speeds of stars and nebulae outward from the center of a galaxy. (page 491)

rotation of the Moon's orbit The slow motion of the major axis of the Moon's orbit due to the gravitational pull of the Sun. (page 43)

RR Lyrae variable One class of pulsating stars with periods less than one day. (pages 426, 484–485)

s process A process of nuclear transformation that occurs when certain isotopes capture neutrons at a slow rate. (page 437)

s-process isotope An isotope that is created only by the s process. (page 437)

saros A particular cycle of similar eclipses that recur about every 18 years. (page 48)

satellite A body that revolves about a larger one.

scarp A line of cliffs formed by the faulting or fracturing of a planet's surface. (page 200)

scattering The deflection of photons or particles by other particles. (page 293)

Schmidt telescope A reflecting telescope invented by Bernard Schmidt that is used to photograph large areas of the sky. (pages 110–111)

Schwarzschild radius The distance from the singularity to the event horizon in a nonrotating black hole. (pages 471–472)

scientific method The basic procedure used by scientists to investigate phenomena. (page 2)

sea-floor spreading The separation of plates under the ocean due to lava emerging in an oceanic rift. (page 156)

second of arc One sixtieth of an arc minute, designated by the symbol ″. (page 7)

sedimentary rock A rock that is formed from material deposited by rain or winds, or on the ocean floor. (page 152)

seeing disk The angular diameter of a star's image. (page 109)

seismic waves Vibrations traveling through a terrestrial planet usually associated with earthquakelike phenomena. (page 153)

seismograph A device used to record and measure seismic waves, such as those produced by earthquakes. (page 153)

seismology The study of earthquakes and related phenomena. (page 153)

self-propagating star formation The process by which the formation of stars in one location in a galaxy stimulates the formation of stars in a neighboring location. (page 492)

semidetached binary A binary in which one star fills its Roche lobe. (page 423)

semimajor axis One-half of the major axis of an ellipse. (page 64)

Seyfert galaxy A spiral galaxy with a bright nucleus whose spectrum exhibits emission lines. (pages 539–540)

shepherd satellite A small satellite that gravitationally confines particles to a ring around a planet. (page 294)

shell helium burning The thermonuclear fusion of helium in a shell surrounding a star's core. (page 431)

shell hydrogen burning The thermonuclear fusion of hydrogen in a shell surrounding a star's core. (pages 416–417, 431)

shell star A star, usually of spectral type A to F, that is surrounded by a shell of gas.

shield volcano A volcano with long, gently sloping sides. (page 218)

shock wave An abrupt, localized region of compressed gas caused by an object traveling through the gas at a speed greater than the speed of sound. (pages 161, 213)

short-period comet A comet that orbits the Sun with a period of less than about 200 years. (page 339)

SI The International System of Units, based on the meter (m), the second (s), and the kilogram (kg). (page 11)

sidereal clock A clock that measures sidereal time. (page 32)

sidereal day The interval between successive meridian passages of the vernal equinox. (page 32)

sidereal month The period of the Moon's revolution about the Earth with respect to the stars. (page 38)

sidereal period The orbital period of one object about another with respect to the stars. (page 60)

sidereal time Time reckoned by the location of the vernal equinox. (page 32)

sidereal year The orbital period of the Earth about the Sun with respect to the stars. (page 31)

siderophile element An element often found with iron. (page 331)

silicate A mineral whose chemical composition is based on the element silicon. (page 151)

silicon burning The thermonuclear fusion of silicon to produce heavier elements, especially iron. (page 436)

single-line spectroscopic binary A spectroscopic binary in which the spectral lines of only one star can be detected. (page 388)

singularity A place of infinite spacetime curvature; the center of a black hole. (pages 470–471)

SMC The Small Magellanic Cloud. (page 507)

solar activity Phenomena that occur in the solar atmosphere such as sunspots, plages, flares, and so forth. (pages 357–361)

solar atmosphere The outer layers of the Sun, consisting of the photosphere, chromosphere, and corona. (page 353)

solar constant The average amount of energy received from the Sun per square meter per second, measured just above the Earth's atmosphere. (page 88)

solar corona Hot, faintly glowing gases seen around the Sun during a total solar eclipse; the uppermost regions of the solar atmosphere. (pages 44–45, 355–356)

solar cycle The semiregular 22-year interval between successive appearances of sunspots at the same latitude and with the same magnetic polarity. (page 362)

solar eclipse An eclipse of the Sun by the Moon; a passage of the Earth through the Moon's shadow. (page 41)

solar flare A violent outburst on the Sun's surface. (pages 162, 361)

solar interior Everything below the solar atmosphere; the inside of the Sun. (pages 351–352)

solar nebula The cloud of gas and dust from which the Sun and solar system formed. (page 138)

solar nebula hypothesis The theory that planetary atmospheres came from gases in the solar nebula. (page 216)

solar nebula wind The theory that planetary atmospheres came from the solar wind. (page 216)

solar seismology The study of the Sun's interior deduced from observable vibrations of its surface. (pages 363–364)

solar system The Sun, planets, their satellites, asteroids, comets, and related objects that orbit the Sun. (pages 3, 128–131)

solar transient A short-lived eruption that moves rapidly outward through the solar corona. (page 356)

solar transit The passage of an object in front of the Sun. (page 195)

solar wind A radial flow of particles (mostly electrons and protons) from the Sun. (page 145)

solstice Either of two points along the ecliptic at which the Sun reaches its maximum distance north or south of the celestial equator. (pages 25–26)

south celestial pole The point directly above the Earth's south pole where the Earth's axis of rotation, if extended, would intersect the celestial sphere. (page 24)

southern lights Aurorae; aurorae australis. (page 162)

special theory of relativity A description of mechanics and electromagnetic theory formulated by Albert Einstein, which explains that measurements of distance, time, and mass are affected by the observer's motion. (pages 75, 470–471)

spectral analysis The identification of chemical substances from the patterns of spectral lines in their spectra. (page 92)

spectral class/type A classification of stars according to the appearance of their spectra. (pages 378–381)

spectrogram The photograph of a spectrum. (page 114)

spectrograph An instrument for photographing a spectrum. (pages 113–114)

spectroheliograph An instrument for photographing the Sun in the monochromatic light of one spectral line.

spectroscope An instrument for directly viewing a spectrum. (page 92)

spectroscopic binary A binary star whose binary nature is deduced from the periodic Doppler shifting of lines in its spectrum. (page 388)

spectroscopic parallax The distance to a star derived by comparing its apparent magnitude to an absolute magnitude inferred from the star's spectrum. (page 385)

spectroscopy The study of spectra. (page 134)

spectrum binary A binary star whose binary nature is deduced from the presence of two sets of incongruous spectral lines. (page 388)

spherical aberration The distortion of an image formed by a telescope due to differing focal lengths of the optical system. (page 110)

spicule A narrow jet of rising gas in the solar chromosphere. (page 355)

spin The intrinsic angular momentum possessed by certain particles. (page 487)

spin–flip transition A transition in the ground state of the hydrogen atom, which occurs when the orientation of the electron's spin changes. (page 487)

spin–orbit coupling The relationship between the rotation of a body to its orbital period. (page 197)

spiral arms Lanes of interstellar gas, dust, and young stars that wind outward in a plane from the central regions of a galaxy. (page 489)

spiral galaxy A flattened, rotating galaxy with pinwheel-like spiral arms winding outward from the galaxy's nucleus. (page 503)

spontaneous symmetry breaking A process by which certain congruities in the mathematics of particle physics are altered to produce new particles and forces. (page 586)

spring tide Ocean tides that occur at new-moon and full-moon phases. (page 180)

standard candle An astronomical object of known intrinsic brightness which can be used to determine extragalactic distances. (page 521)

standing shock wave A shock wave whose position in a supersonic fluid remains fixed. (page 547)

star A self-luminous sphere of gas.

starburst galaxy A galaxy that is experiencing an exceptionally high rate of star formation. (pages 512–513)

starquake A sudden shift in the crust of a neutron star. (page 457)

Stefan–Boltzmann law A relationship between the temperature of a blackbody and the rate at which it radiates energy. (pages 87–88)

Stefan's law *See* Stefan–Boltzmann law.

stellar association A loose grouping of young stars. (page 405)

stellar evolution The changes in size, luminosity, temperature, and so forth that occur as a star ages. (page 397)

stellar model The result of theoretical calculations that give details of physical conditions inside a star. (page 350)

stratosphere A layer in the atmosphere of a planet directly above the troposphere. (pages 159–160)

strong nuclear force The force that binds protons and neutrons together in nuclei. (pages 348, 583–584)

subduction zone A location where colliding tectonic plates cause the Earth's crust to be pulled down into the mantle. (pages 156–157)

subdwarf A star of lower luminosity than main-sequence stars of the same spectral type.

subgiant A star whose luminosity is between that of main-sequence stars and normal giants of the same spectral type. (page 385)

subtend To extend over an angle. (page 7)

summer solstice The point on the ecliptic where the Sun is farthest north of the celestial equator. (page 26)

Sun The star about which the Earth and other planets revolve. (page 346)

Sun-grazing comet A comet that passes quite near the Sun. (page 340)

sunspot A temporary cool region in the solar photosphere. (page 357)

sunspot cycle The semiregular 11-year period with which the number of sunspots fluctuates. (page 358)

sunspot maximum/minimum   That time during the sunspot cycle when the number of sunspots is highest/lowest. (page 358)

supercluster   A collection of clusters of galaxies. (pages 509, 589)

superconductivity   The phenomenon whereby a flowing electric current does not experience any electrical resistance. (page 457)

superfluidity   The phenomenon whereby a fluid flows without experiencing any viscosity. (page 457)

supergiant   A very large, extremely luminous star; stars of luminosity class I. (pages 384–385)

supergrand unified theory   A complete description of all forces and particles, as well as the structure of space and time. (page 585)

supergranule   A large convective feature in the solar atmosphere, usually outlined by spicules. (page 355)

supergravity   A supergrand unified theory that uses higher dimensions to describe the four forces. (page 588)

superior conjunction   The configuration of a planet being behind the Sun as viewed from the Earth. (page 59)

superior planet   A planet that is more distant from the Sun than the Earth is. (page 59)

supermassive black hole   A black hole with a mass in the range of a million to a billion solar masses. (pages 497, 543–549)

supernova   A stellar outburst during which a star suddenly increases its brightness roughly a millionfold. (pages 4, 409, 438–441)

supernova remnant   The gases ejected by a supernova. (page 409)

surface gravity   The weight of a unit mass at the surface of an object such as a planet.

symmetry breaking   A process by which certain congruities in the mathematics of particle physics are altered to produce new particles and forces. (page 585)

synchronous rotation   The rotation of a body with a period equal to its orbital period. (pages 170, 196–197)

synchrotron radiation   The radiation emitted by charged particles moving through a magnetic field; nonthermal radiation. (page 251, 456)

synodic month   The period of revolution of the Moon with respect to the Sun; the length of one cycle of lunar phases. (page 39)

synodic period   The interval between successive occurrences of the same configuration of a planet. (page 60)

T Tauri stars   Young variable stars associated with interstellar matter that show erratic changes in luminosity. (page 400)

T Tauri wind   A flow of particles away from a T Tauri star. (page 145)

tail (of a comet)   Gas and dust particles from a comet's nucleus that have been swept away from the comet's head by the radiation pressure of sunlight and the solar wind. (page 339)

tangential velocity   That portion of an object's velocity perpendicular to the line of sight. (pages 98, 372)

tektites   Rounded glassy objects believed to have a meteoritic origin.

telescope   An instrument for viewing remote objects. (*see* Chapter 6)

temperature (Celsius)   Temperature measured on a scale where water freezes at 0° and boils at 100°. (pages 86–87)

temperature (color)   The temperature of a star determined by comparing the intensity of starlight in two wavelength bands.

temperature (effective)   The temperature of a blackbody that would radiate the same total amount of energy that a particular star does.

temperature (excitation)   The temperature of a star determined from the strengths of various spectral lines that originate in atoms with different stages of excitation.

temperature (Fahrenheit)   Temperature measured on a scale where water freezes at 32° and boils at 212°. (pages 86–87)

temperature (Kelvin)   Absolute temperature measured in units (kelvins, K) equivalent to the degree Celsius. (pages 86–87)

temperature (kinetic)   Temperature directly related to the average speed of atoms or molecules in a substance. (page 138)

terminator   The line dividing day and night on the surface of the Moon or a planet; the line of sunset or sunrise. (page 201)

terra   Cratered lunar highlands. (page 172)

terrestrial planet   Any of the planets Mercury, Venus, Earth, or Mars, and sometimes also including the Galilean satellites and Pluto. (page 132)

theory   A hypothesis that has withstood experimental or observational texts. (page 2)

theory of everything (TOE)   A supergrand unified theory that completely describes all particles and forces as well as the structure of space and time. (page 585)

thermal energy The energy associated with heat stemming from the motions of atoms or molecules in a substance. (page 138)

thermal equilibrium A balance between the input and outflow of heat in a system. (pages 350, 579)

thermal pulses Brief bursts in energy output from the helium-burning shell of an aging low-mass star. (page 431)

thermal radiation The radiation naturally emitted by any object that is not at absolute zero. (page 250)

thermodynamics The branch of physics dealing with heat and the transfer of heat between bodies.

thermonuclear fusion The combining of nuclei under conditions of high temperature in a process that releases substantial energy. (page 348)

thermonuclear reaction A reaction resulting from the high-speed collision of nuclear particles that are moving rapidly because they are at a high temperature. (page 348)

thermosphere A region in the Earth's atmosphere between the mesosphere and the exosphere. (page 160)

third quarter moon The phase of the Moon that occurs when the Moon is 90° west of the Sun. (page 38)

threshold temperature The temperature above which photons spontaneously produce particles and antiparticles of a particular type. (page 580)

tidal force A gravitational force whose strength and/or direction varies over a body and thus tends to deform the body. (pages 179, 287)

time zone A region on the Earth where, by agreement, all clocks have the same time. (pages 30–31)

total eclipse A solar eclipse during which the Sun is completely hidden by the Moon, or a lunar eclipse during which the Moon is completely immersed in the Earth's umbra. (pages 44–45)

total lunar eclipse A lunar eclipse during which the Moon is completely immersed in the Earth's umbra. (page 44)

total solar eclipse A solar eclipse during which the Sun is completely hidden by the Moon. (pages 44–45)

transit The passage of a celestial body across the meridian; the passage of a small object in front of a larger object. (pages 29, 196)

transmission grating A diffraction grating that produces a spectrum when light is shone through it. (page 114)

triple alpha process A sequence of two thermonuclear reactions in which three helium nuclei combine to form one carbon nucleus. (page 418)

triple point The pressure and temperature at which a substance can exist simultaneously as a solid, a liquid, and a gas. (pages 296–297)

Trojan asteroid One of several asteroids that share Jupiter's orbit about the Sun. (page 329)

tropical year The period of revolution of the Earth about the Sun with respect to the vernal equinox. (page 31)

troposphere The lowest level in the Earth's atmosphere. (pages 159–160)

Tully–Fisher relation A correlation between the width of the 21-cm line and the absolute magnitude of a spiral galaxy. (page 524)

turnoff point The point on an H–R diagram where the stars in a cluster are leaving the main sequence. (pages 420–421)

Type I Cepheid A metal-rich Cepheid variable. (page 426)

Type I Seyfert galaxy A Seyfert galaxy whose Balmer lines are significantly broader than any of its other emission lines. (page 539)

Type II Cepheid A metal-poor Cepheid variable. (page 426)

Type II Seyfert galaxy A Seyfert galaxy whose Balmer lines have about the same width as its other emission lines. (page 539)

UBV system A system of stellar magnitude involving measurements of starlight intensity in the ultraviolet, blue, and visible spectral regions. (page 377)

ultraviolet radiation Electromagnetic radiation of wavelengths shorter than those of visible light, but longer than those of X rays. (page 85)

umbra The central, completely dark portion of a shadow. (pages 43–44)

umbra (of a sunspot) The dark, central region of a sunspot. (page 357)

unbounded universe A universe throughout which the average density is less than the critical density. (page 563)

universal constant of gravitation (G) The constant of proportionality in Newton's law of gravitation. (page 69)

universal time Local mean time at the prime meridian. (page 281)

universe All space, along with all the matter and radiation in space.

Van Allen belts Two doughnut-shaped regions around the Earth where many charged particles (protons and electrons) are trapped by the Earth's magnetic field. (page 161)

variable star A star whose luminosity varies.

velocity The speed and direction with which an object moves. (page 68)

vernal equinox The point on the ecliptic where the Sun crosses the celestial equator from south to north. (page 26)

very-long-baseline interferometry (VLBI) A method of connecting widely separated radio telescopes to make observations of very high resolution. (page 115)

virtual pair A particle and antiparticle that exist for such a brief interval that they cannot be observed. (page 570)

visible light Photons detectable by the human eye. (page 85)

visual binary A binary star in which the two components can be resolved through a telescope. (page 386)

void A large volume of space, typically 100 to 400 million light years in diameter, that contains very few galaxies. (page 511)

volatile element An element with low melting and boiling points. (page 183)

vortex A whirling mass of liquid or gas; a whirlpool. (page 259)

waning crescent moon The phase of the Moon that occurs between third quarter and new moon. (page 38)

waning gibbous moon The phase of the Moon that occurs between full moon and third quarter. (page 38)

watt (W) A unit of power; 1 joule per second. (page 87)

wave–particle duality The quantum mechanical notion that particles have wavelike properties and radiation has particlelike properties. (page 558)

wavelength The distance between two successive wave crests. (page 84)

waxing crescent moon The phase of the Moon that occurs between new moon and first quarter. (page 38)

waxing gibbous moon The phase of the Moon that occurs between first quarter and full moon. (page 38)

weak nuclear force The short-range force that is responsible for transforming one particle into another, such as the decay of a neutron into a proton. (pages 348, 583–584)

weight The force with which an object presses downward due to the action of gravity. (page 69)

white dwarf A low-mass star that has exhausted all its thermonuclear fuel and contracted to a size roughly equal to the size of the Earth. (pages 383–384, 434–435)

white hole A black hole from which matter and radiation emerge. (page 572)

Widmanstätten patterns Crystalline structure seen in certain types of meteorites. (page 334)

Wien's law A relationship between the temperature of a blackbody and the wavelength at which it emits the greatest intensity of radiation. (page 89)

winter solstice The point on the ecliptic where the Sun reaches its greatest distance south of the celestial equator. (page 26)

Wolf–Rayet star A class of very hot stars that eject shells of gas at high velocity. (pages 422–423)

X rays Electromagnetic radiation whose wavelength is between that of ultraviolet light and gamma rays. (page 85)

year (yr) The period of revolution of the Earth about the Sun. (pages 30–31)

ZAMS Zero-age main sequence. (page 419)

zap crater A tiny glass-lined crater on a lunar rock presumed to be the result of the impact of a micrometeorite. (page 182)

Zeeman effect A splitting or broadening of spectral lines due to a magnetic field. (pages 359–360)

zenith The point on the celestial sphere opposite to the direction of gravity. (page 26)

zero curvature A surface or space in which parallel lines remain parallel and the sum of the angles of a triangle is exactly 180°. (pages 566–567)

zero-age main sequence The main sequence of young stars that have just begun to burn hydrogen at their cores. (pages 419–420)

zodiac A band of 12 constellations around the sky centered on the ecliptic. (pages 20, 26)

zonal jet Persistent, high-speed winds in Jupiter's atmosphere. (page 257)

zone A light-colored band in Jupiter's atmosphere. (page 247)

Answers to Selected Exercises

Chapter 1: 3 3600″ 7 (a) 10^7 (b) 4×10^5 (c) 6×10^{-2} (d) 1.7×10^{10} 8 3.8 arc sec 9 (a) 8.6 years (b) 8.19×10^{13} km 10 8.92×10^{56} hydrogen atoms 11 3350 km 12 (a) 1.43 m (b) 85.9 m (c) 5.16 km 13 8.3 minutes 14 5500 kg/m^3, which is 5.5 times the density of water 15 The observable universe is about 4×10^{36} times larger than a hydrogen atom. 16 3×10^6 km = $\frac{1}{50}$ AU, which is about eight times the distance between the Earth and the Moon 17 8700 km 18 29 million Suns 19 4.3×10^9 km 20 4 km 21 6×10^{17} s

Chapter 2: 4 56 minutes earlier, at 9:04 pm 11 18:00 12 On the date of the autumnal equinox (September 22) 16 (a) Both sidereal and synodic days would decrease. (b) Both sidereal and synodic days would decrease. (c) The synodic day would be shorter than the sidereal day. 22 $0^{hr}00^{min}00^{sec}$ 23 $26\frac{1}{2}°$ above the southern horizon

Chapter 3: 2 (a) waning crescent (b) full moon (c) waxing crescent (d) at sunrise (e) at noon (f) at sunrise 3 (a) new moon (b) first quarter (c) full moon (d) third quarter 7 One 18 (a) July 22, 2009 in southeast Asia (b) August 12, 2045 19 49 arc sec

Chapter 4: 7 5.2 AU^2 in 1991; 26.0 AU^2 in five years 8 Average distance is 100 AU; maximum distance is 200 AU 9 2.8 years 12 2×10^{20} N, which is about 180,000 times weaker than the force between the Sun and the Earth 13 It would be reduced by a factor of 100. 17 0.24 year 18 It is only possible for a superior planet. Mars is close to having its sidereal period equal its synodic period. 19 (a) 42,365 km 20 (a) 6.11 km/s (b) 3.68 km/s 21 The star is 16 times more massive than the Sun 22 (a) 4.4 m/s (b) 44 m/s (c) 140 m/s 23 You would weigh one-quarter of your Earth weight. 24 You would weigh 38 percent of your Earth weight.

Chapter 5: 3 $7\frac{1}{2}$ trips around the world 4 3.64×10^{14} Hz 5 2.91 m 6 238 nm 7 6520 K 9 10 μm 16 -13.0 km/s 18 5800 K $\approx$ 10,000° F 19 9.35 μm, which is infrared radiation 20 10,000 K 21 2.4×10^{-12} m, which is in the γ-ray region of the electromagnetic spectrum 23 8.75 times that from the Sun's surface 24 16 times brighter than the Sun 25 The star would be $6\frac{1}{4}$ times brighter than the Sun. 26 21 km/s 27 $\frac{2}{3}c \approx$ 86,000 km/s

Chapter 6: 9 Only $\frac{1}{25}$ of the incoming light is obstructed. 10 The Palomar telescope has a light gathering power one million times that of the human eye. 14 The smallest visible features on Jupiter's moons are about 300 km across. Only those features on our Moon larger than about 110 km across can be distinguished with the unaided human eye. 15 (a) 222× (b) 100× (c) 36× 16 No

Chapter 7: 13 8.2×10^{20} J, which is about 10 million times more energy than the bomb that destroyed Hiroshima 14 12 km/s 15 The Sun's escape velocity is 617 km/s. 16 2.74 km/s 18 6.43×10^{23} kg and 3900 kg/m^3 19 $-40°$C = $-40°$F and $574\frac{1}{4}°$F = $574\frac{1}{4}$ K 20 The Earth's mass would have to be about 50 times greater than it is today. 21 Assuming an average density of 1500 kg/m^3, the planet's diameter would be about $1\frac{1}{2}$ times the Earth's diameter.

Chapter 8: 15 2.2×10^8 years 17 17 percent for the core, 82 for the mantle, and 1 percent for the crust 18 The density of the core is about 15,300 kg/m^3. 19 At 55 km, the atmospheric pressure is about 1 millibar.

Chapter 9: 13 The Moon cannot have a significant iron core because, if it did, its average density would be higher than the density of typical crustal rock. 14 28 pounds on the Moon; 176 pounds on the Earth 15 1.5×10^5 kg/yr, which would double the Moon's mass in about 5×10^{17} years 16 552,000 km

Chapter 10: 15 Features larger than about 660 km will be resolved. 16 4.1 μm 17 10^{-3} nm 18 $\mu > 31$ 19 684 N = 176 pounds on Earth; 298 N = 67 pounds on Mercury; 129 N = 29 pounds on the Moon 20 0.61 AU

Chapter 11: 12 3.9 μm 13 6.64×10^{-6} nm 15 The Earth is about $\frac{1}{3}$ as bright as Venus. 17 0.93 km

Chapter 12: 15 450 km for a 1 arc sec resolution; 45 km for the HST 17 6.42×10^{23} kg

Chapter 13: 13 7.43×10^5 km from the Sun's center, which is nearly 50,000 km above the solar surface 14 1.90×10^{27} kg 15 380 pounds 16 Nearly 600 km/hr

Chapter 14: 13 Assuming 1 arc sec seeing, the smallest visible features are about 3000 km in diameter 14 About 60 km 15 64 billion years 16 About 8.1 minutes

Chapter 15: 13 600 km; Tethys, Dione, Rhea, Titan, and Iapetus 14 An approximate 2:3 resonance with Mimas and a 3:1 resonance with Tethys 15 2.6 km/s; $\mu > 12$ 17 16.7 km/s at the outer edge of the A ring; 20.3 km/s at the inner edge of the B ring; the maximum wavelength shift is 0.068 nm

Chapter 16: 12 About 7000 km 13 The HST can resolve the Great Dark Spot on Neptune, but no features on Pluto. 14 2 hours 6 minutes 15 Neptune exerts about 1.6×10^{-4} as much gravitational force on Uranus as does the Sun. 18 The ratio of the diameter of Jupiter to the diameter of Neptune is 2.9 to 1. The ratio of the size of the Great Red Spot to the size of Great Dark Spot is 2.7 to 1.

Chapter 17: 15 678 years; 0.53° per year 16 50 km 17 (a) 35,400 years (b) 1.12 million years (c) 35.4 million years (d) 1.12 billion years 18 The diameter of the iron-rich core of a typical parent asteroid is about $\frac{1}{3}$ of the asteroid's overall diameter.

Chapter 18: 1 1410 kg/m^3 6 (a) 1.8×10^{-9} J (b) 9.0×10^{16}J (c) 5.4×10^{41} J 7 (a) 4.6×10^{-36} s (b) 2.3×10^{-10} s (c) 4.4×10^7 yr 8 1.73×10^{11} kg each second 20 4.9 percent; chemical composition will be 69 percent hydrogen and 30 percent helium 21 photosphere: 500 nm = visible light; chromosphere: 58 nm = ultraviolet light; corona: 1.93 nm = X-rays 22 The energy flux from a sunspot is about 30 percent of that from the undisturbed photosphere.

Chapter 19: 3 When viewed from Jupiter, the Sun is about 3.7 percent as it is from the Earth. 4 (a) 4.85 pc (b) 0.2 arc sec 5 2100 pc 6 55 km/s 15 10 M$_\odot$; 0.1 L$_\odot$ 16 17th magnitude 17 4.31 pc 18 81 km/s 19 (a) 101 km/s (b) 486.29 nm 20 37 AU 22 16 kpc 23 Procyon's absolute magnitude is +2.64 and its luminosity is 7.6 L$_\odot$. 24 0.35 pc, or about one-fifth of its present distance from Earth 26 5 magnitudes 27 After the outburst, its diameter is twice as large as before and its luminosity has increased by a factor of 64. 28 1.48 M$_\odot$ 29 $\frac{3}{4}$M$_\odot$ and $\frac{1}{4}$M$_\odot$ 30 Rigel's diameter is about 28 times larger than the Sun's. 31 temperature = 4700 K; luminosity = 45.7 L$_\odot$; radius = 10.3 R$_\odot$ 32 26 pc

Chapter 20: 14 3.36×10^{10} atoms/m^3, which is much larger than the densities listed in Box 20-1 15 6.3 percent 16 Approximately 1000 times the present diameter of the Sun 18 1.3×10^{31} m^3

Chapter 21: 4 (a) 3 million years, or only about 3×10^{-4} as long as the Sun (b) 140 million years, or about 1.4×10^{-2} as long as the Sun 14 12 percent 15 1.66 M$_\odot$ 16 7.3×10^8 yr 18 620 pc 19 About 100 pc

Chapter 22: 5 13,400 km 13 Its radius is 0.106 R$_\odot$. 15 Nearly 9000 years ago 17 1.83×10^9 kg/m^3; 6450 km/s 19 A magnitude of -12, which is about as bright as the full moon 20 6.3 Mpc

Chapter 23: 1 5466 BC 12 One rotation per 11,000 years 13 The maximum correction is 10^{-4} of the pulsar's period. 14 density of neutron = 4.1×10^{17} kg/m^3; density of neutron star = 1.5×10^{17} kg/m^3 16 The expansion rate is 0.32 arc sec/yr. A change of 1 arc sec requires a wait of about 3.1 years. 17 2.2 kpc; 1300 years ago

Chapter 24: **2** 2.8 m **12** 1.8×10^{17} kg/m^3 **13** 1.4×10^8 M$_\odot$
14 (a) 9 mm; 2×10^{20} kg/m^3 (b) 3 km; 2×10^{19} kg/m^3 (c) 2200
AU; 2 g/m^3 **15** 0.98c **16** 0.99995c **17** 0.87c **18** 2.9 M$_\odot$

Chapter 25: **8** (a) 300 million years (b) 7.4×10^{11} M$_\odot$ **9** 25
times **13** A 10 percent error in radius results in a 10 percent error
in mass. A 10 percent error in velocity results in a 20 percent error
in mass **15** 3.9×10^{11} M$_\odot$ **16** 18,000 ly **17** Once every 90,000
years

Chapter 26: **9** 54 km/s/Mpc **12** 2.5 Mpc **13** Error would be a
factor of 2.5 in the distance **14** 220 Mpc if H$_0$ is 50 km/s/Mpc

15 Orbital period is about 440 million years; mass interior to 20
kpc is about 3.6×10^{11} M$_\odot$

Chapter 27: **8** 13.1 **9** 0.94c **10** 0.96c **11** 1.4×10^8 M$_\odot$
12 2×10^{30} kg/m^3 **13** 20 AU

Chapter 28: **11** 20 billion years, 13 billion years, 10 billion years
12 1.6×10^8 km/s/Mpc **13** 12.6 K **14** 1.9×10^{-26} kg/m^3
17 2.2×10^{-42} m **20** 2.1×10^{70} yr **21** 4.1×10^{22} kg
22 1.35×10^{-43} s **23** 2×10^{11} kg

Chapter 29: **12** 1.0×10^{-26} J **13** 3.5×10^{-25} s **14** 5.9×10^9 K
16 Between 4.2×10^{-35} kg and 1.8×10^{-34} kg

Illustration Credits

ton University; **Fig. 18-14:** Mount Wilson and Las Campanas Observatories; **Fig. 18-17, 18-19, 18-20:** National Optical Astronomy Observatories; **Fig. 18-18:** Naval Research Laboratory; **Fig. 18-23:** National Solar Observatory; **Fig. 18-24:** Kenneth G. Libbrecht, Big Bear Solar Observatory.

Chapter 19 p. 369: Anglo-Australian Observatory; **Fig. 19-3:** Yerkes Observatory; **p. 374:** Robert C. Mitchell, Central Washington University; Figure 19-10: Courtesy of Nancy Houk, Nelson Irvine, and David Rosenbush; **Fig. 19-17:** Yerkes Observatory; **Fig. 19-21:** Lick Observatory.

Chapter 20 p. 396, Figs. 20-1, 20-3, 20-4, 20-7, 20-9, 20-10, 20-14, 20-16, 20-20: Anglo-Australian Observatory; **Figs. 20-2, 20-6b:** Photography by David F. Malin of the Anglo-Australian Observatory from original negatives by the U.K. 1.2-m Schmidt telescope, copyright © 1980 Royal Observatory, Edinburgh; **Fig. 20-6a:** IRAS photograph produced by NLR/ROG, using spline fitted survey data [IRAS was developed and operated by the Netherlands Agency for Aerospace Programs (NIVR), NASA, and the U.K. Science and Engineering Research Council (SERC); **Fig. 20-8:** Bo Reipurth, European Southern Observatory; **p. 404:** Photography by David F. Malin of the Anglo-Australian Observatory from original negatives by the U.K. 1.2-m Schmidt telescope, copyright © 1980 Royal Observatory, Edinburgh; **Fig. 20-11:** U.S. Naval Observatory; **Fig. 20-12a:** Courtesy of Ronald J. Maddalena, Mark Morris, J. Moscowitz, and P. Thaddeus; **Figs. 20-18, 20-19:** Hans Vehrenberg.

Chapter 21 p. 415, Fig. 21-9: Anglo-Australian Observatory; **Fig. 21-4:** U.S. Naval Observatory; **Fig. 21-7:** Lick Observatory photograph; **Fig. 21-8:** National Optical Astronomy Observatories.

Chapter 22 p. 430: Lick Observatory photograph; **Figs. 22-3, 22-4, 22-10:** Anglo-Australian Observatory; **p. 433:** National Optical Astronomy Observatories; **Fig. 22-7:** R. B. Minton; **p. 439:** European Southern Observatory; **Fig. 22-11:** Mount Wilson and Las Campanas Observatories; **Fig. 22-13:** Palomar Observatory, copyright © California Institute of Technology; **Fig. 22-14:** Photography by David F. Malin of the Anglo-Australian Observatory from original negatives by the U.K. Schmidt telescope, copyright © 1980 Royal Observatory, Edinburgh; **Fig. 22-15a:** Stephen S. Murray, Harvard–Smithsonian Institution; **Fig. 22-15b:** National Radio Astronomy Observatory operated by Associated Universities, Inc., under contract with the National Science Foundation (VLA observations by R. J. Tuffs, R. A. Perley, M. T. Brown, S. F. Gull); **Fig. 22-16:** K. S. Luttrell. **Box 22-4:** NASA.

Chapter 23 p. 450: National Optical Astronomy Observatory; **Figs. 23-2, 23-4a:** Palomar Observatory, copyright © California Institute of Technology; **Fig. 23-4b:** Lick Observatory photograph; **Figs. 23-5, 23-7b:** F. R. Harnden, Jr., Harvard–Smithsonian Center for Astrophysics; **Fig. 23-7a:** Photography by David F. Malin of the Anglo-Australian Observatory from original negatives by the U.K. Schmidt telescope, copyright © 1980 Royal Observatory, Edinburgh; **Fig. 23-10:** NASA; **Fig. 23-13:** Jan van Paradijs; **Fig. 23-16:** National Radio Astronomy Observatory operated by Associated Universities, Inc., under contract with the National Science Foundation (VLA observations by R. M. Hjellming, K. J. Johnston); **Fig. 23-17:** Lick Observatory photograph.

Chapter 24 p. 467: D. Norton, Science Graphics; **Fig. 24-6:** National Radio Astronomy Observatory operated by Associated Universities, Inc., under contract with the National Science Foundation (VLA observations by P. E. Greenfield, D. H. Roberts, B. F. Burke); **Fig. 24-7:** Computations by E. Falco (MIT), M. Kurtz, R. Schild, M. Schneps, copyright © 1985 Smithsonian Astrophysical Observatory; **Fig. 24-8:** National Optical Astronomy Observatories; **Fig. 24-9:** National Radio Astronomy Observatory operated by Associated Universities, Inc., under contract with the National Science Foundation (VLA observations by J. N. Hewitt and E. L. Turner; **Fig. 24-10:** Jerome Kristian, Mount Wilson and Las Campanas Observatories; **Figs. 24-12, 24-13:** Larry L. Smarr and John F. Hawley.

Chapter 25 p. 482: Dennis di Cicco; **Fig. 25-1:** Steward Observatory; **Fig. 25-2:** Yerkes Observatory; **Fig. 25-5:** Harvard Observatory; **Fig. 25-6:** National Optical Astronomy Observatories; **Fig. 25-8:** U.S. Naval Observatory; **Fig. 25-11:** Gart Westerhout; **Fig. 25-12a:** Anglo-Australian Observatory; **Fig. 25-12b:** National Radio Astronomy Observatory

operated by Associated Universities, Inc., under contract with the National Science Foundation; **Fig. 25-16:** P. Seiden, D. Elmegreen, B. Elmegreen, and A. Mobarak, IBM Watson Research Center; **Fig. 25-19a:** Dennis di Cicco; **Figs. 25-19b, 25-19c:** NASA; **Fig. 25-20:** Anglo-Australian Telescope; **Fig. 25-21:** National Radio Astronomy Observatory operated by Associated Universities, Inc., under contract with the National Science Foundation [VLA observations by (a) F. Yusef-Zadeh, M. R. Morris, D. R. Chance, (b) Kwok-Yung Lo, M. J. Claussen]; **Fig. 25-22:** Rolf Güsten and Melvyn C. H. Wright.

Chapter 26 p. 500: European Southern Observatory; **Fig. 26-1:** Lund Humphries; **Fig. 26-2:** copyright © D. Elmegreen, B. Elmegreen, and P. Seiden; IBM Watson Research Center; **Fig. 26-3:** Palomar Observatory, copyright © California Institute of Technology; **Fig. 26-4:** Mount Wilson and Las Campanas Observatories; **Fig. 26-5:** (Sa) Rudolph Schild, Harvard–Smithsonian Center for Astrophysics, (Sb, Sc) Philip E. Seiden, IBM Watson Research Center; **Fig. 26-6:** (Sa) National Optical Astronomy Observatory, (Sb, Sc) Rudolph Schild, Harvard–Smithsonian Center for Astrophysics; **Fig. 26-7:** (SBa, SBc) Philip E. Seiden, IBM Watson Research Center, (SBb) Rudolph Schild, Harvard–Smithsonian Center for Astrophysics; **Fig. 26-8:** Yerkes Observatory; **Fig. 26-9:** Rudolph Schild, Harvard–Smithsonian Center for Astrophysics; **Figs. 26-10, 26-12, 26-13, 26-14:** Anglo-Australian Observatory; **Fig. 26-15:** Photography by David F. Malin of the Anglo-Australian Observatory from original negatives by the U.K. Schmidt telescope, copyright © 1987 Royal Observatory, Edinburgh; **Figs. 26-16, 26-17:** Rudolph Schild, Harvard–Smithsonian Center for Astrophysics; **Fig. 26-18:** S. J. Maddox, W. J. Sutherland, G. P. Efstathiou, and J. Loveday, Oxford Astrophysics; **Fig. 26-19:** V. de Lapparent, M. Geller, and J. Huchra; **Fig. 26-20:** C. Jones-Forman, Harvard–Smithsonian Center for Astrophysics; **Fig. 26-22a:** Lick Observatory photograph; **Fig. 26-22b:** Rudolph Schild, Harvard–Smithsonian Center for Astrophysics; **Fig. 26-23:** Joshua Barnes, Canadian Institute for Theoretical Astrophysics: **Fig. 26-24:** Palomar Observatory; **Fig. 26-25:** William C. Keel, University of Alabama; **Fig. 26-26:** Lars Hernquist, Institute for Advanced Study with simulations performed at the Pittsburgh Supercomputing Center; **Fig. 26-30:** Mount Wilson and Las Campanas Observatories; **p. 523:** Martha Haynes, Cornell University; **Fig. 26-32:** Lick Observatory.

Essay p. 528: Josh Grimes.

Chapter 27 p. 530: Courtesy of *Astronomy* Magazine, Kalmbach Publishing Co.; **Figs. 27-1, 27-2:** Palomar Observatory; **Fig. 27-3:** National Optical Astronomy Observatory; **Fig. 27-5:** Copyright © 1960 National Geographic Society–Palomar Sky Survey, reproduced by permission of the California Institute of Technology; **Figs. 27-10a, 27-11:** H. C. Arp; **Fig. 27-10b:** National Optical Astronomy Observatories; **Fig. 27-12:** Palomar Observatory; **Fig. 27-13:** Alan Stockton, Institute for Astronomy, University of Hawaii; **Fig. 27-14:** Palomar Observatory; **Fig. 27-15:** Rudolph Schild, Harvard–Smithsonian Center for Astrophysics; **Fig. 27-16:** National Optical Astronomy Observatories; **Fig. 27-17:** J. E. Grindlay, Harvard–Smithsonian Center for Astrophysics; **Fig. 27-18a:** National Optical Astronomy Observatories; **Fig. 27-18b:** National Radio Astronomy Observatory, operated by Associated Universities, Inc. under contract with the National Science Foundation (VLA observations by J. O. Burns, E. J. Schreier, E. D. Feigelson); **Fig. 27-18c:** E. J. Schreier, Harvard–Smithsonian Center for Astrophysics; **Fig. 27-19:** National Radio Astronomy Observatory, operated by Associated Universities, Inc. under contract with the National Science Foundation (VLA observations by P. A. Scheuer, R. A. Laing, R. A. Perley); **Fig. 27-20:** National Radio Astronomy Observatory, operated by Associated Universities, Inc. under contract with the National Science Foundation (VLA observations by C. P. O'Dea, F. N. Owen); **Fig. 27-21:** T. D. Kinman, Kitt Peak National Observatory; **Fig. 27-23:** National Optical Astronomy Observatories; **Fig. 27-25:** Palomar Observatory, copyright © California Institute of Technology; **Fig. 27-26:** European Southern Observatory; **Fig. 27-27:** Anglo-Australian Observatory; **Fig. 27-28:** H. C. Arp, J. J. Lorre; **Fig. 27-29:** T. Stephenson, Smithsonian Institution; **Figs. 27-31, 27-32:** John F. Hawley and Larry L. Smarr; **p. 552:** Copyright © 1960 National Geographic Society–Palomar Sky Survey, reproduced by permission of the California Institute of Technology.

INDEX

Star Charts

The following set of star charts, one for each month of the year, are from the *Griffith Observer* magazine. To use these charts, first select the chart that best corresponds to the date and time of your observations. Hold the chart vertically and turn it so that the direction you are facing shows at the bottom.

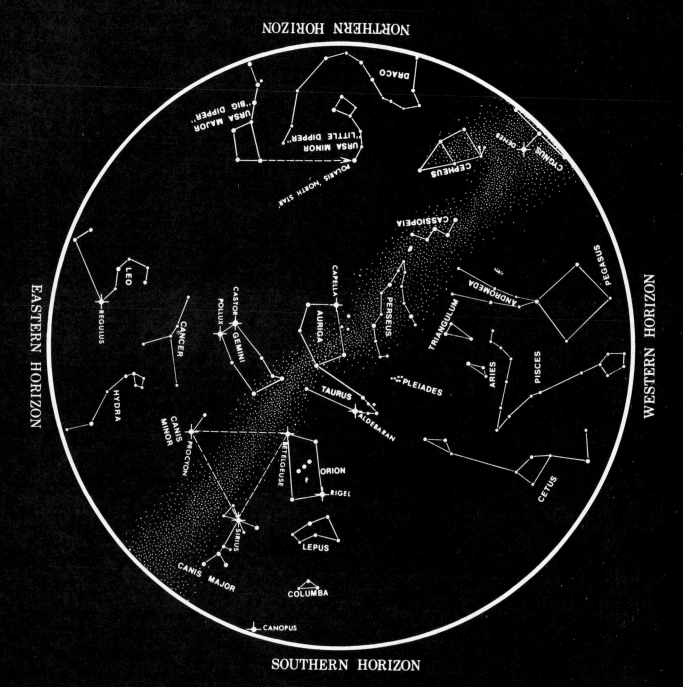

THE NIGHT SKY IN JANUARY

Chart time (Local Standard Time):

10 pm...First of January
9 pm...Middle of January
8 pm...Last of January

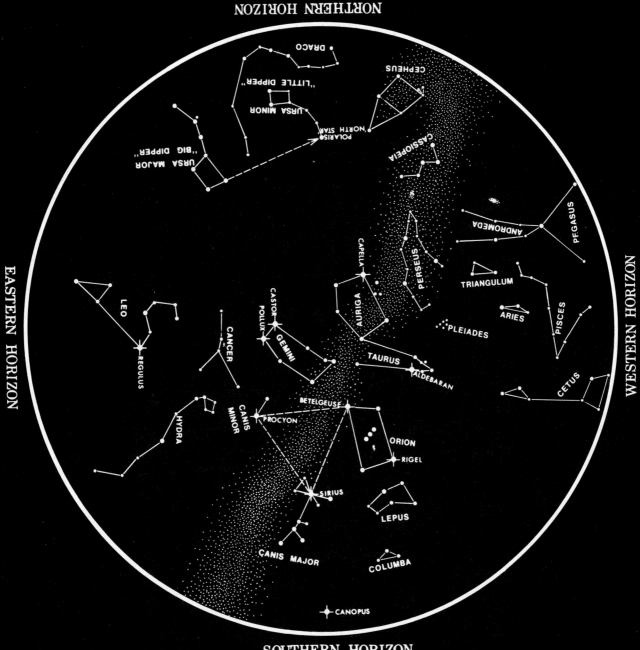

THE NIGHT SKY IN FEBRUARY

Chart time (Local Standard Time):

10 pm...First of February
9 pm...Middle of February
8 pm...Last of February

THE NIGHT SKY IN MARCH

Chàrt time (Local Standard Time):

10 pm...First of March
9 pm...Middle of March
8 pm...Last of March

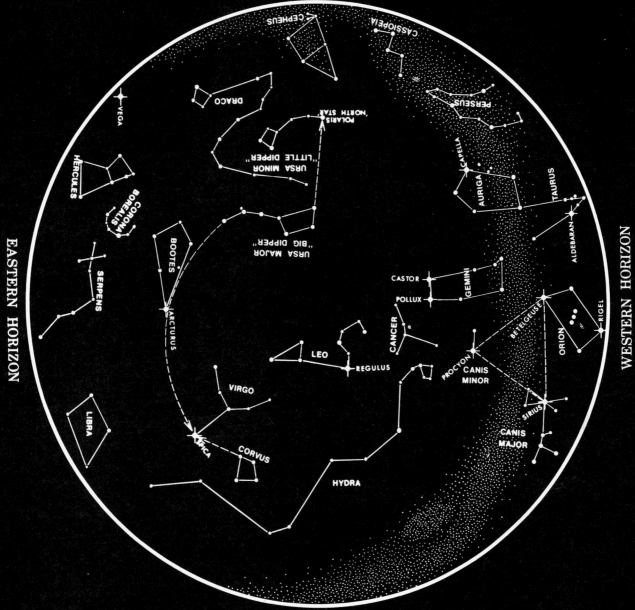

NORTHERN HORIZON

SOUTHERN HORIZON

THE NIGHT SKY IN APRIL

Chart time (Daylight Savings Time):

11 pm...First of April
10 pm...Middle of April
9 pm...Last of April

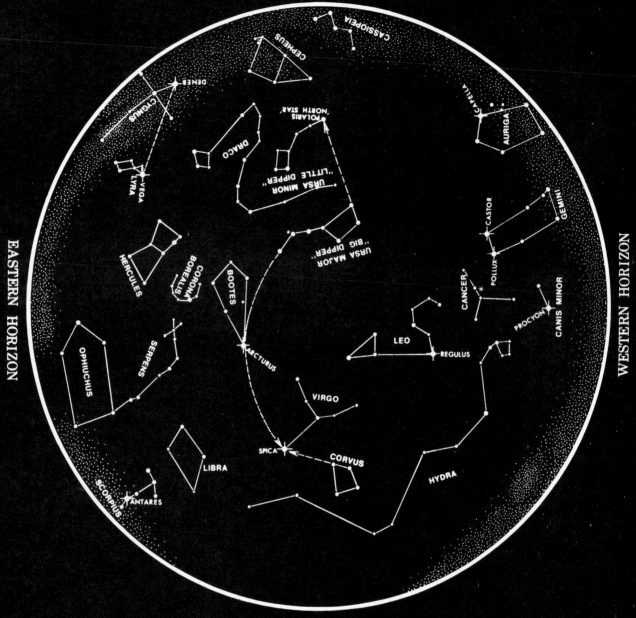

SOUTHERN HORIZON

THE NIGHT SKY IN MAY

Chart time (Daylight Savings Time):

11 pm...First of May

10 pm...Middle of May

9 pm...Last of May

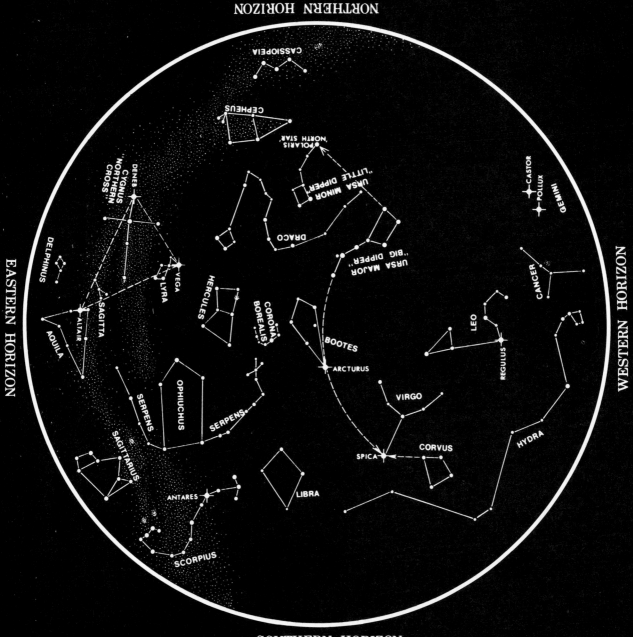

THE NIGHT SKY IN JUNE

Chart time (Daylight Savings Time):

11 pm...First of June
10 pm...Middle of June
9 pm...Last of June

SOUTHERN HORIZON

THE NIGHT SKY IN JULY

Chart time (Daylight Savings Time):

11 pm...First of July
10 pm...Middle of July
9 pm...Last of July

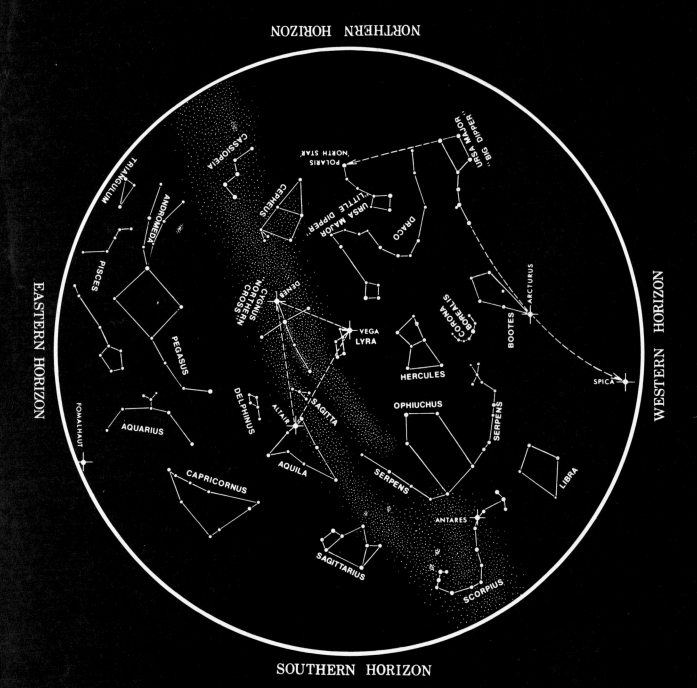

THE NIGHT SKY IN AUGUST

Chart time (Daylight Savings Time):

11 pm ... First of August
10 pm ... Middle of August
9 pm ... Last of August

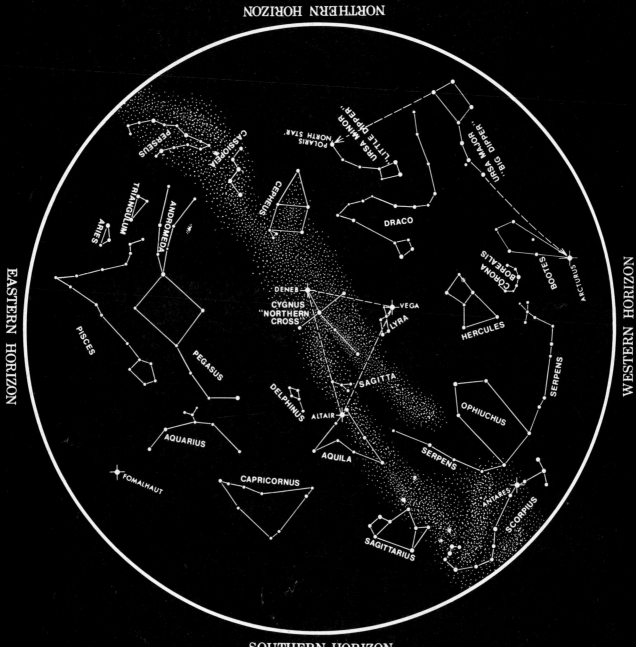

THE NIGHT SKY IN SEPTEMBER

Chart time (Daylight Savings Time):

11 pm...First of September
10 pm...Middle of September
9 pm...Last of September